Sociological Footprints

Introductory Readings in Sociology

TENTH EDITION

LEONARD CARGAN
Wright State University

JEANNE H. BALLANTINE
Wright State University

Australia • Brazil • Canada • Mexico • Singapore • Spain
United Kingdom • United States

Sociological Footprints:
Introductory Readings in Sociology,
Tenth Edition
Leonard Cargan and Jeanne H. Ballantine

Sociology Editor: Robert Jucha
Assistant Editor: Kristen Marrs
Editorial Assistant: Katia Krukowski
Technology Project Manager: Dee Dee Zobian
Marketing Manager: Michelle Williams
Marketing Assistant: Jaren Boland
Marketing Communications Manager: Linda Yip
Project Manager, Editorial Production: Cheri Pamer
Creative Director: Rob Hugel
Print Buyer: Karen Hunt
Permissions Editor: Kiely Sisk
Production Service: Matrix Productions
Photo Researcher: Image Quest
Copy Editor: Victoria Nelson
Illustrator: Integra
Cover Designer: Yvo Riezebos
Cover Image: © Wilfred Krecichwost/Getty Images
Compositor: Integra Software Services
Printer: Webcom

Printed in Canada
2 3 4 5 6 7 10 09 08 07

Library of Congress Control Number: 2005933774

Student Edition ISBN 0-495-00811-7

ISBN- 13: 978-0-495-00811-8
ISBN- 10: 0-495-00811-7

Thomson Higher Education
10 Davis Drive
Belmont, CA 94002-3098
USA

Contents

Preface

THE PRIMARY OBJECTIVE OF THIS ANTHOLOGY is to provide a link between theoretical sociology and everyday life by presenting actual samples of both classical and current sociological studies. If students are to grasp the full meaning of sociological terms and topics, they must be able to translate the jargon of sociology into real and useful concepts that are applicable to everyday life. To this end, *Sociological Footprints* presents viewpoints that demonstrate the broad range of sociological applications and the values of sociological research.

Selecting the readings for the tenth edition involved a number of important steps. As with the previous nine editions, we constantly received feedback from hundreds of students. Feedback was also requested from colleagues who are knowledgeable about the various topics of this anthology. An exhaustive search of the literature was conducted for additional material that was interesting and highly readable, that presented concepts clearly, that represented both recent and classic sociology, and that featured authors of diverse backgrounds. In meeting these criteria we often had to replace popular readings with more comprehensive and up-to-date ones. About one-third of the articles included in this edition are new selections that update all issues and address new concerns. As a final step, we utilized reviewers' comments to make the anthology relevant and useful. In this manner, each edition of *Sociological Footprints* becomes the strongest possible effort in producing a sociologically current, interesting, and highly readable collection.

IN THIS EDITION

The tenth edition builds on the strengths of early editions but also changes with current research and events. Students will gain from reading selections from classical sociology as well as selections from contemporary societal issues. The relevance of this collection is seen in the variety of the topics selected and in our efforts to cover the topics found in introductory sociology courses.

The classic selections cover a wide range of topics and authors. Perhaps the most fitting beginning for this material is Emile Durkheim (2, 47) because he is considered the first scientific sociologist. Next comes a selection by C. Wright Mills from a book long considered the best introduction to sociology as a discipline, *The Sociological Imagination* (5). The second chapter deals with culture and so includes a contribution by one of the fathers of anthropological research, Clyde Kluckholm (6). Another anthropologist, Horace Miner, gives us a typical research examination of a culture with "strange" customs (8). He also shows that research can be presented with a humorous twist. Returning to the "normal," Marvin Harris notes a different kind of culture in India (9).

Chapter 3, on socialization, begins by noting that it is not possible to acquire traits considered human without human contact. Kingsley Davis shows this connection in "A Case of Extreme Isolation" (11). Using a similar idea, the Halls note the importance of interaction in Chapter 4 by noting that interaction takes place between people even when they are silent (17). An important and different type of interaction, bureaucracy, is noted by one of the most important names in sociology, Max Weber (20). Herbert Gans makes a perhaps surprising statement when he claims that our society needs the poor to perform certain tasks; in a stratified society, it is the bottom stratum that does these tasks (25).

Deviance, by definition, concerns those who do not conform to the norm. D. L. Rosenhan turns this supposition around by examining those who are normal in a deviant institution (35). A popular story claims that "in the valley of the blind" the one-eyed person is king. Similarly, Rosenhan asks if the sane are the rulers in the mental institution? A political classic carries a strange title: "If Hitler Asked You to Electrocute a Stranger, Would You? Probably" (60). It is about a notorious study on obedience – something students well understand.

This anthology also covers a wide range of contemporary sociological topics. A perhaps surprising addition is an examination of a virus that has killed and continues to kill millions—AIDS (65). Finding a cure for this virus may not be in the sociologist's realm, but the cause of its spread is because it involves human behavior. Against the prevailing belief that the family has been basically unchanged over the centuries, the family has been a continually changing institution, as such terms as single-parent family, gay marriages, and women's rights in family decisions indicate (37). Sports are a major feature of American life in both the amateur and professional world. Part of this amateur world is school athletics. Thus, the relationship between the role of the school and that of athletics becomes an important question (17). As Gunston notes (74), professional sports are undergoing major changes as they become part of the corporate world—a change that is obvious but not a part of the public's perception.

Most everyone's wallet contains at least one credit card. Credit cards have become a major factor in the economy and an overwhelming influence on individual purchases and activities (54). Another major feature of American life is crime. Not only are a large number of crimes committed, but more people are incarcerated and executed for crimes in America than in any other country in the world. So it may come as a surprise that two of our articles claim that we fear the wrong things as far as crime is concerned (32) and that we do not fight crime intelligently (33). Almost every day newspapers carry stories about domestic violence, yet this is only a limited view of the problem since most domestic violence goes unreported (39). To students and teachers, it probably comes as no surprise that educational inequalities exist on many different levels (42, 43, 44).

Students who are preparing for a future of work are aware that the type of work available is constantly changing; that there will be work available at all is not much of a concern. Yet this is precisely Jeremy Rifkin's theme in his article, "The End of Work" (52). The last two national elections brought about many accusations and investigations of supposed manipulations of the election system. Yet this is only the visible part of such manipulation. Hidden rules of the system mean that 98% of the incumbents get reelected and 40% of the electorate are not registered and so cannot vote; America is only one of three industrialized countries in the world that continues to hold its elections on a workday – a major factor in limiting voter turnout to about 50% of registered voters. These two factors mean that only about 30% of all possible voters actually do – one of the lowest figures in the industrialized world. These are some of the more obvious items affecting election manipulation, but there are others, such as corporate interests, as "Money Changes Everything" reveals (57).

Two major themes running through the events of life today are religious terrorism and gender discrimination. So it comes as no surprise that this anthology has several articles on each topic: two on religious terrorism (47, 48) and four on gender discrimination (13, 14, 15, 30). Gender is such a pervasive topic, in fact, that it can be found in almost every chapter in this anthology.

This brief review reveals that our anthology covers such contemporary sociological topics as AIDS and homosexuality, the changing family, college athletics, the criminal justice system, domestic violence, educational inequality, political election manipulation, the future of work, religious terrorism, and gender discrimination. These are not the only sociological subjects discussed, however. Other topics include TV violence (12), elements of class stratification (21, 22), racial and class discrimination (26, 27, 28, 29), white-collar crime (34), being single in the world of the married (36), fraternity hazing (45), American civil religion (50), aging problems (61, 62, 63), the environment (66) and social change (72, 73, 75).

In sum, *Sociological Footprints* contains a balance of readings in each major chapter that—according to students, instructors, and reviewers—make this collection a valuable excursion into sociology. Although several of the articles have been condensed, the main theme and ideas of each has not been altered. Digressions, repetitions, and detailed descriptions of quantitative data have been omitted in order to emphasize key points.

FEATURES AND ORGANIZATION

We hope this new edition of *Sociological Footprints* will be as valuable to teachers as it is to students—an intention reflected in the book's organization. First, each major part has an introduction that covers the major themes of the topic area, noting how each reading relates to these themes. Second, each reading is also introduced by a comment about the important points in the reading. Third, we provide questions before each reading to guide the reader toward the important points. Fourth, although anthologies do not usually define concepts used in their readings, before many readings we include a glossary of important terms to give students a basic understanding of special terminology.

SUPPLEMENTS

Instructor's Manual with Test Bank

This instructor resource offers instructors teaching suggestions and a correlation grid that correlates each chapter with the chapters of major introductory sociology texts that may be used in conjunction with this reader. The manual also contains a summary analysis of each article in the reader, stating the thesis, findings, and conclusions of each reading. Test items are also provided, including five to ten multiple-choice questions and up to five essay questions for each article.

Companion Website

For additional resources such as Wadsworth's Census 2000: A Student Guide for Sociology, Sociology in Action, NewsEdge daily Sociology articles, and more, see the book's Companion Website at **http://sociology.wadsworth.com/cargan_ ballantine10e.**

Extension: Wadsworth's Sociology Readings Collection Create your own customized reader for your sociology class, drawing from dozens of classic and contemporary articles found on the exclusive Thomson Wadsworth TextChoice database. Using the TextChoice website (**http://www.TextChoice.com**), you can preview articles, select your content, and add your own original material. TextChoice will then produce your materials as a printed supplementary reader for your class.

InfoTrac® College Edition with InfoMarks™

Available as a free option with newly purchased texts, InfoTrac College Edition gives instructors and students four months of free access to an extensive online database of reliable, full-length articles (not just abstracts) from thousands of scholarly and popular publications going back as much as 22 years. Among the journals available 24/7 are *American Journal of Sociology, Social Forces, Social Research,* and *Sociology*. InfoTrac College Edition now also comes with InfoMarks, a tool that allows you to save your research parameters, as well as save links to specific articles. (Available to North American college and university students only; journals are subject to change.)

ACKNOWLEDGMENTS

We wish to thank all those who made this edition of *Sociological Footprints* possible. The reviewers of this edition are Sarah N. Gatson, Texas A&M University; Timothy B. Gongaware, University of Wisconsin—La Crosse; Susan F. Greenwood, University of Maine–Orono; Kristin Marsh, University of Mary Washington; and Jenny M. Stuber, DePauw University. We also thank reviewers of previous editions: Sarah N. Gatson, Texas A&M University; Kristin Marsh, Mary Washington College; Timothy McGettigan, Ph.D., University of Southern Colorado; Mohsen M. Mobasher, Southern Methodist University; and Carol Ray, San Jose State University; Philip Berg, University of Wisconsin-LaCrosse; Kevin J. Christiano, University of Notre Dame; Rodney B. Coates, University of North Carolina at Charlotte; Thomas F. Courtless, George Washington University; Robert W. Duff, University of Portland; Irene Fiala, Baldwin-Wallace College; Michael Goslin, Tallahassee Community College; Susan F. Greenwood, University of Maine; William J. Miller, Ohio University; Martin Monto, University of Portland; Wilbert Nelson, Phoenix College; Dan J. Pence, Southern Utah University; Ralph Peters, Floyd College; Carol Ray, San Jose State University; David R. Rudy, Moorehead State University; George Siefert, Daemen College; Eldon E. Snyder, Bowling Green State University; Larry L. Stealey, Charles Stewart Mott Community College; and Jerry Stockdale, University of North Iowa; Debbie A. Storrs, University of Idaho; Donna Trent, Eckerd College; and Assata Zerai, Syracuse University. Heartfelt thanks, also, go to the many students who took the time to give us their opinions, the departmental secretary Susan Schulthesis and aide Jennifer Meininger Deluca, who helped to assemble and type the material, our proofreader, and to all the good people at Wadsworth who aided in the production of this anthology.

To The Student

THE PURPOSE OF THIS ANTHOLOGY IS to introduce you, the beginning student in sociology, to a wide range of sociological perspectives and to demonstrate their relevance to real-life situations. As you apply sociological perspectives to everyday events, you will begin to realize that sociology is more than jargon, more than dry statistics, more than endless terminology to be memorized. It is an exciting and useful field of study. Unfortunately, no textbook can fully describe the many applications of sociology. This anthology should help to fill the gap by supplying classical readings balanced with contemporary readings on current issues and research.

From our experience in teaching introductory sociology, we know some of the problems that anthologies can present to the student: unexplained terms, readings seemingly unrelated to the text, and different emphases from those of the instructor's lectures. Therefore, to enjoy and benefit fully from *Sociological Footprints*, you should take the following steps:

1. Read and study the related textbook chapter and lecture materials. You must be familiar with the concepts and perspectives before you can clearly observe their daily application.
2. Read the introductions to the assigned sections in the anthology. They are designed to summarize the primary themes of the topic area and relate them to specific readings. In fact, the introductions will not only make the readings easier to understand, they will facilitate your application of the readings to other class materials and real-life situations.
3. Use the glossary that precedes the selection before you read each reading. Knowing the terms will make the reading more interesting and understandable.
4. Read each reading thoroughly. Note the problem or issue being discussed, the evidence the author supplies in support of his or her contentions, and the conclusions drawn from this evidence. Answer the questions posed at the beginning of the piece.
5. Summarize the main ideas of each reading in your own terms, relating them to other material in the course and to your own everyday experiences.

Step 5 is particularly important. Many of the readings address topics of current interest—political issues, population problems, environmental issues, the women's movement, and more. Because these are contemporary problems, you will see related materials in newspapers and magazines and on television. By applying what you have learned from the lectures and this anthology, you should develop a clearer understanding of current issues and of how sociology has aided you in this understanding.

We feel strongly that sociology is a field of study highly relevant to your world and that it can give you a fuller comprehension of day-to-day living. Our aim has been to provide you with a readable, understandable, and enlightening anthology that will convey this relevance.

THE ESSENTIAL WISDOM OF SOCIOLOGY

(Paraphrased from a paper by Earl Babbie from the 1989 ASA annual meeting) "I say this by way of a disclaimer. The essential wisdom of sociology may have twelve or thirteen points, but I'm going to quit at ten."

1. Society has a *sui generis* existence or reality. "You can't fully understand society by understanding individual human beings who comprise it. For example, few people want war, but we have wars all the time."
2. It is possible to study society scientifically. "Society can be more than learned beliefs of 'common sense.' It is actually possible to study society scientifically, just as we study aspects of the physical world."
3. Autopoesis: society creates itself. "Autopoesis (Huberto Maturana's term) might be defined as 'self-creating.' A powerful statement that sociology has to offer is that society is autopoetic: society creates itself."
4. Cultural variations by time and place. "Gaining awareness that differences exist is only the beginning, however . . . Our second task in this regard is to undermine . . . implicit ethnocentrism, offering the possibility of tolerance."
5. Relation of individual and society. "[One might] want to skirt the edge of suggesting that individuals are merely figments of society. Without going quite that far [one might suggest that] individual identity is strongly sociogenetic."
6. System imperatives. "Society is an entity [and] as a system, it has 'needs.'"
7. The inherent conservatism of institutions. "The first function of an institution is institutional survival."
8. Determinism. ". . . We operate with a model that assumes human behavior is determined by forces and factors and circumstances that the individual actors cannot control and/or are unaware of."
9. Paradigms. ". . . paradigms are ways of looking at life, but not life itself. They focus attention so as to reveal things, but they inevitably conceal other things,

rather like microscopes or telescopes, perhaps. They allow us to see things that would otherwise be hidden from us, but they do that at a cost."

10. Sociology is an idea whose time has come. "Finally . . . on a possibly chauvinist note: All the major problems that face us as a society and as a world are to be found within the territory addressed by sociology. I say this in deliberate contrast to our implicit view that most of our problems will be solved by technology."

Introduction: Why Study Sociology?

WHAT IS THIS SUBJECT CALLED SOCIOLOGY? What will I learn from studying sociology? Why should I take sociology? What work do sociologists do? How is sociology useful to me or to the world? If I major in sociology, what can I do when I graduate? These are some of the questions that may be in the back of your mind as you approach your study of sociology. Perhaps you are reading this book because you are curious about the subject, or because sociology is a required course, or because you had sociology in high school and wanted to find out more about it, or because your instructor assigned the book and this article. Whatever the reasons, you will find an introduction to the field of sociology in the discussion that follows.

What you read in the next few pages will only begin to answer the questions just posed. As you learn more about sociology, pieces that at first seemed fragmentary will start to come together like pieces in a puzzle. These pages provide the framework into which those pieces can be placed to answer the opening question: Why study sociology?

WHAT IS THIS SUBJECT CALLED SOCIOLOGY?

First questions first: Sociology is the study of people in groups, of people interacting with one another, even of nations interacting during peace or war. Sociologists' interests are sparked when they see two or more people with a common interest talking or working together. They are interested in how groups work and in how nations of the world relate to one another. When two or more people are interacting, sociologists have the tools to study the process. It could be a married couple in conflict or a teacher and students in a classroom situation; it could be individuals interacting in a work group, sports teams on a playing field, or negotiating teams discussing nuclear disarmament.

Sociology shares a common bond with other social sciences. All are concerned with human behavior in society; they share the perspective of the scientific method and some of the same data-collection methods to study their subject matter. Sociology is the broadest of the social sciences; its main concern is with predicting human group behavior.

"That's a lot to be interested in," you may be saying. In fact, most sociologists specialize. No one sociologist is likely to be an expert in everything, from studies of a few people or small group interaction (microlevel sociology) to large numbers of people in big groups like organizations or nations (macrolevel sociology). Consider the following examples of sociological specializations:

- determining the factors that lead to marital longevity
- identifying effective teachers by classroom observation
- examining public attitudes about the presidency and its policies
- locating satisfaction and problems in certain jobs

The results of these diverse interests lead sociologists into many different areas. Some sociologists specialize in social psychology, a field that considers such questions as how individuals behave in groups, who leaders are and what types of leaders are effective, why some groups accomplish more than other groups, why individuals usually conform to group expectations, and many other topics involving individuals as functioning members of groups. Another area of specialization is political sociology, which studies political power, voting behavior, bureaucracy, and political behavior of individuals and groups. Anthropology examines the culture of different groups; so does sociology. But the methods of study and primary focus differ. Anthropologists often study preliterate groups, whereas sociologists focus primarily on modern groups. Another area that concerns sociologists is social history, which emphasizes the use of history to understand social situations. These are only a few examples of the diverse interests of sociologists and how sociology shares its interests with some other social sciences.

WHAT WILL I LEARN FROM STUDYING SOCIOLOGY?

Consider that in some societies premarital sex is not only allowed but expected; in others, premarital sex is cause for banishment and death. Even though sociologists, like everyone else, have personal opinions, the task of the sociologist is not to judge which social attitude is right or wrong but to understand *why* such divergent practices have evolved. We all have opinions. Usually they come from our experiences, common sense, and family teaching. Some opinions are based on stereotypes or prejudices, some on partial information about an issue. Through systematic scientific study, sociologists gain insight into human behavior in groups, insight not possible through common sense alone. They attempt to understand all sides of an issue; they refrain from making judgments on issues of opinion but try instead to deal objectively with human behavior.

Consider the person who is going through the anguish of a divorce. Self-blame or hostility toward the spouse are often reactions to this personal crisis. Sociology can help us move beyond "individual" explanations to consider the social surroundings that influence the situation: economic conditions, disruptions caused by changing sex roles, and pressures on the family to meet the emotional needs of its members. Thus, sociology teaches us to look beyond individual explanations of our problems to group explanations for behavior; this practice broadens our worldview and gives us a better understanding of why events take place.

A typical college sociology program starts with a basic course introducing the general perspective of sociology; sociological terminology and areas of study; how sociologists get their information, that is, their methods; and the ideas, or theories, that lay the foundations for sociological study. Further sociology courses deal in greater depth with the major components of all societies: family, religion, education, politics, and economics. The sociology department may also offer courses on social processes such as social problems, deviance and corrections, stratification, socialization, and change, or in other areas of social life such as medical, community, urban, sports, or minority sociology.

Family sociology, for instance, usually considers the family social life cycle: young people breaking away from their parents' home, forming a home of their own by selecting a spouse through the courtship process, marrying, selecting a career, making parenting decisions, raising a family, having their children leave home, retiring, and moving into old age.

Students who major in sociology generally take courses in *theory*—the basic ideas of the field—and *methods*—how sociologists approach the social world objectively and do their research. Some sociology departments offer practical experiences where students can use their sociological skills in a job setting.

These are a few examples of what you will learn from the study of sociology and how you will learn it. There is much more to the field of sociology than this, however.

WHY SHOULD I TAKE SOCIOLOGY?

Whether you take a number of sociology courses or only one, you will profit in a number of ways. You will gain personal knowledge, new perspectives, skills needed by employers, background training useful in entering other fields, personal growth and development, new perspectives on the world, and a new way of looking at your relationships with others and your place in society. You will gain tolerance for and fascination with the variety of people in the world around you and their cultural systems. You will be able to understand your interactions with your family and friends better; you will be able to watch the news or read the paper with keener perception. You will have an understanding of how to obtain information to answer questions that you or your boss need answered. And the more sociology you take, the more ability you will have to express your thoughts logically, objectively, and coherently.

It is nice to know that the subjects you take in college will have some personal relevance and professional usefulness. Sociology should provide you with a number of "life skills," such as

1. Ability to view the world more objectively
2. Tools to solve problems by designing studies, collecting data, and analyzing results
3. Ability to understand group dynamics
4. Ability to understand and evaluate problems
5. Ability to understand your personal problems in a broader social context

We know from studies that employers value those applicants with the broad training of such fields as sociology because of the skills they provide. The following are skills employers look for, in order of importance:

1. Ability to work with peers
2. Ability to organize thoughts and information
3. Self-motivation
4. Ability to plan effectively
5. Willingness to adapt to the needs of the organization
6. Ability to interact effectively in group situations
7. Self-confidence about job responsibilities
8. Ability to handle pressure
9. Ability to conceptualize problems clearly
10. Effective problem-solving skills
11. Effective leadership skills
12. Ability to listen to others

Although a college graduate in engineering, computer sciences, or business may enter the job market with a higher salary, the sociology liberal arts major is more likely to rise through the managerial and professional ranks to positions of responsibility and high pay. Businesses and organizations value the skills listed here. In today's rapidly altering society, many of us will change jobs or careers several times during a lifetime. Sociological skills can help us adapt to the expectations of new situations.

Because of the knowledge and skills learned in sociology courses, study in this area provides excellent preparation for other undergraduate and graduate fields. From nursing, business, and education to law and medicine, the knowledge of sociology can be applied to a wide variety of group situations. For instance, a current concern of sociologists who study educational settings is what characteristics make schools effective; by singling out certain characteristics, sociologists can make recommendations to improve schools. Teachers and educational administrators profit from this information.

If we are curious about understanding ourselves and our interactions with others and about why our lives take certain directions, sociology can help us understand. For instance, sociologists are interested in how our social-class standing affects how we think, how we dress, how we speak, what our interests are, whom we are likely to marry, what religion (if any) we belong to, and what our "life chances" are, including how long we will live and what we are likely to do in life. Sociologists have even examined how individuals from different social-class backgrounds raise their children, and implications of child-rearing techniques for our lifestyles. Some use physical punishment and others moral chastisement, but the end result is likely to be a perpetuation of the social class into which we are born.

WHAT WORK DO SOCIOLOGISTS DO?

The most obvious answer is that sociologists *teach*; this is primarily at the higher education level, but high school sociology courses are also offered as part of the social science curriculum. There would be nothing to teach if sociologists were not actively engaged in learning about the social world. Their second major activity is to conduct *research* about questions concerning the social world.

Many sociologists work in business organizations, government agencies, and social service agencies. *Practicing sociologists* are engaged in a variety of activities. Some do family counseling with the whole family group; some conduct market research for companies or opinion polls for news or other organizations; some do surveys for the government to determine what people think or need; some work with juvenile delinquents, prison programs and reforms, and police; some predict how population changes will affect schools and communities.

Applied sociologists use their sociological knowledge to help organizations. They assess organizational needs, plan programs to meet those needs, and evaluate the effectiveness of programs. For instance, a community may want to know how many of its elderly citizens need special services to remain at home rather than be moved to nursing institutions. Sociologists assess this need, help plan programs, and evaluate whether programs are meeting the needs they set out to meet.

The position a sociology major ultimately gets depends in part on the degree he or she holds in sociology. The following are some examples of jobs students have gotten with a B.A. or B.S. degree: director of county group home, research assistant, juvenile probation officer, data processing project director, public administration/district manager, public administration/health coordinator, law enforcement, labor relations/personnel, police commander/special investigations, trucking dispatcher, administrator/social worker, counselor, child caseworker, substance abuse therapist, medical social worker, data programming analyst, activities director at senior citizens center, director of student volunteer program, area sales manager, jury verdict research editor, insurance claims adjustor, employment recruiter, tester for civil service, unemployment office manager, child services houseparent, crisis worker volunteer, advertising copywriter, probation officer, travel consultant, recreation therapist, public TV show hostess, adult education coordinator, research and evaluation specialist, neighborhood youth worker.

Sociologists holding an M.A. or Ph.D. degree are more skilled in sociological theory and methods than B.A. degree holders. They are often involved in research, teaching, or clinical work with families and other clients.

HOW IS SOCIOLOGY USEFUL TO ME AND TO THE WORLD?

Technology is rapidly changing the world. New policies and programs are being implemented in government and private organizations—policies that affect every aspect of our lives. Because sociologists study social processes, they are able to make concrete contributions to the planning of orderly change. Sociological knowledge can also be useful to legislators and courts in making policy decisions. For example, sociologists can assist a juvenile facility to design programs to help young people convicted of crime redirect their energies; how successful such programs are in achieving their goals can be studied by evaluation research.

In summary, sociology is the broadest of the social sciences and, unlike other disciplines, can give us an understanding of the social world. The knowledge and tools of sociology make students of this field valuable in a number of settings, from business to social service to government to education. As you embark on this study, keep in mind that sociology helps us have a deeper understanding of ourselves and our place in the world as well.

Sociology is a study of all people, for all people. To enjoy your encounter with the field and to make the most use of your time in sociology, try to relate the information you read and hear to your own life and relationships with others within the broader context of your social world.

How Will You Spend the 21st Century?

PETER DREIER, OCCIDENTAL COLLEGE

The following article, How Will You Spend the 21st Century?, describes some of the occupations sociology majors have achieved and the many tasks that people with sociology degrees can take on.

. . . I assume some of the parents in the audience today are like my parents were, 30 years ago, when I told them I was going to major in sociology. They weren't quite sure what sociology was, or whether you could get a job with a degree in sociology.

My father was worried that I might become a *social worker*. My mother was worried that I might become a *socialist*.

Well, let me assure you that your sons and daughters will be able to put their sociology degrees to good use. Some of our nation's most outstanding leaders, today and in the past, majored in sociology.

I'm a sports fan, so I've created a Sociology All-Star Team. The team captain is *Regis Philbin*, who majored in sociology at Notre Dame. . . . Other sociology majors in the world of entertainment include comedian *Robin Williams*, actor *Dan Aykroyd, Paul Shaffer* (the band leader on the David Letterman show), and Oscar-nominated actress *Deborah Winger*. And those of you who grew up in the 50s will remember the singer and TV star Dinah Shore, who studied sociology at Vanderbilt. Another sociology major from the world of arts and culture is novelist *Saul Bellow*, winner of the Nobel Prize for literature. The world of sports includes *Alonzo Mourning* (the Miami Heat's All-Star center); *Joe Theisman* (the NFL Hall of Fame quarterback); *Brian Jordan*, the Atlanta Braves' star outfielder; and, from the University of Oregon, *Ahmad Rashad*, the sportscaster and former football star.

Sociology has been a launching pad for people into the world of politics and law. A good example is *Richard Barajas*, chief justice of the Texas Supreme Court, who majored in sociology at Baylor University.

Over the years, quite a few sociology majors have been elected to political office. For example, *Shirley Chisholm*, the first black woman elected to Congress, in 1968, majored in sociology at Brooklyn College. The current Congress includes *Maxine Waters* (the Congresswoman from Los Angeles) and U.S. Senator *Barbara Milkulski* of Maryland.

Quite a few urban majors are on our sociology All-Star team, including *Wellington Webb*, the Democratic mayor of Denver; *Brett Schundler*, the Republican major of Jersey City, New Jersey; and *Annette Strauss*, the former mayor of Dallas.

Who can name a President of the United States who majored in sociology? The answer is *Ronald Reagan*, who has a sociology degree from Eureka College in Illinois.

Sociology is perhaps most well-known as a training ground for social reformers. Whether they go into politics, law, teaching, business, journalism, the arts, urban planning, the clergy, or any other

SOURCE: Reprinted with permission

field, they see their professional careers as a means to improve society and help others. Most sociologists, in other words, are *practical idealists.*

I think playwright George Bernard Shaw had the best understanding of sociology. He said: "Some men see things the way they are and ask: why? Others dream things that never were and ask: why not?" One of those practical idealists was *Saul Alinsky*, the founder of community organizing, who studied sociology at the University of Chicago. Another was *Martin Luther King*, who majored in sociology at Morehouse College in Atlanta.

On the list of great Americans who studied sociology, one of my favorites is Frances Perkins. She may not be a well-known name to many of you, but she was one of the most influential social reformers in American history. Frances Perkins was part of the first generation of women to attend college, entering Mt. Holyoke College in 1898. In one of her courses, students were required to visit a factory and do a survey of its working conditions. Perkins visited several textile mills and paper mills. There she saw the dangerous conditions and low pay that workers endured every day. This project opened her eyes to how "the other half lived."

After graduating in 1903, Perkins got involved in social work among poor immigrants and did extensive sociological research about slum housing and unsafe working conditions in laundries, textile mills, and other industries. She was soon recognized as a national expert in the new field of industrial sociology.

The tragedy of the Triangle Fire galvanized New York City's social reform groups. Perkins became the head of a citizens group called the Committee on Safety. Thanks to this group, within a few years, New York State had enacted 36 new laws protecting workers on the job, limiting the hours of women and children, and compensating victims of on-the-job injuries. Perkins continued this kind of social reform work for the rest of her life. In 1932, President Franklin Roosevelt asked her to become the nation's secretary of labor, the first woman ever to hold a cabinet position, where she became the central figure in the New Deal's efforts to improve the lives of America's poor, unemployed, and elderly. These included the passage of the Social Security Act and of the Fair Labor Standards Act, which established the minimum wage and the eight-hour day. This social legislation forever changed the living and working conditions of most Americans. Frances Perkins was in college 100 years ago. Try to imagine yourselves sitting in a commencement ceremony in 1901.

It is the beginning of a new century. What was America like back then? What kind of society were sociology majors like Frances Perkins about to inherit? In 1901, women didn't have the right to vote. Suffragists, who fought to give women that right, were considered radicals and utopians. Few people could look forward to retirement. Most people worked until they were no longer physically able to do so. And when they could no longer work, they often fell into poverty. A hundred years ago, reformers were calling for "social insurance" for the elderly.

In 1901, lynching was a regular occurrence in the South. Lynching kept black people terrorized. The NAACP was founded back then to fight to outlaw lynching and to abolish laws that denied black people the right to vote.

One hundred years ago, conditions in our factories and our urban housing were incredibly dangerous. Many people were regularly killed or seriously injured on the job. Many apartments were constructed so poorly that they were often fire traps, lacking ventilation. Epidemic diseases like TB were widespread because there were no laws dealing with basic sanitation. Back then, sociologists documented these conditions and worked with reformers for basic changes like government regulations regarding minimal safety standards for factories, schools, and apartment buildings as well as for laws outlawing the exploitation of child labor.

One hundred years ago, these and many other ideas, that today we take for granted—laws protecting consumers from unhealthy and unsafe food; laws regulating air pollution from factories and cars; Pell grants to help students pay college tuition; a minimum wage; government health insurance for the elderly and poor—were considered dangerous, or impractical, or even socialistic.

Each of these ideas has improved the day-to-day lives of Americans. Today, Americans enjoy more rights, better working conditions, better living conditions, and more protection from disease in childhood and old age than anyone could have imagined 100 years ago.

Thanks to Frances Perkins and people like her, America is a much better society than it was 100 years ago.

But that doesn't let you off the hook! There are still many problems and much work to do. Like all agents for social change, whether or not they studied sociology in college, Frances Perkins and Martin Luther King understood the basic point of sociology, that is, to look for the connections between people's everyday personal problems and the larger trends in society.

Things that we experience as personal matters—a woman facing domestic violence, or a low-wage worker who cannot afford housing, or middle-class people stuck in daily traffic jams—are really about how our institutions function. Sociologists hold a mirror up to our society and help us see our society *objectively*. One way to do this is by comparing our own society to others. This sometimes makes us uncomfortable—because we take so much about our society for granted. Conditions that *we* may consider "normal," other societies may consider serious problems.

For example, if we compare the U.S. to other advanced industrial countries like Canada, Germany, France, Sweden, Australia, Holland, and Belgium, we find some troubling things:

- The U.S. has the highest per capita income among those countries. At the same time, the U.S. has, by far, the widest gap between the rich and the poor.
- Almost 30 percent of American workers work full-time, year-round, for poverty-level wages.
- The U.S. has the highest overall rate of poverty. More than 33 million Americans live in poverty.
- Over 12 million of these Americans are children. In fact, one out of six American children is poor. They live in slums and trailer parks, eat cold cereal for dinner, share a bed or a cot with their siblings, and sometimes with their parents, and are often one disaster away from becoming homeless.
- Approximately four million American children under age 12 go hungry.
- Only three out of five children eligible for the Head Start program are enrolled because of the lack of funding.
- About seven million students attend school with life-threatening safety code violations.
- The U.S. has the highest infant mortality rate among the major industrial nations.
- One fifth of all children under two are *not immunized* against serious diseases.
- The U.S. is the only one of these nations without universal health insurance. More than 43 million Americans—including 11 million children—have no health insurance.
- Americans spend more hours stuck in traffic jams than people in any of these other countries. This leads to more pollution, more auto accidents and less time spent with families.
- Finally, the U.S. has a much higher proportion of our citizens in prison than any of these societies . . .

. . . What would you like *your* grandchildren to think about how *you* spent the 21st century? . . . No matter what career you pursue, you have choices about how you will live your lives. As citizens, you can sit on the sidelines and merely be *involved* in your society. Or you can decide to become really *committed* to making this a better world.

What's the different between just being *involved* and really being *committed?* Think about the eggs and bacon that you had for breakfast this morning. The hen was *involved*. But the pig was really *committed*!

Today, there are hundreds of thousands of patriotic Americans committed to making our country live up to its ideals. Some focus on the *environment*, others focus on education, and still others focus on *housing*, or working conditions, or *human rights*, or *global trade*, or *discrimination against women, minorities, and gays and the physically disabled.*

They are asking the same questions that earlier generations of active citizens asked: Why can't our society do a better job of providing equal opportunity, a clean environment, and a decent education for all? They know there are many barriers and obstacles to change, but they want to figure out how to overcome these barriers, and to help build a better society.

So ask yourselves: What are some of the things that *we* take for granted today that need to be changed? What are some ideas for changing things that today might seem "outrageous," but that—25 or 50, or 100 years from now—will be considered common sense?

In fact, your generation has done quite well already. The media stereotypes your generation as being apathetic—but the reality is that a record number of college students today are involved in a wide variety of "community service" activities—such as mentoring young kids in school, volunteering in a homeless shelter, or working in an AIDS hospice.

As a result of this student activism, more than 100 colleges and universities have adopted "anti-sweatshop" codes of conduct for the manufacturers of clothing that bear the names and logos of their institutions.

Positive change *is* possible, but it is *not* inevitable. For about the last decade, America has been holding its breath, trying to decide what kind of country we want to be. I am optimistic that your generation will follow in the footsteps of Frances Perkins and Martin Luther King—not only when you're young, but as a lifelong commitment to positive change.

I know you will *not* be among those who simply "see things the way they are and ask: why?" Instead, you will "dream things that never were and ask: why not?"

PART I

The Discipline of Sociology

Although the term **sociology** *may be familiar, many students are unaware of the areas of study included in the field. Sociology ranges from the study of small groups—perhaps two people—to that of such large entities as corporations and societies. As noted in the following reading, it is the study of interactions within, between, and among groups; and these group interactions encompass all areas of human behavior.*

Leading thinkers in all ages have been concerned about society, human conduct, and the creation of a social order that would bring forth the best man is capable of. But, the study of these problems with the techniques and approaches of science (sociology) is only a little more than a century old.

It was only about 125 years ago that Auguste Comte published his *Cours de philosophie positive* which first included sociology as one of the scientific disciplines. No course in sociology was available in an American university until 1876 at Yale. . . . Before 1900 all the men who identified themselves as professional sociologists were trained originally in other fields such as history, politics, economics, law, and religion. Today, undergraduate students can obtain training in sociology in almost all American four-year liberal arts colleges and in many agricultural colleges and specialized schools; more than 70 schools offer a doctoral program in sociology and many additional schools offer master's degree programs.

Before World War I the opportunities for employment of men and women with professional training in sociology were largely limited to college teaching and research. Besides teaching and research, sociologists today are engaged in more than 25 different kinds of work in professional schools, in local, state, federal, and private agencies, and in business. They work in the fields of education, medicine,

SOURCE: Reprinted by permission of the Sociology Department, University of Kentucky.

law, theology, corrections, agricultural extension, welfare, population study, community development, health, technological change, and the like. In short, sociologists are working on almost all the problems that concern man in relation to his fellow man and the consequences of this relationship for himself and others.

CHAPTER 1

The Sociological Perspective

Is It Just Common Sense?

Alan Mc Evoy

In the first reading of this chapter, Earl Babbie indicates that few people seem to understand the sociologist's true interests and activities despite the common usage of the term *sociology* and extends an offer to learn about the importance of this science. Babbie notes that sociologists are interested in discovering the realities of group interactions. To distinguish their conclusions from everyday, common-sense observations or intuitions, sociologists use scientific data-gathering techniques such as content analysis of documents and other literary items, surveys via questionnaires and interviews, various observation techniques, and laboratory or field experiments. Common sense, in contrast, is the feeling you have about a situation without scientific, systematic analysis; appearances alone are accepted as the criterion for truth. This being the case, what is true for one person may or may not be true for another. Examples of common-sense truths are those found in such advice columns as "Dear Abby" or "Ann Landers." Other types of common-sense truths are found in proverbs such as "Absence makes the heart grow fonder." How true such sayings are is seen when you realize that another saying claims, "Out of sight, out of mind." A number of beliefs about human behavior have the patina of truth because they seem logical and have been around for a long time. For example: "Humans have a natural instinct to mate with the opposite sex"; "a major proportion of those on welfare could work if they wanted to"; "American Catholics are less likely than Protestants to enter interfaith marriages or get divorced and are more likely to oppose birth control"; "a value found in every society is that of romantic love." Despite their seeming logic, all of these statements have been proven false.

As noted, scientists use various scientific techniques to collect and analyze data. This means that they use a procedure that is objective, logical, and systematic to produce reliable knowledge. This procedure can be clarified by examining the components of the previous sentence in more detail.

The first term claims that the procedure is *objective*. This term implies that the social scientist is aware of the difference between social behavior and other types of behavior. In short, we must be aware what activities are social facts. An early and major figure in sociology, Emile Durkheim, notes in the second reading that the rules of the scientific method are an essential part of the discipline of sociology. Durkheim indicates that these rules allow us to recognize what is a social fact and thereby help reduce the influence of personal bias on the research effort. Of course, complete freedom from bias is difficult to achieve because we are products of our society and what we "see" is influenced by this factor.

The next two terms noted are that the procedure must be *logical* and *systematic*. *Logical* refers to the arrangement of the facts collected and their interrelations according to accepted rules of reasoning. Using logic, social scientists construct theories as a guide to possible relationships. A *theory* states the apparent relationship of observed data to other observed data. For example, Durkheim in his classic study of suicide statistics was able to develop three theories on the causes of suicide: altruistic suicide, committed for the good of their group or society; egoistic suicide, committed for personal reasons; and anomic suicide, committed by those who have been overwhelmed by the rapidity of social change. In order to reach these conclusions, Durkheim had to be systematic in the presentation of his data – that is, consistent in the internal order of presentation of his materials.

The next term in the scientific procedure is *method*. This term refers to the techniques that are used to collect data for testing the hypothetical relationships of the variables being examined and that in turn aid in testing the theory. Throughout the readings in this anthology, you will see variations of the major social science techniques noted here: analysis of existing materials, surveys using a questionnaire or by interview, various techniques of observation, and experiment. An excellent way to become familiar with these methods is to see if you can identify the method used to collect the data when it is not indicated in the reading. For example, what was the method used by Horace Miner in his humorous article on the Nacirema (reading 8) and in the perhaps surprising conclusion made by Herbert Gans that poverty will always be with us because the poor perform valuable functions for society (reading 25)? In the third reading of this chapter, Leonard Cargan discusses the various techniques that sociologists use to collect the data that will develop and/or test hypotheses and theories of human behavior.

As noted, methods are the techniques utilized to collect the data, but what guides the search for the data is theoretical paradigms. In the fourth reading, Earl Babbie continues this overview of the scientific procedure by describing some of the theoretical paradigms that guide social research.

The final term used in the definition of a scientific procedure is *reliability*. *Reliable* means that the knowledge produced by the methods used is dependable because a retest made of the observations will produce similar results. It is this factor that allows social science techniques to gather data that make predictions about human behavior. However, unlike mechanical behavior in the physical sciences, human behavior is subject to people's whims and fads. Therefore, predictions in the social sciences always indicate a reliable range in which the behavior is found. Perhaps the most familiar example is found in political polls, which always note that a candidate has a lead of, say, 53% plus or minus 2. This means that the candidate's lead lies in the range of 51–55%.

In the final reading of this chapter, C. Wright Mills attempts to answer the question of what purpose sociology has beyond that of uncovering facts of human behavior. In the field's early days, it was believed that there was a rational explanation for all human behavior and that a scientific study of human behavior could lead to the solution of all social problems. Obviously, this prediction did not come true, and so the goals of sociology evolved to include manifold applications in numerous areas that influence the thinking of all of us. In his classic essay from *The Sociological Imagination*, Mills reaffirms these ideas when he notes that our perceptions are limited by our family, work, and other social experiences. In short, what we observe is bounded by these particular experiences and therefore may lead to biased or commonsense beliefs. As a means of overcoming this limited perspective, Mills suggests the use of what he calls the *sociological imagination*, a perspective that allows us to note relationships that exist between our personal experiences and general social issues.

Mills and the other authors in this chapter contend that adherence to the rules of the scientific method produces reliable knowledge that can be used to prove or disprove commonsense observations. The sociological approach can also expand our perspective beyond the limitations imposed by time and space and in so doing help us deal more efficiently with society's needs. Finally, the publication of the results of scientific investigations may lead to the recognition of other truths—truths that may result in benefits to society. These include the ideas suggested here for aiding family maintenance (reading 41), for dealing with the aftereffects of deregulation (reading 52), for making the criminal justice system more effective (reading 34) and for making the political system more representative (reading 57). As you read this chapter, keep these ideas in mind and apply them to your own common-sense beliefs. In short, challenge your own commonsense beliefs by using your sociological imagination.

1

An Idea Whose Time Has Come

EARL BABBIE

The author claims that this is a time of both unprecedented dangers but also of real achievements. To meet these conditions, we need to learn why people relate to one another sometimes peacefully and sometimes with hostility. This need fits into the role of sociology because it studies interaction and relations among humans, including how humans live together and how rules come into existence. For this reason Babbie claims that sociology is an idea whose time has come.

As you read this selection, consider the following questions as guides:

1. *What do you consider are the dangers now existing in the world and how you would try to resolve them?*
2. *What, if anything, are our politicians doing to resolve these dangers?*
3. *Can you think of any scientific efforts or accomplishments that attempt to deal with these dangers?*
4. *Why do you agree or disagree with Babbie's assertion that sociology is an idea whose time has come?*

There is a more pressing need for sociological insights today than at any time in history. . . .

It is no secret that this generation faces several unprecedented dangers. No sooner has the Cold War seemingly ended than we have come to recognize the danger of localized wars among ethnic groups spilling over into wider, global conflict. This danger is made worse by the possibility that relatively small, impoverished nations could gain access to nuclear weapons, giving terrorists the opportunity to spark an international conflagration. The more we learn about the prospects of a "nuclear winter"—the likely result of the first truly nuclear war—the more evident it is that there would be no real survivors of such a large-scale nuclear exchange.

If, on the other hand, we escape the threat of nuclear extinction, there is a real possibility that we will overpopulate and pollute the planet beyond its carrying capacity. As a single indicator of this problem, some 13 to 18 million people die of starvation around the world every year, three-fourths of them children. Approximately one-fifth of all humans on the planet go to bed hungry every night.

Add to this such persistent problems as crime, inflation, unemployment, prejudice, totalitarianism, and national debts, and you have sufficient grounds for understanding the ancient Chinese curse: "May you live in interesting times."

These are unquestionably interesting times. But the picture is not completely gloomy. These are also the times of great achievements in space: humans

SOURCE: *The Sociological Spirit*, Belmont: Wadsworth, 1994.

landing on the moon, a remote-controlled craft landing on Mars, and others photographing the more distant planets. These are the times when human beings, working cooperatively around the globe, eradicated smallpox—a scourge throughout history. These are the times of an awakening of awareness and commitment to ending world hunger. And the breakup of the former USSR, along with its domination of Eastern Europe, is regarded by most as a positive development. . . .

If it were possible to make comparable lists of the positive and negative aspects of life today, my hunch is that they might be about equal in length. At any rate, both lists would be long ones, indicating that you and I and our fellow human beings face both trying challenges and promising opportunities ahead of us.

Assuming that we'd agree in favoring peace over war, prosperity over hunger and poverty, and so forth, the question you should be asking yourself is: what determines how things turn out? Specifically, what would it take for peace to triumph over the continuing specter of wars large and small? We'd all like the answer to that question.

I suggest there is a prior question you should ask. That is: where should we look for the answer to how peace can prevail over war? Before asking what the answer is, we need to ask where the answer is likely to be found. I suggest that up until now we have tended to look for the answer to how peace can prevail over war within the domain of military technology. Most simply, we have tried to create weapons that would preserve the peace. . . .

The point of this discussion on war and peace is to suggest that what we need to know to establish peace around the world is not likely to arise from military technology. If such an answer is to be found at all, we must look elsewhere: in the study of why people relate to one another as they do—sometimes peacefully, sometimes hostilely. This, as we'll see, lies in the domain of sociology.

THE DOMAIN OF SOCIOLOGY

Sociology involves the study of human beings. More specifically, it is the study of interactions and relations *among* human beings. Whereas psychology is the study of what goes on inside individuals, sociology addresses what goes on between them. Sociology addresses simple, face-to-face interactions such as conversations, dating behavior, and students asking a professor to delay the term paper deadline. Equally, sociology is the study of formal organizations, the functioning of whole societies, and even relations among societies.

Sociology is the study of how human beings live together—in both the good times and the bad. It is no more a matter of how we cooperate and get along than of how we compete and conflict. Both are fundamental aspects of our living together and, hence, of sociology.

You might find it useful to view sociology as the study of our *rules for living together*. Let's take a minute to look at that.

To begin, let's consider some of the things that individuals need or want out of life: food, shelter, companionship, security, satisfaction—the list could go on and on. My purpose in considering such a list is to have us see that the things you and I need or want out of life create endless possibilities for conflict and struggle. When food is scarce, for example, I can only satisfy my need at your expense. Even in the case of companionship—where both people get what they want—you and I may fight over a particular companion.

The upshot of all this is that human beings do not seem to be constructed in a way that ensures cooperation. Bees and ants, by contrast, just seem to be wired that way. As a consequence, human beings *create rules* to establish order in the face of chaos. Sometimes we agree on the rules voluntarily, and other times some people impose the rules on everyone else. In part, sociology is the study of how rules come into existence.

1. Hiram Maxim as quoted in Martin Hellman, *A New Way of Thinking*, Palo Alto, CA: Beyond War, 1985, p. 4.

2. Orville Wright as quoted in Martin Hellman, *A New Way of Thinking*, Palo Alto, CA: Beyond War, 1985, p. 4.

Sociology is also the study of how rules are *organized* and *perpetuated*. It would be worth taking a minute to reflect on the extent and complexity of the rules by which you and I live. There is a rule, for example, that Americans must pay taxes to the government. But it doesn't end there. The rule for paying taxes has been elaborated on by a great many more specific rules indicating how much, when, and to whom taxes are to be paid. In recent years, the index to the IRS tax code has run more than 1,000 pages long, which should give you some idea of the complexity of that set of rules. The much-touted tax simplification of 1986 was 1,855 pages long.

The rules governing our lives are not all legal ones. There are rules about shaking hands when you meet someone, rules about knives and forks at dinner, rules about how long to wear your hair, and rules about what to wear to class, to the symphony, and to mud-wrestling. There are rules of grammar, rules of good grooming, and rules of efficient computer programming.

Many of the rules we've been considering were here long before you and I showed up, and many will still be here after we've left. Moreover, I doubt that you have the experience of having taken part in creating any of the rules I've listed. Nobody asked you to vote on the rules of grammar, for example. But in a critical way, you *did* vote on those rules: you voted by obeying them.

Consider the rule about not going naked in public. Even though you don't recall being asked what you thought about that one, there was a public referendum on that issue this morning—and you voted in favor of clothes. So did I. If this seems silly, by the way, realize that there are other societies in which people voted to accept a different rule this morning.

Sometime today you are likely to be asked to vote on a set of rules about eating. Some of the possibilities are eating spaghetti with a knife, pouring soup on your dessert, and throwing your food against the wall. Let's see how you vote.

The persistence of our rules is largely a function of one generation teaching them to the next generation. We speak of *socialization* as the process of learning the rules, and it becomes apparent that we are all socializing each other all the time through the use of positive and negative *sanctions*—rewards and punishments.

All the rules we've been discussing are fundamentally *arbitrary*—that is, different rules would work just as well. Although Americans have a rule that cars must be driven on the right side of the road, other societies (e.g., England, Japan) manage equally well with people driving on the left side.

Once we've established a rule, however, we tend to add weight to it. We act as though it were better than the other possibilities, that it somehow represents an eternal and universal *truth*. Sociologists often use the term *reification* in reference to the pretense that things are real when they are not, and we often *reify* the rules of society. We make the right side of the road the *right* side for cars to drive on, and we think the British and Japanese strange for not knowing that.

The rules of society take their strongest hold when they are *internalized* by individuals—taken inside ourselves and made our own. Imagine this situation, which you may have actually experienced. It is three o'clock in the morning, and you are driving along a street leading out of a small town. There is no traffic on the street in any direction as you come to a red light at an intersection. You can see there are no cars coming for a mile in every direction. There is no one around. What do you do? There's a good chance that you will sit and wait for the light to change. If someone questioned you about it, you might say, "It just wouldn't feel right."

In the event you would drive through the light and generally regard yourself as above reification and internalization, think again. You have reified and internalized countless rules. How would you feel about having live ants and cockroaches for dinner tonight, for example? Are you willing to give them a try? How do you feel about murder, rape, and child abuse? Are you pretty casual about them, or do you feel they are *really wrong?* If you think about it, you'll find you feel pretty strongly about a lot of our rules.

All of this notwithstanding, sociology is also the study of how we *break* the rules. Some people use bad grammar and pour their soup on the floor, not to mention drive too fast, steal, fix prices, commit murder, and everything in between. Although this may seem like the study of "bad people," beware.

First, the rules of society are so extensive and complex that no one can possibly keep them all. For example, there is probably a street near you that is posted with a twenty-five-mile-per-hour speed limit; that's certainly a rule. And yet if you drive twenty-five miles an hour on that street during rush hour, you may discover you're breaking another rule. Your clue will be the honking horns and shaking fists.

Beyond the impossibility of obeying all the rules, you and I might agree that some of them ought to be broken. Consider the rule, in force a few years ago, that black people had to sit in the back of the bus in some parts of America. The people who finally broke that rule are considered heroes today. By the same token, you might disagree with rules that women can't fix carburetors, that men shouldn't cry, or that professors always know better than students.

The study of how people break the rules is closely related to the study of how the rules *change* over time. Although we are always living in a sea of rules, and many seem to last forever, it is also true that the rules of society are always in a process of change. Rules pertaining to hemlines, hair length, and political views operate a little like yo-yos. Others seem to change in only one direction.

Sociology, then, is an examination of the rules that govern our living together: what they are, how they arise, and how they change. Sociology, however, is a special approach to the rules of social life. As we'll see, there are other approaches.

A SCIENCE OF SOCIETY

. . . Sociology is a science of social life. Like other sciences, sociology has a ***logical/empirical*** basis. This means that, to be accepted, assertions must (1) make sense and (2) correspond to the facts. In this sense, sociology can be characterized by a current buzzword: *critical thinking*. The simple fact is that most of us, most of the time, are uncritical in our thinking. Much of the time we simply believe what we read or hear. Or, when we disagree, we do so on the basis of ideological points of view and prejudices that are not very well thought out.

Suppose you were talking with a friend about the value of going to college. Your friend disagrees: "College is a waste of time. You should get a head start in the job market instead. Most of today's millionaires never went to college, and there are plenty of college graduates pumping gas or out of work altogether." That's the kind of thing people sometimes say, and it can be convincing—especially if it's said with conviction. But does it stand up to logical and empirical testing?

Logically, it doesn't seem to make much sense, since a college education would seem to give a person access to high-paying occupations not open to people with less education. . . .

There isn't any scientific support for the assertion that education is a worthless financial investment—even though there are some individual exceptions to the rule.

It's important to recognize that human beings generally have opinions about everything . . . but it needs saying here that the people you deal with every day have a tendency to express opinions about the way things are—and what they say isn't always so. Consequently, you need to protect yourself from false information. That's what critical thinking is all about, and sociology provides some powerful critical-thinking tools.

I hope these few examples will indicate that sociology is not just something you might study in college and never think about again. Sociology deals with powerful issues that determine the quality of your life. Understanding sociology can empower you to be a more effective participant in the social affairs around you whether you are a conscious player or not. Marriage, employment, prejudice, crime, and politics are only a few of the areas of social life that can be importantly affected by your ability to engage in sociological reasoning.

Let's look at the twin foundations of critical thinking and of science in more detail—seeing how they apply to sociology in particular. . . .

SOCIOLOGICAL QUESTIONS AND ANSWERS

The common image of science is of scientists finding the answers to questions. I will conclude this chapter with a somewhat different view of science and of sociology in particular. Science is sometimes better at raising questions than at finding answers to them. It would be useful for you to regard science as an ongoing inquiry, recognizing that questions initiate new avenues of inquiry, while answers close them down.

Science makes an especially powerful contribution when it calls into question those things that "everyone knows." Everyone used to "know," for example, that blacks were inferior to whites and that women were inferior to men. As you'll see, sociological points of view very often raise questions about things everyone else thought had already been answered. Even when we discover new and seemingly better answers about something, it's important to hold them as tentative.

There is a particular ***recursive*** quality in human life that makes anything we know tentative. Whenever we learn something about ourselves, what we've learned may bring about changes—even to the extent of making what we learned no longer accurate.

Suppose we studied employment opportunities across the country, for example. When our study was complete, we listed the ten cities with the most jobs available. As soon as our findings became widely known, of course, a lot of unemployed people would move to those cities, and soon those cities would not have as many jobs available as before. This is the same thing that happens when a newspaper columnist identifies a local restaurant that has great food, low prices, and no waiting. The "no waiting" part probably won't last another twenty-four hours, and the other two characteristics may disappear, too. In sociology, anything we learn can change things, so no knowledge can be counted on to remain true. Thus, we need to keep asking questions.

Even more fundamentally, sociology deals with a number of questions that will never be answered fully. Who am I? What is a human being? Are we more the result of our genes or of our environment? Is order possible without restricting freedom? . . . and it is unlikely that they will ever be completely answered. As we'll see, however, it can be very useful to keep asking them anyway.

I point all this out to give you an appropriate context for your own inquiry into sociology. Although there are some facts about sociology that are worth learning, it is more important for you to learn to use sociology for your own ongoing critical thinking. If you were studying brain surgery or medieval history, you might not have an opportunity to use what you learned in your day-to-day life. Sociology is very different. Every day you will wake up into a sociological laboratory with a massive experiment underway. We're all subjects in the experiment, and you now have the opportunity to join the researchers.

2

What Is a Social Fact?

EMILE DURKHEIM

In reading this article, be sure to note what behavior is social and why—that is, what is a social fact?

As you read, ask yourself the following questions:

1. *What facts are called social?*
2. *What social facts would you like to see changed?*
3. *What nonsocial facts would you like to see changed?*

GLOSSARY **Aphorism** A general truth embodied in a short saying.

Before inquiring into the method suited to the study of social facts, it is important to know which facts are commonly called "social." This information is all the more necessary since the designation "social" is used with little precision. It is currently employed for practically all phenomena generally diffused within society, however small their social interest. But on that basis, there are, as it were, no human events that may not be called social. Each individual drinks, sleeps, eats, reasons; and it is to society's interest that these functions be exercised in an orderly manner. If, then, all these facts are counted as "social" facts, sociology would have no subject matter exclusively its own, and its domain would be confused with that of biology and psychology.

But in reality there is in every society a certain group of phenomena which may be differentiated from those studied by the other natural sciences. When I fulfil my obligations as brother, husband, or citizen, when I execute my contracts, I perform duties which are defined, externally to myself and my acts, in law and in custom. Even if they conform to my own sentiments and I feel their reality subjectively, such reality is still objective, for I did not create them; I merely inherited them through my education. How many times it happens, moreover, that we are ignorant of the details of the obligations incumbent upon us, and that in order to acquaint ourselves with them we must consult the law and its authorized interpreters! Similarly, the church-member finds the beliefs and practices of his religious life ready-made at birth; their existence prior to his own implies their existence outside of himself. The system of signs I use to express my thought, the system of currency I employ to pay my debts, the instruments of credit I utilize in my commercial relations, the practices followed in my profession, etc., function independently of my own use of them. And these statements can be repeated for each member of society. Here, then, are ways of acting, thinking, and feeling that present the noteworthy property of existing outside the individual consciousness.

These types of conduct or thought are not only external to the individual but are, moreover, endowed with coercive power, by virtue of which they impose themselves upon him, independent of his individual will. Of course, when I fully consent and conform to them, this constraint is felt only slightly, if at all, and is therefore unnecessary. But it is, nonetheless, an intrinsic characteristic of these facts, the proof thereof being that it asserts itself as soon as I attempt to resist it. If I attempt to violate the law, it reacts against me so as to prevent my act before its accomplishment, or to nullify my violation by restoring the damage, if it is accomplished and reparable, or to make me expiate it if it cannot be compensated for otherwise.

In the case of purely moral maxims, the public conscience exercises a check on every act which offends it by means of the surveillance it exercises over the conduct of citizens, and the appropriate penalties at its disposal. In many cases the constraint is less violent, but nevertheless it always exists. If I do not submit to the conventions of society, if in my dress I do not conform to the customs observed in my country and in my class, the ridicule I provoke, the social isolation in which I am kept, produce, although in an attenuated form, the same effects as a punishment in the strict sense of the word. The constraint is nonetheless efficacious for being indirect. I am not obliged to speak French with my fellow-countrymen nor to use the legal currency, but I cannot possibly do otherwise. If I tried to escape this necessity, my attempt would fail miserably. As an industrialist, I am free to apply the technical methods of former centuries; but by doing so, I should invite certain ruin. Even when I free myself from these rules and violate them successfully, I am always compelled to struggle with them. When finally overcome, they make their constraining power sufficiently felt by the resistance they offer. The enterprises of all innovators, including successful ones, come up against resistance of this kind.

Here, then, is a category of facts with very distinctive characteristics: it consists of ways of acting, thinking, and feeling, external to the individual, and endowed with a power of coercion, by reason of which they control him. These ways of thinking could not be confused with biological phenomena, since they consist of representations and of actions; nor with psychological phenomena, which exist only in the individual consciousness and through it. They constitute, thus, a new variety of phenomena; and it is to them exclusively that the term "social" ought to be applied. And this term fits them quite well, for it is clear that, since their source is not in the individual, their substratum can be no other than society, either the political society as a whole or some one of the partial groups it includes, such as religious denominations, political, literary, and occupational associations, etc. On the other hand, this term "social" applies to them exclusively, for it has a distinct meaning only if it designates exclusively the phenomena which are not included in any of the categories of facts that have already been established and classified. These ways of thinking and acting therefore constitute the proper domain of sociology. It is true that, when we define them with this word "constraint," we risk shocking the zealous partisans of absolute individualism. For those who profess the complete autonomy of the individual, man's dignity is diminished whenever he is made to feel that he is not completely self-determinant. It is generally accepted today, however, that most of our ideas and our tendencies are not developed by ourselves but come to us from without. How can they become a part of us except by imposing themselves upon us? This is the whole meaning of our definition. And it is generally accepted, moreover, that social constraint is not necessarily incompatible with the individual personality.[1]

Since the examples that we have just cited (legal and moral regulations, religious faiths, financial systems, etc.) all consist of established beliefs and practices, one might be led to believe that social facts exist only where there is some social organization. But there are other facts without such crystallized form which have the same objectivity and the same ascendency over the individual. These are called "social currents." Thus the great movements of enthusiasm, indignation, and pity in a crowd do not originate in any one of the particular individual consciousness. They come to each one of us from without and can carry us away in spite of ourselves. Of course, it may happen that, in abandoning myself to them unreservedly, I do not feel the pressure they exert upon me. But it is revealed as soon as I try to resist them. Let an individual attempt to oppose one of these collective manifestations, and the emotions that he denies will turn against him. Now, if this power of external coercion asserts itself so clearly in cases of resistance, it must exist also in the first-mentioned cases, although we are unconscious of it. We are then victims of the illusion of having ourselves created that which actually forced itself from without. If the complacency with which we permit ourselves to be carried along conceals the pressure undergone, nevertheless it does not abolish it. Thus, air is no less heavy because we do not detect its weight. So, even if we ourselves have spontaneously contributed to the production of the common emotion, the impression we have received differs markedly from that which we would have experienced if we had been alone. Also, once the crowd has dispersed, that is, once these social influences have ceased to act upon us and we are alone again, the emotions which have passed through the mind appear strange to us, and we no longer recognize them as ours. We realize that these feelings have been impressed upon us to a much greater extent than they were created by us. It may even happen that they horrify us, so much were they contrary to our nature. Thus, a group of individuals, most of whom are perfectly inoffensive, may, when gathered in a crowd, be drawn into acts of atrocity. And what we say of these transitory outbursts applies similarly to those more permanent currents of opinion on religious, political, literary, or artistic matters which are constantly being formed around us, whether in society as a whole or in more limited circles.

To confirm this definition of the social fact by a characteristic illustration from common experience, one need only observe the manner in which children are brought up. Considering the facts as they are and as they have always been, it becomes immediately evident that all education is a continuous effort to impose on the child ways of seeing, feeling, and acting which he could not have arrived at spontaneously. From the very first hours of his life, we compel him to eat, drink, and sleep at regular hours; we constrain him to cleanliness, calmness, and obedience; later we exert pressure upon him in order that he may learn proper consideration for others, respect for customs and conventions, the need for work, etc. If, in time, this constraint ceases to be felt, it is because it gradually gives rise to habits and to internal tendencies that render constraint unnecessary; but nevertheless it is not abolished, for it is still the source from which these habits were derived.... What makes these facts particularly instructive is that the aim of education is, precisely, the socialization of the human being; the process of education, therefore, gives us in a nutshell the historical fashion in which the social being is constituted. This unremitting pressure to which the child is subjected is the very pressure of the social milieu which tends to fashion him in its own image, and of which parents and teachers are merely the representatives and intermediaries.

It follows that sociological phenomena cannot be defined by their universality. A thought which we find in every individual consciousness, a movement repeated by all individuals, is not thereby a social fact. If sociologists have been satisfied with defining them by this characteristic, it is because they confused them with what one might call their reincarnation in the individual. It is, however, the collective aspects of the beliefs, tendencies, and practices of a group that characterize truly social phenomena. As for the forms that the collective states assume when refracted in the individual, these are things of another sort. This duality is clearly demonstrated by the fact that these two orders of phenomena are frequently found dissociated from one another. Indeed, certain of these social manners of acting and thinking acquire, by reason of their repetition, a certain rigidity which on its own account crystallizes them, so to speak, and isolates them from the particular events which reflect them. They thus acquire a body, a tangible form, and constitute a reality in their own right, quite distinct from the individual facts which produce it. Collective habits are inherent not only in the successive acts which they determine but, by a privilege of which we find no example in the biological realm, they are given permanent expression in a formula which is repeated from mouth to mouth, transmitted by education, and fixed even in writing. Such is the origin and nature of legal and moral rules, popular aphorisms and proverbs, articles of faith wherein religious or political groups condense their beliefs, standards of taste established by literary schools, etc. None of these can be found entirely reproduced in the applications made of them by individuals, since they can exist even without being actually applied.

No doubt, this dissociation does not always manifest itself with equal distinctness, but its obvious existence in the important and numerous cases just cited is sufficient to prove that the social fact is a thing distinct from its individual manifestations. Moreover, even when this dissociation is not immediately apparent, it may often be disclosed by certain devices of method. Such dissociation is indispensable if one wishes to separate social facts from their alloys in order to observe them in a state of purity. Currents of opinion, with an intensity varying according to the time and place, impel certain groups either to more marriages, for example, or to more suicides, or to a higher or lower birthrate, etc. These currents are plainly social facts. At first sight they seem inseparable from the forms they take in individual cases. But statistics furnish us with the means of isolating them. They are, in fact, represented with considerable exactness by the rates of births, marriages, and suicides, that is, by the number obtained by dividing the average annual total of marriages, births, suicides, by the number of persons whose ages lie within the range in which marriages, births, and suicides occur.[2] Since each of these figures contains all the individual cases indiscriminately, the individual circumstances which may have had a share in the production of the phenomenon are neutralized and, consequently, do not contribute to its determination. The average, then, expresses a certain state of the group mind. . . .

Such are social phenomena, when disentangled from all foreign matter. As for their individual manifestations, these are indeed, to a certain extent, social, since they partly reproduce a social model. Each of them also depends, and to a large extent, on the organopsychological constitution of the individual and on the particular circumstances in which he is placed. Thus they are not sociological phenomena in the strict sense of the word. They belong to two realms at once; one could call them sociopsychological. They interest the sociologist without constituting the immediate subject matter of sociology. There exist in the interior of organisms similar phenomena, compound in their nature, which form in their turn the subject matter of the "hybrid sciences," such as physiological chemistry, for example. . . .

We thus arrive at the point where we can formulate and delimit in a precise way the domain of sociology. It comprises only a limited group of phenomena. A social fact is to be recognized by the power of external coercion which it exercises or is capable of exercising over individuals, and the presence of this power may be recognized in its turn either by the existence of some specific sanction or by the resistance offered against every individual effort that tends to violate it. One can, however, define it also by its diffusion within the group, provided that, in conformity with our previous remarks, one takes care to add as a second and essential characteristic that its own existence is independent of the individual forms it assumes in its diffusion. This last criterion is perhaps, in certain cases, easier to apply than the preceding one. In fact, the constraint is easy to ascertain when it expresses itself externally by some direct reaction of society, as is the case in law, morals, beliefs, customs, and even fashions. But when it is only indirect, like the constraint which an economic organization exercises, it cannot always be so easily detected. Generality combined with externality may, then, be easier to establish. Moreover, this second definition is but another form of the first; for if a mode of behavior whose existence is external to individual consciousness becomes general, this can only be brought about by its being imposed upon them.

But these several phenomena present the same characteristic by which we defined the others. These "ways of existing" are imposed on the individual precisely in the same fashion as the "ways of acting" of which we have spoken. Indeed, when we wish to know how a society is divided politically, of what these divisions themselves are composed, and how complete is the fusion existing between them, we shall not achieve our purpose by physical inspection and by geographical observations; for these phenomena are social, even when they have some basis in physical nature. It is only by a study of public law that a comprehension of this organization is possible, for it is this law that determines the organization, as it equally determines our domestic and civil relations. This political organization is, then, no less obligatory than the social facts mentioned above. If the population crowds into our cities instead of scattering into the country, this is due to a trend of public opinion, a collective drive that imposes this concentration upon the individuals. We can no more choose the style of our houses than of our clothing—at least, both are equally obligatory. The channels of communication prescribe the direction of internal migrations and commerce, etc., and even their extent. Consequently, at the very most, it should be necessary to add to the list of phenomena which we have enumerated as presenting the distinctive criterion of a social fact only one additional category, "ways of existing"; and, as this enumeration was not meant to be rigorously exhaustive, the addition would not be absolutely necessary.

Such an addition is perhaps not necessary, for these "ways of existing" are only crystallized "ways of acting." The political structure of a society is merely the way in which its component segments have become accustomed to live with one another. If their relations are traditionally intimate, the segments tend to fuse with one another, or, in the contrary case, to retain their identity. The type of habitation imposed upon us is merely the way in which our contemporaries and our ancestors have been accustomed to construct their houses. The methods of communication are merely the channels which the regular currents of commerce and migrations have dug, by flowing in the same direction. To be sure, if the phenomena of a structural character alone presented this permanence, one might believe that they constituted a distinct species. A legal regulation is an arrangement no less permanent than a type of architecture, and yet the regulation is a "physiological" fact. A simple moral maxim is assuredly somewhat more malleable, but it is much more rigid than a simple professional custom or a fashion. There is thus a whole series of degrees without a break in continuity between the facts of the most articulated structure and those free currents of social life which are not yet definitely molded. The differences between them are, therefore, only differences in the degree of consolidation they present. Both are simply life, more or less crystallized. No doubt, it may be of some advantage to reserve the term "morphological" for those social facts which concern the social substratum, but only on condition of not overlooking the fact that they are of the same nature as the others. Our definition will then include the whole relevant range of facts if we say: *A social fact is every way of acting, fixed or not, capable of exercising on the individual an external constraint*; or again, *every way of acting which is general throughout a given society, while at the same time existing in its own right independent of its individual manifestations.*[3]

NOTES

1. We do not intend to imply, however, that all constraint is normal.
2. Suicides do not occur at every age, and they take place with varying intensity at the different ages in which they occur.
3. This close connection between life and structure, organ and function, may be easily proved in sociology because between these two extreme terms there exists a whole series of immediately observable intermediate stages which show the bond between them. Biology is not in the same favorable position. But we may well believe that the inductions on this subject made by sociology are applicable to biology and that, in organisms as well as in societies, only differences in degree exist between these two orders of facts.

3

Doing Social Research

LEONARD CARGAN

According to this article, few people understand what social science actually accomplishes. This accusation may come as a surprise because most of us encounter the products of sociological research every day, from election polls to the positioning of television programs to attract certain audiences to filling out questionnaires (a college application) or having an interview for a job. Perhaps you even engaged in an experiment by trying out similar products in order to see which one was better. This being the case, the author explains what science consists of and in particular what the field of sociology studies.

As you read this selection, consider the following questions as guides:

1. *What areas does social science study?*
2. *What social problem would you study, and how would you go about studying it?*
3. *Do you think that a social scientist can be objective in dealing with a social issue? Consider your answer to question 2 when you answer this one.*

INTRODUCTION: THE SOCIAL SCIENTIFIC METHOD*

Most of the knowledge that we obtain comes through tradition—the culture that we inherit through our own personal experiences or through what we are told by other people. These types of information may be incomplete, incorrect, misleading, and affected by our own particular beliefs. For example, it is possible to hold beliefs that are contradictory in nature. However, no system of logic can prove any belief. A similar kind of error is seen in our personal experiences. Such experiences can lead to *overgeneralizations*—predicting a general pattern of behavior on the observation of a few similar events. This conclusion can, in turn, lead to *selective observation*—focusing on future events that fit the pattern and thereby confirming the original observation. Because selective observation is often used to support one's own biases, it is difficult to make correct decisions with information that is perhaps incorrect.

To overcome these possible inaccuracies, it is necessary to generate the probability that the information being received is in agreement with reality. This, then, is the role of social scientific research: to allow for the probability that information is correct within a predicted range of behavior. It is always within a predicted range of behavior because the

* All of the studies cited in this reading represent the earliest examples of each type of sociological methodology used for the study of social issues. These studies are considered classics because of their pioneering efforts and the subsequent validation of their findings.

social sciences use as their sample aggregates representing groups rather than the individuals themselves. For example, polls will estimate that a politician is leading in the election by 52 to 48 percent plus or minus 3. This prediction means that the range of predicted behavior will be found in 55–45 percent of the population.

Similarly, a brief visual inspection at a mall can determine the popularity of certain clothing styles, but a more accurate assessment can be made by questioning a random sample from a wider population or by doing a long-term systematic observation of shopping habits. To determine which boys in various groups are the leaders, one can ask the boys themselves, but this may lead to misinformation because many boys might nominate their friends as leaders. More accurate information can be obtained by conducting an experiment that records the boys' actions in leadership situations. As these examples indicate, the social scientific method is a means of making observations more deliberate and thereby reducing error. The role played by social research is to provide the means for constantly improving the reliability of the research received. It should be noted, however, that most social behavior changes with changing situations and thus social research cannot provide final answers. The reality of change means that science and especially the social sciences are an unfinished process.

Despite the changing nature of what is true in human behavior, the social sciences are based on the assumption that current human behavior is consistent, amenable to generalization, and predictable. This assumption allows for a more clear-cut definition of what the social sciences do: They apply scientific procedures to the systematic observation and recording of human behavior for the purpose of discovering answers to questions regarding that behavior. By collecting data scientifically rather than by chance, the collected material will be unbiased and logical. Because social research always concerns human behavior, it will always deal with different types of social systems: social relations, social roles, social groups, or societies.

TYPES OF INQUIRY

The overall types of social research are twofold: basic and applied. *Basic research* is designed to add to our knowledge and understanding of the social world, whereas *applied research* is designed to be useful and practical in the future. Social research projects can be of either type or a combination of both. An example of basic research is Emile Durkheim's classic study on suicide, first published in 1897. By examining available suicide statistics from several countries, Durkheim was able to eliminate many beliefs about the cause of suicide, such as bad weather and imitation of others, and to uncover such underlying factors as lack of social cohesion (egoism) and chaotic social systems (anomie) as well as the phenomenon of altruistic suicide.

An applied research study is designed to aid in dealing with a social problem such as prejudice or crime. To accomplish these goals, the social researcher can utilize distinct routes of investigation or several routes in combination. The research effort can be *evaluative*, establishing how successful a particular program is in achieving a desired goal. For example, the goal at one U.S. factory in the 1930s was to increase productivity. The researchers (Roethlisberger & Dickson, 1939) evaluated several experiments for attaining that goal: the degree of lighting was varied, various working schedules were implemented, and direct rewards were given for various productivity goals. In relatively new areas or unstudied topics, the investigative route may be *exploratory*, generating ideas for further research. The primary task in an exploratory effort is to produce analytical or conceptual generalizations that can be tested in further studies; the main emphasis here is on discovery. An example is Robert Bales's (1951) study of small groups in which he aimed to understand how such groups operated and whether they fit into certain patterns. Another investigative route is *description*, portraying accurately the composition of a group, various groups, the attitudes, behaviors, or structures of situations. Examples of descriptive research are seen in population studies and public opinion surveys such as the U.S. Census and election polls. A classic descriptive study of groups and situations is *Street Corner Society*, by William Foote Whyte (1943). Although similar to exploratory studies, the descriptive study draws on prior knowledge about the topic from other available sources, so the findings can be more accurate and precise.

Finally, the research effort may also be *explanatory*. In a sense, all of the investigative routes just discussed are explanatory in nature, but a pure explanatory study focuses on producing empirical generalizations and verifying hypotheses of possible causal relationships between variables. Thus, the study by Durkheim on suicide just noted also uses the exploratory route to uncover statistical variations among groups of individuals who committed suicide. Durkheim examined differences among religious groups, educational levels, the married and unmarried, men and women, the divorced, those not permitted a divorce, and the nondivorced, to name just a few. Interestingly, Durkheim was able to hypothesize three underlying conditions as an explanation for these various groups and their differences in committing suicide.

RESEARCH TECHNIQUES

As these varying routes of investigation indicate, the unit of analysis can include individuals, social groups, social organizations, and/or social artifacts. The unit of analysis can also include numerous types of aggregates, institutions, subcultures, and reoccurring episodes. An example of an aggregate study is the famous investigation by Kinsey and his colleagues (1948/1953), who interviewed thousands of individuals in various aggregates in order to understand sexual behavior in males and females. By statistically manipulating these individual descriptions, they were able to make generalizations about the social groups to which they belonged. An example of a study of a subculture is William Whyte's previously noted *Street Corner Society* (1943). To understand the subculture of gangs, their behavior, structure, and role in the community, Whyte participated in the subcultures of two gangs for over four years. A study of an organization is a study of the typographer's union by Seymour Martin Lipset and his colleagues (*Union Democracy*, 1956). This union was different from all such organizations in that it had a democratic structure; to understand why, the authors had to study all aspects of this organization. A study by Thomas and Znaniecki (1927), *The Polish Peasant in Europe and America*, is an excellent example of the use of social artifacts; the investigators drew on diaries, letters, and newspaper articles to understand the experiences of the Polish emigrant/immigrant in Poland and the United States.

Social scientists have developed several methodological means of collecting the data used in their analysis. To a large extent, the research method used to carry out the investigation depends on the researcher's view of the social world. If the social world is seen as an objective entity that can be described and understood, then such structured techniques as the questionnaire, alone or with an interviewer, will be used. These techniques allow for the quantitative rendering of social phenomena—the counting and measuring of items. During World War II, the army needed information very quickly on a number of items—for example, whether certain base locations would provide higher morale for the various racial groups and whether trainees were ready for combat. Over 500 simple questionnaires were used to collect this vast amount of data in quantifiable form, resulting in *The American Soldier: Adjustment during Army Life* (Stouffer et al., 1949). Kinsey and his colleagues also wanted to collect vast amounts of quantifiable data, but their data on *Sexual Behavior in the Human Male/Female* (1948/1953) was so intertwined with other social factors that it was necessary to use interviewers along with questionnaires in order to be able to explore this topic in depth.

On the other hand, the researcher may view the social world as unstructured and regard reality as not independent of human definition and redefinition. The data collected from this viewpoint are usually qualitative rather than quantitative descriptions because they concern the characteristics, concepts, definitions, descriptions, metaphors, or symbols of things and events. A major method used to collect this type of data is observation. When Roethlisberger and Dickson (*Management and the Worker,* 1939) wanted to find ways to increase productivity in a factory, they began by observing and recording the behavior of workers under several changing conditions. In William Whyte's study of the subcultures of gangs in an Italian neighborhood of Boston (*Street Corner Society*, 1943), his major means of collecting this type of data was direct participant observation. By actually participating in gang life, Whyte was able to understand the gangs' beliefs, values, and organization. A similar methodology is used in most ethnographic studies because it allows for a fuller picture of life in a smaller community. Sometimes it is necessary to conduct an experiment in order to collect this type of data. To understand the ins and outs of friendship, Sherif and his colleagues (*The Robbers Cave Experiment,* 1961) conducted a number of before and after situations for two groups of boys by changing relationships. A final example of a method used to collect this type of data is *The Polish Peasant in Europe and America* (1927), in which the authors used such unobtrusive materials as already available documents like letters and newspapers.

In giving examples of various techniques of social science research, the studies cited were listed under a particular methodology. In reality, most of these studies used several techniques to collect the data necessary for a full understanding of the situation. Multiple techniques are often necessary in order to collect all of the data on an issue because each technique has both strengths and weaknesses in its application. In fact, using several types of research methodology in a scientific enquiry is usually better than using any single one.

THE DIFFICULTIES

Given the complexity of the social world, it is not surprising that social research is not marked by one goal, theoretical model, or methodology. It is also not surprising that regardless of the factors used, social research must follow various scientific criteria in collecting data. This means that the technique used must be an objective, logical, and systematic method of analysis so that the data accumulated consist of reliable knowledge.

Being *objective* indicates that one's private opinions cannot influence the research operation. This means that all aspects of the research design and interpretations of the data collected should be devoid of personal whims and biases. Being objective also means being skeptical of your own findings by checking them out or by submitting the results for scrutiny by others. It is this last factor that confirms the law of objectivity in the social sciences—identical findings should result if two researchers are studying the same phenomenon using the same techniques. Unfortunately, a number of human tendencies influence the way a person sees the world and may thereby affect our objectivity.

A major culprit here is *socialization*—we are all socialized with the values of our culture and of the groups in which we are members. These values act as editors on incoming information and even past information. This effect is easily confirmed if you think about your beliefs concerning family life, religion, country, and other such personal factors. Another factor affecting objectivity is *identification* with the group being studied because the group has similar values to your own. This identity with the group being studied may lead to a misunderstanding of the responses being received, a defense of their responses, a tendency to fill in gaps in information received with what you believe is true, and a difficulty in understanding correctly what is happening. This latter form of bias is particularly true when studying a past time period.

Of course, some biases are beyond the researcher's control if he or she is receiving intentional or unintentional false information. It should also be recognized that your values are even reflected in the selection of a research topic. All these factors taken together mean that it is very difficult to be totally objective and that it would be inane to claim otherwise. Therefore, you should do the best you can to be "objective" by taking extra care in your research efforts.

The next two goals are those of being logical and systematic. The accepted rules of reasoning indicate that being logical and systematic implies internal consistency in the presentation of your data—that is, that there is a consistent arrangement of facts and their interpretation. This means, for example, that your theory and hypotheses are interrelated with each other and are corroboratory. As with being objective, you may find difficulties in this regard because many social concepts are abstract. This being the case, there may be little agreement on how to interpret and measure concepts such as alienation, anomie, and morale.

The final goal of scientific social research is reliability. That is, the methods used to collect the data should produce reliable information. To be reliable means, in turn, that the information obtained can be depended upon in terms of predictability. According to many social scientists, the best means for meeting the criterion of predictability is that the study can be replicated—it can be repeated over and over again with the same results obtained. Unfortunately, although in the physical sciences it is indeed possible to recreate virtually identical conditions, in the social sciences it is often difficult to do so with human subjects. For example, how long it takes for an item to fall to the ground can be determined from such measurements as weight, wind velocity, gravity, and other similar measurements. Social scientists, however, are typically interested in measuring conditions of human behavior that involve abstract concepts such as alienation and anomie and even such commonly used concepts as happiness. Social scientists have difficulty in agreeing on how best to measure these concepts because the meaning often cannot be separated from the social situation being studied.

Another difficulty in dealing with human subjects is the need to be *ethical* by following a code of conduct during the research that involves informed consent from voluntary participants, never injuring participants in the research by safeguarding their concerns and interests via assurances of anonymity and confidentiality, and being honest with the scientific community in analysis of the data collected and reported. Unfortunately, the specific goal of the research may intrude upon one or more of these issues. For example, it would not have been possible for Stanley Milgram (1974) to conduct his famous experiment on obedience to commands to conduct cruel acts had the sample been informed that this was the goal. Similarly, it would not have been possible for Laud Humpreys (1970) to conduct his observations of gay sexual behavior had he informed the participants that this was his goal. Nevertheless, both of these studies were criticized by the scientific community for causing possible harm to the people taking part in the study. The difficulties in dealing with humans requires social scientists to be more skeptical about their findings, which in turn encourages closer examination and testing of the findings.

CONCLUSION

It has been suggested that scientific inquiry is governed by three interacting motives: curiosity (the desire to know more about a subject of concern to humans); intrinsic orderliness (the desire to reduce masses of data to a comprehensible order); and practicality (the desire that one's efforts will be useful) (Chein, Cook, & Harding, 1948). As shown in the referent studies in this paper, these desires have led to inquiries into suicide, understanding the Polish emigrants, finding out the needs of army inductives, uncovering the *sexual behavior of both males and females*, discovering how small groups are organized and operate, and by conducting experiments in an effort to improve factory productivity.

As also was seen, social scientists have a number of tools at their disposal and, despite the difficulties in dealing with human subjects, can produce helpful information.

REFERENCES

Bales, R. A. *Interaction Process Analysis: A Process for the Study of Small Groups*. Reading, MA: Addison-Wesley, 1951.

Chein, I., S. W. Cook, and J. Harding. The Field of Action Research. *American Psychologist* 3 (1948), pp. 43–50.

Durkheim, E. *Suicide: A Study in Sociology*. Trans. J. A. Spaulding and G. Simpson. New York: Free Press, 1964.

Humphreys, L. *Tearoom Trade: Impersonal Sex in Public Places*. Chicago: Aldine, 1970.

Kinsey, A. C., W. B. Pomeroy, & C. L. Martin. *Sexual Behavior in the Human Male/Human Female*. Philadelphia: Saunders, 1948/1953.

Lipset, S. M., M. Trow, & J. S. Coleman. *Union Democracy: the Internal Politics of the International Typographical Union*. Garden City, NY: Doubleday, 1956.

Milgram, S. *Obedience to Authority*. New York: Harper & Row, 1974.

Roethlisberger, F. J., & W. J. Dickson. *Management and the Worker*. Cambridge, MA: Harvard University Press, 1939.

Stouffer, S. A., A. A. Lumsdaine, M. H. Lumsdaine, R. M. Williams, Jr., M. B. Smith, I. L. Janis, S. A. Star, and L. S. Cottrell, Jr. *The American Soldier: Adjustment during Army Life*. Princeton, New Jersey: Princeton University Press, vol. 2, 1949.

Sherif, M., O. J. Harvey, B. J. White, W. R. Hurd, & C. W. Sherif. *Intergroup Conflict and Cooperation: The Robbers Cave Experiment*. Norman, OK: University of Oklahoma Institute of Group Relations, 1961.

Thomas, W. I. & F. Znaniecki. *The Polish Peasant in Europe and America*. New York: Knopf, 1927.

Whyte, W. F. *Street Corner Society: The Social Structure of an Italian Slum*. Chicago: University of Chicago Press, 1943.

4

The Practice of Social Research

EARL BABBIE

The fourth article in the introductory section explains the major theories used in sociology. These theories are the foundation for a scientific approach to the study of social life and together with research methods provide the tools for scientific investigation.

As you read, ask yourself the following questions:

1. *What are the different paradigms used by sociologists in their examination of social data?*
2. *Which of the paradigms noted by Babbie seems most revealing of social behavior? Why?*
3. *What is the relationship between the prior article on doing research and this article?*

GLOSSARY **Theories** A set of logically interrelated statements that attempt to explain or predict social events. **Paradigm** An example, model, or pattern.

INTRODUCTION

There are restaurants in the United States fond of conducting political polls among their diners whenever an election is in the offing. Some take these polls very seriously because of their uncanny history of predicting winners. Some movie theaters have achieved similar success by offering popcorn in bags picturing either donkeys or elephants. Years ago, granaries in the Midwest offered farmers a chance to indicate their political preferences through the bags of grain they selected.

Such idiosyncratic ways of determining trends, though interesting, all follow the same pattern over time: They work for a while, and then they fail. Moreover, we can't predict when or why they will fail.

These unusual polling techniques point to a significant shortcoming of "research findings" based only on the observation of patterns. Unless we can offer logical explanations for such patterns, the regularities we've observed may be mere flukes, chance occurrences. If you flip coins long enough, you'll get ten heads in a row. Scientists might adapt a street expression to describe this situation: "Patterns happen."

Logical explanations are what theories seek to provide. Theories function three ways in research. First, they prevent our being taken in by flukes. If we can't explain why Ma's Diner has been so successful in predicting elections, we run the risk of supporting a fluke. If we know why it has happened, we can anticipate whether or not it will work in the future.

Second, theories make sense of observed patterns in a way that can suggest other possibilities. If we understand the reasons why broken homes produce more juvenile delinquency than do intact homes—lack of supervision, for example—we can take effective action, such as after-school youth programs.

Finally, theories shape and direct research efforts, pointing toward likely discoveries through empirical observation. If you were looking for your lost keys on a dark street, you could whip your flashlight around randomly, hoping to chance upon the errant keys—or you could use your memory of where you had been to limit your search to more likely areas. Theories, by analogy, direct researchers' flashlights where they are most likely to observe interesting patterns of social life. . . .

This reading explores some specific ways theory and research work hand in hand during the adventure of inquiry into social life. We'll begin by looking at some fundamental frames of reference, called *paradigms,* that underlie social theories and inquiry.

SOME SOCIAL SCIENCE PARADIGMS . . .

Macrotheory and Microtheory

Let's begin with a difference concerning focus that stretches across many of the paradigms we'll discuss. Some social theorists focus their attention on society at large, or at least on large portions of it. Topics of study for such **macrotheory** include the struggle between economic classes in a society, international relations, or the interrelations among major institutions in society, such as government, religion, and family. Macrotheory deals with large, aggregate entities of society or even whole societies.

Some scholars have taken a more intimate view of social life. **Microtheory** deals with issues of social life at the level of individuals and small groups. Dating behavior, jury deliberations, and student-faculty interactions are apt subjects for a microtheoretical perspective. Such studies often come close to the realm of psychology, but whereas psychologists typically focus on what goes on inside humans, social scientists study what goes on between them. . . .

Early Positivism

When the French philosopher Auguste Comte (1798–1857) coined the term *sociologie* in 1822, he launched an intellectual adventure that is still unfolding today. Most importantly, Comte identified society as a phenomenon that can be studied scientifically. . . .

Prior to Comte's time, society simply was. To the extent that people recognized different kinds of societies or changes in society over time, religious paradigms generally predominated in explanations of such differences. The state of social affairs was often seen as a reflection of God's will. Alternatively, people were challenged to create a "City of God" on earth to replace sin and godlessness.

Comte separated his inquiry from religion. He felt that religious belief could be replaced with scientific study and objectivity. His "positive philosophy" postulated three stages of history. A "theological stage" predominated throughout the world until about 1300. During the next five hundred years, a "metaphysical stage" replaced God with philosophical ideas such as "nature" and "natural law."

Comte felt he was launching the third stage of history, in which science would replace religion and metaphysics by basing knowledge on observations through the five senses rather than on belief or logic alone. Comte felt that society could be observed and then explained logically and rationally and that sociology could be as scientific as biology or physics. . . .

Social Darwinism . . .

In 1858, when Charles Darwin published his *The Origin of Species,* he set forth the idea of evolution through the process of natural selection. Simply put, the theory states that as a species coped with its environment, those individuals most suited to success would be the most likely to survive long enough to reproduce. Those less well suited would perish. Over time the traits of the survivor would come to dominate the species. As later Darwinians put it, species evolved into different forms through the "survival of the fittest."

As scholars began to study society analytically, it was perhaps inevitable that they would apply Darwin's ideas to changes in the structure of human affairs. The journey from simple hunting-and-gathering tribes to large, industrial civilizations was easily seen as the evolution of progressively "fitter" forms of society. . . .

Conflict Paradigm

. . . Karl Marx (1818–1883) suggested that social behavior could best be seen as the process of conflict: the attempt to dominate others and to avoid being dominated. Marx focused primarily on the struggle among economic classes. Specifically, he examined the way capitalism produced the oppression of workers by the owners of industry. . . .

The conflict paradigm proved to be fruitful outside the realm of purely economic analyses. Georg Simmel (1858–1918) was especially interested in small-scale conflict, in contrast to the class struggle that interested Marx. Simmel noted, for example, that conflicts among members of a tightly knit group tended to be more intense than those among people who did not share feelings of belonging and intimacy.

In a more recent application of the conflict paradigm, when Michael Chossudovsky's (1997) analysis of the International Monetary Fund and World Bank suggested that these two international organizations were increasing global poverty rather than eradicating it, he directed his attention to the competing interests involved in the process. In theory, the chief interest being served should be the poor people of the world or perhaps the impoverished, Third-World nations. The researcher's inquiry, however, identified many other interested parties who benefited: the commercial lending institutions who made loans in conjunction with the IMF and World Bank and multinational corporations seeking cheap labor and markets for their goods, for example. Chossudovsky's analysis concluded that the interests of the banks and corporations tended to take precedence over those of the poor people, who were the intended beneficiaries. Moreover, he found many policies were weakening national economies in the Third World, as well as undermining democratic governments.

Whereas the conflict paradigm often focuses on class, gender, and ethnic struggles, it would be appropriate to apply it whenever different groups have competing interests. For example, it could be fruitfully applied to understanding relations among different departments in an organization, fraternity and sorority rush weeks, or student-faculty-administrative relations, to name just a few.

Symbolic Interactionism . . .

Simmel was one of the first European sociologists to influence the development of U.S. sociology. His focus on the nature of interactions particularly influenced George Herbert Mead (1863–1931), Charles Horton Cooley (1864–1929), and others who took up the cause and developed it into a powerful paradigm for research.

Cooley, for example, introduced the idea of the "primary group," those intimate associates with whom we share a sense of belonging, such as our family, friends, and so forth. Cooley also wrote of the "looking-glass self" we form by looking into the reactions of people around us. If everyone treats us as beautiful, for example, we conclude that we are. Notice how fundamentally the concepts and theoretical focus inspired by this paradigm differ from the society-level concerns of . . . Marx.

Mead emphasized the importance of our human ability to "take the role of the other," imagining how others feel and how they might behave in certain circumstances. As we gain an idea of how people in general see things, we develop a sense of what Mead called the "generalized other."

Mead also showed a special interest in the role of communications in human affairs. Most interactions, he felt, revolved around the process of individuals reaching common understanding through the use of language and other such systems, hence the term *symbolic interactionism*.

This paradigm can lend insights into the nature of interactions in ordinary social life, but it can also help us understand unusual forms of interaction. . . .

Structural Functionalism

Structural functionalism, sometimes also known as "social systems theory," grows out of a notion introduced by Comte and Spencer: A social entity, such as an organization or a whole society, can be viewed as an organism. Like other organisms, a social system is made up of parts, each of which contributes to the functioning of the whole.

By analogy, consider the human body. Each component—such as the heart, lungs, kidneys, skin, and brain—has a particular job to do. The body as a whole cannot survive unless each of these parts does its job, and none of the parts can survive except as a part of the whole body. Or consider an automobile. It is composed of the tires, the steering wheel, the gas tank, the spark plugs, and so forth. Each of the parts serves a function for the whole; taken together, that system can get us across town. None of the individual parts would be very useful to us by itself, however.

The view of society as a social system, then, looks for the "functions" served by its various components. Social scientists using the structural functional paradigm might note that the function of the police, for example, is to exercise social control—encouraging people to abide by the norms of society and bringing to justice those who do not. Notice, though, that they could just as reasonably ask what functions criminals serve in society. Within the functionalist paradigm, we might say that criminals serve as job security for the police. In a related observation, Emile Durkheim (1858–1917) suggested that crimes and their punishment provide an opportunity to reaffirm society's values. By catching and punishing thieves, we reaffirm our collective respect for private property.

To get a sense of the structural-functional paradigm, suppose you were interested in explaining how your college or university works. You might thumb through the institution's catalog and begin assembling a list of the administrators and support staff (such as president, deans, registrar, campus security, maintenance personnel). Then you might figure out what each of them does and relate their roles and activities to the chief functions of your college or university, such as teaching or research. This way of looking at an institution of higher learning would clearly suggest a different line of inquiry than, say, a conflict paradigm, which might emphasize the clash of interests between people who have power in the institution and those who don't.

People often discuss "functions" in everyday conversations. Typically, however, the alleged functions are seldom tested empirically. Some people argue, for example, that welfare, intended to help the poor, actually harms them in a variety of ways. It is sometimes alleged that welfare creates a deviant, violent subculture in society, at odds with the mainstream. From this viewpoint, welfare programs actually result in increased crime rates.

Lance Hannon and James Defronzo (1998) decided to test this last assertion. Working with data drawn from 406 urban counties in the United States, they examined the relationship between levels of welfare payments and crime rates. Contrary to the beliefs of some, their data indicated that higher welfare payments were associated with lower crime rates. In other words, welfare programs have the function of decreasing rather than increasing lawlessness.

Feminist Paradigms . . .

When feminists first began questioning the use of masculine pronouns and nouns whenever gender was ambiguous, their concerns were often viewed as petty, even silly. At most, many felt the issue was one of women having their feelings hurt, their egos bruised . . .

In a similar way, researchers looking at the social world from a feminist paradigm have called attention to aspects of social life that are not revealed by other paradigms. In part, feminist theory and research have focused on gender differences and how they relate to the rest of social organization. These lines of inquiry have drawn attention to the oppression of women in many societies, which in turn has shed light on oppression generally.

Feminist paradigms have also challenged the prevailing notions concerning consensus in society. Most descriptions of the predominant beliefs, values, and norms of a society are written by people representing only portions of society. In the United States, for example, such analyses have typically been written by middle-class white men—not surprisingly, they have written about the beliefs, values, and norms they themselves share. Though George Herbert Mead spoke of the "generalized other" that each of us becomes aware of and can "take the role of," feminist paradigms question whether such a generalized other even exists.

Further, whereas Mead used the example of learning to play baseball to illustrate how we learn about the generalized other, Janet Lever's research suggests that understanding the experience of boys may tell us little about girls.

> Girls' play and games are very different. They are mostly spontaneous, imaginative, and free of structure or rules. Turn-taking activities like jumprope may be played without setting explicit goals. Girls have far less experience with interpersonal competition. The style of their competition is indirect, rather than face to face, individual rather than team affiliated. Leadership roles are either missing or randomly filled.

Social researchers' growing recognition of the general intellectual differences between men and women led the psychologist Mary Field Belenky and her colleagues to speak of *Women's Ways of Knowing*. In-depth interviews with 45 women led the researchers to distinguish five perspectives on knowing that should challenge the view of inquiry as obvious and straightforward:

> *Silence:* Some women, especially early in life, feel themselves isolated from the world of knowledge, their lives largely determined by external authorities.
>
> *Received knowledge:* From this perspective, women feel themselves capable of taking in and holding knowledge originating with external authorities.
>
> *Subjective knowledge:* This perspective opens up the possibility of personal, subjective knowledge, including intuition.
>
> *Procedural knowledge:* Some women feel they have mastered the ways of gaining knowledge through objective procedures.
>
> *Constructed knowledge:* The authors describe this perspective as "a position in which women view all knowledge as contextual, experience themselves as creators of knowledge, and value both subjective and objective strategies for knowing."

"Constructed knowledge" is particularly interesting in the context of paradigms. The positivistic paradigm of Comte would have a place neither for "subjective knowledge" nor for the idea that truth might vary according to its context. The ethnomethodological paradigm, on the other hand, would accommodate these ideas.

Rational Objectivity Reconsidered

We began this discussion of paradigms with Comte's assertion that society can be studied rationally and objectively. Since his time, the growth of science and technology, together with the relative decline of superstition, have put rationality more and more in the center of social life. As fundamental as rationality is to most of us, however, some contemporary scholars have raised questions about it.

For example, positivistic social scientists have sometimes erred in assuming that social reality can be explained in rational terms because humans always act rationally. I'm sure your own experience offers ample evidence to the contrary. Yet many modern economic models fundamentally assume that people will make rational choices in the economic sector: They will choose the highest-paying job, pay the lowest price, and so forth. This assumption ignores the power of tradition, loyalty, image, and other factors that compete with reason and calculation in determining human behavior.

A more sophisticated positivism would assert that we can rationally understand and predict even nonrational behavior. An example is the famous "Asch Experiment." In this experiment, a group of subjects is presented with a set of lines on a screen and asked to identify the two lines that are equal in length.

Imagine yourself a subject in such an experiment. You are sitting in the front row of a classroom in a group of six subjects. A set of lines is projected on the wall in front of you. The experiment asks each of you, one at a time, to identify the line to the right (A, B, or C) that matches the length of line X. The correct answer (B) is pretty obvious to you. To your surprise, however, you find that all the other subjects agree on a different answer!

The experimenter announces that all but one of the group has gotten the correct answer. Since you are the only one who chose B, this amounts to saying that you've gotten it wrong. Then a new set of lines is presented, and you have the same experience. What seems to be the obviously correct answer is said by everyone else to be wrong.

As it turns out, of course, you are the only real subject in this experiment—all the others are working with the experimenter. The purpose of the experiment is to see whether you will be swayed by public pressure to go along with the incorrect answer. In his initial experiments, all of which involved young men, Asch found that a little over one-third of his subjects did just that.

Choosing an obviously wrong answer in a simple experiment is an example of nonrational behavior. But as Asch went on to show, experimenters can examine the circumstances that lead more or fewer subjects to go along with the incorrect answer. For example, in subsequent studies, Asch varied the size of one group and the number of "dissenters" who chose the "wrong" (that is, the correct) answer. Thus, it is possible to study nonrational behavior rationally and scientifically.

More radically, we can question whether social life abides by rational principles at all. . . .

The contemporary challenge to positivism, however, goes beyond the question of whether people behave rationally. In part, the criticism of positivism challenges the idea that scientists can be as objective as the positivistic ideal assumes. Most scientists would agree that personal feelings can and do influence the problems scientists choose to study, what they choose to observe, and the conclusions they draw from their observations.

There is an even more radical critique of the ideal of objectivity. As we glimpsed in the discussions of feminism and ethnomethodology, some contemporary researchers suggest that subjectivity might actually be preferable in some situations. . . .

To begin, all our experiences are inescapably subjective. There is no way out. We can see only through our own eyes, and anything peculiar to our eyes will shape what we see. We can hear things only the way our particular ears and brain transmit and interpret sound waves. You and I, to some extent, hear and see different realities. And both of us experience quite different physical "realities" than, say . . . scientists on the planet Xandu who might develop theories of the physical world based on a sensory apparatus that we humans can't even imagine. Maybe they see X rays or hear colors.

Despite the inescapable subjectivity of our experience, we humans seem to be wired to seek an agreement on what is really real, what is objectively so. Objectivity is a conceptual attempt to get beyond our individual views. It is ultimately a matter of communication, as you and I attempt to find a common ground in our subjective experiences. Whenever we succeed in our search, we say we are dealing with objective reality. . . .

Whereas our subjectivity is individual, our search for objectivity is social. This is true in all aspects of life, not just in science. While you and I prefer different foods, we must agree to some extent on what is fit to eat and what is not, or else there could be no restaurants or grocery stores. The same argument could be made regarding every other form of consumption. Without agreement reality, there could be no movies or television, no sports.

Social scientists as well have found benefits in the concept of a socially agreed-upon objective reality. As people seek to impose order on their experience of life, they find it useful to pursue this goal as a collective venture. What are the causes and cures of prejudice? Working together, social researchers have uncovered some answers that hold up to intersubjective scrutiny. Whatever your subjective experience of things, for example, you can discover for yourself that as education increases, prejudice generally tends to decrease. Because each of us can discover this independently, we say that it is objectively true. . . .

Some say that the ideal of objectivity conceals as much as it reveals. As we saw earlier in years past much of what was regarded as objectivity in Western social science was actually an agreement primarily among white, middle-class European men. Equally real experiences common to women, to ethnic minorities, to non-Western cultures, or to the poor were not necessarily represented in that reality . . . .

Ultimately, we will never be able to distinguish completely between an objective reality and our subjective experience. We cannot know whether our concepts correspond to an objective reality or are simply useful in allowing us to predict and control our environment. So desperate is our need to know what is really real, however, that both positivists and postmodernists are sometimes drawn into the belief that their own view is real and true. There is a dual irony in this. On the one hand, the positivist's belief that science precisely mirrors the objective world must ultimately be based on faith; it cannot be proven by "objective" science since that's precisely what's at issue. And the postmodernists, who say nothing is objectively so and everything is ultimately subjective, do at least feel that that is really the way things are.

Fortunately, as social researchers we are not forced to align ourselves entirely with either of these approaches. Instead, we can treat them as two distinct arrows in our quiver. Each approach compensates for the weaknesses of the other by suggesting complementary perspectives that can produce useful lines of inquiry.

In summary, a rich variety of theoretical paradigms can be brought to bear on the study of social life. With each of these fundamental frames of reference, useful theories can be constructed. . . .

5

The Promise

C. WRIGHT MILLS

Here Mills completes the task of defining what the sociologist does, how he or she does it, and how all subjects of human behavior are in the sociological purview. Mills does this by indicating what the sociological mind helps individuals accomplish—what is going on in the world and happening to themselves.

As you read, ask yourself the following questions:

1. *What does the sociological imagination allow people to accomplish?*
2. *What changes in your personal milieu might be better understood by looking at changes in the society?*
3. *How are the prior articles in this section related to the promise of sociology?*

GLOSSARY **Sociological imagination** The capacity to understand the most impersonal and remote changes in terms of their effect on the human self and to see the relationship between the two. **Personal trouble** A private matter that occurs within the character of an individual and within the range of that individual's immediate relations with others. **Public issue** A matter that transcends the local environment of an individual and the range of that individual's inner life.

Nowadays men* often feel their private lives are a series of traps. They sense that within their everyday worlds, they cannot overcome their troubles, and in this feeling, they are often quite correct: What ordinary men are directly aware of and what they try to do are bounded by the private orbits in which they live; their visions and their powers are limited to the close-up scenes of job, family, neighborhood; in other milieux, they move vicariously and remain spectators. And the more aware they become, however vaguely, of ambitions and of threats which transcend their immediate locales, the more trapped they seem to feel.

Underlying this sense of being trapped are seemingly impersonal changes in the very structure of continent-wide societies. The facts of contemporary history are also facts about the success and the failure of individual men and women. When a society is industrialized, a peasant becomes a worker; a feudal lord is liquidated or becomes a businessman. When classes rise or fall, a man is employed or unemployed; when the rate of investment goes up or down, a man takes new heart or goes broke. When wars happen, an insurance salesman becomes a rocket launcher; a store clerk, a radar man; a wife lives alone; a child grows up without a father. Neither the life of

* This reading was written before it became correct to use gender-neutral terms. Male nouns and pronouns should be understood to refer to males and females alike.

an individual nor the history of a society can be understood without understanding both.

Yet men do not usually define the troubles they endure in terms of historical change and institutional contradiction. The well-being they enjoy, they do not usually impute to the big ups and downs of the societies in which they live. Seldom aware of the intricate connection between the patterns of their own lives and the course of world history, ordinary men do not usually know what this connection means for the kinds of men they are becoming and for the kinds of history-making in which they might take part. They do not possess the quality of mind essential to grasp the interplay of man and society, of biography and history, of self and world. They cannot cope with their personal troubles in such ways as to control the structural transformations that usually lie behind them.

Surely it is no wonder. In what period have so many men been so totally exposed at so fast a pace to such earthquakes of change? That Americans have not known such catastrophic changes as have the men and women of other societies is due to historical facts that are now quickly becoming "merely history." The history that now affects every man is world history. Within this scene and this period, in the course of a single generation, one-sixth of mankind is transformed from all that is feudal and backward into all that is modern, advanced, and fearful. Political colonies are freed; new and less visible forms of imperialism installed. Revolutions occur; men feel the intimate grip of new kinds of authority. Totalitarian societies rise, and are smashed to bits—or succeed fabulously. After two centuries of ascendancy, capitalism is shown up as only one way to make society into an industrial apparatus. After two centuries of hope, even formal democracy is restricted to a quite small portion of mankind. Everywhere in the underdeveloped world, ancient ways of life are broken up, and vague expectations become urgent demands. Everywhere in the overdeveloped world, the means of authority and of violence become total in scope and bureaucratic in form. Humanity itself now lies before us, the super-nation at either pole concentrating its most coordinated and massive efforts upon the preparation of World War III.

The very shaping of history now outpaces the ability of men to orient themselves in accordance with cherished values. And which values? Even when they do not panic, men often sense that older ways of feeling and thinking have collapsed and that newer beginnings are ambiguous to the point of moral stasis. Is it any wonder that ordinary men feel they cannot cope with the larger worlds with which they are so suddenly confronted? That they cannot understand the meaning of their epoch for their own lives? That—in defense of selfhood—they become morally insensible, trying to remain altogether private men? Is it any wonder that they come to be possessed by a sense of the trap?

It is not only information that they need—in this Age of Fact, information often dominates their attention and overwhelms their capacities to assimilate it. It is not only the skills of reason that they need—although their struggles to acquire these often exhaust their limited moral energy.

What they need, and what they feel they need, is a quality of mind that will help them to use information and to develop reason in order to achieve lucid summations of what is going on in the world and of what may be happening within themselves. It is this quality, I am going to contend, that journalists and scholars, artists and publics, scientists and editors are coming to expect of what may be called the sociological imagination.

I

The sociological imagination enables its possessor to understand the larger historical scene in terms of its meaning for the inner life and the external career of a variety of individuals. It enables him to take into account how individuals, in the welter of their daily experience, often become falsely conscious of their social positions. Within that welter, the framework of modern society is sought, and within that framework the psychologies of a variety of men and women are formulated. By such means the personal uneasiness of individuals is focused upon explicit troubles and the indifference of publics is transformed into involvement with public issues.

The first fruit of this imagination—and the first lesson of the social science that embodies it—is the idea that the individual can understand his own experience and gauge his own fate by locating himself within his period; that he can know his own chances in life only by becoming aware of those of all individuals in his circumstances. In many ways it is a terrible lesson; in many ways a magnificent one. We do not know the limits of man's capacities for supreme effort or willing degradation, for agony or glee, for pleasurable brutality or the sweetness of reason. But in our time we have come to know the limits of "human nature" are frighteningly broad. We have come to know that every individual lives, from one generation to the next, in some society; that he lives out a biography, and that he lives it out within some historical sequence. By the fact of his living he contributes, however minutely, to the shaping of this society and to the course of its history, even as he is made by society and by its historical push and shove.

The sociological imagination enables us to grasp history and biography and the relations between the two within society. That is its task and its promise. To recognize this task and this promise is the mark of the classic social analyst. . . . And it is the signal of what is best in contemporary studies of man and society.

No social study that does not come back to the problems of biography, of history, and of their intersections within a society has completed its intellectual journey. Whatever the specific problems of the classic social analysts, however limited or however broad the features of social reality they have examined, those who have been imaginatively aware of the promise of their work have consistently asked three sorts of questions:

1. What is the structure of this particular society as a whole? What are its essential components, and how are they related to one another? How does it differ from other varieties of social order? Within it, what is the meaning of any particular feature for its continuance and for its change?
2. Where does this society stand in human history? What are the mechanics by which it is changing? What is its place within and its meaning for the development of humanity as a whole? How does any particular feature we are examining affect, and how is it affected by, the historical period in which it moves? And this period—what are its essential features? How does it differ from other periods? What are its characteristic ways of history-making?
3. What varieties of men and women now prevail in this society and in this period? And what varieties are coming to prevail? In what ways are they selected and formed, liberated and repressed, made sensitive and blunted? What kinds of "human nature" are revealed in the conduct and character we observe in this society in this period? And what is the meaning of "human nature" of each and every feature of the society we are examining?

Whether the point of interest is a great power state or a minor literary mood, a family, a prison, a creed—these are the kinds of questions the best social analysts have asked. They are the intellectual pivots of classic studies of man in society—and they are the questions inevitably raised by any mind possessing the sociological imagination. For that imagination is the capacity to shift from one perspective to another—from the political to the psychological; from examination of a single family to comparative assessment of the national budgets of the world; from the theological school to the military establishment; from considerations of an oil industry to studies of contemporary poetry. It is the capacity to range from the most impersonal and remote transformations to the most intimate features of the human self—and to see the relations between the two. Back of its use there is always the urge to know the social and historical meaning of the individual in the society and in the period in which he has his quality and his being.

That, in brief, is why it is by means of the sociological imagination that men now hope to grasp what is going on in the world, and to understand what is happening in themselves as minute points of the intersections of biography and history within society. In large part, contemporary man's self-conscious view of himself as at least an outsider, if not a permanent stranger, rests upon an absorbed realization of social relativity and of the transformative power of history. The sociological imagination is the most fruitful form of this self-consciousness. By its use men whose mentalities have swept only a series of limited orbits often come to feel as if suddenly awakened in a house with which they had only supposed themselves to be familiar. Correctly or incorrectly, they often come to feel that they can now provide themselves with adequate summations, cohesive assessments, comprehensive orientations. Older decisions that once appeared sound now seem to them products of a mind unaccountably dense. Their capacity for astonishment is made lively again. They acquire a new way of thinking, they experience a transvaluation of values; in a word, by their reflection and by their sensibility, they realize the cultural meanings of the social sciences.

II

Perhaps the most fruitful distinction with which the sociological imagination works is between "the personal troubles of milieu" and "the public issues of social structure." This distinction is an essential tool of the sociological imagination and a feature of all classic work in social science.

Troubles occur within the character of the individual and within the range of his immediate relations with others; they have to do with his self and with those limited areas of social life of which he is directly and personally aware. Accordingly, the statement and the resolution of troubles properly lie within the individual as a biographical entity and within the scope of his immediate milieu—the social setting that is directly open to his personal experience and to some extent his willful activity. A trouble is a private matter: values cherished by an individual are felt by him to be threatened.

Issues have to do with matters that transcend these local environments of the individual and the range of his inner life. They have to do with the organization of many such milieux into the institutions of an historical society as a whole, with the ways in which various milieux overlap and interpenetrate to form the larger structure of social and historical life. An issue is a public matter: some value cherished by publics is felt to be threatened. Often there is a debate about what that value really is and about what it is that really threatens it. This debate is often without focus if only because it is the very nature of an issue, unlike even widespread trouble, that it cannot very well be defined in terms of the immediate and everyday environment of ordinary men. An issue, in fact, often involves a crisis in institutional arrangements, and often too it involves what Marxists call "contradictions" or "antagonisms."

In these terms, consider unemployment. When, in a city of 100,000, only one man is unemployed, that is his personal trouble, and for its relief we properly look to the character of the man, his skills, and his immediate opportunities. But when in a nation of 50 million employees, 15 million men are unemployed, that is an issue, and we may not hope to find its solution within the range of opportunities open to any one individual. The very structure of opportunities has collapsed. Both the correct statement of the problem and the range of possible solutions require us to consider the economic and political institutions of the society, and not merely the personal situation and character of a scatter of individuals.

Consider war. The personal problem of war, when it occurs, may be how to survive it or how to die in it with honor; how to make money out of it; how to climb into the higher safety of the military apparatus; or how to contribute to the war's termination. In short, according to one's values, to find a set of milieux and within it to survive the war or make one's death in it meaningful. But the structural issues of war have to do with its causes; with what types of men it throws up into command; with its effects upon economic and political, family and religious institutions, with the unorganized irresponsibility of a world of nation-states.

Consider marriage. Inside a marriage a man and a woman may experience personal troubles, but when the divorce rate during the first four years of marriage is 250 out of every 1000 attempts, this is an indication of a structural issue having to do with the institutions of marriage and the family and other institutions that bear upon them.

Or consider the metropolis—the horrible, beautiful, ugly, magnificent sprawl of the great city. For many upper-class people, the personal solution to "the problem of the city" is to have an apartment with a private garage under it in the heart of the city, and forty miles out, a house by Henry Hill, garden by Garrett Eckbo, on a hundred acres of private land. In these two controlled environments—with a small staff at each end and a private helicopter connection—most people could solve many of the problems of personal milieux caused by the facts of the city. But all this, however splendid, does not solve the public issues that the structural fact of the city poses. What should be done with this wonderful monstrosity? Break it all up into scattered units, combining residence and work? Refurbish it as it stands? Or, after evacuation, dynamite it and build new cities according to new plans in new places? What should those plans be? And who is to decide and to accomplish whatever choice is made? These are structural issues; to confront them and to solve them require us to consider political and economic issues that affect innumerable milieux.

Insofar as an economy is so arranged that slumps occur, the problem of unemployment becomes incapable of personal solution. Insofar as war is inherent in the nation-state system and in the uneven industrialization of the world, the ordinary individual in his restricted milieu will be powerless—with or without psychiatric aid—to solve the troubles this system or lack of system imposes upon him. Insofar as the family as an institution turns women into darling little slaves and men into their chief providers and unweaned dependents, the problem of a satisfactory marriage remains incapable of purely private solution. Insofar as the overdeveloped megalopolis and the overdeveloped automobile are built-in features of the overdeveloped society, the issue of urban living will not be solved by personal ingenuity and private wealth.

What we experience in various and specific milieux, I have noted, is often caused by structural changes. Accordingly, to understand the changes of many personal milieux we are required to look beyond them. And the number and variety of such structural changes increase as the institutions within which we live become more embracing and more intricately connected with one another. To be aware of the idea of social structure and to use it with sensibility is to be capable of tracing such linkages among a great variety of milieux. To be able to do that is to possess the sociological imagination.

III

What are the major issues for publics and the key troubles of private individuals in our time? To formulate issues and troubles, we must ask what values are cherished yet threatened, and what values are cherished and supported, by the characterizing trends of our period. In the case both of threat and of support we must ask what salient contradictions of structure may be involved.

When people cherish some set of values and do not feel any threat to them, they experience *well-being*. When they cherish values but *do* feel them to be threatened, they experience a crisis—either as a personal trouble or as a public issue. And if all their values seem involved, they feel the total threat of panic.

But suppose people are neither aware of any cherished values nor experience any threat? That is the experience of *indifference,* which, if it seems to involve all their values, becomes apathy. Suppose, finally, they are unaware of any cherished values, but still are very much aware of a threat? That is the experience of *uneasiness,* of anxiety, which, if it is total enough, becomes a deadly unspecified malaise.

Ours is a time of uneasiness and indifference—not yet formulated in such ways as to permit the work of reason and the play of sensibility. Instead of troubles—defined in terms of values and threats—there is often the misery of vague uneasiness; instead of explicit issues there is often merely the beat feeling that all is somehow not right. Neither the values threatened nor whatever threatens them has been stated; in short, they have not been carried to the point of decision. Much less have they been formulated as problems of social science.

In the thirties there was little doubt—except among certain deluded business circles—that there was an economic issue which was also a pack of personal troubles. In these arguments about "the crisis of capitalism," the formulations of Marx and the many unacknowledged reformulations of his work probably set the leading terms of the issue, and some men came to understand their personal troubles in these terms. The values threatened were plain to see and cherished by all; the structural contradictions that threatened them also seemed plain. Both were widely and deeply experienced. It was a political age.

But the values threatened in the era after World War II are often neither widely acknowledged as values nor widely felt to be threatened. Much private uneasiness goes unformulated; much public malaise and many decisions of enormous structural relevance never become public issues. For those who accept such inherited values as reason and freedom, it is the uneasiness itself that is the trouble; it is the indifference that is the issue. And it is this condition, of uneasiness and indifference, that is the signal feature of our period.

All this is so striking that it is often interpreted by observers as a shift in the very kinds of problems that need now to be formulated. We are frequently told that the problems of our decade, or even the crisis of our period, have shifted from the external realm of economics and now have to do with the quality of individual life—in fact with the question of whether there is soon going to be anything that can properly be called individual life. Not child labor but comic books, not poverty but mass leisure, are at the center of concern. Many great public issues as well as many private troubles are described in terms of "the psychiatric"—often, it seems, in a pathetic attempt to avoid the large issues and problems of modern society. Often this statement seems to rest upon a provincial narrowing of interest to the Western societies, or even to the United States—thus ignoring two-thirds of mankind; often, too, it arbitrarily divorces the individual life from the larger institutions within which that life is enacted, and which on occasion bear upon it more grievously than do the intimate environments of childhood.

Problems of leisure, for example, cannot even be stated without considering problems of work. Family troubles over comic books cannot be formulated as problems without considering the plight of the contemporary family in its new relations with the newer institutions of the social structure. Neither leisure nor its debilitating uses can be understood as problems without recognition of the extent to which malaise and indifference now form the social and personal climate of contemporary American society. In this climate, no problems of "the private life" can be stated and solved without recognition of the crisis of ambition that is part of the very career of men at work in the incorporated economy.

It is true, as psychoanalysts continually point out, that people do often have "the increasing sense of being moved by obscure forces within themselves which they are unable to define." But it is *not* true, as Ernest Jones asserted, that "man's chief enemy and danger is his own unruly nature and the dark forces pent up within him." On the contrary: "Man's chief danger" today lies in the unruly forces of contemporary society itself, with its alienating methods of production, its enveloping techniques of political domination, its international anarchy—in a word, its pervasive transformations of the very "nature" of man and the conditions and aims of his life.

It is now the social scientist's foremost political and intellectual task—for here the two coincide—to make clear the elements of contemporary uneasiness and indifference. It is the central demand made upon him by other cultural workmen—by physical scientists and artists, by the intellectual community in general. It is because of this task and these demands, I believe, that the social sciences are becoming the common denominator of our cultural period, and the sociological imagination our most needed quality of mind.

PART II

Becoming a Member of Society

People survive in group contexts. Most could not survive without other people. Considering the amazing array of ways members of societies around the world carry out the tasks of daily living provides a glimpse into why people differ. Culture, the way of life in each society, provides us with guidelines for living through the values, norms, knowledge, and materials necessary for survival in each particular society.

Differences in cultures stem from historical developments, geographical location, influences from other cultures, and many other factors. Although they vary, each society has cultural expectations, socialization to teach people the culture, and groups to which members belong. Most societies have right and wrong ways for members to fulfill their responsibilities and meet their needs. Newborns begin learning these rules for proper behaviors, or *norms*, from birth. An infant learns quickly that certain actions bring responses from other people. In this way, the transformation of the newborn child into a social being begins. We call this transformation the *socialization process*.

The family is the first agent in this process. It protects and cares for the helpless infant, providing the training the child needs in order to survive and become a participating member of society. As the years pass, the child will come into contact with other agents of the socialization process, including peer groups, schools, and religious organizations.

The socialization process is lifelong. We must continuously learn to deal with the changes that occur throughout our lifetime, in our family circumstances, occupational roles, and in other important aspects of life including aging. What we learn is part of the complex whole known as culture.

Only through our interaction with groups such as the family, peers, schools, and religious organizations do we become social beings. However, this learning process is not always smooth, nor does society always function perfectly. For example, families provide nurturance, but violence may also be part of our family experience. Peers may provide security and a sense of belonging, but they also demand conformity and adherence to strict expectations. Schools train us in the essentials we need to fit into societal positions, but they also function to track students by social class and other factors. Religious organizations may present us with contradictory messages about love and tolerance of those different from ourselves.

Group contact provides needed social interaction and a sense of belonging, but it may also be frustrating. Chance places us in a rich or poor family, which in turn greatly influences our life opportunities. Group membership is our vehicle for carrying out the process of socialization and maintaining the social system, whatever that system might be. Although specifics of each of these three elements—socialization, culture, and groups—vary from society to society, each element is found in every society.

One result of the process of becoming a member of society must be mentioned because it influences how and what we learn. Every society ranks its members according to its own values: This ranking creates differing life opportunities, styles of living, and distribution of power within the society. We call this creation of varying status levels *stratification*. Through socialization, children internalize the class values and beliefs held by their families, thus learning their positions in the stratification hierarchy.

Gender issues, the way we become and behave as women and men in society, are omnipresent in human groups and interactions. Throughout the book articles deal with the idea that gender shapes many of our life experiences. For example, in Chapter 3 on socialization, several readings deal with the effect of gender learning on individuals—how we learn our gender roles and results of these expectations.

As you read this section, consider the processes that influence the infant and turn the child into a productive member of groups within a particular culture. Also consider the different experiences the infant might undergo if reared in another social class or culture.

CHAPTER 2

Culture

Our Way of Life

Alan Mc Evoy

Human behavior is both patterned and orderly because within our society we are taught to follow similar rules of behavior *(norms)* and to cherish similar objects and behaviors *(values)*. These similarities create the culture of a society: *its total way of life*. As Kingsley Davis will note in Chapter 3, the importance of culture is indicated by the fact that most human behavior is learned within a cultural context. It is through *cultural relativism*—looking at other cultures in an objective manner—that we attempt to understand learned cultural patterns and behaviors by considering the functions they serve for society.

In the first reading in this chapter, Clyde Kluckholm sets the stage by presenting a classic definition and discussion of culture, its characteristics, and how culture emerges in groups. An example of cultural variations is presented in "An Indian Father's Plea," outlining some differences in culture that affect Native American children in the U.S. educational system.

Next, Horace Minor gives us an anthropological look at what appears to be a "primitive" group by examining the societal needs served by the unusual attitude of the Nacirema toward the human body. If, as you read this selection, you feel glad to be an American while wondering about the "silly" actions of the Nacirema, then your ethnocentrism is showing. (Read the article carefully, especially the italicized words. You should find it amusing.) The reading by Miner is also a good example of the points made by Mills in Chapter 1. That is, our beliefs are represented by the social system of which we are a part, and they affect subsequent behavior.

Because we are taught the norms, values, language, and beliefs (folklore, legends, proverbs, religion) of our own culture, we frequently find it difficult to see our culture objectively. What we do routinely we accept as right without question, and possibly even without understanding. This *ethnocentrism*—the belief that one's own culture is superior to others—can make it difficult to accept the different ways of others and to change our own ways. It is those two factors of learned behavior and ethnocentrism that lead to cultural constraints on our thoughts and behavior. Ethnocentrism is common despite the fact that many, if not most, of the material items used in any given culture were neither invented nor discovered in that culture but were adopted from other societies through *cultural diffusion*—the spread of cultural behavior and materials from one society to another.

Unfortunately, ethnocentrism can carry over into hostilities toward different groups of people. This hostility exists because groups feel strongly about the rightness of their own cultural beliefs, values, attitudes, and behaviors. A good example of cultural misunderstandings resulting from ethnocentrism is seen in the reading by Marvin Harris, "India's Sacred Cow." It makes the point that cultural practices have origins and can be explained, however strange they seem to outsiders.

Many social scientists feel there is hope for global cooperation despite the ethnocentrism that can result in hostilities and conflicts. Futurist Wendell Bell points out the many important values in human cultures, based on common human needs, that are held around the world. Bell predicts that these shared human values, guideposts to global morality, will emerge into a global ethic. Expanding on the idea of common human values and needs, the final reading by Gary Gardner and Erik Assadourian asks what cultural needs and values make for the "good life." They discuss cultural movements around the world, in rich and poor countries, aimed at achieving a good life. A feeling of belonging and health are key, whether in rich or poor countries—having friends, neighbors, family, a sense of community, and good health. A sense of well-being affects happiness, health, and many other measures of the good life in cultures around the world.

As you read this chapter, ask yourself these questions: Why do the Nacirema have an apparently "pathological" concern with the body, and what are the implications of this concern? Why does our liking of the familiar sometimes lead to violent dislike of groups with dissimilar cultures? How can stereotypes shape our images, sometimes falsely, of cultural patterns? And why do cultural patterns vary so widely, sometimes to the disadvantage of members of society?

6

The Study of Culture

CLYDE KLUCKHOLM

Sociologists generally accept cultural anthropologists' definitions of culture; after all, this is their main field of study. Kluckholm describes what culture is, its characteristics, and briefly how it emerges in groups of humans. Sociologists use and build on these ideas to help understand societies and interactions within and between societies.

As you read, ask yourself the following questions:

1. *How do the elements of culture apply to your society?*
2. *Why is culture a necessary part of society?*
3. *What does it mean to say "culture is omnipresent"?*
4. *Describe explicit and implicit elements of your society.*

GLOSSARY **Culture** Distinctive way of life of a group of people, their complete design for living. **System** Organized design for living.

Culture, as used by American anthropologists, is of course a technical term which must not be confused with the more limited concept of ordinary language and of history and literature. The anthropological term designates those aspects of the total human environment, tangible and intangible, which have been created by men. A "culture" refers to the distinctive way of life of a group of people, their complete "design for living." The Japanese constitute a nation or a society. This entity may be directly observed. "Japanese culture," however, is an abstraction from observed regularities or trends toward regularity in the modes of response of this people.

Recent anthropological research in the United States has by no means been limited to the study of cultures. The community studies of W. Lloyd Warner and other American anthropologists are well known. There have been published some pioneer investigations in quantitative comparative sociology in which the theory is drawn from sociology, psychoanalysis, and behavioristic psychology as well as from anthropology.[1] An increasing number of American anthropologists have been concerned with interrelations between the cultural and the psychological.[2] Others have been developing the interstitial area between biology and anthropology.[3] Still others have concentrated upon the physical environment as a conditioning and limiting factor in cultural development and function.[4]

Nevertheless, culture remains the master concept of American anthropology, with the partial exception of physical anthropology. For ethnologists, folklorists, and anthropological linguists, archaeologists, and social anthropologists, culture

SOURCE: Reprinted from "The Study of Culture" by Clyde Kluckholm, Chapter V of *The Policy Sciences* edited by Daniel Lerner and Harold D. Lasswell with permission of the publishers, Stanford University Press. Publication assisted by a grant from Carnegie Corporation of New York.

is always a point of departure or a point of reference if not invariably the point of central emphasis. During the past fifteen years there have been significant refinements both in the theory of culture and in methods and techniques for the study of cultures.

Many different definitions of culture are current.[5] ... They vary in degree of looseness or precision, in the stressing of one conceptual element as opposed to another. There have also been some recent controversies on epistemological and ontological questions.[6] Neglecting, however, the finer details of terminology and some philosophical nuances, most American anthropologists would agree substantially with the following propositions of Herskovits[7] on the theory of culture:

1. Culture is learned;[8]
2. Culture derives from the biological,[9] environmental, psychological, and historical components of human existence;
3. Culture is structured;
4. Culture is divided into aspects;
5. Culture is dynamic;
6. Culture is variable;
7. Culture exhibits regularities that permit its analysis by the methods of science;
8. Culture is the instrument whereby the individual adjusts to his total setting, and gains the means for creative expression.[10]

A perhaps not unrepresentative brief definition is that of Kluckhohm and Kelly: "A culture is an historically created system of explicit and implicit designs for living, which tends to be shared by all or specially designated members of a group at a specified point in time."[11] Some comments may clarify this definition. Each culture is a precipitate of history from the materials supplied by human biology and the natural environment to which human organisms must make certain minimal adjustments for survival. The selectivity out of the potentialities afforded by human nature and physical surroundings and within the limits set by biological and physical nature is channeled by the historical process. The conventional or arbitrary element (that is, the purely cultural) arises in part out of the accidents of history, including both chance internal events and contacts with other peoples. The word "system" has important implications. The fact that cultures have organization as well as content is now generally recognized. Nor can culture be used as a conceptual instrument for prediction unless due account is taken of this systematic property. The word "tends" warns against reifying an abstraction. One cannot drop a perpendicular from even the most accurate description of a culture or any specific carrier of that culture. No individual thinks, feels, or acts precisely as the "blueprints" which constitute a culture indicate that he will or should. Nor are all the "blueprints" meant by the society to apply to each individual. There are sex differentials, age differentials, occupational differentials, and the like. The best conceptual model of the culture can only state correctly the central tendencies of ranges of variation.

The anthropologist's description of a culture may be compared to a map. A map is obviously not a concrete bit of land but rather an abstract representation of a particular area. If a map is accurate and one can read it, one doesn't get lost. If a culture is correctly portrayed, one will realize the existence of the distinctive features of a way of life and their interrelationships.

Culture is omnipresent; it interposes a double screen between, for example, the psychologist and the native or innate or constitutional personality he is trying to discover and describe. One is tempted to paraphrase Zola's remark that science is nature seen through a temperament and say that personality is a temperament which is both seen through and screened by a culture. Because of the mass of tradition and the complexities of human relationships, even the few simple things that people as animals want have been disguised in cultural patterns. An animal eats when he is hungry—if he can, but the human animal waits for lunch time. Three daily meals are as much an artifact as an automobile. Sneezing at first looks like pure biology. But little customs grow up about it, such as saying "excuse me" or "*Gesundheit.*" People do not sneeze in exactly the same way in different cultures or in various strata of the same society. Sneezing is a biological act caught in a cultural web. It is difficult to point to any activity that is not culturally tailored.

Why do most people, most of the time, adhere to cultural patterns? We cannot give this question the examination it deserves, but two reasons are obvious. First, by following custom one affirms one's solidarity with one's group and escapes a sense of loneliness. Second, patterns are necessary if we are to have a social life, with its attendant division of labor. Imagine people living in the same home and invariably preparing and eating food in different rooms at different times.

The analysis of a culture must encompass both the explicit and the implicit. The explicit culture consists in those regularities in word and deed which may be generalized straight from the evidence of the ear or eye. One has only to observe and to discover the consistencies in one's observations. No arbitrary acts of interpretation on the part of the anthropologist are involved. The implicit culture, however, is an abstraction of the second order. Here the anthropologist infers least common denominators which seem, as it were, to underlie a multiplicity of cultural contents. Only in the most sophisticated and self-conscious of cultures will his attention be called directly to these by carriers of the culture. The implicit culture consists of pure forms. Explicit culture includes both content and structure.

NOTES

1. The most impressive example is G. P. Murdock's *Social Structure* (1949).
2. See, for example, Cora Du Bois, *People of Alor* (1944); and Clyde Kluckhohn and Henry A. Murray (eds.), *Personality in Nature, Society and Culture* (1948).
3. Cf. John Gillin, *The Ways of Men* (1948), pp. 23–175; and Kluckhohn and Murray, op. cit., pp. 107–61 and 377–471.
4. E.g., J. H. Steward, *Basin-Plateau Aboriginal Sociopolitical Groups* (Smithsonian Institution, Bureau of American Ethnology, Bulletin 120 [1938]); A. L. Kroeber, *Cultural and Natural Areas of Native North America* (1939).
5. A. L. Kroeber and C. Kluckhohm, "The Concept of Culture: A Critical Review of Definitions," *Papers of the Peabody Museum* (Harvard University), Vol. XLI (1950). The approximate consensus of these definitions is as follows: "Culture consists in patterned ways of thinking, feeling, and reacting, acquired and transmitted mainly by symbols, constituting the distinctive achievements of human groups, including their embodiments in artifacts; the essential core of culture consists of traditional (i.e., historically derived and selected) ideas and especially their attached values."
6. See D. Bidney, "Human Nature and the Cultural Process," *American Anthropologist*, XLIX, No. 3 (1947), 375–96.
7. Melville J. Herskovits. *Man and His Works* (1940), p. 625.
8. Perhaps it is too obvious to add that while all culture is learned, not all learning is culture: The individual learns a good deal during his own private life-experience which he does not share with others or transmit to others. It might also be commented that some aspects of culture are learned only through the use of symbols, particularly linguistic symbols. Indeed, an argument can be made for R. Bain's definition of culture as "all social behavior which is mediated by symbols."
9. See Claude Lévi-Strauss, *Les Structures Élémentaires de la Parenté* (1949), especially pp. 1–13.
10. Herskovits leaves implicit the fact that participation in a culture or in any part of it is never emotionally neutral. The attitude of the participant may range from hearty acceptance to belligerent revolt, but even what seems to be passive conformance is emotionally tinged.
11. C. Kluckhohm and W. H. Kelly, "The Concept of Culture," in Ralph Linton (ed), *The Science of Man in the World Crisis* (1945), pp. 78–107.

7

An Indian Father's Plea

ROBERT LAKE (MEDICINE GRIZZLYBEAR)

An example of the impact of cultural differences is presented in this letter from an Indian father to his son's teacher. It points out some of the cultural factors that disadvantage his son in the American schools, factors that affect the academic success of his child and potentially all children from different cultural backgrounds than the dominant school culture.

As you read this section, consider the following questions:

1. *On what basis do you think the Indian boy was labeled a "slow learner"?*
2. *How did the means used to socialize the boy through his family and his group disadvantage him in the school setting?*
3. *How would you answer his father when he says the boy is not culturally "disadvantaged," only culturally "different"?*

Dear Teacher,

I would like to introduce you to my son, Wind-Wolf. He is probably what you would consider a typical Indian kid. He was born and raised on the reservation. He has black hair, dark brown eyes, and an olive complexion. And, like so many Indian children his age, he is shy and quiet in the classroom. He is 5 years old, in kindergarten, and I can't understand why you have already labeled him a "slow learner."

He has already been through quite an education compared with his peers in Western society. He was bonded to his mother and to the Mother Earth in a traditional native childbirth ceremony. And he has been continuously cared for by his mother, father, sisters, cousins, aunts, uncles, grandparents, and extended tribal family since this ceremony.

The traditional Indian baby basket became his "turtle's shell" and served as the first seat for his classroom. It is the same kind of basket our people have used for thousands of years. It is specially designed to provide the child with the kind of knowledge and experience he will need to survive in his culture and environment.

Wind-Wolf was strapped in snugly with a deliberate restriction on his arms and legs. Although Western society may argue this hinders motor-skill development and abstract reasoning, we believe it forces the child to first develop his intuitive faculties, rational intellect, symbolic thinking, and five senses. Wind-Wolf was with his mother constantly, closely bonded physically, as she carried him on her back or held him while breast-feeding. She carried him everywhere she went, and every night he slept with both parents. Because of this, Wind-Wolf's educational setting was not only a "secure" environment, but it was also very colorful, complicated, sensitive, and diverse.

As he grew older, Wind-Wolf began to crawl out of the baby basket, develop his motor skills, and explore the world around him. When frightened or sleepy he could always return to the basket, as a

SOURCE: From *Teacher Magazine*, Vol. 2, September 1990, pp. 48–53. Reprinted with permission from *Teacher Magazine*.

turtle withdraws into its shell. Such an inward journey allows one to reflect in privacy on what he has learned and to carry the new knowledge deeply into the unconscious and the soul. Shapes, sizes, colors, texture, sound, smell, feeling, taste, and the learning process are functionally integrated—the physical and spiritual, matter and energy, and conscious and unconscious, individual and social.

It takes a long time to absorb and reflect on these kinds of experiences, so maybe that is why you think my Indian child is a slow learner. His aunts and grandmothers taught him to count and know his numbers while they sorted materials for making abstract designs in native baskets. And he was taught to learn mathematics by counting the sticks we use in our traditional native hand game. So he may be slow in grasping the methods and tools you use in your classroom, ones quite familiar to his white peers, but I hope you will be patient with him. It takes time to adjust to a new cultural system and learn new things.

He is not culturally "disadvantaged," but he is culturally "different."

8

Body Ritual among the Nacirema

HORACE MINER

When sociologists and anthropologists study other cultures, they attempt to be objective in their observations, trying to understand the culture from that culture's point of view. Understanding other cultures can help us gain perspective on our own culture. Yet because individuals are socialized into their own culture's beliefs, values, and practices, other cultures may seem to have strange, even bizarre or immoral activities and beliefs.

As you read Miner's essay, keep in mind these questions:

1. *What practices in the Nacirema culture appear strange to an outsider?*
2. *How does this reading illustrate ethnocentrism and cultural relativity?*
3. *What problems might you encounter in accurately observing other cultures?*
4. *How might someone from a different culture view or interpret practices in your culture?*

GLOSSARY **Body ritual** Ceremonies focusing on the body or body parts.

SOURCE: From *American Anthropologist 58*(3), pp. 503–507, 1956.

The anthropologist has become so familiar with the diversity of ways in which different peoples behave in similar situations that he is not apt to be surprised by even the most exotic customs. In fact, if all of the logically possible combinations of behavior have not been found somewhere in the world, he is apt to suspect that they must be present in some yet undescribed tribe. This point has, in fact, been expressed with respect to clan organization by Murdock (1949:71). In this light, the magical beliefs and practices of the Nacirema present such unusual aspects that it seems desirable to describe them as an example of the extremes to which human behavior can go.

Professor Linton first brought the ritual of the Nacirema to the attention of anthropologists twenty years ago (1936:326), but the culture of this people is still very poorly understood. They are a North American group living in the territory between the Canadian Cree, the Yaqui and Tarahumare of Mexico, and the Carib and Ara-wak of the Antilles. Little is known of their origin, although tradition states that they came from the east. According to Nacirema mythology, their nation was originated by a culture hero, Notgnihsaw, who is otherwise known for two great feats of strength—the throwing of a piece of wampum across the river Pa-To-Mac and the chopping down of a cherry tree in which the Spirit of Truth resided.

Nacirema culture is characterized by a highly developed market economy which has evolved in a rich natural habitat. While much of the people's time is devoted to economic pursuits, a large part of the fruits of these labors and a considerable portion of the day are spent in ritual activity. The focus of this activity is the human body, the appearance and health of which loom as a dominant concern in the ethos of the people. While such a concern is certainly not unusual, its ceremonial aspects and associated philosophy are unique.

The fundamental belief underlying the whole system appears to be that the human body is ugly and that its natural tendency is to debility and disease. Incarcerated in such a body, man's only hope is to avert these characteristics through the use of the powerful influences of ritual and ceremony. Every household has one or more shrines devoted to this purpose. The more powerful individuals in this society have several shrines in their houses and, in fact, the opulence of a house is often referred to in terms of the number of such ritual centers it possesses. Most houses are of wattle and daub construction, but the shrine rooms of the more wealthy are walled with stone. Poorer families imitate the rich by applying pottery plaques to their shrine walls.

While each family has at least one such shrine, the rituals associated with it are not family ceremonies but are private and secret. The rites are normally only discussed with children, and then only during the period when they are being initiated into these mysteries. I was able, however, to establish sufficient rapport with the natives to examine these shrines and to have the rituals described to me.

The focal point of the shrine is a box or chest which is built into the wall. In this chest are kept the many charms and magical potions without which no native believes he could live. These preparations are secured from a variety of specialized practitioners. The most powerful of these are the medicine men, whose assistance must be rewarded with substantial gifts. However, the medicine men do not provide the curative potions for their clients, but decide what the ingredients should be and then write them down in an ancient and secret language. This writing is understood only by the medicine men and by the herbalists who, for another gift, provide the required charm.

The charm is not disposed of after it has served its purpose, but is placed in the charm-box of the household shrine. As these magical materials are specific for certain ills, and the real or imagined maladies of the people are many, the charm-box is usually full to overflowing. The magical packets are so numerous that people forget what their purposes were and fear to use them again. While the natives are very vague on this point, we can only assume that the idea in retaining all the old magical materials is that their presence in the charm-box, before which the body rituals are conducted, will in some way protect the worshipper.

Beneath the charm-box is a small font. Each day every member of the family, in succession, enters the shrine room, bows his head before the charm-box, mingles different sorts of holy water in the font, and proceeds with a brief rite of ablution. The holy waters are secured from the Water Temple of the community, where the priests conduct elaborate ceremonies to make the liquid ritually pure.

In the hierarchy of magical practitioners, and below the medicine men in prestige, are specialists whose designation is best translated "holy-mouth-men." The Nacirema have an almost pathological horror of and fascination with the mouth, the condition of which is believed to have a supernatural influence on all social relationships. Were it not for the rituals of the mouth, they believe that their teeth would fall out, their gums bleed, their jaws shrink, their friends desert them, and their lovers reject them. They also believe that a strong relationship exists between oral and moral characteristics. For example, there is a ritual ablution of the mouth for children which is supposed to improve their moral fiber.

The daily body ritual performed by everyone includes a mouth-rite. Despite the fact that these people are so punctilious about care of the mouth, this rite involves a practice which strikes the uninitiated stranger as revolting. It was reported to me that the ritual consists of inserting a small bundle of hog hairs into the mouth, along with certain magical powders, and then moving the bundle in a highly formalized series of gestures.

In addition to the private mouth-rite, the people seek out a holy-mouth-man once or twice a year. These practitioners have an impressive set of paraphernalia, consisting of a variety of augers, awls, probes, and prods. The use of these objects in the exorcism of the evils of the mouth involves almost unbelievable ritual torture of the client. The holy-mouth-man opens the client's mouth and, using the above-mentioned tools, enlarges any holes which decay may have created in the teeth. Magical materials are put into these holes. If there are no naturally occurring holes in the teeth, large sections of one or more teeth are gouged out so that the supernatural substance can be applied. In the client's view, the purpose of these ministrations is to arrest decay and to draw friends. The extremely sacred and traditional character of the rite is evident in the fact that the natives return to the holy-mouth-man year after year, despite the fact that their teeth continue to decay.

It is to be hoped that, when a thorough study of the Nacirema is made, there will be careful inquiry into the personality structure of these people. One has but to watch the gleam in the eye of a holy-mouth-man, as he jabs an awl into an exposed nerve, to suspect that a certain amount of sadism is involved. If this can be established, a very interesting pattern emerges, for most of the population shows definite masochistic tendencies. It was to these that Professor Linton referred in discussing a distinctive part of the daily body ritual which is performed only by men. This part of the rite involves scraping and lacerating the surface of the face with a sharp instrument. Special women's rites are performed only four times during each lunar month, but what they lack in frequency is made up in barbarity. As part of this ceremony, women bake their heads in small ovens for about an hour. The theoretically interesting point is that what seems to be a preponderantly masochistic people have developed sadistic specialists.

The medicine men have an imposing temple, or *latipso*, in every community of any size. The more elaborate ceremonies required to treat very sick patients can only be performed at this temple. These ceremonies involve not only the thaumaturge but a permanent group of vestal maidens who move sedately about the temple chambers in distinctive costume and headdress.

The *latipso* ceremonies are so harsh that it is phenomenal that a fair proportion of the really sick natives who enter the temple ever recover. Small children whose indoctrination is still incomplete have been known to resist attempts to take them to the temple because "that is where you go to die." Despite this fact, sick adults are not only willing but eager to undergo the protracted ritual purification, if they can afford to do so. No matter how ill the supplicant or how grave the emergency, the guardians of many temples will not admit a client if he cannot give a rich gift to the custodian. Even after one has gained admission and survived the ceremonies, the guardians will not permit the neophyte to leave until he makes still another gift.

The supplicant entering the temple is first stripped of all his or her clothes. In everyday life the Nacirema avoids exposure of his body and its natural functions. Bathing and excretory acts are performed only in the secrecy of the household shrine, where they are ritualized as part of the body-rites. Psychological shock results from the fact that body secrecy is suddenly lost upon entry into the *latipso*. A man, whose own wife has never seen him in an excretory act, suddenly finds himself naked and assisted by a vestal maiden while he performs his natural functions into a sacred vessel. This sort of ceremonial treatment is necessitated by the fact that the excreta are used by a diviner to ascertain the course and nature of the client's sickness. Female clients, on the other hand, find their naked bodies are subjected to the scrutiny, manipulation, and prodding of the medicine men.

Few supplicants in the temple are well enough to do anything but lie on their hard beds. The daily ceremonies, like the rites of the holy-mouth-men, involve discomfort and torture. With ritual precision, the vestals awaken their miserable charges each dawn and roll them about on their beds of pain while performing ablutions, in the formal movements of which the maidens are highly trained. At other times they insert magic wands in the supplicant's mouth or force him to eat substances which are supposed to be healing. From time to time the medicine men come to their clients and jab magically treated needles into their flesh. The fact that these temple ceremonies may not cure, and may even kill the neophyte, in no way decreases the people's faith in the medicine men.

There remains one other kind of practitioner, known as a "listener." This witch-doctor has the power to exorcise the devils that lodge in the heads of people who have been bewitched. The Nacirema believe that parents bewitch their own children. Mothers are particularly suspected of putting a curse on children while teaching them the secret body rituals. The counter-magic of the witch-doctor is unusual in its lack of ritual. The patient simply tells the "listener" all his troubles and fears, beginning with the earliest difficulties he can remember. The memory displayed by the Nacirema in these exorcism sessions is truly remarkable. It is not uncommon for the patient to bemoan the rejection he felt upon being weaned as a babe, and a few individuals even see their troubles going back to the traumatic effects of their own birth.

In conclusion, mention must be made of certain practices which have their base in native esthetics but which depend upon the pervasive aversion to the natural body and its functions. There are ritual fasts to make fat people thin and ceremonial feasts to make thin people fat. Still other rites are used to make women's breasts larger if they are small, and smaller if they are large. General dissatisfaction with breast shape is symbolized in the fact that the ideal form is virtually outside the range of human variation. A few women afflicted with almost inhuman hypermammary development are so idolized that they make a handsome living by simply going from village to village and permitting the natives to stare at them for a fee.

Reference has already been made to the fact that excretory functions are ritualized, routinized, and relegated to secrecy. Natural reproductive functions are similarly distorted. Intercourse is taboo as a topic and scheduled as an act. Efforts are made to avoid pregnancy by the use of magical materials or by limiting intercourse to certain phases of the moon. Conception is actually very infrequent. When pregnant, women dress so as to hide their condition. Parturition takes place in secret, without friends or relatives to assist, and the majority of women do not nurse their infants.

Our review of the ritual life of the Nacirema has certainly shown them to be a magic-ridden people. It is hard to understand how they have managed to exist so long under the burdens which they have imposed upon themselves. But even such exotic customs as these take on real meaning when they are viewed with the insight provided by Malinowski when he wrote (1948: 70):

> Looking from far and above, from our high places of safety in the developed civilization, it is easy to see all the crudity and irrelevance of magic. But without its power and guidance early man could not have mastered his practical difficulties as he has done, nor could man have advanced to the higher stages of civilization.

REFERENCES

Linton, Ralph, 1936. *The Study of Man.* New York: Appleton-Century.

Malinowski, Bronislaw, 1948. *Magic, Science, and Religion.* Glencoe, III.: Free Press.

Murdock, George P., 1949. *Social Structure.* New York: Macmillan.

9

India's Sacred Cow

MARVIN HARRIS

Cultures vary dramatically in their beliefs and practices, yet each cultural practice has evolved with some reason behind it. One of these practices is cow worship. In India, the cultural practice among Hindus is to treat cows with great respect, even in the face of human hunger. Harris discusses this practice, which many find curious.

Consider the following as you read:

1. *Why are cows sacred? What is sacred in your country that might seem strange to others?*
2. *What other practices in different cultures do you find strange or unusual, and what purpose might those practices serve for the culture?*
3. *How and why do cultural traditions evolve?*

GLOSSARY **Untouchables** Lowest group in the stratification (caste) system of India. **Hinduism** Dominant religious belief system in India.

News photographs that came out of India during the famine of the late 1960s showed starving people stretching out bony hands to beg for food while sacred cattle strolled behind undisturbed. The Hindu, it seems, would rather starve to death than eat his cow or even deprive it of food. The cattle appear to browse unhindered through urban markets eating an orange here, a mango there, competing with people for meager supplies of food.

By Western standards, spiritual values seem more important to Indians than life itself. Specialists in food habits around the world like Fred Simoons at the University of California at Davis consider Hinduism an irrational idealogy that compels people to overlook abundant, nutritious foods for scarcer, less healthful foods.

What seems to be an absurd devotion to the mother cow pervades Indian life. Indian wall calendars portray beautiful young women with bodies of fat white cows, often with milk jetting from their teats into sacred shrines.

Cow worship even carries over into politics. In 1966 a crowd of 120,000 people, led by holy men, demonstrated in front of the Indian House of Parliament in support of the All-Party Cow Protection Campaign Committee. In Nepal, the only contemporary Hindu kingdom, cow slaughter is severely punished. As one story goes, the car driven by an official of a United States agency struck and killed a cow. In order to avoid the international incident that would have occurred when the official was arrested for murder, the Nepalese magistrate concluded that the cow had committed suicide.

Many Indians agree with Western assessments of the Hindu reverence for their cattle, the zebu, or *Bos indicus*, a large-humped species prevalent in Asia and Africa. M. N. Srinivas, an Indian anthropologist, states: "Orthodox Hindu opinion regards the killing of cattle with abhorrence, even though the refusal to kill vast number of useless cattle which exist in India today is detrimental to the nation." Even the Indian Ministry of Information formerly maintained that "the large animal population is more a liability than an asset in view of our land resources." Accounts from many different sources point to the same conclusion: India, one of the world's great civilizations, is being strangled by its love for the cow.

The easy explanation for India's devotion to the cow, the one most Westerners and Indians would offer, is that cow worship is an integral part of Hinduism. Religion is somehow good for the soul, even if it sometimes fails the body. Religion orders the cosmos and explains our place in the universe. Religious beliefs, many would claim, have existed for thousands of years and have a life of their own. They are not understandable in scientific terms.

But all this ignores history. There is more to be said for cow worship than is immediately apparent. The earliest Vedas, the Hindu sacred texts from the second millennium B.C., do not prohibit the slaughter of cattle. Instead, they ordain it as part of sacrificial rites. The early Hindus did not avoid the flesh of cows and bulls; they ate it at ceremonial feasts presided over by Brahman priests. Cow worship is a relatively recent development in India; it evolved as the Hindu religion developed and changed.

This evolution is recorded in royal edicts and religious texts written during the last 3,000 years of Indian history. The Vedas from the first millennium B.C. contain contradictory passages, some referring to ritual slaughter and others to a strict taboo on beef consumption. A. N. Bose, in *Social and Rural Economy of Northern India, Cir. 600 B.C.–200 A.D.*, concludes that many of the sacred-cow passages were incorporated into the texts by priests of a later period.

By 200 A.D. the status of Indian cattle had undergone a spiritual transformation. The Brahman priesthood exhorted the population to venerate the cow and forbade them to abuse it or to feed on it. Religious feasts involving the ritual slaughter and the consumption of livestock were eliminated and meat eating was restricted to the nobility.

By 1000 A.D., all Hindus were forbidden to eat beef. Ahimsa, the Hindu belief in the unity of all life, was the spiritual justification for this restriction. But it is difficult to ascertain exactly when this change occurred. An important event that helped to shape the modern complex was the Islamic invasion, which took place in the eighth century A.D. Hindus may have found it politically expedient to set themselves off from the invaders, who were beefeaters, by emphasizing the need to prevent the slaughter of their sacred animals. Thereafter, the cow taboo assumed its modern form and began to function much as it does today.

The place of the cow in modern India is every place—on posters, in the movies, in brass figures, in stone and wood carvings, on the streets, in the fields. The cow is a symbol of health and abundance. It provides the milk that Indians consume in the form of yogurt and ghee (clarified butter), which contribute subtle flavors to much spicy Indian food.

This, perhaps, is the practical role of the cow, but cows provide less than half the milk produced in India. Most cows in India are not dairy breeds. In most regions, when an Indian farmer wants a steady, high-quality source of milk he usually invests in a female water buffalo. In India the water buffalo is the specialized dairy breed because its milk has a higher butterfat content than zebu milk. Although the farmer milks his zebu cows, the milk is merely a by-product.

More vital than zebu milk to South Asian farmers are zebu calves. Male calves are especially valued because from bulls come oxen, which are the mainstay of the Indian agricultural system.

Small, fast oxen drag wooden plows through late-spring fields when monsoons have dampened the dry, cracked earth. After harvest, the oxen break the grain from the stalk by stomping through mounds of cut wheat and rice. For rice cultivation in irrigated fields, the male water buffalo is preferred (it pulls better in deep mud), but for most other crops, including rainfall rice, wheat, sorghum, and millet, and for transporting goods and people to and from town, a team of oxen is preferred. The ox is the Indian peasant's tractor, thresher, and family car combined; the cow is the factory that produces the ox.

If draft animals instead of cows are counted, India appears to have too few domesticated ruminants, not too many. Since each of the 70 million farms in India requires a draft team, it follows that Indian peasants should use 140 million animals in the fields. But there are only 83 million oxen and male water buffalo on the subcontinent, a shortage of 30 million draft teams.

In other regions of the world, joint ownership of draft animals might overcome a shortage, but Indian agriculture is closely tied to the monsoon rains of late spring and summer. Field preparation and planting must coincide with the rain, and a farmer must have his animals ready to plow when the weather is right. When the farmer without a draft team needs bullocks most, his neighbors are all using theirs. Any delay in turning the soil drastically lowers production.

Because of this dependence on draft animals, loss of the family oxen is devastating. If a beast dies, the farmer must borrow money to buy or rent an ox at interest rates so high that he ultimately loses his land. Every year foreclosures force thousands of poverty-stricken peasants to abandon the countryside for the overcrowded cities.

If a family is fortunate enough to own a fertile cow, it will be able to rear replacements for a lost team and thus survive until life returns to normal. If, as sometimes happens, famine leads a family to sell its cow and ox team, all ties to agriculture are cut. Even if the family survives, it has no way to farm the land, no oxen to work the land, and no cows to produce oxen.

The prohibition against eating meat applies to the flesh of cows, bulls, and oxen, but the cow is the most sacred because it can produce the other two. The peasant whose cow dies is not only crying over a spiritual loss but over the loss of his farm as well.

Religious laws that forbid the slaughter of cattle promote the recovery of the agricultural system from the dry Indian winter and from periods of drought. The monsoon, on which all agriculture depends, is erratic. Sometimes, it arrives early, sometimes late, sometimes not at all. Drought has struck large portions of India time and again in this century, and Indian farmers and the zebus are accustomed to these natural disasters. Zebus can pass weeks on end with little or no food and water. Like camels, they store both in their humps and recuperate quickly with only a little nourishment.

During droughts the cows often stop lactating and become barren. In some cases the condition is permanent but often it is only temporary. If barren animals were summarily eliminated, as Western experts in animal husbandry have suggested, cows capable of recovery would be lost along with those entirely debilitated. By keeping alive the cows that can later produce oxen, religious laws against cow slaughter assure the recovery of the agricultural system from the greatest challenge it faces—the failure of the monsoon.

The local Indian governments aid the process of recovery by maintaining homes for barren cows. Farmers reclaim any animal that calves or begins to lactate. One police station in Madras collects strays and pastures them in a field adjacent to the station. After a small fine is paid, a cow is returned to its rightful owner when the owner thinks the cow shows signs of being able to reproduce.

During the hot, dry spring months most of India is like a desert. Indian farmers often complain they cannot feed their livestock during this period. They maintain the cattle by letting them scavenge on the sparse grass along the roads. In the cities the cattle are encouraged to scavenge near food stalls to supplement their scant diet. These are the wandering cattle tourists report seeing throughout India.

Westerners expect shopkeepers to respond to these intrusions with the deference due a sacred animal; instead, their response is a string of curses and the crack of a long bamboo pole across the beast's back or a poke at its genitals. Mahatma Gandhi was well aware of the treatment sacred cows (and bulls and oxen) received in India. "How we bleed her to take the last drop of milk from her. How we starve her to emaciation, how we ill-treat the calves, how we deprive them of their portion of milk, how cruelly we treat the oxen, how we castrate them, how we beat them, how we overload them" [Gandhi, 1954].

Oxen generally receive better treatment than cows. When food is in short supply, thrifty Indian peasants feed their working bullocks and ignore their cows, but rarely do they abandon the cows to die. When the cows are sick, farmers worry over them as they would over members of the family and nurse them as if they were children. When the rains return and when the fields are harvested, the farmers again feed their cows regularly and reclaim their abandoned animals. The prohibition against beef consumption is a form of disaster insurance for all India.

Western agronomists and economists are quick to protest that all the functions of the zebu cattle can be improved with organized breeding programs, cultivated pastures, and silage. Because stronger oxen would pull the plow faster, they could work multiple plots of land, allowing farmers to share their animals. Fewer healthy, well-fed cows could provide Indians with more milk. But pastures and silage require arable land, land needed to produce wheat and rice.

A look at Western cattle farming makes plain the cost of adopting advanced technology in Indian agriculture. In a study of livestock production in the United States, David Pimentel of the College of Agriculture and Life Sciences at Cornell University, found that 91 percent of the cereal, legume, and vegetable protein suitable for human consumption is consumed by livestock. Approximately three-quarters of the arable land in the United States is devoted to growing food for livestock. In the production of meat and milk, American ranchers use enough fossil fuel to equal more than 82 million barrels of oil annually.

Indian cattle do not drain the system in the same way. In a 1971 study of livestock in West Bengal, Stewart Odend'hal [1972] of the University of Missouri found that Bengalese cattle ate only the inedible remains of subsistence crops—rice straw, rice hulls, the tops of sugar cane, and mustard-oil cake. Cattle graze in the fields after harvest and eat the remains of crops left on the ground; they forage for grass and weeds on the roadsides. The food for zebu cattle costs the human population virtually nothing. "Basically," Odend'hal says, "the cattle convert items of little direct human value into products of immediate utility."

In addition to plowing the fields and producing milk, the zebus produce dung, which fires the hearths and fertilizes the fields of India. Much of the estimated 800 million tons of manure produced annually is collected by the farmers' children as they follow the cows and bullocks from place to place. And when the children see the droppings of another farmer's cattle along the road, they pick those up also. Odend'hal reports that the system operates with such high efficiency that the children of West Bengal recover nearly 100 percent of the dung produced by their livestock.

From 40 to 70 percent of all manure produced by Indian cattle is used as fuel for cooking; the rest is returned to the fields as fertilizer. Dried dung burns slowly, cleanly, and with low heat—characteristics that satisfy the household needs of

Indian women. Staples like curry and rice can simmer for hours. While the meal slowly cooks over an unattended fire, the women of the household can do other chores. Cow chips, unlike firewood, do not scorch as they burn.

It is estimated that the dung used for cooking fuel provides the energy-equivalent of 43 million tons of coal. At current prices, it would cost India an extra 1.5 billion dollars in foreign exchange to replace the dung with coal. And if the 350 million tons of manure that are being used as fertilizer were replaced with commercial fertilizers, the expense would be even greater. Roger Revelle of the University of California at San Diego has calculated that 89 percent of the energy used in Indian agriculture (the equivalent of about 140 million tons of coal) is provided by local sources. Even if foreign loans were to provide the money, the capital outlay necessary to replace the Indian cow with tractors and fertilizers for the fields, coal for the fires, and transportation for the family would probably warp international financial institutions for years.

Instead of asking the Indians to learn from the American model of industrial agriculture, American farmers might learn energy conservation from the Indians. Every step in an energy cycle results in a loss of energy to the system. Like a pendulum that slows a bit with each swing, each transfer of energy from sun to plants, plants to animals, and animals to human beings involves energy losses. Some systems are more efficient than others; they provide a higher percentage of the energy inputs in a final, useful form. Seventeen percent of all energy zebus consume is returned in the form of milk, traction, and dung. American cattle raised on Western rangeland return only 4 percent of the energy they consume.

But the American system is improving. Based on techniques pioneered by Indian scientists, at least one commercial firm in the United States is reported to be building plants that will turn manure from cattle feedlots into combustible gas. When organic matter is broken down by anaerobic bacteria, methane gas and carbon dioxide are produced. After the methane is cleansed of the carbon dioxide, it is available for the same purposes as natural gas—cooking, heating, electric generation. The company constructing the biogasification plant plans to sell its product to a gas-supply company, to be piped through the existing distribution system. Schemes similar to this one could make cattle ranches almost independent of utility and gasoline companies, for methane can be used to run trucks, tractors, and cars as well as to supply heat and electricity. The relative energy self-sufficiency that the Indian peasant has achieved is a goal American farmers and industry are now striving for.

Studies like Odend'hal's understate the efficiency of the Indian cow, because dead cows are used for purposes that Hindus prefer not to acknowledge. When a cow dies, an Untouchable, a member of one of the lowest ranking castes in India, is summoned to haul away the carcass. Higher castes consider the body of the dead cow polluting; if they handle it, they must go through a rite of purification.

Untouchables first skin the dead animal and either tan the skin themselves or sell it to a leather factory. In the privacy of their homes, contrary to the teachings of Hinduism, untouchable castes cook the meat and eat it. Indians of all castes rarely acknowledge the existence of these practices to non-Hindus, but most are aware that beefeating takes place. The prohibition against beefeating restricts consumption by the higher castes and helps distribute animal protein to the poorest sectors of the population that otherwise would have no source of these vital nutrients.

Untouchables are not the only Indians who consume beef. Indian Muslims and Christians are under no restriction that forbids them beef, and its consumption is legal in many places. The Indian ban on cow slaughter is state, not national, law and not all states restrict it. In many cities, such as New Delhi, Calcutta, and Bombay, legal slaughterhouses sell beef to retail customers and to restaurants that serve steak.

If the caloric value of beef and the energy costs involved in the manufacture of synthetic leather were included in the estimate of energy, the calculated efficiency of Indian livestock would rise considerably. As well as the system works, experts often claim that its efficiency can be further improved. Alan Heston [et al., 1971], an economist at the University of Pennsylvania, believes that Indians suffer from an overabundance of cows simply because they refuse to slaughter the excess cattle. India could produce at least the same number of oxen and the same quantities of milk and manure with 30 million fewer cows. Heston calculates that only 40 cows are necessary to maintain a population of 100 bulls and oxen. Since India averages 70 cows for every 100 bullocks, the difference, 30 million cows; is expendable.

What Heston fails to note is that sex ratios among cattle in different regions of India vary tremendously, indicating that adjustments in the cow population do take place. Along the Ganges River, one of the holiest shrines of Hinduism, the ratio drops to 47 cows for every 100 male animals. This ratio reflects the preference for dairy buffalo in the irrigated sectors of the Gangetic Plains. In nearby Pakistan, in contrast, where cow slaughter is permitted, the sex ratio is 60 cows to 100 oxen.

Since the sex ratios among cattle differ greatly from region to region and do not even approximate the balance that would be expected if no females were killed, we can assume that some culling of herds does take place; Indians do adjust their religious restrictions to accommodate ecological realities.

They cannot kill a cow but they can tether an old or unhealthy animal until it has starved to death. They cannot slaughter a calf but they can yoke it with a large wooden triangle so that when it nurses it irritates the mother's udder and gets kicked to death. They cannot ship their animals to the slaughterhouse but they can sell them to Muslims, closing their eyes to the fact that the Muslims will take the cattle to the slaughterhouse.

These violations of the prohibition against cattle slaughter strengthen the premise that cow worship is a vital part of Indian culture. The practice arose to prevent the population from consuming the animal on which Indian agriculture depends. During the first millennium B.C., the Ganges Valley became one of the most densely populated regions of the world.

Where previously there had been only scattered villages, many towns and cities arose and peasants farmed every available acre of land. Kingsley Davis, a population expert at the University of California at Berkeley, estimates that by 300 B.C. between 50 million and 100 million people were living in India. The forested Ganges Valley became a windswept semidesert and signs of ecological collapse appeared; droughts and floods became commonplace, erosion took away the rich topsoil, farms shrank as population increased, and domesticated animals became harder and harder to maintain.

It is probable that the elimination of meat eating came about in a slow, practical manner. The farmers who decided not to eat their cows; who saved them for procreation to produce oxen, were the ones who survived the natural disasters. Those who ate beef lost the tools with which to farm. Over a period of centuries, more and more farmers probably avoided beef until an unwritten taboo came into existence.

Only later was the practice codified by the priesthood. While Indian peasants were probably aware of the role of cattle in their society, strong sanctions were necessary to protect zebus from a population faced with starvation. To remove temptation, the flesh of cattle became taboo and the cow became sacred.

The sacredness of the cow is not just an ignorant belief that stands in the way of progress. Like all concepts of the sacred and the profane, this one affects the physical world; it defines the relationships that are important for the maintenance of Indian society.

Indians have the sacred cow, we have the "sacred" car and the "sacred" dog. It would not occur to us to propose the elimination of automobiles and dogs from our society without carefully considering the consequences, and we should not propose the elimination of zebu cattle without first understanding their place in the social order of India.

Human society is neither random nor capricious. The regularities of thought and behavior called culture are the principal mechanisms by which we human beings adapt to the world around us. Practices and beliefs can be rational or irrational, but a society that fails to adapt to its environment is doomed to extinction. Only those societies that draw the necessities of life from their surroundings inherit the earth. The West has much to learn from the great antiquity of Indian civilization, and the sacred cow is an important part of that lesson.

REFERENCES

Gandhi, Mohandas K. 1954. *How to Serve the Cow.* Bombay: Navajivan Publishing House.

Heston, Alan, et al., 1971. "An Approach to the Sacred Cow of India." *Current Anthropology* 12, 191–209.

Odend'hal, Stewart. 1972. "Gross Energetic Efficiency of Indian Cattle in Their Environment." *Journal of Human Ecology* 1, 1–27.

10

Rethinking the Good Life

GARY GARDNER AND ERIK ASSADOURIAN

Around the world, whether in rich or poor cultures, people seek the "good life." Although this term takes on different meanings and changes over time, the common values discussed in the previous article point to some similar aspects of the good life, including security and respect for different faiths, as well as those featured in this final reading, including health and a sense of belonging. Gardner and Assadourian assess human well-being and the meaning of the good life by reviewing studies on what makes people happy and connected. Aspects of life from health to success to social connections are key to well-being.

Think about the following questions as you read their findings:

1. *What factors are key to having "a good life?"*
2. *What do studies show about increasing one's amount of leisure? Productivity?*
3. *Does your community have any plans or programs directed toward increasing well-being?*
4. *What can you do to increase your own well-being?*

GLOSSARY **Good life** Having a feeling of well-being. **GDP** Gross domestic product of a country. **Well-being** Personal and policy goals that promote basics for survival, good health, social relations, security, and freedom.

SOURCE: From the Worldwatch Institute. 2004. *State of the World 2004.* pp. 166. Reprinted by permission of Worldwatch Institute.

THE POWER OF ONE

. . . During the summer of 2003, some 50 million Americans signed up for a government sponsored National Do Not Call Registry designed to prevent commercial telemarketers from phoning them. The outpouring of response to this new government program—in essence an attempt by people to reclaim some of their time and privacy from increasingly aggressive marketing tactics—hints at the frustration many individuals feel when economic forces begin to dominate rather than serve them. Yet a small but growing number of consumers are questioning the way they shop, the amount of "stuff" crowding and complicating their lives and the amount of time they spend at work. These dissatisfied consumers have not yet built a coherent movement, because their actions are mostly private ones occurring in unconnected pockets in many nations. Still, the spontaneous and grassroots nature of these activities may signal a deeply felt desire by many people to build a satisfying life for themselves and their families.[1]

Perhaps the most apparent expression of a desire for a higher quality of life is found in the growing numbers of people who shop with an eye toward well-being. In Europe, for example, demand for organically grown foods drove sales up to $10 billion in 2002, 8 percent above the previous year, as a public bruised by mad cow disease and other food scares increasingly sought assurances of the safety of its food supply. Market analysts estimate that 142 million Europeans are consumers of organics, although a "loyal" core of 20 million accounted for 69 percent of the expenditures on these products in 2001. And 150 million people in Europe are either vegetarians or have reduced their consumption of meat.[2]

Meanwhile, in the United States the group of consumers interested in shopping for better health and a better environment is large enough to have earned recognition by market researchers as a distinct demographic group. Dubbed LOHAS consumers—people who lead Lifestyles of Health and Sustainability—these shoppers buy everything from compact fluorescent lightbulbs and solar cells to fair-trade coffee and chocolate (products that pay a just wage to producers or that have a lighter environmental impact than mainstream purchases do). This group now includes nearly one third of adult Americans and in 2000 accounted for about $230 billion in purchases—some 3 percent of total U.S. consumer expenditures. Although this is a relatively low share of expenditures compared with the number of people identified as LOHAS consumers, this is probably due to the few options for healthy consumption available today.[3]

In many countries, people are joining consumer cooperatives to leverage their market power for a higher quality of life. In Japan, for example, the 250,000-member Seikatsu Club Consumers' Cooperative Union stocks foods free of agricultural chemicals and artificial additives and preservatives, along with household products free of toxins. The club puts its goods in reusable glass jars in order to help reduce the 60 percent of household waste that is packaging. In contrast to many supermarkets that stock tens of thousands of individual items, the Seikatsu Club co-ops carry just 2,000 items, mostly basic foodstuffs. The co-ops typically carry only one or two choices per item, but for members seeking to live a more satisfying life, the high quality, healthy foods, and reduced waste apparently compensate for the somewhat lessened choice. And Seikatsu members are not alone; some 50 million people belong to local co-ops that are affiliated with Consumer Coop International, a global body that helps facilitate training for local consumer co-ops.[4]

In some cases, individuals are turning to organizations for help in greening their consumption. A coalition of organizations in 19 countries known as the Global Action Plan offers training to families on reducing waste, lessening energy use, and switching to eco-friendly products. In the Netherlands, at least 10,000 households are working on redirecting their consumption; after training, these people cut their household waste on average by 28 percent. Six to nine months later, the figure was 39 percent. And in 2003, the French government launched a similar initiative, *la famille durable* (the sustainable family), that offers practical ways for people to live sustainably at home, school, and work and on vacation.[5]

And in the United States, the Center for a New American Dream urges people to live a life of "more fun, less stuff." Through its Turn the Tide program, the Center encourages people to follow a simple nine-step environmental conservation plan, involving such actions as switching to water-efficient faucets and eating less meat. The 14,000 members of this initiative report saving more than 500 million liters of water and preventing over 4 million kilograms of carbon dioxide from being released into the atmosphere.[6]

Beyond a shift in shopping habits, many consumers are trying to simplify their lifestyles in broader ways—a process sometimes called "downshifting." Analyst Cecile Andrews describes the motivation for these individuals: "A lot of people [are] rushed and frenzied and stressed. They have no time for their friends; they snap at their family; they're not laughing very much." Many, she says, "are looking for ways to simplify their lives—to rush less, work less, and spend less. They are beginning to slow down and enjoy life again."[7]

Estimates of the numbers of downshifters are imprecise, but interest in simplifying appears to be growing. In seven European countries, the number of people who have voluntarily reduced their working hours has grown at 5.3 percent each year over the past five years, for example. And the trend toward simplicity is expected to continue. The number of people in these same countries who could at least partially embrace a voluntary simplicity lifestyle is expected to grow from about 7 million in 1997 to at least 13 million in 2007.[8]

Meanwhile, two research surveys in the United States in the mid-1990s suggested that around a quarter of the population were working to simplify their lives, although the extent of course varied greatly from person to person. And the media have registered growing interest in the topic. Articles in U.S. newspapers about simplifying lifestyles grew three to fivefold between 1996 and 1998. In 1997 the Public Broadcasting System aired a documentary called *Affluenza*, which treated consumerism as a contagious disease and offered suggestions for inoculating yourself against it. The program was very popular and was later distributed in 17 countries.[9]

Yet individual initiatives are only part of what is needed to build a society of well being. Individual efforts alone do not necessarily help to build strong, healthy communities (although they can free up time that could lead to greater community involvement), nor can they address the structural obstacles to genuine consumer choice—the lack of organic produce in the supermarket, for instance. Some critics even argue that, pursued in isolation, individual initiatives can be counterproductive. An "individualization of responsibility," as political and environmental scientist Michael Maniates notes, distracts attention from the role that such institutions as business and government play in perpetuating unhealthy consumption. Moreover, to the extent that individuals see their power residing primarily in their pocketbooks, they may neglect their key roles as parents, educators, community members; and citizens in building a society of well-being.[10]

The need for individuals to act collectively to improve their quality of life led a group in Norway in 2000 to launch a campaign entitled 07-06-05. Campaigners are rallying Norwegians to count down to June 7th, 2005, the one-hundredth anniversary of Norway's independence from Sweden, and to once again declare their independence—but this time from the "time poverty" that has accompanied the ascendancy of the consumer culture.[11]

In the United States, an alliance known as the U.S. Simplicity Forum is trying to mobilize the millions of Americans struggling with too much to do and too little time. They organized Take Back Your Time Day on October 24th, 2003, urging Americans to leave work early, arrive late, take longer than usual lunches, or even skip work altogether. Thousands joined events at neighbors' homes, local churches, meeting halls, and universities to discuss the time poverty facing the average American. The date was deliberately chosen—it was nine weeks before the end of the year—to remind Americans that they are some of the most overworked people in the industrial world, putting in 350 hours more on the job (that is, nine work-weeks) each year than the average European.[12]

Organizers hope to use the energy of the American initiative to start a popular movement centered on reclaiming time for a higher quality of life. The campaign would seek to reform national vacation laws, working hours, and other measures that would free up time for the neglected elements of life, such as family, friends, and community. As Take Back Your Time Day coordinator and *Affluenza* producer John de Graaf explains, "The Time Movement is about looking beyond GDP as the measure of a good society and understanding that the real purpose of our economy is not material growth without end, but a balanced, fulfilling, and sustainable life for all."[13]

THE TIES THAT BIND

Humans are social beings, so it is little surprise that good relationships are one of the most important ingredients for a high quality of life. Harvard Professor of Public Policy Robert Putnam notes that "the single most common finding from a half century's research on the correlates of life satisfaction ... is that happiness is best predicted by the breadth and depth of one's social connections." Thus individual efforts to build a satisfying life are more likely to be successful if some of them involve family, friends, or neighbors. Fortunately, individual efforts and community efforts often work hand in hand. The person who works fewer hours each week finds more time for family, friends, and community. And community ties, which are strengthened, for example, when neighbors share tools or babysitting responsibilities, can reduce family expenses and help people lead simpler lives.[14]

People who are socially connected tend to be healthier—often significantly so. More than a dozen long-term studies in Japan, Scandinavia, and the United States show that the chances of dying in a given year, no matter the cause, is two to five times greater for people who are isolated than for socially connected people. For example, one study found that in 1,234 heart attack patients, the rate of a recurring attack within six months was nearly double for those living alone. And a Harvard study of health and mistrust in the United States concluded that moving to a state with a high level of social connections from a state where the level is low would improve a person's health almost as much as quitting smoking.[15]

A particularly impressive example of the relationship between social connectedness and health comes from a study of the town of Roseto, Pennsylvania, which caught the attention of researchers in the 1960s because its rate of heart attacks was less than half the rates in neighboring towns. The usual causes of such an anomaly—diet, exercise, weight, smoking, genetic predisposition, and so on—did not explain the Roseto phenomenon. In fact, people in Roseto scored worse on many of these risk factors than their neighbors. So the researchers looked for other possible explanations and found that the town had a tight-knit social structure that had produced community-initiated sports clubs, churches, a newspaper, and a Scout troop. Extensive informal socializing was the norm. Eventually researchers gave credit to the strong social ties of the residents—most were from the same village in Italy and worked hard to maintain their sense of community in the United States—for the higher levels of health. The sad postscript to the story is that starting in the late 1960s, as social ties weakened in this town and across the United States, the heart attack rate in Roseto rose, eventually surpassing that of a neighboring town.[16]

Researchers offer various explanations for the link between social connectedness and lower risk of health problems. Some are quite practical: connected people have someone to depend on if they run into health problems, thereby reducing the likelihood that sickness will develop into a serious health condition. Social networks may reinforce healthy behaviors; studies show that isolated people are more likely to smoke or drink, for example. And cohesive communities may be more effective at lobbying for medical care. But the connection may run deeper. Social contact may actually stimulate a person's immune system to resist disease and stress. Laboratory animals, for example, are more likely to develop hardening of the arteries when isolated, while animals and humans in isolation both tend to experience decreased immune response and higher blood pressure.[17]

International development professionals also now acknowledge that strong social ties are a major contributor to a country's development. The World Bank, for instance, sees social connectedness as a form of capital—an asset that yields a stream of benefits useful for development. Just as a bank account (financial capital) yields interest, social ties tend to build trust, reciprocity, or information networks, all of which can grease the wheels of economic activity. Trust, for example, facilitates financial transactions by creating a climate of confidence in contractual relationships or in the safety of investments. A World Bank study of social contacts among agricultural traders in Madagascar found that those who are part of an extensive network of traders and can count on colleagues for help in times of trouble have higher incomes than traders with fewer contacts. Indeed, the connected traders say that relationships are more important for their success than many economic factors, including the price of their traded goods or access to credit or equipment.

A lack of social capital also seems to be connected with poor economic growth at the national level. Stephen Knack of the World Bank warns that low levels of societal trust may lock countries in a "poverty trap," in which the vicious circle of mistrust, low investment, and poverty is difficult to break. Knack and his colleagues tested the relationship between trust and economic performance in 29 countries included in the World Values Survey. They found that each 12-point rise in the survey's measure of trust was associated with a 1-percent increase in annual income growth, and that each 7-point rise in trust corresponded to a 1-percent increase in investment's share of GDP.

The role of social glue in facilitating economic transactions is especially evident in microcredit initiatives such as the Grameen Bank of Bangladesh, which provides small loans to very poor women who lack the collateral to borrow from a commercial bank. Participating women organize themselves into borrowing groups of five, and each group applies to the Bank for loans, often of less than $100. The women count on knowledge of their neighbors' dependability when they extend invitations to join the group. This information function—something commercial banks spend money on when they compile an applicant's credit history—is an example of how social capital can lower the costs of financial activity. Social ties are also meant to serve as collateral for the loans. Because women are jointly responsible for repayment, and because a default puts all five in jeopardy of disqualification for future loans, each woman is subject to strong social pressure to repay.

The economic payoff of these types of social connectedness has made microcredit successful in many parts of the world. The Grameen Bank claims that 98 percent of its microcredit loans are repaid, a better record than in most commercial banks. Grameen has inspired the spread of microcredit globally . . . .

Beyond improving health and facilitating economic security, strong social ties are especially helpful in promoting collective consumption, which often has social and environmental advantages. A good example of this is co-housing, a modern form of village living in which 10–40 individual households live in a development designed to stimulate neighborly interaction. Privacy is valued and respected, but residents share key spaces, including a common dining hall, gardens, and recreational space . . . .

CREATING INFRASTRUCTURES OF WELL-BEING

When individuals or communities seek to enhance their quality of life, they may be handcuffed by the set of choices available to them. Organic produce, reusable beverage bottles, or mass transit obviously cannot be bought if they are not offered for sale. The rules and policies that determine the set of choices available, such as oil subsidies that make fossil energy cheaper than wind power, zoning laws that encourage sprawling development, or building codes that frown on the use of recycled building materials, are essentially the "infrastructure of consumption." Creating a higher quality of life requires us all—individuals and communities—to help create new political, physical and cultural "infrastructures of well-being." . . .

GETTING TO THE GOOD LIFE

Lurking beneath growing dissatisfaction with the consumer society is a simple question: What is an economy for? The traditional responses, including prosperity, jobs, and expanded opportunity, seem logical enough—until they become dysfunctional, that is. When prosperity makes us overweight, overwork leaves us exhausted, and a "you can have it all" mindset leads us to neglect family and friends, people start to question more deeply the direction of their lives as well as the system that helps steer them in that direction. The signals emerging in some industrial countries—and some developing ones as well—suggest that many of us are looking for more from life than a bigger house and a new car. People long for something deeper: happy, dignified, and meaningful lives—in a word, well-being. And they expect their economics to be a tool to this end, not an obstacle to it.

NOTES

1. Caroline E. Mayer, "Trade Group to Abide by No-Calls List," *Washington Post*, 29 September 2003.
2. Organic market from "The Global Market for Organic Food & Drink," *Organic Monitor*, July 2003; consumers from Datamonitor, "Organic, Natural, Ethical, & Vegetarian Consumers," report brochure (New York: February 2002).
3. Natural Marketing Institute, "Nearly One-third of Americans Identified as Values-based, Highly Principled Consumers, New Research Shows," press release (Harleysville, PA: 14 May 2002); Amy Cortese, "They Care About the World (and They Shop, Too)," *New York Times*, 20 July 2003; share of expenditures from World Bank, *World Development Indicators Database*, at media.worldbank.org/secure/data/qquery.php, viewed 10 October 2003.
4. Seikatsu Club, "Outline of the Seikatsu Club Consumers' Cooperative Union," at www.seikatsuclub.coop/english/top.html, viewed 21 October 2003; Consumer Coop International, "Action Plan 2002–2003," at www.coop.org/cci/activities/action_plan.htm, viewed 21 October 2003.
5. Global Action Plan at www.globalaction plan.com/index.html, viewed 11 October 2003; Global Action Plan UK, *Annual Report and Accounts 2001–2002* (London: December 2002); Global Action Plan, "The Sustainable Lifestyle Campaign," at www.globalactionplan.org/Files/SLC.htm, viewed 20 October 2003; Minister of the Environment and Sustainable Development, "Week of Sustainable Consumption," press release (Paris: 25 March 2003); *la famille durable* at www.familledurable.com, viewed 10 October 2003.
6. Center for a New American Dream, "Turn the Tide: Nine Actions for the Planet," at www.newdream.org/TurntheTide/default.asp, viewed 10 October 2003.
7. Cecile Andrews cited in Michael Maniates, "In Search of Consumptive Resistance," in Thomas Princen, Michael Maniates, and Ken Conca, eds., *Confronting Consumption* (Cambridge, MA: The MIT Press, 2002), p. 200.
8. Datamonitor, "Simplicity," report brochure (New York: May 2003).
9. Surveys from Michael Maniates, "In Search of Consumptive Resistance," in Princen, Maniates, and Conca, op. cit. note 16, pp. 200–01, from Juliet Schor, *The Overspent American: Why We Want What We Don't Need* (New York: Harper-Perennial, 1998), pp. 113–15, and from The Harwood Group, *Yearning for Balance: Views of Americans on Consumption, Materialism, and the Environment* (Takoma Park, MD: Merck Family Fund, 1995); media and *Affluenza* from Maniates, op. cit. this note, p. 201.
10. Michael Maniates,"Individualization: Plant a Tree, Buy a Bike, Save the World?" in Princen, Maniates, and Conca, op. cit. note 17, p. 45.
11. Information on 07-06-05 at www.07-06-05.com/765/381.htm, viewed 11 October 2003.
12. John de Graaf, Take Back Your Time Day, discussion with Erik Assadourian, 24 October 2003; Take Back Your Time Day, "Take Back Your Time Day Campaign Launch," press release (Seattle, WA: 25 March 2003).

13. De Graff, op. cit. note 12.
14. Robert D. Putnam, *Bowling Alone: The Collapse and Revival of American Community* (New York: Simon & Schuster, 2000), p. 332.
15. Putnam, op. cit. note 14, pp. 327–28; David G. Myers, "Close Relationships and Quality of Life," in D. Kahneman, E. Diener, and N. Schwarz, eds., *Well-Being: The Foundations of Hedonic Psychology* (New York: Russell Sage Foundation, 1999), p. 377.
16. B. Egolf et al., "The Roseto Effect: A 50-Year Comparison of Mortality Rates," *American Journal of Public Health*, August 1992, pp. 1089–92; Putnam. op. cit. note 14, p. 329.
17. Putnam, op. cit. note 14, p. 327.

CHAPTER 3

Socialization and Gender

A Lifelong Learning Process

Alan Mc Evoy

Alan Mc Evoy

How do we become who we are, with our particular values, attitudes, and behaviors? We do so through the process of socialization that takes place within a particular society such as the United States, France, Kenya, or Japan. Each society has a culture or way of life that has developed over time and this dictates appropriate, acceptable behavior. It is within our families that initial socialization takes place. Here we develop a self, learn to be social beings, and develop into members of our culture. From the day we are born, socialization shapes us into social beings, teaching us the behaviors and beliefs that make social existence possible. Through interaction with others, we develop our self-concepts. This process begins at an early age, when young children interact with others in a process called *symbolic interaction*. Very simply, the individual (whether a baby or adult) initiates contact—a cry or words—and receives a response. By interpreting and reacting to that response, the individual learns what brings positive reactions; those actions that receive positive responses and rewards are likely to be continued. Actions receiving negative responses are likely to be dropped.

The process of socialization takes place through interaction with others—interaction that is vital to our social development. The initial *agent,* or transmitter of socialization, is the family. Here we begin to learn our roles for participation in the wider world, an important aspect of which is sex-role socialization. The process continues in educational and religious institutions. When children enter school, they face new challenges and expectations. No longer do they receive unconditional love as they do in most families; now they are judged in a competitive environment, their first introduction to the world outside the protection of home. Socialization typically takes place in a series of developmental stages from birth through old age. Some sociologists focus on childhood stages, others on male or female socialization, and others on middle to old age.

Informal agents of socialization—those whose primary purpose is something other than socialization, such as entertainment—can have a major impact on the process; for instance, the mass media, books, and advertising all send out powerful messages about desirable and appropriate behaviors by presenting role models and lifestyles. Peer groups also affirm or disapprove of behaviors of children. The boy who does not engage in "masculine" activities may be ridiculed, for instance.

You might ask if social development is a natural outgrowth of physical maturity. To answer that question, some social scientists have focused their studies on cases of social isolation. In rare instances, children do not experience early socialization in the family. For example, some cases of physical and mental abuse and neglect and some orphanages provide only minimal care and human contact; children growing up in such environments have been found to show a higher percentage of physical and mental retardation. Only a handful of cases of almost total isolation have been available for study. Kingsley Davis describes the case histories of Anna and Isabelle and considers the impact of social isolation on their mental development. He writes that severe retardation is likely to occur when consistent contact with other human beings is absent, and he concludes that even though socialization can take place after prolonged periods of isolation, some effects of isolation may be permanent.

Although family, educational institutions, and religious organizations have as their stated purposes to socialize their members, we learn from many sources. Informal agents of socialization, mentioned earlier, come in many forms, one of which is television. Television is one of the most powerful informal agents of socialization, especially for children, who may watch several hours a day.

Concern over the impact of TV violence has prompted several studies, including those reported in this chapter. The Kaiser Family Foundation review of studies points out both major findings and methodologies used to study television violence.

The socialization process can be very rigid and difficult to change when it involves ascribed roles. Socialization into gender roles provides examples of both the socialization process and the acquisition and tenacious nature of stereotypes. In the fourth reading in this chapter, William Pollack addresses the socialization experience of men and boys and how they learn masculine gender roles. Each agent of socialization adds to the process of learning attitudes and behaviors: The family provides role models and reinforcement for "proper" behaviors; teachers reward boys and girls for gender-typed behaviors; religious institutions support traditional role behaviors; peer groups pressure boys into acceptable male behaviors. Informal or *nonpurposive* agents of male socialization, including mass media, barbershops, bars, and business meetings, also contribute to the process.

Results of rigid and stereotypic socialization can be destructive for men and women. Consider a problem that afflicts primarily young women—eating disorders, specifically anorexia and bulimia. Diane Taub and Penelope McLorg discuss the social context in which these disorders occur: the societal pressures to achieve and maintain a certain body image, the source of this ideal body image, how women are socialized, and the role models presented to them by the mass media. The result of these socialization influences can be dieting and purging, which cause severe health problems.

Resocialization, radically changing a person's status and role by controlling the physical and social environment, occurs when individuals enter *total institutions* such as prisons, mental institutions, monasteries, and some boarding schools. The final article in this section by Gwynne Dyer addresses the resocialization that takes place in military boot camps.

As you read the selections in this chapter, consider aspects of socialization we have mentioned. What are the effects of isolation from normal human contact? What are the stages of socialization? How does socialization take place through agents of socialization? What are the effects of the socialization process that we take for granted, such as gender stereotyping?

11

Final Note on a Case of Extreme Isolation

KINGSLEY DAVIS

Socialization takes place in stages throughout our lives. We learn through agents of socialization such as family and education, and we are socialized into proper sex roles and other specialized roles as in athletics. Socialization requires contact with other human beings. What would we be like if we were raised in isolation with limited or no contact with other humans? Could socialization take place? Rare cases exist in which humans have grown up in partial or total isolation. This reading discusses two such cases and the results of isolation for these two girls.

As you read, ask yourself the following:

1. *Why do humans need other humans in order to develop "normally"?*
2. *What is missing in the experience of isolated children, and how does this affect the children?*
3. *What happens if a child experiences neglect or abuse during socialization?*
4. *What message does this article provide about what we can do to help children grow up in healthy environments?*

GLOSSARY **Socialization** The process of learning cooperative group living. **Learning stage** The knowledge and ability individuals are expected to have attained at a particular age.

Early in 1940 there appeared . . . an account of a girl called Anna.[1] She had been deprived of normal contact and had received a minimum of human care for almost the whole of her first six years of life. At this time observations were not complete and the report had a tentative character. Now, however, the girl is dead, and with more information available,[2] it is possible to give a fuller and more definitive description of the case from a sociological point of view.

Anna's death, caused by hemorrhagic jaundice, occurred on August 6, 1942. Having been born on March 1 or 6,[3] 1932, she was approximately ten and a half years of age when she died. The previous report covered her development up to the age of almost eight years; the present one recapitulates the earlier period on the basis of new evidence and then covers the last two and a half years of her life.

EARLY HISTORY

The first few days and weeks of Anna's life were complicated by frequent changes of domicile. . . . She was an illegitimate child, the second such child born to her mother, and her grandfather, a widowed

SOURCE: Reprinted from *American Journal of Sociology*, Vol. III, No. 5, March 1947, pp. 432–437 by permission of the author.

farmer in whose house the mother lived, strongly disapproved of this new evidence of the mother's indiscretion. This fact led to the baby's being shifted about.

Two weeks after being born in a nurse's private home, Anna was brought to the family farm, but the grandfather's antagonism was so great that she was shortly taken to the house of one of her mother's friends. At this time a local minister became interested in her and took her to his house with an idea of possible adoption. He decided against adoption, however, when he discovered that she had vaginitis. The infant was then taken to a children's home in the nearest large city. This agency found that at the age of only three weeks she was already in a miserable condition, being "terribly galled and otherwise in very bad shape." It did not regard her as a likely subject for adoption but took her in for a while anyway, hoping to benefit her. After Anna had spent nearly eight weeks in this place, the agency notified her mother to come to get her. The mother responded by sending a man and his wife to the children's home with a view to their adopting Anna, but they made such a poor impression on the agency that permission was refused. Later the mother came herself and took the child out of the home and then gave her to this couple. It was in the home of this pair that a social worker found the girl a short time thereafter. The social worker went to the mother's home and pleaded with Anna's grandfather to allow the mother to bring the child home. In spite of threats, he refused. The child, by then more than four months old, was next taken to another children's home in a nearby town. A medical examination at this time revealed that she had impetigo, vaginitis, umbilical hernia, and a skin rash.

Anna remained in this second children's home for nearly three weeks, at the end of which time she was transferred to a private foster-home. Since, however, the grandfather would not, and the mother could not, pay for the child's care, she was finally taken back as a last resort to the grandfather's house (at the age of five and a half months). There she remained, kept on the second floor in an attic-like room because her mother hesitated to incur the grandfather's wrath by bringing her downstairs.

The mother, a sturdy woman weighing about 180 pounds, did a man's work on the farm. She engaged in heavy work such as milking cows and tending hogs and had little time for her children. Sometimes she went out at night, in which case Anna was left entirely without attention. Ordinarily, it seems, Anna received only enough care to keep her barely alive. She appears to have been seldom moved from one position to another. Her clothing and bedding were filthy. She apparently had no instruction, no friendly attention.

It is little wonder that, when finally found and removed from the room in the grandfather's house at the age of nearly six years, the child could not talk, walk, or do anything that showed intelligence. She was in an extremely emaciated and undernourished condition, with skeletonlike legs and a bloated abdomen. She had been fed on virtually nothing except cow's milk during the years under her mother's care.

Anna's condition when found, and her subsequent improvement, have been described in the previous report. It now remains to say what happened to her after that.

LATER HISTORY

In 1939, nearly two years after being discovered, Anna had progressed, as previously reported, to the point where she could walk, understand simple commands, feed herself, achieve some neatness, remember people, etc. But she still did not speak, and, though she was much more like a normal infant of something over one year of age in mentality, she was far from normal for her age.

On August 30, 1939, she was taken to a private home for retarded children, leaving the country home where she had been for more than a year and a half. In her new setting she made some further progress, but not a great deal. In a report of an examination made November 6 of the same year, the head of the institution pictured the child as follows:

> Anna walks about aimlessly, makes periodic rhythmic motions of her hands, and, at intervals, makes guttural and sucking noises. She regards her hands as if she had seen them for the first time. It was impossible to hold her attention for more than a few seconds at a time—not because of distraction due to external stimuli but because of her inability to concentrate. She ignored the task in hand to gaze vacantly about the room. Speech is entirely lacking. Numerous unsuccessful attempts have been made with her in the hope of developing initial sounds. I do not believe that this failure is due to negativism or deafness but that she is not sufficiently developed to accept speech at this time.... The prognosis is not favorable.

More than five months later, on April 25, 1940, a clinical psychologist, the late Professor Francis N. Maxfield, examined Anna and reported the following: large for her age; hearing "entirely normal"; vision apparently normal; able to climb stairs; speech in the "babbling stage" and "promise for developing intelligible speech later seems to be good." He said further that "on the Merrill-Palmer scale she made a mental score of 19 months. On the Vineland social maturity scale she made a score of 23 months."[4]

Professor Maxfield very sensibly pointed out that prognosis is difficult in such cases of isolation. "It is very difficult to take scores on tests standardized under average conditions of environment and experience," he wrote, "and interpret them in a case where environment and experience have been so unusual." With this warning he gave it as his opinion at that time that Anna would eventually "attain an adult mental level of six or seven years."[5]

The school for retarded children, on July 1, 1941, reported that Anna had reached 46 inches in height and weighed 60 pounds. She could bounce and catch a ball and was said to conform to group socialization, though as a follower rather than a leader. Toilet habits were firmly established. Food habits were normal, except that she still used a spoon as her sole implement. She could dress herself except for fastening her clothes. Most remarkable of all, she had finally begun to develop speech. She was characterized as being at about the two-year level in this regard. She could call attendants by name and bring in one when she was asked to. She had a few complete sentences to express her wants. The report concluded that there was nothing peculiar about her, except that she was "feeble-minded—probably congenital in type."[6]

A final report from the school made on June 22, 1942, and evidently the last report before the girl's death, pictured only a slight advance over that given above. It said that Anna could follow directions, string beads, identify a few colors, build with blocks, and differentiate between attractive and unattractive pictures. She had a good sense of rhythm and loved a doll. She talked mainly in phrases but would repeat words and try to carry on a conversation. She was clean about clothing. She habitually washed her hands and brushed her teeth. She would try to help other children. She walked well and could run fairly well, though clumsily. Although easily excited, she had a pleasant disposition.

INTERPRETATION

Such was Anna's condition just before her death. It may seem as if she had not made much progress, but one must remember the condition in which she had been found. One must recall that she had no glimmering of speech, absolutely no ability to walk, no sense of gesture, not the least capacity to feed herself even when the food was put in front of her, and no comprehension of cleanliness. She was so apathetic that it was hard to tell whether or not she could hear. And all this at the age of nearly six years. Compared with this condition, her capacities at the time of her death seem striking indeed, though they do not amount to much more than a two-and-a-half year mental level. One conclusion therefore seems safe, namely, that her isolation prevented a considerable amount of mental development that was undoubtedly part of her capacity. Just what her original capacity was, of course, is hard to say; but her development after her period of confinement (including the ability to walk and run, to play, to dress, fit into a social situation, and, above all, to speak) shows that she had at least this capacity—capacity that never could have been realized in her original condition of isolation.

A further question is this: What would she have been like if she had received a normal upbringing from the moment of birth? A definitive answer would have been impossible in any case, but even an approximate answer is made difficult by her early death. If one assumes, as was tentatively surmised in the previous report, that it is "almost impossible for any child to learn to speak, think, and act like a normal person after a long period of early isolation," it seems likely that Anna might have had a normal or near-normal capacity, genetically speaking. On the other hand, it was pointed out that Anna represented "a marginal case [because] she was discovered before she had reached six years of age," an age "young enough to allow for some plasticity."[7] While admitting, then, that Anna's isolation *may* have been the major cause (and was certainly a minor cause) of her lack of rapid mental progress during the four-and-a-half years following her rescue from neglect, it is necessary to entertain the hypothesis that she was congenitally deficient.

In connection with this hypothesis, one suggestive though by no means conclusive circumstance needs consideration, namely, the mentality of Anna's forebears. Information on this subject is easier to obtain, as one might guess, on the mother's than on the father's side. Anna's maternal grandmother, for example, is said to have been college educated and wished to have her children receive a good education, but her husband, Anna's stern grandfather, apparently a shrewd, hard-driving, calculating farmowner, was so penurious that her ambitions in this direction were thwarted. Under the circumstances her daughter (Anna's mother) managed, despite having to do hard work on the farm, to complete the eighth grade in a country school. Even so, however, the daughter was evidently not very smart. "A schoolmate of [Anna's mother] stated that she was retarded in school work; was very gullible at this age; and that her morals even at this time were discussed by other students." Two tests administered to her on March 4, 1938, when she was thirty-two years of age, showed that she was mentally deficient. On the Stanford Revision of the Binet-Simon Scale her performance was equivalent to that of a child of eight years, giving her an I.Q. of 50 and indicating mental deficiency of "middle-grade moron type."[8]

As to the identity of Anna's father, the most persistent theory holds that he was an old man about seventy-four years of age at the time of the girl's birth. If he was the one, there is no indication of mental or other biological deficiency, whatever one may think of his morals. However, someone else may actually have been the father.

To sum up: Anna's heredity is the kind that *might* have given rise to innate mental deficiency, though not necessarily.

COMPARISON WITH ANOTHER CASE

Perhaps more to the point than speculation about Anna's ancestry would be a case for comparison. If a child could be discovered who had been isolated about the same length of time as Anna but had achieved a much quicker recovery and a greater mental development, it would be a stronger indication that Anna was deficient to start with.

Such a case does exist. It is the case of a girl found at about the same time as Anna and under strikingly similar circumstances. A full description of the details of this case has not been published, but in addition to newspaper reports, an excellent preliminary account by a speech specialist, Dr. Marie K. Mason, who played an important role in the handling of the child, has appeared.[9] Also the late Dr. Francis N. Maxfield, clinical psychologist at Ohio State University, as was Dr. Mason, has written an as yet unpublished but penetrating analysis of the case.[10] Some of his observations have been included in Professor Zingg's book on feral man.[11] The following discussion is drawn mainly from these enlightening materials. The writer, through the kindness of

Professors Mason and Maxfield, did have a chance to observe the girl in April, 1940, and to discuss the features of her case with them.

Born apparently one month later than Anna, the girl in question, who has been given the pseudonym Isabelle, was discovered in November, 1938, nine months after the discovery of Anna. At the time she was found she was approximately six-and-a-half years of age. Like Anna, she was an illegitimate child and had been kept in seclusion for that reason. Her mother was a deaf-mute, having become so at the age of two, and it appears that she and Isabelle had spent most of their time together in a dark room shut off from the rest of the mother's family. As a result Isabelle had no chance to develop speech; when she communicated with her mother, it was by means of gestures. Lack of sunshine and inadequacy of diet had caused Isabelle to become rachitic. Her legs in particular were affected; they "were so bowed that as she stood erect the soles of her shoes came nearly flat together, and she got about with a skittering gait."[12] Her behavior toward strangers, especially men, was almost that of a wild animal, manifesting much fear and hostility. In lieu of speech she made only a strange croaking sound. In many ways she acted like an infant. "She was apparently utterly unaware of relationships of any kind. When presented with a ball for the first time, she held it in the palm of her hand, then reached out and stroked my face with it. Such behavior is comparable to that of a child of six months."[13] At first it was even hard to tell whether or not she could hear, so unused were her senses. Many of her actions resembled those of deaf children.

It is small wonder that, once it was established that she could hear, specialists working with her believed her to be feeble-minded. Even on nonverbal tests her performance was so low as to promise little for the future. Her first score on the Stanford-Binet was 19 months, practically at the zero point of the scale. On the Vineland social maturity scale her first score was 39, representing an age level of two-and-a-half years.[14] "The general impression was that she was wholly uneducable and that any attempt to teach her to speak, after so long a period of silence, would meet with failure."[15]

In spite of this interpretation, the individuals in charge of Isabelle launched a systematic and skillful program of training. It seemed hopeless at first. The approach had to be through pantomime and dramatization, suitable to an infant. It required one week of intensive effort before she even made her first attempt at vocalization. Gradually she began to respond, however, and, after the first hurdles had at least been overcome, a curious thing happened. She went through the usual stages of learning characteristic of the years from one to six not only in proper succession but far more rapidly than normal. In a little over two months after her first vocalization she was putting sentences together. Nine months after that she could identify words and sentences on the printed page, could write well, could add to ten, and could retell a story after hearing it. Seven months beyond this point she had a vocabulary of 1500–2000 words and was asking complicated questions. Starting from an educational level of between one and three years (depending on what aspect one considers), she had reached a normal level by the time she was eight-and-a-half years old. In short, she covered in two years the stages of learning that ordinarily require six.[16] Or, to put it another way, her I.Q. trebled in a year and a half.[17] The speed with which she reached the normal level of mental development seems analogous to the recovery of body weight in a growing child after an illness, the recovery being achieved by an extra fast rate of growth for a period after the illness until normal weight for the given age is again attained.

When the writer saw Isabelle a year-and-a-half after her discovery, she gave him the impression of being a very bright, cheerful, energetic little girl. She spoke well, walked and ran without trouble, and sang with gusto and accuracy. Today she is over fourteen years old and has passed the sixth grade in a public school. Her teachers say that she participates in all school activities as normally as other children. Though older than her classmates, she has fortunately not physically matured too far beyond their level.[18]

Clearly the history of Isabelle's development is different from that of Anna's. In both cases there was an exceedingly low, or rather blank, intellectual level to begin with. In both cases it seemed that the girl might be congenitally feeble-minded. In both a considerably higher level was reached later on. But the Ohio girl achieved a normal mentality within two years, whereas Anna was still markedly inadequate at the end of four-and-a-half years. This difference in achievement may suggest that Anna had less initial capacity. But an alternative hypothesis is possible.

One should remember that Anna never received the prolonged and expert attention that Isabelle received. The result of such attention, in the case of the Ohio girl, was to give her speech at an early stage, and her subsequent rapid development seems to have been a consequence of that. "Until Isabelle's speech and language development, she had all the characteristics of a feeble-minded child." Had Anna, who, from the standpoint of psychometric tests and early history, closely resembled this girl at the start, been given a mastery of speech at an earlier point by intensive training, her subsequent development might have been much more rapid.[19]

The hypothesis that Anna began with a sharply inferior mental capacity is therefore not established. Even if she were deficient to start with, we have no way of knowing how much so. Under ordinary conditions she might have been a dull normal or, like her mother, a moron. Even after the blight of her isolation, if she had lived to maturity, she might have finally reached virtually the full level of her capacity, whatever it may have been. That her isolation did have a profound effect upon her mentality, there can be no doubt. This is proved by the substantial degree of change during the four-and-a-half years following her rescue.

Consideration of Isabelle's case serves to show, as Anna's case does not clearly show, that isolation up to the age of six, with failure to acquire any form of speech and hence failure to grasp nearly the whole world of cultural meaning, does not preclude the subsequent acquisition of these. Indeed, there seems to be a process of accelerated recovery in which the child goes through the mental stages at a more rapid rate than would be the case in normal development. Just what would be the maximum age at which a person could remain isolated and still retain the capacity for full cultural acquisition is hard to say. Almost certainly it would not be as high as age fifteen; it might possibly be as low as age ten. Undoubtedly various individuals would differ considerably as to the exact age.

Anna's is not an ideal case for showing the effects of extreme isolation, partly because she was possibly deficient to begin with, partly because she did not receive the best training available, and partly because she did not live long enough. Nevertheless, her case is instructive when placed in the record with numerous other cases of extreme isolation. This and the previous article about her are meant to place her in the record. It is to be hoped that other cases will be described in the scientific literature as they are discovered (as unfortunately they will be), for only in these rare cases of extreme isolation is it possible "to observe *concretely separated* two factors in the development of human personality which are always otherwise only analytically separated, the biogenic and the sociogenic factors."[20]

NOTES

1. Kingsley Davis, "Extreme Social Isolation of a Child," *American Journal of Sociology, XLV* (January, 1940), 554–65.
2. Sincere appreciation is due to the officials in the Department of Welfare, commonwealth of Pennsylvania, for their kind cooperation in making available the records concerning Anna and discussing the case frankly with the writer. Helen C. Hubbell, Florentine Hackbusch, and Eleanor Mecklenburg were particularly helpful, as was Fanny L. Matchette. Without their aid neither of the reports on Anna could have been written.
3. The records are not clear as to which day.
4. Letter to one of the state officials in charge of the case.
5. *Ibid.*
6. Progress report of the school.
7. Davis, *op. cit.*, p. 564.
8. The facts set forth here as to Anna's ancestry are taken chiefly from a report of mental tests administered to Anna's mother by psychologists at a state hospital where she was taken for this purpose after the discovery of Anna's seclusion. This excellent report was not available to the writer when the previous paper on Anna was published.

9. Marie K. Mason, "Learning to Speak after Six and One-Half Years of Silence," *Journal of Speech Disorders, VII* (1942), 295–304.
10. Francis N. Maxfield, "What Happens When the Social Environment of a Child Approaches Zero." The writer is greatly indebted to Mrs. Maxfield and to Professor Horace B. English, a colleague of Professor Maxfield, for the privilege of seeing this manuscript and other materials collected on isolated and feral individuals.
11. J. A. L. Singh and Robert M. Zingg, *Wolf-Children and Feral Man* (New York: Harper & Bros., 1941), pp. 248–51.
12. Maxfield, unpublished manuscript cited above.
13. Mason, *op. cit.*, p. 299.
14. Maxfield, unpublished manuscript.
15. Mason, *op. cit.*, p. 299.
16. *Ibid.*, pp. 300–304.
17. Maxfield, unpublished manuscript.
18. Based on a personal letter from Dr. Mason to the writer, May 13, 1946.
19. This point is suggested in a personal letter from Dr. Mason to the writer, October 22, 1946.
20. Singh and Zingg, *op. cit.*, pp. xxi–xxii, in a foreword by the writer.

12

Key Facts on TV Violence

THE HENRY J. KAISER FAMILY FOUNDATION

Socialization takes place through formal agents (family, school, religion) and informal agents whose primary purpose is not socialization but who nonetheless affect the socialization process. One powerful informal agent of socialization is television, which has a major impact on who we are. The following reading presents findings from studies on the effect of violence depicted on television, especially on children. It presents summaries of studies using different methods of data collection: laboratory experiments, field experiments, cross-sectional studies, and longitudinal studies.

Consider the following questions as you read this article:

1. *How does television violence influence the socialization of girls and boys?*
2. *What methods have been used to study violence on TV? How would you conduct such a study?*
3. *What are the implications of this report for children's TV shows and TV viewing?*
4. *What do you think should be done about television violence?*

GLOSSARY **TV violence** Any overt depiction of use of physical force or credible threat of such force intended to harm people physically.

SOURCE: The Henry J. Kaiser Foundation, *Key Facts on TV Violence,* Spring 2003. www.kff.org/entmedia/

Since the advent of television, the effect of TV violence on society has been widely studied and vigorously debated. Based on the cumulative evidence of studies conducted over several decades the scientific and public health communities overwhelmingly conclude that viewing violence poses a harmful risk to children. Critics of the research challenge this conclusion and dispute claims that exposure to TV violence leads to real life aggression. As we move into the digital era with enhanced images and sound, media violence will undoubtedly continue to be a focus of public concern and scientific research.

PREVALENCE OF VIOLENCE ON TV

The National Television Violence Study is the largest content analysis undertaken to date. It analyzed programming over three consecutive TV seasons from 1994 to 1997.[1] Among the findings:

- Nearly 2 out of 3 TV programs contained some violence,[2] averaging about 6 violent acts per hour.[3]
- Fewer than 5% of these programs featured an anti-violence theme or prosocial message emphasizing alternatives to or consequences of violence.[4]
- Violence was found to be more prevalent in children's programming (69%) than in other types of programming (57%). In a typical hour of programming children's shows featured more than twice as many violent incidents (14) than other types of programming (6).[5]
- The average child who watches 2 hours of cartoons a day may see nearly 10,000 violent incidents each year, of which the researchers estimate that at least 500 pose a high risk for learning and imitating aggression and becoming desensitized to violence.[6]
- The number of prime-time programs with violence increased over the three years of the study, from 53% to 67% on broadcast television and from 54% to 64% on basic cable. Premium cable networks have the highest percentage of shows with violence, averaging 92% since 1994.[7]

The UCLA Television Violence Monitoring Report also analyzed three years of programming from 1994 to 1997. This study relied on the qualitative judgments of a team of student monitors and staff researchers rather than a systematic content analysis, to determine whether individual violent depictions "raised concern" for viewers.[8] Among the findings:

- Children's Saturday morning TV shows that feature "sinister combat violence" raised the most serious concerns for these researchers. These are fantasy live-action shows and animated cartoons in which violence is central to the storyline, the villains and superheroes use violence as an acceptable and effective way to get what they want, and the perpetrators are valued for their combatability. Among the most popular shows for children, the number of troubling shows in this genre decreased from seven to four over the three years of the study.[9]
- The number of prime time series that raised frequent concerns about violence steadily declined over the three years, from nine such series in 1994–95 to just two in 1996–97.[10]
- TV specials was the only category that raised new concerns at the end of the three years. In the second year five live-action reality shows featured real or re-created graphic images of animals attacking and sometimes killing people. By the third year the number of such shows had increased again.[11]

SCIENTIFIC STUDIES OF TV VIOLENCE EFFECTS

Researchers hypothesize that viewing TV violence can lead to three potentially harmful effects: increased antisocial or aggressive behavior, desensitization to violence (becoming more accepting of violence in real life and less caring about other people's feelings), or increased fear of becoming a victim of violence.[12] Many researchers believe that children age 7 and younger are particularly vulnerable to the effects of viewing violence because they tend to perceive fantasy and cartoon violence as realistic.[13]

Since the 1960s, a body of research literature has been accumulating on the effects of TV violence. Taken together, the studies conclude that TV violence is one of many factors that contribute to aggressive behavior. Following are examples of the various types of research studies that have been conducted:[14]

Laboratory experiments are conducted in a controlled setting in order to manipulate media exposure and assess the short-term effects. Participants are randomly assigned to view either a violent or nonviolent film clip and their subsequent behavior is observed.

- A series of classic experiments conducted in the 1960s provided the earliest evidence of a link between TV violence and aggression. In these studies, children who were exposed to a TV clip of an actor hitting an inflatable doll were more likely than children who did not see the clip to imitate the action in their play, especially if the aggressive actions in the film clip were rewarded.[15]
- Other laboratory experiments have indicated that exposure to media violence increases children's tolerance for real-life aggression. For example, when third- and fourth-graders were left in charge of two younger children they could see on a TV monitor, the ones who viewed an aggressive film were much more reluctant than those who had not seen the film to ask an adult for help when the younger children began to fight, even though the fight was becoming progressively aggressive.[16]

Field experiments are conducted in a more naturalistic setting. As with the laboratory studies, children are shown video clips and their short-term post-viewing behavior is monitored by researchers. Over the past 30 years, numerous field studies have indicated that some children behave more aggressively after viewing violence.

- In one study, researchers showed children episodes of either *Batman* and *Spiderman* or *Mister Rogers' Neighborhood* over several weeks and then observed their behavior for two weeks afterwards. The children who viewed violent cartoons were more likely to interact aggressively with their peers, while those children who watched *Mister Rogers' Neighborhood* became more cooperative and willing to share toys.[17]
- In another study, researchers exposed children to an episode of *Mighty Morphin Power Rangers* and then observed their verbal and physical aggression in the classroom. Compared to children who had not seen the episode, viewers committed seven times as many aggressive acts such as hitting, kicking, shoving, and insulting a peer.[18]

Cross-sectional studies survey a large and representative sample of viewers at one point in time. Since the 1970s, a large number of these studies have concluded that viewing TV violence is related to aggressive behavior and attitudes.[19] These studies are correlational and do not prove causality; that is, it is difficult to know whether watching violence on TV is causing the increase in aggression or whether viewers who are already aggressive prefer watching violent content.

- In one study, 2,300 junior and senior high school students were asked to list their four favorite programs, which were analyzed for violent content, and to provide a self-reported checklist of activities that ranged from fighting at school to serious delinquency. Researchers found that teens whose favorite programs were violent tended to report a higher incidence in overall aggressive and delinquent behavior.[20]
- A recent study demonstrated a relationship between children's bullying and their exposure to media violence. Third-, fourth-, and fifth-graders who were identified by their peers as being the ones who spread rumors, exclude and insult peers, and behave in ways that hurt others, were more likely to view violence than nonaggressive children.[21]

Longitudinal studies offer the best way to study long-term effects of exposure to violent TV content. These studies survey the same group of individuals at several different times over many years to determine whether viewing violence is related to subsequent aggressive behavior. This method is designed to detect causal relationships and statistically control for environmental, family, and personal characteristics that might otherwise account for aggression.

- One study demonstrated that TV habits of children in the 1960s were a significant predictor of adult aggression, even criminal behavior, regardless of children's initial aggressiveness, IQ, social status, or parenting style. In this study, which spans more than 20 years, boys who preferred and viewed more violent programming at age 8 were more likely to be aggressive as teenagers and have arrests and convictions as adults for interpersonal crimes such as spousal and child abuse, murder, and aggravated assault.[22]
- Television exposure during adolescence has also been linked to subsequent aggression in young adulthood. A 17-year longitudinal study concluded that teens who watched more than one hour of TV a day were almost four times as likely as other teens to commit aggressive acts in adulthood (22% versus 6%), taking into account prior aggressiveness, psychiatric disorders, family income, parental education, childhood neglect, and neighborhood violence.[23]

Meta-analyses use a statistical procedure to combine the results from many different studies.

- The largest meta-analysis on TV violence analyzed 217 studies conducted between 1957 and 1990 and found that viewing violence was significantly linked to aggressive and antisocial behavior, especially among the youngest viewers. The overall effect size was 31, meaning that exposure to TV violence was estimated to account for 10% of the variance in antisocial behavior.[24]

OPPOSING VIEWPOINT

A small number of critics of the scientific evidence have concluded that TV violence does not contribute to real-life aggression. For the most part, they do not base their conclusions on studies with contrary findings, but argue that the studies that have been conducted are flawed.[25]

NOTES

1. Center for Communication and Social Policy, University of California, Santa Barbara (UCSB), National Television Violence Study, Executive Summary, Volume 3, 1998. Commissioned by the National Cable Television Association, the study analyzed almost 10,000 hours of broadcast and cable programming randomly selected from 23 channels over the course of three TV seasons from 1994 to 1997.
2. Ibid., 30. Researchers defined three main types of violent depictions: credible threats, behavioral acts, and harmful consequences.
3. Ibid., 33.
4. Ibid.
5. Barbara Wilson et al., "Violence in Children's Television Programming: Assessing the Risks," *Journal of Communication* 52 (2002): 5–35.
6. Center for Communication and Social Policy, UCSB, 33–34.
7. Ibid., 32.
8. UCLA Center for Communication Policy, UCLA Television Violence Monitoring Report, 1998, <http://ccp.ucla.edu/Webreport96/tableof.htm> (28 September 2002). Commissioned by the National Association of Broadcasters, the researchers analyzed more than 3,000 hours of TV over three consecutive TV seasons from 1994 to 1997. TV shows with violence were divided into four different categories based on the level of concern, ranging from high levels of violence and serious concern to no serious concern because the context is appropriate.
9. Ibid., <http://ccp.ucla.com/Webreport96/network.htm#kids> (28 September 2002).
10. Ibid., <http://ccp.ucla.com/Webreport96/network.htm#Series> (28 September 2002).
11. Ibid., <http://ccp.ucla.com/Webreport96/network.htm#specials> (28 September 2002).

12. American Psychological Association, Report of the American Psychological Association Commission on Violence and Youth, Volume 1, 1993, 33; Stacy L. Smith and Edward Donnerstein, "Harmful Effects of Exposure to Media Violence: Learning of Aggression, Emotional Desensitization, and Fear," *Human Aggression: Theories, Research, and Implications for Social Policy*, eds. R. Geen and E. Donnerstein (New York: Academic Press, 1998), 167–202.
13. Brad Bushman and L. Rowell Huesmann, "Effects of Televised Violence on Aggression," *Handbook of Children and the Media*, eds. D. Singer and J. Singer (Thousand Oaks, CA: Sage Publications, 2001), 223–268.
14. For summaries of these studies, see Bushman and Huesmann, 2001; Victor Strasburger and Barbara Wilson, "Media Violence," *Children, Adolescents & the Media* (Thousand Oaks, CA: Sage Publications, 2002), 73–116; W. James Potter, *On Media Violence* (Thousand Oaks, CA: Sage, 1999).
15. Albert Bandura, Dorothea Ross, and Sheila Ross, "Transmission of Aggression through Imitation of Aggressive Models," *Journal of Abnormal and Social Psychology* 63 (1961): 575–582; Albert Bandura, Dorothea Ross, and Sheila Ross, "Imitation of Film-Mediated Aggressive Models," *Journal of Abnormal and Social Psychology* 66 (1963) 3–11; Albert Bandura, Dorothea Ross, and Sheila Ross, "Vicarious Reinforcement and Imitative Learning," *Journal of Abnormal and Social Psychology* 67 (1963) 601–607.
16. Ronald S. Drabman and Margaret Hanratty Thomas, "Does Media Violence Increase Children's Toleration of Real-Life Aggression?" *Developmental Psychology* 10 (1974): 418–421; Ronald S. Drabman and Margaret Hanratty Thomas, "Does Watching Violence on Television Cause Apathy?" *Pediatrics* 57 (1976): 329–331.
17. Aletha Huston-Stein and L.K. Friedrich, "Television Content and Young Children's Social Behavior," *Television and Social Behavior*, Volume *II, Television and Social Learning,* eds. J. Murray, E. Rubinstein, and G. Comstock (Washington, D.C.: U.S. Government Printing Office, 1972), 207–317.
18. Chris J. Boyatzis, Gina M. Matillo, and Kristen M. Nesbitt, "Effects of the 'Mighty Morphin Power Rangers' on Children's Aggression with Peers," *Child Study Journal* 25: 1 (1995): 45–55.
19. See, for example, Bushman and Huesmann, 2001; Strasburger and Wilson, 2002; Potter, 1999.
20. Jennie McIntyre and James Teevan Jr., "Television Violence and Deviant Behavior," *Television and Social Behavior*, Volume III, *Television and Adolescent Aggressiveness*, eds. G. Comstock and E. Rubinstein (Washington, D.C.: U.S. Government Printing Office, 1972), 383–435.
21. Audrey Buchanan et al., "What Goes in Must Come Out: Children's Media Violence Consumption at Home and Aggressiveness at School," <http.www.mediafamily.org/research/reports/issbd.shtml> (28 September 2002).
22. L. Rowell Huesmann et al., "Stability of Aggression Over Time and Generations," *Developmental Psychology* 20 (1984): 1120–1134. This study was begun in 1960 with a sample of 875 youths in New York state; ten years later 427 were re-interviewed, and twelve years later 409 were interviewed and criminal justice data was collected on 632 of the original subjects.
23. Jeffrey Johnson et al., "Television Viewing and Aggressive Behavior During Adolescence and Adulthood," *Science* 295 (March 29, 2002): 2468–2471. The longitudinal study was conducted over a 17-year time span with a sample of 707 families. Criminal arrest and charge data (assault or physical fights resulting in injury, robbery, threats to injure someone, or weapon used to commit a crime) were obtained.
24. Haejung Paik and George Comstock, "The Effects of Television Violence on Antisocial Behavior: A Meta-Analysis," *Communication Research* 21: 4 (August 1994): 516–546.
25. See, for example, Jonathan Freedman, *Media Violence and Its Effect on Aggression* (Toronto: University of Toronto Press, 2002); this review was commissioned by the Motion Picture Association of America. See also Jib Fowles, *The Case for Television Violence* (Newbury Park, CA: Sage Publications, 1999).

13

Real Boys

Rescuing Our Sons from the Myths of Boyhood

WILLIAM POLLACK

The socialization process teaches us societal expectations. Those expectations restrict our behaviors and can even have some harmful effects as seen in this reading. The Boy Code socializes boys to deny emotional pain and insist everything is all right even when it is not. Pollack analyzes the reasons for the Boy Code and suggests ways to avoid the negative consequences of male socialization.

As you read this article, answer the following questions:

1. *How did the Boy Code evolve and why is it difficult to change?*
2. *How does the Boy Code affect male socialization and behavior?*
3. *What practices reinforce the Boy Code?*
4. *What can be done to alter the effects of the Boy Code? Apply this to someone you know.*

GLOSSARY **Boy Code** Socialization of boys to behave according to ingrained cultural expectations that may have some negative consequences.
Shaming Making boys feel ashamed of their feelings of weakness and fears.

. . . THE GENDER STRAITJACKET

Many years ago, when I began my research into boys, I had assumed that since America was revising its ideas about girls and women, it must have also been reevaluating its traditional ideas about boys, men, and masculinity. But over the years my research findings have shown that as far as boys today are concerned, the old Boy Code—the outdated and constricting assumptions, models, and rules about boys that our society has used since the nineteenth century—is still operating in force. I have been surprised to find that even in the most progressive schools and the most politically correct communities in every part of the country and in families of all types, the Boy Code continues to affect the behavior of all of us—the boys themselves, their parents, their teachers, and society as a whole. None of us is immune—it is so ingrained. I have caught myself behaving in accordance with the code, despite my awareness of its falseness—denying sometimes that I'm emotionally in pain when in fact I am; insisting the everything is all right, when it is not.

The Boy Code puts boys and men into a gender straitjacket that constrains not only them but everyone else, reducing us all as human beings, and eventually making us strangers to ourselves and

SOURCE: From *Real Boys: Rescuing Our Sons from the Myths of Boyhood* by William Pollack, copyright 1998 by William Pollack. Used by permission of Random House, Inc.

to one another—or, at least, not as strongly connected to one another as we long to be. . . .

THE BOY CODE: FOUR INJUNCTIONS

Boys learn the Boy Code in sandboxes, playgrounds, schoolrooms, camps, churches, and hangouts, and are taught by peers, coaches, teachers, and just about everybody else. In the "Listening to Boys' Voices" study, even very young boys reported that they felt they must "keep a stiff upper lip," "not show their feelings," "act real tough," "not act too nice," "be cool," "just laugh and brush it off when someone punches you." These boys were not referring to subtle suggestions about how they "might" comport themselves. Rather, they were invoking strict rules they had absorbed about how they "must" behave, rules that most of them seemed to genuinely fear breaking.

Relying on well-known research, professors Deborah David and Robert Brannon divided these kinds of do-or-die rules, or "injunctions," boys follow into four basic stereotyped male ideals or models of behavior. These four imperatives are at the heart of the Boy Code.

The "Sturdy Oak"

Men should be stoic, stable, and independent. A man never shows weakness. Accordingly, boys are not to share pain or grieve openly. Boys are considered to have broken this guideline, for instance, if they whimper, cry, or complain—or sometimes even if they simply ask for an explanation in a confusing or frightening situation. As one boy in the "Voices" study put it: "If somebody slugs you in the face, probably the best thing you could do is just smile and act like it didn't hurt. You definitely shouldn't cry or say anything." The "sturdy oak" requirement drains boys' energy because it calls upon them to perform a constant "acting job"—to pretend to be confident when they may feel afraid, sturdy when they may feel shaky, independent when they may be desperate for love, attention, and support.

"Give 'em Hell"

This is the stance of some of our sports coaches, of roles played by John Wayne, Clint Eastwood, and Bruce Lee, a stance based on a false self, of extreme daring, bravado, and attraction to violence. This injunction stems largely from the myth that "boys will be boys" . . . —the misconception that somehow boys are biologically wired to act like macho, high-energy, even violent supermen. This is the Boy Code requirement that leads many boys to "dare" each other to engage in risky behaviors and that causes some parents to simply shrug their shoulders if their sons injure themselves or others.

The "Big Wheel"

This is the imperative men and boys feel to achieve status, dominance, and power. Or, understood another way, the "big wheel" refers to the way in which boys and men are taught to avoid shame at all costs, to wear the mask of coolness, to act as though everything is going all right, as though everything is under control, even if it isn't. This Boy Code imperative leads many boys and men to push themselves excessively at academic or career-related work, often in an effort to repress feelings of failure or unhappiness.

"No Sissy Stuff"

Perhaps the most traumatizing and dangerous injunction thrust on boys and men is the literal gender straitjacket that prohibits boys from expressing feelings or urges seen (mistakenly) as "feminine"—dependence, warmth, empathy. According to the ideal of "no sissy stuff," such feelings and behaviors are taboo. Rather than being allowed to explore these emotional states and activities, boys are prematurely forced to shut them out, to become self-reliant. And when boys start to break under the strain, when nonetheless they display "feminine" feelings or behaviors, they are usually greeted not with empathy but with ridicule, with taunts and threats that shame them for their failure to act and feel in stereotypically "masculine" ways. And so boys become determined never to act that way again—they bury those feelings.

And so in several fundamental ways the Boy Code affects the ability of boys and adults to connect.

First, it separates boys from their parents too early, before most boys are actually emotionally prepared for it. When boys encounter some of early childhood's most trying times—when they sleep alone in a crib for the first time, are sent away for two weeks of summer camp, or separate from their parents for the first day of kindergarten—they are often being pushed toward pseudo-independence before they're really ready.

Yet when boys rebel against this push to separate—when they cry, get injured, or tell friends that they'd rather stay at home than go outside and play—society's Boy Code makes them feel ashamed of themselves. Shame haunts many boys all their lives, undermining their core of self-confidence, eroding their fragile self-esteem, leaving them with profound feelings of loneliness, sadness, and disconnection. Moreover, it affects our ability to fully connect with our boys.

Even when boys appear sad or afraid, our culture lets them know in no uncertain terms that they had better toughen up and "tough it out" by themselves. The feelings boys are forced to repress become so troubling that some boys may show the apparent symptoms of attention deficit disorder and serious conduct disorders, become depressed, and—when they're older—turn to alcohol or drugs. Indeed, the same kind of shame that silences adolescent girls from expressing their true voice affects boys at a much younger age—at the age of five or six.

But the good news, I also believe, is that neither boys nor the adults who care for them need to live by these rules. Boys can rebel against them and revise the code for boys and girls so that they can experience a broad range of feelings and behaviors. Parents do not have to resist their deepest feelings for their sons or let myths about boys overwhelm the wisdom of their own instincts. Together we can unlearn the Boy Code. Together we can insist on enjoying close, emotionally rich relationships, based on connection instead of disconnection. . . .

HELPING BOYS RECONNECT: A PRIMER FOR PARENTS

As powerful as the cultural imperatives of the Boy Code may be in pushing boys of all ages to separate from their parents, toughen themselves up, and restrict their emotional lives, there is a lot we can do as adults to help boys overcome these conventional pressures. Here are some basic guidelines I would suggest:

At least once a day, give your boy your undivided attention. This means you're not speaking with someone else, you're not simultaneously trying to cook, clean, read, or do some other task. You're listening closely. He's got your attention. While sometimes he may not want to talk—while he may just want to play a game, get some help on his homework, or complain about having to do chores—showing him this attention, even if he doesn't always soak it up, gives him the message that you're there, that you care, and that he has a daily time and place when he can share things with you. It's not important that he always unload heavy emotions on you. And he may signal that he prefers to talk about things at some later point. He just needs to feel regular loving presence and know that you're eager to know what's happening in his world.

Encourage the expression of a full range of emotions. From the moment a boy is born and throughout his life, it's important he gets the message that all of his emotions are valid. With an infant, this means we need to mirror back all of the feelings the baby expresses. Rather than forcing him to constantly smile or laugh, we also need to show him we're receptive to his sadness, fear, or other painful emotions. So when a young infant begins to frown, yawn, kick, or cry, rather than trying to "cheer him up" or "smooth things over" by making happy faces at him or ignoring his displays of discomfort, show him your empathy, let him know you understand how he's feeling, and show him with your words, facial expressions, and gestures that you respect and understand his genuine feelings. With toddlers and school-age boys, we need to ask questions—"What happened?" "Are you feeling sad about something?" "Tell me what's making you unhappy"—and, again, express our empathy—"Gee, that sounds unfair!" "I'm sorry it hurts so much." We also need to use a broad range of emotion words—happy, sad, tired, disappointed, scared, nervous—rather than limiting our discussion of emotions to words such as "anger" that force boys to channel the gamut of their feelings into one word and one emotion.

In our daily attention-giving time with our sons, we need to pay close attention to what he's saying and how he's acting. If he complains, expresses fears or anxieties, cries, or otherwise shows emotions that reveal he's hurting, ask him what he's going through and let him talk about all that he is experiencing. With an older boy, be sure to ask him questions about his relationships with girls, with other boys, with his siblings, teachers, and other friends and acquaintances. Ask him to share with you not only what's going well in those relationships, but also what's going less well. Ask him what he enjoys about them and what he finds difficult. By probing about both the "positive" and "negative" sides of these relationships, older boys will begin to discuss a broad array of thoughts and feelings.

When a boy expresses vulnerable feelings, avoid teasing or taunting him. While it's natural to want to be playful with our sons, and though showing him a sense of levity and good cheer sometimes helps him to overcome unpleasant feelings or situations, by and large it's important that we not "cut off" his painful emotions by teasing or taunting him. So, for example, when he comes home and complains that his teacher told him he needs a haircut, rather than teasing that he "sure looks like a real fuzz ball," ask him how his teacher's comments made him feel, hear him out, and tell him that you too don't appreciate what the teacher said. Or if your teenage son announces despondently that his sweetheart just "dumped him," rather than joking that it must have been his bad breath that got to her or that his heart "must just be totally broken," instead ask him if he'd like to talk about it and, if so, listen to what he'd like to share with you and try to mirror back in an empathic way the feelings you sense he's trying to convey. Teasing and taunting rarely heal the boy. Empathy, however, goes miles to help him learn how to express and cope with a broad range of feelings.

Avoid using shaming language in talking with a boy. Research, as well as everyday observation, reveals that parents often—although unintentionally—use shaming language with their male children that they do not use with girls. It's important to find ways to talk with boys that do not shame them, and that they can respond to. If a boy does something that surprises or concerns you, a natural reaction is to ask, "How could you do that?" But that implies that the act, whatever it was, was wrong and casts the boy in the role of the evil perpetrator. Rather, you might ask, "What's going on?" or "What happened?" which suggests that you have not formed a judgment about the situation under discussion.

If a boy comes home with a less than stellar report card, a parent—understandably concerned—may challenge him and deliver an ultimatum, "You're going to have to work harder than this. These grades won't get you into a good college." Undoubtedly, the boy knows he is not performing as well as others or as well as he would like. The better parental response might be "You're still struggling with math, aren't you. What could we do to help?"

Or suppose a boy declines an invitation to visit a friend or go to a party. Rather than say, "It would do you good to get out of the house. Besides, that boy is really nice," you could try to find out why the boy no longer wants to be with his friend—"Has something happened between you two guys?"—or what it is about the party that doesn't appeal to him—"Will somebody be there you don't get along with?"

Such language carries tremendous power to make a boy feel shame and to reinforce his own conception that he is somehow toxic.

Look behind anger, aggression, and rambunctiousness. In so many cases, a boy who seems angry, displays a lot of aggression, or is constantly rambunctious is indirectly asking for our help. If you notice a boy who's acting in such ways, try to create a setting where he'll feel comfortable talking with you and then ask him how things are with him. With a young boy, you might not be able to ask him a lot of direct questions—and he may not yet be able to talk about feelings in a clear way—but try your best to get a sense of what he's feeling. For instance, if you notice that your son has seemed angry a lot lately, you might say, "Gosh, you've seemed upset a lot. Is everything OK? Have things been rough for you lately?" Or, if you're a schoolteacher and you notice a boy who's constantly roughing up and provoking other kids, rather than chastising the boy, ask him how things are at home. Ask him if he's upset about something. Try to get a sense of whether there might be deeper, more vulnerable feelings that are motivating his anger or rowdy behavior. You might even tell him that sometimes when we act irritably or show aggression, we might be feeling sadness or other upset feelings.

Express your love and empathy openly and generously. Despite all the messages you might receive about "letting go" of your son, of not staying too attached to him, of not "babying" him, you simply can never show him too much love or empathy. Cutting off your affection and support, to let him "stand on his own," as we've discussed in this chapter, can actually traumatize him.

Tell your boy that you love him as often as you like. Give him hugs. Tell him you're proud of him and that you care about him. Stay involved in his emotional life. See opportunities to connect with him for moments of playful closeness and emotional sharing. If he asks you to let him alone, give him the space he needs, but let him know that you love him very much and that when he's ready to spend time together, you'll be up for it. You cannot "spoil" your son with too much love or attention. You will not make him "girl-like" or "feminine" by maintaining a close relationship. There's simply no such thing as too much love!

Let boys know that they don't need to be "sturdy oaks." So many boys, even at a very young age, feel that they need to act like a "sturdy oak." When there are problems at home, when he suffers his own failures or disappointments, or when there's a need for somebody who's physically or emotionally "strong" for others to lean on and he feels he has to be that support, the boy is often pushed to "act like a man," to be the one who is confident and unflinching. No boy should be called upon to be the tough one. No boy should be hardened in this way. So through thick and thin, let your boy know that he doesn't have to act like a "sturdy oak." Talk to him honestly about your own fears and vulnerabilities and encourage him to do the same. The more genuine he feels he can be with you, the more he'll be free to express his vulnerability and the stronger he will become.

Create a model of masculinity for him that is broad and inclusive. Despite all the narrow messages about "being a guy" that they may get at school, on television, or elsewhere, you can help boys to create their own model of masculinity. Try to help them develop a model that is broad and inclusive. Try to do for them what we have done for girls by valuing them as people before evaluating them as a distinct (and therefore restricted) gender. This means encouraging boys in all their interests, relationships, and activities. It means letting them know that "big guys *do* cry." It also means exposing boys to people who bend society's strict gender rules—to men who are nurses, women who are plumbers, girls who are "jocks," boys who cook, and so on. Boys especially benefit from getting to know adult male "role models" who exude masculinity in a genuine and expansive way. When you give your son a sense that there's no one single way of being "manly," you're helping him develop confidence about who he really is. You're letting him know that no matter what he enjoys doing, whom he likes spending time with, and what sorts of feelings he experiences, he's a "real boy" on his way to being a "real man."

14

Influences of Gender Socialization and Athletic Involvement on the Occurrence of Eating Disorders

DIANE E. TAUB AND PENELOPE A. MCLORG

Not everything we learn through the process of socialization has positive results. Consider eating disorders that result from the preoccupation many people, especially young women, have with body shape and weight. As Taub and McLorg note, in order to reduce this prevalent problem, we must consider the source of the negative body images and how to change the emphasis in the socialization process on ideal images. The authors also discuss the role of athletics in eating disorders.

As you read, think about the following:

1. *What messages do women and men receive about ideal body types, and from where?*
2. *What role do athletics play in contributing to or discouraging eating disorders?*
3. *What could be done to change the negative results of socialization, such as abuse, neglect, and unrealistic body images?*

GLOSSARY **Anorexia and bulimia nervosa** Eating disorders that involve (1) self-starving and (2) binging and purging behaviors. **Internalization** The process of making ideas and behavior patterns an integral part of one's repertoire of behaviors. **Gender socialization** Learning the gender roles expected in society. **Agents of gender socialization** Ways gender expectations are passed on.

The eating disorders of anorexia nervosa (self-starvation) and bulimia nervosa (binge-purge syndrome) are a major health and social problem, with the reported occurrence increasing steadily over the past 30 years (Gordon, 1988; Harrison and Cantor, 1997; Wiseman and others, 1992). As a risk group, females are much more likely than males to be affected, comprising approximately 90 percent of reported cases (American Psychiatric Association, 1994; Haller, 1992). This gender difference can be clarified by examining factors of gender socialization that relate to physical appearance.

Traditionally, more emphasis has been placed on the appearance of females than of males (Lovejoy, 2001; Thornton and Maurice, 1997). Women show awareness of this focus by being

SOURCE: Original article. Reprinted by permission of the authors, 2006.

more concerned with their appearance than are men (Pliner, Chaiken, and Flett, 1990; Thompson and Heinberg, 1999). Physical appearance is also more crucial to self-concept among females than among males (Lovejoy, 2001; Rodin, Silberstein, and Striegel-Moore, 1985). Whereas in men self-image is associated with skill and achievement, among women it is linked to physical characteristics (Hesse-Biber, Clayton-Matthews, and Downey, 1987). In terms of bodily appeal, the physical attractiveness of males is related to physical abilities, with their bodies valued for being active and functional. In contrast, the bodies of females are judged on the basis of beauty (Rodin and others, 1985; Thompson and Heinberg, 1999).

One important consideration in appearance is body shape. Concerns about body shape expressed by women range from mild weight consciousness at one extreme to eating disorders at the other (Gordon, 1988; Rodin and others, 1985; Twamley and Davis, 1999). As part of the socialization in the "cult of thinness" (Hesse-Biber and others, 1987, p. 512), females accept an ideal body shape and a corresponding need for weight control. Dissatisfaction from a failure to meet slim appearance standards and subsequent dieting behavior have been identified as risk factors for the development of eating disorders (Drewnowski and Yee, 1987; Polivy and Herman, 1986; Striegel-Moore, Silberstein, and Rodin, 1986; Twamley and Davis, 1999).

This reading examines the relationship between eating disorders and gender socialization. Our purpose is to demonstrate the important contribution female and male socialization makes to the social context of anorexia nervosa and bulimia nervosa, including the gender difference in occurrence. To illustrate the connections between gender socialization and eating disorders, we use the following framework: (1) ideal body shape, as a representation of gender norms, (2) role models and mass media messages, as agents of gender socialization, and (3) dieting, as an expression of gender socialization. Ideal body shape affects agents of socialization, which in turn reinforce ideal body shape. Acceptance of ideal body shape norms and exposure to agents of gender socialization are expressed through dieting behavior; eating disorders are an extreme response. Also discussed in this reading is the influence of athletic involvement on eating disorders. Participation in certain sport activities converges with female body norms and heightens females' vulnerability to eating disorders.

IDEAL BODY SHAPE

Current appearance expectations specify thinness for women (Garner and others, 1980; Lovejoy, 2001; Thompson and Heinberg, 1999). Slim bodies are regarded as the most beautiful and worthy ones; overweight is seen as not only unhealthy but also offensive and disgusting (Harrison and Cantor, 1997; Schwartz, Thompson, and Johnson, 1982). Although both males and females are socialized to devalue fatness, women are more exposed to the need to be thin (Raudenbush and Zellner, 1997; Rodin and others, 1985).

In contrast, males are socialized to be muscular and not skinny or weak (Harrison, 2000; Leon and Finn, 1984). In ratings by preadolescent, adolescent, and college-aged males, the mesomorphic or muscular male body type is associated with socially favorable behaviors and personality traits (Raudenbush and Zellner, 1997). Compared with endomorphic (plump) and ectomorphic (slender) individuals, mesomorphs are judged more likely to assume leadership and be assertive, smart, and most wanted as a friend. The devaluing of a thin body for males is also reflected in the frequent desire of preadolescent and teenage boys as well as college males to gain weight and/or size (Collins, 1991; Harrison, 2000; Raudenbush and Zellner, 1997; Striegel-Moore and others, 1986).

Among females, the orientation toward slimness is so established that even when they are not overweight, they frequently perceive themselves as such (Connor-Greene, 1988; Raudenbush and Zellner, 1997; Wiseman, Harris, and Halmi, 1998). For example, although over 60 percent of college females believe that they are overweight, only 2 percent actually are (Connor-Greene, 1988). In addition, college females underestimate the occurrence of being underweight. While 31 percent of college women are measured as underweight, only 13 percent think they weigh below weight norms (Connor-Greene, 1988). Further, over four out of five college women report that they want to lose weight (Hesse-Biber and others, 1987). Even among college women whose weight is within the normal range, three-fourths want to be thinner (Raudenbush and Zellner, 1997).

Other results indicate that the ideal body shape of college women is significantly thinner than both their actual body type (Fallon and Rozin, 1985; Raudenbush and Zellner, 1997) and the body they perceive as most attractive to males (Fallon and Rozin, 1985). In addition, Raudenbush and Zellner (1997) report that women who engage in disordered eating behavior desire a body shape that is thinner than what they believe is attractive to males. In a study of families, Rozin and Fallon (1988) show that mothers and daughters both want slimmer bodies than they currently have. Furthermore, the shape these females believe is most attractive to males is thinner than what males actually prefer.

Collins (1991) demonstrates that gender-based ideas of attractive bodies develop in children as young as six or seven years. First- through third-grade girls select illustrations of their ideal figures that are significantly thinner than their current figures; this pattern is found across all levels of actual weight. Moreover, girls choose significantly slimmer figures than boys do for the ideal girl, ideal female adult, and ideal male adult (Collins, 1991).

Among females, learning to desire thinness begins at an early age. In general, females want to be slim and are critical of their weight, regardless of their actual body size and weight. A similar concern for thinness in male bodies is not common among males (Collins, 1991; Connor-Greene, 1988; Fallon and Rozin, 1985; Harrison, 2000; Hesse-Biber and others, 1987; Rozin and Fallon, 1988). The inaccuracy with which females of all ages perceive their body shapes (Rodin and others, 1985) parallels the distorted body images held by individuals with eating disorders.

AGENTS OF GENDER SOCIALIZATION

Images of ideal body shape affect agents of gender socialization, such as role models and mass media. In turn, these influences support expectations of body size. Reflecting gender socialization, traditional female role models and mass media messages express thinness norms for females.

Role Models

Examining patterns of ideal body shape, Garner and colleagues (1980) study the measurements of Miss America contestants over the 20-year span from 1959 to 1978. Mazur (1986) and Wiseman and colleagues (1992) conduct similar analyses of contestants' dimensions covering the period of 1979 to the mid-1980s. Both Garner and coworkers (1980) and Mazur (1986) report decreases in bust and hip measurements of Miss America contestants over the study periods. However, waist dimensions demonstrate periods of increase, suggesting a less hourglass standard. Further, weight for height of Miss America contestants progressively declines (Garner and others, 1980; Mazur, 1986), with a trend from 1970 to 1978 for pageant winners to be thinner than the average contestant (Garner and others, 1980). Analyzing the weight of contestants in relation to expected weight for their height and age, Wiseman and colleagues (1992) additionally find a significant decrease in the women's percentage of expected weight from 1979 to 1985.

Garner and colleagues (1980), Mazur (1986), Wiseman and coworkers (1992), and Katzmarzyk and Davis (2001) also examine the beauty ideal represented by *Playboy* centerfolds. As with Miss America contestants, bust and hip dimensions decline and waist measurements rise from 1959 to 1978 (Garner and others, 1980). During the early 1980s, bust, waist, and hip dimensions of centerfolds decrease (Mazur, 1986). Centerfolds also show declines in weight for height between 1959 and the early 1980s (Garner and others, 1980; Mazur, 1986). Further, *Playboy* centerfolds continue the diminished body size through the 1990s, with 70 percent of centerfolds from 1978 to 1998 being classified as underweight (Katzmarzyk and Davis, 2001).

The trends of slenderization exhibited by both *Playboy* centerfolds and Miss America contestants illustrate the slimness norm. In fact, from 1979 through the mid-1980s, approximately two-thirds of these ideals of female beauty weighed 15 percent or more below their expected weight (Wiseman and others, 1992). The pattern of at least two-thirds of *Playboy* centerfolds weighing 15 percent below expected weight norms persists through the 1990s (Katzmarzyk and Davis, 2001). Maintaining such a weight level is one of the criteria for anorexia nervosa (Wiseman and others, 1992). Thus, the declining size of female figures considered admirable represents a body size reflective of an eating disorder.

Other female role models, such as movie stars and magazine models, have become less curvaceous over the latter half of the 20th century (Silverstein and others, 1986). Hence, portrayals of females in media geared toward women as well as men demonstrate the thinness norm. In a study that examines cover models appearing on the four most popular American fashion magazines for the years 1959–1999, overall body size decreases significantly during the 1980s and 1990s (Sypeck, Gray, and Ahrens, 2004). From the 1960s to the 1990s, there is a dramatic increase in the frequency with which the covers depict the models' upper bodies and hips. Further, in comparison with the earliest period investigated, the latest covers reveal more of the models' bodies (Sypeck and others, 2004). Thus, over recent decades, figures of magazine cover models have become thinner and more emphasized and revealed.

The "anorectic body type" of models in major women's fashion magazines illustrates "an idealized standard of beauty and high fashion" (Gordon, 1988, p. 157). As preferred bodies, fashion models set an example of slimness that is unrealistic for most women (Thompson and Heinberg, 1999; Wiseman and others, 1998). College women viewing pictures of catalog models who typify the idealized thin female figure exhibit lowered self-esteem and increased self-consciousness, anxiety about their body, and body dissatisfaction (Thornton and Maurice, 1997). In general, through exposure to female role models, the majority of women are continually reminded of their inadequacy (Pliner and others, 1990) and encouraged to obsess about thinness and beauty (Lovejoy, 2001).

Further, observing thin female celebrities and models pictured in magazines is related to anorexic symptoms in girls between the ages of 11 and 18, and to bulimic tendencies among girls between 15 and 18 years old (Harrison, 2000). For college women, reading fashion and fitness magazines is associated with eating disorder attitudes and behavior (Harrison, 1997; Harrison and Cantor, 1997). The relationship between exposure to these media sources and disordered eating patterns is maintained whether or not the women have an initial interest in fashion and fitness. Fitness magazines are particularly associated with anorexic rather than bulimic tendencies because they provide examples of diet and exercise behavior characteristic of anorexics, but do not offer illustrations of the binging and purging behavior of bulimics (Harrison, 1997; Harrison and Cantor, 1997).

Beyond simple exposure to magazines, the attraction of college women to role models in magazines relates to eating disorder symptoms. Factors such as feeling similar to, wanting to be like, or liking thin figures increases both anorexic and bulimic behavior and attitudes (Harrison, 1997). Images of thinness in magazines are more likely to be associated with body dissatisfaction and eating disturbance among college women who accept the thin body ideal (Twamley and Davis, 1999). In general, internalization of the thin body an as ideal has been shown to be a causal risk factor for body-image and eating disturbances. This risk increases when the internalization coexists with other risk factors, such as body dissatisfaction, dieting, and negative emotions (Thompson and Stice, 2001).

Research on newspaper and magazine advertisements indicates that male figures are generally portrayed as bigger than female figures (Goffman,

1979). Such representation of the size of men symbolizes their "social weight," in power, authority, or rank (Goffman, 1979, p. 28), as well as the positive valuing of a larger body in males (Raudenbush and Zellner, 1997). In contrast, females in print advertisements provide additional exposure to the thin female standard.

Women's socialization to be slim is also demonstrated by role models on television over the past 25 years. For example, in prime-time, top-10 Nielson-rated television shows and their commercials during the 1970s, females are more likely to be thin than heavy and to be thinner than males (Kurman, 1978). A related study (Silverstein and others, 1986) of most-watched television programs demonstrates that 69 percent of the actresses and only 17.5 percent of the actors are slim. In addition, 5 percent of the women are evaluated as heavy, while over a quarter of the men are rated as such. These contrasts remain over a range of ages of the performers (Silverstein and others, 1986). Most female television characters are thinner than the average woman, while fewer that 10 percent of the women on television are overweight (Thompson and Heinberg, 1999).

In a recent study of girls and boys between 11 and 18 years old, watching television shows with overweight characters is associated with bulimic-like behavior and attitudes among girls. For boys, only the youngest demonstrate a relationship between exposure to overweight characters on television and body dissatisfaction (Harrison, 2000). Research on college women reports that viewing television shows with overweight characters is related to body dissatisfaction, while exposure to television shows with thin characters is associated with drive for thinness (Harrison, 1997; Harrison and Cantor, 1997). Similar to findings with magazine figures, identification with thin characters on television is additionally associated with both anorexic and bulimic behavior and attitudes (Harrison, 1997). Further, college women who accept the thin body ideal represented by television characters are more likely to demonstrate body dissatisfaction and eating disturbance (Twamley and Davis, 1999).

Mass Media Messages

In addition to displaying slim female role models for imitation, agents of gender socialization promote consciousness of weight and diet. Mass media messages encourage virtually uniform standards of beauty (Mazur, 1986); messages directed toward females emphasize the thin ideal (Thompson and Heinberg, 1999). The presence of electronic as well as print media has increased the accessibility of cultural standards of appearance (Thompson and Heinberg, 1999).

Among children and adolescents, media messages about physical beauty and the thin body ideal influence self-perceptions, body image, weight concerns, and eating behaviors (Morris and Katzman, 2003). In an analysis of 25 children's videos, messages emphasize the importance of physical attractiveness as well as support body stereotypes. Video characters with thin or muscular body figures are found to possess desirable traits, whereas obese characters have negative traits. Compared with videos, however, 20 children's books for ages 4–8 do not exhibit as many body-related messages (Herbozo and others, 2004).

Surveying major women's magazines, Garner and colleagues (1980) find a significant increase in diet articles from 1959 to 1978. From 1979 to 1988, the same women's magazines show a leveling off in number of diet articles but an increase in articles on exercise as a strategy for weight loss (Wiseman and others, 1992). In addition, dieting and weight control listings in the *Reader's Guide to Periodical Literature* almost double from 1977 to 1986 (Hesse-Biber and others, 1987).

Media emphasis on a slim body standard for females is also illustrated in comparisons of female- and male-directed magazines. An analysis (Silverstein and others, 1986) of the most popular women's and men's magazines indicates significant differences in content of articles and advertisements. In women's magazines, ads for diet foods, and articles and advertisements dealing with body shape or size, appear 63 times and 12 times more often, respectively, than in men's magazines (Silverstein and others, 1986).

Similarly, Andersen and DiDomenico (1992) find that the most popular magazines among women aged 18 to 24 contain 10 times more articles and advertisements on dieting or losing weight than do the most popular magazines among men aged 18 to 24. The focus on weight control in young women's magazines may have particular importance, as adolescence is the primary period of onset for both anorexia nervosa and bulimia nervosa (American Psychiatric Association, 1994; Haller, 1992; Harrison, 2000).

In addition to advertisements and articles on weight control, women's magazines surpass men's magazines in material concerning food. Articles on food and ads for food (excluding those for diet foods) in women's magazines exceed those in men's magazines by 71 to 1 (Silverstein and others, 1986). Thus, through this printed medium, females are being presented with conflicting messages. While food advertisements and articles encourage the consumption and enjoyment of food, diet aids and body shape ads and articles reinforce control of eating and weight. Popular magazines effectively maintain women's weight control preoccupation through their dual messages of eat and stay slim (McLorg and Taub, 1987). Moreover, although exposure to magazine advertising may be similar for individuals with and without eating disorders, anorexics and bulimics are especially likely to believe that advertisements promote the desirability of slimness (Peterson, 1987).

As agents of socialization, mass media and role models present a consistent portrayal of thin females in material directed toward both female and male audiences. Media preference for slimness in women is additionally shown in messages encouraging the weight control efforts of females. Through their selective representation of thin women, these sources not only reflect the slim ideal body shape for females, but also strengthen this gender norm for appearance. The influence of mass media and role models is affected by the extent to which females identify with the role models (Harrison, 1997) and accept the messages about thinness (Thompson and Heinberg, 1999; Twamley and Davis, 1999). Role models and mass media effectively reinforce the weight consciousness of females. The impact of these influences is to encourage and perpetuate women's repeated attempts to conform to the thin standard. A similar promotion or expression of a slimness ideal is not an aspect of the socialization experience of males (Andersen and DiDomenico, 1992; Drewnowski and Yee, 1987; Harrison, 2000; Silverstein and others, 1986).

DIETING

Concerns about thin body size reflect gender socialization of females. With the role obligation of being visually attractive, women alter their bodies to conform to an appearance ideal. Dieting can be viewed as a response to the gender norm of slimness in females. Worrying about weight and weight-loss efforts are so common among females that they have become normative (Rodin and others, 1985; Striegel-Moore and others, 1986). Nasser (1988, p. 574) terms dieting a "cultural preoccupation" among females, with concerns about weight and weight control persisting even into women's elderly years (Pliner and others, 1990). Further, females' continual efforts toward the thinness ideal are usually unsuccessful (Silverstein and others, 1986; Wiseman and others, 1998).

Frequency of dieting is related to actual body size, as well as to ideal body size and the emphasis a woman places on the importance of attractiveness (Silverstein and Perdue, 1988). A history of dieting, beginning with the teen years, is common among anorexics and bulimics (McLorg and Taub, 1987; Wiseman and others, 1998). In fact, researchers consider dieting a "precondition" (Polivy and Herman, 1986, p. 328) or a "chief risk factor" (Drewnowski and Yee, 1987, p. 633) of eating disorders.

As shown in a Nielson survey over 20 years ago, 56 percent of all women aged 24 to 54 dieted during the course of the year; 76 percent did so for cosmetic, rather than health, reasons (Schwartz and others, 1982). Among high school females, over half had dieted by the time they entered high school; and nearly 40 percent were currently dieting (Johnson and others, 1983). More recently, over half of women report that they are dieting (Haller, 1992). Another study demonstrates an even greater occurrence of dieting in preadolescent girls. Half of 9-year-olds and nearly 80 percent of 10- and 11-year-olds indicate that they have dieted (Stein, 1986). With dieting so common among girls and women (Lovejoy, 2001), it is not surprising that "serious dieting" in females is seen as "normal" (Leon and Finn, 1984, p. 328).

Compared with young females, young males are much less likely to diet. In one sample, 10 percent of boys versus 80 percent of girls had been on a diet before the age of 13 (Hawkins and others, cited in Striegel-Moore and others, 1986). Similarly, 64 percent of first-year college women, but only 29 percent of first-year males, followed a reduced-calorie diet in the previous month (Drewnowski and Yee, 1987). Compared with their male classmates, high school senior females are much more likely to use prescription amphetamines without a physician's orders for the purpose of losing weight (Taub, 1986). Further, among a sample of high school females, 34 percent had fasted to control weight, 16 percent had used diet pills, and 10 percent had taken diuretics during the past year, with 7 percent indicating daily use of diet pills and diuretics (Taub and Blinde, 1994).

Overall, studies demonstrate that females of differing ages frequently diet or engage in other weight loss behavior. In addition, females are much more likely than males to pursue weight loss through various methods. The extent and persistence of the dieting efforts of females indicate acceptance of the thin ideal. As an expression of gender socialization, dieting also reveals exposure to both role model and mass media promotion of slimness. Anorexia nervosa and bulimia nervosa represent extreme responses to female socialization toward thinness, with dieting usually preceding an eating disorder.

THE FACTOR OF ATHLETIC INVOLVEMENT

As noted, eating disorders are health problems in which social and cultural factors are very important. Body norms for women are shown to play a significant role in the genesis and maintenance of eating disorders. These cultural influences include the thinness norm, role models, mass media messages, and dieting/weight reducing behavior. The risk of eating disorders is increased when gender role expectations and athletic involvement interact. In particular, athletes are more prone to eating disorders than nonathletes when they participate in organized physical activities that have an aesthetic focus or that emphasize body shape and control (Benson and Taub, 1993; Hausenblas and Carron, 2002; Smolak, Murnen, and Ruble, 2000). Athletes not only internalize gender socialization about the thin body ideal, but they also hear from coaches and fellow participants that excessive leanness is essential for superior athletic performance. A high percentage of body fat is assumed to slow movement and hinder performance (Thornton, 1990). The sports most mentioned for increasing vulnerability to eating disorders are swimming, diving, figure skating, cross country running, dance, and gymnastics (Benson and Taub, 1993; Combs, 1982; Hausenblas and Carron, 2002; Smolak and others, 2000; Thornton, 1990).

In general, particular features of sport have been found to be associated with susceptibility to eating disorders. For example, eating disorders tend to be more prevalent in sports that are individual rather than team in nature, in which individual performance is more discernible. In addition, the risk of eating disorders is heightened in sports in which appearance is highlighted in the overall performance evaluation. Thirdly, sports that involve airborne movements exacerbate the need for thinness and increase vulnerability to eating disorders. And lastly, eating disorders tend to be more common in sports that involve a competitive uniform that exposes body size and contours (Combs, 1982; Hausenblas and Carron, 2002; Smolak and others,

2000; Thornton, 1990). Disordered eating is promoted when athletes experience public weighing or measuring, assessments of weight or percent fat at team practice, or pressure from coaches to lose weight (Benson and Taub, 1993; Thornton, 1990). Overall, occurrence of eating disorders increases as the competitive level of the sport environment increases (Smolak and others, 2000).

CONCLUSION

The influence of gender socialization in anorexia nervosa and bulimia nervosa is suggested by the gender distribution of the syndromes, with occurrence at least 10 times higher among females than males. In their connections with appearance expectations, eating disorders illustrate normative elements of female socialization. For example, weight loss efforts of females can be attributed to the greater importance of appearance in evaluations of women than of men; women's figures are more emphasized and more critically assessed (Lovejoy, 2001; Rodin and others, 1985).

As agents of gender socialization, role models and mass media are affected by, and support, notions of ideal body shape. Beauty queens, *Playboy* centerfolds, fashion models, and female television and movie characters are predominantly slender (e.g., Silverstein and others, 1986; Sypeck and others, 2004; Wiseman and others, 1992). Such role models serve as ideals for the female body shape. In addition to promoting the slimness standard, nude layouts and beauty contests epitomize the viewing of women as objects, with women's bodies judged according to narrow beauty standards (Lovejoy, 2001).

Also supporting the expectation of thinness in females are the numerous articles and advertisements on diet aids and body size in women's magazines (Andersen and DiDomenico, 1992; Silverstein and others, 1986). These media messages especially promote dieting behavior when accompanied by ample food articles and advertisements that encourage individuals to eat. Although females may enjoy forbidden food, they are continually reminded of the need to be thin. This double-edged message of enjoy eating but control your weight reinforces dieting (McLorg and Taub, 1987).

A low rate of success for dieting (Silverstein and others, 1986; Wiseman and others, 1998), combined with consistent pressure to be slim, results in repeated weight-loss efforts by females. Such manipulations of eating and body shape illustrate the tendency of females to view their bodies as objects, subject to modification for an attractiveness standard (Lovejoy, 2001; Rodin and others, 1985). While dieting reflects acceptance of gender norms for body shape, anorexia nervosa and bulimia nervosa represent extreme examples of gender socialization toward slimness.

Women's preoccupation with body shape ranges from mild weight consciousness to fully developed eating disorders (Gordon, 1988; Rodin and others, 1985; Twamley and Davis, 1999).

Individuals with eating disorders exemplify extreme concern about one's weight (Twamley and Davis, 1999), and can be viewed as extensions of the slim body ideal for females (Nasser, 1988; Rodin and others, 1985; Thompson and Heinberg, 1999). Female athletes are especially at risk of eating disorders if their athletic participation includes an aesthetic focus or emphasis on body shape and control (Benson and Taub, 1993; Smolak and others, 2000). Understanding of the gender distribution of anorexia nervosa and bulimia nervosa can be expanded by examining linkages with elements of gender socialization. Analysis of these factors is crucial for explaining the social context in which eating disorders occur.

REFERENCES

American Psychiatric Association, 1994. *Diagnostic and Statistical Manual of Mental Disorders*, 4th ed. Washington, DC: Author.

Andersen, Arnold E., and Lisa DiDomenico, 1992. Diet vs. shape content of popular male and female magazines: A dose-response relationship to the incidence of eating disorders? *International Journal of Eating Disorders* 11: 283–287.

Benson, RoseAnn and Diane E. Taub, 1993. Using the PRECEDE model for causal analysis of bulimic tendencies among elite women swimmers. *Journal of Health Education* 24: 360–368.

Collins, M. Elizabeth, 1991. Body figure perceptions and preferences among preadolescent children. *International Journal of Eating Disorders 10*: 199–208.

Combs, Margaret R., 1982. By food possessed. *Women's Sports 4(2)*: 12–13, 16–17.

Connor-Greene, Patricia Anne, 1988. Gender differences in body weight perception and weight-loss strategies of college students. *Women and Health* 14(2): 27–42.

Drewnowski, Adam, and Doris K. Yee, 1987. Men and body image: Are males satisfied with their body weight? *Psychosomatic Medicine* 49: 626–634.

Fallon, April E., and Paul Rozin, 1985. Sex differences in perceptions of desirable body shape. *Journal of Abnormal Psychology* 94: 102–105.

Garner, David M., Paul E. Garfinkel, Donald Schwartz, and Michael Thompson, 1980. Cultural expectations of thinness in women. *Psychological Reports* 47: 483–491.

Goffman, Erving, 1979. *Gender Advertisements*. New York: Harper & Row.

Gordon, Richard A., 1988. A sociocultural interpretation of the current epidemic of eating disorders. In Barton J. Blinder, Barry F. Chaitin, and Renee S. Goldstein (eds.), *The Eating Disorders: Medical and Psychological Bases of Diagnosis and Treatment*. New York: PMA.

Haller, Ellen, 1992. Eating disorders: A review and update. *Western Journal of Medicine* 157: 658–662.

Harrison, Kristen, 1997. Does interpersonal attraction to thin media personalities promote eating disorders? *Journal of Broadcasting & Electronic Media* 41: 478–500.

Harrison, Kristen, 2000. The body electric: Thin-ideal media and eating disorders in adolescents. *Journal of Communication* 50: 119–143.

Harrison, Kristen, and Joanne Cantor, 1997. The relationship between media consumption and eating disorders. *Journal of Communication* 47: 40–67.

Hausenblas, Heather A. and Albert V. Carron, 2002. Assessing eating disorder symptoms in sport groups: A critique with recommendations for future research. *International Sports Journal* 6: 65–74.

Herbozo, Sylvia, Stacey Tantleff-Dunn, Jessica Gokee-Larose, and J. Kevin Thompson, 2004. Beauty and thinness messages in children's media: A content analysis. *Eating Disorders* 12: 21–34.

Hesse-Biber, Sharlene, Alan Clayton-Matthews, and John A. Downey, 1987. The differential importance of weight and body image among college men and women. *Genetic, Social, and General Psychology Monographs* 113: 511–528.

Johnson, Craig L., Chris Lewis, Susan Love, Marilyn Stuckey, and Linda Lewis, 1983. A descriptive survey of dieting and bulimic behavior in a female high school population. In *Understanding Anorexia Nervosa and Bulimia: Report of the Fourth Ross Conference on Medical Research*. Columbus, OH: Ross Laboratories.

Katzmarzyk, Peter T. and Caroline Davis, 2001. Thinness and body shape of *Playboy* centerfolds from 1978 to 1998. *International Journal of Obesity* 25: 590–592.

Kurman, Lois, 1978. An analysis of messages concerning food, eating behaviors, and ideal body image on prime-time American network television. *Dissertation Abstracts International 39*: 1907A–1908A.

Leon, Gloria R., and Stephen Finn, 1984. Sex-role stereotypes and the development of eating disorders. In Cathy Spatz Widom (ed.), *Sex Roles and Psychopathology*. New York: Plenum.

Lovejoy, Meg, 2001. Disturbances in the social body: Differences in body image and eating problems among African American and white women. *Gender & Society* 15: 239–261.

Mazur, Allan, 1986. U.S. trends in feminine beauty and overadaptation. *Journal of Sex Research* 22: 281–303.

McLorg, Penelope A., and Diane E. Taub, 1987. Anorexia nervosa and bulimia: The development of deviant identities. *Deviant Behavior* 8: 177–189.

Morris, Anne M. and Debra K. Katzman, 2003. The impact of the media on eating disorders in children and adolescents. *Paediatrics & Child Health* 8: 287–289.

Nasser, Mervat, 1988. Culture and weight consciousness. *Journal of Psychosomatic Research* 32: 573–577.

Peterson, Robin T., 1987. Bulemia and anorexia in an advertising context. *Journal of Business Ethics* 6: 495–504.

Pliner, Patricia, Shelly Chaiken, and Gordon L. Flett, 1990. Gender differences in concern with body weight and physical appearance over the life span. *Personality and Social Psychology Bulletin* 16: 263–273.

Polivy, Janet, and C. Peter Herman, 1986. Dieting and binging reexamined: A response to Lowe. *American Psychologist* 41: 327–328.

Raudenbush, Bryan, and Debra A. Zellner, 1997. Nobody's satisfied: Effects of abnormal eating behaviors and actual and perceived weight status on body image satisfaction in males and females. *Journal of Social and Clinical Psychology* 16: 95–110.

Rodin, Judith, Lisa Silberstein, and Ruth Striegel-Moore, 1985. Women and weight: A normative discontent. In Theo B. Sonderegger (ed.), *Nebraska Symposium on Motivation. Vol. 32: Psychology and Gender.* Lincoln: University of Nebraska.

Rozin, Paul, and April Fallon, 1988. Body image, attitudes to weight, and misperceptions of figure preferences of the opposite sex: A comparison of men and women in two generations. *Journal of Abnormal Psychology* 97: 342–345.

Schwartz, Donald M., Michael G. Thompson, and Craig L. Johnson, 1982. Anorexia nervosa and bulimia: The socio-cultural context. *International Journal of Eating Disorders* 1(3): 20–36.

Silverstein, Brett, and Lauren Perdue, 1988. The relationship between role concerns, preferences for slimness, and symptoms of eating problems among college women. *Sex Roles* 18: 101–106.

Silverstein, Brett, Lauren Perdue, Barbara Peterson, and Eileen Kelly, 1986. The role of the mass media in promoting a thin standard of bodily attractiveness for women. *Sex Roles* 14: 519–532.

Smolak, Linda, Sarah K. Murnen, and Anne E. Ruble, 2000. Female athletes and eating problems: A meta-analysis. *International Journal of Eating Disorders* 27: 371–380.

Stein, Jeannine, 1986. Why girls as young as 9 fear fat and go on diets to lose weight. *Los Angeles Times*, October 29, Part v: 1, 10.

Striegel-Moore, Ruth, Lisa R. Silberstein, and Judith Rodin, 1986. Toward an understanding of risk factors for bulimia. *American Psychologist* 41: 246–263.

Sypeck, Mia Foley, James J. Gray, and Anthony H. Ahrens, 2004. No longer just a pretty face: Fashion magazines' depictions of ideal female beauty from 1959 to 1999. *International Journal of Eating Disorders* 36: 342–347.

Taub, Diane E., 1986. Amphetamine usage among high school senior women, 1976–1982: An evaluation of social bonding theory. Unpublished doctoral dissertation, Lexington: University of Kentucky.

Taub, Diane E. and Elaine M. Blinde, 1994. Disordered eating and weight control among adolescent female athletes and performance squad members. *Journal of Adolescent Research* 9: 483–497.

Thompson, J. Kevin, and Leslie J. Heinberg, 1999. The media's influence on body image disturbance and eating disorders: We've reviled them, now can we rehabilitate them? *Journal of Social Issues* 55: 339–353.

Thompson, J. Kevin and Eric Stice, 2001. Thin-ideal internalization: Mounting evidence for a new risk factor for body-image disturbance and eating pathology. *Current Directions in Psychological Science* 10:(5) 181–183.

Thornton, Bill, and Jason Maurice, 1997. Physique contrast effect: Adverse impact of idealized body images for women. *Sex Roles* 37: 433–439.

Thornton, James S., 1990. Feast or famine: Eating disorders in athletes. *Physician and Sportsmedicine* 18(4): 116, 118–122.

Twamley, Elizabeth W., and Mary C. Davis, 1999. The sociocultural model of eating disturbance in young women: The effects of personal attributes and family environment. *Journal of Social and Clinical Psychology* 18: 467–489.

Wiseman, Claire V., James J. Gray, James E. Mosimann, and Anthony H. Ahrens, 1992. Cultural expectations of thinness in women: An update. *International Journal of Eating Disorders* 11: 85–89.

Wiseman, Claire V., Wendy A. Harris, and Katherine A. Halmi, 1998. Eating disorders. *Medical Clinics of North America* 82: 145–159.

15

Anybody's Son Will Do

GWYNNE DYER

During our lifetime we may chose or be forced to undergo the process of resocialization—casting off an "old self" and replacing it with a new self. This process can take place in boarding schools, religious convents, mental hospitals, prisons, and even after a major life transition such as becoming a parent or getting divorced.

Resocialization into the military is the focus of this article. Note that although the author acknowledges women in the military, the focus is on young males. In excerpts from Dyer's 1985 book, War, *she explains the means of socializing men out of the civilian role and into the soldier/killer role that has become institutionalized as a result of centuries of experience.*

As you read about military socialization, ask yourself the following questions:

1. *Why is it desirable to recruit young men and women under 20?*
2. *What are the steps in the resocialization process?*
3. *How does the military resocialization process work?*
4. *Why do countries need to resocialize recruits in order to train them?*

GLOSSARY **Resocialization** Process of discarding former behavior patterns and accepting new ones as part of a transition in one's life. **Ideology** Cultural beliefs that justify certain social arrangements, including patterns of inequality.

All soldiers belong to the same profession, no matter what country they serve, and it makes them different from everybody else. They have to be different, for their job is ultimately about killing and dying, and those things are not a natural vocation for any human being. Yet all soldiers are born civilians. The method for turning young men into soldiers—people who kill other people and expose themselves to death—is basic training. It's essentially the same all over the world, and it always has been, because young men everywhere are pretty much alike.

Human beings are fairly malleable, especially when they are young, and in every young man there are attitudes for any army to work with: the inherited values and postures, more or less dimly recalled, of the tribal warriors who were once the model for every young boy to emulate. Civilization did not involve a sudden clean break in the way people behave, but merely the progressive distortion and redirection of all the ways in which people in the old tribal societies used to behave, and modern definitions of maleness still contain a great deal of the old warrior ethic. The anarchic machismo of the primitive warrior is not what modern armies really need in their soldiers, but it does provide them with promising raw material for the transformation they must work in their recruits.

SOURCE: Gwynne Dyer, "Anybody's Son Will Do." From *War: Past, Present, and Future* by Gwynne Dyer.

Just how this transformation is wrought varies from time to time and from country to country. In totally militarized societies—ancient Sparta, the samurai class of medieval Japan, the areas controlled by organizations like the Eritrean People's Liberation Front today[1]—it begins at puberty or before, when the young boy is immersed in a disciplined society in which only the military values are allowed to penetrate. In more sophisticated modern societies, the process is briefer and more concentrated, and the way it works is much more visible. It is, essentially, a conversion process in an almost religious sense—and as in all conversion phenomena, the emotions are far more important than the specific ideas . . . .

. . . The rhetoric of military patriotism . . . is virtually irrelevant so far as the actual job of soldiering is concerned. Soldiers are not just robots; they are ordinary human beings with national and personal loyalties, and many of them do feel the need for some patriotic or ideological justification for what they do. But which nation, which ideology, does not matter: men will fight as well and die as bravely for the Khmer Rouge as for "God, King, and Country." Soldiers are the instruments of politicians and priests, ideologues and strategists, who may have high national or moral purposes in mind, but the men down in the trenches fight for more basic motives. The closer you get to the frontline, the fewer abstract nouns you hear.

Armies know this. It is their business to get men to fight, and they have had a long time to work out the best way of doing it. All of them pay lip service to the symbols and slogans of their political masters, though the amount of time they must devote to this activity varies from country to country. It is less in the United States than in the Soviet Union, and it is still less in a country like Israel, which actually fights frequent wars. Nor should it be thought that the armies are hypocritical—most of their members really do believe in their particular national symbols and slogans. But their secret is that they know these are not the things that sustain men in combat.

What really enables men to fight is their own self-respect, and a special kind of love that has nothing to do with sex or idealism. Very few men have died in battle, when the moment actually arrived, for the United States of America or for the sacred cause of Communism, or even for their homes and families; if they had any choice in the matter at all, they chose to die for each other and for their own vision of themselves . . . .

The way armies produce this sense of brotherhood in a peacetime environment is basic training: a feat of psychological manipulation on the grand scale which has been so consistently successful and so universal that we fail to notice it as remarkable. In countries where the army must extract its recruits in their late teens, whether voluntarily or by conscription, from a civilian environment that does not share the military values, basic training involves a brief but intense period of indoctrination whose purpose is not really to teach the recruits basic military skills, but rather to change their values and their loyalties. "I guess you could say we brainwash them a little bit," admitted a U.S. Marine drill instructor, "but you know they're good people." . . .

It's easier if you catch them young. You can train older men to be soldiers; it's done in every major war. But you can never get them to believe that they like it, which is the major reason armies try to get their recruits before they are twenty. There are other reasons too, of course, like the physical fitness, lack of dependents, and economic dispensability of teenagers, that make armies prefer them, but the most important qualities teenagers bring to basic training are enthusiasm and naiveté. Many of them actively want the discipline and the closely structured environment that the armed forces will provide, so there is no need for the recruiters to deceive the kids about what will happen to them after they join.

> There is discipline. There is drill . . . . When you are relying on your mates and they are relying on you, there's no room for slackness or sloppiness. If you're not prepared to accept the rules, you're better off where you are.
>
> — *British army recruiting advertisement, 1976*

> People are not born soldiers, they become soldiers.... And it should not begin at the moment when a new recruit is enlisted into the ranks, but rather much earlier, at the time of the first signs of maturity, during the time of adolescent dreams.
>
> —Red Star *(Soviet army newspaper), 1973*

Young civilians who have volunteered and have been accepted by the Marine Corps[2] arrive at Parris Island, the Corps's East Coast facility for basic training, in a state of considerable excitement and apprehension: most are aware that they are about to undergo an extraordinary and very difficult experience. But they do not make their own way to the base; rather, they trickle in to Charleston airport on various flights throughout the day on which their training platoon is due to form, and are held there, in a state of suppressed but mounting nervous tension, until late in the evening. When the buses finally come to carry them the seventy-six miles to Parris Island, it is often after midnight—and this is not an administrative oversight. The shock treatment they are about to receive will work most efficiently if they are worn out and somewhat disoriented when they arrive.

The basic training organization is a machine, processing several thousand young men every month, and every facet and gear of it has been designed with the sole purpose of turning civilians into Marines as efficiently as possible. Provided it can have total control over their bodies and their environment for approximately three months, it can practically guarantee converts. Parris Island provides that controlled environment, and the recruits do not set foot outside it again until they graduate as Marine privates eleven weeks later....

For the young recruits, basic training is the closest thing their society can offer to a formal rite of passage,[3] and the institution probably stands in an unbroken line of descent from the lengthy ordeals by which young males in precivilized groups were initiated into the adult community of warriors. But in civilized societies it is a highly functional institution whose product is not anarchic warriors, but trained soldiers.

Basic training is not really about teaching people skills; it's about changing them, so that they can do things they wouldn't have dreamt of otherwise. It works by applying enormous physical and mental pressure to men who have been isolated from their normal civilian environment and placed in one where the only right way to think and behave is the way the Marine Corps wants them to. The key word the men who run the machine use to describe this process is motivation.

> I can motivate a recruit and in third phase, if I tell him to jump off the third deck, he'll jump off the third deck. Like I said before, it's a captive audience and I can train that guy; I can get him to do anything I want him to do.... They're good kids and they're out to do the right thing. We get some bad kids, but you know, we weed those out. But as far as motivation—here, we can motivate them to do anything you want, in recruit training.
>
> —*USMC drill instructor, Parris Island*

The first three days the raw recruits spend at Parris Island are actually relatively easy, though they are hustled and shouted at continuously. It is during this time that they are documented and inoculated, receive uniforms, and learn the basic orders of drill that will enable young Americans (who are not very accustomed to this aspect of life) to do everything simultaneously in large groups. But the most important thing that happens in "forming" is the surrender of the recruits' own clothes, their hair—all the physical evidence of their individual civilian identities.

During a period of only seventy-two hours, in which they are allowed little sleep, the recruits lay aside their former lives in a series of hasty rituals (like being shaven to the scalp) whose symbolic significance is quite clear to them even though they are quite deliberately given absolutely no time for reflection, or any hint that they might have the option of turning back from their commitment. The men in charge of them know how delicate a tightrope they are walking, though, because at this stage the recruits are still newly caught civilians who have not yet made their ultimate inward submission to the discipline of the Corps. . . .

. . . The frantic bustle of forming is designed to give the recruit no time to think about resisting what is happening to him. And so the recruits emerge from their initiation into the system, stripped of their civilian clothes, shorn of their hair, and deprived of whatever confidence in their own identity they may previously have had as eighteen-year-olds, like so many blanks ready to have the Marine identity impressed upon them.

The first stage in any conversion process is the destruction of an individual's former beliefs and confidence, and his reduction to a position of helplessness and need. It isn't really as drastic as all that, of course, for three days cannot cancel out eighteen years; the inner thoughts and the basic character are not erased. But the recruits have already learned that the only acceptable behavior is to repress any unorthodox thoughts and to mimic the character the Marine Corps wants. Nor are they, on the whole, reluctant to do so, for they want to be Marines. From the moment they arrive at Parris Island, the vague notion that has been passed down for a thousand generations that masculinity means being a warrior becomes an explicit article of faith, relentlessly preached: to be a man means to be a Marine.

There are very few eighteen-year-old boys who do not have highly romanticized ideas of what it means to be a man, so the Marine Corps has plenty of buttons to push. And it starts pushing them on the first day of real training: the officer in charge of the formation appears before them for the first time, in full dress uniform with medals, and tells them how to become men.

> The United States Marine Corps has 205 years of illustrious history to speak for itself. You have made the most important decision in your life . . . by signing your name, your life, pledge to the Government of the United States, and even more importantly, to the United States Marine Corps—a brotherhood, an elite unit. In 10.3 weeks you are going to become a member of that history, those traditions, this organization—if you have what it takes. . . .
>
> —*Captain Pingree, USMC*

The recruits, gazing at him with awe and adoration, shout in unison, "Yes sir!" just as they have been taught. They do it willingly, because they are volunteers—but even conscripts tend to have the romantic fervor of volunteers if they are only eighteen years old. Basic training, whatever its hardships, is a quick way to become a man among men, with an undeniable status, and beyond the initial consent to undergo it, it doesn't even require any decisions.

> I had just dropped out of high school and I wasn't doing much on the street except hanging out, as most teenagers would be doing. So they gave me an opportunity—a recruiter picked me up, gave me a good line, and said that I could make it in the Marines, that I have a future ahead of me. And since I was living with my parents, I figured that I could start my own life here and grow up a little.
>
> —*USMC recruit, 1982*

The training, when it starts, seems impossibly demanding physically for most of the recruits—and then it gets harder week by week. There is a constant barrage of abuse and insults aimed at the recruits, with the deliberate purpose of breaking down their pride and so destroying their ability to resist the transformation of values and attitudes that the Corps intends them to undergo. At the same time the demands for constant alertness and for instant obedience are continuously stepped up, and the standards by which the dress and behavior of the recruits are judged become steadily more unforgiving. But it is all carefully calculated by the men who run the machine, who think and talk in terms of the stress they are placing on the recruits: "We take so many c.c.'s of stress and we administer it to each man—they should be a little bit scared and they should be unsure, but they're adjusting." The aim is to keep the training arduous but just within most of the recruits' capability to withstand. One of the most striking achievements of the drill instructors is to create and maintain the illusion that basic training is an extraordinary challenge, one that will set those who graduate apart from others, when in fact almost everyone can succeed.

There has been some preliminary weeding out of potential recruits even before they begin training, to eliminate the obviously unsuitable minority, and some people do "fail" basic training and get sent home, at least in peacetime. The standards of acceptable performance in the U.S. armed forces, for example, tend to rise and fall in inverse proportion to the number and quality of recruits available to fill the forces to the authorized manpower levels. (In 1980, about 15 percent of Marine recruits did not graduate from basic training.) But there are very few young men who cannot be turned into passable soldiers if the forces are willing to invest enough effort in it. Not even physical violence is necessary to effect the transformation, though it has been used by most armies at most times....

There is, indeed, a good deal of fine-tuning in the roles that the men in charge of training any specific group of recruits assume. At the simplest level, there is a sort of "good cop–bad cop" manipulation of the recruits' attitudes toward those applying the stress. The three younger drill instructors with a particular serial are quite close to them in age and unremittingly harsh in their demands for ever higher performance, but the senior drill instructor, a man almost old enough to be their father, plays a more benevolent and understanding part and is available for individual counseling. And generally offstage, but always looming in the background, is the company commander, an impossibly austere and almost godlike personage.

At least these are the images conveyed to the recruits, although of course all these men cooperate closely with an identical goal in view. It works: in the end they become not just role models and authority figures, but the focus of the recruits' developing loyalty to the organization.

> I imagine there's some fear, especially in the beginning, because they don't know what to expect.... I think they hate you at first, at least for a week or two, but it turns to respect.... They're seeking discipline, they're seeking someone to take charge, 'cause at home they never got it.... They're looking to be told what to do and then someone is standing there enforcing what they tell them to do, and it's kind of like the father-and-son game, all the way through. They form a fatherly image of the DI[4] whether they want to or not.
>
> —*Sergeant Carrington, USMC*

Just the sheer physical exercise, administered in massive doses, soon has the recruits feeling stronger and more competent than ever before. Inspections, often several times daily, quickly build up their ability to wear the uniform and carry themselves like real Marines, which is a considerable source of pride. The inspections also help to set up the pattern in the recruits of unquestioning submission to military authority: standing stock-still, staring straight ahead, while somebody else examines you closely for faults is about as extreme a ritual act of submission as you can make with your clothes on.

But they are not submitting themselves merely to the abusive sergeant making unpleasant remarks about the hair in their nostrils. All around them are deliberate reminders—the flags and insignia displayed on parade, the military music, the marching formations and drill instructors' cadenced calls—of the idealized organization, the "brotherhood" to which they will be admitted as full members if they submit and conform. Nowhere in the armed forces are the military courtesies so elaborately observed, the staffs' uniforms so immaculate (some DIs change several times a day), and the ritual aspects of military life so highly visible as on a basic training establishment.

Even the seeming inanity of close-order drill has a practical role in the conversion process. It has been over a century since mass formations of men were of any use on the battlefield, but every army in the world still drills its troops, especially during basic training, because marching in formation, with every man moving his body in the same way at the same moment, is a direct physical way of learning two things a soldier must believe: that orders have to be obeyed automatically and instantly, and that you are no longer an individual, but part of a group.

The recruits' total identification with the other members of their unit is the most important lesson of all, and everything possible is done to foster it. They spend almost every waking moment together—a recruit alone is an anomaly to be looked into at once—and during most of that time they are enduring shared hardships. They also undergo collective punishments, often for the misdeed or omission of a single individual (talking in the ranks, a bed not swept under during barracks inspection), which is a highly effective way of suppressing any tendencies toward individualism. And, of course, the DIs place relentless emphasis on competition with other "serials" in training: there may be something infinitely pathetic to outsiders about a marching group of anonymous recruits chanting, "Lift your heads and hold them high, 3313 is a-passin' by," but it doesn't seem like that to the men in the ranks.

Nothing is quite so effective in building up a group's morale and solidarity, though, as a steady diet of small triumphs. Quite early in basic training, the recruits begin to do things that seem, at first sight, quite dangerous: descend by ropes from fifty-foot towers, cross yawning gaps hand-over-hand on high wires (known as the Slide for Life, of course), and the like. The common denominator is that these activities are daunting but not really dangerous: the ropes will prevent anyone from falling to his death off the rappelling tower, and there is a pond of just the right depth—deep enough to cushion a falling man, but not deep enough that he is likely to drown—under the Slide for Life. The goal is not to kill recruits, but to build up their confidence as individuals and as a group by allowing them to overcome apparently frightening obstacles....

If somebody does fail a particular test, he tends to be alone, for the hurdles are deliberately set low enough that most recruits can clear them if they try. In any large group of people there is usually a goat: someone whose intelligence or manner or lack of physical stamina marks him for failure and contempt. The competent drill instructor, without deliberately setting up this unfortunate individual for disgrace, will use his failure to strengthen the solidarity and confidence of the rest. When one hapless young man fell off the Slide for Life into the pond, for example, his drill instructor shouted the usual invective—"Well, get out of the water. Don't contaminate it all day"—and then delivered the payoff line: "Go back and change your clothes. You're useless to your unit now."

"Useless to your unit" is the key phrase, and all the recruits know that what it means is "I'm useless in battle." The Marine drill instructors at Parris Island know exactly what they are doing to the recruits, and why. They are not rear-echelon people filling comfortable jobs, but the most dedicated and intelligent NCOs[5] the Marine Corps can find: even now, many of them have combat experience. The Corps has a clear-eyed understanding of precisely what it is training its recruits for—combat—and it ensures that those who do the training keep that objective constantly in sight.

The DIs "stress" the recruits, feed them their daily ration of synthetic triumphs over apparent obstacles, and bear in mind all the time that the goal is to instill the foundations for the instinctive, selfless reactions and the fierce group loyalty that is what the recruits will need if they ever see combat. They are arch-manipulators, fully conscious of it, and utterly unashamed. These kids have signed up as Marines, and they could well see combat; this is the way they have to think if they want to live....

Combat is the ultimate reality that Marines—or any other soldiers, under any flag—have to deal with. Physical fitness, weapons training, battle drills, are all indispensable elements of basic training, and it is absolutely essential that the recruits learn the attitudes of group loyalty and interdependency which will be their sole hope of survival and success in combat. The training inculcates or fosters all of those things, and even by the halfway point in the eleven-week course, the recruits are generally responding with enthusiasm to their tasks.

But there is nothing in all this (except the weapons drill) that would not be found in the training camp of a professional football team. What sets soldiers apart is their willingness to kill. But it is not a willingness that comes easily to most men—even young men who have been provided with uniforms, guns, and official approval to kill those whom their government has designated as enemies. They will, it is true, fall very readily into the stereotypes of the tribal warrior group. Indeed, most of them have had at least a glancing acquaintance in their early teens with gangs (more or less violent, depending on, among other things, the neighborhood), the modem relic of that ancient institution.

And in many ways what basic training produces is the uniformed equivalent of a modern street gang: a bunch of tough, confident kids full of bloodthirsty talk. But gangs don't actually kill each other in large numbers. If they behaved the way armies do, you'd need trucks to clean the bodies off the streets every morning. They're held back by the civilian belief—the normal human belief—that killing another person is an awesome act with huge consequences.

There is aggression in all of us—men, women, children, babies. Armies don't have to create it, and they can't even increase it. But most of us learn to put limits on our aggression, especially physical aggression, as we grow up. . . .

Armies had always assumed that, given the proper rifle training, the average man could kill in combat with no further incentive than the knowledge that it was the only way to defend his own life. After all, there are no historical records of Roman legionnaires refusing to use their swords, or Marlborough's infantrymen[6] refusing to fire their muskets against the enemy. But then dispersion hit the battlefield, removing each rifleman from the direct observation of his companions—and when U.S. Army Colonel S. L. A. Marshall finally took the trouble to inquire into what they were doing in 1943–45, he found that on average only 15 percent of trained combat riflemen fired their weapons at all in battle. The rest did not flee, but they would not kill—even when their own position was under attack and their lives were in immediate danger. . . .

But the question naturally arises: if the great majority of men are not instinctive killers, and if most military killing these days is in any case done by weapons operating from a distance at which the question of killing scarcely troubles the operators—then why is combat an exclusively male occupation? The great majority of women, everyone would agree, are not instinctive killers either, but so what? If the remote circumstances in which the killing is done or the deliberate conditioning supplied by the military enable most men to kill, why should it be any different for women?

My own guess would be that it probably wouldn't be different; it just hasn't been tried very extensively . . . war has moved a very long way from its undeniably warrior male origins, and human behavior, male or female, is extremely malleable. Combat of the sort we know today, even at the infantryman's level let alone the fighter pilot's—simply could not occur unless military organizations put immense effort into reshaping the behavior of individuals to fit their unusual and exacting requirements. The military institution, for all its imposing presence, is a highly artificial structure that is maintained only by constant endeavor. And if ordinary people's behavior is malleable in the direction the armed forces require, it is equally open to change in other directions. . . .

NOTES

1. Eritrea, an Italian colony from 1885 to 1941, was annexed by Ethiopia in 1962. After a 30-year civil war, Eritrea gained its independence in 1992.
2. Something you might not know if you have no military experience and don't watch war movies or attend the ballet is that the word *corps* is pronounced "core" (from the Latin *corpus*, meaning "body").
3. The concept of rite of passage (or *rites de passage*) is discussed in *The Practical Skeptic*, chapter 10.
4. Drill Instructor.
5. Noncommissioned officers.
6. John Churchill (1650–1722), first Duke of Marlborough, British General, supreme commander of the British forces in the War of the Spanish Succession.

CHAPTER 4

Social Interaction, Groups, and Bureaucracy

Life Is with People

Alan Mc Evoy

Groups are the setting for socialization, the lifelong process of learning the ways of society. This process takes place first in the family, then in play groups, followed by more formal group associations in institutions of education and religion, and as adults in economic and political institutions. As individuals grow, so do their group contacts and roles within these groups. Familial relations expand to other primary groups such as play and peer groups, then to more formal groups such as the school. When individuals become adults, many of their interactions take place within secondary groups or organizations characterized by more formal relations.

Individuals make choices about some of their group affiliations; other memberships are obligatory. In the first reading, Peter Blau discusses the classic *exchange theory*—the rewards, costs and reciprocal expectations of individuals in groups. When the costs of group membership, be it a dyad (two people) or more, outweigh the rewards, the group may fail.

A major part of our interaction with other individuals or within groups takes place through nonverbal communication. Edward and Mildred Hall describe the silent part of language—the nonverbal communication of unconscious body movements. Body language involves physical cues—glances, movement, facial expressions, and so on—as well as social distance and personal space. Because of its subtleties and the vast range of possible nonverbal rules, silent language is more difficult to master than verbal speech. To complicate matters, as Hall and Hall show, similar movements may have different meanings in different situations or cultures.

Most of our interactions take place in groups. A large body of sociological literature deals with group relations, from role relationships, leadership, and decision making to the tendency toward conformity in group situations because individuals strive to get along with others. In a classic discussion of in-groups and out-groups, William Graham Sumner points out that some groups bind members together because of

> relation to each other (kin, neighborhood, alliance, connubium and commercium) which draws them together and differentiates them from others. Thus a differentiation arises between ourselves, the we-group, or in-group, and everybody else, or the others-groups, out-groups. The insiders in a we-group are in a relation of peace, order, law, government, and industry to each other. Their relation to all outsiders, or others-groups, is one of war and plunder, except so far as agreements have modified it.[1]

1. William Graham Sumner, *Folkways* (New York: Ginn and Company, 1904), secs. 12, 13.

Understanding group dynamics involves us in everything from patterns of communication to roles we play within groups, as discussed in the next reading. The Adlers analyze the process of learning and carrying out the college athlete role. The intense pressure for performance, adulation from crowds and the press, and intimidation of classroom expectations combine to provide an example of group socialization into rigid role expectations.

Although the case of athletes presents a stressful kind of group for the members, Philip Zimbardo asks what happens when the situation facilitates or even encourages negative group behavior. He considers prison abuse in which bad behavior is allowed and even encouraged by analyzing the abuse scandals at Abu Ghraib prison in Iraq in the context of group norms and the setting.

One famous observer of groups and society was Max Weber. Writing at the turn of the century, he saw a modern form of groups emerging as societies industrialized. As businesses grew larger and more technical, owners no longer hired friends and relatives, but rather hired and fired people on the basis of their training, skills, and performance. Bureaucracies as Weber described them were organized as hierarchies with sets of rules and regulations, and individuals were formally contracted to carry out certain tasks in exchange for compensation. Today it would seem strange if organizations did not function in this manner. Characteristics of bureaucratic organizations include the formal contractual relationships, rules and regulations, and hiring and firing based on merit and competence, as discussed in the last reading.

Since Weber first wrote about bureaucracy, there have been changes in the way bureaucracies function. Technology, for instance, has created new relationships between organizations and individuals. At the same time, individual participation in groups has been changing. Today there is less personal involvement in groups and more technological involvement, which tends to isolate individuals.

As you read the selections in this chapter, consider the definitions of primary and secondary groups and reflect on examples of these groups in modern society. What dynamics do we need to understand in order to interact and communicate effectively in groups? Are some problems in group dynamics outcomes of our need for belonging and acceptance? How are organizations changing? How is individual participation in groups changing?

16

The Exchange of Social Rewards

PETER M. BLAU

Our actions and choices have positive or negative consequences. Much satisfaction comes from social life, but so does human suffering. An act may make some people happy and others sad; rewards for one may mean disappointment or loss to another. In this classic reading, Blau explains the dynamic of interaction that forms the base of exchange theory: rewards, costs, and reciprocal expectations that create human bonds. If costs in a relationship are too high compared to rewards, we may end that relationship.

As you read, consider the following questions:

1. *Think of examples of what motivates people to act.*
2. *When have you felt rewarded or disappointed by the actions of others?*
3. *When do you feel an obligation to reciprocate?*

GLOSSARY **Altruism** Benefitting or being kind to others. **Egoism** Motivation by expectation of social reward.

Most human pleasures have their roots in social life. Whether we think of love or power, professional recognition or sociable companionship, the comforts of family life or the challenge of competitive sports, the gratifications experienced by individuals are contingent on actions of others. The same is true for the most selfless and spiritual satisfactions. To work effectively for a good cause requires making converts to it. Even the religious experience is much enriched by communal worship. Physical pleasures that can be experienced in solitude pale in significance by comparison. Enjoyable as a good dinner is, it is the social occasion that gives it its luster. Indeed, there is something pathetic about the person who derives his major gratification from food or drink as such, since it reveals either excessive need or excessive greed; the pauper illustrates the former, the glutton, the latter. To be sure, there are profound solitary enjoyments—reading a good book, creating a piece of art, producing a scholarly work. Yet these, too, derive much of their significance from being later communicated to and shared with others. The lack of such anticipation makes the solitary activity again somewhat pathetic: the recluse who has nobody to talk to about what he reads; the artist or scholar whose works are completely ignored, not only by his contemporaries but also by posterity.

Much of human suffering as well as much of human happiness has its source in the actions of other human beings. One follows from the other, given the facts of group life, where pairs do not exist in complete isolation from other relations. The same human acts that cause pleasure to some typically cause displeasure to others. For one boy to enjoy the love of a girl who has committed herself to be his steady date, other boys who had gone out with her must suffer the pain of having been rejected. The satisfaction a

SOURCE: Reprinted from *Exchange and Power in Social Life* by Peter M. Blau, John Wiley, New York, 1964, pp. 14–17, by permission of the publishers.

man derives from exercising power over others requires that they endure the deprivation of being subject to his power. For a professional to command an outstanding reputation in his field, most of his colleagues must get along without such pleasant recognition, since it is the lesser professional esteem of the majority that defines his as outstanding. The joy the victorious team members experience has its counterpart in the disappointment of the losers. In short, the rewards individuals obtain in social associations tend to entail a cost to other individuals. This does not mean that most social associations involve zero-sum games in which the gains of some rest on the losses of others. Quite the contrary, individuals associate with one another because they all profit from their association. But they do not necessarily all profit equally, nor do they share the cost of providing the benefits equally, and even if there are no direct costs to participants, there are often indirect costs born by those excluded from the association, as the case of the rejected suitors illustrates.

Some social associations are intrinsically rewarding. Friends find pleasure in associating with one another, and the enjoyment of whatever they do together—climbing a mountain, watching a football game—is enhanced by the gratification that inheres in the association itself. The mutual affection between lovers or family members has the same result. It is not what lovers do together but their doing it *together* that is the distinctive source of their special satisfaction—not seeing a play but sharing the experience of seeing it. Social interaction in less intimate relations than those of lovers, family members, or friends, however, may also be inherently rewarding. The sociability at a party or among neighbors or in a work group involves experiences that are not especially profound but are intrinsically gratifying. In these cases, all associates benefit simultaneously from their social interaction, and the only cost they incur is the indirect one of giving up alternative opportunities by devoting time to the association.

Social associations may also be rewarding for a different reason. Individuals often derive specific benefits from social relations because their associates deliberately go to some trouble to provide these benefits for them. Most people like helping others and doing favors for them—to assist not only their friends but also their acquaintances and occasionally even strangers, as the motorist who stops to aid another with his stalled car illustrates. Favors make us grateful and our expressions of gratitude are social rewards that tend to make doing favors enjoyable, particularly if we express our appreciation and indebtedness publicly and thereby help establish a person's reputation as a generous and competent helper. Besides, one good deed deserves another. If we feel grateful and obligated to an associate for favors received, we shall seek to reciprocate his kindness by doing things for him. He in turn is likely to reciprocate, and the resulting mutual exchange of favors strengthens, often without explicit intent, the social bond between us.

A person who fails to reciprocate favors is accused of ingratitude. This very accusation indicates that reciprocation is expected, and it serves as a social sanction that discourages individuals from forgetting their obligations to associates. Generally, people are grateful for favors and repay their social debts, and both their gratitude and their repayment are social rewards for the associate who has done them favors.[1] The fact that furnishing benefits to others tends to produce these social rewards is, of course, a major reason why people often go to great trouble to help their associates and enjoy doing so. We would not be human if these advantageous consequences of our good deeds were not important inducements for our doing them.[2] There are, to be sure, some individuals who selflessly work for

1. "We rarely meet with ingratitude, so long as we are in a position to confer favors." François La Rochefoucauld, *The Maxims*, London: Oxford University Press, 1940, p. 101 (#306).

2. Once a person has become emotionally committed to a relationship, his identification with the other and his interest in continuing the association provide new independent incentives for supplying benefits to the other. Similarly, firm commitments to an organization lead members to make recurrent contributions to it without expecting reciprocal benefits in every instance. The significance of these attachments is further elaborated in subsequent chapters.

others without any thought of reward and even without expecting gratitude, but these are virtually saints, and saints are rare. The rest of us also act unselfishly sometimes, but we require some incentive for doing so, if it is only the social acknowledgment that we are unselfish.

An apparent "altruism" pervades social life; people are anxious to benefit one another and to reciprocate for the benefits they receive. But beneath this seeming selflessness an underlying "egoism" can be discovered; the tendency to help others is frequently motivated by the expectation that doing so will bring social rewards. Beyond this self-interested concern with profiting from social associations, however, there is again an "altruistic" element or, at least, one that removes social transactions from simple egoism or psychological hedonism. A basic reward people seek in their associations is social approval, and selfish disregard for others makes it impossible to obtain this important reward.[3]

The social approval of those whose opinions we value is of great significance to us; but its significance depends on its being genuine. We cannot force others to give us their approval, regardless of how much power we have over them, because coercing them to express their admiration or praise would make these expressions worthless. "Action can be coerced, but a coerced show of feeling is only a show."[4] Simulation robs approval of its significance, but its very importance makes associates reluctant to withhold approval from one another and, in particular, to express disapproval, thus introducing an element of simulation and dissimulation into their communications. As a matter of fact, etiquette prescribes that approval be simulated in disregard of actual opinions under certain circumstances. One does not generally tell a hostess, "Your party was boring," or a neighbor, "What you say is stupid." Since social conventions require complimentary remarks on many occasions, these are habitually discounted as not reflecting genuine approbation, and other evidence that does reflect it is looked for such as whether guests accept future invitations or whether neighbors draw one into further conversations.

3. Bernard Mandeville's central theme is that private vices produce public benefits because the importance of social approval prompts men to contribute to the welfare of others in their own self-interest. As he put it tersely at one point, "Moral Virtues Are the Political Offspring Which Flattery Begot upon Pride." *The Fable of the Bees*, Oxford: Clarendon, 1924, Vol. I, 51; see also pp. 63–80.

4. Erving Goffman, *Asylums*. Chicago: Aldine, 1962. p. 115.

17

The Sounds of Silence

EDWARD T. HALL AND MILDRED REED HALL

A crucial element in human interaction and group behavior is language, both verbal and nonverbal. Nonverbal communication is quite as important in communicating as verbal language. When we study foreign language, we may learn the words, but the gestures and facial expressions are difficult to master unless we have been raised in the culture. The Halls discuss nonverbal communication patterns and differences across cultures.

Think about the following questions as you read this selection:

1. *What is meant by "the sounds of silence"?*
2. *How do individuals learn nonverbal language?*
3. *Can you think of times you have been uncomfortable because of misunderstandings in nonverbal communication?*
4. *What might be the consequences of misunderstood nonverbal language cues between men and women?*

GLOSSARY **Silent language** Communication that takes place through gestures, facial expressions, and other body movements. **Machismo** Assertive masculinity.

BOB LEAVES HIS APARTMENT at 8:15 A.M. and stops at the corner drugstore for breakfast. Before he can speak, the counterman says, "The usual?" Bob nods yes. While he savors his Danish, a fat man pushes onto the adjoining stool and overflows into his space. Bob scowls and the man pulls himself in as much as he can. Bob has sent two messages without speaking a syllable.

Henry has an appointment to meet Arthur at 11 o'clock; he arrives at 11:30. Their conversation is friendly, but Arthur retains a lingering hostility. Henry has unconsciously communicated that he doesn't think the appointment is very important or that Arthur is a person who needs to be treated with respect.

George is talking to Charley's wife at a party. Their conversation is entirely trivial, yet Charley glares at them suspiciously. Their physical proximity and the movements of their eyes reveal that they are powerfully attracted to each other.

José Ybarra and Sir Edmund Jones are at the same party, and it is important for them to establish a cordial relationship for business reasons. Each is trying to be warm and friendly, yet they will part with mutual distrust and their business transaction will probably fall through. José, in Latin fashion, moved closer and closer to Sir Edmund as they spoke, and this movement was miscommunicated as pushiness to Sir Edmund, who kept backing away from this intimacy, and this was miscommunicated to José as coldness. The silent languages of Latin and English cultures are more difficult to learn than their spoken languages.

In each of these cases, we see the subtle power of nonverbal communication. The only language used throughout most of the history of humanity (in evolutionary terms, vocal communication is relatively recent), it is the first form of communication you learn. You use this preverbal language, consciously and unconsciously, every day to tell other people how you feel about yourself and them. This language includes your posture, gestures, facial expressions, costume, the way you walk, even your treatment of time and space and material things. All people communicate on several different levels at the same time but are usually aware of only the verbal dialog and don't realize that they respond to nonverbal messages. But when a person says one thing and really believes something else, the discrepancy between the two can usually be sensed. Nonverbal communication systems are much less subject to the conscious deception that often occurs in verbal systems. When we find ourselves thinking, "I don't know what it is about him, but he doesn't seem sincere," it's usually this lack of congruity between a person's words and his behavior that makes us anxious and uncomfortable.

Few of us realize how much we all depend on body movement in our conversation or are aware of the hidden rules that govern listening behavior. But we know instantly whether or not the person we're talking to is "tuned in" and we're very sensitive to any breach in listening etiquette. In white middle-class American culture, when someone wants to show he is listening to someone else, he looks either at the other person's face, or specifically, at his eyes, shifting his gaze from one eye to the other.

If you observe a person conversing, you'll notice that he indicates he's listening by nodding his head. He also makes little "Hmm" noises. If he agrees with what's being said, he may give a vigorous nod. To show pleasure or affirmation, he smiles; if he has some reservations, he looks skeptical by raising an eyebrow or pulling down the corners of his mouth. If a participant wants to terminate the conversation, he may start shifting his body position, stretching his legs, crossing or uncrossing them, bobbing his foot or diverting his gaze from the speaker. The more he fidgets, the more the speaker becomes aware that he has lost his audience. As a last measure, the listener may look at his watch to indicate the imminent end of the conversation.

Talking and listening are so intricately intertwined that a person cannot do one without the other. Even when one is alone and talking to oneself, there is part of the brain that speaks while another part listens. In all conversations, the listener is positively or negatively reinforcing the speaker all the time. He may even guide the conversation without knowing it, by laughing or frowning or dismissing the argument with a wave of his hand.

The language of the eyes—another age-old way of exchanging feelings—is both subtle and complex. Not only do men and women use their eyes differently but there are class, generation, regional, ethnic, and national cultural differences. Americans often complain about the way foreigners stare at people or hold a glance too long. Most Americans look away from someone who is using his eyes in an unfamiliar way because it makes them self-conscious. If a man looks at another man's wife in a certain way, he's asking for trouble, as indicated earlier. But he might not be ill-mannered or seeking to challenge the husband. He might be a European in this country who hasn't learned our visual mores. Many American women visiting France or Italy are acutely embarrassed because, for the first time in their lives, men really look at them—their eyes, hair, nose, lips, breasts, hips, legs, thighs, knees, ankles, feet, clothes, hairdo, even their walk. . . .

Analyzing the mass of data on the eyes, it is possible to sort out at least three ways in which the eyes are used to communicate: dominance versus submission, involvement versus detachment, and positive versus negative attitude. In addition, there are three levels of consciousness and control, which can be categorized as follows: (1) conscious use of the eyes to communicate, such as the flirting blink and the intimate nose-wrinkling squint; (2) the very extensive category of unconscious but learned behavior governing where the eyes are directed and when (this unwritten set of rules dictates how and under what circumstances the sexes, as well as people of all status categories, look at each other); and (3) the response of the eye itself, which is completely outside both awareness and control—changes in the cast (the sparkle) of the eye and the pupillary reflex.

The eye is unlike any other organ of the body, for it is an extension of the brain. The unconscious pupillary reflex and the cast of the eye have been known by people of Middle Eastern origin for years—although most are unaware of their knowledge. Depending on the context, Arabs and others look either directly at the eyes or deeply *into* the eyes of their interlocutor. We became aware of this in the Middle East several years ago while looking at jewelry. The merchant suddenly started to push a particular bracelet at a customer and said, "You buy this one." What interested us was that the bracelet was not the one that had been consciously selected by the purchaser. But the merchant, watching the pupils of the eyes, knew what the purchaser really wanted to buy. Whether he specifically knew *how* he knew is debatable.

A psychologist at the University of Chicago, Eckhard Hess, was the first to conduct systematic studies of the pupillary reflex. His wife remarked one evening, while watching him reading in bed, that he must be very interested in the text because his pupils were dilated. Following up on this, Hess slipped some pictures of nudes into a stack of photographs that he gave to his male assistant. Not looking at the photographs but watching his assistant's pupils, Hess was able to tell precisely when the assistant came to the nudes. In further experiments, Hess retouched the eyes in a photograph of a woman. In one print, he made the pupils small, in another, large; nothing else was changed. Subjects who were given the photographs found the woman with the dilated pupils much more attractive. Any man who has had the experience of seeing a woman look at him as her pupils widen with reflex speed knows that she's flashing him a message.

The eye-sparkle phenomenon frequently turns up in our interviews of couples in love. It's apparently one of the first reliable clues in the other person that love is genuine. To date, there is no scientific data to explain eye sparkle; no investigation of the pupil, the cornea, or even the white sclera of the eye shows how the sparkle originates. Yet we all know it when we see it.

One common situation for most people involves the use of the eyes in the street and in public. Although eye behavior follows a definite set of rules, the rules vary according to the place, the needs and feelings of the people, and their ethnic background. For urban whites, once they're within definite recognition distance (16–32 feet for people with average eyesight), there is mutual avoidance of eye contact—unless they want something specific: a pickup, a handout, or information of some kind. In the West and in small towns generally, however, people are much more likely to look at and greet one another, even if they're strangers.

It's permissible to look at people if they're beyond recognition distance; but once inside this sacred zone, you can only steal a glance at strangers. You *must* greet friends, however; to fail to do so is insulting. Yet, to stare too fixedly even at them is considered rude and hostile. Of course, all of these rules are variable. . . .

[A] very basic difference between people of different ethnic backgrounds is their sense of territoriality and how they handle space. This is the silent communication, or miscommunication, that caused friction between Mr. Ybarra and Sir Edmund Jones in our earlier example. We know from research that everyone has around himself an invisible bubble of space that contracts and expands depending on several factors: his emotional state, the activity he's performing at the time, and his cultural background. This bubble is a kind of mobile territory that he will defend against intrusion. If he is accustomed to close personal distance between himself and others, his bubble will be smaller than that of someone who's accustomed to greater personal distance. People of North European heritage—English, Scandinavian, Swiss, and German—tend to avoid contact. Those whose heritage is Italian,

French, Spanish, Russian, Latin American, or Middle Eastern like close personal contact.

People are very sensitive to any intrusion into their spatial bubble. If someone stands too close to you, your first instinct is to back up. If that's not possible, you lean away and pull yourself in, tensing your muscles. If the intruder doesn't respond to these body signals, you may then try to protect yourself, using a briefcase, umbrella, or raincoat. . . . As a last resort, you may move to another spot and position yourself behind a desk or a chair that provides screening. Everyone tries to adjust the space around himself in a way that's comfortable for him; most often, he does this unconsciously.

Emotions also have a direct effect on the size of a person's territory. When you're angry or under stress, your bubble expands and you require more space. New York psychiatrist Augustus Kinzel found a difference in what he calls Body-Buffer Zones between violent and nonviolent prison inmates. Dr. Kinzel conducted experiments in which each prisoner was placed in the center of a small room and then Dr. Kinzel slowly walked toward him. Nonviolent prisoners allowed him to come quite close, while prisoners with a history of violent behavior couldn't tolerate his proximity and reacted with some vehemence.

Apparently, people under stress experience other people as looming larger and closer than they actually are. Studies of schizophrenic patients have indicated that they sometimes have a distorted perception of space, and several psychiatrists have reported patients who experience their body boundaries as filling up an entire room. For these patients, anyone who comes into the room is actually inside their body, and such an intrusion may trigger a violent outburst.

Unfortunately, there is little detailed information about normal people who live in highly congested urban areas. We do know, of course, that the noise, pollution, dirt, crowding, and confusion of our cities induce feelings of stress in most of us, and stress leads to a need for greater space. [People who are] packed into a subway, jostled in the street, crowded into an elevator, and forced to work all day in a bull pen or in a small office without auditory or visual privacy [are] going to be very stressed at the end of [the] day. They need places that provide relief from constant overstimulation. . . . Stress from overcrowding is cumulative and people can tolerate more crowding early in the day than later; note the increased bad temper during the evening rush hour as compared with the morning melee. Certainly one factor in people's desire to commute by car is the need for privacy and relief from crowding (except, often, from other cars); it may be the only time of day when nobody can intrude.

In crowded public places, we tense our muscles and hold ourselves stiff, and thereby communicate to others our desire not to intrude on their space and, above all, not to touch them. We also avoid eye contact and the total effect is that of someone who has "tuned out." Walking along the street, our bubble expands slightly as we move in a stream of strangers, taking care not to bump into them. In the office, at meetings, in restaurants, our bubble keeps changing as it adjusts to the activity at hand.

Most white middle-class Americans use four main distances in their business and social relations: intimate, personal, social, and public. Each of these distances has a near and a far phase and is accompanied by changes in the volume of the voice. Intimate distance varies from direct physical contact with another person to a distance of six to eighteen inches and is used for our most private activities—caressing another person or making love. At this distance, you are overwhelmed by sensory inputs from the other person—heat from the body, tactile stimulation from the skin, the fragrance of perfume, even the sound of breathing—all of which literally envelop you. Even at the far phase, you're still within easy touching distance. In general, the use of intimate distance in public between adults is frowned on. It's also much too close for strangers, except under conditions of extreme crowding.

In the second zone—personal distance—the close phase is one and a half to two and a half feet; it's at this distance that wives usually stand from their husbands in public. If another woman moves into this zone, the wife will most likely be disturbed. The far phase—two and a half to four feet—is the distance used to "keep someone at arm's length" and is the most common spacing used by people in conversation.

The third zone—social distance—is employed during business transactions or exchanges with a clerk or repairman. People who work together tend to use close social distance—four to seven feet. This is also the distance for conversation at social gatherings. To stand at this distance from one who is seated has a dominating effect (for example, teacher to pupil, boss to secretary). The far phase of the third zone—seven to twelve feet—is where people stand when someone says, "Stand back so I can look at you." This distance lends a formal tone to business or social discourse. In an executive office, the desk serves to keep people at this distance.

The fourth zone—public distance—is used by teachers in classrooms or speakers at public gatherings. At its farthest phase—25 feet and beyond—it is used for important public figures. Violations of this distance can lead to serious complications. During his 1970 U.S. visit, the president of France, Georges Pompidou, was harassed by pickets in Chicago, who were permitted to get within touching distance. Since pickets in France are kept behind barricades a block or more away, the president was outraged by this insult to his person, and President Nixon was obliged to communicate his concern as well as offer his personal apologies.

It is interesting to note how American pitchmen and panhandlers exploit the unwritten, unspoken conventions of eye and distance. Both take advantage of the fact that once explicit eye contact is established, it is rude to look away, because to do so means to brusquely dismiss the other person and his needs. Once having caught the eye of his mark, the panhandler then locks on, not letting go until he moves through the public zone, the social zone, the personal zone, and finally, into the intimate sphere, where people are most vulnerable.

Touch also is an important part of the constant stream of communication that takes place between people. A light touch, a firm touch, a blow, a caress are all communications. In an effort to break down barriers among people, there's been a recent upsurge in group-encounter activities, in which strangers are encouraged to touch one another. In special situations such as these, the rules for not touching are broken with group approval and people gradually lose some of their inhibitions.

Although most people don't realize it, space is perceived and distances are set not by vision alone but with all the senses. Auditory space is perceived with the ears, thermal space with the skin, kinesthetic space with the muscles of the body, and olfactory space with the nose. And, once again, it's one's culture that determines how his senses are programmed—which sensory information ranks highest and lowest. The important thing to remember is that culture is very persistent. In this country, we've noted the existence of culture patterns that determine distance between people in the third and fourth generations of some families, despite their prolonged contact with people of very different cultural heritages.

Whenever there is great cultural distance between two people, there are bound to be problems arising from differences in behavior and expectations. An example is the American couple who consulted a psychiatrist about their marital problems. The husband was from New England and had been brought up by reserved parents who taught him to control his emotions and to respect the need for privacy. His wife was from an Italian family and had been brought up in close contact with all the members of her large family, who were extremely warm, volatile, and demonstrative.

When the husband came home after a hard day at the office, dragging his feet and longing for peace and quiet, his wife would rush to him and smother him. Clasping his hands, rubbing his brow, crooning over his weary head, she never left him alone. But when the wife was upset or anxious about her day, the husband's response was to withdraw completely and leave her alone. No comforting, no affectionate embrace, no attention—just solitude. The woman became convinced her husband didn't love her and, in desperation, she consulted a psychiatrist. Their problem wasn't basically psychological but cultural.

Why [have people] developed all these different ways of communicating messages without words? One reason is that people don't like to spell out certain kinds of messages. We prefer to find other ways of showing our feelings. This is especially true in relationships as sensitive as courtship.... We work out subtle ways of encouraging or discouraging each other that save face and avoid confrontations....

If a man sees a woman whom he wants to attract, he tries to present himself by his posture and stance as someone who is self-assured. He moves briskly and confidently. When he catches the eye of the woman, he may hold her glance a little longer than normal. If he gets an encouraging smile, he'll move in close and engage her in small talk. As they converse, his glance shifts over her face and body. He, too, may make preening gestures—straightening his tie, smoothing his hair, or shooting his cuffs.

How do people learn body language? The same way they learn spoken language—by observing and imitating people around them as they're growing up. Little girls imitate their mothers or an older female. Little boys imitate their fathers or a respected uncle or a character on television. In this way, they learn the gender signals appropriate for their sex. Regional, class, and ethnic patterns of body behavior are also learned in childhood and persist throughout life.

Such patterns of masculine and feminine body behavior vary widely from one culture to another. In America, for example, women stand with their thighs together. Many walk with their pelvis tipped sightly forward and their upper arms close to their body. When they sit, they cross their legs at the knee or cross their ankles. American men hold their arms away from their body, often swinging them as they walk. They stand with their legs apart (an extreme example is the cowboy, with legs apart and thumbs tucked into his belt). When they sit, they put their feet on the floor with legs apart and, in some parts of the country, they cross their legs by putting one ankle on the other knee.

Leg behavior indicates sex, status, and personality. It also indicates whether or not one is at ease or is showing respect or disrespect for the other person. Young Latin American males avoid crossing their legs. In their world of *machismo*, the preferred position for young males when with one another (if there is no old dominant male present to whom they must show respect) is to sit on the base of the spine with their leg muscles relaxed and their feet wide apart. Their respect position is like our military equivalent: spine straight, heels and ankles together—almost identical to that displayed by properly brought up young women in New England in the early part of this century.

American women who sit with their legs spread apart in the presence of males are *not* normally signaling a come-on—they are simply (and often unconsciously) sitting like men. Middle-class women in the presence of other women to whom they are very close may on occasion throw themselves down on a soft chair or sofa and let themselves go. This is a signal that nothing serious will be taken up. Males, on the other hand, lean back and prop their legs up on the nearest object.

The way we walk, similarly, indicates status, respect, mood, and ethnic or cultural affiliation.... To white Americans, some French middle-class males walk in a way that is both humorous and suspect. There is a bounce and looseness to the French walk, as though the parts of the body were somehow unrelated. Jacques Tati, the French movie actor, walks this way; so does the great mime, Marcel Marceau....

All over the world, people walk not only in their own characteristic way but have walks that communicate the nature of their involvement with whatever it is they're doing. The purposeful walk of North Europeans is an important component of proper behavior on the job. Any male who has been in the military knows how essential it is to walk properly.... The quick shuffle of servants in the Far East in the old days was a show of respect. On the island of Truk, when we last visited, the inhabitants even had a name for the respectful walk that one used when in the presence of a chief or when walking past a chief's house. The term was *sufan*, which meant to be humble and respectful.

The notion that people communicate volumes by their gestures, facial expressions, posture and walk is not new; actors, dancers, writers, and psychiatrists have long been aware of it. Only in recent years, however, have scientists begun to make systematic observations of body motions. Ray L. Birdwhistell of the University of Pennsylvania is one of the pioneers in body-motion research and coined the term *kinesics* to describe this field. He developed an elaborate notation system to record both facial and body movement, using an approach similar to that of the linguist, who studies the basic elements of speech. Birdwhistell and other kinesicists such as Albert Shellen, Adam Kendon, and William Condon take movies of people interacting. They run the film over and over again, often at reduced speed for frame-by-frame analysis, so that they can observe even the slightest body movements not perceptible at normal interaction speeds. These movements are then recorded in notebooks for later analysis. . . .

Several years ago in New York City, there was a program for sending children from predominantly black and Puerto Rican low income neighborhoods to summer school in a white upper-class neighborhood on the East Side. One morning, a group of young black and Puerto Rican boys raced down the street, shouting and screaming and overturning garbage cans on their way to school. A doorman from an apartment building nearby chased them and cornered one of them inside a building. The boy drew a knife and attacked the doorman. This tragedy would not have occurred if the doorman had been familiar with the behavior of boys from low-income neighborhoods, where such antics are routine and socially acceptable and where pursuit would be expected to invite a violent response.

The language of behavior is extremely complex. Most of us are lucky to have under control one subcultural system—the one that reflects our sex, class, generation, and geographic region within the United States. Because of its complexity, efforts to isolate bits of nonverbal communication and generalize from them are in vain; you don't become an instant expert on people's behavior by watching them at cocktail parties. Body language isn't something that's independent of the person, something that can be donned and doffed like a suit of clothes.

Our research and that of our colleagues have shown that, far from being a superficial form of communication that can be consciously manipulated, nonverbal-communication systems are interwoven into the fabric of the personality and, as sociologist Erving Goffman has demonstrated, into society itself. They are the warp and woof of daily interactions with others, and they influence how one expresses oneself, how one experiences oneself as a man or a woman.

Nonverbal communications signal to members of your own group what kind of person you are, how you feel about others, how you'll fit into and work in a group, whether you're assured or anxious, the degree to which you feel comfortable with the standards of your own culture, as well as deeply significant feelings about the self, including the state of your own psyche. For most of us, it's difficult to accept the reality of another's behavioral system. And, of course, none of us will ever become fully knowledgeable of the importance of every nonverbal signal. But as long as each of us realizes the power of these signals, this society's diversity can be a source of great strength rather than a further—and subtly powerful—source of division.

18

Backboards & Blackboards

College Athletes and Role Engulfment

PATRICIA A. ADLER AND PETER ADLER

We all carry out roles in groups to which we belong. We are socialized into school, work, and other roles. Sometimes this socialization is intense and influences all other roles. Such is the case with the achieved role of college athletes. The Adlers provide an example from the world of NCAA basketball, describing the process of socialization, role engulfment, and abandonment of former roles, and formation of team cliques that reflect earlier socialization experiences. The athletes work hard to achieve their goals and in the process are shaped by pressures from adoring fans and intimidating professors. Their roles in the group are shaped by all these pressures.

Consider the following as you read:

1. *What are the pressures faced by high-profile athletes?*
2. *How do the roles of student athletes differ from roles of most other students?*
3. *What would you recommend doing to reduce role engulfment?*
4. *What role responsibilities have you had that involve role engulfment and abandonment of former roles?*

GLOSSARY **Role engulfment** Demands and rewards of athletic role supersede other roles. **Role domination** Process by which athletes become engulfed in athletic roles to exclusion of other roles. **Role abandonment** Detachment from investment in other roles, letting go of alternative goals and priorities. **Statuses** Positions in organized groups related to other positions by set of normative expectations. **Roles** Activities of people of given status.

It was a world of dreams. They expected to find fame and glory, spotlights and television cameras. There was excitement and celebrity, but also hard work and discouragement, a daily grind characterized by aches, pains, and injuries, and an abundance of rules, regulations, and criticism. Their lives alternated between contacts with earnest reporters, adoring fans, and fawning women, and with intimidating professors, demanding boosters, and unrelenting coaches. There was secrecy and intrigue, drama and adulation, but also isolation and alienation, loss of freedom and personal autonomy, and overwhelming demands. These conflicts and dualisms are the focus of this reading. This is a study of the socialization of college athletes.

For five years we lived in and studied the world of elite NCAA (National Collegiate Athletic Association) college basketball. Participant-observers,

SOURCE: *Backboards & Blackboards: College Athletes and Role Engulfment* by Patricia A. Adler and Peter Adler.

we fit ourselves into the setting by carving out evolving roles that integrated a combination of team members' expectations with our interests and abilities. Individually and together, we occupied a range of different positions including friend, professor, adviser, confidant, and coach. We observed and interacted with all members of the team, gaining an intimate understanding of the day-to-day and year-to-year character of this social world. From behind the scenes of this secretive and celebrated arena, we document the experiences of college athletes, focusing on changes to their selves and identities over the course of their college years. . . .

THEORETICAL APPROACH

This is not only a study of college athletes, but also a study in the social psychology of the self. Our observations reveal a significant pattern of transformation experienced by all our subjects: *role engulfment.* Many of the individuals we followed entered college hoping to gain wealth and fame through their involvement with sport. They did not anticipate, however, the cost of dedicating themselves to this realm. While nearly all conceived of themselves as athletes first, they possessed other self-images that were important to them as well. Yet over the course of playing college basketball, these individuals found the demands and rewards of the athletic role overwhelming and became engulfed by it. However, in yielding to it, they had to sacrifice other interests, activities, and, ultimately, dimensions of their selves. They immersed themselves completely in the athletic role and neglected or abandoned their identities lodged in these other roles. They thus became extremely narrow in their focus. In this work we examine *role domination*, the process by which athletes became engulfed in their athletic role as it ascended to a position of prominence. We also examine the concomitant process of *role abandonment*, where they progressively detached themselves from their investment in other areas and let go of alternative goals or priorities. We analyze the changes this dual process of self-engulfment had on their self-concepts and on the structure of their selves. . . .

Role theory focuses on the systems, or institutions, into which interaction fits. According to its tenets, *statuses* are positions in organized groups or systems that are related to other positions by a set of normative expectations. Statuses are not defined by the people that occupy them, but rather they are permanent parts of those systems. Each status carries with it a set of role expectations specifying how persons occupying that status should behave. *Roles* consist of the activities people of a given status are likely to pursue when following the normative expectations for their positions. *Identities* (or what McCall and Simmons 1978, call role-identities), are the self-conceptions people develop from occupying a particular status or enacting a role. The *self* is the more global, multirole, core conception of the real person.

Because in modern society we are likely to be members of more than one group, we may have several statuses and sets of role-related behavioral expectations. Each individual's total constellation of roles forms what may be termed a *role-set*, characterized by a series of relationships with role-related others, or role-set members (Merton 1957). Certain roles or role-identities may be called to the fore, replacing others, as people interact with individuals through them. Individuals do not invest their core feelings of identity or self in all roles equally, however. While some roles are more likely to be called forth by the expectations of others, other roles are more salient to the individual's core, or "real self," than others. They are arranged along a *hierarchy of salience* from peripheral roles to those that "merge with the self" (Turner 1978), and their ranking may be determined by a variety of factors. Role theory enhances an understanding of both the internal structure of the self and the relation between self and society; it sees this relation as mediated by the concept of role and its culturally and structurally derived expectations.

The interpretive branch of symbolic interactionism focuses on examining agency, process, and change. One of the concepts most critical to symbolic interactionism is the self. Rather than merely looking at roles and their relation to society, symbolic interactionism looks at the individuals filling those roles and the way they engage not only in role-taking, but also in active, creative role-making (Turner 1962). The self is the thinking and feeling being connecting the various roles and identities individuals put forth in different situations (Cooley 1962; Mead 1934). Symbolic interactionism takes a dynamic view of individuals in society, believing that they go beyond merely reproducing existing roles and structures to collectively defining and interpreting the meaning of their surroundings. These subjective, symbolic assessments form the basis for the creation of new social meanings, that then lead to new, shared patterns of adaptation (Blumer 1969). In this way individuals negotiate their social order as they experience it (Strauss 1978). They are thus capable of changing both themselves and the social structures within which they exist. Symbolic interactionism enhances understandings of the dynamic processes characterizing human group life and the reciprocal relation between those processes and changes in the self.

In integrating these two perspectives we show how the experiences of college athletes are both bounded and creative, how athletes integrate structural, cultural, and interactional factors, and how they change and adapt through a dynamic process of action and reaction, forging collective adaptations that both affirm and modify existing structures.

THE SETTING

We conducted this research at a medium-size (6000 students), private university (hereafter referred to as "the University") in the south-western portion of the United States. Originally founded on the premise of a religious affiliation, the University had severed its association with the church several decades before, and was striving to make a name for itself as one of the finer private, secular universities in the region. For many years it had served the community chiefly as a commuter school, but had embarked on an aggressive program of national recruiting over the past five to ten years that considerably broadened the base of its enrollment. Most of the students were white, middle class, and drawn from the suburbs of the South, Midwest, and Southwest. Academically, the University was experimenting with several innovative educational programs designed to enhance its emerging national reputation. Sponsored by reforms funded by the National Endowment for the Humanities, it was changing the curriculum, introducing a more interdisciplinary focus, instituting a funded honors program, increasing the general education requirements, and, overall, raising academic standards to a fairly rigorous level.

Within the University, the athletic program overall had considerable success during the course of our research: the University's women's golf team was ranked in the top three nationally, the football team won their conference each season, and the basketball program was ranked in the top forty of Division I NCAA schools, and in the top twenty for most of two seasons. The basketball team played in post-season tournaments every year, and in the five complete seasons we studied them, they won approximately four times as many games as they lost. In general, the basketball program was fairly representative of what Coakley (1986) and Frey (1982a) have termed "big-time" college athletics. Although it could not compare to the upper echelon of established basketball dynasties or to the really large athletic programs that wielded enormous recruiting and operating budgets, its success during the period we studied it compensated for its size and lack of historical tradition. The University's basketball program could thus best be described as "up and coming." Because the basketball team (along with the athletic department more generally) was ranked nationally and sent graduating members into the professional leagues, the entire athletic milieu was imbued with a sense of seriousness and purpose.

The team's professionalism was also enhanced by the attention focused on it by members of the community. Located in a city of approximately 500,000 with no professional sports teams, the University's programs served as the primary source of athletic entertainment and identification for the local population. When the basketball program meteorically rose to prominence, members of the city embraced it with fanatical support. Concomitant with the team's rise in fortunes, the region—part of the booming oil and sun belts—was experiencing increased economic prosperity. This surging local pride and financial windfall cast its glow over the basketball team, as it was the most charismatic and victorious program in the city's history, and the symbol of the community's newfound identity. Interest and support were therefore lavished on the team members.

THE PEOPLE

Over the course of our research, we observed 39 players and seven coaches. Much like Becker et al. (1961), we followed the players through their recruitment and entry into the University, keeping track of them as they progressed through school. We also watched the coaches move up their career ladders and deal with the institutional structures and demands.

Players, like students, were recruited primarily from the surrounding region. Unlike the greater student population, though, they generally did not hail from suburban areas. Rather, they predominantly came from the farming and rural towns of the prairies or southlands, and from the ghetto and working class areas of the mid-sized cities.

Demographics

Over the course of our involvement with the team, two-thirds of the players we studied were black and one-third white. White middle class players accounted for approximately 23 percent of the team members. They came from intact families where fathers worked in such occupations as wholesale merchandising, education, and sales. Although several were from suburban areas, they were more likely to come from mid- to larger-sized cities or exurbs. The remaining white players (10 percent) were from working class backgrounds. They came from small factory towns or cities and also from predominantly (although not exclusively) intact families. Some of their parents worked in steel mills or retail jobs.

The black players were from middle, working, and lower class backgrounds. Those from the middle class (15 percent) came from the cities and small towns of the Midwest and South. They had intact families with fathers who worked as police chiefs, ministers, high school principals, or in the telecommunications industry. A few came from families in which the mothers also worked. Several of these families placed a high premium on education; one player was the youngest of five siblings who had all graduated from college and gone on for professional degrees, while another's grandparents had received college educations and established professional careers. The largest group of players (33 percent) were blacks from working class backgrounds. These individuals came from some small Southern towns, but more often from the mid- to larger-sized cities of the South, Midwest, and Southwest. Only about half came from intact families; the rest were raised by their mothers or extended families. Many of them lived in the ghetto areas of these larger cities, but their parents held fairly steady jobs in factories, civil service, or other blue-collar or less skilled occupations. The final group (18 percent) was composed of lower class blacks. Nearly all of these players came from broken homes. While the majority lived with their mothers, one came from a foster family, another lived with his father and sister, and a third was basically reared by his older brothers. These individuals came from the larger cities in the Southwest, South, and Midwest. They grew up in ghetto areas and were street smart and tough. Many of their families subsisted on welfare; others lived off menial jobs such as domestic service. They were poor, and had the most desperate dreams of escaping.

Cliques

Moving beyond demographics, the players fell into four main coalitions that served as informal social groups. Not every single member of the team neatly fit into one of these categories or belonged to one of these groups, but nearly all players who stayed on the team for at least a year eventually drifted into a camp. At the very least, individuals associated with the various cliques displayed many behavioral characteristics we will describe. Players often forged friendship networks within these divisions, because of common attitudes, values, and activities. Once in a clique, no one that we observed left it or shifted into another one. In presenting these cliques, we trace a continuum from those with the most "heart" (bravery, dedication, willingness to give everything they had to the team or their teammates), a quality highly valued by team members, to those perceived as having the least.

Drawing on the vernacular shared by players and coaches, the first group of players were the "*bad niggas*." All of these individuals were black, from the working or lower classes, and shared a ghetto upbringing. Members of what Edwards (1985) has called the underclass (contemporary urban gladiators), they possessed many of the characteristics cited by Miller (1958) in his study of delinquent gangs' lower class culture: trouble, toughness, smartness, excitement, fate, and autonomy. They were street smart and displayed an attitude that bespoke their defiance of authority. In fact, their admiration for independence made it hard for them to adjust to domination by the coach (although he targeted those with reform potential as pet "projects"). Fighters, they would not hesitate to physically defend their honor or to jump into the fray when a teammate was in trouble. They worked hard to eke the most out of their athletic potential, for which they earned the respect of their teammates; they had little desire to do anything else. These were the players with the most heart. They may not have been "choir boys," but when the competition was fierce on the court, these were the kind of players the coach wanted out there, the kind he knew he could count on. Their on-court displays of physical prowess contributed to their assertions of masculinity, along with sexual conquests and drug use. They were sexually promiscuous and often boasted about their various "babes." With drugs, they primarily used marijuana, alcohol (beer), and cocaine. Their frequency of use varied from daily to occasional, although who got high and how often was a significant behavioral difference dividing the cliques. This type of social split, and the actual amounts of drugs team members used, is no different from the general use characteristic of a typical college population (see Moffatt 1989).

Tyrone was one of the bad niggas. He came from the ghetto of a mid-sized city in the Southwest, from an environment of outdoor street life, illicit opportunity, and weak (or absent) parental guidance. He was basically self-raised: he saw little of his mother, who worked long hours as a maid, and he had never known his father. Exceptionally tall and thin, he walked with a swagger (to show his "badness"). His speech was rich with ghetto expressions and he felt more comfortable hanging around with "brothers" than with whites. He often promoted himself boastfully, especially in speaking about his playing ability, future professional chances, and success with women. His adjustment to life at the University was difficult, although after a year or so he figured out how to "get by"; he became acclimated to dorm life, classes, and the media and boosters. When it came to common sense street-smarts, he was one of the brightest people on the team. He neither liked nor was favored by many boosters, but he did develop a solid group of friends within the ranks of the other bad niggas.

Apollo was another bad nigga. His family upbringing was more stable than Tyrone's, as his parents were together and his father was a career government worker (first in the military, then in the postal service). They had never had much money though, and scraped by as best they could. He was the youngest of six children, the only boy, and was favored by his father because of this. Tall and handsome, he sported an earring (which the coach made him remove during the season) and a gold tooth. One of the most colorful players, Apollo had a charismatic personality and a way with words. He was a magnetic force on the team, a leader who related emotionally to his teammates to help charge and arouse them for big games. He spoke in the common street vernacular of a ghetto "brother," although he could converse in excellent "White English" when it was appropriate. He was appealing to women, and enjoyed their attention, even though he had a steady girlfriend on the women's basketball team. Like Tyrone, Apollo was intelligent and articulate; he was able to express his perspective on life in a way that was insightful, entertaining, and outrageous. He was eager to explore and experience the zest of life, traveling the world, partying heavily, and seizing immediate gratification. He disdained the boring life of the team's straighter members, and generally did not form close relations with them. Yet he managed to enjoy his partying and playing and still graduate in four years. He would never have thought about college except for basketball, and had to overcome several debilitating knee injuries, but he ended up playing professionally on four continents and learning a foreign language.

A second group of players were the "*candy-asses.*" These individuals were also black, but from the middle and working classes. Where the bad niggas chafed under the authority of the coach, the candy-asses craved his attention and approval; they tended to form the closest personal ties with both him and his family. In fact, their strong ties to the coach made them the prime suspects as "snitches," those who would tell the coaches when others misbehaved. They "browned up" to the coaches and to the boosters and professors as well. The candy-asses were "good boys," the kind who projected the public image of conscientious, religious, polite, and quiet individuals. Several of them belonged to the Fellowship of Christian Athletes. They could be counted on to stay out of trouble. Yet although they projected a pristine image, they did not live like monks. They had girlfriends, enjoyed going to discos, and occasionally drank a few beers for recreation. They enjoyed parties, but, responsibly, moderated their behavior. The candy-asses enjoyed a respected position on the team because, like the bad niggas, they were good athletes and had heart. As much as they sought to be well rounded and attend to the student role, they did not let this interfere with their commitment to the team. They cared about the team first, and would sacrifice whatever was necessary—playing in pain, coming back too soon from an injury, relinquishing personal statistics to help the team win, diving to get the ball—for its benefit. Above all, they could be counted on for their loyalty to the coach, the team, and the game.

Rob was one such player from a large, extended working class family in a sizable Southern city. He was friendly and easy-going, with a positive attitude that came out in most of his activities. Although he was black and most of his close friends on the team were as well, Rob interacted much more easily with white boosters and students than did Tyrone. Rob transferred to the University to play for the coach because he had competed against him and liked both his reputation and style of play. Once there, he devoted himself to the coach, and was adored by the coach's wife and children. He often did favors for the family; one summer he painted the house in his spare time, and ate many of his meals there during the off-season. Rob's family was very close-knit, and both his mother and brother moved to town to be near him while he was at the University. They grew close with the coach's family, and often did things together. They also became regulars at the games, and were courted by many boosters who wanted to feel as if they knew Rob. Rob kept his academic and social life on an even keel during his college years; he worked hard in class (and was often the favorite of the media when they wanted to hype the image of the good student-athlete), and had a steady girlfriend. She was also very visible, with her young child from a previous boyfriend, in the basketball stadium. Like one or two of the others on the team, Rob spent one summer traveling with a Christian group on an around-the-world basketball tour.

Another typical candy-ass was Darian. He lived next door to Rob for two of the years they overlapped at the University, and the two were very close. Darian came from a middle class family in a nearby state and his family came to town for most of the home games. He was much more serious about life than Rob, and worked hard in everything he did. He was recruited by the team at the last minute (he had health problems that many thought would keep him from playing), yet he devoted himself to improvement. By his senior year he was one of the outstanding stars and had dreams of going pro. He wanted to make the most of his college education as well, and spent long hours in his room trying to study. Most people on the team looked up to him because he did what they all intended to do—work hard, sacrifice, and make the most of their college opportunity. His closest friends, then, were others with values like his, who were fairly serious, respectful, and who deferred their gratification in hopes of achieving a future career.

The "*whiners*" constituted a third category of players. Drawn from the middle and working classes and from both races, this group was not as socially cohesive as the previous two, yet it contained friendship cliques of mixed class and race. These individuals had neither the athletic prowess of the candy-asses nor the toughness of the bad niggas, yet they wanted respect. In fact the overriding trait they shared was their outspoken belief that they deserved more than they were getting: more playing time, more deference, more publicity. In many ways they envied and aspired to the characteristics of the two other groups. They admired the bad niggas' subculture, their independence, toughness, and disrespect for authority, but they were not as "bad." They wanted the attention (and perceived favoritism) the candy-asses received, but they could not keep themselves out of trouble. Like the bad niggas, they enjoyed getting high. While their athletic talent varied, they did not live up to their potential; they were not willing to devote themselves to basketball. They were generally not the kind of individuals who would get into fights, either on the court or on the street, and they lacked the heart of the previous two groups. Therefore, despite their potential and their intermittent complaining, they never enjoyed the same kind of respect among their teammates as the bad niggas, nor did they achieve the same position of responsibility on the team as the candy-asses.

Buck fell into this category. A young black from a rural, Southern town, Buck came from a broken home. He did not remember his father, and his mother, who worked in a factory, did not have the money to either visit him at the University or attend his games. Yet he maintained a close relationship with her over the phone. Buck had gone to a primarily white school back home and felt more comfortable around white people than most of the black players; in fact, several of his best friends on the team were white and he frequently dated several white girls. He had a solid academic background and performed well in his classes, although he was not as devoted to the books as some of the candy-asses. He liked to party, and occasionally got in trouble with the coach for breaking team rules. He spent most of his time hanging around with other whiners and with some of the bad niggas, sharing the latter's critical attitude toward the coach's authoritarian behavior. He felt that the coach did not recognize his athletic potential, and that he did not get the playing time he deserved. He had been warned by the coach about associating too much with some of the bad niggas who liked to party and not study hard. Yet for all their partying, the bad niggas were fiercer on the court than Buck, and could sometimes get away with their lassitude through outstanding play. He could not, and always had the suspicion that he was on the coach's "shit list." Like the candy-asses, he wanted to defer his gratification, do well in school, and get a good job afterward, yet he was not as diligent as they were and ended up going out more. He redeemed himself in the eyes of his peers during his senior year by playing the whole season with a debilitating chronic back injury that gave him constant pain.

James was another player in this clique, who also fell somewhere in between the bad niggas and the candy-asses. Like Buck, he socialized primarily with whiners and with some of the bad niggas, although he was white. He came from a middle class family in a small town near the school, and was recruited to play along with his brother (who was a year younger). James came to the University enthusiastic about college life and college basketball. He threw himself into the social whirl and quickly got into trouble with the coach for both his grades and comportment. He readily adapted to the predominantly black ambience of the players' peer subculture, befriending blacks and picking up their jargon. He occasionally dated black women (although this was the cause of one major fight between him and some of the football players, since black women were scarce on this campus). He had heart, and was willing to commit his body to a fight. The most famous incident erupted during a game where he wrestled with an opponent over territorial advantage on the court. Yet he had neither the speed nor the size of some other players, and only occasionally displayed flashes of the potential the coaches had seen in him.

The final group was known to their teammates as the "*L-7s*" (a "square," an epithet derived from holding the thumbs and forefingers together to form the "square" sign). Members of this mixed group were all middle class, more often white than black. They were the most socially isolated from other team members, as they were rejected for their squareness by all three other groups, and even among themselves seldom made friends across racial lines (the white and black L-7s constituted separate social groups). They came from rural, suburban, or exurban backgrounds and stable families. They were fairly moralistic, eschewing drinking, smoking, and partying. They attempted to project a studious image, taking their books with them on road trips and speaking respectfully to their professors. Compared with other players, they had a stronger orientation to the student population and booster community. They were, at heart, upwardly mobile, more likely to consider basketball a means to an end than an end in itself. Because of this orientation, several of the white players landed coveted jobs, often working at boosters' companies, at the end of their playing careers. They tended to be good technical players bred on the polished courts of their rural and exurban high schools rather than on the street courts of the cities; they knew how to play the game, but it did not occupy their full attention. They had varying (usually lesser) degrees of athletic ability, but they were even less likely than the whiners to live up to their potential. In contrast to the other groups, they had decidedly the least commitment, least loyalty, and least heart.

Mark was a dirty-blond-haired white boy from the West Coast with a strong upper body, built up from surfing. He was clean cut, respectful, and a favorite of the boosters. His sorority girlfriend was always on his arm, and helped create his desired image of a future businessman. He consciously worked to nurture this image, ostentatiously carrying books to places where he would never use them, and dressing in a jacket and tie whenever he went out in public. It was very important to him to make it financially, because he came from a working class family without much money. He had arrived at the University highly touted, but his talent never materialized to the degree the coaches expected. He was somewhat bitter about this assessment, because he felt he could have "made it" if given more of a chance. Like Buck, part of his problem may have lain in the difference between his slow-down style and the fast-paced style favored by the coach. He roomed and associated with other L-7s on the road and at home, but he also spent a lot of his time with regular students. After graduation he got a job from a booster working for a life insurance company.

Constellation of Role-Set Members

Basketball players generally interacted within a circle that was largely determined by their athletic environment. Due to the obligations of their position, these other role-set members fell into three main categories: athletic, academic, and social. Within the athletic realm, in addition to their teammates, athletes related primarily to the coaching staff, secretaries, and athletic administrators. The coaching staff consisted of the head coach, his first assistant (recruiting, playing strategy), the second assistant (recruiting, academics, some scouting), a part-time assistant (scouting, tape exchange with other teams, statistics during games), a graduate assistant (running menial aspects of practices, monitoring study halls, tape analysis), the trainer (injuries, paramedical activities), and the team manager (laundry, equipment). . . .

Secondary members of the athletic role-set included boosters, fans, athletic administrators (the Athletic Director, Sports Information Director, and their staffs), and members of the media. . . .

Within the academic arena, athletes' role-set members consisted of professors, tutors, and students in their classes, and, to a lesser extent, academic counselors and administrators. The players also tended to regard their families as falling primarily into this realm, although family members clearly cared about their social lives and athletic performance as well.

Socially, athletes related to girlfriends, local friends, and other students (non-athletes), but most especially to other college athletes: the teammates and dormmates (football players) who were members of their peer subculture.

REFERENCES

Becker, Howard, Blanche Geer, Everett Hughes, and Anselm Strauss, 1961. *Boys in White*. Chicago: University of Chicago Press.

Blumer, Herbert, 1969. *Symbolic Interactionism*. Englewood Cliffs, N.J.: Prentice-Hall.

Coakley, Jay J., 1986. *Sport in Society*. Third Edition. St. Louis: Mosby [Second edition, 1982.]

Cooley, Charles H., 1962. *Social Organization*. New York: Scribners.

Edwards, Harry, 1985. Beyond symptoms: Unethical behavior in American collegiate sport and the problem of the color line: *Journal of Sport and Social Issues* 9:3–11.

Frey, James H., 1982a. Boosterism, scarce resources and institutional control: The future of American intercollegiate athletics. *International Review of Sport Sociology* 17:53–70.

McCall, George J. and Jerry L. Simmons, 1978. *Identities and Interaction*. New York: Free Press.

Mead, George H., 1934. *Mind, Self and Society*. Chicago: University of Chicago Press.

Merton, Robert K., 1957. The role-set: Problems in sociological theory. *British Journal of Sociology* 8:106–20.

Miller, Walter B., 1958. Lower class culture as a generating milieu of gang delinquency. *Journal of Social Issues* 14:5–19.

Moffatt, Michael, 1989. *Coming of Age in New Jersey*. New Brunswick, N.J.: Rutgers University Press.

Strauss, Anselm, 1978. *Negotiations*. San Francisco: Jossey-Bass.

Turner, Ralph H., 1962. Role taking: Process versus conformity. In A. M. Rose, ed., *Human Behavior and Social Processes*, pp. 20–40. Boston: Houghton-Mifflin.

———1978. The role and the person. *American Journal of Sociology* 84:1–23.

19

You Can't Be a Sweet Cucumber in a Vinegar Barrel

PHILIP ZIMBARDO

Why do good people do bad things? This question has guided the work of social psychologist Zimbardo, founder of the National Center for the Psychology of Terrorism, for several decades. In this article he discusses questions underlying his research studies (especially his famous prison studies) and the applicability of his findings to the prison scandals at Abu Ghraib prison in Iraq and Guantanamo prison in Cuba. Zimbardo discusses why otherwise "nice kids" could do such awful things; social group and setting contribute greatly to behaviors, he says.

Consider the following as you read the selection:

1. *What were the key findings of Zimbardo's studies?*
2. *How can we apply Zimbardo's findings to current events?*
3. *Why do good kids do terrible things?*
4. *Have you ever been tempted to do something out of character for you? What were the circumstances?*

GLOSSARY **Dehumanization** Defining individuals as inferior, worthless. **Social modeling** Someone takes the lead in an activity, providing "permission" for others to follow. **Deindividualization** Sense of anonymity.

For years I've been interested in a fundamental question concerning what I call the psychology of evil: Why is it that good people do evil deeds? I've been interested in that question since I was a little kid. Growing up in the ghetto in the South Bronx, I had lots of friends who I thought were good kids, but for one reason or another they ended up in serious trouble. They went to jail, they took drugs, or they did terrible things to other people. My whole upbringing was focused on trying to understand what could have made them go wrong.

When you grow up in a privileged environment, you want to take credit for the success you see all around, so you become a dispositionalist. You look for character, genes, or family legacy to explain things, because you want to say your father did good things, you did good things, and your kid will do good things. Curiously, if you grow up poor, you tend to emphasize external situational factors when trying to understand unusual behavior. When you look around and you see that your father's not working, and you have friends who are

SOURCE: Edited from an interview with Philip Zimbardo at http://www.edge.org/3rd_culture/zimbardo05/zimbardo05_index.html. Reprinted by permission of Philip Zimbardo.

selling drugs or their sisters are in prostitution, you don't want to say it's because there's something inside them that makes them do it, because then there's a sense in which it's in your line. Psychologists and social scientists that focus on situations more often than not come from relatively poor, immigrant backgrounds. That's where I came from.

Over the years I've asked that question in more and more refined ways. I began to investigate what specific kinds of situational variables or processes could make someone step across that line between good and evil. We all like to think that the line is impermeable—that people who do terrible things like commit murder, treason, or kidnapping are on the other side of the line—and we could never get over there. We want to believe that we're with the good people. My work began by saying, no, that line is permeable. The reason some people are on the good side of the line is that they've never really been tested. They've never really been put in unusual circumstances where they were tempted or seduced across that line. My research over the last 30 years has created situations in the laboratory or in field settings in which we take good, normal, average, healthy people—more often than not healthy college students—and expose them to these kinds of settings. . . .

To investigate this I created an experiment. We took women students at New York University and made them anonymous. We put them in hoods, put them in the dark, took away their names, gave them numbers, and put them in small groups. And sure enough, within half an hour those sweet women were giving painful electric shocks to other women within an experimental setting. We also repeated that experiment on deindividuation with the Belgian military, and in a variety of formats, with the same outcomes. Any situation that makes you anonymous and gives permission for aggression will bring out the beast in most people. That was the start of my interest in showing how easy it is to get good people to do things they say they would never do.

I also did research on vandalism. When I was a teacher at NYU, I noticed that there were hundreds and hundreds of vandalized cars on the streets throughout the city. I lived in Brooklyn and commuted to NYU in the Bronx, and I'd see a car in the street. I'd call the police and say, "You know, there's a car demolished on 167th and Sedgwick Avenue. Was it an accident?" When he told me it was vandals, I said, "Who were the vandals? I'd like to interview them." He told me that they were little black or Puerto Rican kids who come out of the sewers, smash everything, paint graffiti on the walls, break windows and disappear.

So I created what ethologists would call "releaser cues." I bought used cars, took off license plates, and put the hood up, and we photographed what happened. It turns out that it wasn't little black or Puerto Rican kids, but white, middle-class Americans who happened to be driving by. We had a car near NYU in the Bronx. Within ten minutes the driver of the first car that passed by jacked it up and took a tire. Ten minutes later a little family would come. The father took the radiator, the mother emptied the trunk, and the kid took care of the glove compartment. In 48 hours we counted 23 destructive contacts with that car. In only one of those were kids involved. We did a comparison in which we set out a car a block from Palo Alto, where Stanford University is. The car was out for a week, and no one touched it until the last day when it rained and somebody put the hood down. God forbid that the motor should get wet.

This gives you a sense of what a community is. A sense of community means people are as concerned about any property or people on their turf because there's a sense of reciprocal concern. The assumption is that I am concerned because you will be concerned about me and my property. In an anonymous environment nobody knows who I am and nobody cares, and I don't care to know about anyone else. The environment can convey anonymity externally, or it can be put on like a Ku Klux Klan outfit.

And so I and other colleagues began to do research on dehumanization. What are the ways in which, instead of changing yourself and becoming the aggressor, it becomes easier to be hostile against other people by changing your psychological conception of them? You think of them as worthless animals. That's the killing power of stereotypes.

I put that all together with other research I did 30 years ago during the Stanford prison experiment. The question there was, what happens when you put good people in an evil place? We put good, ordinary college students in a very realistic, prison-like setting in the basement of the psychology department at Stanford. We dehumanized the prisoners, gave them numbers, and took away their identity. We also deindividuated the guards, calling them Mr. Correctional Officer, putting them in khaki uniforms, and giving them silver reflecting sunglasses like in the movie *Cool Hand Luke*. Essentially, we translated the anonymity of *Lord of the Flies* into a setting where we could observe exactly what happened from moment to moment.

What's interesting about that experiment is that it is really a study of the competition between institutional power versus the individual will to resist. The companion piece is the study by Stanley Milgram, who was my classmate at James Monroe High School in the Bronx. (Again, it is interesting that we are two situationists who came from the same neighborhood.) His study investigated the power of an individual authority: Some guy in a white lab coat tells you to continue to shock another person even though he's screaming and yelling. That's one way that evil is created as blind obedience to authority. But more often than not, somebody doesn't have to tell you to do something. You're just in a setting where you look around and everyone else is doing it. Say you're a guard and you don't want to harm the prisoners—because at some level you know they're just college students—but the two other guards on your shift are doing terrible things. They provide social models for you to follow if you are going to be a team player.

In this experiment we selected normal, healthy, good kids that we found through ads in the paper.... within a few days, if they were assigned to the guard role, they became abusive, red-necked prison guards.

Every day the level of hostility, abuse, and degradation of the prisoners became worse and worse and worse. Within 36 hours the first prisoner had an emotional breakdown, crying, screaming, and thinking irrationally. We had to release him, and each day after that we had to release another prisoner because of extreme stress reactions. The study was supposed to run for two weeks, but I ended it after six days because it was literally out of control. Kids we chose because they were normal and healthy were breaking down. Kids who were pacifists were acting sadistically, taking pleasure in inflicting cruel, evil punishment on prisoners....

There are stunning parallels between the Stanford prison experiment and what happened at Abu Ghraib, where some of the visual scenes that we have seen include guards stripping prisoners naked, putting bags over heads, putting them in chains, and having them engage in sexually degrading acts. And in both prisons the worst abuses came on the night shift. Our guards committed very little physical abuse. There was a prisoner riot on the second day, and the guards used physical abuse....

These are exact parallels between what happened in this basement at Stanford 30 years ago and at Abu Ghraib, where you see images of prisoners stripped naked, wearing hoods or masks as guards get them to simulate sodomy. The question is whether what we learned about the psychological mechanisms that transformed our good volunteers into these creatively evil guards can be used to understand the transformation of good American Army reservists into the people we see in these trophy photos in Abu Ghraib. And my answer is, yes, there are very direct parallels....

These terrible deeds form an interesting analog in America, because there are two things we are curious to understand about Abu Ghraib. First, how did the soldiers get so far out of hand? And secondly, why would the soldiers take pictures of themselves in positions that make them legally culpable? The ones that are on trial now are the ones in those pictures, although obviously there are many more people involved in various ways. We can understand why they did so not only by applying the basic social-psychological processes from the Stanford prison study, but also by analyzing what was unique in Abu Ghraib....

At Abu Ghraib you had the social modeling in which somebody takes the lead in doing something. You had the dehumanization, the use of labels of the other as inferior, as worthless. There was a diffusion of responsibility such that nobody was personally accountable.... The other thing, of course, is that you had low-level army reservists who had no "mission-specific" training in how to do this difficult new job. There was little or no supervision of them on the night and there was literally no accountability. This went on for months in which the abuses escalated over time.... And then there is the hidden factor of boredom. One of the main contributors to evil, violence, and hostility in all prisons that we underplay is the boredom factor. In fact, the worst things that happened in our prisons occurred during the night shifts....

Dehumanization also occurred because the prisoners often had no prison clothes available, or were forced to be naked as a humiliation tactic by the military police and higher-ups. There were too many of them; in a few months the number soared from 400 to over a thousand. They didn't have regular showers, did not speak English, and they stank. Under these conditions it's easy for guards to come to think of the prisoners as animals, and dehumanization processes set in.

When you put that set of horrendous work conditions and external factors together, it creates an evil barrel. You could put virtually anybody in it and you're going to get this kind of evil behavior. The Pentagon and the military say that the Abu Ghraib scandal is the result of a few bad apples in an otherwise good barrel. That's the dispositional analysis. The social psychologist in me, and the consensus among many of my colleagues in experimental social psychology, says that's the wrong analysis. It's not the bad apples, it's the bad barrels that corrupt good people. Understanding the abuses at this Iraqi prison starts with an analysis of both the situational and systematic forces operating on those soldiers working the night shift in that "little shop of horrors."

Coming from New York, I know that if you go by a delicatessen, and you put a sweet cucumber in the vinegar barrel, the cucumber might say, "No, I want to retain my sweetness." But it's hopeless. The barrel will turn the sweet cucumber into a pickle. You can't be a sweet cucumber in a vinegar barrel. My sense is that we have the evil barrel of war, into which we've put this evil barrel of this prison—it turns out actually all of the military prisons have had similar kinds of abuses—and what you get is the corruption of otherwise good people.

20

Characteristics of Bureaucracy

MAX WEBER

We've all stood in lines to transact some business, only to find after our wait that we were missing a necessary form or receipt. Bureaucracy can be frustrating, but it is also an efficient way to conduct business, relative to other systems of patronage or favoritism. Weber's classic work on bureaucracy analyzes the rise of this form of organization. In the following brief excerpt, Weber outlines the key components that define bureaucracy.

As you read, consider the following:

1. *How do the characteristics of bureaucracy discussed by Weber compare with the characteristics of an organization with which you are familiar?*
2. *Which parts of bureaucracy have you experienced?*
3. *Can you design a more efficient form of organization than that described by Weber?*

GLOSSARY **Bureaucracy** Formal organization in which rules and hierarchical rankings are used to achieve efficiency.

Modern officialdom functions in the following specific manner:

I. There is the principle of fixed and official jurisdictional areas, which are generally ordered by rules, that is, by laws or administrative regulations.

1. The regular activities required for the purposes of the bureaucratically governed structure are distributed in a fixed way as official duties.
2. The authority to give the commands required for the discharge of these duties is distributed in a stable way and is strictly delimited by rules concerning the coercive means, physical, sacerdotal, or otherwise, which may be placed at the disposal of officials.
3. Methodical provision is made for the regular and continuous fulfilment of these duties and for the execution of the corresponding rights; only persons who have the generally regulated qualifications to serve are employed.

In public and lawful government these three elements constitute "bureaucratic authority." In private economic domination, they constitute bureaucratic "management." Bureaucracy, thus understood, is fully developed in the private economy, only in the most advanced institutions of capitalism. Permanent and public office authority, with fixed jurisdiction, is not the historical rule, but rather the exception. This is so even in large political structures such as those of the ancient Orient,

SOURCE: From *Max Weber: Essays in Sociology*, pp. 196–204, edited by H. H. Gerth and C. Wright Mills, copyright 1946, by Oxford University Press. Reprinted by permission.

the Germanic and Mongolian empires of conquest, or of many feudal structures of state. In all these cases, the ruler executes the most important measures through personal trustees, table-companions, or court-servants. Their commissions and authority are not precisely delimited and are temporarily called into being for each case.

II. The principles of office hierarchy and of levels of graded authority mean a firmly ordered system of super- and subordination in which there is a supervision of the lower offices by the higher ones. Such a system offers the governed the possibility of appealing the decision of a lower office to its higher authority in a definitely regulated manner. With the full development of the bureaucratic type, the office hierarchy is monocratically organized. The principle of hierarchical office authority is found in all bureaucratic structures: in state and ecclesiastical structures as well as in large party organizations and private enterprises. It does not matter for the character of bureaucracy whether its authority is called "private" or "public."

When the principle of jurisdictional "competency" is fully carried through, hierarchical subordination—at least in public office—does not mean that the "higher" authority is simply authorized to take over the business of the "lower." Indeed, the opposite is the rule. Once established and having fulfilled its task, an office tends to continue in existence and be held by another incumbent.

III. The management of the modern office is based upon written documents ("the files"), which are preserved in their original or draught form. There is, therefore, a staff of subaltern officials and scribes of all sorts. The body of officials actively engaged in a "public" office, along with the respective apparatus of material implements and the files, make up a "bureau." In private enterprise, "the bureau" is often called "the office."

In principle, the modern organization of the civil service separates the bureau from the private domicile of the official, and, in general, bureaucracy segregates official activity as something distinct from the sphere of private life. Public monies and equipment are divorced from the private property of the official. . . .

IV. Office management, at least all specialized office management—and such management is distinctly modern—usually presupposes thorough and expert training. This increasingly holds for the modern executive and employee of private enterprises, in the same manner as it holds for the state official.

V. When the office is fully developed, official activity demands the full working capacity of the official, irrespective of the fact that his obligatory time in the bureau may be firmly delimited. In the normal case, this is only the product of a long development, in the public as well as in the private office. Formerly, in all cases the normal state of affairs was reversed: official business was discharged as a secondary activity.

VI. The management of the office follows general rules, which are more or less stable, more or less exhaustive, and which can be learned. Knowledge of these rules represents a special technical learning which the officials possess. It involves jurisprudence, or administrative or business management.

PART III

Inequalities between Groups

The world can be kind or cruel to its inhabitants. Some people live privileged lives with modern comforts, plenty of food, nice houses, access to health care, communication and transportation. Others barely scrape by with each day a struggle for survival—and many of these do *not* survive, thanks to starvation, disease, and war. Why are you in college while millions of young people around the world will be lucky to get an elementary school education?

The chapters in this section explore some of the issues that separate rich from poor, elites from powerless, dominant groups from minorities. Several sociological theories provide explanations for why these differences between individuals and groups develop. Explanations range from "survival of the fittest" (those who are most able get ahead) to "the rich get richer."

Many recent sociological theories consider the disadvantages faced by those holding minority status because of their race, class, gender, religion, age, or other distinguishing characteristics. The arguments often point to prejudice and discrimination that limit opportunities for these members of society.

To set the stage for understanding inequalities, Chapter 5 on stratification focuses on how people come to be layered in societies, why some are rich and some poor, and why poverty persists within and between societies. The introduction to the chapter on stratification notes that the U.S. Declaration of Independence declares that "all men are created equal." Thus, this chapter begins by noting that there is a gender difference in the way our founding "fathers" considered their citizens; women of the country would be denied the vote for almost one hundred years. This is but one example of factors that stratify members of society; articles in Chapter 5 examine the causes and consequences of stratification.

The intermingling of the world's peoples has produced group relations that sometimes lead to stereotyping, prejudice, and discrimination based on cultural or physical differences between people, and may result in poverty for some groups. The readings in Chapter 6 on inequality discuss what is happening to some specific race, class, and gender groups who are attempting to change their experiences and situations.

To function smoothly, society requires normative behavior of its members. The socialization process generally results in individuals who follow the norms of society. However, norms are generally determined by the dominant groups in society, and some theorists argue that minority groups are at a disadvantage, either being judged as deviant as a group or finding that their actions are defined as deviant according to the dominant group. Our prisons are filled with miscreants who have violated the norms of society. A disproportionate number of these criminals are from the lower classes and minority groups. In fact, wealthy criminals, those who have evaded taxes or cooked the books, tend to make the news because we don't expect to see corporate giants committing deviant acts; meanwhile common criminals are forgotten.

Some people do bad things! However, deviance theory does imply that some individuals in society are labeled as different, suspicious, or bad; are apprehended more readily because they are easily recognized or suspected of wrongdoing; and lack the equal opportunity to get ahead in legitimate ways because avenues for success have been limited, providing a partial explanation for their deviance.

The inequalities discussed in this section are important to understand because they are the basis for differences in the experiences and opportunities of individuals and groups around the world.

CHAPTER 5

Stratification

Some Are More Equal than Others

The Declaration of Independence states that "all men are created equal"—a statement that is more hopeful than factual. Although we are well aware that upper, middle, and lower social classes exist in this country, this fact is often covered up with beliefs to the contrary. In reality, the United States has a stratified class structure. Like most societies, this class ranking is based on family background, wealth, authority, occupation, age, and gender. The presence of a class system in a society that professes equality for all raises two questions. First, what are the effects of a class system on the various segments of the population? Why do the citizens of the country tolerate such a system?

In answer to the first question, sociological research has found that social class significantly affects almost all facets of our lives. It has also found that inequality begins at the very start of life, with chances of survival during the birth process. The process continues with unequal opportunities for health, health care, and education. Class status also affects the supposed guarantees of a democratic society of equal protection and justice under the law. When the United States had a military draft, inequality also meant that you had a greater chance of being drafted and being wounded or killed if you were from the lower rankings of the army. These findings indicate that class status determines your control over your own life choices. In the first reading Andrew Hacker acknowledges the existence of a class system and its unequal balance in numbers. He shows the influence of those who have money on those who don't.

As for the second question—why a country that believes in equality for all tolerates an unequal system—one reason for the existence of an unequal social class structure may lie in the meaning of the term *doublethink*. In his futuristic novel *1984*, George Orwell defined doublethink as holding two contradictory ideas at the same time and believing in both of them. It is a kind of thinking that can only be done unconsciously and simply means that we often do not examine our beliefs closely enough to note that they are, perhaps, contradictory. For example, Americans strongly believe in equal justice under the law while recognizing that wealth affects the dispensing of justice. Another reason for maintaining a social class system is the underlying factor involved in unequal benefits—some people simply profit from the system as it is. This is the thesis behind the reading by Harold Kerbo, who describes how the wealthy maintain the class system for their own advantage.

The continued existence of an unequal class system also relies on the dubious belief that it is possible to climb the social class ladder—if not in reality, then at least in imitation. This imitation of the wealthy gives the illusion of success and its resulting pleasure. In the third reading Juliet Schor describes this process as "Keeping up with the Trumps."

Our social class affects our beliefs and behaviors. Richard Kahlenberg notes in the fourth reading that the influence of class may override other influences in our lives. Class appeals, he notes, may cause people to vote against their best interests.

The last reading provides a final reason for the existence of an unequal class sytem—a functional reason. Herbert Gans notes that lower classes exist because they perform various important tasks for society. In short, we maintain inequality because it is good for society as a whole if not for some.

As you read the selections in this chapter, ask yourself realistically what class you belong to. Also ask yourself what your opportunities are for mobility. Finally, ask yourself what, if anything, you can do to increase your opportunities.

21

Money and the World We Want

ANDREW HACKER

Two issues are covered in this reading: First, is there a class structure, and what is its breakdown? Second, what are the consequences of this class structure?

As you read, ask yourself the following questions:

1. *What problems result from the way the United States distributes income?*
2. *What changes would you make to help all people to be fully Americans?*
3. *Do you believe that Americans are becoming more equal? Why or why not?*

GLOSSARY **Corollary** A natural result or easily drawn conclusion. **Esoteric** Understood by those with special knowledge. **Skepticism** Doubt, unbelief.

The three decades spanning 1940 to 1970 were the nation's most prosperous years. Indeed, so far as can happen in America's kind of economy, a semblance of redistribution was taking place. During this generation, the share of national income that went to the bottom fifth of families rose by some fractions of a point to an all-time high, while that received by the richest fifth fell to its lowest level. This is the very period of shared well-being that many people would like to re-create in this country.

Yet in no way is this possible. It was an atypical era, in which America won an adventitious primacy because of a war it delayed entering and its geographic isolation which spared it the ravages that the other combatants suffered.

America's postwar upsurge lasted barely three decades. The tide began to turn in the early 1970s. Between 1970 and 1980, family income rose less than 7 percent. And that small increase resulted entirely from the fact that additional family members were joining the workforce. Indeed, the most vivid evidence of decline is found in the unremitting drop in men's earnings since 1970. Averages and medians, however, conceal important variations, such as that some American households have done quite well for themselves during the closing decades of the century. Those with incomes of $1 million or more have reached an all-time high, as are families and individuals making over $100,000. By all outward appearances, there is still plenty of money around, but it is landing in fewer hands. Yet it is by no means apparent that people are being paid these generous salaries and options and fees because their work is adding much of substance to the nation's output. Indeed, the coming century will test whether an economy can flourish by exporting most notably action movies and flavored water.

The term *upper class* is not commonly used to describe the people in America's top income tier since it connotes a hereditary echelon that passes on its holdings from generation to generation. Only a few American families have remained at the very top for more than two or three generations, and even when they do, as have the du Ponts and

SOURCE: Reprinted with permission of Scribner, a division of Simon & Schuster, Inc., from *Money: Who Has How Much and Why*, by Andrew Hacker. Copyright 1997 by Andrew Hacker.

How Families Fare: Three Tiers

Tier	Income Range	Number of Families/Percentage	
Comfortable	Over $75,000	12,961,000	18.6%
Coping	$25,000–$75,000	36,872,000	53.0%
Deprived	Under $25,000	19,764,000	28.4%

Rockefellers, successive descendants slice the original pie into smaller and smaller pieces. So if America does not have a "class" at its apex, what should we call the people who have the most money? The answer is to refer to them as we usually do, as being "wealthy" or "rich."

The rich, members of the 68,064 households who in 1995 filed federal tax returns that declared a 1994 income of $1 million or more, have varied sources of income. . . . For present purposes, wealth may be considered holdings that would yield you an income of $1 million a year without your having to put in a day's work. Assuming a 7 percent return, it would take income-producing assets of some $15 million to ensure that comfort level.

Unfortunately, we have no official count of how many Americans possess that kind of wealth. It is measurably less than 68,064, since those households report that the largest segment of their incomes come from salaries. Indeed, among the chairmen of the one hundred largest firms, the median stock holding is only $8.4 million. A liberal estimate of the number of wealthy Americans would be about thirty thousand households, one-thirtieth of one percent of the national total.

This still puts almost 90 percent of all Americans between the very poor and the wealthy and the rich. One way to begin to define this majority is by creating a more realistic bottom tier than merely those people who fall below the official poverty line. Since Americans deserve more than subsistence, we may set $25,000 a year for a family of three as a minimum for necessities without frills. And even this is a pretty bare floor. Indeed, only about 45 percent of the people questioned in the Roper-Starch poll felt that their households could "get by" on $25,000 a year. In 1995, almost 20 million families—28.4 percent of the total—were living below that spartan standard. This stratum is a varied group. . . . For over a third, all of their income comes from sources other than employment, most typically Social Security and public assistance. Over 43.3 percent of the households have only one earner, who at that income level is usually a woman and the family's sole source of support. The remainder consists of families where two or more members have had jobs of some kind, which suggests sporadic employment at close to the minimum wage. Whatever designation we give to these households—poor or just getting by—their incomes leave them deprived of even the more modest acquisitions and enjoyments available to the great majority of Americans.

The question of who belongs to America's middle class requires deciding where to place its upper and lower boundaries. While it is difficult to set precise boundaries for this stratum, it is accurate to say that, by one measure or another, most Americans fit into a middle class.

While it is meaningless to classify the households in the middle of America's income distribution because this stratum is so substantial and it encompasses such a wide range of incomes, meaningful divisions can be drawn in our country's overall income distribution.

This reading has sought to make clear that the prominent place of the rich tells us a great deal about the kind of country we are, as does the growing group of men and women with $100,000 salaries and households with $100,000 incomes. The same stricture applies to the poor, who, while not necessarily increasing in number, are too often permanently mired at the bottom. This noted, a three-tier division can be proposed, with the caveat that how you fare on a given income can depend on local costs and social expectations.

. . .

That the rich have become richer would seem to bear out Karl Marx's well-known prediction. The nation's greatest fortunes are substantially larger than those of a generation ago. Households with incomes exceeding $1 million a year are also netting more in real purchasing power. At a more mundane level, the top 5 percent of all households in 1975 averaged $122,651 a year; by 1995, in inflation-adjusted dollars, their average annual income had ascended to $188,962.

But Marx did not foresee that the number of rich families and individuals would actually increase over time. Between 1979 and 1994, the number of households declaring incomes of $1 million or more rose from 13,505 to 68,064, again adjusting for inflation. In 1996, *Forbes*'s 400 richest Americans were all worth at least $400 million. In 1982, the year of the magazine's first list, only 110 people in the 400 had holdings equivalent to the 1996 cutoff figure. In other words, almost three-quarters of 1982's wealthiest Americans would not have made the 1996 list. And families with incomes over $100,000—once more, in constant purchasing power—increased almost threefold between 1970 and 1995, rising from 3.4 percent to 9 percent of the total. During the same period, the group of men making more than $50,000 rose from 12 percent of the total to 17 percent.

It is one thing for the rich to get richer when everyone is sharing in overall economic growth, and it is quite another for the better off to prosper while others are losing ground or standing still. But this is what has been happening. Thus 1995 found fewer men earning enough to place them in the $25,000 to $50,000 tier compared with twenty-five years earlier. And the proportion in the bottom bracket has remained essentially unchanged. But this is not necessarily a cause for cheer. In more halcyon times, it was assumed that each year would bring a measure of upward movement for people at the bottom of the income ladder and a diminution of poverty.

The wage gap of our time reflects both the declining fortunes of many Americans and the rise of individuals and households who have profited from recent trends. Economists generally agree on what brought about static wages and lowered living standards. In part, well-paying jobs are scarcer because goods that were once produced here, at American wage scales, are now made abroad and then shipped here for sale. A corollary cause has been the erosion of labor unions, which once safeguarded generous wages for their members. Between 1970 and 1996, the portion of the workforce represented by unions fell from 27 percent to 15 percent. Today, the most highly organized occupations are on public payrolls, notably teachers and postal employees. Only 11 percent of workers in the private sector belong to unions. For most of the other 89 percent this means that their current paychecks are smaller than they were in the past.

Analysts tend to differ on the extent to which the paltriness of the minimum wage has lowered living standards. Despite the 1996 increase, the minimum wage produces an income that is still below the poverty line. Even more contentious is the issue of to what degree immigrants and aliens have undercut wages and taken jobs once held by people who were born here. We can all cite chores that Americans are unwilling to do, at least at the wages customarily paid for those jobs. Scouring pots, laundering clothes, herding cattle, and caring for other people's children are examples of such tasks. At the same time, employers frequently use immigrants to replace better-paid workers, albeit by an indirect route. The most common practice is to remove certain jobs from the firm's payroll and then to hire outside contractors, who bring in their own staffs, which are almost always lower paid and often recently arrived in this country.

Most economists agree that the primary cause of diverging earnings among American workers has been the introduction of new technologies. These new machines and processes are so esoteric and complex that they require sophisticated skills that call for premium pay. In this view, the expanded stratum of Americans earning over $50,000 is made up largely of men and women who are adept at current techniques for producing goods, organizing information, and administering personnel. New technologies have reduced the number of people who are needed as telephone operators, tool and die makers, and aircraft mechanics. Between 1992 and 1996, Delta Airlines was scheduling the same number of flights each year, even though it was discharging a quarter of its employees. Closer to the top, there is a strong demand for individuals who are skilled at pruning payrolls. And this cadre has been doing its work well. . . . [I]n 1973, the five hundred largest industrial firms employed some 15.5 million men and women. By 1993, these firms employed only 11.5 million people. But this reduction in the industrial workforce amounted to more than a loss of 4 million positions. Given the increase in production that took place over this twenty-year period, it meant a comparable output could be achieved in 1993 with half as many American workers as were needed in 1973. In fact, American workers account for an even smaller share of the output, since many of the top five hundred industrial firms are having more of their production performed by overseas contractors and subsidiaries.

So what special skills do more highly educated workers have that make them eligible for rising salaries? In fact, such talents as they may display have only marginal ties to technological expertise. The years at college and graduate school pay off because they burnish students' personalities. The time spent on a campus imparts cues and clues on how to conduct oneself in corporate cultures and professional settings. This demeanor makes for successful interviews and enables a person to sense what is expected of him during the initial months on a job.

Does America's way of allocating money make any sense at all? Any answer to this question requires establishing a rationale. The most common explanation posits that the amounts people get are set in an open market. Thus, in 1995, employers offered some 14.3 million jobs that paid between $20,000 and $25,000, and were able to find 14.3 million men and women who were willing to take them. The same principle applied to the 1.7 million positions pegged at $75,000 to $85,000, and to the dozen or so corporate chairmen who asked for or were given more than $10 million. By the same token, it can be argued that market forces operate at the low end of the scale. Wal-Mart and Pepsico's Pizza Hut cannot force people to work for $6.50 an hour, but those companies and others like them seem to attract the workers they need by paying that wage.

The last half century gave many groups a chance to shield themselves from the labor auction.

But many, if not most, of these protections are no longer being renewed. The up-and-coming generations of physicians, professors, and automobile workers are already finding that they must settle for lower pay and fewer safeguards and benefits. The most graphic exception to this new rule has been in the corporate world, where boards of directors still award huge salaries to executives, without determining whether such compensation is needed to keep their top people from leaving or for any other reason. They simply act as members of an inbred club who look after one another. Only rarely do outside pressures upset these arrangements, which is why they persist.

A market rationale also presumes that those receiving higher offers will have superior talents or some other qualities that put them in demand. Some of the reasons why one person makes more than another make sense by this standard. Of course, a law firm will pay some of its members more if they bring in new business or satisfy existing clients. Two roommates have just received master's degrees with distinction, one in education and the other in business administration. The former's first job is teaching second-graders and will pay $23,000. The latter, at an investment firm, will start at $93,000. About all that can be said with certainty is that we are unlikely to arrive at a consensus on which roommate will be contributing more to the commonwealth.

Would America be a better place to live, and would Americans be a happier people, if incomes were more evenly distributed? Even as the question is being posed, the answers can be anticipated.

One side will respond with a resounding "Yes!" After which will come a discourse on how poverty subverts the promise of democracy, while allowing wealth in so few hands attests to our rewarding greed and selfishness. There would be far less guilt and fear if the rich were not so rich and no Americans were poor. But the goal, we will be told, is not simply to take money from some people and give it to others. Rather, our goal should be to create a moral culture where citizens feel it is right to have no serious disparities in living standards. Other countries that also have capitalist economies have shown that this is possible.

Those who exclaim "No way!" in response to the same question will be just as vehement. To exact taxes and redistribute the proceeds is an immoral use of official power since it punishes the productive and rewards the indolent. And do we want the government telling private enterprises what wages they can offer? The dream of economic equality has always been a radical's fantasy. Apart from some primitive societies, such a system has never worked. If you want efficiency and prosperity, and almost everyone does, then variations in incomes are part of the equation.

These different responses arise in part from disparate theories of human nature. Since the earliest days of recorded history, philosophers have disagreed over whether our species is inherently competitive or cooperative.

Also at issue is whether greater economic equality can only be achieved by giving oppressive powers to the state, either to limit incomes through heavy taxes or by setting levels of earnings.

Of course, the price that America pays for economic inequality is its persisting poverty.

In one way or another all Americans will pay the high costs of poverty. California now spends more on its prison system than it does on higher education, and other states will soon be following suit. Bolstering police forces is hardly cheap: upward of $75,000 per officer, when overtime and benefits are added in. Being poor means a higher chance of being sick or being shot, or bearing low-birth-weight babies, all of which consume medical resources and have helped to make Medicaid one of the costliest public programs. In addition, poor Americans now represent the fastest-growing group of AIDS victims: mainly drug users and the women and children they infect. Generally, $100,000 worth of medical treatment is spent on each person dying of AIDS. The poor also have more of their children consigned to "special education" classes, which most never leave and where the tab can reach three times the figure for regular pupils.

Additional expenses are incurred by families who put as much distance as possible between themselves and the poor. Doing so often entails the upkeep of gated communities, security systems, and privately supplied guards. Yet these are expenses better-off Americans readily bear. They are willing to foot the bills for more prisons and police, as well as the guns they keep in their bedrooms and the alarms for their cars. Indeed, their chief objection is to money given to non-married mothers who want to be at home with their pre-school-age children. Virtually every such penny is begrudged, followed by the demand that these mothers take jobs even before their children go to kindergarten. While the imposition of work may make some taxpayers cheerier, it will not do much to close the income gap.

How disparities in income affect the nation's well-being has long been debated by economists. Much of the argument centers on what people do with their money. The poor and those just trying to cope devote virtually all of what they have to necessities plus the few extras they can afford. If their incomes were raised, they would obviously spend more, which would create more demand and generate more jobs. It should not be forgotten that the year 1929, which was noted for a severe imbalance in incomes, gave us a devastating economic depression that lasted a decade.

The traditional reply to critics of economic inequality has been that we need not only the rich, but also a comfortably off class, who are able to put some of their incomes into investments. In other words, disparities give some people more than they "need," which allows them to underwrite the new enterprises that benefit everyone. While there is obvious validity to this argument, it should be added that much of this outlay now goes to paper contrivances, which have only a remote connection with anything productive, if any at all. Nor are the rich as necessary as they may once have been, since institutions now supply most invested capital. Metropolitan Life, Merrill Lynch, Bank of America, and the California Public Employees Pension Fund put substantially more into new production than do the 68,064 families with $1 million incomes.

In one sphere, the income gap comes closer to home. Most young Americans will not live as well as their parents did. Indeed, in many instances this is already occurring. A generation ago, many men had well-paid blue-collar jobs; now their sons are finding that those jobs are no longer available. In that earlier era, college graduates could enter a growing managerial stratum; today, firms view their payrolls as bloated and are ending the security of corporate careers. . . .

The patterns of decline prevail if the entire nation is viewed as the equivalent of family. Federal programs now award nine times as much to retirees as they do to the nation's children, so senior citizens as a group fare better than younger Americans. Twice as many children as Americans over the age of sixty live in households below the poverty line. (And as death approaches, the government is more generous: almost 30 percent of the total Medicare budget is spent on the terminal year of elderly patients' lives.) Many retired persons have come to view a comfortable life as their entitlement, and have concluded that they no longer have obligations to repay. Grandparents tend to support campaigns for the rights of the elderly, not for school bonds and bigger education budgets.

In the end, the issue may be simply stated: what would be required for all Americans—or at least as many as possible—to make the most of their lives? . . .

Poverty takes its greatest toll in the raising of children. With a few exceptions, being poor consigns them to schools and surroundings that do little to widen their horizons. The stark fact is that we have in our midst millions of bright and talented children whose lives are fated to be a fraction of what they might be. And by any moderate standard, deprivation extends above the official poverty line. In most of the United States, families with incomes of less than $25,000 face real limits to the opportunities open to their children. Less than one child in ten from these households now enters and graduates from college. The statistics are apparent to outside observers, and to the children themselves, who very early on become aware of the barriers they face. And from this realization results much of the behavior that the rest of the society deplores. The principal response from solvent Americans has been to lecture the poor on improving their ways.

No one defends poverty, but ideologies differ on what can or should be done to alleviate it. Conservatives generally feel it is up to the individual: those at the bottom should take any jobs they can find and work hard to pull themselves up. Hence the opposition to public assistance, which is seen as eroding character and moral fiber. Indeed, conservatives suggest that people will display character and moral fiber if they are made to manage on their own.

Not many voting Americans favor public disbursements for the poor or even for single working mothers who cannot make ends meet. Most American voters have grown weary of hearing about the problems of low-income people. Yet even those who are unsettled by the persistence of income imbalances no longer feel that government officials and experts know how to reduce the disparities.

Of course, huge redistributions occur everyday. Funds for Social Security are supplied by Americans who are currently employed, providing their elders with pensions that now end up averaging $250,000 above what their own contributions would have warranted. Agricultural subsidies give farmers enough extra cash to ensure that they will have middle-class comforts. The same subventions furnish farms owned by corporations with generous profit margins. . . .

In contrast, there is scant evidence that public programs have done much for the bottom tiers of American society. Despite the New Deal and the Great Society, including public works and public assistance, since 1935, the share of income going to the poorest fifth of America's households has remained between 3.3 percent and 4.3 percent.

Thus, if many elderly Americans have been raised from poverty, it is clear that younger people are now taking their places. . . .

All parts of the population except the richest fifth have smaller shares of the nation's income than they did twenty years ago. The gulf between the best-off and the rest shows no signs of diminishing, and by some political readings this should mean increased tensions between the favored fifth and everyone else. But declines in living standards have not been so severe or precipitous as to lead many people to question the equity of the economic system. The economy has ensured that a majority of Americans remain in moderate comfort and feel able to count their lives a reasonable success. Airline reservationists making $14,000 do not consider themselves "poor," and no one tells them that they are. Thus a majority of Americans still see themselves as middle class, and feel few ties or obligations to the minority with incomes less than their own.

Given this purview, why should the way America distributes its income be considered a problem? At this moment, certainly, there is scant sentiment for imposing further taxes on the well-to-do and doing more for the poor. As has been observed, there is little resentment felt toward the rich; if anything, greater animus is directed toward families receiving public assistance. Nor is it regarded as untoward if the well-off use their money to accumulate luxuries while public schools must cope with outdated textbooks and leaking roofs. Although this reading is about money, about why some have more and others less, it should not be read as a plea for income redistribution. The reason is straightforward: if people are disinclined to share what they have, they will not be persuaded by a reproachful tone. Rather, its aim is to enhance our understanding of ourselves, of the forces that propel us, and the shape we are giving to the nation of which we are a part.

How a nation allocates its resources tells us how it wishes to be judged in the ledgers of history and morality. America's chosen emphasis has been on offering opportunities to the ambitious, to those with the desire and the drive to surpass. America has more self-made millionaires and more men and women who have attained $100,000 than any other country.

But because of the upward flow of funds, which has accelerated in recent years, less is left for those who lack the opportunities or the temperament to succeed in the competition. The United States now has a greater percentage of its citizens in prison or on the streets, and more neglected children, than any of the nations with which it is appropriately compared. Severe disparities—excess alongside deprivation—sunder the society and subvert common aims. With the legacy we are now creating, millions of men, women, children are prevented from being fully American, while others pride themselves on how much they can amass.

22

Upper-Class Power

HAROLD R. KERBO

In class-based stratification systems the upper classes or elites of society dominate the corporate structure and the political structure. Kerbo analyzes the means by which the upper class dominates the corporate structure and the political structure.

As you read, ask yourself the following questions:

1. *What tactics do members of the upper class use in order to dominate?*
2. *What changes would you make to adjust the power of the upper class?*
3. *Is it really possible to lessen the power of the upper class? Why?*

Upper-Class Economic Power

If we have an upper class in this country that, because of its power, can be described as a governing class, by what means does it govern or dominate? . . . We will examine first how the upper class is said to have extensive influence over the economy through stock ownership, then turn to the question of economic power through extensive representation in major corporate offices.

Stock Ownership As some argue, the most important means of upper-class economic power lies in its ownership of the primary means of production. The upper class has power over our economy because of its control of the biggest corporations through stock ownership. . . .

Legally, the ultimate control of corporations is found not with top corporate executives, but with major stockholders. In a sense, top corporate executives and boards of directors are charged with managing these corporations for the real owners—the stockholders. Stockholders have the authority to elect corporate directors who are to represent stockholder interests. These directors are then responsible for general corporate policy, including the task of filling top executive positions. The day-to-day management of the corporation is in turn the responsibility of the executive officers, who must generally answer to stockholders and their representatives on the board of directors.

Assuming for now that corporate authority actually operates this way . . . the largest stockholder or stockholders in a corporation should be in control. Thus, if we find that upper-class families have extensive stock ownership and that this stock is in major corporations, we can say that upper-class families dominate the American economy.

It is clear . . . that wealth is very unequally distributed in this country—more so even than family or personal income. One of the most important categories of wealth . . . is corporate stock. . . . 1 percent of the people in this country owned *56.5 percent* of the privately held corporate stock, and only 0.5 percent of the people owned *49.3 percent* of the privately held corporate stock in the United States. Thus, from 1 to 0.5 percent of the people in this country . . . hold most of the privately owned corporate stock.

This concentration of private stock ownership is even more striking when we find that most of the remaining stock is controlled by large financial corporations (see U.S. Senate Committee on Governmental Affairs 1978a, 1980; Kerbo and Della Fave 1983, 1984). To the degree that the upper class also has a lot of influence over these financial corporations (such as banks with large amounts of stock control in other big corporations), the actual stock control of the upper class is much greater. . . .

In the early stages of industrialization in this country the control of corporations was fairly easy to estimate. Most corporations were owned, and thus controlled, by specific families. We knew, for example, that the Rockefeller family controlled Standard Oil, the McCormick family controlled International Harvester, the Mellon family controlled the Aluminum Company of America, and the Morgan family controlled Morgan Bank by virtue of their extensive stock ownership of these companies. But this concentration of stock ownership by specific families in one or a few corporations has changed greatly in recent decades. Few clearly family-controlled corporations such as the Ford Motor Company (with the Ford family owning 40 percent of the stock) are found today.

Because of the wide distribution of stockholders in most corporations, government agencies and researchers agree that 5 to 10 percent ownership in a particular company by a family will often result in control of that company by the family.

A government study, however, found only 13 of the top 122 corporations in this country to be clearly controlled by one family (see U.S. Senate Committee on Governmental Affairs 1978a:252). But we must emphasize clearly controlled. One of the problems in determining control is that the ownership of stock in specific corporations is often hidden. For example, the owner of stock may be listed under meaningless names (such as street names) or under trusts and foundations (Zeitlin, 1974). To make the situation even more complex, corporations (especially banks) control stock in other corporations.

Consider the following situation: A family owns about 2 percent of the stock in corporation A, with other families also owning about 2 percent each. In turn, this original family owns, say, 5 percent of the stock in corporation B (a bank) and 6 percent in corporation C (an insurance company). We find upon further investigation that company B (the bank) controls 4 percent of the stock in corporation A, and corporation C (the insurance company) controls 7 percent of the stock in corporation A. Who controls corporation A?

It *may* be that our original family does, through its stock in corporation A, as well as B and C. But other families own stock in A who in addition have much stock in corporations D and E. And (you are probably ahead of me), corporations D and E also control stock in corporation A! This example is not an exaggeration, as anyone will see in examining the data on stock ownership published in a Senate study (U.S. Senate Committee on Governmental Affairs 1978a, 1980). In the face of this complexity of wide stockholdings, many researchers simply conclude that top managers of a corporation control by default (for example, Berle 1959; Galbraith 1971; Bell 1976). But, as we will see in the following, this generalization also has many drawbacks.

Upper-Class Backgrounds of Economic Elites

Aside from actual stock ownership, there is another possible means of upper-class leverage over the economy. After the authority of stockholders in a corporation, we find the board of directors and top executive officers. We will call these people *economic elites*. The family backgrounds of these economic elites may be important in how they think, whom they trust, and what group interests they serve while making decisions in their positions of authority in the corporate world. Ruling-class theorists such as Domhoff believe that these economic elites often come from, or have backgrounds in, upper-class families. Thus, even if upper-class families may not own enough stock to control many corporations, their people are there in important positions of authority. . . .

Domhoff (1967) examined the directors from many top corporations. He found . . . that of the top twenty industrial corporations, 54 percent of the board members were from the upper class; of the top fifteen banks, 62 percent were upper-class members; of the top fifteen insurance companies, 44 percent were upper-class members; of the top fifteen transportation companies, 53 percent were upper-class members; and of the top fifteen utility corporations, 30 percent were upper-class members. Clearly we find much overrepresentation by the upper class on these boards of directors when it is noted that the upper class accounts for only about 0.5 percent of the population. . . .

In another study Soref (1976) took a random sample of board members from the top 121 corporations in the United States. Using Domhoff's definition of upper class, he found upper-class board members had more board positions in other companies (average of 3.49 for upper-class directors, 2.0 for others), and were more often members of board subcommittees that made important long-range decisions in the company. . . .

In conclusion, we find some evidence supporting the argument that the upper class is able to dominate the economy through positions of authority in major corporations. But this evidence is far from conclusive. . . .

There is also the question of whether upper-class members act exclusively to protect the interests of the upper class when in positions of corporate authority. In part, this second reservation pertains to the strength of upper-class unity and consciousness discussed earlier. It is clear that corporate elite membership in social clubs and interlocking directorates through multiple board memberships help unify the structure of large corporations today. However, the question of whose interests (an upper class or corporate elites themselves) are served by this unified corporate structure remains inadequately answered.

Upper-Class Political Power

The next questions of importance for ruling-class or governing-class theorists are the degree and means of political power exercised by the upper class. The significance of the state, and especially the federal government, on domestic affairs in this nation has increased rapidly since the 1930s. We find today a federal government with an annual budget well over $1.5 trillion, with programs designed for such purposes as regulating the economy as well as its traditional job of managing foreign affairs.

The potential impact of the federal government upon upper-class interests is clear. If the upper class is to maintain a position of dominance in the nation, it is imperative that it have influence over the state as well as the economy. In this section we will consider evidence suggesting upper class influence over the government through (1) direct participation by the upper class in politics, (2) the selection of government leaders, (3) the activities of lobby organizations, and (4) organizations established to shape the development of government policy.

Upper-Class Participation in Government

Research on direct participation by the upper class in government is focused heavily on the president's cabinet. Cabinet members are under the direction of the president, but because of the president's many concerns and lack of time in gathering all the needed information in making policy, the president must rely heavily upon cabinet members for advice and information. If these cabinet members represent the interests of the upper class, they can provide the president with information to guide his policy decisions in a way that will ensure that upper-class interests are maintained. . . .

Using his definition of upper-class membership outlined earlier, Domhoff (1967:97–99; see also Kerbo and Della Fave 1979:7) examined the backgrounds of secretaries of state, the treasury, and defense between 1932 and 1964. He found that 63 percent of the secretaries of state, 62 percent of the secretaries of defense, and 63 percent of the secretaries of the treasury could be classified as members of the upper class before assuming office. . . . As Domhoff admits, the preceding represents only a small part of the cabinet for a period of little more than thirty years. But with these positions we find the upper class represented in proportions far greater than their 0.5 percent of the population would suggest.

Since Domhoff's earlier work, an extensive study of cabinet members has been conducted by Beth Mintz (1975). Using Domhoff's indicators of upper-class membership, Mintz (1975, along with

Peter Freitag, 1975) undertook the massive job of examining the backgrounds of all cabinet members (205 people) serving between 1897 and 1973. Her most interesting finding at this point is that *66 percent* of these cabinet members could be classified as members of the upper class before obtaining their cabinet positions.... Also interesting is that the number of cabinet members coming from the upper class was fairly consistent between 1897 and 1973.... And in case anyone believes that the wealthy and upper class strongly favor Republicans over Democrats, Mintz's data show that Republican presidents chose over 71 percent of their cabinet members from the upper class, while Democratic presidents chose over 60 percent from the upper class.

In her background research on these cabinet members, Mintz also included information pertaining to the previous occupations of these people. Along with Freitag (1975), she reports that over 76 percent of the cabinet members were associated with big corporations before or after their cabinet position, 54 percent were from *both* the upper class and top corporate positions, and *90 percent* either came from the upper class or were associated with big corporations. Focusing on corporate ties of cabinet members, Freitag (1975) shows that these ties have not changed much over the years, and vary only slightly by particular cabinet position. In fact, even most secretaries of labor have been associated with big corporations in the capacity of top executives, board members, or corporate lawyers.

Most ruling-class or governing-class theorists consider the cabinet to be the most important position for direct government participation by the upper class. The cabinet allows easy movement into government and then back to top corporate positions. As might be expected, Mintz (1975) found most cabinet members between 1897 and 1973 coming from outside of government, rather than working their way up within government bureaucracies. The United States and England are unique in this aspect of top government elite recruitment. Putnam (1976:48–49) has found that in most other Western industrial societies the top political elites (with positions comparable to those in the U.S. cabinet) are more likely to come from within government bureaucracies, working their way to the top in a line of career promotions. In the United States and England, this atypical method of political elite recruitment affords the upper class and corporate elite opportunities for political influence lacking in these other industrial nations.

In a massive three-volume work by Burch (1981) examining elites throughout American history, we find that the rich and corporate elite have always dominated the top federal government positions. From 1789 to 1861, 96 percent of the cabinet and diplomatic appointees "were members of the economic elite, with a great many landowners, lawyers, and merchants in the group" (Domhoff 1983:142). Then from 1861 to 1933 the proportion was 84 percent, with more of these people now coming from major corporations that did not exist before 1861....

Political Campaign Contributions Today it costs money, lots of money, to obtain a major elective office. In 1972, for example, Richard Nixon spent $60 million to win reelection, while his opponent spent $30 million. Since this time limits have been placed on presidential campaigns, but House and Senate campaigns have not been so restricted. In the 1978 U.S. congressional elections, special-interest groups alone contributed $35 million to candidates. This figure increased to $55 million in 1980, and to $150 million in 1988! The average Senate campaign in 1988 cost $4 million.

In his famous work on the power elite just a little over thirty years ago, C. Wright Mills had relatively little to say about campaign contributions. But the subject can no longer be neglected. Especially in an age when political campaigns are won more through presenting images than through issues, the image-creating mass media are extremely important and costly. Most presidents and congressional office-holders are wealthy, but they are not superrich. With a few rare exceptions, they cannot afford to finance their own political campaigns. Who, then, pays for these campaigns? Thousands of contributors send $25 or $50 to favored candidates. For the most part, however, the money comes from corporations and the wealthy.

With the nationwide reaction against Watergate and the many illegal campaign contributions to Nixon's reelection committee in 1972, some election reforms were undertaken by Congress in 1974. Among these reforms was the creation of a voluntary $1-per-person campaign contribution from individual income tax reports. A Presidential Election Campaign Fund was established to distribute this money to the major parties and candidates during an election year. In addition, a Federal Election Commission was established to watch over campaign spending, and people were limited to a $1000 contribution in any single presidential election, with organizations limited to $5000. Perhaps these reforms contributed to less spending in the 1976 presidential election (Dye 1979:90); Carter and Ford spent a combined total of about $50 million in 1976, compared with Nixon and McGovern's $90 million in 1972. But $50 million continues to be a substantial investment, and requires large contributors.

An interesting outcome of the campaign reform law of 1974 is that much of the illegal activity in Nixon's 1972 campaign was *made legal* as long as correct procedures are followed. For example, organizations are limited to $5000 in political contributions per election. However, if there are more organizations, more money can be contributed. And this is precisely what happened by 1976. There was an explosion in the number of political action committees (PACs) established by large corporations and their executives, an increase far outnumbering those established by any other group, such as labor unions (Domhoff 1983:125). By the 1980 congressional elections, 1585 corporate, health industry, and other business PACs contributed $36 million to candidates, while $13 million was contributed by 240 labor union PACs.

Campaign contributions, therefore, continue to be an important means of political influence. The wealthy are not assured that their interests will be protected by those they help place in office, but they obviously consider the gamble worth taking. Usually, it is hoped that these campaign contributions are placing people in office who hold political views that lead to the defense of privilege when unforeseen challenges to upper-class interests occur along the way.

For example, it seems that campaign contributions by oil companies to congressional candidates paid off in 1979 when 95 percent of the people receiving this campaign money voted for the bill sponsored by the oil industry challenging President Carter's windfall profits tax on oil companies. The oil industry investments in campaign contributions will bring a sizable return in increased profits over the years.

Since the early 1970s, a number of studies have been done on this subject (Mintz 1989). For example, Allen and Broyles (1989) examined data pertaining to the campaign contributions of 100 of the most wealthy families (629 individuals) in the United States. They found that about one-half of these individuals made large contributions. And it was the more "visible" and active rich who made these large contributions. By this they mean that the rich were more likely to make contributions if they were corporate directors or executives, were listed in *"Who's Who,"* and/or were directors of nonprofit foundations. These people were more likely to contribute to Republicans, and this was especially so with the new rich, non-Jews, and people with extensive oil stocks. In a similar manner, Burris (1987) found a split between the old rich (Yankees) and the new rich in who they supported with their money (the new rich were more likely to support Republicans). Burris also found that the rich were more likely to make large campaign contributions if their company did business with the federal government (such as a defense contractor) or was in some way regulated by the government.

In another study, Boies (1989) examined the top 500 companies in the United States between 1976 and 1980 to determine what explains the varied amount of money they contribute through PACs. As we might expect from what has already been discussed, he found that companies with more material interests in the outcomes of government policy were most likely to contribute huge amounts through PACs. Specifically, they were more likely to contribute large amounts if the company was a top defense contractor, if they were trying to get the government to approve a new corporate merger, or if they were the subject of some kind of criminal investigation by the government. And in another study of PAC contributions by big corporations, Clawson, Newstadtl, and Bearden (1986) found extensive consensus by big corporations on the "best" business candidates to support: When looking at individual campaigns, they found that in 75 percent of the cases one candidate received nine times more of the corporate PAC money compared with the political opponent. And in a large study of corporate campaign contributions, Mizruchi (1992) found that corporations grouped together through various ties (such as interlocking directors...) are more likely to support the same politicians and political issues even if many corporations in the group have no individual interests in doing so.

Congressional Lobbying If the interests of the wealthy are not ensured by their direct participation in government, and if those the wealthy helped put in office seem to be forgetting their debtors, a third force can be brought into action. The basic job of a lobbyist is to make friends among congressional leaders, provide them with favors such as trips, small gifts, and parties, and, most importantly, provide these leaders with information and arguments favoring their employers' interests and needs. All of this requires a large staff and lots of money.

In one of the first empirical studies of the effects of certain characteristics of corporations on government policies toward these corporations (such as tax policies), Salamon and Siegfried (1977) found that the size of the corporation showed a strong inverse relation to the amount of taxes paid by the corporation. And this inverse relation between size of the corporation and the corporation's tax rate was especially upheld when examining the oil companies and including their state as well as federal taxes paid (Salamon and Siegfried 1977:1039). Thus, the bigger the corporation, the less it tends to pay in corporate taxes.

Later studies have confirmed this relationship between size (and power) and corporate tax rates. Jacobs (1988), however, measured the concentration of powerful corporations within each type of industry. The findings were similar: The more corporate concentration (meaning the size of the firms in the industry and their dominance in the industry), the lower the taxes for the corporations in that industry. In examining how this is done in the oil industry and health-care industry, Laumann, Knoke, and Kim (1985) studied 166 government policy decisions relating to these industries and interviewed 458 lobbyists for these industries. They found that there are leading corporations in these industries that have a reputation for being most politically active in influencing government for the overall industry, and that this reputation is very accurate when measuring their lobbying activity....

Lobby organizations, therefore, can be of major importance in ensuring that the special interests of a wealthy upper-class and corporate elite are served. If special favors are not completely ensured through direct participation in the cabinet and campaign contributions, the powerful lobby organizations may then move into action. The upper class and big business are not the only groups that maintain lobby organizations in Washington. The American Medical Association, the National Rifle Association, the Milk Producers Association, and many others have maintained successful lobby organizations. But when considering the big issues such as how to deal with inflation, tax policy, unemployment, foreign affairs, and many others that broadly affect the lives of people in this country, the corporate and upper-class lobbies are most important....

Shaping Government Policy Of the various means of upper-class and corporate political influence, the type least recognized by the general public is referred to as the *policy-forming process* (see Domhoff 1979:61–128; 1983:98–112; 1990; Dye 1990:250–270). As scholars believe, in the long run this means of political influence is perhaps one of the most important. The basic argument is this: The federal government is faced with many national problems for which there are many possible alternative solutions. For example, consider the problem of unemployment. The possible government means of dealing with this problem are varied, and a key is that different solutions to the problem may favor different class interests. Some possible solutions (such as stimulating the economy with low interest rates and restricting imports) are believed to favor the working class and help create new jobs, and thus are pushed by labor unions. Other possible solutions (such as the "Reagonomics" idea of cutting taxes for the rich and corporations) favor the interests of corporations and the upper class. . . . One important means of ensuring that the federal government follows a policy that is favorable to your class interests is to convince the government through various types of research data that one line of policy is the overall best policy. Generating the needed information and spelling out the exact policy required take a lot of planning, organization, personnel, and resources. And there must be avenues for getting this policy information to the attention of government leaders. It is no surprise, ruling-class theorists argue, that the upper class and its corporations are able to achieve this and guide government policy in their interests. . . .

At the heart of this process are (1) upper-class and corporate *money and personnel* (2) that fund and guide *research* on important questions through foundations and universities, (3) then process the information through *policy-planning groups* sponsored by the upper class (4) that make direct recommendations to government, and (5) influence the opinion-making centers, such as the media and government commissions, which in turn influence the population and government leaders in favoring specific policy alternatives. . . .

Many writers in sociology, political science, and economics have come to stress the increased importance of information and ideas generated through research in guiding the economy and government in advanced or postindustrial societies (see Galbraith 1971; Bell 1976). As a consequence, some writers have argued that an upper-class or wealthy elite is no longer in control of the economy or political system because the ideas and specialized knowledge are in the hands of a new group of elites—strategic elites or technocrats (see Galbraith 1971; Keller 1963).

Others, especially ruling-class theorists, counter by charging that the knowledge and information behind the operation of the economy and government today are not always neutral. Much of this knowledge is generated through upper-class sponsorship, and thus favors its class interests. Increasingly, knowledge needed by corporations and government is generated through research conducted at major universities. Scientific research requires a lot of time, money, and personnel. The upper class and corporations, it is argued, influence the research process through funding and authority positions in major research-oriented universities.

A major source of funds for research is large foundations. These foundations possess extensive wealth that is given to fund research projects (see Lundberg 1968:498–505) that the foundation directors judge to be important in generating information needed in guiding political and economic decisions. In most cases, these large foundations were established by wealthy people as a means of reducing taxes. But these families often maintain control of the foundations, influencing their funding policies. . . .

In a study of top foundations (those with over half the assets held by all foundations), Dye (1983:143) found that 50 percent of the directors were members of upper-class social clubs.

Also important in the research process are the major universities in which much of this research is conducted. . . . In these universities faculty are often released from their teaching responsibilities to devote most of their time to conducting research sponsored by large corporations and foundations, as well as the federal government. One means of upper-class and corporate influence, therefore, is through guiding what type of research is conducted by the faculty.

We have only a few studies of the exact impact of funding sources on this research, but what we do have is enlightening. For example, Useem, Hoops, and Moore . . . have found that there is a relationship between members from the upper class on a university's board of trustees and obtaining more funds from corporations and foundations. . . . A majority of these professors admitted that their research plans (what type of research they could do) were influenced by the policies of funding agencies (in this case, the federal government). In other words, they were doing the research the funding agency thought was most important, not what they or their scientific disciplines thought most important. . . .

Finally, there is the more general influence over university policy that may be exercised by corporations and the upper class. Most universities are governed much like a corporation. Above the executive officers (such as presidents and vice presidents) is the board of trustees (resembling a board of directors in a corporation). This board of trustees has broad authority in governing the university and its general policies. Thus, it is often deemed important to understand the outside interests of the university board of trustees.

In Dye's (1995) study of elites he examined the backgrounds of presidents and trustees of the top twelve private universities. . . . He found that *62 percent* of these university trustees were members of just 37 exclusive upper-class clubs. Much of the research sponsored by corporations and the upper class is in state-sponsored universities not included in Dye's sample. But other research indicates that the trustees of these universities are also dominated by corporations and the upper class (Domhoff 1967:77–79; Dye 1983:157–160).

In this policy-forming process the next important link is through what has been called *policy-planning groups*. The corporate elites and upper class come together in these groups, discuss policy, publish and disseminate research, and, according to Dye and Domhoff, arrive at some consensus about what should be done in the nation. The most important of the policy groups are sponsored directly by the upper class for the purpose of linking the research information discussed earlier to specific policy alternatives and making certain these policy alternatives find their way to government circles.

Perhaps the most has been written about the Council on Foreign Relations (CFR) and the Committee on Economic Development (CED). Both groups are clearly upper-class institutions (as defined by Domhoff), the former specializing in foreign policy and the latter specializing in domestic issues. The CFR was established shortly after World War I by upper-class members with the direct intent of influencing the U.S. government with respect to their business interests overseas (see Shoup 1975). Among the CFR's early successes were to have the government define Japan as an economic threat by 1940, to establish some of the ideas behind the development of the World Bank and the United Nations, and to motivate the government to define certain parts of the world as important areas of economic interest (multinational corporation interests). Membership in the CFR is limited to 1400 people, half of whom are members of the upper class. (Domhoff 1983). The CED emerged out of the Business Advisory Council in 1942 to continue the input into government by the upper class that began with this earlier organization in the 1930s (Domhoff 1970:123–128; 1979:67–69; Collins 1977).

We do not have studies of the overall membership of the CED, as we do with the CFR; Dye . . . , however, has traced the backgrounds of directors of both the CED (n = 61) and the CFR (n = 22), finding them to be strongly upper-class. There were an average of 4.1 corporate directorships among CED directors (3.2 for the CFR) and an average of 1.0 government offices held (such as cabinet, 3.0 for the CFR); 72 percent belonged to upper-class clubs (64 percent for the CFR), and 62 percent went to elite universities (. . . 82 percent for the CFR). With respect to research on major universities, 67 percent of CED directors and 68 percent of CFR directors were also trustees of major universities. . . .

Various government commissions are established from time to time to make recommendations on such issues as civil disorders, the conduct of the CIA, and energy. . . . These commissions make public the recommendations of these upper-class policy groups and provide their members with a semiofficial position in the government.

As for the national news media, they are often said to have a liberal slant in their views and to possess much power in shaping public and government opinion. . . . But the moderate conservative wing of the upper class and corporate leaders most influences the media. The major television networks, magazines, and newspapers are highly concentrated and tied to corporations. In terms of the backgrounds of top leaders in the national media, Dye (1995) found 33 percent had previous careers in big corporations and 44 percent were members of upper-class clubs. Their upper-class backgrounds are not as extensive as those of corporate leaders, but neither are the top media leaders from humble origins.

The backgrounds of media directors, the extensive corporate influence through ownership, and the huge funding from advertising all contribute to making mass media organizations cautious in presenting views that may be overly critical of the upper class and corporate interests. Information critical of these interests is not ignored by the mass media. The upper class might not even want this information to be ignored, for corrective action is often needed to prevent economic problems and corporate abuse from getting out of hand and requiring more drastic solutions that may harm more general corporate interests. The news media are in part a policing agency involved in calling attention to problems that need correction. It is not what is criticized as much as how it is criticized, where blame is placed, and what possible solutions are offered.

One final point requires emphasis. Few theorists writing in this area (on the upper class or, more specifically, on upper-class influence in the mass media) suggest that the upper-class or corporate elites completely control the mass media in the country. Neither do most writers in this area believe that there is some kind of upper-class secret conspiracy to control the mass media—or anything else in the country, for that matter. Rather, they are trying to call attention to an economic structure that allows more influence (in the many ways outlined previously) to fall into the hands of groups like the upper-class and corporate elites. Each class or economic interest group tends to have a world-view or way of perceiving reality that has been shaped by its own economic and political interests. When one group has more influence over the major means of conveying information, its view of reality often comes to be accepted by more people.

In summarizing the total policy-forming process, we find an underground network in this country that is highly influenced by corporate and upper-class institutions. The federal government and Congress have the authority to adopt or reject the policy recommendations flowing from this process, but most often they accept them, leaving the government to decide exactly how the policy will be carried out (Dye 1995).

THE UPPER CLASS: A CONCLUSION

. . . We have found some support for the existence of upper-class unity through interaction patterns in prep schools, social clubs, policy-formation organizations, and multiple corporate board positions. We have also found evidence of upper-class influence in the economy through stock ownership and membership on corporate boards. And we have found evidence of upper-class political influence through direct participation in government, campaign contributions, lobby organizations, and a policy-formation process. Upper-class interests are said to be maintained through these means of influence in the economy and the political system . . . .

Editors' Note: For documentation, see the source note on page 140.

23

Keeping Up with the Trumps

JULIET B. SCHOR

This reading deals with the sociological concept of the reference group. *According to this concept, the group that we choose to compare ourselves to determines whether we feel positive or negative about ourselves in financial or other ways. Schor notes the effect of class beliefs on our own assumptions about class and spending habits.*

As you read this selection, consider the following questions as guides:

1. *Do the television programs you watch affect the amount of money you spend on consumer goods or types of products?*
2. *Based on the material in this reading, what do you believe is your class ranking? Why?*
3. *Many people are overloaded with debt on their credit cards and are only able to pay off the minimum each month. Why do you think this is so? What can be done about it?*

For most of us, social space begins with relatives, friends, and co-workers. These are the people whose spending patterns we know and care most about. They are the people against whom we judge our own material lifestyles, and with whom we try and keep up. The comparisons we make between ourselves and them matter deeply to us. And we act upon that deep feeling. According to the research I have done, how people stack up financially against the group with which they most often compare themselves, or their reference group, has an enormous impact on their overall spending. I began my study by asking all the respondents at the company where I was conducting my research to identify their primary reference group. I then asked them, "How does your financial status compare to that of most of the members of the reference group you have chosen?"...

I elicited this information in order to figure out whether Americans really do keep up with others. I reasoned that a person who is trying to associate or identify with a group above himself or herself will spend more, all other things being equal, than someone who has chosen a comparison group of people with less money....

The idea is that where you stand relative to those with whom you compare yourself has a significant impact on your spending....

What is that evidence? I estimated statistical equations that explained the amount of saving and spending each person did in the year of the survey. I included a wide range of factors likely to affect spending, such as the respondent's age, number of dependents, household income in that year, long-term expected income (or what economists call permanent income), and so on. These are the standard variables that economists typically use to explain variations in spending propensities across the population. Then I added my own comparative variable—how the respondent stacked up financially compared to his or her reference group.

As it turns out, this variable has a very large impact. In the savings equation, each step a respondent moved down the scale (from much better off than the reference group to better off) reduced the amount saved by $2,953 a year. Moving down two steps reduced saving by twice that. The sheer magnitude of this effect can be appreciated when we remember that the average employee at the company where I conducted my research saved only $10,450 per year, including retirement savings. According to these estimates, disaster ensues as a person slides down the reference group scale. Moving from the top to the bottom would lead you to save $15,000 less each year—or more likely, to take on some of that amount in debt.

SCREEN WITH ENVY

While television has long been suspected as a promoter of consumer desire, there has been little hard evidence to support that view, at least for adult spending. After all, there's not an obvious connection. Many of the products advertised on television are everyday low-cost items such as aspirin, laundry detergent, and deodorant. Those TV ads are hardly a spur to excessive consumerism. Leaving aside other kinds of ads for the moment (for cars, diamonds, perfume) there's another counter to the argument that television causes consumerism: TV is a substitute for spending. One of the few remaining free activities, TV is a popular alternative to costly recreational spending such as movies, concerts, and restaurants. If it causes us to spend, that effect must be powerful enough to overcome its propensity to save us money.

Apparently it is. My research shows that the more TV a person watches, the more he or she spends. The likely explanation for the link between television and spending is that what we see on TV inflates our sense of what's normal. The lifestyles depicted on television are far different from the average American's: With a few exceptions, TV characters are upper-middle class, or even rich.

Studies by the consumer researchers Thomas O'Guinn and L. J. Shrum confirm this upward situation. The more people watch television, the more they think American households have tennis courts, private planes, convertibles, car telephones, maids, and swimming pools. Heavy watchers also overestimate the portion of the population who are millionaires, have had cosmetic surgery, and belong to a private gym, as well as those suffering from dandruff, bladder control problems, gingivitis, athlete's foot, and hemorrhoids (the effect of all those ads for everyday products). What one watches also matters. Dramatic shows—both daytime soap operas and prime time—have a stronger impact on viewer perceptions than other kinds of programs (say news, sports, or weather).

Heavy watchers are not the only ones, however, who tend to overestimate standards of living. Almost everyone does. (And almost everyone watches TV.) In one study, ownership rates for 22 of 27 consumer products were generally overstated. Your own financial position also matters. Television inflates standards for lower-, average-, and above-average-income students, but it does the reverse for really wealthy ones. (Among those raised in a financially rarefied atmosphere, TV is almost a reality check.) Social theories of consumption hold that the inflated sense of consumer norms promulgated by the media raises people's aspirations and leads them to buy more. . . .

Television also affects norms by giving us real information about how other people live and what they have. It allows us to be voyeurs, opening the door to the "private world" inside the homes and lives of others. . . . As O'Guinn and Shrum note, television has replaced personal contact as our source of information about "what members of other social classes have and how they consume, even behind their closed doors."

Another piece of evidence for the TV-spending link is the apparent correlation between debt and excessive TV viewing. In the Merck Family Fund poll, the fraction responding that they "watch too much TV" rose steadily with indebtedness. More than half (56 percent) of all those who reported themselves "heavily" in debt also said they watched too much TV.

It is partly because of television that the top 20 percent of the income distribution, and even the top 5 percent within it, has become so important in setting and escalating consumption standards for more than just the people immediately below them. Television lets everyone see what these folks have and allows viewers to want it in concrete, product-specific ways. Let's not forget that television programming and movies are increasingly filled with product placements—the use of identifiable brands by characters. TV shows and movies are more and more like running ads. We've become so inured to this practice that it's hard to remember that a can of soda in TV show was once labeled "soda" rather than "Coke" or "Pepsi."...

Part of what keeps the see-want-borrow-and-buy sequence going is lack of attention. Americans live with high levels of denial about their spending patterns. We spend more than we realize, hold more debt than we admit to, and ignore many of the moral conflicts surrounding our acquisitions.... What is not well understood is that the spending of many normal consumers is also predicated on denial....

Nowhere is denial so evident as with credit cards. Contrary to economists' usual portrayal of credit card debtors as fully rational consumers who use the cards to smooth out temporary shortfalls in income, the finding of the University of Maryland economist Larry Ausubel was that people greatly underestimate the amount of debt they hold on their cards—1992's actual $182 billion in debt was thought to be a mere $70 billion. Furthermore, most people do not expect to use their cards to borrow, but, of course, they do. Eighty percent end up paying finance charges within any given year, with just under half (47 percent) always holding unpaid balances.

Not paying attention to what we spend is also very common....

FEAR OF FALLING

Fifty years ago, most people just wanted to secure their place in the American middle class, doing whatever it took to stay there. At one time, that was acquiring a houseful of "decencies," the status symbols of the middle class, situated between the necessities of the poor and the luxuries of the rich. Today, in a world where being middle-class is not good enough for many people and indeed that social category seems like an endangered species, securing a place means going upscale. But when everyone's doing it, upscaling can mean simply keeping up. Even when we are aiming high, there's a strong defensive component to our comparisons. We don't want to fall behind or lose the place we've carved out for ourselves. We don't want to get stuck in the "wrong" lifestyle cluster. How we spend has become a crucial part of our self-image, personal identity, and social network.

The historical record highlights the fact that beneath—indeed driving—our system of competitive consumption are deep class inequalities. The classless-society and end-of-ideology literature of 25 years ago turns out to have been wishful thinking. Ironically, inequality began to arise soon after these ideas appeared. The dirty little secret of American society is that not everyone did become middle-class. We still have rich and poor and gradations in between. Class background and income level affect not only the obvious—if and where you go to college, the quality of your children's elementary school, the kind of job you get—but also your likelihood of getting heart disease, the way you talk, and how respectfully you're treated by others. At all levels of the class structure, we are motivated, as Barbara Ehrenreich put it some years ago, by "fear of falling".... At all levels, a structure of inequality injects insecurity and fear into our psyches. The penalties of dropping down are perhaps the most powerful psychological hooks that keep us keeping up, even as the heights get dizzying.

However rational it may be for individuals to keep up with the upscaling of consumer standards, it can be deeply irrational for society as a whole.... The more our consumer satisfaction is tied into social comparisons—whether upscaling, just keeping up, or not falling too far behind—the less we achieve when consumption grows, because the people we compare ourselves to are also experiencing rising consumption. Our relative position does not change. Jones' delight at being able to afford the Honda Accord is dampened when he sees Smith's new Camry. Both must put in long hours to make the payments, suffer with congested highways and dirty air, and have less in the bank at the end of the day. And both remain frustrated when they think about the Land Cruiser down the street.

Of course, relative positions do change. Some people get promotions or pay raises that place them higher up in the hierarchy. Others fall behind. But these random changes cancel each other out. Of more interest is how the broad social groupings that make up the major comparison groups fare. From the end of the Second World War until the mid-1970s, growth was relatively equally distributed. The rough doubling in living standards was experienced by most Americans, including the poor. In fact, the income distribution was even compressed, as people at the bottom gained some ground relative to those at the top. Since then, however, and particularly since the 1980s, the income groups have diverged.

Middle-class Americans began to experience themselves falling behind as their slow-growing wages and salaries lagged behind those of the groups above them. Their anxiety grew, and it became commonplace that it was no longer possible to achieve a middle-class standard of living on one salary. At the same time, increasing numbers began to lose completely the respectability that defined their class. Below them, a segment of downwardly mobile working people found that their reduced job prospects and declining wages had placed them in the ranks of the working poor. And the nonemployed poor fell even further as their numbers grew and their average income fell.

Thus, relative position has worsened for most people, making it increasingly difficult to keep up. The excitement, convenience, or joy that households may have experienced through the billions in additional spending between 1979 and the present seems to have been overshadowed by feelings of deprivation. Among the upper echelons, all those personal computers, steam showers, Caribbean vacations, and piano lessons have not been sufficient to offset the anxieties inherent in a rapidly upscaling society.

The current mood has led to nostalgia about the older, simpler version of the American dream. There is a palpable sense of unease, a yearning for the less expansive, and less expensive, aspirations of our parents. In the words of one young man, "My dream is to build my own house. When my parents grew up, they weren't so much 'I want this, I've got to have that.' They just wanted to be comfortable. Now we're more—I know I am—'I need this.' And it's not really a need."

24

Class, Not Race

RICHARD KAHLENBERG

In dealing with the problem of poverty, racial attitudes tend to intrude. Thus, affirmative action is considered racially biased. According to Kahlenberg, the means for correcting this bias is to realize that poverty is at one end of a class spectrum of rich and poor; therefore affirmative action should be based on class, not race.

As you read, ask yourself the following questions:

1. *Why does the author claim that the problem is not between black and white but between rich and poor?*
2. *What would you include in a "bill of rights" for the disadvantaged?*
3. *Do you foresee a situation when the poor of all colors will unite in their common cause? Why or why not?*

GLOSSARY **Deus ex machina** An artificial intervention intended to settle an involved situation.

In an act that reflected panic as much as cool reflection, Bill Clinton said recently that he is reviewing all federal affirmative action programs to see "whether there is some other way we can reach [our] objective without giving a preference by race or gender." As the country's mood swings violently against affirmative action, . . . the whole project of legislating racial equality seems suddenly in doubt. . . .

There is, as Clinton said, a way "we can work this out." But it isn't the "*Bakke* straddle," which says yes to affirmative action (race as a factor) but no to quotas. It isn't William Julius Wilson's call to "emphasize" race-neutral social programs, while downplaying affirmative action. The days of downplaying are gone. . . . The way out—an idea Clinton hinted at—is to introduce the principle of race neutrality and the goal of aiding the disadvantaged into affirmative action preference programs themselves; to base preferences in education, entry-level employment and public contracting, on class, not race. . . .

But despite its association with conservatives . . . the idea of class-based affirmative action should in fact appeal to the left as well. After all, its message of addressing class unfairness and its political potential for building cross-racial coalitions are traditional liberal staples.

For many years, the left argued not only that class was important, but also that it was more important than race. This argument was practical, ideological and political. An emphasis on class inequality meant Robert Kennedy riding in a motorcade through cheering white and black sections of racially torn Gary, Indiana, in 1968, with black Mayor Richard Hatcher on one side, and white working-class boxing hero Tony Zale on the other.

Ideologically, it was clear that with the passage of the Civil Rights Act of 1964, class replaced caste as

SOURCE: From *The New Republic*, April 3, 1995, pp. 21–27. Reprinted by permission of the publisher.

the central impediment to equal opportunity. Martin Luther King Jr. moved from the Montgomery Boycott to the Poor People's Campaign, which he described as "his last, greatest dream," and "something bigger than just a civil rights movement for Negroes." RFK told David Halberstam that "it was pointless to talk about the real problem in America being black and white, it was really rich and poor, which was a much more complex subject."

Finally, the left emphasized class because to confuse class and race was seen not only as wrong but as dangerous. This notion was at the heart of the protest over Daniel Patrick Moynihan's 1965 report, *The Negro Family: The Case for National Action*, in which Moynihan depicted the rising rates of illegitimacy among poor blacks. While Moynihan's critics were wrong to silence discussion of illegitimacy among blacks, they rightly noted that the title of the report, which implicated all blacks, was misleading, and that fairly high rates of illegitimacy also were present among poor whites —a point which Moynihan readily endorses today. (In the wake of the second set of L.A. riots in 1992, Moynihan rose on the Senate floor to reaffirm that family structure "is not an issue of race but of class. . . . It is class behavior.")

The irony is that affirmative action based on race violates these . . . liberal insights. . . . It says that despite civil rights protections, the wealthiest African American is more deserving of preference than the poorest white. It relentlessly focuses all attention on race.

In contrast, Lyndon Johnson's June 1965 address to Howard University, in which the concept of affirmative action was first unveiled, did not ignore class. In a speech drafted by Moynihan, Johnson spoke of the bifurcation of the black community, and, in his celebrated metaphor, said we needed to aid those "hobbled" in life's race by past discrimination. This suggested special help for disadvantaged blacks, not all blacks; for the young Clarence Thomas, but not for Clarence Thomas's son. Johnson balked at implementing the thematic language of his speech. His Executive Order 11246, calling for "affirmative action" among federal contractors, initially meant greater outreach and required hiring without respect to race. In fact, LBJ rescinded his Labor Department's proposal to provide for racial quotas in the construction industry in Philadelphia. It fell to Richard Nixon to implement the "Philadelphia Plan," in what Nixon's aides say was a conscious effort to drive a wedge between blacks and labor. . . .

The ironies were compounded by the Supreme Court. In the 1974 case *DeFunis* v. *Odegaard*, in which a system of racial preferences in law school admissions was at issue, it was the Court's liberal giant, William O. Douglas, who argued that racial preferences were unconstitutional, and suggested instead that preferences be based on disadvantage. Four years later, in the *Bakke* case, the great proponent of affirmative action as a means to achieve "diversity" was Nixon appointee Lewis F. Powell Jr. . . .

Today, the left pushes racial preferences, even for the most advantaged minorities, in order to promote diversity and provide role models for disadvantaged blacks—an argument which, if it came from Ronald Reagan, the left would rightly dismiss as trickle-down social theory. Today, when William Julius Wilson argues the opposite of the Moynihan report—that the problems facing the black community are rooted more in class than race—it is Wilson who is excoriated by civil rights groups. The left can barely utter the word "class," instead resorting to euphemisms such as "income groups," "wage earners" and "people who play by the rules."

For all of this, the left has paid a tremendous price. On a political level, with a few notable exceptions, the history of the past twenty-five years is a history of white, working-class Robert Kennedy Democrats turning first into Wallace Democrats, then into Nixon and Reagan Democrats and ultimately into today's Angry White Males. Time and again, the white working class votes its race rather than its class, and Republicans win. The failure of the left to embrace class also helps turn poor blacks, for whom racial preferences are, in Stephen Carter's words, "stunningly irrelevant," toward Louis Farrakhan.

On the merits, the left has committed itself to a goal—equality of group results—which seems highly radical, when it is in fact rather unambitious. To the extent that affirmative action, at its ultimate moment of success, merely creates a self-perpetuating black elite along with a white one, its goal is modest—certainly more conservative than real equality of opportunity, which gives blacks and whites and other Americans of all economic strata a fair chance at success.

The priority given to race over class has inevitably exacerbated white racism. Today, both liberals and conservatives conflate race and class because it serves both of their purposes to do so. Every year, when SAT scores are released, the breakdown by race shows enormous gaps between blacks on the one hand and whites and Asians on the other. The NAACP cites these figures as evidence that we need to do more. Charles Murray cites the same statistics as evidence of intractable racial differences. We rarely see a breakdown of scores by class, which would show enormous gaps between rich and poor, gaps that would help explain the differences in scores by race.

On the legal front, it once made some strategic sense to emphasize race over class. But when states moved to the remedial phrase—and began trying to address past discrimination—the racial focus became a liability. The strict scrutiny that struck down Jim Crow is now used, to varying degrees, to curtail racial preferences. Class, on the other hand, is not one of the suspect categories under the Fourteenth Amendment, which leaves class-based remedies much less assailable.

If class-based affirmative action is a theory that liberals should take seriously, how would it work in practice? In this magazine, Michael Kinsley has asked, "Does Clarence Thomas, the sharecropper's kid, get more or fewer preference points than the unemployed miner's son from Appalachia?" Most conservative proponents of class-based affirmative action have failed to explain their idea with any degree of specificity. Either they're insincere—offering the alternative only for tactical reasons—or they're stumped.

The former is more likely. While the questions of implementation are serious and difficult, they are not impossible to answer. At the university level, admissions committees deal every day with precisely the type of apples-and-oranges question that Kinsley poses. Should a law school admit an applicant with a 3.2 GPA from Yale or a 3.3 from Georgetown? How do you compare those two if one applicant worked for the Peace Corps but the other had slightly higher LSATs?

In fact, a number of universities already give preferences for disadvantaged students in addition to racial minorities. Since 1989 Berkeley has granted special consideration to applicants "from socioeconomically disadvantaged backgrounds . . . regardless of race or ethnicity." Temple University Law School has, since the 1970s, given preference to "applicants who have overcome exceptional and continuous economic depravation." And at Hastings College of Law, 20 percent of the class is set aside for disadvantaged students through the Legal Equal Opportunity Program. Even the U.C. Davis medical program challenged by Allan Bakke was limited to "disadvantaged" minorities, a system which Davis apparently did not find impossible to administer.

Similar class-based preference programs could be provided by public employers and federal contractors for high school graduates not pursuing college, on the theory that at that age their class-based handicaps hide their true potential and are not at all of their own making. In public contracting, government agencies could follow the model of New York City's old class-based program, which provided preferences based not on the ethnicity or gender of the contractor, but to small firms located in New York City which did part of their business in depressed areas or employed economically disadvantaged workers.

The definition of class or disadvantage may vary according to context, but if, for example, the government chose to require class-based affirmative action from universities receiving federal funds, it is possible to devise an enforceable set of objective standards for deprivation. If the aim of class-based affirmative action is to provide a system of genuine equality of opportunity, a leg up to promising students who have done well despite the odds, we have a wealth of sociological data to devise an obstacles test. While some might balk at the very idea of reducing disadvantage to a number, we currently reduce intellectual promise to numbers—SATs and GPAs—and adding a number for disadvantage into the calculus just makes deciding who gets ahead and who does not a little fairer.

There are three basic ways to proceed: with a simple, moderate or complex definition. The simple method is to ask college applicants their family's income and measure disadvantage by that factor alone, on the theory that income is a good proxy for a whole host of economic disadvantages (such as bad schools or a difficult learning environment). This oversimplified approach is essentially the tack we've taken with respect to compensatory race-based affirmative action. For example, most affirmative action programs ask applicants to check a racial box and sweep all the ambiguities under the rug. Even though African Americans have, as Justice Thurgood Marshall said in *Bakke*, suffered a history "different in kind, not just degree, from that of other ethnic groups," universities don't calibrate preferences based on comparative group disadvantage (and, in the Davis system challenged by Bakke, two-thirds of the preferences went to Mexican-Americans and Asians, not blacks). We also ignore the question of when an individual's family immigrated in order to determine whether the family was even theoretically subject to the official discrimination in this country on which preferences are predicated.

"Diversity" was supposed to solve all this by saying we don't care about compensation, only viewpoint. But, again, if universities are genuinely seeking diversity of viewpoints, they should inquire whether a minority applicant really does have the "minority viewpoint" being sought. Derrick Bell's famous statement—"the ends of diversity are not served by people who look black and think white"—is at once repellent and a relevant critique of the assumption that all minority members think alike. In theory, we need some assurance from the applicant that he or she will in fact interact with students of different backgrounds, lest the cosmetic diversity of the freshman yearbook be lost to the reality of ethnic theme houses.

The second way to proceed, the moderately complicated calculus of class, would look at what sociologists believe to be the Big Three determinants of life chances: parental income, education and occupation. Parents' education, which is highly correlated with a child's academic achievement, can be measured in number of years. And while ranking occupations might seem hopelessly complex, various attempts to do so objectively have yielded remarkably consistent results—from the Barr Scale of the early 1920s to Alba Edwards' Census rankings of the 1940s to the Duncan Scores of the 1960s.

The third alternative, the complex calculus of disadvantage, would count all the factors mentioned, but might also look at net worth, the quality of secondary education, neighborhood influences and family structure. An applicant's family wealth is readily available from financial aid forms, and provides a long-term view of relative disadvantage, to supplement the "snapshot" picture that income provides. We also know that schooling opportunities are crucial to a student's life chances, even controlling for home environment. Some data suggest that a disadvantaged student at a middle-class school does better on average than a middle-class student at a school with high concentrations of poverty. Objective figures are available to measure secondary school quality—from per student expenditure, to the percentage of students receiving free or reduced-price lunches, to a school's median score on standardized achievement tests. Neighborhood influences, measured by the concentration of poverty within Census tracts or zip codes, could also be factored in, since numerous studies have found that living in a low-income community can adversely affect an individual's life chances above and beyond family income. Finally, everyone from Dan Quayle to Donna Shalala agrees that children growing up in single-parent homes have a tougher time. This factor could be taken into account as well.

The point is not that this list is the perfect one, but that it *is* possible to devise a series of fairly objective and verifiable factors that measure the degree to which a teenager's true potential has been hidden. (As it happens, the complex definition is the one that disproportionately benefits African Americans. Even among similar income groups, blacks are more likely than whites to live in concentrated poverty, go to bad schools, and live in single-parent homes.) It's just not true that a system of class preferences is inherently harder to administer than a system based on race. Race only seems simpler because we have ignored the ambiguities. And racial preferences are just as easy to ridicule. To paraphrase Kinsley, does a new Indian immigrant get fewer or more points than a third-generation Latino whose mother is Anglo?

Who should benefit? Mickey Kaus, in "Class Is In," . . . argued that class preferences should be reserved for the underclass. But the injuries of class extend beyond the poorest. The offspring of the working poor and the working class lack advantages, too, and indeed SAT scores correlate lockstep with income at every increment. Unless you believe in genetic inferiority, these statistics suggest unfairness is not confined to the underclass. As a practical matter, a teenager who emerges from the underclass has little chance of surviving at an elite college. At Berkeley, administrators found that using a definition of disadvantaged, under which neither parent attended a four-year college and the family could not afford to pay $1,000 in education expenses, failed to bring in enough students who were likely to pass.

Still, there are several serious objections to class-based preferences that must be addressed.

1. *We're not ready to be color-blind because racial discrimination continues to afflict our society.* Ron Brown said affirmative action "continues to be needed not to redress grievances of the past, but the current discrimination that continues to exist." This is a relatively new theory, which conveniently elides the fact that preferences were supposed to be temporary. It also stands logic on its head. While racial discrimination undoubtedly still exists, the Civil Rights Act of 1964 was meant to address prospective discrimination. Affirmative action—discrimination in itself—makes sense only to the extent that there is a current-day legacy of *past* discrimination which new prospective laws cannot reach back and remedy.

   In the contexts of education and employment, the Civil Rights Act already contains powerful tools to address intentional and unintentional discrimination. The Civil Rights Act of 1991 reaffirmed the need to address unintentional discrimination—by requiring employers to justify employment practices that are statistically more likely to hurt minorities—but it did so without crossing the line to required preferences. This principle also applies to Title VI of the Civil Rights Act, so that if, for example, it can be shown that the SAT produces an unjustified disparate impact, a university can be barred from using it. In addition, "soft" forms of affirmative action, which require employers and universities to broaden the net and interview people from all races are good ways of ensuring positions are not filled by word of mouth, through wealthy white networks.

   We have weaker tools to deal with discrimination in other areas of life—say, taxi drivers who refuse to pick up black businessmen—but how does a preference in education or employment remedy that wrong? By contrast, there is nothing illegal about bad schools, bad housing and grossly stunted opportunities for the poor. A class preference is perfectly appropriate.

2. *Class preferences will be just as stigmatizing as racial preferences.* Kinsley argues that "any debilitating self-doubt that exists because of affirmative action is not going to be mitigated by being told you got into Harvard because of your 'socioeconomic disadvantage' rather than your race."

   But class preferences are different from racial preferences in at least two important respects. First, stigma—in one's own eyes and the eyes of others—is bound up with the question of whether an admissions criterion is accepted as legitimate. Students with good grades aren't seen as getting in "just because they're smart." And there appears to be a societal consensus—from Douglas to

Scalia—that kids from poor backgrounds deserve a leg up. Such a consensus has never existed for class-blind racial preferences.

Second, there is no myth of inferiority in this country about the abilities of poor people comparable to that about African Americans. Now, if racial preferences are purely a matter of compensatory justice, then the question of whether preferences exacerbate white racism is not relevant. But today racial preferences are often justified by social utility (bringing different racial groups together helps dispel stereotypes) in which case the social consequences are highly relevant. The general argument made by proponents of racial preferences—that policies need to be grounded in social reality, not ahistorical theory—cuts in favor of the class category. Why? Precisely because there is no stubborn historical myth for it to reinforce.

Kaus makes a related argument when he says that class preferences "will still reward those who play the victim." But if objective criteria are used to define the disadvantaged, there is no way to "play" the victim. Poor and working-class teenagers are the victims of class inequality not of their own making. Preferences, unlike, say, a welfare check, tell poor teenagers not that they are helpless victims, but that we think their long-run potential is great, and we're going to give them a chance—if they work their tails off—to prove themselves.

3. *Class preferences continue to treat people as members of groups as opposed to individuals.* Yes. But so do university admissions policies that summarily reject students below a certain SAT level. It's hard to know what treating people as individuals means. (Perhaps if university admissions committees interviewed the teachers of each applicant back to kindergarten to get a better picture of their academic potential, we'd be treating them more as individuals.) The question is not whether we treat people as members of groups—that's inevitable—but whether the group is a relevant one. And in measuring disadvantage (and hidden potential) class is surely a much better proxy than race.

4. *Class-based affirmative action will not yield a diverse student body in elite colleges.* Actually, there is reason to believe that class preferences will disproportionately benefit people of color in most contexts—since minorities are disproportionately poor. In the university context, however, class-based preferences were rejected during the 1970s in part because of fear that they would produce inadequate numbers of minority students. The problem is that when you control for income, African American students do worse than white and Asian students on the SAT—due in part to differences in culture and linguistic patterns, and in part to the way income alone as a measurement hides other class-based differences among ethnic groups.

   The concern is a serious and complicated one. Briefly, there are four responses. First, even Murray and Richard Herrnstein agree that the residual racial gap in scores has declined significantly in the past two decades, so the concern, though real, is not as great as it once was. Second, if we use the sophisticated definition of class discussed earlier—which reflects the relative disadvantage of blacks vis-à-vis whites of the same income level—the racial gap should close further. Third, we can improve racial diversity by getting rid of unjustified preferences—for alumni kids or students from underrepresented geographic regions—which disproportionately hurt people of color. Finally, if the goal is to provide genuine equal opportunity, not equality of group result, and if we are satisfied that a meritocratic system which corrects for class inequality is the best possible approximation of that equality, then we have achieved our goal.

5. *Class-based affirmative action will cause as much resentment among those left out as race-based affirmative action.* Kinsley argues that the rejected applicant in the infamous Jesse Helms commercial from 1990 would feel just as angry for losing out on a class-based as a race-based

> preference, since both involve "making up for past injustice." The difference, of course, is that class preferences go to the actual victims of class injury, mooting the whole question of intergenerational justice. In the racial context, this was called "victim specificity." Even the Reagan administration was in favor of compensating actual victims of racial discrimination.

The larger point implicit in Kinsley's question is a more serious one: that any preference system, whether race- or class-based, is "still a form of zero-sum social engineering." Why should liberals push for class preferences at all? Why not just provide more funding for education, safer schools, better nutrition? The answer is that liberals should do these things; but we cannot hold our breath for it to happen. . . . Cheaper alternatives, such as preferences, must supplement more expensive strategies of social spending. Besides, to the extent that class preferences help change the focus of public discourse from race to class, they help reforge the coalition needed to sustain the social programs liberals want.

Class preferences could restore the successful formula on which the early civil rights movement rested: morally unassailable underpinnings and a relatively inexpensive agenda. It's crucial to remember that Martin Luther King Jr. called for special consideration based on class, not race. After laying out a forceful argument for the special debt owed to blacks, King rejected the call for a Negro Bill of Rights in favor of a Bill of Rights for the Disadvantaged. It was King's insight that there were nonracial ways to remedy racial wrongs, and that the injuries of class deserve attention along with the injuries of race. . . .

25

No, Poverty Has Not Disappeared

HERBERT J. GANS

As noted in the prior readings in this chapter, we tend to blame the cause of poverty mostly on personal attributes rather than the restrictions of reality. Gans indicates another reason for the persistence of poverty: Poor people are fundamentally needed.

As you read, ask yourself the following questions:

1. *The author believes that poverty will continue until functional alternatives are developed. What are these alternatives?*
2. *Do you believe that the alternatives suggested or your own will eliminate poverty? Why?*
3. *Regardless of any efforts, poverty will always be with us. Why?*

SOURCE: Reprinted from *Social Policy*, July–August 1971, pp. 20–24, published by Social Policy Corporation, New York, NY 10036.

GLOSSARY **Functional analysis** A means of analysis that examines the objective consequences of an action, a law, or the like. **Latent function** A consequence that is not readily apparent. **Dysfunction** An objective consequence that hinders the fulfillment of a goal. **Negative income tax** Receipt of tax dollars when income falls below a set figure. **Family assistance plan** Various programs for aiding needy families. **Vicarious** Participating in another person's experience through the imagination.

Some twenty years ago Robert K. Merton applied the notion of functional analysis[1] to explain the continuing though maligned existence of the urban political machine: if it continued to exist, perhaps it fulfilled latent—unintended or unrecognized—positive functions. Clearly it did. Merton pointed out how the political machine provided central authority to get things done when a decentralized local government could not act, humanized the services of the impersonal bureaucracy for fearful citizens, offered concrete help (rather than abstract law or justice) to the poor, and otherwise performed services needed or demanded by many people but considered unconventional or even illegal by formal public agencies.

Today, poverty is more maligned than the political machine ever was; yet it, too, is a persistent social phenomenon. Consequently, there may be some merit in applying functional analysis to poverty, in asking whether it also has positive functions that explain its persistence.

Merton defined functions as "those observed consequences [of a phenomenon] which make for the adaptation or adjustment of a given [social] system." I shall use a slightly different definition; instead of identifying functions for an entire social system, I shall identify them for the interest groups, socioeconomic classes, and other population aggregates with shared values that "inhabit" a social system. I suspect that in a modern heterogeneous society, few phenomena are functional or dysfunctional for the society as a whole, and that most result in benefits to some groups and costs to others. Nor are any phenomena indispensable; in most instances, one can suggest what Merton calls "functional alternatives" or equivalents for them, in other words, other social patterns or policies that achieve the same positive functions but avoid the dysfunctions.[2]

Associating poverty with positive functions seems at first glance to be unimaginable. Of course, the slumlord and the loan shark are commonly known to profit from the existence of poverty, but they are viewed as evil men, so their activities are classified among the dysfunctions of poverty. However, what is less often recognized, at least by the conventional wisdom, is that poverty also makes possible the existence or expansion of respectable professions and occupations, for example, penology, criminology, social work, and public health. More recently, the poor have provided jobs for professional and paraprofessional "poverty warriors," and for journalists and social scientists, this author included, who have supplied the information demanded by the revival of public interest in poverty.

Clearly, then, poverty and the poor may well satisfy a number of positive functions for many nonpoor groups in American society. I shall describe thirteen such functions—economic, social, and political—that seem to me most significant.

THE FUNCTIONS OF POVERTY

First, the existence of poverty ensures that society's "dirty work" will be done. Every society has such work: physically dirty or dangerous, temporary, dead-end and underpaid, undignified and menial jobs. Society can fill these jobs by paying higher wages than for "clean" work, or it can force people who have no other choice to do the dirty work—and at low wages. In America, poverty functions to provide a low-wage labor pool that is willing—or, rather, unable to be *un*willing—to perform dirty work at low cost. Indeed, this function of the poor is so important that in some Southern states, welfare payments have been cut off during the summer months when the poor are needed to work in the fields. Moreover, much of the debate about the Negative Income Tax and the Family Assistance Plan has concerned their impact on the work incentive, by which is actually meant the incentive of the poor to do the needed dirty work if the wages therefrom are no larger than the income grant. Many economic activities that involve dirty work depend on the poor for their existence: restaurants, hospitals, parts of the garment industry, and "truck farming," among others, could not persist in their present form without the poor.

Second, because the poor are required to work at low wages, they subsidize a variety of economic activities that benefit the affluent. For example, domestics subsidize the upper middle and upper classes, making life easier for their employers and freeing affluent women for a variety of professional, cultural, civic, and partying activities. Similarly, because the poor pay a higher proportion of their income in property and sales taxes, among others, they subsidize many state and local governmental services that benefit more affluent groups. In addition, the poor support innovation in medical practice as patients in teaching and research hospitals and as guinea pigs in medical experiments.

Third, poverty creates jobs for a number of occupations and professions that serve or "service" the poor, or protect the rest of society from them. As already noted, penology would be minuscule without the poor, as would the police. Other activities and groups that flourish because of the existence of poverty are the numbers game, the sale of heroin and cheap wines and liquors, pentecostal ministers, faith healers, prostitutes, pawn shops, and the peacetime army, which recruits its enlisted men mainly from among the poor.

Fourth, the poor buy goods others do not want and thus prolong the economic usefulness of such goods—day-old bread, fruit and vegetables that would otherwise have to be thrown out, second-hand clothes, and deteriorating automobiles and buildings. They also provide incomes for doctors, lawyers, teachers, and others who are too old, poorly trained, or incompetent to attract more affluent clients.

In addition to economic functions, the poor perform a number of social functions.

Fifth, the poor can be identified and punished as alleged or real deviants in order to uphold the legitimacy of conventional norms. To justify the desirability of hard work, thrift, honesty, and monogamy, for example, the defenders of these norms must be able to find people who can be accused of being lazy, spendthrift, dishonest, and promiscuous. Although there is some evidence that the poor are about as moral and law-abiding as anyone else, they are more likely than middle-class transgressors to be caught and punished when they participate in deviant acts. Moreover, they lack the political and cultural power to correct the stereotypes that other people hold of them and thus continue to be thought of as lazy, spendthrift, and so on, by those who need living proof that moral deviance does not pay.

Sixth, and conversely, the poor offer vicarious participation to the rest of the population in the uninhibited sexual, alcoholic, and narcotic behavior in which they are alleged to participate and which, being freed from the constraints of affluence, they are often thought to enjoy more than the middle classes. Thus many people, some social scientists included, believe that the poor not only are more given to uninhibited behavior (which may be true, although it is often motivated by despair more than lack of inhibition) but derive more pleasure from it than affluent people (which research by Lee Rainwater, Walter Miller, and others shows to be patently untrue). However, whether the poor actually have more sex and enjoy it more is irrelevant; so long as middle-class people believe this to be true, they can participate in it vicariously when instances are reported in factual or fictional form.

Seventh, the poor also serve a direct cultural function when culture created by or for them is adopted by the more affluent. The rich often collect artifacts from extinct folk cultures of poor people; and almost all Americans listen to the blues, Negro spirituals, and country music, which originated among the Southern poor. Recently they have enjoyed the rock styles that were born, like the Beatles, in the slums; and in the last year, poetry written by ghetto children has become popular in literary circles. The poor also serve as culture heroes, particularly, of course, to the left, but the hobo, the cowboy, the hipster, and the mythical prostitute with a heart of gold perform this function for a variety of groups.

Eighth, poverty helps to guarantee the status of those who are not poor. In every hierarchical society someone has to be at the bottom; but in American society, in which social mobility is an important goal for many and people need to know where they stand, the poor function as a reliable and relatively permanent measuring rod for status comparisons. This is particularly true for the working class, whose politics is influenced by the need to maintain status distinctions between themselves and the poor, much as the aristocracy must find ways of distinguishing itself from the *nouveaux riches*.

Ninth, the poor also aid the upward mobility of groups just above them in the class hierarchy. Thus a goodly number of Americans have entered the middle class through the profits earned from the provision of goods and services in the slums, including illegal or nonrespectable ones that upper-class and upper-middle-class businessmen shun because of their low prestige. As a result, members of almost every immigrant group have financed their upward mobility by providing slum housing, entertainment, gambling, narcotics, etc., to later arrivals—most recently to blacks and Puerto Ricans.

Tenth, the poor help to keep the aristocracy busy, thus justifying its continued existence. "Society" uses the poor as clients of settlement houses and beneficiaries of charity affairs; indeed, the aristocracy must have the poor to demonstrate its superiority over other elites who devote themselves to earning money.

Eleventh, the poor, being powerless, can be made to absorb the costs of change and growth in American society. During the nineteenth century, they did the backbreaking work that built the cities; today, they are pushed out of their neighborhoods to make room for "progress." Urban renewal projects to hold middle-class taxpayers in the city and expressways to enable suburbanites to commute downtown have typically been located in poor neighborhoods, since no other group will allow itself to be displaced. For the same reason, universities, hospitals, and civic centers also expand into land occupied by the poor. The major costs of the industrialization of agriculture have been borne by the poor, who are pushed off the land without recompense; and they have paid a large share of the human cost of the growth of American power overseas, for they have provided many of the foot soldiers for Vietnam and other wars.

Twelfth, the poor facilitate and stabilize the American political process. Because they vote and participate in politics less than other groups, the political system is often free to ignore them. Moreover, since they can rarely support Republicans, they often provide Democrats with a captive constituency that has no other place to go. As a result, the Democrats can count on their votes, and be more responsive to voters—for example, the white working class—who might otherwise switch to the Republicans.

Thirteenth, the role of the poor in upholding conventional norms (see the fifth point, above) also has a significant political function. An economy based on the ideology of laissez faire requires a deprived population that is allegedly unwilling to work or that can be considered inferior because it must accept charity or welfare in order to survive. Not only does the alleged moral deviancy of the poor reduce the moral pressure on the present political economy to eliminate poverty but socialist alternatives can be made to look quite unattractive if those who will benefit most from them can be described as lazy, spendthrift, dishonest, and promiscuous.

THE ALTERNATIVES

I have described thirteen of the more important functions poverty and the poor satisfy in American society, enough to support the functionalist thesis that poverty, like any other social phenomenon, survives in part because it is useful to society or some of its parts. This analysis is not intended to suggest that because it is often functional, poverty *should* exist, or that it *must* exist. For one thing, poverty has many more dysfunctions than functions; for another, it is possible to suggest functional alternatives.

For example, society's dirty work could be done without poverty, either by automation or by paying "dirty workers" decent wages. Nor is it necessary for the poor to subsidize the many activities they support through their low-wage jobs. This would, however, drive up the costs of these activities, which would result in higher prices to their customers and clients. Similarly, many of the professionals who flourish because of the poor could be given other roles. Social workers could provide counseling to the affluent, as they prefer to do anyway; and the police could devote themselves to traffic and organized crime. Other roles would have to be found for badly trained or incompetent professionals now relegated to serving the poor, and someone else would have to pay their salaries. Fewer penologists would be employable, however. And pentecostal religion could probably not survive without the poor—nor would parts of the second- and third-hand-goods market. And in many cities, "used" housing that no one else wants would then have to be torn down at public expense.

Alternatives for the cultural functions of the poor could be found more easily and cheaply. Indeed, entertainers, hippies, and adolescents are already serving as the deviants needed to uphold traditional morality and as devotees of orgies to "staff" the fantasies of vicarious participation.

The status functions of the poor are another matter. In a hierarchical society, some people must be defined as inferior to everyone else with respect to a variety of attributes, but they need not be poor in the absolute sense. One could conceive of a society in which the "lower class," though last in the pecking order, received 75 percent of the median income, rather than 15–40 percent, as is now the case. Needless to say, this would require considerable income redistribution.

The contribution the poor make to the upward mobility of the groups that provide them with goods and services could also be maintained without the poor's having such low incomes. However, it is true that if the poor were more affluent, they would have access to enough capital to take over the provider role, thus competing with, and perhaps rejecting, the "outsiders." (Indeed, owing in part to anti-poverty programs, this is already happening in a number of ghettos, where white storeowners are being replaced by blacks.) Similarly, if the poor were more affluent, they would make less willing clients for upper-class philanthropy, although some would still use settlement houses to achieve upward mobility, as they do now. Thus "society" could continue to run its philanthropic activities.

The political functions of the poor would be more difficult to replace. With increased affluence the poor would probably obtain more political power and be more active politically. With higher incomes and more political power, the poor would be likely to resist paying the costs of growth and change. Of course, it is possible to imagine urban renewal and highway projects that properly reimbursed the displaced people, but such projects would then become considerably more expensive, and many might never be built. This, in turn, would reduce the comfort and convenience of those who now benefit from urban renewal and expressways. Finally, hippies could serve also as more deviants to justify the existing political economy—as they already do. Presumably, however, if poverty were eliminated, there would be fewer attacks on that economy.

In sum, then, many of the functions served by the poor could be replaced if poverty were eliminated, but almost always at higher costs to others, particularly more affluent others. Consequently, a functional analysis must conclude that poverty persists not only because it fulfills a number of positive functions but also because many of the functional alternatives to poverty would be quite dysfunctional for the affluent members of society. A functional analysis thus ultimately arrives at much the same conclusion as radical sociology, except that radical thinkers treat as manifest what I describe as latent; that social phenomena that are functional for affluent or powerful groups and dysfunctional for poor or powerless ones persist; that when the elimination of such phenomena through functional alternatives would generate dysfunctions for the affluent or powerful, they will continue to persist; and that phenomena like poverty can be eliminated only when they become dysfunctional for the affluent or powerful, or when the powerless can obtain enough power to change society.

NOTES

1. "Manifest and Latent Functions," in *Social Theory and Social Structure* (Glencoe, III.: Free Press, 1949), p. 71.
2. I shall henceforth abbreviate positive functions as functions and negative functions as dysfunctions. I shall also describe functions and dysfunctions, in the planner's terminology, as benefits and costs.

CHAPTER 6

Race, Class, and Gender

The Problems of Inequality

Alan Mc Evoy

The process of stratification results in inequalities between individuals and groups. This chapter explores major sources of inequality and results of inequality.

How easy it is to judge by generalization and stereotypes rather than fact, and to maintain our ethnocentric attitude favoring the "we," the in-group, *our* group. If categorization serves as the primary means of ordering our lives, is that bad? Potentially, yes. In the process of categorizing, we accept some people and ideas, and rule out others—often by using arbitrary stereotypes that discriminate against individuals who belong to certain groups.

In the first reading Michael Omi and Howard Winant discuss the concept of *social race*, based partly on stereotypes and how they affect the dominant group's reactions toward those who differ.

The United States, land of opportunity, has proven to be anything but that for segments of the population. The country was founded on principles of freedom and justice, but certain groups, including Africans, Hispanics, and Asians, were left out of the formula. Africans came as slaves and have a long road to travel before the remnants of their suppression—prejudice, discrimination, racism—are overcome. Asians often supplied needed labor for building railroads, but when economic times were hard they were no longer welcomed by the dominant group. Native Americans, the indigenous peoples of the Americas, were labeled "savages," their cultures destroyed, and tribes systematically decimated. Media images, usually negative, reinforce stereotypes of minority groups, as is the case with Native Americans. Similar patterns of race relations between dominant and subordinate groups occur in countries around the world. In this section we consider the status of two minority groups in the United States, Latinos and Asians.

What has been referred to as the "black underclass"[1] is due mainly to historical discrimination and economic trends. These problems have causes that lie in the economic and social structure of society; they are difficult to change, and in fact are increasing in many urban areas. Many frustrated, disaffected youth who see no opportunities and face daily hostile environments find belonging, acceptance, and protection in gang membership. Latinos, the fastest-growing minority group in the United States, experience prejudice and discrimination because of differences in culture and language. Moore and Pinderhughes trace the history of these groups and their current isolation and stigmatization in society, factors that will not make change easy. These authors outline economic and social problems faced by Latinos living in the United States today and discuss the issue of underclass as it relates to Latinos.

Asian Americans have been called the "model minority" because of the educational and business success of certain groups, especially Japanese and Chinese immigrants from certain periods of history; most of the success stories are from families that were middle class or higher to begin with. However, many Asian Americans are not so fortunate, entering the United States as refugees with few skills to cope with U.S. society. The reading by Min Zhou discusses the variety of Asian groups in America and some of the stereotypes and obstacles they face.

The inequality experienced by these two groups in the United States does not represent an isolated case. Inequality is a worldwide problem. Eduardo

1. William Julius Wilson, "The Black Underclass," *Wilson Quarterly* (Spring 1984).

Bonilla-Silva and Mary Hovsepian discuss the problems that occur when workers from developing countries seek opportunities in western Europe and the United States and the rise of hostilities that greets these workers and their families, especially when economic conditions in the host countries are poor. Generally these workers are from different racial and ethnic groups and many international workers are women, setting them up for prejudice and discrimination.

A pervasive example of the importing of international workers is seen in the sex trade; in fact, many of these "workers" have been bought and sold into the trade. Keven Bales discusses this topic that illustrates inequality based on race, class and gender.

If inequality—stemming from racism, classism, sexism, prejudice, stereotyping, and discrimination—is to be reduced, then individuals and groups must recognize the effects of prejudice and discrimination embedded in the societal structure on their own beliefs and actions. Only in this manner can they overcome their own stereotypes and ethnocentrism and view others from a cultural relativist's perspective. As you consider these readings, think of ways that this can be done in our society. Consider why some minorities have had more success than others. Also note other examples you have observed of groups segregating themselves for self-protection, of false stereotyping, and of economic conflicts between groups that lead to discrimination.

26

Racial Formations

MICHAEL OMI AND HOWARD WINANT

Race—a much discussed and debated topic! Sociologists focus on the effects of social race, *or using race to categorize people into groups against which prejudice and discrimination may be directed. Omi and Winant open this section with a discussion of this social concept and of racial ideology and identity that results from how people view race.*

As you read, think about the following:

1. *The idea of race is relatively new in human history. How did it develop?*
2. *What are examples of social race? Of hypo-descent?*
3. *How does the idea of race affect* all *groups in the United States?*

GLOSSARY **Social race** Categories defining social relations between ethnic groups based on historical contexts and events. **Hypo-descent** Defining anyone with one drop of blood (African, Asian, Native American) as a marker of the ethnic group. **Passing** Attempting to be seen or recognized as being a member of a different group, usually white.

WHAT IS RACE?

Race consciousness, and its articulation in theories of race, is largely a modern phenomenon. When European explorers in the New World "discovered" people who looked different than themselves, these "natives" challenged then existing conceptions of the origins of the human species, and raised disturbing questions as to whether *all* could be considered in the same "family of man."[1] Religious debates flared over the attempt to reconcile the Bible with the existence of "racially distinct" people. Arguments took place over creation itself, as theories of polygenesis questioned whether God had made only one species of humanity ("monogenesis"). Europeans wondered if the natives of the New World were indeed human beings with redeemable souls. At stake were not only the prospects for conversion, but the types of treatment to be accorded them. The expropriation of property, the denial of political rights, the introduction of slavery and other forms of coercive labor, as well as outright extermination, all presupposed a worldview which distinguished Europeans—children of God, human beings, etc.—from "others." Such a worldview was needed to explain why some should be "free" and others enslaved, why some had rights to land and property while others did not. Race, and the interpretation of racial differences, was a central factor in that worldview.

In the colonial epoch science was no less a field of controversy than religion in attempts to comprehend the concept of race and its meaning. Spurred on by the classificatory scheme of living organisms devised by Linnaeus in *Systema Naturae*, many scholars in the eighteenth and nineteenth centuries dedicated themselves to

SOURCE: From Michael Omi and Howard Winant, eds., *Racial Formation in the United States*, second edition, pp. 3-13.

the identification and ranking of variations in humankind. Race was thought of as a *biological* concept, yet its precise definition was the subject of debates which, as we have noted, continue to rage today. Despite efforts ranging from Dr. Samuel Morton's studies of cranial capacity[2] to contemporary attempts to base racial classification on shared gene pools,[3] the concept of race has defied biological definition. . . .

Attempts to discern the *scientific meaning* of race continue to the present day. Although most physical anthropologists and biologists have abandoned the quest for a scientific basis to determine racial categories, controversies have recently flared in the area of genetics and educational psychology. For instance, an essay by Arthur Jensen arguing that hereditary factors shape intelligence not only revived the "nature or nurture" controversy, but raised highly volatile questions about racial equality itself.[4] Clearly the attempt to establish a *biological* basis of race has not been swept into the dustbin of history, but is being resurrected in various scientific arenas. All such attempts seek to remove the concept of race from fundamental social, political, or economic determination. They suggest instead that the truth of race lies in the terrain of innate characteristics, of which skin color and other physical attributes provide only the most obvious, and in some respects most superficial, indicators.

RACE AS A SOCIAL CONCEPT

The social sciences have come to reject biologistic notions of race in favor of an approach which regards race as a *social* concept. Beginning in the eighteenth century, this trend has been slow and uneven, but its direction clear. In the nineteenth century Max Weber discounted biological explanations for racial conflict and instead highlighted the social and political factors which engendered such conflict.[5] The work of pioneering cultural anthropologist Franz Boas was crucial in refuting the scientific racism of the early twentieth century by rejecting the connection between race and culture, and the assumption of a continuum of "higher" and "lower" cultural groups. Within the contemporary social science literature, race is assumed to be a variable which is shaped by broader societal forces.

Race is indeed a pre-eminently *socio-historical* concept. Racial categories and the meaning of race are given concrete expression by the specific social relations and historical context in which they are embedded. Racial meanings have varied tremendously over time and between different societies.

In the United States, the black/white color line has historically been rigidly defined and enforced. White is seen as a "pure" category. Any racial intermixture makes one "nonwhite." In the movie *Raintree County*, Elizabeth Taylor describes the worst of fates to befall whites as "havin' a little Negra blood in ya'—just one little teeny drop and a person's all Negra."[6] This thinking flows from what Marvin Harris has characterized as the principle of *hypo-descent:*

> By what ingenious computation is the genetic tracery of a million years of evolution unraveled and each man [sic] assigned his proper social box? In the United States, the mechanism employed is the rule of hypo-descent. This descent rule requires Americans to believe that anyone who is known to have had a Negro ancestor is a Negro. We admit nothing in between. . . . "Hypo-descent" means affiliation with the subordinate rather than the super-ordinate group in order to avoid the ambiguity of intermediate identity. . . . The rule of hypo-descent is, therefore, an invention, which we in the United States have made in order to keep biological facts from intruding into our collective racist fantasies.[7] . . .

By contrast, a striking feature of race relations in the lowland areas of Latin America since the abolition of slavery has been the relative absence of sharply defined racial groupings. No such rigid descent rule characterizes racial identity in many Latin American societies. Brazil, for example, has historically had less rigid conceptions of race, and thus a variety of "intermediate" racial categories exist. Indeed, as Harris notes, "One of the most striking consequences of the Brazilian system of racial identification is that parents and children and even brothers and sisters are frequently accepted as representatives of quite opposite racial types."[8] Such a possibility is incomprehensible within the logic of racial categories in the US.

To suggest another example: the notion of "passing" takes on new meaning if we compare various American cultures' means of assigning racial identity. In the United States, individuals who are actually "black" by the logic of hypo-descent have attempted to skirt the discriminatory barriers imposed by law and custom by attempting to "pass" for white.[9] Ironically, these same individuals would not be able to pass for "black" in many Latin American societies.

Consideration of the term "black" illustrates the diversity of racial meanings which can be found among different societies and historically within a given society. In contemporary British politics the term "black" is used to refer to all nonwhites. Interestingly this designation has not arisen through the racist discourse of groups such as the National Front. Rather, in political and cultural movements, Asian as well as Afro-Caribbean youth are adopting the term as an expression of self-identity.[10] The wide-ranging meanings of "black" illustrate the manner in which racial categories are shaped politically.[11]

The meaning of race is defined and contested throughout society, in both collective action and personal practice. In the process, racial categories themselves are formed, transformed, destroyed and reformed. We use the term *racial formation* to refer to the process by which social, economic and political forces determine the content and importance of racial categories, and by which they are in turn shaped by racial meanings. Crucial to this formulation is the treatment of race as a *central axis* of social relations which cannot be subsumed under or reduced to some broader category or conception.

RACIAL IDEOLOGY AND RACIAL IDENTITY

The seemingly obvious, "natural" and "common sense" qualities which the existing racial order exhibits themselves testify to the effectiveness of the racial formation process in constructing racial meanings and racial identities.

One of the first things we notice about people when we meet them (along with their sex) is their race. We utilize race to provide clues abut *who* a person is. This fact is made painfully obvious when we encounter someone whom we cannot conveniently racially categorize—someone who is, for example, racially "mixed" or of an ethnic/racial group with which we are not familiar. Such an encounter becomes a source of discomfort and momentarily a crisis of racial meaning. Without a racial identity, one is in danger of having no identity.

Our compass for navigating race relations depends on preconceived notions of what each specific racial group looks like. Comments such as, "Funny, you don't look black," betray an underlying image of what black should be. We also become disoriented when people do not act "black," "Latino," or indeed "white." The content of such stereotypes reveals a series of unsubstantiated beliefs about who these groups are and what "they" are like.[12]

In US society, then, a kind of "racial etiquette" exists, a set of interpretative codes and racial meanings which operate in the interactions of daily life. Rules shaped by our perception of race in a comprehensively racial society determine the "presentation of self,"[13] distinctions of status, and appropriate modes of conduct. "Etiquette" is not mere universal adherence to the dominant group's rules, but a more dynamic combination of these rules with the values and beliefs of subordinated groupings. This racial "subjection" is quintessentially ideological. Everybody learns some combination, some version, of the rules of racial classification, and of their own racial identity, often without obvious teaching or conscious inculcation. Race becomes "common sense"—a way of comprehending, explaining and acting in the world.

Racial beliefs operate as an "amateur biology," a way of explaining the variations in "human nature."[14] Differences in skin color and other obvious physical characteristics supposedly provide visible clues to differences lurking underneath. Temperament, sexuality, intelligence, athletic ability, aesthetic preferences and so on are presumed to be fixed and discernible from the palpable mark of race. Such diverse questions as our confidence and trust in others (for example, clerks or salespeople, media figures, neighbors), our sexual preferences and romantic images, our tastes in music, films, dance, or sports, and our very ways of talking, walking, eating and dreaming are ineluctably shaped by notions of race. Skin color "differences" are thought to explain perceived differences in intellectual, physical and artistic temperaments, and to justify distinct treatment of racially identified individuals and groups.

The continuing persistence of racial ideology suggests that these racial myths and stereotypes cannot be exposed as such in the popular imagination. They are, we think, too essential, too integral, to the maintenance of the US social order. Of course, particular meanings, stereotypes and myths can change, but the presence of a *system* of racial meanings and stereotypes, of racial ideology, seems to be a permanent feature of US culture.

Film and television, for example, have been notorious in disseminating images of racial minorities which establish for audiences what people from these groups look like, how they behave, and "who they are."[15] The power of the media lies not only in their ability to reflect the dominant racial ideology, but in their capacity to shape that ideology in the first place. D. W. Griffith's epic *Birth of a Nation*, a sympathetic treatment of the rise of the Ku Klux Klan during Reconstruction, helped to generate, consolidate and "nationalize" images of blacks which had been more disparate (more regionally specific, for example) prior to the film's appearance.[16] In US television, the necessity to define characters in the briefest and most condensed manner has led to the perpetuation of racial caricatures, as racial stereotypes serve as shorthand for scriptwriters, directors and actors, in commercials, etc. Television's tendency to address the "lowest common denominator" in order to render programs "familiar" to an enormous and diverse audience leads it regularly to assign and reassign racial characteristics to particular groups, both minority and majority.

These and innumerable other examples show that we tend to view race as something fixed and immutable—something rooted in "nature." Thus we mask the historical construction of racial categories, the shifting meaning of race, and the crucial role of politics and ideology in shaping race relations. Races do not emerge full-blown. They are the results of diverse historical practices and are continually subject to challenge over their definition and meaning.

NOTES

1. Thomas F. Gossett notes:

   > Race theory . . . had up until fairly modern times no firm hold on European thought. On the other hand, race theory and race prejudice were by no means unknown at the time when the English colonists came to North America. Undoubtedly, the age of exploration led many to speculate on race differences at a period when neither Europeans nor Englishmen were prepared to make allowances for vast cultural diversities. Even though race theories had not then secured wide acceptance or even sophisticated formulation, the first contacts of the Spanish with the Indians in the Americas can now be recognized as the beginning of a struggle between conceptions of the nature of primitive peoples which has not yet been wholly settled. (Thomas F. Gossett, *Race: The History of an Idea in America* (New York: Schocken Books, 1965), p. 16.)

   Winthrop Jordan provides a detailed account of early European colonialists' attitudes about color and race in *White over Black: American Attitudes Toward the Negro, 1550–1812* (New York: Norton, 1977 [1968]), pp. 3–43.
2. Pro-slavery physician Samuel George Morton (1799–1851) compiled a collection of 800 crania from all parts of the world which formed the sample for his studies of race. Assuming that the larger the size of the cranium translated into greater

intelligence, Morton established a relationship between race and skull capacity. Gossett reports that:

> In 1849, one of his studies included the following results: The English skulls in his collection proved to be the largest, with an average cranial capacity of 96 cubic inches. The Americans and Germans were rather poor seconds, both with cranial capacities of 90 cubic inches. At the bottom of the list were the Negroes with 83 cubic inches, the Chinese with 82, and the Indians with 79. (Ibid., p. 74.)

On Morton's methods, see Stephen J. Gould, "The Finagle Factor," *Human Nature* (July 1978).

3. Definitions of race founded upon a common pool of genes have not held up when confronted by scientific research which suggests that the differences *within* a given human population are greater than those between populations. See L. L. Cavalli-Sforza, "The Genetics of Human Populations," *Scientific American*, September 1974, pp. 81–89.
4. Arthur Jensen, "How Much Can We Boost IQ and Scholastic Achievement?", *Harvard Educational Review 39* (1969):1–123.
5. Ernst Moritz Manasse, "Max Weber on Race," *Social Research 14* (1947):191–221.
6. Quoted in Edward D. C. Campbell, Jr., *The Celluloid South: Hollywood and the Southern Myth* (Knoxville: University of Tennessee Press, 1981), pp. 168–70.
7. Marvin Harris, *Patterns of Race in the Americas* (New York: Norton, 1964), p. 56.
8. Ibid., p. 57.
9. After James Meredith had been admitted as the first black student at the University of Mississippi, Harry S. Murphy announced that he, and not Meredith, was the first black student to attend "Ole Miss." Murphy described himself as black but was able to pass for white and spent nine months at the institution without attracting any notice (ibid., p. 56).
10. A. Sivanandan, "From Resistance to Rebellion: Asian and Afro-Caribbean Struggles in Britain," *Race and Class* 23(2–3) (Autumn–Winter 1981).
11. Consider the contradictions in racial status which abound in the country with the most rigidly defined racial categories—South Africa. There a race classification agency is employed to adjudicate claims for upgrading of official racial identity. This is particularly necessary for the "coloured" category. The apartheid system considers Chinese as "Asians" while the Japanese are accorded the status of "honorary whites." This logic nearly detaches race from any grounding in skin color and other physical attributes and nakedly exposes race as a juridical category subject to economic, social and political influences. (We are indebted to Steve Talbot for clarification of some of these points.)
12. Gordon W. Allport, *The Nature of Prejudice* (Garden City, NY: Doubleday, 1958), pp. 184–200.
13. We wish to use this phrase loosely, without committing ourselves to a particular position on such social psychological approaches as symbolic interactionism, which are outside the scope of this study. An interesting study on this subject is S. M. Lyman and W. A. Douglass, "Ethnicity: Strategies of Individual and Collective Impression Management," *Social Research* 40(2) (1973).
14. Michael Billig, " Patterns of Racism: Interviews with National Front Members," *Race and Class* 20(2) (Autumn 1978):161-79
15. "Miss San Antonio USA Lisa Fernandez and other Hispanics auditioning for a role in a television soap-opera did not fit the Hollywood image of real Mexicans and had to darken their faces before filming." Model Aurora Garza said that their faces were bronzed with powder because they looked too white. " 'I'm a real Mexican [Garza said] and very dark anyway. I'm even darker right now because I have a tan. But they kept wanting me to make my face darker and darker' " (*San Francisco Chronicle*, 21 September 1984). A similar dilemma faces Asian American actors who feel that Asian character lead roles inevitably go to white actors who make themselves up to be Asian. Scores of Charlie Chan films, for example, have been made with white leads (the last one was the 1981 *Charlie Chan and the Curse of the Dragon Queen*). Roland Winters, who played in six Chan features, was asked by playwright Frank Chin to explain the logic of casting a white man in the role of Charlie Chan: " 'The only thing I can think of is, if you want to cast a homosexual in a show, and get a homosexual, it'll be awful. It won't be funny . . . and maybe there's something there . . .' " (Frank Chin, "Confessions of the Chinatown Cowboy," *Bulletin of Concerned Asian Scholars* 4(3) (Fall 1972)).
16. Melanie Martindale-Sikes, "Nationalizing 'Nigger' Imagery Through 'Birth of a Nation'," paper prepared for the 73rd Annual Meeting of the American Sociological Association, 4–8 September 1978, San Francisco.

27

In the Barrios

Latinos and the Underclass Debate

JOAN MOORE AND RAQUEL PINDERHUGHES

In an important work by sociologist William Julius Wilson called The Truly Disadvantaged, *the term* underclass *was introduced to mean persistent poverty due largely to economic restructuring. The use of this term has been debated by scholars. Some see it as blaming the victim, the poor, for their condition because of their values and behaviors. Others see it as a debate over who is responsible for the poor—the individuals themselves or society? Behavioral pathology or economic structure?*

In this discussion by Moore and Pinderhughes, the concept of underclass is considered as it applies to Latinos.

As you read, consider the following:

1. *Does the term* underclass *apply to the Latino population?*
2. *To what does "Latino population" refer?*
3. *What makes the Latino population unique as a minority group in the United States?*
4. *What might be done to alleviate problems for Latinos?*

GLOSSARY **Underclass** Meaning is debated, but it often refers to the poorest of the poor in the United States. **Polarization of the labor market** High- and low-level jobs but few in the middle. **Rustbelt** Area of the country (Midwest) where jobs are being lost. **Sunbelt** Area of the country (mostly south) where jobs are increasing. **Informal economic activities** Outside government control, small-scale.

In the publication *The Truly Disadvantaged*, William Julius Wilson's seminal work on persistent, concentrated poverty in Chicago's black neighborhoods, Wilson used the term "underclass" to refer to the new face of poverty, and traced its origins to economic restructuring. He emphasized the impact of persistent, concentrated poverty not only on individuals but on communities.

. . . The term "Hispanic" is used particularly by state bureaucracies to refer to individuals who reside in the United States who were born in, or trace their ancestry back to, one of twenty-three Spanish-speaking nations. Many of these individuals prefer to use the term "Latino." . . .

No matter what the details, when one examines the history of the term underclass among

SOURCE: Joan Moore and Raquel Pinderhughes, eds. *In the Barrios: Latinos and the Underclass Debate.* New York: Russell Sage Foundation, 1993. Excerpts from pp. xi to xxxix.

sociologists, it is clear that Wilson's 1987 work seriously jolted the somewhat chaotic and unfocused study of poverty in the United States. He described sharply increasing rates of what he called "pathology" in Chicago's black ghettos. By this, Wilson referred specifically to female headship, declining marriage rates, illegitimate births, welfare dependency, school dropouts, and youth crime. The changes in the communities he examined were so dramatic that he considered them something quite new.

Two of the causes of this new poverty were particularly important, and his work shifted the terms of the debate in two respects. First, Wilson argued effectively that dramatic increases in joblessness and long-term poverty in the inner city were a result of major economic shifts—economic restructuring. "Restructuring" referred to changes in the global economy that led to deindustrialization, loss and relocation of jobs, and a decline in the number of middle-level jobs—a polarization of the labor market. Second, he further fueled the debate about the causes and consequences of persistent poverty by introducing two neighborhood-level factors into the discussion. He argued that the outmigration of middle- and working-class people from the urban ghetto contributed to the concentration of poverty. These "concentration effects" meant that ghetto neighborhoods showed sharply increased proportions of very poor people. This, in turn, meant that residents in neighborhoods of concentrated poverty were isolated from "mainstream" institutions and role models. As a result, Wilson postulates, the likelihood of their engaging in "underclass behavior" was increased. Thus the social life of poor communities deteriorated because poverty intensified. . . .

THE LATINO POPULATION—SOME BACKGROUND

American minorities have been incorporated into the general social fabric in a variety of ways. Just as Chicago's black ghettos reflect a history of slavery, Jim Crow legislation, and struggles for civil and economic rights, so the nation's Latino barrios reflect a history of conquest, immigration, and a struggle to maintain cultural identity.

In 1990 there were some 22 million Latinos residing in the United States, approximately 9 percent of the total population. Of these, 61 percent were Mexican in origin, 12 percent Puerto Rican, and 5 percent Cuban. These three groups were the largest, yet 13 percent of Latinos were of Central and South American origin and another 9 percent were classified as "other Hispanics." Latinos were among the fastest-growing segments of the American population, increasing by 7.6 million, or 53 percent, between 1980 and 1990. There are predictions that Latinos will outnumber blacks by the twenty-first century. If Latino immigration and fertility continue at their current rate, there will be over 54 million Latinos in the United States by the year 2020.

This is an old population: as early as the sixteenth century, Spanish explorers settled what is now the American Southwest. In 1848, Spanish and Mexican settlers who lived in that region became United States citizens as a result of the Mexican-American War. Although the aftermath of conquest left a small elite population, the precarious position of the masses combined with the peculiarities of southwestern economic development to lay the foundation for poverty in the current period (see Barrera 1979; Moore and Pachon 1985).

In addition to those Mexicans who were incorporated into the United States after the Treaty of Guadalupe Hidalgo, Mexicans have continually crossed the border into the United States, where they have been used as a source of cheap labor by U.S. employers. The volume of immigration from Mexico has been highly dependent on fluctuations in certain segments of the U.S. economy. This dependence became glaringly obvious earlier in this century. During the Great Depression of the 1930s state and local governments "repatriated" hundreds of thousands of unemployed Mexicans, and just a few years later World War II labor shortages reversed the process as Mexican contract-laborers (*braceros*) were eagerly sought. A little later, in the

1950s, massive deportations recurred when "operation Wetback" repatriated hundreds of thousands of Mexicans. Once again, in the 1980s, hundreds of thousands crossed the border to work in the United States, despite increasingly restrictive legislation.

High levels of immigration and high fertility mean that the Mexican-origin population is quite young—on the average, 9.5 years younger than the non-Latino population—and the typical household is large, with 3.8 persons, as compared with 2.6 persons in non-Latino households (U.S. Bureau of the Census 1991b). Heavy immigration, problems in schooling, and industrial changes in the Southwest combine to constrain advancement. The occupational structure remains relatively steady, and though there is a growing middle class, there is also a growing number of very poor people. . . .

Over the past three decades the economic status of Puerto Ricans dropped precipitously. By 1990, 38 percent of all Puerto Rican families were below the poverty line. A growing proportion of these families were concentrated in poor urban neighborhoods located in declining industrial centers in the Northeast and Midwest, which experienced massive economic restructuring and diminished employment opportunities for those with less education and weaker skills. The rising poverty rate has also been linked to a dramatic increase in female-headed households. Recent studies show that the majority of recent migrants were not previously employed on the island. Many were single women who migrated with their young children (Falcon and Gurak 1991). Currently, Puerto Ricans are the most economically disadvantaged group of all Latinos. As a group they are poorer than African Americans.

Unlike other Latino migrants, who entered the United States as subordinate workers and were viewed as sources of cheap labor, the first large waves of Cuban refugees were educated middle- and upper-class professionals. Arriving in large numbers after Castro's 1959 revolution, Cubans were welcomed by the federal government as bona fide political refugees fleeing communism and were assisted in ways that significantly contributed to their economic well-being. Cubans had access to job-training programs and placement services, housing subsidies, English-language programs, and small-business loans. Federal and state assistance contributed to the growth of a vigorous enclave economy (with Cubans owning many of the businesses and hiring fellow Cubans) and also to the emergence of Miami as a center for Latin American trade. Cubans have the highest family income of all Latino groups. Nevertheless, in 1990, 16.9 percent of the Cuban population lived below the poverty line.

In recent years large numbers of Salvadorans and Guatemalans have come to the United States in search of refuge from political repression. But unlike Cubans, few have been recognized by the U.S. government as bona fide refugees. Their settlement and position in the labor market have been influenced by their undocumented (illegal) status. Dominicans have also come in large numbers to East Coast cities, many also arriving as undocumented workers. Working for the lowest wages and minimum job security, undocumented workers are among the poorest in the nation.

Despite their long history and large numbers, Latinos have been an "invisible minority" in the United States. Until recently, few social scientists and policy analysts concerned with understanding stratification and social problems in the United States have noticed them. Because they were almost exclusively concerned with relations between blacks and whites, social scientists were primarily concerned with generating demographic information on the nation's black and white populations, providing almost no information on other groups. Consequently, it has been difficult, sometimes impossible, to obtain accurate data about Latinos.

Latinos began to be considered an important minority group when census figures showed a huge increase in the population. By 1980 there were significant Latino communities in almost every metropolitan area in the nation. As a group, Latinos have low education, low family incomes, and are more clustered in low-paid, less-skilled occupations. Most Latinos live in cities, and poverty has become an increasing problem. On the whole, Latinos are more likely to live in poverty than the general U.S. population: poverty is widespread for all Latino subgroups except Cubans. They were affected by structural factors that influenced the socioeconomic status of all U.S. workers. In 1990, 28 percent were poor as compared with 13 percent of all Americans and 32 percent of African Americans (U.S. Bureau of the Census 1991b). Puerto Ricans were particularly likely to be poor....

THE IMPORTANCE OF ECONOMIC RESTRUCTURING

The meaning of economic restructuring has shaped the debate about the urban underclass....

First, there is the "Rustbelt in the Sunbelt" phenomenon. Some researchers have argued that deindustrialization has been limited to the Rustbelt, and that the causal chain adduced by Wilson therefore does not apply outside that region. But the fact is that many Sunbelt cities developed manufacturing industries, particularly during and after World War II. Thus Rustbelt-style economic restructuring—deindustrialization, in particular—has also affected them deeply. In the late 1970s and early 1980s cities like Los Angeles experienced a major wave of plant closings that put a fair number of Latinos out of work (Morales 1985).

Second, there has been significant reindustrialization and many new jobs in many of these cities, a trend that is easily overlooked. Most of the expanding low-wage service and manufacturing industries, like electronics and garment manufacturing, employ Latinos (McCarthy and Valdez 1986; Muller and Espenshade 1986), and some depend almost completely on immigrant labor working at minimum wage (Fernandez-Kelly and Sassen 1991). In short, neither the Rustbelt nor the Sunbelt has seen uniform economic restructuring.

Third, Latinos are affected by the "global cities" phenomenon, particularly evident in New York and Chicago. This term refers to a particular mix of new jobs and populations and an expansion of both high- and low-paid service jobs (see Sassen-Koob 1984). When large multinational corporations centralize their service functions, upper-level service jobs expand. The growing corporate elite want more restaurants, more entertainment, more clothing, and more care for their homes and children, but these new consumer services usually pay low wages and offer only temporary and part-time work. The new service workers in turn generate their own demand for low-cost goods and services. Many of them are Latino immigrants and they create what Sassen calls a "Third World city... located in dense groupings spread all over the city": this new "city" also provides new jobs (1989, p. 70).

Los Angeles... has experienced many of these patterns. The loss of manufacturing jobs has been far less visible than in New York or Chicago, for although traditional manufacturing declined, until the 1990s high-tech manufacturing did not. Moreover, Los Angeles' international financial and trade functions flourished (Soja 1987). The real difference between Los Angeles on the one hand and New York and Chicago on the other was that more poor people in Los Angeles seemed to be working. In all three cities internationalization had similar consequences for the *structure* of jobs for the poor. More of the immigrants pouring into Los Angeles were finding jobs, while the poor residents of New York and Chicago were not.

Fourth, even though the deindustrialization framework remains of overarching importance in understanding variations in the urban context of Latino poverty, we must also understand that economic restructuring shows many different faces. It is different in economically specialized cities. Houston, for example, has been called "the oil capital of the world," and most of the devastating economic shifts in that city were due to "crisis and reorganization in the world oil-gas industry" (Hill and Feagin 1987, p. 174). Miami is another special case. The economic changes that have swept Miami have little to do with deindustrialization, or with Europe or the Pacific Rim, and much to do with the overpowering influence of its Cuban population, its important "enclave economy," and its "Latino Rim" functions (see Portes and Stepick 1993).

Finally, economic change has a different effect in peripheral areas. Both Albuquerque and Tucson are regional centers in an economically peripheral area. Historically, these two cities served the ranches, farms, and mines of their desert hinterlands. Since World War II, both became military centers, with substantial high-tech defense industrialization. Both cities are accustomed to having a large, poor Latino population, whose poverty is rarely viewed as a crisis. In Tucson, for example, unemployment for Mexican Americans has been low, and there is stable year-round income. But both cities remain marginal to the national economy, and this means that the fate of their poor depends more on local factors.

Laredo has many features in common with other cities along the Texas border, with its substantial military installations, and agricultural and tourist functions. All of these cities have been affected by general swings in the American and Texan economy. These border communities have long been the poorest in the nation, and their largely Mexican American populations have suffered even more from recent economic downturns. They are peripheral to the U.S. economy, but the important point is that their economic well-being is intimately tied to the Mexican economy. They were devastated by the collapse of the peso in the 1980s. They are also more involved than most American cities in international trade in illicit goods, and poverty in Laredo has been deeply affected by smuggling. Though Texas has a long history of discrimination against Mexican Americans, race is not an issue within Laredo itself, where most of the population—elite as well as poor—is of Mexican descent. . . .

THE INFORMAL AND ILLICIT ECONOMIES

The growth of an informal economy is part and parcel of late twentieth-century economic restructuring. Particularly in global cities, a variety of "informal" economic activities proliferates—activities that are small-scale, informally organized, and largely outside government regulations (cf. Portes, Castells, and Benton 1989). Some low-wage reindustrialization, for example, makes use of new arrangements in well-established industries (like home work in the garment industry, as seamstresses take their work home with them). Small-scale individual activities such as street vending and "handyman" house repairs and alterations affect communities in peripheral as well as global cities. . . . These money-generating activities are easily ignored by researchers who rely exclusively on aggregate data sources: they never make their way into the statistics on labor-market participation, because they are "off the books." But they play a significant role in the everyday life of many African American neighborhoods as well as in the barrios.

And, finally, there are illicit activities—most notoriously, a burgeoning drug market. There is not much doubt that the new poverty in the United States has often been accompanied by a resurgence of illicit economic activities. It is important to note that most of the Latino communities . . . have been able to contain or encapsulate such activities so that they do not dominate neighborhood life. But in most of them there is also little doubt that illicit economic activities form an "expanded industry." They rarely provide more than a pittance for the average worker: but for a very small fraction of barrio households they are part of the battery of survival strategies.

Researchers often neglect this aspect of the underclass debate because it is regarded as stigmatizing. However, some . . . make it clear that the neglect of significant income-generating activities curtails our understanding of the full range of survival strategies in poor communities. At the worst (as in Laredo) it means that we ignore a significant aspect of community life, including its ramifications in producing yet more overpolicing of the barrios. Even more important, many of these communities have been able to encapsulate illicit economic activities so that they are less disruptive. This capacity warrants further analysis.

IMMIGRATION

Immigration—both international and from Puerto Rico—is of major significance for poor Latino communities in almost every city in every region of the country. Further, there is every reason to believe that immigration will continue to be important.

First, it has important economic consequences. Immigration is a central feature of the economic life of global cities: for example, Los Angeles has been called the "capital of the Third World" because of its huge Latino and Asian immigration (Rieff 1991). In our sample, those cities most bound to world trends (New York, Los Angeles, Chicago, Houston, and Miami) experienced massive Latino immigration in the 1980s. In the Los Angeles, Houston, and Miami communities . . . immigration is a major factor in the labor market, and the residents of the "second settlement" Puerto Rican communities described in New York and Chicago operate within a context of both racial and ethnic change and of increased Latino immigration. The restructured economy provides marginal jobs for immigrant workers, and wage scales seem to drop for native-born Latinos in areas where immigration is high. This is a more complicated scenario than the simple loss of jobs accompanying Rustbelt deindustrialization. Immigrants are ineligible for most government benefits, are usually highly motivated, and are driven to take even the poorest-paying jobs. They are also more vulnerable to labor-market swings.

These may be construed as rather negative consequences, but in addition, immigrants have been a constructive force in many cities. For example, these authors point to the economic vitality of immigrant-serving businesses. Socially and culturally, there are references . . . to the revival of language and of traditional social controls, the strengthening of networks, and the emergence of new community institutions. Recent research in Chicago focuses on the "hard work" ethos of many Mexican immigrants and the extensive resource base provided by kinship networks, a pattern that is echoed and amplified. . . . Most of Tucson's Chicano poor—not just immigrants—are involved in such helping networks.

Though immigrants have been less important in the peripheral cities of Albuquerque, Laredo, and Tucson, each of these cities is special in some way. Albuquerque has attracted few Mexican immigrants, but it draws on a historical Latino labor pool—English-speaking rural *Manitos*—who are as economically exploitable as are Spanish-speaking immigrants from Mexico. Until recently Tucson was also largely bypassed by most Mexican immigrants. Instead, there is an old, relatively self-contained set of cross-border networks, with well-established pathways of family movement and mutual aid. Similar networks also exist in Laredo. Laredo's location on the border means that many of its workers are commuters—people who work in Laredo but live in Mexico.

In recent years, immigration has not been very significant in most African American communities, and as a consequence it is underemphasized in the underclass debate. It is also often interpreted as wholly negative. This is partly because the positive effects can be understood only by researchers who study immigrant communities themselves, partly because in some places large numbers of immigrants have strained public resources, and partly because immigrants have occasionally become a source of tension among poor minority populations. Though the specific contouring of immigration effects varies from place to place, in each city . . . immigration is a highly significant dimension of Latino poverty, both at the citywide level and also in the neighborhoods. It is an issue of overriding importance for the understanding of Latino poverty, and thus for the understanding of American urban poverty in general. . . .

The concentration of poverty comes about not only because of market forces or the departure of the middle classes for better housing; in Houston, Rodriguez shows that restructuring in real estate had the effect of concentrating poverty. Concentrated poverty can also result from government planning. Chicago's decision decades ago to build a concentration of high-rise housing projects right next to one another is a clear case in point. Another is in New York's largely Latino South Bronx, where the city's ten-year-plan created neighborhoods in which the least enterprising of the poor are concentrated, and in which a set of undesirable "Not-In-My-Back-Yard" institutions, such as drug-treatment clinics and permanent shelters for the homeless, were located. These neighborhoods are likely to remain as pockets of unrelieved poverty for many generations to come (Vergara 1991). It was not industrial decline and the exodus of stable working people that created these pockets: the cities of Chicago and New York chose to segregate their problem populations in permanent buildings in those neighborhoods. . . .

In addition, studies demonstrate that it is not just poverty that gets concentrated. Most immigrants are poor, and most settle in poor communities, thus further concentrating poverty. But, as Rodriguez shows, immigrant communities may be economically, culturally, and socially vital. Social isolation early in the immigration process, he argues, can strengthen group cohesion and lead to community development, rather than to deterioration. Los Angeles portrays institution-building among immigrants in poor communities, and institutional "resilience" characterizes many of the communities . . . especially New York and Chicago. Analysis of poverty in Tucson points to the overwhelming importance of "funds of knowledge" shared in interdependent household clusters. Although a priori it makes sociological sense that concentrated poverty should destroy communities, these studies offer evidence that a different pattern emerges under certain circumstances. To use Grenier and Stepick's term, "social capital" also becomes concentrated.

In short, the concentration of poverty need not plunge a neighborhood into disarray. . . . This line of reasoning raises other issues. If it isn't just demographic shifts that weaken neighborhoods, then what is it? These questions strike at the heart of the underclass debate. The old, rancorous controversy about the usefulness of the "culture of poverty" concept questioned whether the poor adhered to a special set of self-defeating values, and if so, whether those values were powerful enough to make poverty self-perpetuating. That argument faded as research focused more effectively on the situational and structural sources of poverty. We do not intend to revive this controversy. It is all too easy to attribute the differences between Latino and black poverty to "the culture." This line can be invidious, pitting one poor population against another in its insinuation that Latino poverty is somehow "better" than black poverty. (Ironically, this would reverse another outdated contention—i.e., that Latinos are poor *because* of their culture.) . . .

OTHER ASPECTS OF URBAN SPACE . . .

Where a poor neighborhood is located makes a difference.

First, some are targets for "gentrification." This is traditionally viewed as a market process by which old neighborhoods are revitalized and unfortunate poor people displaced. But there is a different perspective. Sassen (1989) argues that gentrification is best understood in the context of restructuring, globalization, and politics. It doesn't happen everywhere . . . gentrification, along with downtown revitalization and expansion, affects Latino neighborhoods in Chicago, Albuquerque, New York, and west side Los Angeles. In Houston, a variant of "gentrification" is documented. Apartment owners who were eager to rent to Latino immigrants when a recession raised their vacancy rates were equally eager to "upgrade" their tenants when the economy recovered and the demand for housing rose once again. Latinos were "gentrified" out of the buildings.

Second, Latinos are an expanding population in many cities, and they rub up against other populations. Most of the allusions to living space center on ethnic frictions accompanying the expansion of Latino areas of residence. Ethnic succession is explicit in Albuquerque and in Chicago. . . . It is implicit in East Los Angeles, with the Mexicanization of Chicano communities, and in Houston, with the immigration of Central Americans to Mexican American neighborhoods and the manipulated succession of Anglos and Latinos. In Albuquerque and East Los Angeles, Latinos are "filling-in" areas of the city, in a late phase of ethnic succession. Ethnic succession is *not* an issue in Laredo because the city's population is primarily of Mexican origin. It is crucial in Miami, where new groups of immigrants are establishing themselves within the Latino community: newer immigrants tend to move into areas vacated by earlier Cuban arrivals, who leave for the suburbs. In Brooklyn, a different kind of urban ecological function is filled by the Puerto Rican barrio—that of an ethnic buffer between African American and Anglo communities. Los Angeles' Westlake area is most strongly affected by its location near downtown: it is intensely involved in both gentrification and problems of ethnic succession. Here the Central Americans displaced a prior population, and, in turn, their nascent communities are pressured by an expanding Koreatown to the west and by gentrification from the north and from downtown.

These details are important in themselves, but they also have implications for existing theories of how cities grow and how ethnic groups become segregated (and segregation is closely allied to poverty). Most such theories take the late nineteenth-century industrial city as a point of departure—a city with a strong central business district and clearly demarcated suburbs. In these models, immigrants initially settle in deteriorating neighborhoods near downtown. Meanwhile, earlier generations of immigrants, their predecessors in those neighborhoods, leapfrog out to "areas of second settlement," often on the edge of the city. . . .

Thus it is no surprise that the "traditional" Rustbelt pattern of ethnic location and ethnic succession fails to appear in most cities discussed in this volume. New Latino immigrants are as likely to settle initially in communities on the edge of town (near the new jobs) as they are to move near downtown; or their initial settlement may be steered by housing entrepreneurs, as in Houston. The new ecology of jobs, housing, and shopping malls has made even the old Rustbelt cities like Chicago less clearly focused on a central downtown business district.

Housing for the Latino poor is equally distinctive. Poor communities in which one-third to one-half of the homes are owner-occupied would seem on the face of it to provide a different ambience from public housing—like the infamous phalanx of projects on Chicago's South Side that form part of Wilson's focus. . . .

Finally, space is especially important when we consider Mexican American communities on the border. Mexican Americans in most border communities have important relationships with kin living across the border in Mexico, and this is certainly the case in Tucson and Laredo. But space is also important in economic matters. Shopping, working, and recreation are conditioned by the proximity of alternative opportunities on both sides of the border. And in Laredo the opportunities for illicit economic transactions also depend on location. The Laredo barrios in which illicit activities are most concentrated are located right on the Rio Grande River, where cross-border transactions are easier.

In sum, when we consider poor minority neighborhoods, we are drawn into a variety of issues that go well beyond the question of how poverty gets concentrated because middle-class families move out. We must look at the role of urban policy in addition to the role of the market. We must look at the factors that promote and sustain segregation. We must look at how housing is allocated, and where neighborhoods are located within cities. And, finally, we must look at how the location of a neighborhood facilitates its residents' activity in licit and illicit market activities.

REFERENCES

Editors' Note: For documentation, see the source note on page 174.

AFL-CIO Industrial Union Department 1986. *The Polarization of America*. Washington, DC: AFL-CIO Industrial Union Department.

Auletta, Ken, 1982. *The Underclass*. New York: Random House.

Barrera, Mario, 1979. *Race and Class in the Southwest*. Notre Dame, IN: University of Notre Dame Press.

Bluestone, Barry, and Bennett Harrison, 1982. *The Deindustrialization of America*. New York: Basic Books.

Chenault, Lawrence Royce, 1938. *The Puerto Rican Migrant in New York*. New York: Columbia University Press.

Clark, Margaret, 1959. *Health in the Mexican American Culture*. Berkeley: University of California Press.

Crawford, Fred, 1961. *The Forgotten Egg*. San Antonio, TX: Good Samaritan Center.

Edmundson, Munro S., 1957. *Los Manitos: A Study of Institutional Values*. New Orleans: Tulane University, Middle American Research Institute.

Ellwood, David T., 1988. *Poor Support: Poverty in the American Family*. New York: Basic Books.

Falcon, Luis, and Douglas Gurak, 1991. "Features of the Hispanic Underclass: Puerto Ricans and Dominicans in New York." Unpublished manuscript.

Fernandez-Kelly, Patricia, and Saskia Sassen, 1991. "A Collaborative Study of Hispanic Women in the Garment and Electronics Industries: Executive Summary." New York: New York University, Center for Latin American and Caribbean Studies.

Galarza, Ernesto, 1965. *Merchants of Labor*. San Jose, CA: The Rosicrucian Press Ltd.

Goldschmidt, Walter, 1947. *As You Sow*. New York: Harcourt, Brace.

Gosnell, Patricia Aran, 1949. *Puerto Ricans in New York City*. New York: New York University Press.

Handlin, Oscar, 1959. *The Newcomers: Negroes and Puerto Ricans*. Cambridge, MA: Harvard University Press.

Hill, Richard Child, and Joe R. Feagin, 1987. "Detroit and Houston: Two Cities in Global Perspective." In Michael Peter Smith and Joe R. Feagin, eds. *In The Capitalist City*, pp. 155–177. New York: Basil Blackwell.

Kluckhohn, Florence, and Fred Strodtbeck, 1961. *Variations in Value Orientations*. Evanston, IL: Row, Peterson.

Leonard, Olen, and Charles Loomis, 1938. *Culture of a Contemporary Rural Community: El Cerito, NM.* Washington, DC: U.S. Department of Agriculture.

Levy, Frank, 1977. "How Big Is the Underclass?" Working Paper 0090-1. Washington, DC: Urban Institute.

Maldonado-Denis, Manuel, 1972. *Puerto Rico: A Sociohistoric Interpretation*. New York: Random House.

Massey, Douglas, and Mitchell Eggers, 1990. "The Ecology of Inequality: Minorities and the Concentration of Poverty." *American Journal of Sociology* 95:1153–1188.

Matza, David, 1966. "The Disreputable Poor." In Reinhardt Bendix and Seymour Martin Lipset, eds. *Class, Status and Power*, pp. 289–302. New York: Free Press.

McCarthy, Kevin, and R.B. Valdez, 1986. *Current and Future Effects of Mexican Immigration in California*. Santa Monica, CA: Rand Corporation.

McWilliams, Carey, 1949. *North From Mexico*. New York: J. B. Lippincott.

Menefee, Seldon, and Orin Cassmore, 1940. *The Pecan Shellers of San Antonio*. Washington: WPA, Division of Research.

Mills, C. Wright, Clarence Senior, and Rose K. Goldsen, 1950. *The Puerto Rican Journey*. New York: Harper.

Montiel, Miguel, 1970. "The Social Science Myth of the Mexican American Family." *El Grito* 3:56–63.

Moore, Joan, 1989. "Is There a Hispanic Underclass?" *Social Science Quarterly* 70:265–283.

Moore, Joan, and Harry Pachon, 1985. *Hispanics in the United States*. Englewood Cliffs, NJ: Prentice Hall.

Morales, Julio, 1986. *Puerto Rican Poverty and Migration: We Just Had to Try Elsewhere*. New York: Praeger.

Morales, Rebecca, 1985. "Transitional Labor: Undocumented Workers in the Los Angeles Automobile Industry." *International Migration Review* 17:570–96.

Morris, Michael, 1989. "From the Culture of Poverty to the Underclass: An Analysis of a Shift in Public Language." *The American Sociologist* 20:123–133.

Muller, Thomas, and Thomas J. Espenshade, 1986. *The Fourth Wave*. Washington, DC: Urban Institute Press.

Murray, Charles, 1984. *Losing Ground*. New York: Basic Books.

Padilla, Elena, 1958. *Up From Puerto Rico*. New York: Columbia University Press.

Perry, David, and Alfred Watkins, 1977. *The Rise of the Sunbelt Cities*. Beverly Hills, CA: Sage.

Portes, Alejandro, Manuel Castells, and Lauren A. Benton, 1989. *The Informal Economy*. Baltimore: Johns Hopkins University Press.

Portes, Alejandro, and Alex Stepick, 1993. *City on the Edge: The Transformation of Miami*. Berkeley: University of California Press.

Rand, Christopher, 1958. *The Puerto Ricans*. New York: Oxford University Press.

Ricketts, Erol, and Isabel V. Sawhill, 1988. "Defining and Measuring the Underclass." *Journal of Policy Analysis and Management* 7:316–325.

Rieff, David, 1991. *Los Angeles: Capital of the Third World*. New York: Simon and Schuster.

Rodriguez, Clara, 1989. *Puerto Ricans: Born in the U.S.A.* Boston: Unwin Hyman.

Romano-V, Octavio I, 1968. "The Anthropology and Sociology of the Mexican Americans." *El Grito* 2:13–26.

Russell, George, 1977. "The American Underclass." *Time Magazine* 110 (August 28):14–27.

Sanchez, George, 1940. *Forgotten People: A Study of New Mexicans*. Albuquerque: University of New Mexico Press.

Sassen, Saskia, 1989. "New Trends in the Sociospatial Organization of the New York City Economy." In Robert Beauregard, ed. *Economic Restructuring and Political Response*. Newberry Park, CA.

Sassen-Koob, Saskia, 1984. "The New Labor Demand in Global Cities." In Michael Smith, ed. *Cities in Transformation*. Beverly Hills, CA: Sage.

Saunders, Lyle, 1954. *Cultural Differences and Medical Care*. New York: Russell Sage Foundation.

Senior, Clarence Ollson, 1965. *Our Citizens from the Caribbean*. New York: McGraw Hill.

Soja, Edward, 1987. "Economic Restructuring and the Internationalization of the Los Angeles Region." In Michael Peter Smith and Joe R. Feagin, eds. *The Capitalist City*, pp. 178–198. New York: Basil Blackwell.

Stevens Arroyo, Antonio M., 1974. *The Political Philosophy of Pedro Abizu Campos: Its Theory and Practice*. Ibero American Language and Area Center. New York: New York University Press.

Sullivan, Mercer L., 1989a. *Getting Paid: Youth Crime and Work in the Inner City*. Ithaca: Cornell University Press.

Taylor, Paul, 1928. *Mexican Labor in the U.S.: Imperial Valley*. Berkeley: University of California Publications in Economics.

———, 1930. *Mexican Labor in the U.S.: Dimit County, Winter Garden District, South Texas*. Berkeley: University of California Publications in Economics.

———, 1934. *An American-Mexican Frontier*. Chapel Hill, NC: University of North Carolina Press.

U.S. Bureau of the Census, 1991b. *The Hispanic Population in the United States: March 1991*. Current Population Reports, Series P-20, No. 455. Washington, DC: U.S. Government Printing Office.

Vaca, Nick, 1970. "The Mexican American in the Social Sciences." *El Grito* 3:17–52.

Vergara, Camilo Jose, 1991. "Lessons Learned: Lessons Forgotten: Rebuilding New York City's Poor Communities." *The Livable City* 15:3–9.

Wagenheim, Kal, 1975. *A Survey of Puerto Ricans on the U.S. Mainland*. New York: Praeger.

Wakefield, Dan, 1959. *Island in the City*. New York: Corinth Books.

Wilson, William Julius, 1987. *The Truly Disadvantaged: The Inner City, the Underclass, and Public Policy*. Chicago: The University of Chicago Press.

———, 1990. "Social Theory and Public Agenda Research: The Challenge of Studying Inner-city Social Dislocations." Paper presented at Annual Meeting of the American Sociological Association.

28

Varieties of Asian Americans

MIN ZHOU

Sometimes it is a burden to be considered a member of the "model minority." Min Zhou discusses issues faced by Asian Americans, including that of lumping all Asians from many diverse cultures into one category and assuming all groups fit the model.

As you read, consider the following:

1. *Why might it be a burden to be considered a member of a model minority?*
2. *What problems result from the label "Asian American"?*
3. *What changes take place for Asian Americans who have been in the United States more than one generation?*
4. *What stereotypes affect the lives of Asian Americans?*

GLOSSARY **Honorary white** Considered "white" in accomplishments, but not allowed certain privileges of whites in the United States. **Model minority** Overcoming barriers and succeeding in United States society.

Privately, few Americans of Asian ancestry would spontaneously identify themselves as Asian, and fewer still as Asian American. They instead link their identities to specific countries of origin, such as China, Japan, Korea, the Philippines, India or Vietnam. In a study of Vietnamese youth in San Diego, for example, 53 percent identified themselves as Vietnamese, 32 percent as Vietnamese American, and only 14 percent as Asian American. But they did not take these labels lightly; nearly 60 percent of these youth considered their chosen identity as very important to them.

Some Americans of Asian ancestry have family histories in the United States longer than many Americans of Eastern or Southern European origin. However, Asian-origin Americans became numerous only after 1970, rising from 1.4 million to 11.9 million (4 percent of the total U.S. population), in 2000. Before 1970, the Asian-origin population was largely made up of Japanese, Chinese and Filipinos. Now, Americans of Chinese and Filipino ancestries are the largest subgroups (at 2.8 million and 2.4 million, respectively), followed by Indians, Koreans, Vietnamese and Japanese (at more than one million). Some 20 other national-origin groups, such as Cambodians, Pakistanis, Laotians, Thai, Indonesians and Bangladeshis, were officially counted in government statistics only after 1980; together they amounted to more than two million Americans in 2000.

The sevenfold growth of the Asian-origin population in the span of 30-odd years is primarily due to accelerated immigration following the Hart-Celler Act of 1965, which ended the national origins quota system, and the historic resettlement of Southeast Asian refugees after the Vietnam War. Currently, about 60 percent of the Asian-origin population is foreign-born (the first generation), another 28 percent are U.S.-born of foreign-born parents (the second generation), and just 12 percent were born to U.S.-born parents (the third generation and beyond).

Unlike earlier immigrants from Asia or Europe, who were mostly low-skilled laborers looking for work, today's immigrants from Asia have more varied backgrounds and come for many reasons, such as to join their families, to invest their money in the U.S. economy, to fill the demand for highly skilled labor, or to escape war, political or religious persecution and economic hardship. For example, Chinese, Taiwanese, Indian, and Filipino Americans tend to be overrepresented among scientists, engineers, physicians and other skilled professionals, but less educated, low-skilled workers are more common among Vietnamese, Cambodian, Laotian, and Hmong Americans, most of whom entered the United States as refugees. While middle-class immigrants are able to start their American lives with high-paying professional careers and comfortable suburban lives, low-skilled immigrants and refugees often have to endure low-paying menial jobs and live in inner-city ghettos.

Asian Americans tend to settle in large metropolitan areas and concentrate in the West. California is home to 35 percent of all Asian Americans. But recently, other states such as Texas, Minnesota and Wisconsin, which historically received few Asian immigrants, have become destinations for Asian American settlement. Traditional ethnic enclaves, such as Chinatown, Little Tokyo, Manilatown, Koreatown, Little Phnom Penh, and Thaitown, persist or have emerged in gateway cities, helping new arrivals to cope with cultural and linguistic difficulties. However, affluent and highly-skilled immigrants tend to bypass inner-city enclaves and settle in suburbs upon arrival, belying the stereotype of the "unacculturated" immigrant. Today, more than half of the Asian-origin population is spreading out in suburbs surrounding traditional gateway cities, as well as in new urban centers of Asian settlement across the country.

Differences in national origins, timing of immigration, affluence and settlement patterns profoundly inhibit the formation of a pan-ethnic identity. Recent arrivals are less likely than those born or raised in the United States to identify as Asian American. They are also so busy settling in that they have little time to think about being Asian or Asian American, or, for that matter, white. Their diverse origins include drastic differences in languages and dialects, religions, cuisines and customs. Many national groups also bring to America their histories of conflict (such as the Japanese colonization of Korea and Taiwan, Japanese attacks on China, and the Chinese invasion of Vietnam).

Immigrants who are predominantly middle-class professionals, such as the Taiwanese and Indians, or predominantly small business owners, such as the Koreans, share few of the same concerns and priorities as those who are predominantly uneducated, low-skilled refugees, such as Cambodians and Hmong. Finally, Asian-origin people living in San Francisco or Los Angeles among many other Asians and self-conscious Asian Americans develop a stronger ethnic identity than those living in predominantly Latin Miami or predominantly European Minneapolis. A politician might get away with calling Asians "Oriental" in Miami but get into big trouble in San Francisco. All of these differences create obstacles to fostering a cohesive pan-Asian solidarity. As Yen Le Espiritu shows, pan-Asianism is primarily a political ideology of U.S.-born, American-educated, middle-class Asians rather than of Asian immigrants, who are conscious of their national origins and overburdened with their daily struggles for survival.

UNDERNEATH THE MODEL MINORITY: "WHITE" OR "OTHER"

The celebrated "model minority" image of Asian Americans appeared in the mid-1960s, at the peak of the civil rights and the ethnic consciousness movements, but before the rising waves of immigration and refugee influx from Asia. Two articles in 1966—"Success Story, Japanese-American Style," by William Petersen in the *New York Times Magazine*, and "Success of One Minority Group in U.S.," by the *US News & World Report* staff—marked a significant departure from how Asian immigrants and their descendants had been traditionally depicted in the media. Both articles congratulated Japanese and Chinese Americans on their persistence in overcoming extreme hardships and discrimination to achieve success, unmatched even by U.S.-born whites, with "their own almost totally unaided effort" and "no help from anyone else." (The implicit contrast to other minorities was clear.) The press attributed their winning wealth and respect in American society to hard work, family solidarity, discipline, delayed gratification, non-confrontation and eschewing welfare.

This "model minority" image remains largely unchanged even in the face of new and diverse waves of immigration. The 2000 U.S. Census shows that Asian Americans continue to score remarkable economic and educational achievements. Their median household income in 1999 was more than $55,000—the highest of all racial groups, including whites—and their poverty rate was under 11 percent, the lowest of all racial groups. Moreover, 44 percent of all Asian Americans over 25 years of age had at least a bachelor's degree, 18 percentage points more than any other racial group. Strikingly, young Asian Americans, including both the children of foreign-born physicians, scientists, and professionals and those of uneducated and penniless refugees, repeatedly appear as high school valedictorians and academic decathlon winners. They also enroll in the freshman classes of prestigious universities in disproportionately large numbers. In 1998, Asian Americans, just 4 percent of the nation's population, made up more than 20 percent of the undergraduates at universities such as Berkeley, Stanford, MIT and Cal Tech. Although some ethnic groups, such as Cambodians, Lao, and Hmong, still trail behind other East and South Asians in most indicators of achievement, they too show significant signs of upward mobility. Many in the media have dubbed Asian Americans the "new Jews." Like the second-generation Jews of the past, today's children of Asian immigrants are climbing up the ladder by way of extraordinary educational achievement.

One consequence of the model-minority stereotypes is that it reinforces the myth that the United States is devoid of racism and accords equal opportunity to all, fostering the view that those who lag behind do so because of their own poor choices and inferior culture. Celebrating "model minorities" can help impede other racial minorities' demands for social justice by pitting minority groups against each other. It can also pit Asian Americans against whites. On the surface, Asian Americans seem to be on their way to becoming white, just like the offspring of earlier European immigrants. But the model-minority image implicitly casts Asian Americans as different from whites. By placing Asian Americans above whites, this image still sets them apart from other Americans, white or nonwhite, in the public mind.

There are two other less obvious effects. The model-minority stereotype holds Asian Americans to higher standards, distinguishing them from average Americans. "What's wrong with being a model minority?" a black student once asked, in a class I taught on race, "I'd rather be in the model minority than in the downtrodden minority that nobody respects." Whether people are in a model minority or a downtrodden minority, they are still judged by standards different from average Americans. Also, the model-minority stereotype places particular expectations on members of the group so labeled, channeling them to specific avenues of success, such as science and engineering. This, in turn, makes it harder for Asian Americans to pursue careers outside these designated fields. Falling into this trap, a Chinese immigrant father gets upset when his son tells him he has changed his major from engineering to English. Disregarding his son's talent for creative writing, such a father rationalizes his concern, "You have a 90 percent chance of getting a decent job with an engineering degree, but what chance would you have of earning income as a writer?" This thinking represents more than typical parental concern; it constitutes the self-fulfilling prophecy of a stereotype.

The celebration of Asian Americans rests on the perception that their success is unexpectedly high. The truth is that unusually many of them, particularly among the Chinese, Indians and Koreans, arrive as middle-class or upper middle-class immigrants. This makes it easier for them and their children to succeed and regain their middle-class status in their new homeland. The financial resources that these immigrants bring also subsidize ethnic businesses and services, such as private after-school programs. These, in turn, enable even the less fortunate members of the groups to move ahead more quickly than they would have otherwise.

NOT SO MUCH BEING "WHITE" AS BEING AMERICAN

Most Asian Americans seem to accept that "white" is mainstream, average and normal, and they look to whites as a frame of reference for attaining higher social position. Similarly, researchers often use non-Hispanic whites as the standard against which other groups are compared, even though there is great diversity among whites, too. Like most immigrants to the United States, Asian immigrants tend to believe in the American Dream and measure their achievements materially. As a Chinese immigrant said to me in an interview, "I hope to accomplish nothing but three things: to own a home, to be my own boss, and to send my children to the Ivy League." Those with sufficient education, job skills and money manage to move into white middle-class suburban neighborhoods immediately upon arrival, while others work intensively to accumulate enough savings to move their families up and out of inner-city ethnic enclaves. Consequently, many children of Asian ancestry have lived their entire childhood in white communities, made friends with mostly white peers, and grown up speaking only English. In fact, Asian Americans are the most acculturated non-European group in the United States. By the second generation, most have lost fluency in their parents' native languages (see "English-Only Triumphs, but the Costs are High," *Contexts*, Spring 2002). David Lopez finds that in Los Angeles, more than three-quarters of second-generation Asian Americans (as opposed to one-quarter of second-generation Mexicans) speak only English at home. Asian Americans also intermarry extensively with whites and with members of other minority groups. Jennifer Lee and Frank Bean find that more than one-quarter of married Asian Americans have a partner of a different racial background, and 87 percent of those marry whites; they also find that 12 percent of all Asian Americans claim a multiracial background, compared to 2 percent of whites and 4 percent of blacks.

Even though U.S.-born or U.S.-raised Asian Americans are relatively acculturated and often intermarry with whites, they may be more ambivalent about becoming white than their immigrant parents. Many only cynically agree that "white" is synonymous with "American." A Vietnamese high school student in New Orleans told me in an interview, "An American is white. You often hear people say, hey, so-and-so is dating an 'American.' You know she's dating a white boy. If he were black, then people would say he's black." But while they recognize whites as a frame of reference, some reject the idea of becoming white themselves: "It's not so much being white as being American," commented a Korean-American student in my class on the new second generation. This aversion to becoming white is particularly common among second-generation college students who have taken ethnic studies courses, and among Asian-American community activists. However, most of the second generation continues to strive for the privileged status associated with whiteness, just like their parents. For example, most U.S.-born or U.S.-raised Chinese-American youth end up studying engineering, medicine, or law in college, believing that these areas of study guarantee a middle-class life.

Second-generation Asian Americans are also more conscious of the disadvantages associated with being nonwhite than their parents, who as immigrants tend to be optimistic about overcoming the disadvantages of this status. As a Chinese-American woman points out from her own experience, "The truth is, no matter how American you think you are or try to be, if you have almond-shaped eyes, straight black hair, and a yellow complexion, you are a foreigner by default.... You can certainly be as good as or even better than whites, but you will never become accepted as white." This remark echoes a commonly-held frustration among second-generation, U.S.-born Asians who detest being treated as immigrants or foreigners. Their experience suggests that whitening has more to do with the beliefs of white America, than with the actual situation of Asian Americans. Speaking perfect English, adopting mainstream cultural values, and even intermarrying members of the dominant group may help reduce this "otherness" for particular individuals, but it has little effect on the group as a whole. New stereotypes can emerge and unwhiten Asian Americans, no matter how "successful" and "assimilated" they have become. For example, Congressman David Wu once was invited by the Asian-American employees of the U.S. Department of Energy to give a speech in celebration of Asian-American Heritage Month. Yet, he and his Asian-American staff were not allowed into the department building, even after presenting their congressional Identification, and were repeatedly asked about their citizenship and country of origin. They were told that this was standard procedure for the Department of Energy and that a congressional ID card was not a reliable document. The next day, a congressman of Italian descent was allowed to enter the same building with his congressional ID, no questions asked.

The stereotype of the "honorary white" or model minority goes hand-in-hand with that of the "forever foreigner." Today, globalization and U.S.–Asia relations, combined with continually high rates of immigration, affect how Asian Americans are perceived in American society. Many historical stereotypes, such as the "yellow peril" and "Fu Manchu" still exist in contemporary American life, as revealed in such highly publicized incidents as the murder of Vincent Chin, a Chinese American mistaken for Japanese and beaten to death by a disgruntled while auto worker in the 1980s; the trial of Wen Ho Lee, a nuclear scientist suspected of spying for the Chinese government in the mid-1990s; the 1996 presidential campaign finance scandal, which implicated Asian Americans in funneling foreign contributions to the Clinton campaign; and most recently, in 2001, the Abercrombie & Fitch t-shirts that depicted Asian cartoon characters in stereotypically negative ways, with slanted eyes, thick glasses and heavy Asian accents. Ironically, the ambivalent, conditional nature of their acceptance by whites prompts many Asian Americans to organize pan-ethnically to fight back—which consequently heightens their racial distinctiveness. So becoming white or not is beside the point. The bottom line is: Americans of Asian ancestry still have to constantly prove that they truly are loyal Americans.

REFERENCES

Liu, Eric. 1988. *The Accidental Asian.* New York: Random House.

29

"This Is a White Country"

The Racial Ideology of the Western Nations of the World-System

EDUARDO BONILLA-SILVA AND MARY HOVSEPIAN

Racial inequalities and racism are found in many countries. Bonilla-Silva and Hovsepian provide evidence that the world-system of capitalist economies, including rich and poor nations, has resulted in increased racial tensions. As citizens of poorer countries seek job opportunities in the Western world, for instance, racist backlash increases.

Note the following as you read this selection:

1. *What is causing a racial backlash between rich and poor countries?*
2. *What is the result of the globalization of race relations?*
3. *What is the "new racism" and how does it differ from past practices?*
4. *What can be done to reduce racial backlash in the Western world?*

GLOSSARY **Racism** The belief that one category of people is superior or inferior to another. **GNP** A nation's gross national product. **Neonazis** Modern-day white supremacist groups.

INTERNATIONAL CONTEXT: THE INTERNATIONALIZATION OF THE ECONOMY AND THE GLOBALIZATION OF RACE RELATIONS

The national capitalist economies of the world have formed a "world-system" for over 600 years (Braudel 1979; Wallerstein 1996). The extension of that system into Africa, the Americas, and Asia in the 16th century involved the racialization of the peoples of the entire world (Balibar and Wallerstein 1991; Rodney 1981). In order to dominate the "new world," European nations developed a structure of knowledge-meaning that created the notion of the "West" (Hopkins and Wallerstein 1996). This intellectual construction facilitated the expansion of the world-system by *racializing* the inhabitants of peripheral and core nations (Rodney 1981). The concept of the West crystallized a set of binary oppositions that defined the peoples of Western and of non-Western nations: human/subhuman, developed/underdeveloped, civilized/barbarian, rational/instinctive, Christian/heathen, superior/inferior, and clean/unclean (Markus 1994). By defining non-Western nations in this fashion, core nations were able to conquer, exploit, and massacre Indian, African, and Asiatic peoples without much guilt and to use their natural resources to advance their own social, economic, and political interests—including the development of democratic regimes

SOURCE: From *Sociological Footprints,* 8th ed., 2000. Used by permission of the author.

with extensive citizenship rights for all (white) citizens (Berkhoffer 1979; Gunder-Frank 1978; Hopkins and Wallerstein 1996; Rodney 1981).

This Western discourse was not—and is not—just a set of ideas revolving in the heads of Europeans. This discourse was an essential component in the structuration of various kinds of social relations of domination and subordination between "Western" and non-Western peoples, between whites and nonwhites in the world-system (Balibar and Wallerstein 1991; Bonilla-Silva 1997; Spoonley 1988). Racism (racial ideology), as I have suggested elsewhere, is not a free-floating ideology. Racism is always anchored in real practices and it reinforces social relations among racialized subjects in a social order, that is, it supports a racialized social structure (Bonilla-Silva 1997). Thus, for example, the racial ideology of Canada, Australia, and the United States is the direct product of their own racial situations. Even Western countries that did not have historical racial minorities, such as the Netherlands, France, or England, established racial structures in their colonies which have shaped the way in which they have dealt with "colonial immigrants" and other immigrants of color. For instance, although France had by 1930 the highest level of foreigners of any country in the world with seven percent, Arabs, who were neither the largest immigrant group nor the last arrivals, were the object of the most severe antipathies and found themselves at the bottom of the occupational structure (Stora 1996).

Since the mid-1960s, the capitalist world-system has experienced a systemic transformation or, properly speaking, a crisis, that has produced a dramatic restructuration, the famous "globalization" that we hear about almost every day (Amin 1992). The central features of this transformation are the "decline in the importance of territorially based mass production, the globalization of finance and technology, and the increased specialization and diversity of markets" (Kaldor 1996). Each of these elements is a result of the serious world-systemic crisis of accumulation in the late 1960s and early 1970s that produced drastic shifts in the loci of production (from center to peripheries), investment (from productive to financial), and the countries spending a significant portion of their GNP on military expenditures (by incorporating peripheral and semi-peripheral nation-states as central actors in the military race). Although advocates of capitalism interpret these various changes as progressive and speak of a "global village," this new stage in the world-system should be characterized as "the empire of chaos" (Amin 1992).

The chaos produced by the restructuration of the world-system has had local (plant relocations), national (downsizing of the labor force of large multinational companies in the core and "shock therapies" in the periphery), and international repercussions (NAFTA, new world-level economic and political arrangements, etc.). The dislocations caused by these changes and labor recruitment policies by some core nation-states have led to monumental migrations of people from the Third World into core states and the deterioration of the status of workers in the Western world (Cohen 1997). Although a substantial part of this migration is legal and even sponsored by the core states, increasingly since the 1970s the migration has been illegal (Wallerstein 1996).

This new international order has led to the globalization of race and race relations and the intensification and diversification of the numbers of racial Others in the Western world. Although race has fractured countries such as the United States, Australia, New Zealand, and Canada since their inception, it was until recently a marginal social category in most Western nations. Today, as a direct result of the international movement of peoples, all Western nations have interiorized the Other, colonial and otherwise (Winant 1994). In European nations such as Luxembourg, Belgium, Austria, the Netherlands, France, and Germany, the geographical distance between the "uncivilized" and the "civilized" has been "bridged" through what Balibar calls the "interiorization of the exterior" (Balibar and Wallerstein 1991). Accordingly, today immigrants and minorities of color in Europe constitute anywhere between 1.4 percent of the population, as in Italy, to 27.5 percent, as in Luxembourg (Castles and Miller 1993: 80).

Although many analysts conceive these immigrants as basically workers who have been racialized as an "underclass" (Castles and Miller 1993; Cohen 1997; Loomis 1990) I contend that the racialized character of their experience is deeper and in line with 500 years of Western history (Potts 1990; Jayasuriya 1996). For example, in England, although European (white) workers were viewed as easily assimilable, a clear stigma was attached to Caribbean workers whose absorption into the social body was deemed "very difficult." In France, even before the development of the fascist National Front Party, French workers had racist views and feelings toward "black" (Algerians and Caribbean) workers and were among the first to oppose immigration. Finally, since immigration is not a new phenomenon in these countries, and in many, a substantial proportion of the immigrants are white (two-thirds of those in Europe and most of those in England), the "immigrants" that matter are those defined as "black," "non-Western," "unchristian." Accordingly, for example, although Belgium has over half a million French and Italian "foreigners," it targets its 250,000 Arabs and the "blacks" as the objects of scapegoating. It is also significant that studies of the various "immigrants" show that darker immigrants (Caribbeans, Arabs, and Southern Europeans) are viewed and treated much worse than white immigrants. In England, although immigration restrictions were imposed on all groups, political leaders have said that immigration from "Canada, Australia, and New Zealand formed no part of the [immigration] problem" (Saggar 1992: 105) and that earlier migrations of Irish, in contrast to those of Jews and blacks (19th century until 1960), did not produce major reactions from the body politic (Solomos 1989).

Despite the different legal status of these people of color in Western nations (guest workers, asylum seekers, "aliens," or "citizens"), they have a number of similarities. First, in economic terms, all experience a racialized class status characterized by segmented labor market experiences—even segmentation in middle class occupations, overrepresentation in manual and "underclass" locations in the class structure, and significantly higher levels of unemployment (Berrier 1985; Castles and Miller 1993). They also experience very little occupational mobility even among second-generation "immigrants." Second, they tend to live in ethnic quarters or ghettos and are more likely to rent rather than to own their houses (Loomis 1990). This is partly due to discrimination in the housing markets (Loomis 1990). Finally, all people of color in Europe (Turks, Arabs, Native peoples, blacks from the Caribbean and Africa, etc.), whether immigrant or not, experience as "expressive exploitation" or the psychological aspects of depreciation—derogatory attitudes, stereotyping and related behavior, and racially motivated violence. In short, people of color in the historically white countries of the West experience a status and are treated as second-class citizens, a status that resembles that of the historical racial minorities in Western nations such as the United States, Canada, Australia, and New Zealand. . . .

CONCLUSIONS

In the postmodern world no one is racist except for Nazis and neonazis and members of white supremacist groups. Yet racial minorities and immigrants of color are experiencing a racial backlash all over the Western world. That backlash is evident in attacks on affirmative action-type policies, the growth in racial violence, the increase in electoral support for populist racial parties, and the move to the right by mainstream parties on racially-perceived matters such as immigration. Faced with economic insecurity, restructuring, transnationalism, and new political alignments, whites in the Western world are struggling—ideologically and practically—to maintain what they regard as their "rights" to cultural, social, political, economic, and psychological advantages as white, "civilized," and "Christian" citizens over racial minorities, immigrants, or any representative of the Other.

The apparent contradiction between a racial backlash and a Western world that pretends to be cosmopolitan, multicultural, and raceless is explained by the fact that contemporary racial ideology combines abstract and technocratic liberalism with ethnonational and culturalist elements. Laissez-faire racism, which ideologically equalizes the races ("We are all equal!") although in fact they remain unequal, provides the ammunition for whites to feel *moral indignation, anger, resentment,* and even *hate* toward minorities and the programs viewed as providing "preferential" treatment to them. Therefore, this new racial ideology allows whites in the West to defend their racial privilege without appearing to be "racist." Contemporary racial struggle is waged with a new racial language and new racial ideas. Instead of the biologically based racism of the past, the new racial ideology allows even racists such as Enoch Powell to express racial resentment—evident in statements such as the one below—in a way that is acceptable to most whites in the West.

> The spectacle which I cannot help seeing . . . is that of Britain which has lost, quite suddenly, in the space of less than a generation, all consciousness and conviction of being a nation: the web which binds it to its past has been torn asunder, and what has made the spectacle the more impressive has been the indifference, not to say levity, with which the change has been greeted (Enoch Powell's statement to *The Guardian*, cited in Saggar 1992: 176).

This new racism is not a hangover from the past, an articulation of the New Right, a simple case of scapegoating, or something affecting only workers. I suggest that the new racism is world-systemic and affects all Western nations although its specific articulations vary by locality. The reason why racism in all Western nations has a similar macro-racial discourse is because these nations share a history of racial imperialism and the notion of the West, have real but differing racial structures (Bonilla-Silva 1997), and have a significant presence of the Other, either through immigration, as in most European nations, or through their history of constitution as nations, as in New World nations. . . .

Although anti-racist organizations have surfaced in all Western nations (e.g., *Lichterketten* in Germany, SOS Racism in Britain, etc.), they have not been able to mount a counteroffensive rooted in the recognition of the materiality of racialized discourse and behavior. In too many places anti-racist campaigns have been highly ritualistic (candlelights, commercials, etc.) or very narrowly defined (against certain fascist groups) with little concern for developing a broader political agenda. Unless anti-racist organizations understand the centrality and meaning of race and the new racism, they will not be able to develop a progressive agenda around a reconceptualized notion of citizenship that includes both the idea of equality of rights and the equality of status (Jayasuriya 1996; Tlati 1996). Failure to do so, regardless of talks about racial reconciliation (as President Clinton has proposed in the United States), liberal views on "cosmopolitanism" (as many Europeans suggest), or programs based on an abstract liberal universalism, will maintain people of color in the West as denizens.

REFERENCES

Editors' Note: Complete notes can be found in the article in *Sociological Footprints*, 8th ed., 2000.

Amin, Samir. 1992. Empire of Chaos. New York: Monthly Review Press.

Balibar, Etienne, and Immanuel Wallerstein. 1991. *Race, Nation, Class: Ambiguous Identities*. London and New York: Verso.

Berkhoffer, Robert F. Jr., 1979. *The White Man's Indian*. New York: Vintage Books.

Berrier, Robert J. 1985. "The French Textile Industry: A Segmented Labor Market." In *Guests Come to Stay: The Effects of European Labor Migration on Sending and Receiving Countries*, edited by Rosemarie Rogers, pp. 51–68. Boulder and London: Westview Press.

Bonilla-Silva, Eduardo. 1997. "Rethinking Racism: Toward a Structural Interpretation." *American Sociological Review*, Vol. 62, No. 3, pp. 465–480.

Braudel, Fernand. 1979. *The Perspective of the World: Civilization and Capitalism, 15th–18th Century*, Vol. III. New York: Harper & Row.

Castles, Stephen. 1994. "Democracy and Multicultural Citizenship. Australian Debates and their Relevance for Western Europe." In *From Aliens to Citizens: Redefining the Status of Immigrants in Europe*, edited by Rainer Baubock, pp. 3–27. Germany: Avebury.

Castles, Stephen, and Mark J. Miller. 1993. *The Age of Migration: International Population Movements in the Modern World*. Hong Kong: MacMillan.

Cohen, Robin. 1997. *Global Diasporas: An Introduction*. Seattle: University of Washington Press.

Hopkins, Terrence K., and Immanuel Wallerstein. 1996. "The World System: Is There a Crisis?" In *The Age of Transition: Trajectory of the World-System, 1945–2025*, pp. 1–12. Lechhardt, Australia: Pluto Press.

Gunder-Frank, Andre. 1978. *World Accumulation: 1492–1789*. New York: Monthly Review Press.

Jayasuriya, Laksiri. 1996. "Immigration and Settlement in Australia: An Overview and Critique of Multiculturalism." In *Immigration and Integration in Post-Industrial Societies: Theoretical Analysis and Policy-Related Research*, edited by Naomi Cameron, pp. 206–226. London: St. Martin's Press.

Kaldor, Mary. 1996. "Cosmopolitanism Versus Nationalism: The New Divide." In *Europe's New Nationalism: States and Minorities in Conflict*, edited by Richard Caplan and John Feffer, pp. 42–58. New York: Oxford University Press.

Loomis, Terrence. 1990. *Pacific Migrant Labour, Class and Racism in New Zealand: Fresh Off the Boat*. Aldershot, England: Avebury.

Markus, Andrew. 1994. *Australian Race Relations, 1788–1993*. St. Leonards, Australia: Allen & Unwin.

Potts, Lydia. 1990. *The World Labor Market: A History of Immigration*. London: Zed Books.

Rodney, Walter. 1981 [1972]. *How Europe Underdeveloped Africa*. Washington, D.C: Howard University Press.

Saggar, Shamit. 1992. *Race and Politics in Britain*. London: Harvester/Wheatsheaf.

Schoenbaum, David, and Elizabeth Pond. 1996. *The German Question and Other German Questions*. New York: St. Martin's Press.

Solomos, John. 1989. *Race and Racism in Contemporary Britain*. London: MacMillan.

Spoonley, Paul. 1988. *Racism and Ethnicity*. Auckland, New Zealand: Oxford University Press.

Stora, Benjamin. 1996. "Locate, Isolate, Place under Surveillance: Algerian Migration to France in the 1930s." In *Franco-Arab Encounters: Studies in Memory of David C. Gordon*, edited by L. Carl Brown and Matthew S. Gordon, pp. 373–391. Beirut, Lebanon: American University of Beirut.

Tlati, Soraya. 1996. "French Nationalism and the Issue of North African Immigration." In *Franco-Arab Encounters: Studies in Memory of David C. Gordon*, edited by L. Carl Brown and Matthew S. Gordon, pp. 392–414. Beirut, Lebanon: American University of Beirut.

Wallerstein, Immanuel. 1996. "The Global Picture, 1945–90." In *The Age of Transition: Trajectory of the World-System, 1945–2025*, edited by Terence K. Hopkins and Immanuel Wallerstein, pp. 209–225. Lechhardt, Australia: Pluto Press.

Winant, Howard. 1994. *Racial Conditions: Politics, Theory, Comparisons*. Minneapolis, Minnesota: University of Minnesota Press.

Zainu'ddin, Ailsa. 1968. *A Short History of Indonesia*. Melbourne, Australia: Cassell.

30

Because She Looks like a Child

KEVIN BALES

Finally, Kevin Bales discusses a topic that crosses race, class and *gender lines—sex trafficking. Who are the targets? Poor, young minority women who are helpless to resist enslavement. They are bought by traders and sold by poor families, or sometimes just kidnapped. The experiences of these young women are horrific and illustrate the deep divisions in humanity.*

Think about the following ideas as you read this selection:

1. *Who are the girls engaged in the sex trade who serve Thai and foreign men? What are their backgrounds?*
2. *How can families and governments allow sex trafficking to happen? Who is benefitting and how?*
3. *What should the world community do about cross-border trafficking, if anything?*
4. *Are you aware of any trafficking in your area?*

GLOSSARY **Brothel** Place where men pay for sex with prostitutes (in this case, often sex slaves).

When Siri wakes it is about noon.[1] In the instant of waking she knows exactly who and what she has become. As she explained to me, the soreness in her genitals reminds her of the fifteen men she had sex with the night before. Siri is fifteen years old. Sold by her parents a year ago, she finds that her resistance and her desire to escape the brothel are breaking down and acceptance and resignation are taking their place.

In the provincial city of Ubon Ratchathani, in northeastern Thailand, Siri works and lives in a brothel. About ten brothels and bars, dilapidated and dusty buildings, line the side street just around the corner from a new Western-style shopping mall. Food and noodle vendors are scattered between the brothels. The woman behind the noodle stall outside the brothel where Siri works is also a spy, warder, watchdog, procurer, and dinner lady to Siri and the other twenty-four girls and women in the brothel.

The brothel is surrounded by a wall, with iron gates that meet the street. Within the wall is a dusty yard, a concrete picnic table, and the ubiquitous spirit house, a small shrine that stands outside all Thai buildings. A low door leads into a windowless concrete room that is thick with the smell of cigarettes, stale beer, vomit, and sweat. This is the "selection" room (*hong du*). On one side of the room are stained and collapsing tables and booths; on the other side is a narrow elevated platform with a bench that runs the length of the room. Spotlights pick out this bench, and at night the girls and women sit here under the

SOURCE: Bales, Kevin. "Because She Looks like a Child." In Barbara Ehrenreich and Arlie Russell Hochschild, eds. *Global Woman: Nannies, Maids, and Sex Workers in the New Economy*. New York: Henry Holt and Company. 2002.

glare while the men at the tables drink and choose the one they want.

Passing through another door, at the far end of the bench, the man follows the girl past a window, where a bookkeeper takes his money and records which girl he has selected. From there he is led to the girl's room. Behind its concrete front room, the brothel degenerates even further, into a haphazard shanty warren of tiny cubicles where the girls live and work. A makeshift ladder leads up to what may have once been a barn. The upper level is now lined with doors about five feet apart, which open into rooms of about five by seven feet that hold a bed and little else.

Scraps of wood and cardboard separate one room from the next, and Siri has plastered her walls with pictures of teenage pop stars cut from magazines. Over her bed, as in most rooms, there also hangs a framed portrait of the king of Thailand; a single bare lightbulb dangles from the ceiling. Next to the bed a large tin can holds water; there is a hook nearby for rags and towels. At the foot of the bed, next to the door, some clothes are folded on a ledge. The walls are very thin, and everything can be heard from the surrounding rooms; a shout from the bookkeeper echoes through all of them, whether their doors are open or closed.

After rising at midday, Siri washes herself in cold water from the single concrete trough that serves the brothel's twenty-five women. Then, dressed in a T-shirt and skirt, she goes to the noodle stand for the hot soup that is a Thai breakfast. Through the afternoon, if she does not have any clients, she chats with the other girls and women as they drink beer and play cards or make decorative handicrafts together. If the pimp is away the girls will joke around, but if not they must be constantly deferential and aware of his presence, for he can harm them or use them as he pleases. Few men visit in the afternoon, but those who do tend to have more money and can buy a girl for several hours if they like. Some will even make appointments a few days in advance.

At about five, Siri and the other girls are told to dress, put on their makeup, and prepare for the night's work. By seven the men will be coming in, purchasing drinks, and choosing girls; Siri will be chosen by the first of the ten to eighteen men who will buy her that night. Many men choose Siri because she looks much younger than her fifteen years. Slight and round faced, dressed to accentuate her youth, she could pass for eleven or twelve. Because she looks like a child, she can be sold as a "new" girl at a higher price, about $15, which is more than twice that charged for the other girls.

Siri is very frightened that she will get AIDS. Long before she understood prostitution she knew about HIV, as many girls from her village returned home to die from AIDS after being sold into the brothels. Every day she prays to Buddha, trying to earn the merit that will preserve her from the disease. She also tries to insist that her clients use condoms, and in most cases she is successful, because the pimp backs her up. But when policemen use her, or the pimp himself, they will do as they please; if she tries to insist, she will be beaten and raped. She also fears pregnancy, but like the other girls she receives injections of the contraceptive drug Depo-Provera. Once a month she has an HIV test. So far it has been negative. She knows that if she tests positive she will be thrown out to starve.

Though she is only fifteen, Siri is now resigned to being a prostitute. The work is not what she had thought it would be. Her first client hurt her; and at the first opportunity she ran away. She was quickly caught, dragged back, beaten, and raped. That night she was forced to take on a chain of clients until the early morning. The beatings and the work continued night after night, until her will was broken. Now she is sure that she is a very bad person to have deserved what has happened to her. When I comment on how pretty she looks in a photograph, how like a pop star, she replies, "I'm no star; I'm just a whore, that's all." She copes as best she can. She takes a dark pride in her higher price and the large number of men who choose her. It is the adjustment of the concentration camp, an effort to make sense of horror.

In Thailand prostitution is illegal, yet girls like Siri are sold into sex slavery by the thousands. The brothels that hold these girls are but a small part of a much wider sex industry. How can this wholesale trade in girls continue? What keeps it working?

The answer is more complicated than we might think. Thailand's economic boom and its social acceptance of prostitution contribute to the pressures that enslave girls like Siri.

RICE IN THE FIELD. FISH IN THE RIVER. DAUGHTERS IN THE BROTHEL.

Thailand is blessed with natural resources and sufficient food. The climate is mild to hot, there is dependable rain, and most of the country is a great plain, well watered and fertile. The reliable production of rice has for centuries made Thailand a large exporter of grains, as it is today. Starvation is exceedingly rare in its history, and social stability very much the norm. An old and often-repeated saying in Thai is "There is always rice in the fields and fish in the river." And anyone who has tried the imaginative Thai cuisine knows the remarkable things that can be done with those two ingredients and the local chili peppers.

One part of Thailand that is not so rich in necessities of life is the mountainous north. In fact, that area is not Thailand proper; originally the kingdom of Lanna, it was integrated into Thailand only in the late nineteenth century. The influence of Burma here is very strong—as are the cultures of the seven main hill tribes, which are distinctly foreign to the dominant Thai society. Only about a tenth of the land of the north can be used for agriculture, though what can be used is the most fertile in the country. The result is that those who control good land are well-off; those who live in the higher elevations, in the forest, are not. In another part of the world this last group might be called hillbillies, and they share the hardscrabble life of mountain dwellers everywhere.

The harshness of this life stands in sharp contrast to that on the great plain of rice and fish. Customs and culture differ markedly as well, and one of those differences is a key to the sexual slavery practiced throughout Thailand today. For hundreds of years many people in the north, struggling for life, have been forced to view their own children as commodities. A failed harvest, the death of a key breadwinner, or any serious debt incurred by a family might lead to the sale of a daughter (never a son) as a slave or servant. In the culture of the north it was a life choice not preferred but acceptable and one that was used regularly. In the past these sales fed a small, steady flow of servants, workers, and prostitutes south into Thai society.

ONE GIRL EQUALS ONE TELEVISION

The small number of children sold into slavery in the past has become a flood today. This increase reflects the enormous changes in Thailand over the past fifty years as the country has gone through the great transformation of industrialization—the same process that tore Europe apart over a century ago. If we are to understand slavery in Thailand, we must understand these changes as well, for like so many other parts of the world, Thailand has always had slavery, but never before on this scale.

The economic boom of 1977 to 1997 had a dramatic impact on the northern villages. While the center of the country, around Bangkok, rapidly industrialized, the north was left behind. Prices of food, land, and tools all increased as the economy grew, but the returns for rice and other agriculture were stagnant, held down by government policies guaranteeing cheap food for factory workers in Bangkok. Yet visible everywhere in the north is a flood of consumer goods—refrigerators, televisions, cars and trucks, rice cookers, air conditioners—all of which are extremely tempting. Demand for these goods is high as families try to join the ranks of the prosperous. As it happens, the cost of participating in this consumer boom can be met from an old source that has become much more profitable: the sale of children.

In the past, daughters were sold in response to serious family financial crises. Under threat of losing its mortgaged rice fields and facing destitution, a family might sell a daughter to redeem its debt, but for the most part daughters were worth about as much at home as workers as they would realize when sold. Modernization and economic growth have changed all that. Now parents feel a great pressure to buy consumer goods that were unknown even twenty years ago; the sale of a daughter might easily finance a new television set. A recent survey in the northern provinces found that of the families who sold their daughters, two-thirds could afford not to do so but instead preferred to buy color televisions and video equipment.[2] And from the perspective of parents who are willing to sell their children, there has never been a better market.

The brothels' demand for prostitutes is rapidly increasing. The same economic boom that feeds consumer demand in the northern villages lines the pockets of laborers and workers in the central plain. Poor economic migrants from the rice fields now work on building sites or in new factories earning many times what they did on the land. Possibly for the first time in their lives, these laborers can do what more well-off Thai men have always done: go to a brothel. The purchasing power of this increasing number of brothel users strengthens the call for northern girls and supports a growing business in their procurement and trafficking.

Siri's story was typical. A broker, a woman herself from a northern village, approached the families in Siri's village with assurances of well paid work for their daughters. Siri's parents probably understood that the work would be as a prostitute, since they knew that other girls from their village had gone south to brothels. After some negotiation they were paid 50,000 baht (US$2,000) for Siri, a very significant sum for this family of rice farmers.[3] This exchange began the process of debt bondage that is used to enslave the girls. The contractual arrangement between the broker and the parents requires that this money be paid by the daughter's labor before she is free to leave or is allowed to send money home. Sometimes the money is treated as a loan to the parents, the girls being both the collateral and the means of repayment. In such cases the exorbitant interest charged on the loan means there is little chance that a girl's sexual slavery will ever repay the debt.

Siri's debt of 50,000 baht rapidly escalated. Taken south by the broker, Siri was sold for 100,000 baht to the brothel where she now works. After her rape and beating Siri was informed that the debt she must repay to the brothel equaled 200,000 baht. In addition, Siri learned of the other payments she would be required to make, including rent for her room, at 30,000 baht per month, as well as charges for food and drink, fees for medicine, and fines if she did not work hard enough or displeased a customer.

The total debt is virtually impossible to repay, even at Siri's higher rate of 400 baht. About 100 baht from each client is supposed to be credited to Siri to reduce her debt and pay her rent and other expenses; 200 goes to the pimp and the remaining 100 to the brothel. By this reckoning, Siri must have sex with three hundred men a month just to pay her rent, and what is left over after other expenses barely reduces her original debt. For girls who can charge only 100 to 200 baht per client, the debt grows even faster. This debt bondage keeps the girls under complete control as long as the brothel owner and the pimp believe they are worth having. Violence reinforces the control and any resistance earns a beating as well as an increase in the debt. Over time if the girl becomes a good and cooperative prostitute, the pimp may tell her she has paid off the debt and allow her to send small sums home. This "paying off" of the debt usually has nothing to do with an actual accounting of earnings but is declared at the discretion of the pimp, as a means to extend the brothel's profits by making the girl more pliable. Together with rare visits home, money sent back to the family operates to keep her at her job.

Most girls are purchased from their parents, as Siri was, but for others the enslavement is much more direct. Throughout Thailand agents travel to villages, offering work in factories or as domestics. Sometimes they bribe local officials to vouch for them, or they befriend the monks at the local temple to gain introductions. Lured by the promise of good jobs and the money that the daughters will send back to the village, the deceived families dispatch their girls with the agent, often paying for the privilege. Once they arrive in a city, the girls are sold to a brothel, where they are raped, beaten, and locked in. Still other girls are simply kidnapped. This is especially true of women and children who have come to visit relatives in Thailand from Burma or Laos. At bus and train stations, gangs watch for women and children who can be snatched or drugged for shipment to brothels.

Direct enslavement by trickery or kidnapping is not really in the economic interest of the brothel owners. The steadily growing market for prostitutes, the loss of girls to HIV infection, and the especially strong demand for younger and younger girls make it necessary for brokers and brothel owners to cultivate village families so that they can buy more daughters as they come of age. In Siri's case this means letting her maintain ties with her family and ensuring that after a year or so she send a monthly postal order for 10,000 baht to her parents. The monthly payment is a good investment, since it encourages Siri's parents to place their other daughters in the brothel as well. Moreover, the young girls themselves become willing to go when their older sisters and relatives returning for holidays bring stories of the rich life to be lived in the cities of the central plain. Village girls lead a sheltered life, and the appearance of women only a little older than themselves with money and nice clothes is tremendously appealing. They admire the results of this thing called prostitution with only the vaguest notion of what it is. Recent research found that young girls knew that their sisters and neighbors had become prostitutes, but when asked what it means to be a prostitute their most common answer was "wearing Western clothes in a restaurant."[4] Drawn by this glamorous life, they put up little opposition to being sent away with the brokers to swell an already booming sex industry.

By my own conservative estimate there are perhaps thirty-five thousand girls like Siri enslaved in Thailand. Remarkably, this is only a small proportion of the country's prostitutes. In the mid-1990s the government stated that there were 81,384 prostitutes in Thailand—but that official number is calculated from the number of registered (though still illegal) brothels, massage parlors, and sex establishments. One Thai researcher estimated the total number of prostitutes in 1997 to be around 200,000.[5] Every brothel, bar, and massage parlor we visited in Thailand was unregistered, and no one working with prostitutes believes the government figures. At the other end of the spectrum are the estimates put forward by activist organizations such as the Center for the Protection of Children's Rights. These groups assert that there are more than 2 million prostitutes. I suspect that this number is too high in a national population of 60 million. My own reckoning, based on information gathered by AIDS workers in different cities, is that there are between half à million and 1 million prostitutes.

Of this number, only about one in twenty is enslaved. Most become prostitutes voluntarily, through some start out in debt bondage. Sex is sold everywhere in Thailand: barbershops, massage parlors, coffee shops and cafes, bars and restaurants, nightclubs and karaoke bars, brothels, hotels, and even temples traffic in sex. Prostitutes range from the high-earning "professional" women who work with some autonomy, through the women working by choice as call girls or in massage parlors, to the enslaved rural girls like Siri. Many women work semi-independently in bars, restaurants, and night clubs—paying a fee to the owner, working when they choose, and having the power to decide whom to take as a customer. Most bars and clubs cannot use an enslaved prostitute like Siri, as the women are often sent out on call and their clients expect a certain amount of cooperation and friendliness. Enslaved girls serve the lowest end of the market: the laborers, students, and workers who can afford only the 100 baht per half hour. It is low-cost sex in volume, and the demand is always there. For a Thai man, buying a woman is much like buying a round of drinks. But the reasons why such large numbers of Thai men use prostitutes are much more complicated and grow out of their culture, their history, and a rapidly changing economy.

"I DON'T WANT TO WASTE IT, SO I TAKE HER"

Until it was officially disbanded in 1910, the king of Thailand maintained a harem of hundreds of concubines, a few of whom might be elevated to the rank of "royal mother" or "minor wife." This form of polygamy was closely imitated by status-hungry nobles and emerging rich merchants of the nineteenth century. Virtually all men of any substance kept at least a mistress or a minor wife. For those with fewer resources, prostitution was a perfectly acceptable option, as renting took the place of out-and-out ownership.

Even today everyone in Thailand knows his or her place within a very elaborate and precise status system. Mistresses and minor wives continue to enhance any man's social standing, but the consumption of commercial sex has increased dramatically.[6] If an economic boom is a tide that raises all boats, then vast numbers of Thai men have now been raised to a financial position from which they can regularly buy sex. Nothing like the economic growth in Thailand was ever experienced in the West, but a few facts show its scale: in a country the size of Britain, one-tenth of the workforce moved from the land to industry in just the three years from 1993 to 1995; the number of factory workers doubled from less than 2 million to more than 4 million in the eight years from 1988 to 1995; and urban wages doubled from 1986 to 1996. Thailand is now the world's largest importer of motorcycles and the second-largest importer of pickup trucks, after the United States. Until the economic downturn of late 1997, money flooded Thailand, transforming poor rice farmers into wage laborers and fueling consumer demand.

With this newfound wealth, Thai men go to brothels in increasing numbers. Several recent studies show that between 80 and 87 percent of Thai men have had sex with a prostitute. Most report that their first sexual experience was with a prostitute. Somewhere between 10 and 40 percent of married men have paid for commercial sex within the past twelve months, as have up to 50 percent of single men. Though it is difficult to measure, these reports suggest something like 3 to 5 million regular customers for commercial sex. But it would be wrong to imagine millions of Thai men sneaking furtively on their own along dark streets lined with brothels; commercial sex is a social event, part of a good night out with friends. Ninety-five percent of men going to a brothel do so with their friends, usually at the end of a night spent drinking. Groups go out for recreation and entertainment, and especially to get drunk together. That is a strictly male pursuit, as Thai women usually abstain from alcohol. All-male groups out for a night on the town are considered normal in any Thai city, and whole neighborhoods are devoted to serving them. One man interviewed in a recent study explained, "When we arrive at the brothel; my friends take one and pay for me to take another. It costs them money; I don't want to waste it so I take her."[7] Having one's prostitute paid for also brings an informal obligation to repay in kind at a later date. Most Thais, men and women, feel that commercial sex is an acceptable part of an ordinary outing for single men, and about two-thirds of men and one-third of women feel the same about married men.[8]

For most married women, having their husbands go to prostitutes is preferable to other forms of extramarital sex. Most wives accept that men naturally want multiple partners, and prostitutes are seen as less threatening to the stability of the family.[9] Prostitutes require no long-term commitment or emotional involvement. When a husband uses a prostitute he is thought to be fulfilling a male role, but when he takes a minor wife or mistress, his wife is thought to have failed. Minor wives are usually bigamous second wives, often married by law in a district different than that of the men's first marriage (easily done, since no national records are kept). As wives, they require upkeep, housing, and regular support, and their offspring have a claim on inheritance; so they present a significant danger to the well-being of the major wife and her children. The potential disaster for the first wife is a minor wife who convinces the man to leave his first family, and this happens often enough to keep first wives worried and watchful.

For many Thai men, commercial sex is a legitimate form of entertainment and sexual release. It is not just acceptable: it is a clear statement of status and economic power. Such attitudes reinforce the treatment of women as mere markers in a male game of status and prestige. Combined with the new economy's relentless drive for profits, the result for women can be horrific. Thousands more must be found to feed men's status needs, thousands more must be locked into sexual slavery to feed the profits of investors. And what are the police, government, and local authorities doing about slavery? Every case of sex slavery involves many crimes—fraud, kidnap, assault, rape, sometimes murder. These crimes are not rare or random; they are systematic and repeated in brothels thousands of times each month. Yet those with the power to stop this terror instead help it continue to grow and to line the pockets of the slaveholders.

MILLIONAIRE TIGER AND BILLIONAIRE GEESE

Who are these modern slaveholders? The answer is anyone and everyone—anyone, that is, with a little capital to invest. The people who *appear* to own the enslaved prostitutes—the pimps, madams, and brothel keepers—are usually just employees. As hired muscle, pimps and their helpers provide the brutality that controls women and makes possible their commercial exploitation. Although they are just employees, the pimps do rather well for themselves. Often living in the brothel, they receive a salary and add to that income by a number of scams; for example, food and drinks are sold to customers at inflated prices, and the pimps pocket the difference. Much more lucrative is their control of the price of sex. While each woman has a basic price, the pimps size up each customer and pitch the fee accordingly. In this way a client may pay two or three times more than the normal rate, and all of the surplus goes to the pimp. In league with the bookkeeper, the pimp systematically cheats the prostitutes of the little that is supposed to be credited against their debt. If they manage the sex slaves well and play all of the angles, pimps can easily make ten times their basic wage—a great income for an ex-peasant whose main skills are violence and intimidation, but nothing compared to the riches to be made by the brokers and the real slaveholders.

The brokers and agents who buy girls in the villages and sell them to brothels are only short-term slaveholders. Their business is part recruiting agency, part shipping company, part public relations, and part kidnapping gang. They aim to buy low and sell high while maintaining a good flow of girls from the villages. Brokers are equally likely to be men or women, and they usually come from the regions in which they recruit. Some are local people dealing in girls in addition to their jobs as police officers, government bureaucrats, or even schoolteachers. Positions of public trust are excellent starting points for buying young girls. In spite of the character of their work, they are well respected. Seen as job providers and sources of large cash payments to parents, they are well known in their communities. Many of the women brokers were once sold themselves; some spent years as prostitutes and now, in their middle age, make their living by supplying girls to the brothels. These women are walking advertisements for sexual slavery. Their lifestyle and income, their Western clothes and glamorous, sophisticated ways promise a rosy economic future for the girls they buy. That they have physically survived their years in the brothel may be the exception—many more young women come back to the village to die of AIDS—but the parents tend to be optimistic.

Whether these dealers are local people or traveling agents, they combine the business of procuring with other economic pursuits. A returned prostitute may live with her family, look after her parents, own a rice field or two, and buy and sell girls on the side. Like the pimps, they are in a good business, doubling their money on each girl within two or three weeks; but also like the pimps, their profits are small compared to those of the long-term slaveholders.

The real slaveholders tend to be middle-aged businessmen. They fit seamlessly into the community, and they suffer no social discrimination for what they do. If anything, they are admired as successful, diversified capitalists. Brothel ownership is normally only one of many business interests for the slaveholder. To be sure, a brothel owner may have some ties to organized crime, but in Thailand organized crime includes the police and much of the government. Indeed, the work of the modern slaveholder is best seen not as aberrant criminality but as a perfect example of disinterested capitalism. Owning the brothel that holds young girls in bondage is simply a business matter. The investors would say that they are creating jobs and wealth. There is no hypocrisy in their actions, for they obey an important social norm: earning a lot of money is good enough reason for anything.

The slaveholder may in fact be a partnership, company, or corporation. In the 1980s, Japanese investment poured into Thailand, in an enormous migration of capital that was called "Flying Geese."[10] The strong yen led to buying and building across the country, and while electronics firms built television factories, other investors found that there was much, much more to be made in the sex industry. Following the Japanese came investment from the so-called Four Tigers (South Korea, Hong Kong, Taiwan, and Singapore), which also found marvelous opportunities in commercial sex. (All five of these countries further proved to be strong import markets for enslaved Thai girls, as discussed below.) The Geese and the Tigers had the resources to buy the local criminals, police, administrators, and property needed to set up commercial sex businesses. Indigenous Thais also invested in brothels as the sex industry boomed; with less capital, they were more likely to open poorer, working-class outlets.

Whether they are individual Thais, partnerships, or foreign investors, the slaveholders share many characteristics. There is little or no racial or ethnic difference between them and the slaves they own (with the exception of the Japanese investors). They feel no need to rationalize their slaveholding on racial grounds. Nor are they linked in any sort of hereditary ownership of slaves or of the children of their slaves. They are not really interested in their slaves at all, just in the bottom line on their investment.

To understand the business of slavery today we have to know something about the economy in which it operates. Thailand's economic boom included a sharp increase in sex tourism tacitly backed by the government. International tourist arrivals jumped from 2 million in 1981 to 4 million in 1988 to over 7 million in 1996.[11] Two-thirds of tourists were unaccompanied men; in other words, nearly 5 million unaccompanied men visited Thailand in 1996. A significant proportion of these were sex tourists.

The recent downturn in both tourism and the economy may have slowed, but not dramatically altered, sex tourism. In 1997 the annual illegal income generated by sex workers in Thailand was roughly $10 billion, which is more than drug trafficking is estimated to generate.[12] According to ECPAT, an organization working against child prostitution, the economic crisis in Southeast Asia may have increased the exploitation of young people in sex tourism:

> According to Professor Lae Dilokvidhayarat from Chulalongkorn University, there has been a 10 percent decrease in the school enrollment at primary school level in Thailand since 1996. Due to increased unemployment, children cannot find work in the formal sector, but instead are forced to "disappear" into the informal sector. This makes them especially vulnerable to sexual exploitation. Also, a great number of children are known to travel to tourist areas and to big cities hoping to find work.
>
> We cannot overlook the impact of the economic crisis on sex tourism, either. Even though travelling costs to Asian countries are approximately the same as before mid 1997, when the crisis began, the rates for sexual services in many places are lower due to increased competition in the business. Furthermore, since there are more children trying to earn money, there may also be more so called situational child sex tourists, i.e. those who do not necessarily prefer children as sexual partners, but who may well choose a child if the situation occurs and the price is low."[13]

In spite of the economic boom, the average Thai's income is very low by Western standards. Within an industrializing country, millions still live in rural poverty. If a rural family owns its house and has a rice field, it might survive on as little as 500 baht ($20) per month. Such absolute poverty means a diet of rice supplemented with insects (crickets, grubs, and maggots are widely eaten), wild plants, and what fish the family can catch. If a family's standard of living drops below this level, which can be sustained only in the countryside, it faces hunger and the loss of its house or land. For most Thais, an income of 2,500 to 4,000 baht per month ($100 to $180) is normal. Government figures from December 1996 put two-thirds of the population at this level. There is no system of welfare or health care, and pinched budgets allow no space for saving. In these families, the 20,000 to 50,000 baht ($800 to $2,000) brought by selling a daughter provides a year's income. Such a vast sum is a powerful inducement that often blinds parents to the realities of sexual slavery.

DISPOSABLE BODIES

Girls are so cheap that there is little reason to take care of them over the long term. Expenditure on medical care or prevention is rare in the brothels, since the working life of girls in debt bondage is fairly short—two to five years. After that, most of the profit has been drained from the girl and it is more cost-effective to discard her and replace her with someone fresh. No brothel wants to take on the responsibility of a sick or dying girl.

Enslaved prostitutes in brothels face two major threats to their physical health and to their lives: violence and disease. Violence—their enslavement enforced through rape, beatings, or threats—is always present. It is a girl's typical introduction to her new status as a sex slave. Virtually every girl interviewed repeated the same story: after she was taken to the brothel or to her first client as a virgin, any resistance or refusal was met with beatings and rape. A few girls reported being drugged and then attacked; others reported being forced to submit at gunpoint. The immediate and forceful application of terror is the first step in successful enslavement. Within hours of being brought to the brothel, the girls are in pain and shock. Like other victims of torture they often go numb, paralyzed in their minds if not in their bodies. For the youngest girls, who understand little of what is happening to them, the trauma is overwhelming. Shattered and betrayed, they often have few clear memories of what occurred.

After the first attack, the girl has little resistance left, but the violence never ends. In the brothel, violence and terror are the final arbiters of all questions. There is no argument; there is no appeal. An unhappy customer brings a beating, a sadistic client brings more pain; in order to intimidate and cheat them more easily, the pimp rains down terror randomly on the prostitutes. The girls must do anything the pimp wants if they are to avoid being beaten. Escape is impossible. One girl reported that when she was caught trying to escape, the pimp beat her and then took her into the viewing room; with two helpers he then beat her again in front of all the girls in the brothel. Afterward she was locked into a room for three days and nights with no food or water. When she was released she was immediately put to work. Two other girls who attempted escape told of being stripped naked and whipped with steel coat hangers by pimps. The police serve as slave catchers whenever a girl escapes; once captured, girls are often beaten or abused at the police station before being sent back to the brothel. For most girls it soon becomes clear that they can never escape, that their only hope for release is to please the pimp and to somehow pay off their debt.

In time, confusion and disbelief fade, leaving dread, resignation, and a break in the conscious link between mind and body. Now the girl does whatever it takes to reduce the pain, to adjust mentally to a life that means being used by fifteen men a day. The reaction to this abuse takes many forms: lethargy, aggression, self-loathing, suicide attempts, confusion, self-abuse, depression, full-blown psychoses, and hallucinations. Girls who have been freed and taken into shelters exhibit all of these disorders. Rehabilitation workers report that the girls suffer emotional instability; they are unable to trust or to form relationships, to readjust to the world outside the brothel, or to learn and develop normally. Unfortunately, psychological counseling is virtually unknown in Thailand, as there is a strong cultural pressure to keep mental problems hidden. As a result, little therapeutic work is done with girls freed from brothels. The long-term impact of their experience is unknown.

The prostitute faces physical dangers as well as emotional ones. There are many sexually transmitted diseases, and prostitutes contract most of them. Multiple infections weaken the immune system and make it easier for other infections to take hold. If the illness affects a girl's ability to have sex, it may be dealt with, but serious chronic illnesses are often left untreated. Contraception often harms the girls as well. Some slaveholders administer contraceptive pills themselves, continuing them without any break and withholding the monthly placebo pills so that the girls can work more nights of the month. These girls stop menstruating altogether.

Not surprisingly, HIV/AIDS is epidemic in enslaved prostitutes. Thailand now has one of the highest rates of HIV infection in the world. Officially, the government admits to 800,000 cases, but health workers insist there are at least twice that many. Mechai Veravaidya, a birth-control campaigner and expert who has been so successful that *mechai* is now the Thai word for condom, predicts there will be 4.3 million people infected with HIV by 2001.[14] In some rural villages from which girls are regularly trafficked, the infection rate is over 60 percent. Recent research suggests that the younger the girl, the more susceptible she is to HIV, because her protective vaginal mucous membrane has not fully developed. Although the government distributes condoms, some brothels do not require their use.

BURMESE PROSTITUTES

The same economic boom that has increased the demand for prostitutes may, in time, bring an end to Thai sex slavery. Industrial growth has also led to an increase in jobs for women. Education and training are expanding rapidly across Thailand, and women and girls are very much taking part. The ignorance and deprivation on which the enslavement of girls depends are on the wane, and better-educated girls are much less likely to fall for the promises made by brokers. The traditional duties to family, including the debt of obligation to parents, are also becoming less compelling. As the front line of industrialization sweeps over northern Thailand, it is bringing fundamental changes. Programs on the television bought with the money from selling one daughter may carry warning messages to her younger sisters. As they learn more about new jobs, about HIV/AIDS, and about the fate of those sent to the brothels, northern Thai girls refuse to follow their sisters south. Slavery functions best when alternatives are few, and education and the media are opening the eyes of Thai girls to a world of choice.

For the slaveholders this presents a serious problem. They are faced with an increase in demand for prostitutes and a diminishing supply. Already the price of young Thai girls is spiraling upward. The slaveholders' only recourse is to look elsewhere, to areas where poverty and ignorance still hold sway. Nothing, in fact, could be easier: there remain large, oppressed, and isolated populations desperate enough to believe the promises of the brokers. From Burma to the west and Laos to the east come thousands of economic and political refugees searching for work; they are defenseless in a country where they are illegal aliens. The techniques that worked so well in bringing Thai girls to brothels are again deployed, but now across borders. Investigators from Human Rights Watch, which made a special study of this trafficking in 1993, explain:

> The trafficking of Burmese women and girls into Thailand is appalling in its efficiency and ruthlessness. Driven by the desire to maximize profit and the fear of HIV/AIDS, agents acting on behalf of brothel owners infiltrate ever more remote areas of Burma seeking unsuspecting recruits. Virgin girls are particularly sought after because they bring a higher price and pose less threat of exposure

> to sexually transmitted disease. The agents promise the women and girls jobs as waitresses or dishwashers, with good pay and new clothes. Family members or friends typically accompany the women and girls to the Thai border, where they receive a payment ranging from 10,000 to 20,000 baht from someone associated with the brothel. This payment becomes the debt, usually doubled with interest, that the women and girls must work to pay off, not by waitressing or dishwashing, but through sexual servitude.[15]

Once in the brothels they are in an even worse situation than the enslaved Thai girls: because they do not speak Thai their isolation is increased, and as illegal aliens they are open to even more abuse. The pimps tell them repeatedly that if they set foot outside the brothel, they will be arrested. And when they are arrested, Burmese and Lao girls and women are afforded no legal rights. They are often held for long periods at the mercy of the police, without charge or trial. A strong traditional antipathy between Thais and Burmese increases the chances that Burmese sex slaves will face discrimination and arbitrary treatment. Explaining why so many Burmese women were kept in brothels in Ranong, in southern Thailand, the regional police commander told a reporter for the *Nation*: "In my opinion it is disgraceful to let Burmese men [working in the local fishing industry] frequent Thai prostitutes. Therefore I have been flexible in allowing Burmese prostitutes to work here."[16]

A special horror awaits Burmese and Lao women once they reach the revolving door at the border. If they escape or are dumped by the brothel owners, they come quickly to the attention of the police, since they have no money for transport and cannot speak Thai. Once they are picked up, they are placed in detention, where they meet women who have been arrested in the periodic raids on brothels and taken into custody with only the clothes they are wearing. In local jails, the foreign women might be held without charge for as long as eight months while they suffer sexual and other abuse by the police. In time, they might be sent to the Immigrant Detention Center in Bangkok or to prison. In both places, abuse and extortion by the staff continue, and some girls are sold back to the brothels from there. No trial is necessary for deportation, but many women are tried and convicted of prostitution or illegal entry. The trials take place in Thai without interpreters, and fines are charged against those convicted. If they have no money to pay the fines, and most do not, they are sent to a factory-prison to earn it. There they make lightbulbs or plastic flowers for up to twelve hours a day; the prison officials decide when they have earned enough to pay their fine. After the factory-prison the women are sent back to police cells or the Immigrant Detention Center. Most are held until they can cover the cost of transportation (illegal aliens are required by law to pay for their own deportation); others are summarily deported.

The border between Thailand and Burma is especially chaotic and dangerous. Only part of it is controlled by the Burmese military dictatorship; other areas are in the hands of tribal militias or warlords. After arriving at the border, the deportees are held in cells by immigration police for another three to seven days. Over this time, the police extort money and physically and sexually abuse the inmates. The police also use this time to make arrangements with brothel owners and brokers, notifying them of the dates and places of deportation. On the day of deportation, the prisoners are driven in cattle trucks into the countryside along the border, far from any village, and then pushed out. Abandoned in the jungle, miles from any major road, they are given no food or water and have no idea where they are or how to proceed into Burma. As the immigration police drive away, the deportees are approached by agents and brokers who followed the trucks from town by arrangement with the police. The brokers offer work and transportation back into Thailand. Abandoned in the jungle, many women see the offer as their only choice. Some who don't are attacked and abducted. In either case, the cycle of debt bondage and prostitution begins again.

If they do make it into Burma, the women face imprisonment or worse. If apprehended by Burmese border patrols they are charged with "illegal departure" from Burma. If they cannot pay the fine, and most cannot, they serve six months' hard labor. Imprisonment applies to all those convicted—men, women, and children. If a girl or woman is suspected of having been a prostitute, she can face additional charges and long sentences. Women found to be HIV-positive have been imprisoned and executed. According to Human Rights Watch, there are consistent reports of "deportees being routinely arrested, detained, subjected to abuse and forced to porter for the military. Torture, rape and execution have been well documented by the United Nations bodies, international human rights organizations and governments."[17]

The situation on Thailand's eastern border with Laos is much more difficult to assess. The border is more open, and there is a great deal of movement back and forth. Lao police, government officials, and community leaders are involved in the trafficking, working as agents and making payments to local parents. They act with impunity, as it is very difficult for Lao girls to escape back to their villages; those who do find it dangerous to speak against police or officials. One informant told me that if a returning girl did talk, no one would believe her *and* she would be branded as a prostitute and shunned. There would be no way to expose the broker and no retribution; she would just have to resign herself to her fate. It is difficult to know how many Lao women and girls are brought into Thailand. In the northeast many Thais speak Lao, which makes it difficult to tell whether a prostitute is a local Thai or has actually come from Laos. Since they are illegal aliens, Lao girls will always claim to be local Thais and will often have false identity cards to prove it. In the brothels their lives are indistinguishable from those of Thai women.

TO JAPAN, SWITZERLAND, GERMANY, THE UNITED STATES

Women and girls flow in both directions over Thailand's borders.[18] The export of enslaved prostitutes is a robust business, supplying brothels in Japan, Europe, and America. Thailand's Ministry of Foreign Affairs estimated in 1994 that as many as 50,000 Thai women were living illegally in Japan and working in prostitution. Their situation in these countries parallels that of Burmese women held in Thailand. The enticement of Thai women follows a familiar pattern. Promised work as cleaners, domestics, dishwashers, or cooks, Thai girls and women pay large fees to employment agents to secure jobs in rich, developed countries. When they arrive, they are brutalized and enslaved. Their debt bonds are significantly larger than those of enslaved prostitutes in Thailand, since they include airfares, bribes to immigration officials, the costs of false passports, and sometimes the fees paid to foreign men to marry them and ease their entry.

Variations on sex slavery occur in different countries. In Switzerland girls are brought in on "artist" visas as exotic dancers. There, in addition to being prostitutes, they must work as striptease dancers in order to meet the carefully checked terms of their employment. The brochures of the European companies that have leaped into the sex-tourism business leave the customer no doubt about what is being sold:

> Slim, sunburnt, and sweet, they love the white man in an erotic and devoted way. They are masters of the art of making love by nature, an art that we Europeans do not know. (Life Travel, Switzerland) [M]any girls from the sex world come from the poor north-eastern region of the country and from the slums of Bangkok. It has become a custom that one of the nice looking daughters goes into the business in order to earn money for the poor family... [Y]ou can get the feeling that taking a girl here is as easy as buying a package of cigarettes... little slaves who give real Thai warmth. (Kanita Kamha Travel, the Netherlands)[19]

In Germany they are usually bar girls, and they are sold to men by the bartender or bouncer. Some are simply placed in brothels or apartments controlled by pimps. After Japanese sex tours to Thailand began in the 1980s, Japan rapidly became the largest importer of Thai women. The fear of HIV in Japan has also increased the demand for virgins. Because of their large disposable incomes, Japanese men are able to pay considerable sums for young rural girls from Thailand. Japanese organized crime is involved throughout the importation process, sometimes shipping women via Malaysia or the Philippines. In the cities, the Japanese mob maintains bars and brothels that trade in Thai women. Bought and sold between brothels, these women are controlled with extreme violence. Resistance can bring murder. Because the girls are illegal aliens and often enter the country under false passports, Japanese gangs rarely hesitate to kill them if they have ceased to be profitable or if they have angered their slaveholders. Thai women deported from Japan also report that the gangs will addict girls to drugs in order to manage them more easily.

Criminal gangs, usually Chinese or Vietnamese, also control brothels in the United States that enslave Thai women. Police raids in New York, Seattle, San Diego, and Los Angeles have freed more than a hundred girls and women.[20] In New York, thirty Thai women were locked into the upper floors of a building used as a brothel. Iron bars sealed the windows and a series of buzzer-operated armored gates blocked exit to the street. During police raids, the women were herded into a secret basement room. At her trial, the brothel owner testified that she'd bought the women outright, paying between $6,000 and $15,000 for each. The women were charged $300 per week for room and board; they worked from 11:00 A.M until 4:00 A.M. and were sold by the hour to clients. Chinese and Vietnamese gangsters were also involved in the brothel, collecting protection money and hunting down escaped prostitutes. The gangs owned chains of brothels and massage parlors through which they rotated the Thai women in order to defeat law enforcement efforts. After being freed from the New York brothel, some of the women disappeared—only to turn up weeks later in similar circumstances three thousand miles away, in Seattle. One of the rescued Thai women, who had been promised restaurant work and then enslaved, testified that the brothel owners "bought something and wanted to use it to the full extent and they didn't think those people were human beings."[21]

OFFICIAL INDIFFERENCE AND A GROWTH ECONOMY

In many ways, Thailand closely resembles another country, one that was going through rapid industrialization and economic boom one hundred years ago. Rapidly shifting its labor force off the farm, experiencing unprecedented economic growth, flooded with economic migrants, and run by corrupt politicians and a greedy and criminal police force, the United States then faced many of the problems confronting Thailand today. In the 1890s, political machines that brought together organized crime with politicians and police ran the prostitution and protection rackets, drug sales, and extortion in American cities. Opposing them were a weak and disorganized reform movement and a muckraking press. I make this comparison because it is important to explore why Thailand's government is so ineffective when faced with the enslavement of its own citizens, and also to remember that conditions *can* change over time. Discussions with Thais about the horrific nature of sex slavery often end with their assertion that "nothing will ever change this . . . the problem is just too big . . . and those with power will never allow change." Yet the social and economic underpinnings of slavery in Thailand are always changing, sometimes for the worse and sometimes for the better. No society can remain static, particularly one undergoing such upheavals as Thailand.

As the country takes on a new Western-style materialist morality, the ubiquitous sale of sex sends a clear message: women can be enslaved and exploited for profit. Sex tourism helped set the stage for the expansion of sexual slavery.

Sex tourism also generates some of the income that Thai men use to fund their own visits to brothels. No one knows how much money it pours into the Thai economy, but if we assume that just one-quarter of sex workers serve sex tourists and that their customers pay about the same as they would pay to use Siri, then 656 billion baht ($26.2 billion) a year would be about right. This is thirteen times more than the amount Thailand earns by building and exporting computers, one of the country's major industries, and it is money that floods into the country without any concomitant need to build factories or improve infrastructure. It is part of the boom raising the standard of living generally and allowing an even greater number of working-class men to purchase commercial sex.

Joining the world economy has done wonders for Thailand's income and terrible things to its society. According to Pasuk Phongpaichit and Chris Baker, economists who have analyzed Thailand's economic boom,

> Government has let the businessmen ransack the nation's human and natural resources to achieve growth. It has not forced them to put much back. In many respects, the last generation of economic growth has been a disaster. The forests have been obliterated. The urban environment has deteriorated. Little has been done to combat the growth in industrial pollution and hazardous wastes. For many people whose labour has created the boom, the conditions of work, health, and safety are grim.
>
> Neither law nor conscience has been very effective in limiting the social costs of growth. Business has reveled in the atmosphere of free-for-all. The machinery for social protection has proved very pliable. The legal framework is defective. The judiciary is suspect. The police are unreliable. The authorities have consistently tried to block popular organizations to defend popular rights.[22]

The situation in Thailand today is similar to that of the United States in the 1850s; with a significant part of the economy dependent on slavery, religious and cultural leaders are ready to explain why this is all for the best. But there is also an important difference: this is the new slavery, and the impermanence of modern slavery and the dedication of human-rights workers offer some hope.

NOTES

1. Siri is, of course, a pseudonym; the names of all respondents have been changed for their protection. I spoke with them in December 1996.
2. "Caught in Modern Slavery: Tourism and Child Prostitution in Thailand," Country Report Summary prepared by Sudarat Sereewat-Srisang for the Ecumenical Consultation held in Chiang Mai in May 1990.
3. Foreign exchange rates are in constant flux. Unless otherwise noted, dollar equivalences for all currencies reflect the rate at the time of the research.
4. From interviews done by Human Rights Watch with freed child prostitutes in shelters in Thailand, reported in Jasmine Caye, *Preliminary Survey on Regional Child Trafficking for Prostitution in Thailand* (Bangkok: Center for the Protection of Children's Rights, 1996), p. 25.
5. Kulachada Chaipipat, "New Law Targets Human Trafficking," *Bangkok Nation*, November 30, 1997.
6. Thais told me that it would be very surprising if a well-off man or a politician did not have at least one mistress. When I was last in Thailand there was much public mirth over the clash of wife and mistress outside the hospital room of a high government official who had suffered a heart attack, as each in turn barricaded the door.
7. Quoted in Mark Van Landingham, Chanpen Saengtienchai, John Knodel, and Anthony Pramualratana, *Friends, Wives, and Extramarital Sex in Thailand* (Bangkok: Institute of Population Studies, Chulalongkorn University, 1995), p. 18.
8. Van Landingham et al., 1995, pp. 9–25.
9. Van Landingham et al., 1995, p. 53.
10. Pasuk Phongpaichit and Chris Baker, *Thailand's Boom* (Chiang Mai: Silkworm Books, 1996), pp. 51–54.
11. Center for the Protection of Children's Rights, *Case Study Report on Commercial Sexual Exploitation of Children in Thailand* (Bangkok, October 1996), p.37.
12. David Kyle and John Dale, "Smuggling the State Back In: Agents of Human Smuggling Reconsidered," in *Global Human Smuggling: Comparative Perspectives*, ed. David Kyle and Rey Koslowski (Baltimore: Johns Hopkins University Press, 2001).
13. "Impact of the Asian Economic Crisis on Child Prostitution," *ECPAT International Newsletter 27*

(May 1, 1999), found at http://www.ecpat.net/eng/Ecpat_inter/IRC/articles.asp?articleID=143&NewsID=21.

14. Mechai Veravaidya, address to the International Conference on HIV/AIDS, Chiang Mai, September 1995. See also Gordon Fairclough, "Gathering Storm," *Far Eastern Review*, September 21, 1995, pp. 26–30.
15. Human Rights Watch, *A Modern Form of Slavery*, p. 3.
16. "Ranong Brothel Raids Net 148 Burmese Girls," *Nation* (July 16 1993), p. 12.
17. Dorothy O. Thomas, ed., *A Modern Form of Slavery: Trafficking of Burmese Women and Girls into Brothels in Thailand* (New York: Human Rights Watch, 1993), p. 112.
18. *International Report on Trafficking in Women (Asia-Pacific Region)* (Bangkok: Global Alliance Against Traffic in Women, 1996); Sudarat Sereewat, *Prostitution: Thai-European Connection* (Geneva: Commission on the Churches' Participation in Development, World Council of Churches, n.d.). Women's rights and antitrafficking organizations in Thailand have also published a number of personal accounts of women enslaved as prostitutes and sold overseas. These pamphlets are disseminated widely in the hope of making young women more aware of the threat of enslavement. Good examples are Siriporn Skrobanek, *The Diary of Prang* (Bangkok: Foundation for Women, 1994); and White Ink (pseud.), *Our Lives, Our Stories* (Bangkok: Foundation for Women, 1995). They follow the lives of women "exported," the first to Germany and the second to Japan.
19. The brochures are quoted in Truong, *Sex, Money, and Morality: Prostitution and Tourism in Southeast Asia* (London: Zed Books, 1990), p. 178.
20. Carey Goldberg, "Sex Slavery, Thailand to New York," *New York Times* (September 11, 1995), p. 81.
21. Quoted in Goldberg.
22. Phongpaichit and Baker, 1996, p. 237.

CHAPTER 7

Deviance

Violating the Norms of Society

Alan Mc Evoy

Deviance is one of the most myth laden of social issues. Most often, myths are simplistic explanations based on an individual's own set of biases. The main problem in dealing with myths is that they often contain a grain of truth and this makes it difficult to convince people that their belief is really a myth. For example, the very definition of the term *deviance* implies that it is abnormal behavior, but as Randal Collins explains in the first reading of this chapter, crime is so prevalent that it should be considered a normal behavior of society.

A more recognizable myth-belief is that crime is genetic. This is the excuse people make when they assert that there is a relationship between one's genes and one's behavior. This myth explains to its believers why there are higher crime rates for certain physical types or ethnic groups or racial groups. However, like all human behavior, crime is not limited to any particular physical type, ethnic or racial group.

Another myth attempts to explain the nature of crime. Most crime, it asserts, is committed against people. FBI reports show, however, that most crime is committed against property. This myth-belief comes about because the mass media tends to concentrate on what they believe is of greatest interest to their readers/viewers: murder and rape rather than burglary.

This type of reporting leads to another myth about crime rates. If only certain types of crimes are reported by the media, then there is the mistaken impression that the rates of these crimes are increasing. Thus, frequent reporting of drug busts leads to the impression that drug abuse is pandemic. However, crime rates are usually a reflection of the system used in collecting the data. Thus, any system listing all persons charged with a crime will show a much higher rate than a system that lists only those convicted of a crime. Still, the media convinces us that crime rates are climbing even though they have been falling for several years; drug abuse is on the same downward trend. These types of myth-beliefs lead Barry Glassner to claim in the second reading that "Americans Fear the Wrong Things."

Related to the myth that crime rates are climbing is the belief that the only way to reduce these numbers is through punishment and incarceration. Adding impetus to this belief is the idea of revenge—revenge against those who would dare violate the norms of society. Politicians, recognizing the popularity of the public's desire for vengeance, are encouraged to promote tougher sentencing programs. Adding to the desire for longer prison sentences is the feeling that people are safer from criminals while the latter are in prison. Thus, the increase in the number of persons imprisoned is a reflection of tougher sentencing laws and not, as the public may believe, of increasing crime rates. Yet when one considers the high rates of repeated criminal behavior despite severe punishment, it must be concluded that there is little or no relationship between incarceration and crime reduction. It is this factor of repeated criminal behavior that leads Jeffrey Reiman in the third reading to declare that the U.S. system of crime control is a failure and, worse, that we have learned nothing from this failure.

This failure to learn from the poor results of our incarceration policy has, as noted, led to large increases in our prison population. Despite the large increase in the building of prisons to accommodate the burgeoning numbers of those imprisoned, many prison systems are running out of space. This results in the need to open up space for the newcomers by giving early release to some of those imprisoned earlier. Some believe that a better way to meet this need for more prison space is to decriminalize those behaviors labeled as crimes that have no victims. According to this theory, such crimes as prostitution and pot smoking only affect those doing this behavior and thus "revenge" by society is unnecessary. What do you think?

On the other hand, other crimes have countless victims but are not widely recognized as a social problem. White-collar crime is indeed criminal behavior because it fits the legal definition of a socially harmful act or an act for which the law provides a penalty. Richard Ball in the fourth reading describes the pervasiveness of this little-known criminal behavior. For example, in the 1990s there were 38 environmental crimes, 20 antitrust crimes, 13 fraud crimes, seven campaign finance crimes, and six food and drug crimes among the top 100 corporate crimes. Interestingly enough, six of these top corporate criminals were recidivists. Perhaps a question that is just as important to ask is: Why is there so little known about this widely occurring, repeated criminal behavior? Possibly the best-known reason for the public's lack of knowledge about white collar crime is the fact that the news of such criminal behavior and trials appears on the business pages and almost never on other media sources despite the fact that this criminal behavior may affect, and even kill, hundreds of people.

A related question is: Why do these corporations repeatedly commit these crimes? Two reasons are usually given for this recidivism: One, the perpetrators are considered to be the corporations, not the individuals that led the corporation into the criminal behavior, and so individuals rarely go to jail. In fact, they may get bonuses for raising the corporation's profits. If they do go to jail, it is usually for a relatively short time at an upscale prison. Second, the corporation pays the fine, which can legally be deducted as a business expense. How do you think white-collar criminals should be treated? Considering its ubiquity, its repetitiveness, and the widespread harm it does, why do you think corporate crime is not considered a social problem? Is this another aspect of the corporate political power noted in Chapter 12?

In a sense, the readings in this chapter deal with the issue of *labeling*. D. L. Rosenhan discusses the labeling of people as deviant in the final reading. The label, once applied, can seemingly be never removed even if the person is found innocent of deviant behavior. For example, a number of child-care workers were accused of being child molesters. Despite the fact that many were found innocent, the so-called perpetrators were never able to again obtain jobs or continue working in the child-care field. Thus the act of labeling a person causes them to be identified with that label whether the description is true or not. Just as important, once labeled a certain way, that person will never be considered to be otherwise despite the absence of the labeled behavior. In short, the label defines the person in society's eyes rather than the person himself or herself. After reading this selection, think about other forms of deviance that may be an "artifact of a label." What can or should be done about this type of situation? What other areas should be labeled as deviant or, for that matter, not be labeled at all?

31

The Normalcy of Crime

RANDALL COLLINS

The author discusses three all-encompassing explanations for crime: conservative, liberal, and radical. These explanations ignore the possibility that crime is normal and even functional.

As you read this selection, consider the following questions as guides:

1. *Do you believe that the broad reasons given for committing crime are too simplified? Why, or why not?*
2. *Defend or dispute the author's claim that crime is normal and functional.*
3. *If, indeed, crime is normal and functional, what should be done about it?*
4. *If crime is normal and functional, can anything be done about it?*

GLOSSARY **Altruistic** Unselfishly concerned.

There have been several widely accepted views about crime. The more obvious explanations begin at the level of common sense. The trouble with common sense, though, is that there are usually opposite opinions on any subject, both of which are equally commonsensical to those who believe in them. These views have generally corresponded to popular political beliefs. Roughly speaking, we may refer to them as the conservative and liberal views on crime. . . .

The most sophisticated and least obvious theory of crime, I will suggest, goes back to Durkheim. The problem has not turned out to be just what we once thought it was. We may have to face a paradox: crime exists because it is built into the structure of society itself. This does not mean that nothing can be done about it, but the social costs of controlling crime may involve more difficult change than we have been aware of.

CONSERVATIVE EXPLANATIONS OF CRIME

One view of crime is that criminals are simply bad people; the only way to deal with them is to punish them. The more crime there is, the harder we should crack down on it. This position has been held for many centuries, and it keeps on being restated today. The trouble is that it has never really worked. In Europe during the 1600s and 1700s, punishments were as severe as one could imagine. People were hung for stealing a loaf of bread; others were branded or had their ears cut off. Some offenders, especially people accused of religious or political crimes, were tortured to death. All these punishments were public spectacles. A crowd would gather around to watch a good execution, while vendors sold refreshments and people made

SOURCE: From *Sociological Insight: An Introduction to Nonobvious Sociology* 2/e by Randall Collins.© 1992 by Oxford University Press, Inc. Used by permission of the publisher.

bets on how long the criminal would yell while he or she was burning at the stake. People today who advocate severe punishments as a deterrent for crime would have been delighted by the situation.

But the brutal punishments did not work. Crime kept right on occurring at a tremendous rate for hundreds of years, despite the hangings and the mutilations. . . .

The same kind of situation still can be found in some parts of the world today. In Saudi Arabia and some other Muslim countries, theft is punished by cutting off a hand, and many other offenses by death. Executions are carried out in public, with the whole community required to attend. But the results are the same as they were in medieval Europe. . . .

We can begin to see, therefore, that the philosophy of punishing criminals as violently as possible is not really a policy that people advocate because it has proven effective. It is a political position, or what comes to the same thing, a moral philosophy, which declares it is good to be tough and even brutal or malicious to offenders. Just why people hold this position is itself a question for sociology to explain, since they must hold it for some other reason than its practical effects. The holders of this position doubtless consider it rational, but here again we see that their rationality has a nonrational foundation. They do not bother to look at the evidence for whether severe deterrents work but already "know" their policy is right. This sense of rightness is the mark of a partisan position, in this case political conservatism.

A somewhat more scientific version of this political position has tried to tie crime to biology. Today some assert that criminals have bad genes; their propensities to crime are inborn and, hence, nothing can be done about them. Society could only pick them out at an early age by appropriate testing and then presumably get rid of them in some way. Just how this is to be done is not yet worked out: whether the police would hold a complete dossier on all people with bad genes, or whether such people would be locked up for life, or be sterilized, or even exterminated. The issue hasn't really gotten to this point because the position so far is completely theoretical. No one knows how to make a test for bad genes, and there is no real comparative evidence that such genes are causes of crime. The modern genetic theory of crime is another version of conservative political ideology. This is easy to see, since the same arguments about criminals are also applied to welfare recipients and other social types who are anathema in conservative thinking. . . .

LIBERAL EXPLANATIONS

If there is a conservative version of common sense about crime, there is a liberal common sense as well. The liberal position makes an effort to understand what it is like being in the criminal's shoes. Why would someone enter a life of crime, and what can be done to help them out of it? There have been several answers to these questions.

One is that criminals are people who have gotten in with the wrong crowd. Youths hang around with a delinquent gang and start to pick up delinquent values themselves. Soon they are committing petty thefts, small acts of vandalism, and the like. This moves them more and more into the delinquent culture, and eventually they move on to serious crimes and become full-fledged criminals.

A similar type of explanation is that criminals come from broken homes and run-down neighborhoods. These childhood stresses and strains make people hostile and insecure, and lead them to a life of crime. Growing up in an area of poverty and disillusionment, these youths have no reason to be attached to normal society. They feel that society has no use for them, and they have every reason to take revenge in any way they can.

Sometimes this argument is taken one step further to propose that it isn't just their background that makes some people become criminals, but also the lack of opportunities to change their social condition. If children from poor families or racial minorities had a chance to rise in the world, they would become normal, productive members of society. It is because they are trapped by the lack of opportunities to get ahead that they turn to crime. It is proposed, moreover, that the social atmosphere of the United States makes this feeling particularly strong. For the U.S. is an achievement-oriented culture, where people are expected to make a success on their own. . . .

Some of these arguments, we can see, get to be rather complicated. Nevertheless, they all share the notion that crime is not really the fault of the criminal. He (or she–although in fact the great majority of criminals are male) would rather not be a criminal if he could help it. It is only the adverse social conditions that force them into a criminal career.

This type of explanation certainly has the appeal of sounding altruistic, and it has given rise to a great many efforts at reform and rehabilitation to set criminals back on the path to normal social participation. . . .

In this way, all of the various social causes that are believed to account for crime are to be counteracted by an appropriate social reform. If it is a delinquent milieu that starts youth on their evil ways, we provide youth services and group workers to try to lure the gangs off the streets and onto supervised playgrounds. For broken homes and run-down neighborhoods, there are social workers and urban renewal projects. For blocked mobility opportunities, there are various efforts to improve the life chances of the disadvantaged, to keep them in school longer, to provide remedial services, and the like.

As I said, all of this is very altruistic, but it has one big drawback. It simply has not worked very well. . . .

These sorts of facts are a fairly serious indictment of the liberal theories of crime and its prevention, but this hasn't entirely convinced the proponents of these theories that they are wrong. They can continue to argue, for example, that the proper counteractive measures have not been applied vigorously enough. We need more youth group workers, they may reply, or a more extensive attack on the existence of poverty and racial discrimination, or a more serious effort to create career mobility opportunities for deprived youth and ex-convicts alike. This has some plausibility since it is certainly true that much more could be done in this altruistic direction. But the suspicion has been growing that the underlying theories just may not be accurate.

Take the broken-family-and-blighted-neighborhood hypothesis about crime. This explanation seems to fit our commonsense view of the world: stress and deprivation lead to crime. But the evidence does not exactly bear this out. Not everyone from a divorced family becomes a criminal; in fact, most such children do not. This is especially apparent today, when divorce has become a normal and accepted part of otherwise quite average families. Nor is it fair to say that everyone who lives in a poor neighborhood is a criminal: again, it is only a minority within this area who are. Hence it cannot be poverty per se that causes crime but some other factor. This becomes even clearer when we realize that by no means are all criminals poor or from racial minorities. Delinquent youths are found in middle-class areas as well as poor ones. Rich boys at fraternity parties commit acts of vandalism, too, as well as violence, rape, theft, and all the rest of it, although they are not always charged with these crimes. The same thing is true among adults. It is not just the poorer social classes that commit crimes. So-called white-collar crime is also a major problem, ranging from passing bad checks to embezzling business funds or conspiring to bribe government officials or to evade legal regulations.

The altruistic, liberal theories of crime are just not adequate to deal with these phenomena. What looked at first glance like a realistic sociological explanation of crime turns out on closer examination not to fit the facts very well at all. There is less crime in the deprived areas of society than the theory would predict and more crime elsewhere in society where these conditions do not hold. It is no wonder, one might conclude, that the liberal methods for preventing crime and rehabilitating criminals have not had much success.

RADICAL EXPLANATIONS OF CRIME

In recent sociology there has been an upsurge of theorizing that rejects the more traditional kinds of theories in favor of a radically new look at the crime issue. Here the theories enter the realm of the nonobvious and even the paradoxical.

The basic turn in the argument has been to shift attention away from the criminal side and to look critically instead at the agents of law-enforcement. For example, it is sometimes argued that increases in the crime rate have nothing to do with how many crimes are actually committed. All that has changed, it is suggested, is that more crimes are being reported. Sometimes a newspaper will create a crime wave by running crime stories more prominently on the front pages–perhaps for political purposes, to attack a city administration or make an issue of the crime problem for an up-coming election. The police, too, it is charged, inflate the crime rate by improving their record-keeping capabilities. Unsolved crimes that formerly were left unreported are now included. This makes a good argument for police appeals that they need an increased budget.

It does appear to be true that some alleged shifts in crime rates are produced in this way. Newspapers in particular are not a very reliable source of information on social trends, and official police statistics are also subject to biases due to shifts in reporting methods. Whenever one sees a rapid jump in crime rates over a space of a year, it is often due to a purely administrative change in the statistical accounting system. At the same time, it has to be said that not all of the shifts in crime rates can be attributed to causes of this sort.

But there is a much more radical sense in which it is proposed that crimes are created by the law-enforcement side. This is referred to as the labeling theory. The argument goes like this. All sorts of youths violate the laws. They engage in petty thefts and acts of vandalism. They get into fights, drink illegally, have illicit sex, smoke dope or use drugs, and so on. This is widespread and almost normal behavior at a certain age. What is crucial, though, is that some of these young people get caught. They are apprehended by the authorities for one thing or another. Now even at this point there is a possibility of heading off the negative social consequences. Some of these youths get off with a warning, because their school principal likes them, say, or because their parents intervene, or because the police are sympathetic to them. If so, then they have escaped going down into a long funnel at the end of which lies a full-fledged criminal identity.

If a young offender is actually arrested, charged with a crime, convicted, and all the rest, this has a crucial effect upon the rest of his or her career. This happens in several ways. One effect is psychological: those who had previously regarded themselves as more or less like anyone else, just goofing off perhaps, now are someone special. They are now labeled an offender, a juvenile delinquent, a criminal; they are caught up in a network of criminal-processing organizations. Every step along the way reinforces the sense that they have become someone different from the normal. They acquire a criminal identity. . . .

In this way, a self-perpetuating chain of criminal activities builds up. The key point in the whole sequence is right at the beginning, where the labeling process begins. It is the first, dramatic confrontation with the law that makes all the difference, deciding which way the individuals will go. Either they will get by with a bit of normal goofing off, or they are embarked on a career of crime in which everything that is done to prevent it actually makes it all the more inevitable.

This is a rather psychological way of describing the dynamics of the labeling process. I could fill in the process from a different angle, one that does not stress so much the shift that takes place within the novice "criminal's" mind but within the organization of the law-enforcement world itself. Sociologists who have studied the police point out that the police constitute an organization, with administrative problems just like any other organization. A business organization needs to keep up its sales; a police organization needs to keep apprehending criminals and solving crimes. This is by no means an easy thing to do. Some crimes are relatively easy to solve, such as murder. But these make up only a small percentage of total crimes. The most common crimes, and those which most widely affect the public, are burglary, auto theft, and other types of larceny. These are hard to solve precisely because there are so many of them. There usually is little evidence left at the scene of the crime, and there are rarely any witnesses. . . . How, then, do the police try to control this large category of crime?

The best strategy they can follow is to try to get confessions from the criminals that they do apprehend. So whenever someone is arrested with goods from a burglary, say, a great deal of pressure is put on them to confess to other burglaries. . . . The most effective sort of pressure, though, is usually in the form of a bargain. The accused criminals are encouraged to confess to a list of unsolved thefts; in return for this, they are allowed to plead guilty to some restricted charge, e.g., one or two counts of burglary, or even some lesser offense. This is a typical plea bargain. . . .

All this has a powerful effect in reinforcing the "labeling" process that keeps people going in criminal careers. The way police can make their system work is to keep tabs on people whom it is easiest to arrest. . . . The easiest people to arrest are people who have been arrested before. So one way police can "solve" a round of burglaries is to pay a surprise call on formerly convicted criminals in the area who are out on parole. One of the conditions of parole often is that the ex-convict should be subject to search. So the police arrive, look for stolen property, illegal drugs, or other violations. Often it is not hard to find these, especially since drugs of one kind or another are generally a part of the criminal culture. (Which is not to say that these same drugs may not also be part of the life-style of people who are not in the criminal world.)

So the police then are able to set the bargaining process in motion. . . .

Thus, the chain of events that starts when someone is labeled a criminal for some initial offense, can end up as a kind of invisible prison in its own right. Once someone becomes known to the police, they are subject to organizational pressures that will send them through the system over and over again. Whether they come to strongly identify themselves personally with a criminal identity or not, the police will tend to do so, and that makes it all the harder to get out. Ex-convicts are trapped in a machine that constantly reprocesses them because they are its easiest materials to reach.

The labeling theory declares that crime is actually created by the process of getting caught. Unlike the previous types of theories that we looked at, the personal characteristics of the individuals, or their social class or ethnic or neighborhood background, is not a crucial point. It is assumed that all sorts of people violate the law. But only some of them get caught, are prosecuted, labeled and all the rest, thereby becoming full-fledged criminals. If criminals who go through the courts and the prisons are so often likely to be disproportionately poor, black, or otherwise fit someone's idea of "social undesirables" or the "socially deprived," it is because these are the types of people who are most likely to be apprehended and prosecuted. The fraternity boys stealing a college monument or raping the sorority girls at a party are let off with a reprimand because these are labelled "college pranks." The poor black youth who does the same sort of thing gets sent to juvenile court and a start on a career of serious crime.

There is an even stronger version of the radical approach to crime. This argues that it is not simply the police who create the criminals but the law itself. To cite an obvious example: possession of drugs such as narcotics was not a crime until laws were passed making private possession of them a felony. In the 1800s, the use of opium and opiate-based preparations such as laudanum was not illegal, and it was fairly widespread. The drugs could be bought over the counter at a pharmacy. Many people used them in patent medicines. Others used them for pain-killers, escape, or because they liked the sensations they produced. The same was true of hashish and marijuana, or of coca and cocaine, which were used in greater or lesser quantities by various kinds of people. In the early 1900s, the public use of opium and its derivatives was outlawed in the United States, and under a series of international agreements, by most of the modern states around the world. Other laws followed, outlawing cocaine and cannabinols.

These laws suddenly created a new category of crime. People who had previously been engaging in a purely private act were now breaking a fairly serious law. This had a great many social ramifications.

For one thing, the labeling processes outlined above, both psychological and organizational, were set in motion. . . .

The illegalizing of drugs, moreover, had an important economic effect. When drugs were sold on the open market, their cost was relatively low because they are relatively inexpensive to produce and transport. But when drugs became illegal, the whole business was greatly restricted. As one can see from a simple application of the economics of supply and demand, restricting the supply raised the price. Whereas a modest supply of opium in early nineteenth-century England cost a shilling (the equivalent of perhaps $25 today), heroin (a twentieth century derivative of opium) now costs some $2000 an ounce. Drug dealers and smugglers incur much greater expenses, keeping their activities hidden as much as possible, paying out bribes, and also paying for the inevitable legal fees when they are caught. So the illegalizing of drugs, by raising the prices, ramified into many other crimes that had formerly been unconnected with the drug market. Smuggling and bribery expanded, of course, but so did burglary and robbery. Most drug addicts, unable to pay for the expense of supporting a costly opiate habit, turned to theft as a main way of keeping the money coming in. From the initial decision to outlaw drugs, then, many other crimes followed.

The same kind of analysis has been applied to many other sorts of crimes. The national prohibition of alcohol in the United States, which held sway between 1919 and 1933, for example, created a whole illegal culture of speak-easies, stills, alcohol smugglers, and an organized crime network to "protect" these operations.. . .

The radical approach to the analysis of crime turns up a great many ironic interconnections between crime and the social structure. Actions taken by citizens in the name of morality and law-abidingness add up to vastly increasing the amount of criminality. Some sociologists have argued that an explanation of crime really boils down to an explanation of how certain things came to be defined as crimes. It has been suggested that crimes are manufactured by "moral entrepreneurs," people who try to create a morality and enforce it upon others. Other sociologists have gone farther, to look for the economic and organizational interests or the social movements that create crimes in this way. It may be suggested, for example, that the outlawing of drugs in the early twentieth century was part of the efforts of the medical profession to monopolize control over all drugs for itself. The prohibitionist movement has been explained as a last-ditch effort of rural Anglo-American Protestants to try to head off what they saw as the degenerate alcoholic culture of the immigrants in the big cities. An analysis along these lines could be applied to current movements that are attempting to create new definitions of crime, such as the antiabortion movement.

At this point, one might step back and ask a question. The examples given have all been of the type of activities that offend some people's moral sense as to what is proper. Drug taking, drinking, gambling–one could add prostitution, pornography, homosexuality and other sexual practices–all involve people who willingly consent to these actions. These actions offend only outsiders. They are what are called "victimless crimes." Here the idea that society creates these crimes in a fairly arbitrary sense, just by passing a law against them, has a good deal of plausibility. But what about "real" crimes, such as robbery, murder, assault, rape, and all other actions that hurt someone's life, body, or property? One could well maintain that these actions would not be considered licit by most people, even if there were no laws prohibiting them. These seem to be "natural," rather than "artificial" categories of crime, and people would want to stop them without the necessity of some kind of moral crusade trying to have laws passed to outlaw them.

However, the most radical position in sociological theory attempts to show that these crimes, too, are socially created. For example, the crime of robbery is only a crime because of the system of property. . . .

This is certainly a theory worth thinking about. It has the merit of seeing that "real" crimes are a matter of conflict between people in a stratified society, and especially that economic crimes are part of the system of economic stratification in general. Since economic crimes like robbery and auto theft make up the largest proportion of all crimes, this kind of theory can potentially explain a great deal.

Nevertheless, we cannot immediately jump to the conclusion that crime is class struggle of exactly the same sort as usually featured in the Marxian model. For one thing, when we look at who are the victims of crimes, we find a rather surprising pattern. The poorer classes are much more likely to be robbed or burglarized than the wealthier classes. And this is true, in the United States, for both whites and blacks. In fact, blacks with the lowest incomes are the most likely of all to be victims of crimes of virtually all sorts, including murder and rape as well as property crime.

Clearly, then, there is a stratified pattern of crime, but it is not primarily the poor robbing (and murdering and raping) the rich. Criminals are not Robin Hoods. What appears to be going on, rather, is that crime is mainly *local*. People rob, burglarize, murder, and rape in their own neighborhoods above all. The reason is fairly simple: these are the easiest opportunities, especially for teenagers, who commit the majority of all crimes.

The end result is that there is a social-class pattern in crime after all, but it comes out in the fact that neighborhoods tend to be segregated by social class, as well as by race and ethnicity. Hence it is the least privileged people who commit the largest number of crimes, but their victims are primarily people like themselves. It is mainly the poor robbing the poor. . . .

There has been a lot of controversy over the death penalty in recent decades. If we leave aside the moral questions involved in this, and concentrate only on the research that has been done, we can see some interesting patterns. Some states in the U.S. have the death penalty, while others have abolished it. If we compare states that are similar in their social characteristics, it turns out that they have about the same murder rates, whether they have the death penalty or not. That implies that people do not decide to commit murder or not according to whether they expect to risk a severe penalty for it. Murders do not seem to be related to any social calculus. By the same token, none of the sociological theories given above seem to explain murder very well.

I mentioned earlier that murders are relatively easy for the police to solve. Why is this? It is because the large majority of murders are committed by people who know their victims personally. For that matter, the largest single category of murders happens within the family, especially one spouse murdering the other. Hence to solve a murder is not particularly difficult. The police need only look for someone who knew the victim and who had some motive to be especially angry with them. So if you are thinking about killing your husband or your wife, forget it; you will automatically be the number one suspect.

All this adds up to a picture in which crimes divide into quite different sorts. There are victimless crimes, very much created by social movements that define them as criminal; people who become labeled as criminal because of these sorts of offenses usually become involved in networks of other sorts of criminality as a result of the law-enforcement process. There are also property crimes, which have some relevance to the way in which individuals make their careers as criminals, but which would by no means disappear if laws stopped being enforced. And there are crimes of passion, which seem to be of a much more personal nature, and which do not seem to be related to any of the factors we have considered here.

Is there any perspective that encompasses all of this? Yes, I believe there is. But it is the most nonobvious of all, and one which does not resonate any too well in the hearts of either conservatives, liberals, or radicals. It is a perspective that declares that crime is a normal, and even necessary, feature of all societies.

THE SOCIAL NECESSITY OF CRIME

This perspective, like so many of the nonobvious ideas in sociology, traces back to Émile Durkheim. In this view, crime and its punishment are a basic part of the rituals that uphold any social structure. Suppose it is true that the process of punishing or reforming criminals is not very effective. The courts, the police, the parole system–none of these very effectively deter criminals from going on to a further life of crime. This would not surprise Durkheim very much. It can be argued that the social purpose of these punishments is not to have a real effect upon the criminal, but to enact a ritual for the benefit of society.

Recall that a ritual is a standardized, ceremonial behavior, carried out by a group of people. It involves a common emotion, and it creates a symbolic belief that binds people closer to the group. Carrying out rituals over and over again is what serves to keep the group tied together. Now in the case of punishing criminals, the group that is held together is not the criminals' group. It is the rest of society, the people who punish the criminals. The criminal is neither the beneficiary of the ritual nor a member of the group that enacts the ritual, but only the raw material out of which the ritual is made. . . .

The main object of a crime-punishment ritual, then, is not the criminal but the society at large. The trial reaffirms belief in the laws, and it creates the emotional bonds that tie the members of society together again. From this point of view, exactly how the criminal reacts to all this is irrelevant. The criminal is an outsider, an object of the ritual, not a member of it. He or she is the necessary material for this solidarity-producing machine, not the recipient of its benefits. It is the dramatics of the trial that counts, the moments when it is before the public eye. Afterwards, it may all come unravelled. The conviction may be reversed on appeal for some technical error. Criminals may go to an overcrowded prison where they make new criminal contacts and acquire a deeper commitment to the criminal role. Sooner than expected, the parole board may decide to relieve crowding in the prison by releasing them, and they are back out on parole and into the routine of police checks and parole officers and all the rest of an ongoing criminal career. If we look at the criminal justice system from the point of view of somehow doing something to deter the criminal, it appears ineffective, even absurd. It makes more sense once we realize that all the social pressure falls upon dramatizing the initiation of punishment, and that this is done to convince society at large of the validity of the rules, not necessarily to convince the criminal.

An even more paradoxical conclusion follows from this. Society needs crime, says Durkheim, if it is to survive; without crimes, there would be no punishment rituals. The rules could not be ceremonially acted out and would decay in the public consciousness. The moral sentiments that are aroused when the members of society feel a common outrage against some heinous violation would no longer be felt. If a society went too long without crimes and punishments, its own bonds would fade away and the group would fall apart.

For this reason, Durkheim explained, society is in the business of manufacturing crimes, if they do not already exist in sufficient abundance. Just what would count as a crime may vary a great deal, relative to what type of society it is. Even a society of saints would find things to make crimes out of: any little matters of falling off into less saintliness than the others would do. To put it another way, the saints, too, would have their central, especially sacred rules, and those who did not respect them as intensely as the others would be singled out for punishment rituals that served to dramatize and elevate the rules all the more. . . .

Punishment rituals hold society together in a certain sense: they hold together the structure of domination. They do this partly by mobilizing emotional support for politicians and the police. Above all, they increase the feelings of solidarity within the privileged classes and enable them to feel superior to those who do not follow their own ideals. Outrage about crime legitimates the social hierarchy. The society that is held together by the ritual punishment of crime is the stratified society.

In this sense, crime is built into the social structure. Whatever resources the dominant group uses for control will have corresponding crimes attached to them. Since there is an ongoing struggle among groups over domination, some groups will violate other groups' standards. And those individuals who are least integrated into any groups will pursue their own individual aims without regard for the morality held by others. Therefore, there is usually no shortage of actions that are offensive to many groups in a society. And these violations are to a certain extent welcome by the dominant groups. Crime gives them an occasion for putting on ceremonies of punishment that dramatize the moral feelings of the community, which bolsters their group domination.

This means that every type of society will have its own special crimes. What is constant in all societies is that somehow the laws will be set in such a way that crimes and punishments do occur. A tribal society has its taboos, the violation of which calls down ferocious punishment. The Puritans of the New England colonies, with all their intense moral pressures, believed in the crime of witchcraft, Capitalist societies have endless definitions of criminality relating to property. Socialist societies have their crimes as well, especially political crimes of disloyalty to the state, as well as the individualistic crimes of failing to participate whole-heartedly in the collective. The ritual perspective finds that all societies manufacture their own types of crime. It may be possible to shift from one type of crime to another, but not to do away with crime altogether.

Crime is not simply a matter of poverty and social disorganization, nor of particularly evil or biologically defective individuals. The labeling theory is closer to the truth, but the processes are much wider than merely social-psychological occurrences within the minds of offenders. Criminals are only part of a larger system, which encompasses the whole society.

THE LIMITS OF CRIME

If the whole social structure is producing crime, we might wonder if there is any limit to how much crime it produces. If crime helps hold society together, doesn't it follow, paradoxically, that the more crime there is the better integrated the society will be? Obviously, there must come a point at which the amount of crime is too great. There would be no one left to enforce the laws, and society would fall apart.

Nevertheless, this does not usually happen. If we look further into the matter, the reasons turn out to be not so much that the law-enforcement side effectively controls crime, but that crime tends to limit itself. Look at what happens when crime becomes more and more successful. Individual criminals can do only so much. They are much more effective at stealing, embezzling, or whatever if they are organized. Individual thieves give way to gangs, and gangs to organized crime syndicates. But notice; organized crime now becomes a little society of its own. It creates its own hierarchy, its own rules, and it attempts to enforce these rules upon its own members. Organized crime tends toward regularity and normalcy. It begins to deplore unnecessary violence and strife. The more successful it is, the more it approximates an ordinary business. The very success of crime, then, tends to make it more law-abiding and less criminal. The same thing can be seen historically.

At some points in history, political power consisted of little more than marauding gangs of warriors or robber barons that plundered whoever came their way. The very success of some of these well-armed criminals, if we may call them that, meant that they had to take more responsibility for maintaining social order around them. At a minimum, the violent gang of warriors had to maintain discipline among themselves if it was to operate effectively in plundering others. The more successful a robber-baron became, the more he turned into an enforcer of laws. The state arose from a type of criminality but was forced to create a morality just to survive.

If social life creates crime, then crime also tends to create its own antithesis. Crime tends to drive out crime. It is not so easy, after all, to be a successful criminal. If you start out today to be a thief, let us say, how do you go about it? In many ways it is like learning any other occupation. You need to learn the tricks of the trade: how to break into a house, how to open a locked car. You need to know where to acquire the proper tools: where to get guns, if you want to be an armed robber. And you need to learn how to dispose of the loot once you have stolen it; it doesn't do you much good to steal a lot of television sets and stereos if you have no way of selling them for cash. And the more expensive the stolen goods, the more difficult it is to dispose of them profitably. To realize very much when stealing jewelry or artwork, for example, one needs both special training in how to recognize objects of value and special connections for getting rid of them. Stolen cars, too, because of the elaborate regulations of licensing and serial numbers, can only be profitably gotten rid of by tying in with a smoothly functioning criminal organization.

Any new criminal starting out on a life of crime has a lot to learn and many connections to make. Most novice criminals cannot make it very far in the crime world for exactly the same sort of reasons that most people in legitimate business never make it to the level of corporation executive. The average robbery nets less than $100, which is not exactly a fast way to get rich. Crime is a competitive world, too, as soon as one goes into it seriously in order to make a good living from it. Part of this is a kind of market effect, a process of sheer supply and demand. The more stolen goods show up at the fence, the less will be paid for them. The more criminals involved in any particular racket, the less take there will be for any one of them. Established criminals have no reason to want to help just anyone who wants to learn the trade and acquire the necessary connections. Hence, many novice criminals are simply "flunked out"; there isn't enough room for them in the world of crime.

Perhaps it is for this reason that crime rates peak for the youth population between ages fifteen and eighteen, and drop off rapidly thereafter. Youths at this age are not seriously committed to crime; they do not know much about the criminal world. They don't have much money of their own, or very much sense of what one can do with money. Small robberies may seem like an easy way to get a few luxuries. Auto theft, for example, is especially high at this age. But teenagers have little sense of how to market a stolen car; they are more likely to joy-ride around in it for a while and then abandon it. Obviously one can't make much of a living out of this sort of thing. If the crime rate starts dropping off in the late teens, and reaches a fairly low level by the age of thirty, it is not so much because of the effectiveness of the law enforcement system but simply because most youthful criminals wash out of a career in crime. (Again, as I mentioned, most crimes are committed by males, and that is the occupational pattern to pay attention to here.) Crime simply doesn't bring in enough income for them and they are forced to turn to something else to make their way in the adult world.

In the final analysis, the problem of crime, and its solution as well, is built much more deeply into the social structure than common sense would lead us to believe. Crime is so difficult to control because it is produced by large-scale social processes. The police, the courts, the prisons, the parole system are not very effective in counteracting criminality, and their very ineffectiveness seems foreordained by their largely ritualistic nature. Yet on the other side, crime has its own limitations. It works best the more it is organized, but the more organized it becomes, the more it becomes law-abiding and self-disciplining in its own ways. Individual criminals get squeezed out by the competitiveness of the world of crime itself, forced back into the world of ordinary society and its laws, whether they like it or not. Crime and society sway back and forth on this dialectic of opposing ironies.

32

The Culture of Fear

Why Americans Fear the Wrong Things

BARRY GLASSNER

The author notes a number of factors that many Americans fear. He also explains why these fears are based on falsehoods and, therefore, are unnecessary.

As you read this selection, consider the following questions as guides:

1. *Consider each fear in turn and indicate the reasons that you agree or disagree with the author's contention that it is groundless.*
2. *Think of items that should be added to the list of real fears. List other false fears.*
3. *If a fear is false, what should or can be done about the problem to which it refers?*

Why are so many fears in the air, and so many of them unfounded? Why, as crime rates plunged throughout the 1990s, did two-thirds of Americans believe they were soaring? How did it come about that by mid-decade 62 percent of us described ourselves as "truly desperate" about crime–almost twice as many as in the late 1980s, when crime rates were higher? Why, on a survey in 1997, when the crime rate had already fallen for a half dozen consecutive years, did more than half of us disagree with the statement "This country is finally beginning to make some progress in solving the crime problem"?[1]

In the late 1990s the number of drug users had decreased by half compared to a decade earlier; almost two-thirds of high school seniors had never used any illegal drugs, even marijuana. So why did a majority of adults rank drug abuse as the greatest danger to America's youth? Why did nine out of ten believe the drug problem is out of control, and only one in six believe the country was making progress?[2]

Give us a happy ending and we write a new disaster story. In the late 1990s the unemployment rate was below 5 percent for the first time in a quarter century. People who had been pounding the pavement for years could finally get work. Yet pundits warned of imminent economic disaster. They predicted inflation would take off, just as they had a few years earlier–also erroneously–when the unemployment rate dipped below 6 percent.[3]

INTRODUCTION

We compound our worries beyond all reason. Life expectancy in the United States has doubled during the twentieth century. We are better able to cure and control diseases than any other civilization in history. Yet we hear that phenomenal numbers of us are dreadfully ill. . . . The scope of our health fears seems limitless. Besides worrying

SOURCE: From *The Culture of Fear: Why Americans Are Afraid of the Wrong Things* by Barry Glassner. New York: Basic Books, 1998, pp. x-xxv.

disproportionately about legitimate ailments and prematurely about would-be diseases, we continue to fret over already refuted dangers. . . .

KILLER KIDS

When we are not worrying about deadly diseases we worry about homicidal strangers. Every few months for the past several years it seems we discover a new category of people to fear: government thugs in Waco, sadistic cops on Los Angeles freeways and in Brooklyn police stations, mass-murdering youths in small towns all over the country. A single anomalous event can provide us with multiple groups of people to fear. After the 1995 explosion at the federal building in Oklahoma City first we panicked about Arabs. "Knowing that the car bomb indicates Middle Eastern terrorists at work, it's safe to assume that their goal is to promote free-floating fear and a measure of anarchy, thereby disrupting American life," a *New York Post* editorial asserted. "Whatever we are doing to destroy Mideast terrorism, the chief terrorist threat against Americans, has not been working," wrote A. M. Rosenthal in the *New York Times*.[4]

When it turned out that the bombers were young white guys from middle America, two more groups instantly became spooky: right-wing radio talk show hosts who criticize the government–depicted by President Bill Clinton as "purveyors of hatred and division"–and members of militias. No group of disgruntled men was too ragtag not to warrant big, prophetic news stories.[5]

We have managed to convince ourselves that just about every young American male is a potential mass murderer–a remarkable achievement, considering the steep downward trend in youth crime throughout the 1990s. Faced year after year with comforting statistics, we either ignore them–adult Americans estimate that people under eighteen commit about half of all violent crimes when the actual number is 13 percent–or recast them as "The Lull Before the Storm" (*Newsweek* headline).[6] . . .

The more things improve the more pessimistic we become. Violence-related deaths at the nation's schools dropped to a record low during the 1996–97 academic year (19 deaths out of 54 million children), and only one in ten public schools reported *any* serious crime. Yet *Time* and *U.S. News & World Report* both ran headlines in 1996 referring to "Teenage Time Bombs." In a nation of "Children Without Souls" (another *Time* headline that year), "America's beleaguered cities are about to be victimized by a paradigm shattering wave of ultraviolent, morally vacuous young people some call 'the superpredators,'" William Bennett, the former Secretary of Education, and John DiIulio, a criminologist, forecast in a book published in 1996.[7]

Instead of the arrival of superpredators, violence by urban youths continued to decline. So we went looking elsewhere for proof that heinous behavior by young people was "becoming increasingly more commonplace in America" (CNN). After a sixteen-year-old in Pearl, Mississippi, and a fourteen-year-old in West Paducah, Kentucky, went on shooting sprees in late 1997, killing five of their classmates and wounding twelve others, these isolated incidents were taken as evidence of "an epidemic of seemingly depraved adolescent murderers" (Geraldo Rivera). Three months later in March 1998 all sense of proportion vanished after two boys ages eleven and thirteen killed four students and a teacher in Jonesboro, Arkansas. No longer, we learned in *Time*, was it "unusual for kids to get back at the world with live ammunition." When a child psychologist on NBC's "Today" show advised parents to reassure their children that shootings at schools are rare, reporter Ann Curry corrected him. "But this is the fourth case since October," she said.[8]

Over the next couple of months young people failed to accommodate the trend hawkers. None committed mass murder. . . . Yet given what had happened in Mississippi, Kentucky, Arkansas, and Oregon, could anyone doubt that today's youths are "more likely to pull a gun than make a fist," as Katie Couric declared on the "Today" show?[9]

ROOSEVELT WAS WRONG

We had better learn to doubt our inflated fears before they destroy us. Valid fears have their place; they cue us to danger. False and over-drawn fears only cause hardship.

Even concerns about real dangers, when blown out of proportion, do demonstrable harm. Take the fear of cancer. Many Americans overestimate the prevalence of the disease, underestimate the odds of surviving it, and put themselves at greater risk as a result. . . .[10]

Still more ironic, if harder to measure, are the adverse consequences of public panics. Exaggerated perceptions of the risks of cancer at least produce beneficial by-products, such as bountiful funding for research and treatment of this leading cause of death. When it comes to large-scale panics, however, it is difficult to see how potential victims benefit from the frenzy. Did panics a few years ago over sexual assaults on children by preschool teachers and priests leave children better off? Or did they prompt teachers and clergy to maintain excessive distance from children in their care, as social scientists and journalists who have studied the panics suggest? How well can care givers do their jobs when regulatory agencies, teachers' unions, and archdioceses explicitly prohibit them from any physical contact with children, even kindhearted hugs?[11] Was it a good thing for children and parents that male day care providers left the profession for fear of being falsely accused of sex crimes? . . .

We all pay one of the costs of panics: huge sums of money go to waste. Hysteria over the ritual abuse of children cost billions of dollars in police investigations, trials, and imprisonments. Men and women went to jail for years "on the basis of some of the most fantastic claims ever presented to an American jury," as Dorothy Rabinowitz of the *Wall Street Journal* demonstrated in a series of investigative articles for which she became a Pulitzer Prize finalist in 1996. Across the nation expensive surveillance programs were implemented to protect children from fiends who reside primarily in the imaginations of adults.[12]

The price tag for our panic about overall crime has grown so monumental that even law-and-order zealots find it hard to defend. The criminal justice system costs Americans close to $100 billion a year, most of which goes to police and prisons. In California we spend more on jails than on higher education. Yet increases in the number of police and prison cells do not correlate consistently with reductions in the number of serious crimes committed. Criminologists who study reductions in homicide rates, for instance, find little difference between cities that substantially expand their police forces and prison capacity and others that do not.[13]

The turnabout in domestic public spending over the past quarter century, from child welfare and antipoverty programs to incarceration, did not even produce reductions in *fear* of crime. Increasing the number of cops and jails arguably has the opposite effect: it suggests that the crime problem is all the more out of control.[14]

Panic-driven public spending generates over the long term a pathology akin to one found in drug addicts. The more money and attention we fritter away on our compulsions, the less we have available for our real needs, which consequently grow larger. While fortunes are being spent to protect children from dangers that few ever encounter, approximately 11 million children lack health insurance, 12 million are malnourished, and rates of illiteracy are increasing.[15]

I do not contend, as did President Roosevelt in 1933, that "the only thing we have to fear is fear itself." My point is that we often fear the wrong things. . . .

One of the paradoxes of a culture of fear is that serious problems remain widely ignored even though they give rise to precisely the dangers that the populace most abhors . . . .

TWO EASY EXPLANATIONS

In the following discussion I will try to answer two questions: Why are Americans so fearful lately, and why are our fears so often misplaced? To both questions the same two-word answer is commonly given by scholars and journalists: premillennial tensions. The final years of a millennium and the early years of a new millennium provoke mass anxiety and ill reasoning, the argument goes. So momentous does the calendric change seem, the populace cannot keep its wits about it.... In a classic study thirty years ago Alan Kerckhoff and Kurt Back pointed out that "the belief in a tangible threat makes it possible to explain and justify one's sense of discomfort."[16]...

Another popular explanation blames the news media. We have so many fears, many of them off-base, the argument goes, because the media bombard us with sensationalistic stories designed to increase ratings. This explanation, sometimes called the media-effects theory, is less simplistic than the millennium hypothesis and contains sizable kernels of truth. When researchers from Emory University computed the levels of coverage of various health dangers in popular magazines and newspapers they discovered an inverse relationship: much less space was devoted to several of the major causes of death than to some uncommon causes. The leading cause of death, heart disease, received approximately the same amount of coverage as the eleventh-ranked cause of death, homicide. They found a similar inverse relationship in coverage of risk factors associated with serious illness and death. The lowest-ranking risk factor, drug use, received nearly as much attention as the second-ranked risk factor, diet and exercise.[17]

Disproportionate coverage in the news media plainly has effects on readers and viewers.... Asked in a national poll why they believe the country has a serious crime problem, 76 percent of people cited stories they had seen in the media. Only 22 percent cited personal experience.[18]

When professors Robert Blendon and John Young of Harvard analyzed forty-seven surveys about drug abuse conducted between 1978 and 1997, they too discovered that the news media, rather than personal experience, provide Americans with their predominant fears.[19]...

Television news programs survive on scares. On local newscasts, where producers live by the dictum "if it bleeds, it leads," drug, crime, and disaster stories make up most of the news portion of the broadcasts. Evening newscasts on the major networks are somewhat less bloody, but between 1990 and 1998, when the nation's murder rate declined by 20 percent, the number of murder stories on network newscasts increased 600 percent (*not* counting stories about O. J. Simpson).[20]

After the dinnertime newscasts the networks broadcast newsmagazines, whose guiding principle seems to be that no danger is too small to magnify into a national nightmare.... A wide array of groups, including businesses, advocacy organizations, religious sects, and political parties, promote and profit from scares. News organizations are distinguished from other fear-mongering groups because they sometimes bite the scare that feeds them.

A group that raises money for research into a particular disease is not likely to negate concerns about that disease. A company that sells alarm systems is not about to call attention to the fact that crime is down. News organizations, on the other hand, periodically allay the very fears they arouse to lure audiences....

Several major newspapers parted from the pack in other ways. *USA Today* and the *Washington Post*, for instance, made sure their readers knew that what should worry them is the availability of guns. *USA Today* ran news stories explaining that easy access to guns in homes accounted for increases in the number of juvenile arrests for homicide in rural areas during the 1990s.... *USA Today* ran an op-ed piece proposing legal parameters for gun ownership akin to those for the use of alcohol and motor vehicles. And the paper published its own editorial in support of laws that require gun owners to lock their guns or keep them in locked containers. Adopted at that time by only fifteen states, the laws had reduced the number of deaths among children in those states by 23 percent.[21]

The *Washington Post*, meanwhile, published an excellent investigative piece by reporter Sharon Walsh showing that guns increasingly were being marketed to teenagers and children. . . . "Seems like only yesterday that your father brought you here for the first time," reads the copy beside a photo of a child aiming a handgun, his father by his side. "Those sure were the good times–just you, dad and his Smith & Wesson."[22]

As a social scientist I am impressed and somewhat embarrassed to find that journalists, more often than media scholars, identify the jugglery involved in making small hazards appear huge and huge hazards disappear from sight. Take, for example, the scare several years ago over the Ebola virus. . . . A report by *Dateline NBC* on deaths in Zaire, for instance, interspersed clips from *Outbreak*, a movie whose plot involves a lethal virus that threatens to kill the entire U.S. population. Alternating between Dustin Hoffman's character exclaiming, "We can't stop it!" and real-life science writer Laurie Garrett, author of *The Coming Plague*, proclaiming that "HIV is not an aberration . . . it's part of a trend," *Dateline*'s report gave the impression that swarms of epidemics were on their way. . . .[23]

"It is one of the ironies of the analysis of alarmists such as Preston that they are all too willing to point out the limitations of human beings, but they neglect to point out the limitations of microscopic life forms," Gladwell notes. . . .[24]

Among my personal favorites is an article published in 1996 titled "Fright by the Numbers," in which reporter Cynthia Crossen rebuts a cover story in *Time* magazine on prostate cancer. One in five men will get the disease, *Time* thundered. "That's scary. But it's also a lifetime risk–the accumulated risk over some 80 years of life," Crossen responds. A forty-year-old's chance of coming down with (not dying of) prostate cancer in the next ten years is 1 in 1,000, she goes on to report. His odds rise to 1 in 100 over twenty years. Even by the time he's seventy, he has only a 1 in 20 chance of *any* kind of cancer, including prostate.[25]

In the same article Crossen counters other alarmist claims as well, such as the much-repeated pronouncement that one in three Americans is obese. The number actually refers to how many are overweight, a less serious condition. Fewer are *obese* (a term that is less than objective itself), variously defined as 20 to 40 percent above ideal body weight as determined by current standards. . . .[26]

MORALITY AND MARKETING

From a psychological point of view extreme fear and outrage are often projections. Consider, for example, the panic over violence against children. By failing to provide adequate education, nutrition, housing, parenting, medical services, and child care over the past couple of decades we have done the nation's children immense harm. Yet we project our guilt onto a cavalcade of bogeypeople– pedophile preschool teachers, preteen mass murderers, and homicidal au pairs, to name only a few. . . .[27]

Diverse groups used the ritual-abuse scares to diverse ends. Well-known feminists such as Gloria Steinem and Catharine MacKinnon took up the cause, depicting ritually abused children as living proof of the ravages of patriarchy and the need for fundamental social reform.[28] This was far from the only time feminist spokeswomen have mongered fears about sinister breeds of men who exist in nowhere near the high numbers they allege. . . .

Within public discourse fears proliferate through a process of exchange. It is from crosscurrents of scares and counterscares that the culture of fear swells ever larger. . . . Samuel Taylor Coleridge was right when he claimed, "In politics, what begins in fear usually ends up in folly." Political activists are more inclined, though, to heed an observation from Richard Nixon: "People react to fear, not love. They don't teach that in Sunday school, but it's true." That principle, which guided the late president's political strategy throughout his career, is the sine qua non of contemporary political campaigning. Marketers of products and services ranging from car alarms to TV news programs have taken it to heart as well.[29]

The short answer to why Americans harbor so many misbegotten fears is that immense power and money await those who tap into our moral insecurities and supply us with symbolic substitutes. . . .

NOTES

1. Crime data here and throughout are from reports of the Bureau of Justice Statistics unless otherwise noted. Fear of crime: Esther Madriz, *Nothing Bad Happens to Good Girls* (Berkeley: University of California Press, 1997), Ch. 1; and Richard Morin, "As Crime Rate Falls, Fears Persist," *Washington Post* National Edition, 16 June 1997, p. 35; David Whitman, "Believing the Good News," *U.S. News & World Report*, 5 January 1998, pp. 45–46.
2. Eva Bertram, Morris Blachman et al., *Drug War Politics* (Berkeley: University of California Press, 1996), p. 10; Mike Males, *Scapegoat Generation* (Monroe, ME: Common Courage Press, 1996), ch. 6; Karen Peterson, "Survey: Teen Drug Use Declines," *USA Today*, 19 June 1998, p. A6; Robert Blendon and John Young, "The Public and the War on Illicit Drugs," *Journal of the American Medical Association* 279 (18 March 1998): 827–32. In presenting these statistics and others I am aware of a seeming paradox: I criticize the abuse of statistics by fearmongering politicians, journalists, and others but hand down precise-sounding numbers myself. Yet to eschew all estimates because some are used inappropriately or do not withstand scrutiny would be as foolhardy as ignoring all medical advice because some doctors are quacks. Readers can be assured I have interrogated the statistics presented here as factual. As notes throughout the book make clear, I have tried to rely on research that appears in peer-reviewed scholarly journals. Where this was not possible or sufficient, I traced numbers back to their sources, investigated the research methodology utilized to produce them, or conducted searches of the popular and scientific literature for critical commentaries and conflicting findings.
3. Bob Herbert, "Bogeyman Economics," *New York Times*, 4 April 1997, p. A15; Doug Henwood, "Alarming Drop in Unemployment," *Extra*, September 1994, pp. 16–17; Christopher Shea, "Low Inflation and Low Unemployment Spur Economists to Debate 'Natural Rate' Theory," *Chronicle of Higher Education*, 24 October 1997, p. A13.
4. Jim Naureckas, "The Jihad That Wasn't," *Extra*, July 1995, pp. 6–10, 20 (contains quotes). See also Edward Said, "A Devil Theory of Islam," *Nation*, 12 August 1996, pp. 28–32.
5. Lewis Lapham, "Seen but Not Heard," *Harper's*, July 1995, pp. 29–36 (contains Clinton quote). See also Robin Wright and Ronald Ostrow, "Illusion of Immunity Is Shattered," *Los Angeles Times*, 20 April 1995, pp. A1, 18; Jack Germond and Jules Witcover, "Making the Angry White Males Angrier," column syndicated by Tribune Media Services, May 1995; and articles by James Bennet and Michael Janofsky in the *New York Times*, May 1995.
6. Tom Morganthau, "The Lull Before the Storm?" *Newsweek*, 4 December 1995, pp. 40–42; Mike Males, "Wild in Deceit," *Extra*, March 1996, pp. 7–9; *Progressive*, July 1997, p. 9 (contains Clinton quote); Robin Templeton, "First, We Kill All the 11-Year-Olds," *Salon*, 27 May 1998.
7. Statistics from "Violence and Discipline Problems in U.S. Public Schools: 1996-97," National Center on Education Statistics, U.S. Department of Education Washington, DC, March 1998; CNN, "Early Prime," 2 December 1997; and Tamar Lewin, "Despite Recent Carnage, School Violence Is Not on Rise," *New York Times*, 3 December 1997, p. A14. Headlines: *Time*, 15 January 1996; *U.S. News & World Report*, 25 March 1996; Margaret Carlson, "Children Without Souls," *Time*, 2 December 1996, p. 70. William J. Bennett, John J. DiIulio, and John Walters, *Body Count* (New York: Simon & Schuster, 1996).
8. CNN, "Talkback Live," 2 December 1997; CNN, "The Geraldo Rivera Show," 11 December 1997; Richard Lacayo, "Toward the Root of Evil," *Time*, 6 April 1998, pp. 38–39; NBC, "Today," 25 March 1998. See also Rick Bragg, "Forgiveness, After 3 Die in Shootings in Kentucky," *New York Times*, 3 December 1997, p. A14; Maureen Downey, "Kids

and Violence," 28 March 1998, *Atlanta Journal and Constitution*, p. A12.

9. Jocelyn Stewart, "Schools Learn to Take Threats More Seriously," *Los Angeles Times*, 11 May 1998, pp. A1, 17; "Kindergarten Student Faces Gun Charges," *New York Times*, 11 May 1998, p. A11; Rick Bragg, "Jonesboro Dazed by Its Darkest Day" and "Past Victims Relive Pain as Tragedy Is Repeated," *New York Times*, 18 April 1998, p. A7, and idem, 25 May 1998, p. A8. Remaining quotes are from Tamar Lewin, "More Victims and Less Sense in Shootings," *New York Times*, 22 May 1998, p. A20; NPR, "All Things Considered," 22 May 1998; NBC, "Today," 25 March 1998. See also Mike Males, "Who's Really Killing Our Schoolkids," *Los Angeles Times*, 31 May 1998, pp. M1, 3; Michael Sniffen, "Youth Crime Fell in 1997, Reno Says," Associated Press, 20 November 1998.
10. Overestimation of breast cancer: Willam C. Black et al., "Perceptions of Breast Cancer Risk and Screening Effectiveness in Women Younger Than 50," *Journal of the National Cancer Institute* 87 (1995): 720–31; B. Smith et al., "Perception of Breast Cancer Risk Among Women in Breast and Family History of Breast Cancer," *Surgery* 120 (1996): 297–303. Fear and avoidance: Steven Berman and Abraham Wandersman, "Fear of Cancer and Knowledge of Cancer," *Social Science and Medicine* 31 (1990): 81–90; S. Benedict et al., "Breast Cancer Detection by Daughters of Women with Breast Cancer," *Cancer Practice* 5 (1997): 213–19; M. Muir et al., "Health Promotion and Early Detection of Cancer in Older Adults," *Cancer Oncology Nursing Journal* 7 (1997): 82–89. For a conflicting finding see Kevin McCaul et al., "Breast Cancer Worry and Screening," *Health Psychology* 15 (1996): 430–33.
11. Philip Jenkins, *Pedophiles and Priests* (New York: Oxford University Press, 1996), see esp. Ch. 10; Debbie Nathan and Michael Snedeker, *Satan's Silence* (New York: Basic Books, 1995), see esp. Ch. 6; Jeffrey Victor, "The Danger of Moral Panics," *Skeptic* 3 (1995): 44–51. See also Noelle Oxenhandler, "The Eros of Parenthood," *Family Therapy Networker* (May 1996): 17–19.
12. Dorothy Rabinowitz, "A Darkness in Massachusetts," *Wall Street Journal*, 30 January 1995, p. A20 (contains quote); "Back in Wenatchee" (unsigned editorial), *Wall Street Journal*, 20 June 1996, p. A18; Dorothy Rabinowitz, "Justice in Massachusetts," *Wall Street Journal*, 13 May 1997, p. A19. See also Nathan and Snedeker, *Satan's Silence*; James Beaver, "The Myth of Repressed Memory," *Journal of Criminal Law and Criminology*, 86 (1996): 596–607; Kathryn Lyon, *Witch Hunt* (New York: Avon, 1998); Pam Belluck, " 'Memory' Therapy Leads to a Lawsuit and Big Settlement," *New York Times*, 6 November 1997, pp. A1, 10.
13. Elliott Currie, *Crime and Punishment in America* (New York: Metropolitan, 1998); Tony Pate et al., *Reducing Fear of Crime in Houston and Newark* (Washington, DC: Police Foundation, 1986); Steven Donziger, *The Real War on Crime* (New York: HarperCollins, 1996); Christina Johns, *Power, Ideology and the War on Drugs* (New York: Praeger, 1992); John Irwin et al., "Fanning the Flames of Fear," *Crime and Delinquency* 44 (1998): 32–48.
14. Steven Donziger, "Fear, Crime and Punishment in the U.S.," *Tikkun* 12 (1996): 24–27, 77.
15. Peter Budetti, "Health Insurance for Children," *New England Journal of Medicine* 338 (1998): 541–42; Eileen Smith, "Drugs Top Adult Fears for Kids' Well-being," *USA Today*, 9 December 1997, p. D1. Literacy statistic: Adult Literacy Service.
16. Alan Kerckhoff and Kurt Back, *The June Bug* (New York: Appleton-Century-Crofts, 1968), see esp. pp. 160-61.
17. Karen Frost, Erica Frank et al., "Relative Risk in the News Media," *American Journal of Public Health* 87 (1997): 842–45. Media-effects theory: Nancy Signorielli and Michael Morgan, eds., *Cultivation Analysis* (Newbury Park, CA: Sage, 1990); Jennings Bryant and Dolf Zillman, eds., *Media Effects* (Hillsdale, NJ: Erlbaum, 1994); Ronald Jacobs, "Producing the News, Producing the Crisis," *Media, Culture and Society* 18 (1996): 373–97.
18. Madriz, *Nothing Bad Happens to Good Girls*, see esp. pp. 111–14; David Whitman and Margaret Loftus, "Things Are Getting Better? Who Knew," *U.S. News & World Report*, 16 December 1996, pp. 30–32.
19. Blendon and Young, "War on Illicit Drugs," See also Ted Chiricos et al., "Crime, News and Fear of Crime," *Social Problems* 44 (1997). 342–57.
20. Steven Stark, "Local News: The Biggest Scandal on TV," *Washington Monthly* (June 1997): 38–41; Barbara Bliss Osborn, "If It Bleeds, It Leads," *Extra*, September–October 1994, p. 15; Jenkins, *Pedophiles*

and Priests, pp. 68–71; "It's Murder," *USA Today*, 20 April 1998, p. D2; Lawrence Grossman, "Does Local TV News Need a National Nanny?" *Columbia Journalism Review* (May 1998): 33.

21. "Licensing Can Protect," *USA Today*, 7 April 1998, p. A11; Jonathan Kellerman, "Few Surprises When It Comes to Violence," *USA Today*, 27 March 1998, p. A13; Gary Fields, "Juvenile Homicide Arrrest Rate on Rise in Rural USA," *USA Today*, 26 March 1998, p. A11; Karen Peterson and Glenn O'Neal, "Society More Violent, So Are Its Children," *USA Today*, 25 March 1998, p. A3; Scott Bowles, "Armed, Alienated and Adolescent," *USA Today*, 26 March 1998, p. A9. Similar suggestions about guns appear in Jonathan Alter, "Harnessing the Hysteria," *Newsweek*, 6 April 1998, p. 27.
22. Sharon Walsh, "Gun Sellers Look to Future–Children," *Washington Post*, 28 March 1998, pp. A1, 2.
23. John Schwartz, "An Outbreak of Medical Myths," *Washington Post* National Edition, 22 May 1995, p. 38.
24. Richard Preston, *The Hot Zone* (New York: Random House, 1994); Malcolm Gladwell, "The Plague Year," *New Republic*, 17 July 1995, p. 40.
25. Erik Larson, "A False Crisis: How Workplace Violence Became a Hot Issue," *Wall Street Journal*, 13 October 1994, pp. A1, 8; Cynthia Crossen, "Fright By the Numbers," *Wall Street Journal*, 11 April 1996, pp. B1, 8. See also G. Pascal Zachary, "Junk History," *Wall Street Journal*, 19 September 1997, pp. A1, 6.
26. On variable definitions of obesity see also Werner Cahnman, "The Stigma of Obesity," *Sociological Quarterly* 9 (1968): 283–99; Susan Bordo, *Unbearable Weight* (Berkeley: University of California Press, 1993); Joan Chrisler, "Politics and Women's Weight," *Feminism and Psychology* 6 (1996): 181–84.
27. See Marina Warner, "Peroxide Mug-shot," *London Review of Books*, 1 January 1998, pp. 10–11.
28. Nathan and Snedeker, *Satan's Silence* (quote from p. 240). See also David Bromley, "Satanism: The New Cult Scare," in James Richardson et al., eds., *The Satanism Scare* (Hawthorne, NY: Aldine de Gruyter, 1991), pp. 49–71.
29. Henry Nelson Coleridge, ed., *Specimens of the Table Talk of the Late Samuel Taylor Coleridge* (London: J. Murray, 1935), entry for 5 October 1830. Nixon quote cited in William Safire, *Before the Fall* (New York: Doubleday, 1975), Prologue.

33

Crime Control in America

Nothing Succeeds like Failure

JEFFREY REIMAN

The author of this reading implies that crime control is a purposeful failure since it is known what would be more successful. The purposefulness is tied in with Durkheim's theory of the necessity of crime. What do you think of the means for dealing with crime? Why do you think they are not instigated?

As you read, ask yourself the following questions:

1. *The author claims that our assaults on the crime problem are a failure. Why?*
2. *What changes would you make in the criminal justice system? Defend your changes.*
3. *Would executing all murderers within six months of conviction lessen the murder rate? Why or why not?*

GLOSSARY **FBI Crime Index** Federal Bureau of Investigation report on criminal offenses "against the person" and "against property." **UCR** Uniform Crime Reports, an FBI measure of crime.

DESIGNED TO FAIL

The plain fact is that virtually no student of the crime problem believes we can arrest and imprison our way out of the crime problem. To be sure, we have seen an enormous increase in the number of Americans behind bars. Between 1980 and 1996 the number of persons incarcerated in state and federal prisons nearly quadrupled, growing from 329,000 to nearly 1.2 million. Including those who are locked up in jails, there are currently more than 1.8 million people behind bars in the United States—a number more than twice the population of San Francisco![1] The Justice Policy Institute estimated that by February 15, 2000, "America's prison and jail population will top 2 million [!]"[2] The Bureau of Justice Statistics reports that if current incarceration rates remain unchanged, 9 percent of men and 28 percent of black men can expect to serve time in prison during their lifetime![3]

And what are the results? Violent crimes have declined since 1992, but they still are not down to where they were in 1985, when crime was still very high. For example, in its *Uniform Crime Reports (UCR)* for 1992, the FBI reported 1,932,270 crimes of violence, and a violent crime rate of 758 per 100,000 persons in the population. In 1998 the FBI reported a decline in the number of violent crimes to 1,531,044, and in the crime rate to 566 per 100,000. In 1985, however, the FBI counted 1,328,800 violent crimes with a rate per

SOURCE: From *The Rich Get Richer and the Poor Get Prison*, 6th edition, by Jeffrey Reiman. Copyright 2001 by Allyn & Bacon. Reprinted by permission.

100,000 of 557.[4] And the Justice Department's *National Crime Victimization Survey (NCVS)*, which picks up many crimes not reported to the police, found 8,116,000 violent crimes in 1998 (3,790 per 100,000)—higher than in 1992, when survey respondents reported 6,621,000 victimizations; and higher than in 1985, when 5,823,000 violent crimes were reported.[5] The recent reductions in crime are little more than a leveling off from the rises in crimes that preceded them, and they have come down to rates that existed when far fewer of our fellows were being locked up. In short, the crime reductions for which our leaders are now claiming credit are actually no more than reductions from *very, very high* crime rates to rates that are merely *very high*....

A less comforting view appeared in a December 5, 1999, article in *The Washington Post* under the headline: "Despite Rhetoric, Violent Crime Climbs." Begins the article:

> Rosy assessments of the nation's declining crime rate wrongly focus on the short-term drops from crime peaks early in the decade and ignore the overall rise of violence since the 1960s, according to a new report.
>
> The 30-year update of a landmark study by the National Commission on the Causes and Prevention of Violence found that violent crime in major cities reported to the FBI has risen by 40 percent since 1969.
>
> The new study is intended as a counterpoint to the drumbeat of optimistic reports describing the current drop in crime, and it offers a sober reminder that the United States still suffers from a historically high level of violence.[6]

This new study was conducted by the Milton S. Eisenhower Foundation.... The foundation study noted the strikingly higher rates of violent crime in the United States compared to other industrialized nations: "In 1995, handguns were used to kill 2 people in New Zealand, 15 in Japan, 30 in Great Britain, 106 in Canada, 213 in Germany, and 9,390 in the United States."[7] "The most optimistic view after looking at this," said foundation president Lynn A. Curtis, who also worked on the 1969 violence report, "is that we are in roughly the same ballpark now in the late 1990s as we were in the late 1960s, when everyone said crime is so bad we need a national commission to study it."[8] The difference is that in 1969 there were 197,136 individuals in state and federal prisons, but by 1997 that number had grown *nearly 600 percent* to 1,130,465—growth that has cost us billions of dollars, given prison records to huge numbers of nonviolent criminals, and torn up inner-city communities, but that has not made much of a difference in the amount of crime we have and leaves Americans significantly more fearful of crime than they were in 1969.[9]...

Moreover, it is not hard to figure out why this unprecedented imprisonment binge has produced such meager effects. First of all, since American jurisdictions have always been highly likely to imprison violent offenders, an increase in the rate of imprisonment necessarily means that we are imprisoning more less-serious criminals than the criminals already in prison. The result is that:

> Violent offenders are now a considerably smaller proportion of the prison population than they were in the recent past, and non-violent offenders are therefore correspondingly greater. Inmates convicted of violent crimes were 57 percent of the state and federal prison population in 1980 and 44 percent in 1995; drug offenders rose from just 8 percent to 26 percent in the same period.[10]

... This is not to deny that we have recently seen some significant declines in crime rates. The point is that very little of these declines can be attributed to the enormous increase in our prison population. Indeed, numerous students of the crime problem attribute the recent declines to factors other than criminal justice policies. For example, the United States currently has an unemployment rate of slightly above 4 percent, the lowest rate in over 30 years—but none of the officials who are claiming credit for reducing crime, from the president on down, mentions this as a cause of lower crime rates. Moreover, the recent declines in crime come after a period in which much violent crime was attributed to turf wars between inner-city drug gangs. Now that the wars have been fought, more or less stable turf boundaries exist, and the rate of violence subsides—not because the police have succeeded, but in fact because the (surviving) drug dealers have succeeded in turning their trade into a stable inner-city business. Says Temple University professor of criminology and former New York City Police Captain Jim Fyfe: "When a new illegal and profitable substance comes along, there is fighting and scratching for control.... Then dealers kill each other off, and the market stabilizes and the amount of violence decreases."[11] And *U.S. News & World Report* points out that, "contrary to popular impression, turf wars among [drug] gangs are increasingly rare. A staple of the late 1980s and early 1990s..., such battles have now largely succumbed to what criminologist James Lynch of Washington [D.C.]'s American University terms the "routinization of the drug trade."[12]

Other experts attribute at least some of the decline in violent crime to "the fact that the nation's population of young men—the group most likely to commit violent crimes—is smaller now because of the aging of the baby-boomers."[13] The number of people aged 15 to 29 in our population declined steadily from 1976 through 1994, a demographic trend to which many criminologists attribute not only the recent decline but the larger decline that marked the years 1979 through 1984. Now this age group is slowly on the increase again, bringing with it dire predictions of dramatic increases in crime to come. Commenting on these demographic trends, Northeastern University criminologist James Alan Fox says: "This is the calm before the crime storm."[14]...

And Elliot Currie suggests that some of the reduction in violent crime is due to the fact that:

> ... a significant number of those at highest risk of being either perpetrators or victims have been removed from the picture—through death, disease, or disability.... Overall, young black men aged fifteen to twenty-four were 66 percent more likely to die in 1993 than in 1985—a stunning reversal of decades of general improvement in life expectancy.[15]

So, while politicians claim credit for the recent declines in crime, the real story appears to be this: While the enormous growth in our prison population over the last decade coupled with questionable police tactics may have contributed in some measure to the decline, much if not most of the decline can be attributed to factors beyond the criminal justice system: the reduction in unemployment, the stabilization of the drug trade, and the decrease in the number of young men in the crime-prone years. For all that, new growth in crime—"another tidal wave" *Time* calls it, "a ticking time bomb" says *The New York Times*—is right around the corner.[16]

The recent decline in the crime rates does not represent dramatic success in dealing with the U.S. crime problem. It represents rather a reduction from extremely high crime rates to rates that are slightly less high, *but high nonetheless*. In short, crime is still rampant, and, for all their crowing and claiming credit, neither politicians nor criminal justice policy makers have come close to changing this fact. The criminal justice system may win the occasional skirmish, but it is still losing the war against crime.

In 1960, the average citizen had less than a 1-in-50 chance of being a victim of one of the crimes on the FBI Index (murder, forcible rape, robbery, aggravated assault, burglary, larceny, or auto theft). In 1970, that chance grew to 1 in 25. In 1986, the FBI reported nearly 5,500 Index crimes per 100,000 citizens, a further increase in the likelihood of victimization to a 1-in-18 chance. And by 1991, this had reached 5,898 per 100,000 citizens—a better than 1-in-17 chance. The FBI reported slight declines in 1992 and 1993, with the rate for 1993 at 5,483 per 100,000, roughly where it was in 1986. Most of the decline was accounted for by a drop in property crimes. Even with these declines, the FBI said in its 1993 report: "Every American now has a realistic chance of murder victimization in view of the random nature the crime has assumed."[17] And for 1997, the FBI reports a further decline to 4,615

Index crimes per 100,000 in the U.S. population, a decrease in the chance of victimization to roughly 1 in 22—still higher than the rates and risks of Index crimes for 1970, and much higher than those for 1960, before our enormous investment in prisons began and while a far smaller number of Americans were behind bars.[18]

Moreover, American violent crime rates are still far higher than those in other countries. Writes criminologist Elliot Currie:

> By the mid-1990s . . . , a young American male was 37 times as likely to die by deliberate violence as his English counterpart—and 12 times as likely as a Canadian youth, 20 times as likely as Swede, 26 times as likely as a young Frenchman, and over 60 times as likely as a Japanese.[19]

In sum, when we look behind the politicians' claims to have turned the tide against violence, the fact remains that criminal justice policy is failing to make our lives substantially safer. How are we to comprehend this failure? . . .

One way to understand this failure is to look at the *excuses* that are offered for it. This we will do—but mainly to show that they do not hold up!

One commonly heard excuse is that we can't reduce crime because our laws and our courts are too lenient. *Translation*: We are failing to reduce crime because we don't have the heart to do what has to be done.

Other excuses point to some feature of modern life, such as urbanization or population growth (particularly the increase in the number of individuals in the crime-prone ages of 15 to 24), and say that this feature is responsible for the growth in crime. This means that crime cannot be reduced unless we are prepared to return to horse-and-buggy days or to abolish adolescence. *Translation*: We are failing to reduce crime because it is impossible to reduce crime.

Some try to excuse our failure by claiming that we simply do not know how to reduce crime. *Translation*: Even though we are doing our best, we are failing to reduce crime because our knowledge of the causes of crime is still too primitive to make our best good enough.

These excuses simply do not pass muster. There is no evidence that we are too lenient on crime . . . . Our rates of incarceration are as high as or higher than those of other modern nations, and we are the only one that still imposes the death penalty.

Moreover, although increasing urbanization and a growing youth population account for some of our high level of crime, they by no means account for all of it, and certainly not for the impossibility of reducing crime. Crime rates vary widely (and wildly) when we compare cities of similar population size and density . . . . This means that growing crime is not a simple, unavoidable consequence of increasing urbanization . . . . Similarly, the crime rate has increased far more rapidly than the youth population, both in absolute numbers and as a fraction of the total population. This means that growing crime is not a simple, unavoidable consequence of a growing youth population . . . .

On the other hand, the excuse that we do not know how to reduce crime also does not hold up. The bald truth is that we *do* know some of the sources of crime and *we obstinately refuse to remedy them!* . . .

. . . So the question "How are we to comprehend our failure to reduce crime?" still stares us in the face. Examination of the excuses and then of policies that could reduce crime suggest that our failure is avoidable. What has to be explained is not why *we cannot* reduce crime but why *we will not!* Oddly enough, this paradoxical result points us in the direction of an answer to our question.

KNOWN SOURCES OF CRIME . . .

Poverty

Those youngsters who figure so prominently in arrest statistics are not drawn equally from all economic strata. Although there is much reported and even more unreported crime among middle-class youngsters, the street crime attributed to this age group that makes our city streets a perpetual war zone is largely the work of poor ghetto youth. This is the group at the lowest end of the economic spectrum. This is a group among whom unemployment approaches 50 percent, with underemployment (the rate of persons either jobless or with part-time, low-wage jobs) still higher. This is a group with no realistic chance (for any but a rare individual) to enter college or amass sufficient capital (legally) to start a business or to get into the union-protected, high-wage, skilled job markets. We know that poverty is a *source* of crime, even if we do not know how it *causes* crime—and yet we do virtually nothing to improve the life chances of the vast majority of the inner-city poor. They are as poor as ever and are facing cuts in welfare and other services. . . .

An analysis of data issued by the Congressional Budget Office indicates that "Among the bottom fifth of households, average after-tax income is anticipated to *fall* nine percent from 1977 to 1999." The same report concludes that:

> Income disparities have widened to such a degree that in 1999, the richest one percent of the population is projected to receive as much after-tax income as the bottom 38 percent combined. That is, the 2.7 million Americans with the largest incomes are expected to receive as much after-tax income as the 100 million Americans with the lowest incomes.[20]

. . . In fact, poverty contributes to crime by creating need, while—at the other end of the spectrum—wealth can contribute to crime by unleashing greed. Some criminologists have argued that economic inequality itself worsens crimes of the poor and of the well off by increasing the opportunities for the well off and increasing the humiliation of the poor.[21] And inequality has worsened in recent years. . . .

Almost a decade later, an article in *Business Week*, looking back at the 1980s, confirms the charge in retrospect: "At the uppermost end of the income scale, tax cuts made aftertax income surge even higher than pretax income. And at the low end of the distribution scale, cuts in income transfers hurt the poor." The article notes also "the extraordinarily high level of child poverty in America today. One in five children under the age of 15 lives in poverty, and a staggering 50% of all black children under the age of six live in poverty."[22] And Edward Wolff writes that the "Equalizing trends of the 1930s–1970s reversed sharply in the 1980s. The gap between the haves and have-nots is greater now than at any time since 1929. The sharp increase in inequality since the late 1970s has made wealth distribution in the United States more unequal than in what used to be perceived as the class-ridden societies of north-western Europe.[23]

Moreover, as unemployment has gone up and down over the past decades, unemployment at the bottom of society remains strikingly worse than the national average. For example, over the past 25 years black unemployment has remained slightly more than twice the rate of white unemployment. In 1967, when 3.4 percent of white workers were unemployed, 7.4 percent of black workers were jobless. By 1997, 4.2 percent of white workers were unemployed and 10 percent of blacks were. Among those in the crime-prone ages of 16 to 19, 12.7 percent of white youngsters and 31.6 percent (more than one of every three) black youngsters were jobless.[24]

Writes Todd Clear, professor of criminal justice at Rutgers University, "Let's start investing in things that really reduce crime: good schools, jobs and a future for young parents and their children."[25] Why don't we?

Prison

There is more. We know that prison produces more criminals than it cures. We know that more than 70 percent of the inmates in the nation's prisons or jails are not there for the first time. We know that prison inmates are denied autonomy and privacy and subjected to indignities, mortifications, and acts of violence as regular features of their confinement—all of which is heightened by overcrowding. As of the last day of 1998, state prison systems were operating at between 13 percent and 22 percent over their reported capacity; the federal prison system was operating at 27 percent above capacity.[26] The predictable result, as delineated by Robert Johnson and Hans Toch in *The Pains of Imprisonment*, "is that the prison's survivors become tougher, more pugnacious, and less able to feel for themselves and others, while its nonsurvivors become weaker, more susceptible, and less able to control their lives."[27] Prisoners are thus bereft of both training and capacity to handle daily problems in competent and socially constructive ways, inside or outside of prison. Once on the outside, burdened with the stigma of a prison record and rarely trained in a marketable skill, they find few opportunities for noncriminal employment open to them. . . .

Should we then really pretend that we do not *know* why ex-cons turn to crime? Can we honestly act as if we do not know that our prison system (combined with our failure to ensure a meaningful postrelease noncriminal alternative for the ex-con) is a *source* of crime? Recidivism does not happen because ex-cons miss their alma mater. In fact, if prisons are built to deter people from crime, one would expect that ex-prisoners would be the most deterred because the deprivations of prison are more real to them than to the rest of us. Recidivism is thus a doubly poignant testimony to the job that prison does in preparing its graduates for crime—and yet we do little to change the nature of prisons or to provide real services to ex-convicts.

Guns

We know that it is about as difficult to obtain a handgun in the United States as a candy bar. . . .

. . . Gary Kleck estimates that, by 1990, the civilian stock of guns in the United States had passed the 200 million mark.[28] This estimate is corroborated by a 1993 report from the Bureau of Alcohol, Tobacco and Firearms, which estimated 200 million guns, about 1 percent of which were assault rifles. They also note that the "number of large caliber pistols produced annually increased substantially after 1986."[29] Nearly one-half of U.S. households have at least one gun. And about a quarter have at least one handgun. Half of handgun owners surveyed said that their guns were currently loaded.[30]

The President's Crime Commission reported that, in 1965, "5,600 murders, 34,700 aggravated assaults and the vast majority of the 68,400 armed robberies were committed by means of firearms. All but 10 of 278 law enforcement officers murdered during the period 1960–65 were killed with firearms." The commission concluded almost 30 years ago that

> . . . more than one-half of all willful homicides and armed robberies, and almost one-fifth of all aggravated assaults, involves use of firearms. *As long as there is no effective gun-control legislation, violent crimes and the injuries they inflict will be harder to reduce than they might otherwise be.*[31]

The situation has worsened since the commission's warning. The FBI reported that the "proportion of violent crimes committed with firearms has increased in recent years"—from being employed in the commission of 26 percent of violent offenses in 1987 to 32 percent in 1993. The FBI writes: "In 1975, 66 percent of murders of persons (aged 15 to 19) were attributable to guns, while in 1992 the figure rose to 85 percent. This increase supports the theory that today's high-school-aged youths are exposed to an environment that includes guns."[32] The Office of Juvenile Justice reports that "By 1997, the homicide rate for 15- to 24-year-olds was 15.2 per 100,000, which is higher than the combined total homicide rate of eleven industrialized nations," and goes on to point out that "Firearms were the weapons of choice in nearly two-thirds of all murders."[33] . . .

Can we believe that our leaders sincerely want to cut down on violent crime and the injuries it produces when they oppose even as much as *registering* guns or *licensing* gun owners, much less actually restricting the sale and movement of guns as a matter of national policy? . . . Are we to believe that the availability of guns does not contribute to our soaring crime rate? Zimring's study indicates that areas with a high number of privately owned guns have more crimes involving guns than do areas with lower numbers of privately owned firearms. His data also indicate that cities that experience an increase in legal gun sales also experience an increase in gun-related suicides, accidents, and crimes.[34]

This is hardly more than what common sense would lead us to expect. Can we really believe that if guns were less readily available, violent criminals would simply switch to other weapons to commit the same number of crimes and do the same amount of damage? Is there a weapon other than the handgun that works as quickly, that allows its user so safe a distance, or that makes the criminal's physical strength (or speed or courage for that matter) as irrelevant? Could a bank robber hold a row of tellers at bay with a switchblade? Would an escaping felon protect himself from a pursuing police officer with a hand grenade? Zimring's studies also indicate that if gun users switched to the next deadliest weapon—the knife—and attempted the same number of crimes, we could still expect *80 percent fewer fatalities* because the fatality rate of the knife is roughly one-fifth that of the gun. Another researcher found that family and intimate assaults involving firearms were 12 times more likely to result in death than those that did not.[35] In other words, even if guns were eliminated and crimes not reduced, we could expect to save as many as four out of every five persons who are now the victims of firearm homicide, and maybe more!

Drugs

Finally, the United States has an enormous drug abuse and addiction problem. There is considerable evidence, however, that our attempts to cure it are worse than the disease itself. Consider first heroin. Some people think this drug is out of fashion and no longer widely used. Far from it! Its use is widespread and persistent. . . . A U.S. government report covering 1994 states that "[g]rowing evidence indicated that domestic heroin consumption was on the rise." And further that "[e]stimates suggested that there may be 600,000 hardcore drug users who report heroin as their principal drug of abuse," and that "heroin was readily available to addicts in all major metropolitan areas. . . .

As shocking as these numbers may be, it must be at least as shocking to discover that there is little evidence proving that heroin is a *dangerous* drug. There is no evidence conclusively establishing a link between heroin and disease or tissue degeneration such as that established for tobacco and alcohol. James Q. Wilson, a defender of the prohibition on heroin and other drugs, admits that "there are apparently no specific pathologies—serious illnesses or physiological deterioration—that are known to result from heroin use per se."[36] On the basis of available scientific evidence, there is every reason to suspect that we do our bodies more damage, more *irreversible* damage, by smoking cigarettes and drinking liquor than by using heroin. Most of the physical damage associated with heroin use is probably attributable to the trauma of withdrawal—and this is a product not so much of heroin as of its occasional unobtainability. . . .

Suffice it to say, then, at the very least, our attitudes about heroin are inconsistent and irrational, and there is reason to believe they are outrageous and hypocritical. Even if this were not so, even if we could be much more certain that heroin addiction is a disease worth preventing, the fact would remain that the "cure" we have chosen is worse than the disease. We *know* that treating the possession of heroin as a criminal offense produces more crime than it prevents. . . .

. . . Professor Blumstein agrees, "you need money to buy drugs, so the higher the price of the drug, the greater the incentive to commit the crime."[37] The result is a recipe for large-scale and continual robbery and burglary, which would not exist if the drug were available legally. A recent study by Anglin and Speckart of the relationship between narcotics use and crime concludes that there is "strong evidence that there is a strong causal relationship, at least in the United States, between addiction to narcotics and property crime levels."[38] . . .

Looking at heroin, we must recognize it is not the "disease" of heroin addiction that leads to property crime. There is, writes Trebach, "nothing in the pharmacology, or physical and psychological impact, of the drug that would propel a user to crime."[39] Nor is there anything about heroin itself that makes it extremely costly. The heroin for which an addict pays $100 or more a day could be produced legally at a cost of a few cents for a day's supply. Thus, it is not the "disease" of heroin addiction but its "cure" that leads to property crime. . . .

In sum, we have an antidrug policy that is failing at its own goals and succeeding only in adding to crime. First, there are the heroin and crack addicts, who must steal to support their habits. Then, there are the drug merchants who are offered fabulous incentives to provide illicit substances to a willing body of consumers. This in turn contributes to the high rate of inner-city murders and other violence as drug gangs battle for the enormous sums of money available. Next, there are the law enforcement officials who, after risking their lives for low salaries, are corrupted by nearly irresistible amounts of money. Finally, there are the otherwise law-abiding citizens who are made criminals because they use cocaine, a drug less harmful than tobacco, and those who are made criminals because they use marijuana, a drug that is safer than alcohol and less deadly than aspirin. Much of the recent dramatic growth in our prison population . . . is the result of the hardening of drug enforcement policy in the Reagan years: In 1968 there were 162,000 drug arrests nationwide, in 1977 there were 569,000, and in 1989 there were 1,150,000 drug arrests.[40] In 1997 there were 1.6 million drug arrests, 200,000 more than in 1995.[41] And the Bureau of Justice Statistics reports that 63 percent of federal inmates were serving sentences for drug violations, an increase of 3 percent since 1993, and of 28 percent since 1980.[42] Since numerous studies show that arrested drug dealers in inner-city neighborhoods are quickly replaced, it was apparent from the start that this policy would have little success in reducing the availability of illicit drugs.[43]

And all this is occurring at a time when there is increasing evidence that what does work to reduce substance abuse is public education. Because this has succeeded in reducing alcohol and tobacco consumption and, in some cases, marijuana and cocaine consumption as well, it's time that we take the money we are wasting in the "war on drugs" and spend it on public education instead. . . .

In the face of all this, it is hard to believe that we do not know how to reduce crime at all. It is hard not to share the frustration expressed by Norval Morris, former dean of the University of Chicago Law School: "It is trite but it remains true that the main causes of crime are social and economic. The question arises whether people really care. The solutions are so obvious. *It's almost as if America wished for a high crime rate*."[44] If this is so, then *the system's failure is only in the eye of the victim: For those in control, it is a roaring success!*

WHAT WORKS TO REDUCE CRIME

Surveying the programs that might contribute to reducing crime, criminologist Elliot Currie concludes that "four priorities seem especially critical: preventing child abuse and neglect, enhancing children's intellectual and social development, providing support and guidance to vulnerable adolescents, and working extensively with juvenile offenders. . . .

In short, there is a growing body of knowledge showing that early childhood intervention and drug treatment programs can work to reduce crime. As Professor Blumstein observed, "If you intervene early, you not only save the costs of incarceration, you also save the costs of crime and gain the benefits of an individual who is a taxpaying contributor to the economy."[45] But, as Peter Greenwood, author of the Rand Corporation Study, *Diverting Children from a Life of Crime*, says, "The big policy question is, Who will act on this?"[46] . . .

SUMMARY

In this chapter, I have tried to establish the first part of the Pyrrhic defeat theory, namely, that the war on crime is a failure and an avoidable one: The

American criminal justice system—by which I mean the entire process from lawmaking to law enforcing—has done little to reduce the enormous amount of crime that characterizes our society and threatens our citizens. Over the last several decades, crime has generally risen, although in recent years it has declined. No doubt demographic changes, most significantly the growth followed by the decrease in the number of youngsters in the crime-prone years, have played a role in this. This in itself suggests that criminal justice policy and practice cannot be credited with the recent occasional declines. At the same time, however, neither can it be thought on this basis that public policy cannot reduce the crime we have. To support this, I have shown that crime is neither a simple and unavoidable consequence of the number of youngsters nor of the degree of urbanization. I have suggested that there are a number of policies we have good reason to believe would succeed in reducing crime—effective gun control, decriminalization of illicit drugs, and, of course, amelioration of poverty—that we refuse to implement. I concluded the chapter by showing that the Pyrrhic defeat theory shares, with the Durkheim-Erikson view of the functional nature of crime, the idea that societies may promote behavior that they seem to want to eliminate. My theory differs from their view in insisting that the failure to stamp out crime doesn't simply reflect an existing consensus but contributes to creating one, one that is functional for only a certain part of our society.

NOTES

Editors Note: The original chapter from which this selection was taken has extensive references that could not be listed here. For more documentation, see source on p. 231.

1. Elliot Currie, *Crime and Punishment in America* (New York: Metropolitan Books, Henry Holt & Company, 1998), p. 12; and James Lynch and William Sobol, *Did Getting Tough on Crime Pay?* (Washington, D.C.: The Urban Institute, 1997), p. 1; BJS, *Prisoners in 1998* (NCJ175687).
2. Reported in William Raspberry, "2 Million and Counting." *The Washington Post*, December 13, 1999, p. A25.
3. BJS, *Criminal Offender Statistics*, at http://www.ojp.usdoj.gov/bjs/crimoff.htm (last revised December 5, 1999).
4. *UCR—1998*, p. 10; Joseph D. Davey, *The Politics of Prison Expansion* (Westport, Conn.: Praeger, 1998), pp. 118–19.
5. *Sourcebook—1998*, p. 2, Table 1.1; Davey, *Politics of Prison Expansion*, p. 122.
6. David Vise and Lorraine Adams, "Despite Rhetoric, Violent Crime Climbs," *The Washington Post*, December 5, 1999, p. A3.
7. Quoted in ibid.
8. Quoted in ibid.
9. *Sourcebook—1998*, p. 502, Table 6.48. See also Vise and Adams, "Despite Rhetoric, Violent Crime Climbs."
10. Currie, *Crime and Punishment in America*, p. 60.
11. Quoted in Pierre Thomas, "In a Reversal, U.S. Homicide Numbers Fall," *Washington Post*, December 31, 1995, p. A8.
12. "Popgun Politics," *U.S. News & World Report*, September 30, 1996, p. 33.
13. Thomas, "In a Reversal, U.S. Homicide Numbers Fall," p. A8.
14. Richard Lacayo, "Law and Order," *Time*, January 15, 1996, pp. 48–54.
15. Currie, *Crime and Punishment in America*, p. 26.
16. Lacayo, "Law and Order," p. 50; Fox Butterfield, "Experts on Crime Warn of a Ticking Time Bomb," *The New York Times*, January 6, 1996, p. 6.
17. *UCR—1992*, p. 58; *UCR—1993*, p. 5; the quote is on p. 287.
18. *UCR—1997*, p. 5.
19. Currie, *Crime and Punishment in America*, pp. 24–25.
20. Isaac Shapiro and Robert Greenstein, *The Widening Income Gulf* (Washington, D.C.: Center on Budget and Policy Priorities, September 4, 1999), pp. 1, 2.
21. See, for example, John Braithwaite, "Poverty, Power, and White-Collar Crime," in Kip Schlegel and David Weisburd, eds., *White-Collar Crime Reconsidered* (Boston: Northeastern University Press, 1992), pp. 78–107.
22. Karen Pennar, "The Rich Are Richer—And America May Be the Poorer," *Business Week*, November 18, 1991, pp. 85, 88.

23. Edward Wolff, *Top Heavy: A Study of the Increasing Inequality of Wealth in America* (New York: The Twentieth Century Fund Press, 1995), p. 2.
24. *StatAbst—1992*, p. 399, Table 635; *StatAbst–1998*, p. 51, Table 51; p. 407, Table 651. See also "*Racial Gulf*: Blacks' Hopes, Raised by '68 Kerner Report, Are Mainly Unfulfilled," *The Wall Street Journal*, February 26, 1988, pp. 1, 9; *StatAbst*–1992, p. 80, Table 109; "Today's Native Sons," *Time*, December 1, 1986, pp. 26–29; and *StatAbst–1988*, p. 75, Table 113.
25. Todd R. Clear, " 'Tougher' Is Dumber," *The New York Times*, December 4, 1993, p. 21.
26. BJS. *Prisoners in 1998* (NCI 175687), p. 1.
27. Robert Johnson and Hans Toch, "Introduction," in Johnson and Toch, eds., *The Pains of Imprisonment* (Beverly Hills, Calif.: Sage, 1982), pp. 19–20.
28. Gary Kleck, *Point Blank: Guns and Violence in America* (New York: Aldine de Gruyter: 1991), p. 17. Kleck's estimate of the number of guns is supported by the Bureau of Alcohol, Tobacco and Firearms, which calculated 200 million in 1990 (cited in Albert Reiss and Jeffrey Roth, eds., *Understanding and Preventing Violence* [Washington D.C.: National Academy Press, 1993], p. 256).
29. BJS, *Guns Used in Crime*, July 1995 (NCJ-148201), pp. 3, 6.
30. Kleck, *Point Blank: Guns and Violence in America*, p. 54. Kleck's claim is based in part on a 1989 Gallup poll that found 47 percent of households had firearms and the *Time*/CNN poll of the same year that found 48 percent of households with firearms. Also, see Douglas Weil and David Henenway, "Loaded Guns in the Home: Analysis of a National Random Survey of Gun Owners," *Journal of the American Medical Association* 267, no. 22 (June 10, 1992). They report that 46 percent of households had firearms, including 25 percent that had a handgun, half of which were currently loaded (p. 3033).
31. *Challenge*, p. 239 (emphasis added).
32. *UCR—1995*, p. 36.
33. Office of Juvenile Justice and Delinquency Prevention, Fact Sheet, February 1999, no. 93 (available online at *www.ncjrs.org/jjfact.htm*).
34. Zimring, *Firearms and Violence in American Life*, Chap. 11.
35. Linda Saltzman et al., "Weapon Involvement and Injury Outcomes in Family and Intimate Assaults," *Journal of the American Medical Association*, 276, no. 22 (June 10, 1992), p. 3043.
36. Quoted in Doug Bandow, "War on Drugs or War on America?" p. 246.
37. "A LEN interview with Professor Alfred Blumstein of Carnegie Mellon University," p. 11.
38. M. Douglas Anglin and George Speckart, "Narcotics Use and Crime: A Multisample, Multimethod Analysis," *Criminology* 26, No. 2 (1988), p. 226.
39. Trebach, *The Heroin Solution*, p. 246.
40. Bandow, "War on Drugs or War on America?" p. 243; McCoy and Block, "U.S. Narcotics Policy: An Anatomy of Failure,"p. 6.
41. Christopher Mumola, "Substance Abuse and Treatment, State and Federal Prisoners, 1997," BJS Special Report (NCJ 172871), January 1997.
42. Ibid.
43. Michael Tonry, "Racial Politics, Racial Disparities, and the War on Crime," *Crime & Delinquency* 40, No. 4 (October 1994), p. 487.
44. *Time*, June 30, 1975, p. 17 (emphasis added).
45. Quoted in Butterfield, "Intervening Early Costs Less . . .," p. A24.
46. Ibid. For an extensive list of promising programs aimed at reducing crime, see Donziger, ed., *The Real War on Crime*, app. B.

34

Top 10 Corporate Criminals of the Decade

RUSSELL MOKHIBER

Mokhiber brings to our attention a most pervasive type of crime, one that is seemingly unknown by the public. White collar crime *is crime committed in the pursuance of one's occupation; it actually costs the public more in both physical injuries and financial cost than crimes such as robbery and burglary. According to the author, the majority of these types of crime are committed by the operators of the major corporations in the country. He also notes that they are often repeat offenders.*

As you read this selection, consider the following questions as guides:

1. *What reasons do you believe lie behind the public's lack of knowledge of white collar crimes?*
2. *Would corporations be less likely to commit these crimes if their penalties (usually fines) were not considered business expenses and also deductible form taxes? Why or why not?*
3. *Do you believe that the officers who actually direct the corporations into the crimes committed would be less likely to commit crimes if they received longer prison terms and were fined any bonuses received for the so-called increase in corporate profits? Why or why not?*

INTRODUCTION

Every year, the major business magazines put out their annual surveys of big business in America. The point of these surveys is simple – to identify and glorify the biggest and most profitable corporations. The list contained in this report, *The Top 100 Corporate Criminals of the Decade* focuses public attention on a wave of corporate criminality that has swamped prosecutors offices around the country. This is the dark underside of the marketplace that is given little sustained attention and analysis by politicians and news outlets.

To compile The Top 100 Corporate Criminals of the 1990s, we used the most narrow and conservative of definitions – corporations that have pled guilty or no contest to crimes and have been criminally fined. The 100 corporate criminals fell into *14 categories of crime*: Environmental (38), antitrust (20), fraud (13), campaign finance (7), food and drug (6), financial crimes (4), false statements (3), illegal exports (3), illegal boycott (1), worker death (1), bribery (1), obstruction of justice (1) public corruption (1), and tax evasion (1).

There are millions of Americans who care about morality in the marketplace. But few Americans realize that when they buy Exxon stock, or when they fill up at an Exxon gas station, they are in fact supporting a criminal recidivist corporation. And few Americans realize that when they take a ride on a cruise ship owned by Royal Caribbean Cruise Lines, they are riding on a ship owned by a criminal recidivist corporation. Six corporations that made

SOURCE: From *Top 100 Corporate Criminals of the Decade*, by Russell Mokhiber. Corporate Crime Reporter, pp. 1–5. Reprinted with permission.

the list of the Top 100 Corporate Criminals were criminal recidivist companies during the 1990s. In addition to Exxon and Royal Caribbean, Rockwell International, Warner-Lambert, Teledyne, and United Technologies each pled guilty to more than one crime during the 1990s.

A few caveats about this report

Caveat one Big companies that are criminally prosecuted represent only the tip of a very large iceberg of corporate wrongdoing. For every company convicted of health care fraud, there are hundreds of others who get away with ripping off Medicare and Medicaid, or face only mild slap-on-the-wrist fines and civil penalties when caught. For every company convicted of polluting the nation's waterways, there are many others who are not prosecuted because their corporate defense lawyers are able to offer up a low-level employee to go to jail in exchange for a promise from prosecutors not to touch the company or high-level executives. For every corporation convicted of bribery or of giving money directly to a public official in violation of federal law, there are thousands who give money legally through political action committees to candidates and political parties. They profit from a system that effectively has legalized bribery. For every corporation convicted of selling illegal pesticides, there are hundreds more who are not prosecuted because their lobbyists have worked their way in Washington to ensure that dangerous pesticides remain legal. For every corporation convicted of reckless homicide in the death of a worker, there are hundreds of others that don't even get investigated for reckless homicide when a worker is killed on the job. Only a few district attorneys across the country (Michael McCann, the DA in Milwaukee County, Wisconsin, being one) regularly investigate workplace deaths as homicides.

Caveat two Corporations define the laws under which they live. For example, the automobile industry over the past 30 years has worked its will on Congress to block legislation that would impose criminal sanctions on knowing and willful violations of the federal auto safety laws. Now, if an auto company is caught violating the law, and if the cops are not asleep at the wheel, only a civil fine is imposed.

Caveat three Because of their immense political power, big corporations have the resources to defend themselves in courts of law and in the court of public opinion. Few prosecutors are willing to subject themselves to the constant legal and public relations barrage that a corporation's well connected and high-priced legal talent can inflict. It is a testament to the tenacity of a few dedicated federal prosecutors that Royal Caribbean Cruise Lines, for example, was criminally convicted of polluting the oceans. In the criminal prosecution of Royal Caribbean Cruise Lines the company was facing a team of two federal criminal prosecutors. To defend itself, Royal Caribbean hired Judson Starr and Jerry Block, both of whom have served as head of the Justice Department's Environmental Crimes Section, and former Attorney General Benjamin Civiletti. Also representing Royal Caribbean were former federal prosecutors Kenneth C. Bass III, and Norman Moscowitz. Donald Carr of Winthrop & Stimson also joined the defense team. Hired on as experts on international law issues were former Attorney General Eliot Richardson, University of Virginia law professor John Norton Moore, former State Department officials Terry Leitzell and Bernard Oxman, and four retired senior admirals. As the case proceeded to trial, Royal Caribbean engaged in a massive public relations campaign, taking out ads during the Super Bowl, putting former Environmental Protection Agency (EPA) Administrators on its board of directors, and donating thousands of dollars to environmental groups. Federal prosecutors overcame this legal and public relations barrage and convicted the company. But that was an unusual prosecution and unusually determined prosecutors.

While the 1990s was a decade of booming markets and booming profits, it was also a decade of rampant corporate criminality. There is an emerging consensus among corporate criminologists. Corporate crime and violence inflicts far more damage on society than all street crime combined. The FBI estimates, for example, that burglary and robbery – street crimes – costs the nation $3.8 billion a year. Compare this to the hundreds of billions of dollars stolen from Americans as a result of corporate and white-collar fraud. Health care fraud alone costs Americans $100 billion to $400 billion a year.

The savings and loan fraud – which former Attorney General Dick Thornburgh called "the biggest white collar swindle in history" – cost us anywhere from $300 billion to $500 billion. And then you have your lesser frauds: auto repair fraud, $40 billion a year, securities fraud, $15 billion a year – and on down the list. Recite this list of corporate frauds and people will immediately say to you: but you can't compare street crime and corporate crime – corporate crime is not violent crime.

Unfortunately, corporate crime is often violent crime. The FBI estimates that 19,000 Americans are murdered every year. Compare this to the 56,000 Americans who die every year on the job or from occupational diseases such as black lung and asbestosis and the tens of thousands of other Americans who fall victim to the silent violence of pollution, contaminated foods, hazardous consumer products, and hospital malpractice. These deaths are often the result of criminal recklessness. They are sometimes prosecuted as homicides or as criminal violations of federal laws.

And environmental crimes often result in death, disease and injury. In 1998, for example, a Tampa, Florida company and the company's plant manager were found guilty of violating a federal hazardous waste law. Those illegal acts resulted in the deaths of two nine-year-old boys who were playing in a dumpster at the company's facility.

This report is only a tiny step in an effort to fill a great void in corporate crime research. The Justice Department has the information and should get the budget to begin putting out yearly reports on corporate crime. Every year, the Justice Department puts out an annual report titled "Crime in the United States." But by "Crime in the United States," the Justice Department means "street crime in the United States." So, in "Crime in the United States" document you will read about burglary, robbery and theft. There is nothing in it about price-fixing, corporate fraud, pollution, or public corruption. A yearly Justice Department report on Corporate Crime in the United States is long overdue.

The Top 10 Corporate Criminals of the 1990s

1. **F. Hoffmann-La Roche Ltd.**
   Type of Crime: Antitrust
   Criminal Fine: $500 million
   12 Corporate Crime Reporter 21(1), May 24, 1999
2. **Daiwa Bank Ltd.**
   Type of Crime: Financial
   Criminal Fine: $340 million
   10 Corporate Crime Reporter 9(3), March 4, 1996
3. **BASF Aktiengesellschaft**
   Type of Crime: Antitrust
   Criminal Fine: $225 million
   12 Corporate Crime Reporter 21(1), May 24, 1999
4. **SGL Carbon Aktiengesellschaft (SGL AG)**
   Type of Crime: Antitrust
   Criminal Fine: $135 million
   12 Corporate Crime Reporter 19(4), May 10, 1999
5. **Exxon Corporation and Exxon Shipping**
   Type of Crime: Environmental
   Criminal Fine: $125 million
   5 Corporate Crime Reporter 11(3), March 18, 1991
6. **UCAR International, Inc.**
   Type of Crime: Antitrust
   Criminal Fine: $110 million
   12 Corporate Crime Reporter 15(6), April 13, 1998
7. **Archer Daniels Midland**
   Type of Crime: Antitrust
   Criminal Fine: $100 million
   10 Corporate Crime Reporter 40(1), October 21, 1996
8. **(tie) Banker's Trust**
   Type of Crime: Financial
   Criminal Fine: $60 million
   12 Corporate Crime Reporter 11(1), March 15, 1999

9. **(tie) Sears Bankruptcy Recovery Management Services**
Type of Crime: Fraud
Criminal Fine: $60 million
13 Corporate Crime Reporter 7(1), February 15, 1999

10. **Haarman & Reimer Corp.**
Type of Crime: Antitrust
Criminal fine: $50 million
11 Corporate Crime Reporter 5(4), February 3, 1997

35

On Being Sane in Insane Places

D. L. ROSENHAN

Rosenhan reveals a push toward deviance not noted by most theorists—the labeling of behavior as not normal. The question being raised in this reading is, "What is normality as opposed to abnormality?"

As you read, ask yourself the following questions:

1. *What were the main factors affecting the treatment of the patients?*
2. *What other institutions are affected by labeling? Why do you say this?*
3. *What activities would you remove from those considered abnormal?*

GLOSSARY **Pseudopatient** A pretend patient. **Type 1 error** A false-negative error; for example, diagnosing a sick person as healthy. **Type 2 error** A false-positive error; for example, diagnosing a healthy person as sick. **Depersonalization** Removing a sense of the individual from treatment. **Prima facie evidence** Presumption of fact. **Veridical** Truthful.

If sanity and insanity exist, how shall we know them?

The question is neither capricious nor itself insane. However much we may be personally convinced that we can tell the normal from the abnormal, the evidence is simply not compelling. It is commonplace, for example, to read about murder trials wherein eminent psychiatrists for the defense are contradicted by equally eminent psychiatrists for the prosecution on the matter of the defendant's sanity. More generally, there are a great deal of conflicting data on the reliability, utility, and meaning of such terms as "sanity," "insanity," "mental illness," and "schizophrenia."[1] Finally, as early as 1934, Benedict suggested that normality and abnormality are not universal.[2] What is viewed as normal in one culture may be seen as quite aberrant in another. Thus, notions of normality and abnormality may not be quite as accurate as people believe they are.

To raise questions regarding normality and abnormality is in no way to question the fact that some behaviors are deviant or odd. Murder is deviant. So, too, are hallucinations. Nor does raising such questions deny the existence of the personal anguish that is often associated with "mental illness." Anxiety and depression exist. Psychological suffering exists. But normality and abnormality, sanity and insanity, and the diagnoses that flow from them may be less substantive than many believe them to be.

At its heart, the question of whether the sane can be distinguished from the insane (and whether degrees of insanity can be distinguished from each other) is a simple matter: do the salient characteristics that lead to diagnoses reside in the patients themselves or in the environments and contexts in which observers find them? From Bleuler, through Kretchmer, through the formulators of the recently revised *Diagnostic and Statistical Manual* of the American Psychiatric Association, the belief has been strong that patients present symptoms, that those symptoms can be categorized, and, implicitly, that the sane are distinguishable from the insane. More recently, however, this belief has been questioned. Based in part on theoretical and anthropological considerations, but also on philosophical, legal, and therapeutic ones, the view has grown that psychological categorization of mental illness is useless at best and downright harmful, misleading, and pejorative at worst. Psychiatric diagnoses, in this view, are in the minds of the observers and are not valid summaries of characteristics displayed by the observed.

Gains can be made in deciding which of these is more nearly accurate by getting normal people (that is, people who do not have, and have never suffered, symptoms of serious psychiatric disorders) admitted to psychiatric hospitals and then determining whether they were discovered to be sane and, if so, how. If the sanity of such pseudopatients were always detected, there would be prima facie evidence that a sane individual can be distinguished from the insane context in which he is found. Normality (and presumably abnormality) is distinct enough that it can be recognized wherever it occurs, for it is carried within the person. If on the other hand, the sanity of the pseudopatients were never discovered, serious difficulties would arise for those who support traditional modes of psychiatric diagnosis. Given that the hospital staff was not incompetent, that the pseudopatient had been behaving as sanely as he had been outside of the hospital, and that it had never been previously suggested that he belonged in a psychiatric hospital, such an unlikely outcome would support the view that psychiatric diagnosis betrays little about the patient but much about the environment in which an observer finds him.

This reading describes such an experiment. Eight sane people gained secret admission to 12 different hospitals. Their diagnostic experiences constitute the data of the first part of this article; the remainder is devoted to a description of their experiences in psychiatric institutions. Too few psychiatrists and psychologists, even those who have worked in such hospitals, know what the experience is like. They rarely talk about it with former patients, perhaps because they distrust information coming from the previously insane. Those who have worked in psychiatric hospitals are likely to have adapted so thoroughly to the settings that they are insensitive to the impact of that experience. And while there have been occasional reports of researchers who submitted themselves to psychiatric hospitalization,[3] these researchers have commonly remained in the hospitals for short periods of time, often with the knowledge of the hospital staff. It is difficult to know the extent to which they were treated like patients or like research colleagues. Nevertheless, their reports about the inside of the psychiatric hospital have been valuable. This reading extends those efforts.

PSEUDOPATIENTS AND THEIR SETTINGS

The eight pseudopatients were a varied group. One was a psychology graduate student in his twenties. The remaining seven were older and "established." Among them were three psychologists, a pediatrician, a psychiatrist, a painter, and a housewife. Three pseudopatients were women, five were men. All of them employed pseudonyms, lest their alleged diagnoses embarrass them later. Those who were in mental health professions alleged another occupation in order to avoid the special attentions that might be accorded by staff, as a matter of courtesy or caution, to ailing colleagues.[4] With the exception of myself (I was the first pseudopatient and my presence was known to the hospital administrator and chief psychologist and, so far as I can tell, to them alone), the presence of pseudopatients and the nature of the research program was not known to the hospital staff.[5]

The settings were similarly varied. In order to generalize the findings, admission into a variety of hospitals was sought. The 12 hospitals in the sample were located in five different states on the East and West coasts. Some were old and shabby, some were quite new. Some were research-oriented, others not. Some had good staff-patient ratios, others were quite understaffed. Only one was a strictly private hospital. All of the others were supported by state or federal funds or, in one instance, by university funds.

After calling the hospital for an appointment, the pseudopatient arrived at the admissions office complaining that he had been hearing voices. Asked what the voices said, he replied that they were often unclear, but as far as he could tell they said "empty," "hollow," and "thud." The voices were unfamiliar and were of the same sex as the pseudopatient. The choice of these symptoms was occasioned by their apparent similarity to existential symptoms. Such symptoms are alleged to arise from painful concerns about the perceived meaninglessness of one's life. It is as if the hallucinating person were saying, "My life is empty and hollow." The choice of these symptoms was also determined by the *absence* of a single report of existential psychoses in the literature.

Beyond alleging the symptoms and falsifying name, vocation, and employment, no further alterations of person, history, or circumstances were made. The significant events of the pseudopatient's life history were presented as they had actually occurred. Relationships with parents and siblings, with spouse and children, with people at work and in school, consistent with the aforementioned exceptions, were described as they were or had been. Frustrations and upsets were described along with joys and satisfactions. These facts are important to remember. If anything, they strongly biased the subsequent results in favor of detecting sanity, since none of their histories or current behaviors were seriously pathological in any way.

Immediately upon admission to the psychiatric ward, the pseudopatient ceased simulating *any* symptoms of abnormality. In some cases, there was a brief period of mild nervousness and anxiety, since none of the pseudopatients really believed that they would be admitted so easily. Indeed, their shared fear was that they would be immediately exposed as frauds and greatly embarrassed. Moreover, many of them had never visited a psychiatric ward; even those who had, nevertheless, had some genuine fears about what might happen to them. Their nervousness, then, was quite appropriate to the novelty of the hospital setting, and it abated rapidly.

Apart from that short-lived nervousness, the pseudopatient behaved on the ward as he "normally" behaved. The pseudopatient spoke to patients and staff as he might ordinarily. Because there is uncommonly little to do on a psychiatric ward, he attempted to engage others in conversation. When asked by staff how he was feeling, he indicated that he was fine, that he no longer experienced symptoms. He responded to instructions from attendants, to calls for medication (which was not swallowed), and to dining-hall instructions. Beyond such activities as were available to him on the admissions ward, he spent his time writing down his observations about the ward, its patients, and the staff. Initially these notes were written "secretly," but as it soon became clear that no one much cared, they were subsequently written on standard tablets of paper in such public places as the day-room. No secret was made of these activities.

The pseudopatient, very much as a true psychiatric patient, entered a hospital with no foreknowledge of when he would be discharged. Each was told that he would have to get out by his own devices, essentially by convincing the staff that he was sane. The psychological stresses associated with hospitalization were considerable, and all but one of the pseudopatients desired to be discharged immediately after being admitted. They were, therefore, motivated not only to behave sanely, but to be paragons of cooperation. That their behavior was in no way disruptive is confirmed by nursing reports, which have been obtained on most of the patients. These reports uniformly indicate that the patients were "friendly," "cooperative," and "exhibited no abnormal indications."

THE NORMAL ARE NOT DETECTABLY SANE

Despite their public "show" of sanity, the pseudopatients were never detected. Admitted, except in one case, with a diagnosis of schizophrenia,[6] each was discharged with a diagnosis of schizophrenia "in remission." The label "in remission" should in no way be dismissed as a formality, for at no time during any hospitalization had any question been raised about any pseudopatient's simulation. Nor are there any indications in the hospital records that the pseudopatient's status was suspect. Rather, the evidence is strong that, once labeled schizophrenic, the pseudopatient was stuck with that label. If the pseudopatient was to be discharged, he must naturally be "in remission"; but he was not sane, nor, in the institution's view, had he ever been sane.

The uniform failure to recognize sanity cannot be attributed to the quality of the hospitals, for, although there were considerable variations among them, several are considered excellent. Nor can it be alleged that there was simply not enough time to observe the pseudopatients. Length of hospitalization ranged from 7 to 52 days, with an average of 19 days. The pseudopatients were not, in fact, carefully observed, but this failure clearly speaks more to traditions within psychiatric hospitals than to lack of opportunity.

Finally, it cannot be said that the failure to recognize the pseudopatients' sanity was due to the fact that they were not behaving sanely. While there was clearly some tension present in all of them, their daily visitors could detect no serious behavioral consequences—nor, indeed, could other patients. It was quite common for the patients to "detect" the pseudopatients' sanity. During the first three hospitalizations, when accurate counts were kept, 35 of a total of 118 patients on the admissions ward voiced their suspicions, some vigorously. "You're not crazy. You're a journalist, or a professor [referring to the continual notetaking]. You're checking up on the hospital." While most of the patients were reassured by the pseudopatient's insistence that he had been sick before he came in but was fine now, some continued to believe that the pseudopatient was sane throughout his hospitalization.[7] The fact that the patients often recognized normality when staff did not raises important questions.

Failure to detect sanity during the course of hospitalization may be due to the fact that physicians operate with a strong bias toward what statisticians call the type 2 error. This is to say that physicians are more inclined to call a healthy person sick (a false positive, type 2) than a sick person healthy (a false negative, type 1). The reasons for this are not hard to find: it is clearly more dangerous to misdiagnose illness than health. Better to err on the side of caution, to suspect illness even among the healthy.

But what holds for medicine does not hold equally well for psychiatry. Medical illnesses, while unfortunate, are not commonly pejorative. Psychiatric diagnoses, on the contrary, carry with them personal, legal, and social stigmas.[8] It was therefore important to see whether the tendency toward diagnosing the sane insane could be reversed. The following experiment was arranged at a research and teaching hospital whose staff had heard these findings but doubted that such an error could occur in their hospital. The staff was informed that at some time during the following 3 months, one or more pseudopatients would attempt to be admitted into the psychiatric hospital. Each staff member was asked to rate each patient who presented himself at admission or on the ward according to the likelihood that the patient was a pseudopatient. A 10-point scale was used, with a 1 and 2 reflecting high confidence that the patient was a pseudopatient.

Judgments were obtained on 193 patients who were admitted for psychiatric treatment. All staff who had had sustained contact with or primary responsibility for the patient—attendants, nurses, psychiatrists, physicians, and psychologists—were asked to make judgments. Forty-one patients were alleged, with high confidence, to be pseudopatients by at least one member of the staff. Twenty-three were considered suspect by at least one psychiatrist. Nineteen were suspected by one psychiatrist *and* one other staff member. Actually, no genuine pseudopatient (at least from my group) presented himself during this period.

The experiment is instructive. It indicates that the tendency to designate sane people as insane can be reversed when the stakes (in this case, prestige and diagnostic acumen) are high. But what can be said of the 19 people who were suspected of being "sane" by one psychiatrist and another staff member? Were these people truly "sane," or was it rather the case that in the course of avoiding the type 2 error the staff tended to make more errors of the first sort—calling the crazy "sane"? There is no way of knowing. But one thing is certain: any diagnostic process that lends itself so readily to massive errors of this sort cannot be a very reliable one.

THE STICKINESS OF PSYCHODIAGNOSTIC LABELS

Beyond the tendency to call the healthy sick—a tendency that accounts better for diagnostic behavior on admission than it does for such behavior after a lengthy period of exposure—the data speak to the massive role of labeling in psychiatric assessment. Having once been labeled schizophrenic, there is nothing the pseudopatient can do to overcome the tag. The tag profoundly colors others' perceptions of him and his behavior.

From one viewpoint, these data are hardly surprising, for it has long been known that elements are given meaning by the context in which they occur. Gestalt psychology made this point vigorously, and Asch[9] demonstrated that there are "central" personality traits (such as "warm" versus "cold") which are so powerful that they markedly color the meaning of other information in forming an impression of a given personality.[10] "Insane," "schizophrenic," "manic-depressive," and "crazy" are probably among the most powerful of such central traits. Once a person is designated abnormal, all of his other behaviors and characteristics are colored by that label. Indeed, that label is so powerful that many of the pseudopatients' normal behaviors were overlooked entirely or profoundly misinterpreted. Some examples may clarify this issue.

Earlier I indicated that there were no changes in the pseudopatient's personal history and current status beyond those of name, employment, and, where necessary, vocation. Otherwise, a veridical description of personal history and circumstances was offered. Those circumstances were not psychotic. How were they made consonant with the diagnosis of psychosis? Or were those diagnoses modified in such a way as to bring them into accord with the circumstances of the pseudopatient's life, as described by him?

As far as I can determine, diagnoses were in no way affected by the relative health of the circumstances of a pseudopatient's life. Rather, the reverse occurred: the perception of his circumstances was shaped entirely by the diagnosis. A clear example of such translation is found in the case of a pseudopatient who had had a close relationship with his mother but was rather remote from his father during his early childhood. During adolescence and beyond, however, his father became a close friend, while his relationship with his mother cooled. His present relationship with his wife was characteristically close and warm. Apart from occasional angry exchanges, friction was minimal. The children had rarely been spanked. Surely there is nothing especially pathological about such a history. Indeed, many readers may see a similar pattern in their own experiences, with no markedly deleterious consequences. Observe, however, how such a history was translated in the psychopathological context, this from the case summary prepared after the patient was discharged.

> This white 39-year-old male . . . manifests a long history of considerable ambivalence in close relationships, which begins in early childhood. A warm relationship with his mother cools during his adolescence. A distant relationship to his father is described as becoming very intense. Affective stability is absent. His attempts to control emotionality with his wife and children are punctuated by angry outbursts and, in the case of the children, spanking. And while he says that he has several good friends, one senses considerable ambivalence embedded in those relationships also . . .

The facts of the case were unintentionally distorted by the staff to achieve consistency with a popular theory of the dynamics of a schizophrenic reaction.[11] Nothing of an ambivalent nature had been described in relations with parents, spouse, or friends. To the extent that ambivalence could be inferred, it was probably not greater than is found in all human relationships. It is true the pseudopatient's relationships with his parents changed over time, but in the ordinary context that would hardly be remarkable—indeed, it might very well be expected. Clearly, the meaning ascribed to his verbalizations (that is, ambivalence, affective instability) was determined by the diagnosis: schizophrenia. An entirely different meaning would have been ascribed if it were known that the man was "normal."

All pseudopatients took extensive notes publicly. Under ordinary circumstances, such behavior would have raised questions in the minds of observers, as, in fact, it did among patients. Indeed, it seemed so certain that the notes would elicit suspicion that elaborate precautions were taken to remove them from the ward each day. But the precautions proved needless. The closest any staff member came to questioning these notes occurred when one pseudopatient asked his physician what kind of medication he was receiving and began to write down the response. "You needn't write it," he was told gently. "If you have trouble remembering, just ask me again."

If no questions were asked of the pseudopatients, how was their writing interpreted? Nursing records for three patients indicate that the writing was seen as an aspect of their pathological behavior. "Patient engages in writing behavior" was the daily nursing comment on one of the pseudopatients who was never questioned about his writing. Given that the patient is in the hospital, he must be psychologically disturbed. And given that he is disturbed, continuous writing must be a behavioral manifestation of that disturbance, perhaps a subset of the compulsive behaviors that are sometimes correlated with schizophrenia.

One tacit characteristic of psychiatric diagnosis is that it locates the sources of aberration within the individual and only rarely within the complex of stimuli that surrounds him. Consequently, behaviors that are stimulated by the environment are commonly misattributed to the patient's disorder. For example, one kindly nurse found a pseudopatient pacing the long hospital corridors. "Nervous, Mr. X?" she asked. "No, bored," he said.

The notes kept by pseudopatients are full of patient behaviors that were misinterpreted by well-intentioned staff. Often enough, a patient would go "berserk" because he had, wittingly or unwittingly, been mistreated by, say, an attendant. A nurse coming upon the scene would rarely inquire even cursorily into the environmental stimuli of the patient's behavior. Rather, she assumed that his upset derived from his pathology, not from his present interactions with other staff members. Occasionally, the staff might assume that the patient's family (especially when they had recently visited) or other patients had stimulated the outburst. But never were the staff found to assume that one of themselves or the structure of the hospital had anything to do with a patient's behavior. One psychiatrist pointed to a group of patients who were sitting outside the cafeteria entrance half an hour before lunchtime. To a group of young residents he indicated that such behavior was characteristic of the oral-acquisitive nature of the syndrome. It seemed not to occur to him that there were very few things to anticipate in a psychiatric hospital besides eating.

A psychiatric label has a life and an influence of its own. Once the impression has been formed that the patient is schizophrenic, the expectation is that he will continue to be schizophrenic. When a sufficient amount of time has passed, during which the patient has done nothing bizarre, he is considered to be in remission and available for discharge. But the label endures beyond discharge, with the unconfirmed expectation that he will behave as a schizophrenic again. Such labels, conferred by mental health professionals, are as influential on the patient as they are on his relatives and friends, and it should not surprise anyone that the diagnosis acts on all of them as a self-fulfilling prophecy. Eventually, the patient himself accepts the diagnosis, with all of its surplus meanings and expectations, and behaves accordingly.

The inferences to be made from these matters are quite simple. Much as Zigler and Phillips have demonstrated that there is enormous overlap in the symptoms presented by patients who have been variously diagnosed,[12] so there is enormous overlap in the behaviors of the sane and the insane. The sane are not "sane" all of the time. We lose our tempers "for no good reason." We are occasionally depressed or anxious, again for no good reason. And we may find it difficult to get along with one or another person—again for no reason that we can specify. Similarly, the insane are not always insane. Indeed, it was the impression of the pseudopatients while living with them that they were sane for long periods of time—that the bizarre behaviors upon which their diagnoses were allegedly predicated constituted only a small fraction of their total behavior. If it makes no sense to label ourselves permanently depressed on the basis of an occasional depression, then it takes better evidence than is presently available to label all patients insane or schizophrenic on the basis of bizarre behaviors or cognitions. It seems more useful, as Mischel[13] has pointed out, to limit our discussion to *behaviors*, the stimuli that provoke them, and their correlates.

It is not known why powerful impressions of personality traits, such as "crazy" or "insane," arise. Conceivably, when the origins of and stimuli that give rise to a behavior are remote or unknown, or when the behavior strikes us as immutable, trait labels regarding the *behavior* arise. When, on the other hand, the origins and stimuli are known and available, discourse is limited to the behavior itself. Thus, I may hallucinate because I am sleeping, or I may hallucinate because I have ingested a peculiar drug. These are termed sleep-induced hallucinations, or dreams, and drug-induced hallucinations, respectively. But when the stimuli to my hallucinations are unknown, that is called craziness, or schizophrenia—as if that inference were somehow as illuminating as the others.

THE EXPERIENCE OF PSYCHIATRIC HOSPITALIZATION

The term "mental illness" is of recent origin. It was coined by people who were humane in their inclinations and who wanted very much to raise the station of (and the public's sympathies toward) the psychologically disturbed from that of witches and "crazies" to one that was akin to the physically ill. And they were at least partially successful, for the treatment of the mentally ill *has* improved considerably over the years. But while treatment has improved, it is doubtful that people really regard the mentally ill in the same way that they view the physically ill. A broken leg is something one recovers from, but mental illness allegedly endures forever.[14] A broken leg does not threaten the observer, but a crazy schizophrenic? There is by now a host of evidence that attitudes toward the mentally ill are characterized by fear, hostility, aloofness, suspicion, and dread.[15] The mentally ill are society's lepers.

That such attitudes infect the general population is perhaps not surprising, only upsetting. But that they affect the professionals—attendants, nurses, physicians, psychologists, and social workers—who treat and deal with the mentally ill is more disconcerting, both because such attitudes are self-evidently pernicious and because they are unwitting. Most mental health professionals would insist that they are sympathetic toward the mentally ill, that they are neither avoidant nor hostile. But it is more likely that an exquisite ambivalence characterizes their relations with psychiatric patients, such that their avowed impulses are only part of their entire attitude. Negative attitudes are there too and can easily be detected. Such attitudes should not surprise us. They are the natural offspring of the labels patients wear and the places in which they are found.

Consider the structure of the typical psychiatric hospital. Staff and patients are strictly segregated. Staff have their own living space, including their dining facilities, bathrooms, and assembly places. The glassed quarters that contain the professional staff, which the pseudopatients came to call "the cage," sit out on every dayroom. The staff emerge primarily for care-taking purposes—to give medication, to conduct a therapy or group meeting, to instruct or reprimand a patient. Otherwise, staff keep to themselves, almost as if the disorder that afflicts their charges is somehow catching.

So much is patient-staff segregation the rule that, for four public hospitals in which an attempt was made to measure the degree to which staff and patients mingle, it was necessary to use "time out of the staff cage" as the operational measure. While it was not the case that all time spent out of the cage was spent mingling with patients (attendants, for example, would occasionally emerge to watch television in the dayroom), it was the only way in which one could gather reliable data on time for measuring.

The average amount of time spent by attendants outside of the cage was 11.3 percent (range, 3 to 52 percent). This figure does not represent only time spent mingling with patients, but also includes time spent on such chores as folding laundry, supervising patients while they shave, directing ward cleanup, and sending patients to off-ward activities. It was the relatively rare attendant who spent time talking with patients or playing games with them. It proved impossible to obtain a "percent mingling time" for nurses, since the amount of time they spent out of the cage was too brief. Rather, we counted instances of emergence from the cage. On the average, daytime nurses emerged from the cage 11.5 times per shift, including instances when they left the ward entirely (range, 4 to 39 times). Late afternoon and night nurses were even less available, emerging on the average 9.4 times per shift (range, 4 to 41 times). Data on early morning nurses, who arrived usually after midnight and departed at 8 A.M., are not available because patients were asleep during most of this period.

Physicians, especially psychiatrists, were even less available. They were rarely seen on the wards. Quite commonly, they would be seen only when they arrived and departed, with the remaining time being spent in their offices or in the cage. On the average, the physicians emerged on the ward 6.7 times per day (range, 1 to 17 times). It proved difficult to make an accurate estimate in this regard, since physicians often maintained hours that allowed them to come and go at different times.

The hierarchical organization of the psychiatric hospital has been commented on before,[16] but the latent meaning of that kind of organization is worth noting again. Those with the most power have least to do with patients, and those with the least power are most involved with them. Recall, however, that the acquisition of role-appropriate behaviors occurs mainly through the observation of others, with the most powerful having the most influence. Consequently, it is understandable that attendants not only spend more time with patients than do any other members of the staff—that is required by their station in the hierarchy—but also, insofar as they learn from their superiors' behavior, spend as little time with patients as they can. Attendants are seen mainly in the cage, which is where the models, the action, and the power are.

I turn now to a different set of studies, these dealing with staff response to patient-initiated contact. It has long been known that the amount of time a person spends with you can be an index of your significance to him. If he initiates and maintains eye contact, there is reason to believe that he is considering your requests and needs. If he pauses to chat or actually stops and talks, there is added reason to infer that he is individuating you. In four hospitals, the pseudopatient approached the staff member with a request which took the following form: "Pardon me, Mr. [or Dr. or Mrs.] X, could you tell me when I will be eligible for grounds privileges?" (or ". . . when I will be presented at the staff meeting?" or ". . . when I am likely to be discharged?"). While the content of the question varied according to the appropriateness of the target and the pseudopatient's (apparent) current needs the form was always a courteous and relevant request for information. Care was taken never to approach a particular member of the staff more than once a day, lest the staff member become suspicious or irritated. In examining these data, remember that the behavior of the pseudopatients was neither bizarre nor disruptive. One could indeed engage in good conversation with them. . . .

. . . Minor differences between these four institutions were overwhelmed by the degree to which staff avoided continuing contacts that patients had initiated. By far, their most common response consisted of either a brief response to the question, offered while they were "on the move" and with head averted, or no response at all.

The encounter frequently took the following bizarre form: (pseudopatient) "Pardon me, Dr. X. Could you tell me when I am eligible for grounds privileges?" (physician) "Good morning, Dave. How are you today?" (Moves off without waiting for a response.)

It is instructive to compare these data with data recently obtained at Stanford University. It has been alleged that large and eminent universities are characterized by faculty who are so busy that they have no time for students. For this comparison, a young lady approached individual faculty members who seemed to be walking purposefully to some meeting or teaching engagement and asked them the following six questions.

1. "Pardon me, could you direct me to Encina Hall?" (at the medical school: ". . . to the Clinical Research Center?")
2. "Do you know where Fish Annex is?" (there is no Fish Annex at Stanford).
3. "Do you teach here?"
4. "How does one apply for admission to the college?" (at the medical school: ". . . to the medical school?")
5. "Is it difficult to get in?"
6. "Is there financial aid?"

Without exception . . . all of the questions were answered. No matter how rushed they were, all respondents not only maintained eye contact, but stopped to talk. Indeed, many of the respondents went out of their way to direct or take the questioner to the office she was seeking, to try to locate "Fish Annex," or to discuss with her the possibilities of being admitted to the university.

Similar data . . . were obtained in the hospital. Here too, the young lady came prepared with six questions. After the first question, however, she remarked to 18 of her respondents . . . "I'm looking for a psychiatrist," and to 15 others . . . , "I'm looking for an internist." Ten other respondents received no inserted comment. . . . The general degree of cooperative responses is considerably higher for these university groups than it was for pseudopatients in psychiatric hospitals. Even so, differences are apparent within the medical school setting. Once having indicated that she was looking for a psychiatrist, the degree of cooperation elicited was less than when she sought an internist.

POWERLESSNESS AND DEPERSONALIZATION

Eye contact and verbal contact reflect concern and individuation: their absence, avoidance, and depersonalization. The data I have presented do not do justice to the rich daily encounters that grow up around matters of depersonalization and avoidance. I have records of patients who were beaten by staff for the sin of having initiated verbal contact. During my own experience, for example, one patient was beaten in the presence of other patients for having approached an attendant and told him,

"I like you." Occasionally, punishment meted out to patients for misdemeanors seemed so excessive that it could not be justified by the most radical interpretations of psychiatric canon. Nevertheless, they appeared to go unquestioned. Tempers were often short. A patient who had not heard a call for medication would be roundly excoriated, and the morning attendants would often wake patients with, "Come on, you m—f—s, out of bed!"

Neither anecdotal nor "hard" data can convey the overwhelming sense of powerlessness which invades the individual as he is continually exposed to the depersonalization of the psychiatric hospital. It hardly matters *which* psychiatric hospital—the excellent public ones and the very plush private hospital were better than the rural and shabby ones in this regard, but, again, the features that psychiatric hospitals had in common overwhelmed by far their apparent differences.

Powerlessness was evident everywhere. The patient is deprived of many of his legal rights by dint of his psychiatric commitment.[17] He is shorn of credibility by virtue of his psychiatric label. His freedom of movement is restricted. He cannot initiate contact with the staff, but may only respond to such overtures as they make. Personal privacy is minimal. Patient quarters and possessions can be entered and examined by any staff member, for whatever reason. His personal history and anguish is available to any staff member (often including the "gray lady" and "candy striper" volunteer) who chooses to read his folder, regardless of their therapeutic relationship to him. His personal hygiene and waste evacuation are often monitored. The water closets may have no doors.

At times, depersonalization reached such proportions that pseudopatients had the sense that they were invisible, or at least unworthy of account. Upon being admitted, I and other pseudopatients took the initial physical examinations in a semipublic room, where staff members went about their own business as if we were not there.

On the ward, attendants delivered verbal and occasionally serious physical abuse to patients in the presence of other observing patients, some of whom (the pseudopatients) were writing it all down. Abusive behavior, on the other hand, terminated quite abruptly when other staff members were known to be coming. Staff are credible witnesses. Patients are not.

A nurse unbuttoned her uniform to adjust her brassiere in the presence of an entire ward of viewing men. One did not have the sense that she was being seductive. Rather, she didn't notice us. A group of staff persons might point to a patient in the dayroom and discuss him animatedly, as if he were not there.

One illuminating instance of depersonalization and invisibility occurred with regard to medications. All told, the pseudopatients were administered nearly 2100 pills, including Elavil, Stelazine, Compazine, and Thorazine, to name but a few. (That such a variety of medications should have been administered to patients presenting identical symptoms is itself worthy of note.) Only two were swallowed. The rest were either pocketed or deposited in the toilet. The pseudopatients were not alone in this. Although I have no precise records on how many patients rejected their medications, the pseudopatients frequently found the medications of other patients in the toilet before they deposited their own. As long as they were cooperative, their behavior and the pseudopatients' own in this matter, as in other important matters, went unnoticed throughout.

Reactions to such depersonalization among pseudopatients were intense. Although they had come to the hospital as participant observers and were fully aware that they did not "belong," they nevertheless found themselves caught up in and fighting the process of depersonalization. Some examples: a graduate student in psychology asked his wife to bring his textbooks to the hospital so he could "catch up on his homework"—this despite the elaborate precautions taken to conceal his professional association. The same student, who had trained for quite some time to get into the hospital, and who had looked forward to the experience, "remembered" some drag races that he had wanted to see on the weekend and insisted that he be discharged by that time. Another pseudopatient attempted a romance with a nurse. Subsequently he informed the staff that he was applying for admission to graduate school in psychology and was very likely to be admitted, since a graduate professor was one of his regular hospital visitors. The same person began to engage in psychotherapy with other patients—all of this as a way of becoming a person in an impersonal environment.

THE SOURCES OF DEPERSONALIZATION

What are the origins of depersonalization? I have already mentioned two. First are attitudes held by all of us toward the mentally ill—including those who treat them—attitudes characterized by fear, distrust, and horrible expectations on the one hand, and benevolent intentions on the other. Our ambivalence leads, in this instance as in others, to avoidance.

Second, and not entirely separate, the hierarchical structure of the psychiatric hospital facilitates depersonalization. Those who are at the top have least to do with patients, and their behavior inspires the rest of the staff. Average daily contact with psychiatrists, psychologists, residents, and physicians combined ranged from 3.9 to 25.1 minutes, with an overall mean of 6.8 (six pseudopatients over a total of 129 days of hospitalization). Included in this average are time spent in the admissions interview, ward meetings in the presence of a senior staff member, group and individual psychotherapy contacts, case presentation conferences, and discharge meetings. Clearly, patients do not spend much time in interpersonal contact with doctoral staff. And doctoral staff serve as models for nurses and attendants.

There are probably other sources. Psychiatric installations are presently in serious financial straits. Staff shortages are pervasive, staff time at a premium. Something has to give, and that something is patient contact. Yet while financial stresses are realities, too much can be made of them. I have the impression that the psychological forces that result in depersonalization are much stronger than the fiscal ones and that the addition of more staff would not correspondingly improve patient care in this regard. The incidence of staff meetings and the enormous amount of record-keeping on patients, for example, have not been as substantially reduced as has patient contact. Priorities exist, even during hard times. Patient contact is not a significant priority in the traditional psychiatric hospital, and fiscal pressures do not account for this. Avoidance and depersonalization may.

Heavy reliance upon psychotropic medication tacitly contributes to depersonalization by convincing staff that treatment is indeed being conducted and that further patient contact may not be necessary. Even here, however, caution needs to be exercised in understanding the role of psychotropic drugs. If patients were powerful rather than powerless, if they were viewed as interesting individuals rather than diagnostic entities, if they were socially significant rather than social lepers, if their anguish truly and wholly compelled our sympathies and concerns, would we not *seek* contact with them, despite the availability of medications? Perhaps for the pleasure of it all?

THE CONSEQUENCES OF LABELING AND DEPERSONALIZATION

Whenever the ratio of what is known to what needs to be known approaches zero, we tend to invent "knowledge" and assume that we understand more than we actually do. We seem unable to acknowledge that we simply don't know. The needs for diagnosis and remediation of behavioral and emotional problems are enormous. But rather than acknowledge that we are just embarking on understanding, we continue to label patients "schizophrenic," "manic-depressive," and "insane," as if in those words we had captured the essence of understanding. The facts of the matter are that we have known for a long time that diagnoses are often not useful or reliable, but we have nevertheless continued to use them. We now know that we cannot distinguish insanity from sanity. It is depressing to consider how that information will be used.

Not merely depressing, but frightening. How many people, one wonders, are sane but not recognized as such in our psychiatric institutions? How many have been needlessly stripped of their privileges of citizenship, from the right to vote and drive to that of handling their own accounts? How many have feigned insanity in order to avoid the criminal consequences of their behavior, and, conversely, how many would rather stand trial than live interminably in a psychiatric hospital—but are wrongly thought to be mentally ill? How many have been stigmatized by well-intentioned, but nevertheless erroneous, diagnoses? On the last point, recall again that a "type 2 error" in psychiatric diagnosis does not have the same consequences it does in medical diagnosis. A diagnosis of cancer that has been found to be in error is cause for celebration. But psychiatric diagnoses are rarely found to be in error. The label sticks, a mark of inadequacy forever.

Finally, how many patients might be "sane" outside the psychiatric hospital but seem insane in it—not because craziness resides in them, as it were, but because they are responding to a bizarre setting, one that may be unique to institutions which harbor neither people? Goffman calls the process of socialization to such institutions "mortification"—an apt metaphor that includes the processes of depersonalization that have been described here. And while it is impossible to know whether the pseudopatients' responses to these processes are characteristic of all inmates—they were, after all, not real patients—it is difficult to believe that these processes of socialization to a psychiatric hospital provide useful attitudes or habits of response for living in the "real world."

SUMMARY AND CONCLUSIONS

It is clear that we cannot distinguish the sane from the insane in psychiatric hospitals. The hospital itself imposes a special environment in which the meanings of behavior can easily be misunderstood. The consequences to patients hospitalized in such an environment—the powerlessness, depersonalization, segregation, mortification, and self-labeling—seem undoubtedly countertherapeutic.

I do not, even now, understand this problem well enough to perceive solutions. But two matters seem to have some promise. The first concerns the proliferation of community mental health facilities, of crisis intervention centers, of human potential movement, and of behavior therapies that, for all of their own problems, tend to avoid psychiatric labels, to focus on specific problems and behaviors, and to retain the individual in a relatively nonpejorative environment. Clearly, to the extent that we refrain from sending the distressed to insane places, our impressions of them are less likely to be distorted. (The risk of distorted perceptions, it seems to me, is always present, since we are much more sensitive to an individual's behaviors and verbalizations than we are to the subtle contextual stimuli that often promote them. At issue here is a matter of magnitude. And, as I have shown, the magnitude of distortion is exceedingly high in the extreme context that is a psychiatric hospital.)

The second matter that might prove promising speaks to the need to increase the sensitivity of mental health workers and researchers to the *Catch 22* position of psychiatric patients. Simply reading materials in this area will be of help to some such workers and researchers. For others, directly experiencing the impact of psychiatric hospitalization will be of enormous use. Clearly, further research into the social psychology of such total institutions will both facilitate treatment and deepen understanding.

I and other pseudopatients in the psychiatric setting had distinctly negative reactions. We do not pretend to describe the subjective experiences of true patients. Theirs may be different from ours, particularly with the passage of time and the necessary process of adaptation to one's environment. But we can and do speak to the relatively more objective indices of treatment within the hospital. It could be a mistake, and a very unfortunate one, to consider that what happened to us derived from malice or stupidity on the part of the staff. Quite the contrary, our overwhelming impression of them was of people who really cared, who were committed, and who were uncommonly intelligent. Where they failed, as they sometimes did painfully, it would be more accurate to attribute those failures to the environment in which they, too, found themselves than to personal callousness. Their perceptions and behavior were controlled by the situation, rather than being motivated by a malicious disposition. In a more benign environment, one that was less attached to global diagnosis, their behaviors and judgments might have been more benign and effective.

NOTES

1. . . . For an analysis of these artifacts and summaries of the disputes, see J. Zubin, *Annu. Rev. Psychol. 18*, 373 (1967); L. Phillips and J. G. Draguns, *ibid. 22*, 447 (1971).
2. R. Benedict, *J. Gen. Psychol. 10*, 59 (1934).
3. A. Barry, *Bellevue Is a State of Mind* (Harcourt Brace Jovanovich, New York, 1971); I. Belknap, *Human Problems of a State Mental Hospital* (McGraw-Hill, New York, 1956); W. Caudill, F. C. Redlich, H. R. Gilmore, E. B. Brody, *Amer. J. Orthopsychiat. 22*, 314 (1952); A. R. Goldman, R. H. Bohr, T. A. Steinberg, *Prof. Psychol. 1*, 427 (1970); unauthored, *Roche Report 1* (No. 13), 8 (1971).
4. Beyond the personal difficulties that the pseudopatient is likely to experience in the hospital, there are legal and social ones that, combined, require considerable attention before entry. . . .
5. However distasteful such concealment is, it was a necessary first step to examining these questions. Without concealment, there would have been no way to know how valid these experiences were; nor was there any way of knowing whether whatever detections occurred were a tribute to the diagnostic acumen of the staff or to the hospital's rumor network. . . .
6. Interestingly, of the 12 admissions, 11 were diagnosed as schizophrenic and one, with the identical symptomatology, as manic-depressive psychosis. This diagnosis has a more favorable prognosis, and it was given by the only private hospital in our sample. On the relations between social class and psychiatric diagnosis, see A. deB. Hollingshead and F. C. Redlich, *Social Class and Mental Illness: A Community Study* (Wiley, New York, 1958).
7. It is possible, of course, that patients have quite broad latitudes in diagnosis and therefore are inclined to call many people sane, even those whose behavior is patently aberrant. However, although we have no hard data on this matter, it was our distinct impression that this was not the case. In many instances, patients not only singled us out for attention, but came to imitate our behaviors and styles.
8. J. Cumming and E. Cumming, *Community Ment. Health 1*, 135 (1965); A. Farina and K. Ring, *J. Abnorm. Psychol. 70*, 47 (1965); H. E. Freeman and O. G. Simmons, *The Mental Patient Comes Home* (Wiley, New York, 1963); W. J. Johannsen, *Ment. Hygiene 53*, 218 (1969); A. S. Linsky, *Soc. Psychiat.* 5, 1966 (1970).
9. S. E. Asch, *J. Abnorm. Soc. Psychol. 41*, 258 (1946); *Social Psychology* (Prentice-Hall, New York, 1952).
10. See also I. N. Mensh and J. Wishner, *J. Personality 16*, 188 (1947); J. Wishner, *Psychol. Rev. 67*, 96 (1960); J. S. Bruner and R. Tagiuri, in *Handbook of Social Psychology*, G. Lindzey, Ed. (Addison-Wesley, Cambridge Mass., 1954), vol. 2, pp. 634–654; J. S. Bruner, D. Shapiro, R. Tagiuri, in *Person Perception and Interpersonal Behavior*, R. Tagiuri and L. Petrullo, Eds. (Stanford Univ. Press, Stanford, Calif., 1958), pp. 277–288.
11. For an example of a similar self-fulfilling prophecy, in this instance dealing with the "central" trait of intelligence, see R. Rosenthal and L. Jacobson, *Pygmalion in the Classroom* (Holt, Rinehart & Winston, New York, 1968).
12. E. Zigler and L. Phillips, *J. Abnorm. Soc. Psychol. 63*, 69 (1961). See also R. K. Freudenberg and J. P. Robertson, *A.M.A. Arch. Neurol. Psychiatr. 76*, 14 (1956).
13. W. Mischel, *Personality and Assessment* (Wiley, New York, 1968).
14. The most recent and unfortunate instance of this tenet is that of Senator Thomas Eagleton.
15. T. R. Sarbin and J. C. Mancuso, *J. Clin. Consult. Psychol. 35*, 159 (1970); T. R. Sarbin, *ibid. 31*, 447 (1967); J. C. Nunnally, Jr., *Popular Conceptions of Mental Health* (Holt, Rinehart & Winston, New York, 1961).
16. A. H. Stanton and M. S. Schwartz, *The Mental Hospital: A Study of Institutional Participation in Psychiatric Illness and Treatment* (Basic, New York, 1954).
17. D. B. Wexler and S. E. Scoville, *Ariz. Law Rev. 13*, 1 (1971).

PART IV

Major Institutions in Society

An institution is a formal relationship organized around common values to meet basic needs within the society. When a behavior pattern becomes fixed and expected, it can be said to have become part of an institution. Although long-standing normative patterns are difficult to change, new behavior patterns are occurring constantly because of innovations in the material culture (for example, the invention of the automobile) or challenges to expected behavior (the availability of the automobile removed parental supervision of dating from the home). As the new behavior is adopted, it becomes part of the institutionalized, normative expectation. Then, through the socialization function of that institution, the new behavior is passed on to the rest of the society.

All societies have certain functional prerequisites that are necessary for survival. To fulfill these prerequisites, societies have developed major institutions. They are economics, education, the family, politics, and religion. Although their forms may differ greatly, each of these institutions can be found in every society. Before we begin our discussion of major social institutions, we should review the prerequisites that social institutions fulfill. According to Talcott Parsons, there are two: (1) The social system must be relatively compatible with both the individual members of the society and the cultural system as a whole, and (2) the social system requires the support of the other systems around it.

It is the second of Parsons's prerequisites that leads us to expect social institutions to be interdependent and interrelated. What happens in one institution will affect all others. Suppose, for example, a recession in the economy occurs. Family members may lose jobs, churches may receive fewer and smaller donations, and politicians may have a hard time getting reelected.

These ideas of Parsons's indicate that institutions are essential in helping us develop and deal with our wants and needs but that institutions can also be dysfunctional—the means and ends can mean distortion and negative interrelations, with negative results. The effects of socialization mean that transforming our institutions will be difficult and that the interrelationships among institutions will make it harder for them to accomplish their original ends.

Each of us is involved at all times with the major institutions of our society. The effects of family, education, religion, economics, and politics on our lives are constant. Illustrations of such effects are seen in the ways family morals or beliefs and our school experiences affect our thinking—and even the nonbeliever will be affected by religious beliefs embodied in laws or reform efforts. Similarly, the economy, with its cycles of inflation and unemployment, has an effect on prices and taxes, and politics affects our lives through the passage of laws and expenditure of taxes.

As you read through the various selections in Part IV, think about the changes that our major social institutions are undergoing and need to undergo—for that is the major theme of each chapter in this section.

CHAPTER 8

Marriage and Family

Diversity and Change

Alan Mc Evoy

Societies consist of five major institutions: family, education, religion, economic, and political. Some would argue that the most important of these is the family because it is the one with which we have the most contact, the one from which we traditionally receive the most emotional support. It is also the one from which we receive our introduction into the world, the one that supplies our early nurturance, and the one on which society depends to socialize our young. Considering the importance of these factors, it may seem strange that we begin this chapter on families with an article on single people. What does this growing trend imply? Is it just a fad that will reverse itself down the years? Will it continue to grow and thus become a threat to the institution of the family as we know it? If single people become the majority, what will replace the family in performing its important societal functions? It may not be too early to begin thinking about these questions.

Despite the fact that the family today is still a major social institution that performs important societal functions, there is a great deal of misunderstanding about its nature and those functions. As the Cargan article implies, this apparently staid institution is undergoing rapid social change: A rising delay in the ages that mark people's entry into marriage and childbearing has brought about significant changes in the stages of family life, and this change has caused age-graded recreational activities to replace family enterprises. There has been a continuing movement toward earlier and more permissive sexual standards, with the accompanying risk of such diseases as AIDS; there has been a continuing movement toward equality within the female-male relationship that has huge ramifications for both personal and social interactions, especially in the home and the workplace. Along with the increase in numbers of both parents working outside the home has come what many are calling the disappearance of childhood, as children are forced to face problems not previously encountered. Longer life spans will have increasing and serious implications for the way people live with their families, especially their elder members. Finally, changes in the concept of family have come about as divorce rather than death has become the major interrupter of family life, and we still have not come to grips with the effects of globalization on the family.

What all these changes indicate is that our traditional concept of marriage and family is too narrow. As the chapter title suggests, there are many types of marriages and many types of families. The rising divorce rate has created the dual institutions of first marriages and remarriages. With the greater freedom accorded gay people has come gay marriage. Although they are restricted in this country, at least two types of multimate marriages are now found around the world.

Similarly, the concept of family now has different meanings: first, the newlywed family; with the birth of children, the parental family; with both parents working outside the home, the dual-earner family; with divorce, the single-parent family; in its aftermath, a remarriage and blended family; and finally, with greater longevity, the postparental family and the extended family. Along with these changes in the types of marriage and types of families has come a change in what was once considered the model family. The American stereotype of the working father and mother at home taking care of two or three children fits only about 15% of all families. Finally, we noted that entry into marriage is occurring at later and later ages. This effect has led to the myth-belief that Americans are rejecting marriage altogether. However, the facts reveal that the United States is still one of the most marrying societies in the world, with over 90% of Americans marrying at least once.

The second reading by Arlene and Jerome Skolnick examines in more detail our beliefs concerning the family. They conclude that the changes occurring in society are not changing the family as much as they are exposing the fact that beliefs formerly held about the family were, surprisingly, not based on reality in the first place. The Skolnicks do not claim that change does not affect the family, only that the many beliefs held about the family are not always true. Examining just one area of change readily proves that change has a major impact on the family. The third reading by Robert Hughes, Jr., and Jason Hans shows this effect by examining the Internet and its impact on the family.

If wider societal changes are altering the institution of the family, then it should not be surprising that they are also affecting other social institutions. All social institutions are interrelated, and changes in one institution also impact the others. For example, reforms in education (Chapter 9) are demanding greater family involvement in educational issues at the very same time that economic conditions (Chapter 11) require both parents to work outside the home and for longer hours. Relief might be available in the form of reasonably priced day care, availability of family leave at crucial times, flex time to fit home needs, or even part-time work. Unfortunately, the nation's political institutions (Chapter 12) have not seen fit to deal with these economic and family problems. In some situations, however, solving the problem seems relatively simple. For example, many countries have allowed their schools to function as full-time day-care centers as well. Because the buildings already exist and have to be heated or cooled for students and staff, the cost of providing day care could be less costly than day care provided by private providers. At the same time, it would bring much needed income to the schools. The institution of religion (Chapter 10) is also having an important impact on the family. The demands by some religious groups to prevent sexual education and/or the distribution of condoms can lead to unwanted pregnancies with resulting hardships for the affected families.

The fourth reading in this chapter examines domestic violence, which has always existed. Those who would label family violence as something recent that resulted from working wives or economic strains on nonworking wives are merely trying to find an excuse for the myth that all families are supposed to be loving. Emery and Laumann-Billings discuss this and other aspects related to family violence.

Obviously, the government has policies that affect the family—on education, health care, taxes, and so forth—but it does not have a policy specifically designed for the family. Yet the changes noted and the problems that have resulted call out for the development of a systematic family policy. What elements should such a policy contain? For example, should such a policy promote traditional family values in regards to issues such as family duties and sexual activity? Should the policy promote parental and government responsibility in such areas as family planning and aid in various family expenses? In the final reading George Martin describes the four cornerstones that such a policy should contain.

As you read this chapter, consider the significance of the family for each of us and what the outcomes are when this institution is unable to fulfill its societal functions, for us individually and for society as a whole. Also consider the means available for dealing with this situation. Are the family policies noted here legitimate, and should there be others as well? Are the proposed policies realistic?

36

Being Single on Noah's Ark

LEONARD CARGAN

The author describes what it is like to be an unmarried person in a world consisting of a majority of married people. By not marrying by a certain age, these unmarried people are violating the expected pattern of adult life. This deviation from the expected pattern leads to imagined reasons for their not following the pattern. Since little is really known about these deviants, their believed behavior is described in stereotypes such as "swingers" and "loneliness". Despite these negative applications, the singles population has been growing since the "golden age" of marriage in the 1950s due to an age delay in getting married and a rising divorce rate.

As you read this article, consider the following questions as guides:

1. *Do you believe that the situation described is temporary, will stay about the same, or will continue to grow in numbers? Why?*
2. *The increase in numbers of those who delay marriage or do not marry at all is also seen among men. Is it for the same reasons as women? If not, what reasons do men have for postponing marriage or staying single permanently?*
3. *Do you consider that the delayed marriage and nonmarriage phenomenon is a problem? Why or why not?*

All social systems must have norms and values. Norms are agreed-to rules of behavior that allow the group, the organization, or the community to attain its goals – those values that are considered important to achieve. However, if everyone pursued these valued items without regard to norms, there would soon be chaos. In America, a major value is that of individualism – being responsible for our own actions and ideas. An example of such values is seen in this country's great emphasis on competition. This emphasis on competition usually means that government proposed or run programs are usually considered inferior to those run by private enterprise. Another prized value is that of individuality and so we are encouraged to think for ourselves. In seeking these values it is necessary to follow norms – certain rules of behavior. For example, there are numerous norms that must be followed in order to drive safely and prevent accidents. Similarly, there are other norms to protect the public from those who do not follow the norms in seeking valued possessions. The importance of norms and values does not mean that they do not undergo change. For example, the value of a kingdom was exchanged for the value of a republic in the revolutionary war, the value of slavery was exchanged for freedom for all in the civil war, and the value of equal political rights was given to women when given the right to vote.

As America has grown larger in population and urbanized in the twentieth century, it has also become more of an organizational society. These changes have transformed all aspects of social life and required the country to put a greater emphasis on norms, on conforming to the rules whether written or simply understood as the way things are done. In regard to the prized value of individuality it means that it cannot be carried beyond certain limits without being restrained by the norms of conformity. To do so would bring about penalties either directly by, perhaps, going to jail or indirectly via the application of negative stereotypes regarding that behavior and a resulting shame.

This emphasis on conformity can lead to contradictions and even problems since values and norms change and may change rapidly due to rapidly changing technology. It appears that the penalties of such contradictions in the norms are more likely to be felt by minority groups since by definition they are different from the majority in the way things should be done. Thus, ethnic and racial groups may be discriminated against because they are different in their language, dress, food, and values. In a system of rapidly changing values and norms, it is possible for other groups to become recognized as minorities and subsequently discriminated against. For example, it really has only been recently that such groups as women, the poor, and the handicapped were recognized as minorities that are being discriminated against. Although the discrimination was obvious, it apparently was accepted since it was believed that little could be done about it by a minority group. Thus, it has only been relatively recently that efforts have been undertaken to lessen and perhaps eliminate discrimination against women and the handicapped. Obviously, such discrimination changes are slow. The result is that the unmarried are considered a minority in a society of mostly married and are discriminated against by that majority. This discrimination is seen in the development of beliefs designed to perpetuate the dominant system – couples. Thus, singleness is seen as a negative choice as compared to being married – a positive choice. As a means of encouraging this positive choice, singles have been discriminated against via acts – lack of promotions and beliefs – stereotypes concerning their behavior – a bunch of "old maids."

CHANGING FUNCTIONS

An important question to ask at this point is why? Why bother to discriminate against what seems to be a harmless group who mostly will eventually join the majority? Approximately 90 percent of all adult singles will eventually marry. To answer this question, it is necessary to examine the functions that marriage provides and that society believes are valuable and, therefore, should be perpetuated. Despite the loss of many functions in regard to education, protection, recreation, and work, marriage is still seen as the means for aiding and maintaining life: providing for such personal needs as affection and security, socializing the infant with human qualities, and providing the "matrix for the development of personality" (Ackerman, 1972: 16). In addition, the values sought in marriage are among the best needed and desired and for which there are no reliable substitutes: love, loyalty, and stability (O'Brien, 1974: 51). It is for these reasons that marriage is considered a valued part of the natural order of life's progression and that those who do not conform to this order are believed to threaten it. This being the belief, it becomes necessary to encourage conformity to this order via discrimination and stereotypical beliefs. Thus, marriage is presented as the normal and as the only healthy solution to life's dilemmas. This is the value presented despite the facts that show marriage to be an institution experiencing a number of problems. As a result, singleness has been relegated to the margins of society and being single is seen as a temporary position. There is also the added pressure of the family and society to get married. All of these factors contribute to the single feeling deviant and viewing life through this lens. Despite these factors, there has been a growing population of singles.

People are single for a number of reasons. Due to changing social conditions, these reasons are changing and as a result the numbers of people who are single are spending more time in this category. As noted, a major reason for this time spent as a single is the delay in age for a first marriage. In 1960, the number of unmarried men at age 24 was 53 percent but by 1990, it had risen by 25 percent to 78 percent. Similarly, the rates for unmarried women had spurted from 28 percent to 61 percent – a 33 percent increase. The average age for both men and women at first marriage is now at an all time high. The main reason for this jump in age at first marriage appears to be a value change requiring increasing education to obtain work, especially "good" work.

Another factor adding to the population growth of singles is the rapid growth in those divorced. Apparently, divorce is no longer a rare and stigmatizing event. Currently, there are 1.2 million divorces a year. This figure has led to the mistaken belief that the divorce rate is a staggering 50 percent. This belief is based on the fact that there are about 2.4 million marriages per year. Hence, 50 percent of this figure is 1.2 million – the number of divorces occurring each year. This belief would be true if all these divorces came only from those 2.4 million marriages. However, they do not. These divorces come from all existing marriages and so the correct divorce rate is the number of divorces per thousands of existing marriages or 20 percent. Surprisingly, this is actually a lower percentage than that which occurred in the 1920s but much higher than what was believed to be the golden age for marriage – the 1950s. Another good question to ask at this point is why there is a growing rate of those being divorced? The factors believed to be involved include: a declining influence in the role of religion, the liberalization of divorce laws, and the increased participation of women in the labor market. This latter item is believed to have made women less dependent on men. Thus, 60.7 percent of divorces filed were by wives versus about half that amount (32.5 percent) by husbands. Some think the divorce rate is not as bad as it seems since 75 percent of those getting a divorce will remarry. This may be wishful thinking since half of the remarried will re-divorce.

The changes noted that have made being single more common have also made it somewhat more acceptable but only up to a certain age bracket. Beyond 30, singleness remains uncomfortable for both men and women. The truth is that we still live in a world of couples and families. Singleness is still defined only in its relation to marriage – it is the absence of marriage. The status of being single is still seen as a transitional stage as one moves toward marriage. It is marriage that indicates intimacy and the sanctioned way of expressing that intimacy. Marriage is the norm for the community and also indicates the approved means for having children and becoming a part of the family-oriented community. As noted, this means that most people want to marry and do marry, usually by the age of 25 for men and 23 for women, and even remarry after a divorce.

Turning to those who remained single after the seemingly approved age of 30, an opposite picture is revealed. The single person is seen as one who is moving against the approved value that defines marriage as the most desirable state adults can attain. They are often perceived as one who, lacking a partner, are alone and not complete. They are, therefore, "unfit and deviant" (Bell, 1972:89).

THE STEREOTYPES

To not marry requires stereotypical beliefs regarding this failure to follow this approved path: you are socially inadequate and so failed in the dating game or are an alcoholic; you are immature and unwilling to assume responsibility; you are a homosexual and hostile to the opposite sex; you are overfocused on economics in that you believe you are too poor to marry or that marriage is a threat to the pursuance of your career; even if that career is a theological one, your single status is due to your being a religious fanatic or a recluse; or you are single due to geographic or occupational limitations. If none of these excuses for not marrying applies, then the failure not to marry must have been an oversight. The listing of why people do not marry is then a reflection of the belief that every one should marry and that everyone can if they really want to. For the stereotypes, singles are one of the fastest growing segments of American households – a record 40 million compose this category.

Freedom, Workaholic, Affluent

These three stereotypes are so interrelated that it is easier to deal with them as a whole rather than separately. Since the single is not burdened with a wife and family, they are perceived as being freer to have more time to do as they wish. This freedom also implies that the single is less responsible. They can, if they wish, be only concerned with having fun and looking after their own interests. Their lack of responsibilities and community ties also means that they are irresponsible and immature. Freedom is actually a feature that both the single and non-single population share – the freedom to divide their time between office work, housework, social responsibilities, and fun objectives. As noted, with their lesser obligations to deal with a wife and family, the single is freer to spend more time at work in an effort to gain promotions and affluence or more time having fun. Unfortunately, the stereotypes concerning singles provide a hindrance to his work ambition – it is the married who usually get the promotions since they are seen as more mature and responsible, they are the ones restrained by the obligations of the family. Thus, it is the married that may become workaholics since they have greater obligations.

Happiness and Loneliness

This is a mixed set of stereotypes. On the one hand, the single is pictured as happier because they have more freedom to make choices, to get out more, and subsequently have more fun. This also means that they have fewer worries. All these beliefs indicate that the single should be happier than the non-single. On the other hand, there is the likelihood of usually living alone and having no permanent person with whom to share. Therefore, the single person is lonelier and unhappy. In reply to these absolutes regarding single life, singles note that life is not usually a one-way street for singles or the married. There are times when they are enjoying themselves and so are happy and there are times when they wished that they had company but also glad to have, at times, solitude. Thus, loneliness may be a problem at times but it is a problem willingly accepted for the perceived advantages of solitude and being single.

Deviance

Perhaps the single most approbation leveled at singles is that they are deviant. Since marriage is seen as the normal transition process for adults between certain ages, they should be married. Therefore, this charge is true – they are deviating from the norm of marriage. At the same time, singles are and always have been a substantial proportion of the population and it seems strange to label such a large minority as deviant. Population trends also indicate that this minority is rapidly growing due to a later age for a first marriage and a slowly increasing divorce rate. It is possible that before long the minority will be the majority. Still it is doubtful that the beliefs associated with being a minority will shift to the married.

Given these stereotypes, we are socialized with the belief that to be truly happy and fulfilled, it is necessary to marry and take on the responsibilities of a family. This norm is seen in the efforts by others to have you meet the right person and the joy displayed by parents when an engagement is announced and a wedding takes place. In these ways, marriage is proclaimed as society's norm. Despite all these beliefs and married, it appears beyond belief that remaining single might have been a rational choice.

The result of these beliefs are descriptions designed to degrade the single and their lifestyle: the single person is a "poseur, a squanderer, a narcissist, a wastrel." His lifestyle consists of dancing "the hustle in the apartment's house party room" and "loafing on his plastic horse in the . . . swimming pool." He "lives for lotions, balms, and sprays." He is also a "non-stop lover, drinker, laugher, and more (or less)" (Rosenblatt, 1977:14). Even when so-called positive images of singles are presented, they must be degraded. Thus, before marrying, Joe Namath is presented as a positive image of a single – the happy, swinging single yet, at the same time, it implies overtones of being immature, selfish, lechery, and social irresponsibility (Libby, 1978:165; Stein, 1978:2).

The fact that the preceding description of a single described a male does not mean that such stereotypes do not apply to females. According to the beliefs regarding single females, it is she who is "saved" by marriage since this is the only way that she can find love and sexual fulfillment. Thus, it is important for them to marry lest they face emotional and physical deterioration. Single women are such because they are unattractive, handicapped, or incompetent (Deegan, 1969:3) and so they are people to be pitied, ridiculed, disliked, and ascribed a low status except when needed such as in times of war. The single young woman may be intriguing and even challenging but prolonged singleness brings forth suspicions since it seems hard for married people to believe that these women are single because they do not want to be wives.

As a means of better understanding these stereotypical beliefs, it is necessary to sort them out into more specific groupings since their large number results in stereotypes that are entangled, overlapping, and contradictory. In order to be fair, a reply to these stereotypes will be attempted by those being condemned – the singles.

Immaturity

Since marriage is seen as a normal stage of adult development, the "failure" to marry must reflect some kind of immaturity. A man is seen as being tied to his mother's apron strings whereas the woman, being a spinster has failed to experience life's adventures. The sign of immaturity may be a reflection of their lack of altruism – they are selfish and unable to share. To contradict such beliefs, one merely need to point to the many singles like Isaac Newton and Ralph Nader who were experienced, selfless, and obviously mature.

Sexual Deviant

One of the most prevalent stereotypes of the single is seen in the approbation that one is a "swinger" – a lecher hopping from bed to bed. Or their sexual needs may be satisfied by an act of selfishness such as masturbation, an act of sexual inversion which is considered as unnatural as abstinence. What is being ignored in these allegations is that such behaviors have also been found to be quite extensive in the married population as well (Kinsey, 1948).

REFERENCES

Ackerman, Nathan W. in Harrold Hart (ed), *Marriage For and Against*. New York: Hart, 1972, pp. 10–26.

Bell, Robert, in Harrold Hart (ed), *Marriage For and Against*. New York: Hart, 1972, pp. 79–99.

Deegan, Dorothy Y. *The Stereotypes of the Single Woman in American Novels: A Social Study With Implications for the Education of Women*. New York: Octagon, 1969.

Kinsey, A. C., W. B. Pomeroy, & C. E. Martin. *Sexual Behavior in the Human Male*. Philadelphia: W.B. Saunders, 1948.

Libby, Roger W. "Creative Singlehood as a Sexual Lifestyle: Beyond Marriage as a Rite of Passage" in Bernard I. Murster (ed), *Exploring Intimate Lifestyles*. New York: Springer, 1978.

OBrien, Patricia. *The Woman Alone*. New York: Quadrangle, 1974.

Rosenblatt, Roger. *The Self as a Sybarite*. Harpers (March, 1977), pp. 12–14.

Stein, Peter. "The Lifestyle and Life Chances of the Never-Married." *Marriage and Family Review* (July/August, 1978) 1–11.

37

Family in Transition

ARLENE S. SKOLNICK AND JEROME H. SKOLNICK

The authors note that the seeming staid institution of family is actually undergoing a constant and sometimes dramatic transformation.

As you read this article, consider the following questions as guides:

1. *What are some of the surprising changes that have taken place since the so-called "golden age" of family life in the 1950s?*
2. *Why will the institution of family never disappear despite greater options as well as delay and decline in marriage?*
3. *The previous reading discussed the surge in the singles population among the over-30 age group. Does this increase mean that this age group has rejected marriage?*
4. *Does the increase in the numbers of singles in the over 30-age group pose a threat to the institution of marriage?*

GLOSSARY **Cloning** Asexual reproduction of a group of organisms by a single individual. **Flapper** A woman free from social and moral restraints in the post–World War I decade of the 1920s. **Genetic engineering** Practical application of pure science to reproduction. **Introspective** Inclined to examine one's own mental and emotional state.

Even before the attacks on September 11, 2001, it was clear that we had entered a period of profound and unsettling transformations. "Every once in a great while," states the introduction to a *Business Week* issue on the twenty-first century, "the established order is overthrown. Within a span of decades, technological advances, organizational innovations, and new ways of thinking transform economies" (Farrell, 1994, p. 16).

Whether we call the collection of changes *globalization, the information age,* or *post-industrial society,* they are affecting every area of the globe and every aspect of life, including the family. Although the state of the U.S. family has been the subject of great public attention in recent years, the discussion of family has been strangely disconnected from talk of the other transformations we are living through. Instead, family talk is suffused with nostalgia, confusion, and anxiety—what happened, we wonder, to the fifties family of Ozzie and Harriet?. . . .

Of course family life has changed massively in recent decades, and too many children and families are beset by serious stresses and troubles. But we can't understand these changes and problems without understanding the impact of large-scale changes in society on the small worlds of everyday family life. All the industrialized nations, and many of the emerging ones, have experienced the same changes the

SOURCE: Excerpted with permission from *Family in Transition*, 11th edition by Arlene S. Skolnick and Jerome H. Skolnick. Copyright 2001 by Allyn & Bacon. Excerpts from pp. 1–13.

United States has—a transformation in women's roles, rising divorce rates, lower marriage and birth rates, and an increase in single-parent families. In no other country, however, has family change been so traumatic and divisive as in the United States.

For example, the two-earner family has replaced the breadwinner/housewife family as the norm in the United States, even when there are young children in the home. In the mid-nineties, more than 60 percent of married women with children under six were in the paid labor force (Han and Moen, 1999). Yet the question of whether mothers "should" work or stay home is still a hotly debated issue (except if the mother is on welfare).

A range of other, once-"deviant" lifestyles—from living with a partner while unmarried, to remaining single or childless, or even having a child while single (à la Murphy Brown)—has become both acceptable and controversial. It was once unthinkable that gay and lesbian families could be recognized or even tolerated, but despite persisting stigma and the threat of violence, they have been. Local governments and some leading corporations have granted these families increasing recognition as domestic partnerships entitled to spousal benefits.

The changes of recent decades have affected more than the forms of family life; they have been psychological as well. A major study of U.S. attitudes over two decades revealed a profound shift in how people think about family life, work, and themselves (Veroff, Douvan, and Kulka, 1981). In 1957 four-fifths of respondents thought that a man or woman who did not want to marry was sick, immoral, and selfish. By 1976 only one-fourth of respondents thought that choice was bad. Summing up many complex findings, the authors conclude that the United States underwent a "psychological revolution" in the two decades between surveys.

Ever since the 1970s, the mass media have been serving up stories and statistics that seemingly show the family is disintegrating, falling apart, or on the verge of disappearing. . . .

A sudden blizzard of newspaper columns, magazine articles, and talk show "experts" warned that divorce and single parenthood are inflicting serious damage on children and on society in general. This family structure, they argued, is the single biggest problem facing the country, because it is the root cause of all the rest—poverty, crime, drugs, school failure, youth violence, and other social ills.

The proposed solution? Restore the "traditional" family. Make divorce and single parenthood socially unacceptable once again. Do away with welfare and make divorces more difficult to obtain. These arguments flooded the media as welfare reform was being debated in Congress: both sides of the debate cited social science evidence to bolster their case. . . .

It's not surprising then that public debate about the family often sinks to the level of a "food fight" as it lurches from one hot topic to another—single mothers, divorce, gay marriage, nannies, and working mothers. Each issue has only two sides: Are you for or against the two-parent family? Is divorce bad or good for children? Should mothers of young children work or not? Is the family "in decline" or not?. . .

When one extreme position debates the opposite extreme position, it becomes difficult to realistically discuss the issues and problems facing the country. It doesn't describe the range of views among family scholars, and it doesn't fit the research evidence. For example, if someone takes the position that "divorce is damaging to children," the argument culture leads us to assume that there are people on the other "side" who will argue just the opposite—in other words, that divorce is "good," or at least not harmful. But as researcher Paul Amato suggests, the right question to ask is "Under what circumstances is divorce harmful or beneficial to children?" (1994). In most public debates about divorce, however, that question is never asked, and the public never hears the useful information they should.

Still another problem with popular discourse about the family is that it exaggerates the extent of change. For example, we sometimes hear that the traditional nuclear family no longer exists, or has shrunk to a tiny percentage of the population. But that statement depends on a very narrow definition of family—two biological parents, in their first marriage, with a full-time breadwinner husband and a full-time homemaker wife, and two or three children under the age of 18. Of course that kind of family has declined—for the simple reason that most wives and mothers now work outside the home. It has also declined because there are more married couples with grown children than there used to be.

Similarly, we hear that divorce rates have shot up since the 1950s, but we are not told that the trend toward higher divorce rates started in the nineteenth century, with more marital breakups in each succeeding generation. Nor do we hear that despite the current high divorce rates (actually down from 1979), the United States has the highest marriage rates in the industrial world. About 90 percent of Americans marry at some point in their lives, and virtually all who do either have, or want to have, children. Further, surveys repeatedly show that family is central to the lives of most Americans. They find family ties their deepest source of satisfaction and meaning, as well as the source of their greatest worries (Mellman, Lazarus, and Rivlin, 1990). In sum, family life in the United States is a complex mixture of both continuity and change.

While the transformations of the past three decades do not mean the end of family life, they have brought a number of new difficulties. For example, most families now depend on the earnings of wives and mothers, but the rest of society has not caught up to the new realities. There is still an earnings gap between men and women. Employed wives and mothers still bear most of the workload in the home. For both men and women, the demands of the job are often at odds with family needs.

UNDERSTANDING THE CHANGING FAMILY

During the same years in which the family was becoming the object of public anxiety and political debate, a torrent of new research on the family was pouring forth.... Ironically, much of the new scholarship is at odds with the widespread assumption that the family had a long, stable history until hit by the social "earthquake" of the 1960s and 1970s. We have learned from historians that the "lost" golden age of family happiness and stability we yearn for never actually existed....

Part of the confusion surrounding the current status of the family arises from the fact that the family is a surprisingly problematic area of study; there are few if any self-evident facts, even statistical ones. Researchers have found, for example, that when the statistics of family life are plotted for the entire twentieth century, or back into the nineteenth century, a surprising finding emerges: Today's young people—with their low marriage, high divorce, and low fertility rates—appear to be behaving in ways consistent with long-term historical trends (Cherlin, 1981; Masnick and Bane, 1980). The recent changes in family life only appear deviant when compared to what people were doing in the 1940s and 1950s. But it was the postwar generation that married young, moved to the suburbs, and had three, four, or more children that departed from twentieth-century trends. As one study put it, "Had the 1940s and 1950s not happened, today's young adults would appear to be behaving normally" (Masnick and Bane, 1980, p. 2).

Thus, the meaning of change as a particular indicator of family life depends on the time frame in which it is placed. If we look at trends over too short a period of time—say ten or twenty years—we may think we are seeing a marked change, when, in fact, an older pattern may be reemerging. For some issues, even discerning what the trends are can be a problem.

For example, whether we conclude that there is an "epidemic" of teenage pregnancy depends on how we define adolescence and what measure of illegitimacy we use. Contrary to the popular notion of skyrocketing teenage pregnancy, teenaged childbearing has actually been on the decline during the past two decades. It is possible for the *ratio* of illegitimate births to all births to go up at the same time as there are declines in the *absolute number* of births and in the likelihood that an individual will bear an illegitimate child. This is not to say that concern about teenage pregnancy is unwarranted; but the reality is much more complex than the simple and scary notion an "epidemic" implies. Given the complexities of interpreting data on the family, it is little wonder that, as Joseph Featherstone observes (1979), the family is a "great intellectual Rorschach blot. (p. 37)"

1. The Myth of Universality

To say that the family is the same everywhere is in some sense true. Yet families vary in organization, membership, life cycles, emotional environments, ideologies, social and kinship networks, and economic and other functions. Although anthropologists have tried to come up with a single definition of family that would hold across time and place, they generally have concluded that doing so is not useful (Geertz, 1965; Stephens, 1963).

Biologically, of course, a woman and a man must unite sexually to produce a child—even if only sperm and egg meet in a test tube. But no social kinship ties or living arrangements flow inevitably from biological union. Indeed, the definition of marriage is not the same across cultures. Although some cultures have weddings and notions of monogamy and permanence, many cultures lack one or more of these attributes. In some cultures, the majority of people mate and have children without legal marriage and often without living together. In other societies, husbands, wives, and children do not live together under the same roof.

In our own society, the assumption of universality has usually defined what is normal and natural both for research and therapy and has subtly influenced our thinking to regard deviations from the nuclear family as sick or perverse or immoral. As Suzanne Keller (1971) once observed:

> The fallacy of universality has done students of behavior a great disservice. By leading us to seek and hence to find a single pattern, it has blinded us to historical precedents for multiple legitimate family arrangements.

2. The Myth of Family Harmony

To question the idea of the happy family is not to say that love and joy are not found in family life or that many people do not find their deepest satisfactions in their families. Rather, the happy-family assumption omits important, if unpleasant, aspects of family life. Intimate relations inevitably involve antagonism as well as love. This mixture of strong positive and negative feelings sets close relationships apart from less intimate ones. . . .

In recent years, family scholars have been studying family violence such as child abuse and wife beating to better understand the normal strains of family life. Long-known facts about family violence have recently been incorporated into a general analysis of the family. More police officers are killed and injured dealing with family fights than in dealing with any other kind of situation; of all the relationships between murderers and their victims, the family relationship is most common. Studies of family violence reveal that it is much more widespread than had been assumed, cannot easily be attributed to mental illness, and is not confined to the lower classes. Family violence seems to be a product of psychological tensions and external stresses that can affect all families at all social levels.

The study of family interaction has also undermined the traditional image of the happy, harmonious family. . . .

3. The Myth of Parental Determinism

The kind of family a child grows up in leaves a profound, lifelong impact. But a growing body of studies shows that early family experience is not the all-powerful, irreversible influence it has sometimes been thought to be. An unfortunate childhood does not doom a person to an unhappy adulthood. Nor does a happy childhood guarantee a similarly blessed future (Emde and Harmon, 1984; Macfarlane, 1964; Rubin, 1996).

First, children come into this world with their own temperamental and other individual characteristics. As parents have long known, child rearing is not like molding clay or writing on a blank slate.

Rather, it's a two-way process in which both parent and child shape each other. Further, children are active perceivers and interpreters of the world. Finally, parents and children do not live in a social vacuum; children are also influenced by the world around them and the people in it—the kin group, the neighborhood, other children, the school, and the media. . . .

4. The Myth of a Stable, Harmonious Past

Laments about the current state of decay of the family imply some earlier era when the family was more stable and harmonious. But unless we can agree what earlier time should be chosen as a baseline and what characteristics of the family should be specified, it makes little sense to speak of family decline. Historians have not, in fact, located a golden age of the family.

Indeed, they have found that premarital sexuality, illegitimacy, generational conflict, and even infanticide can best be studied as a part of family life itself rather than as separate categories of deviation. . . .

The most shocking finding of recent years is the prevalence of infanticide throughout European history. Infanticide has long been attributed to primitive peoples or assumed to be the desperate act of an unwed mother. It now appears that infanticide provided a major means of population control in all societies lacking reliable contraception, Europe included, and that it was practiced by families on legitimate children (Hrdy, 1999).

Rather than being a simple instinctive trait, having tender feelings toward infants—regarding a baby as a precious individual—seems to emerge only when infants have a decent chance of surviving and adults experience enough security to avoid feeling that children are competing with them in a struggle for survival. Throughout many centuries of European history, both of these conditions were lacking.

Another myth about the family is that of changelessness—the belief that the family has been essentially the same over the centuries, until recently, when it began to come apart. Family life has always been in flux; when the world around them changes, families change in response. At periods when a whole society undergoes some major transformation, family change may be especially rapid and dislocating.

In many ways, the era we are living through today resembles two earlier periods of family crisis and transformation in U.S. history (see Skolnick, 1991). The first occurred in the early nineteenth century, when the growth of industry and commerce moved work out of the home. Briefly, the separation of home and work disrupted existing patterns of daily family life, opening a gap between the way people actually lived and the cultural blueprints for proper gender and generational roles (Ryan, 1981). In the older pattern, when most people worked on farms, a father was not just the head of the household, but also boss of the family enterprise. Mother and children and hired hands worked under his supervision. But when work moved out, father—along with older sons and daughters—went with it, leaving behind mother and the younger children. These dislocations in the functions and meaning of family life unleashed an era of personal stress and cultural confusion.

Eventually, a new model of family emerged that not only reflected the new separation of work and family, but also glorified it. No longer a workplace, the household now became idealized as "home sweet home," an emotional and spiritual shelter from the heartless world outside. Although father remained the head of the family, mother was now the central figure in the home. The new model celebrated the "true woman's" purity, virtue, and selflessness. Many of our culture's most basic ideas about the family in U.S. culture, such as "a woman's place is in the home," were formed at this time. In short, the family pattern we now think of as traditional was in fact the first version of the modern family.

Historians label this model of the family "Victorian" because it became influential in England and Western Europe as well as in the United States during the reign of Queen Victoria. It reflected, in idealized form, the nineteenth-century middle-class family. However, the Victorian model became the prevailing cultural definition of family. Few families could live up to the ideal in all its particulars; working-class, black, and ethnic families, for example, could not get by without the economic contributions of wives, mothers, and daughters. And even for middle-class families, the Victorian ideal prescribed a standard of perfection that was virtually impossible to fulfill (Demos, 1986).

Eventually, however, social change overtook the Victorian model. Beginning around the 1880s, another period of rapid economic, social, and cultural change unsettled Victorian family patterns, especially their gender arrangements. Several generations of so-called new women challenged Victorian notions of femininity. They became educated, pursued careers, became involved in political causes—including their own—and created the first wave of feminism. This ferment culminated in the victory of the women's suffrage movement. It was followed by the 1920s' jazz-age era of flappers and flaming youth—the first, and probably the major, sexual revolution of the twentieth century.

To many observers at the time, it appeared that the family and morality had broken down. Another cultural crisis ensued, until a new cultural blueprint emerged—the companionate model of marriage and the family. The new model was a revised, more relaxed version of the Victorian family; companionship and sexual intimacy were now defined as central to marriage.

This highly abbreviated history of family and cultural change forms the necessary backdrop for understanding the family upheavals of the late twentieth and early twenty-first centuries. As in earlier times, major changes in the economy and society have destabilized an existing model of family life and the everyday patterns and practices that have sustained it.

We have experienced a triple revolution: first, the move toward a postindustrial service and information economy; second, a life course revolution brought about by the reductions in mortality and fertility; and third, a psychological transformation rooted mainly in rising educational levels.

Although these shifts have profound implications for everyone in contemporary society, women have been the pacesetters of change. Most women's lives and expectations over the past three decades, inside and outside the family, have departed drastically from those of their own mothers. Men's lives today also are different from their fathers' generation, but to a much lesser extent.

THE TRIPLE REVOLUTION

The Postindustrial Family

The most obvious way the new economy affects the family is in its drawing women, especially married women, into the workplace. A service and information economy produces large numbers of jobs that, unlike factory work, seem suitable for women. Yet as Jessie Bernard (1982) once observed, the transformation of a housewife into a paid worker outside the home sends tremors through every family relationship. It creates a more "symmetrical" family, undoing the sharp contrast between men's and women's roles that marks the breadwinner/housewife pattern. It also reduces women's economic dependence on men, thereby making it easier for women to leave unhappy marriages.

Beyond drawing women into the workplace, shifts in the nature of work and a rapidly changing globalized economy have unsettled the lives of individuals and families at all class levels. The well-paying industrial jobs that once enabled a blue-collar worker to own a home and support a family are no longer available. The once secure jobs that sustained the "organization men" and their families in the 1950s and 1960s have been made shaky by downsizing, an unstable economy, corporate takeovers, and a rapid pace of technological change.

The new economic climate has also made the transition to adulthood increasingly problematic. The uncertainties of work is in part responsible for young adults' lower fertility rates and for women flooding the workplace. Further, the family formation patterns of the 1950s are out of step with the increased educational demands of today's post-industrial society. In the postwar years, particularly in the United States, young people entered adulthood in one giant step—going to work, marrying young and moving to a separate household from their parents, and having children quickly. Today, few young adults can afford to marry and have children in their late teens or early twenties. In an economy where a college degree is necessary to earn a living wage, early marriage impedes education for both men and women.

Those who do not go on to college have little access to jobs that can sustain a family. Particularly in the inner cities of the United States, growing numbers of young people have come to see no future for themselves at all in the ordinary world of work. In middle-class families, a narrowing opportunity structure has increased anxieties about downward mobility for offspring, and parents as well. The "incompletely launched young adult syndrome" has become common: Many young adults deviate from their parents' expectations by failing to launch careers and become successfully independent adults, and many even come home to crowd their parents' empty nest (Schnaiberg and Goldenberg, 1989).

The Life Course Revolution

The demographic transformations of the twentieth century were no less significant than the economic ones. We cannot hope to understand current predicaments of family life without understanding how radically the demographic and social circumstances of U.S. culture have changed. In earlier times, mortality rates were highest among infants, and the possibility of death from tuberculosis, pneumonia, or other infectious diseases was an ever-present threat to young and middle-aged adults. Before the turn of the twentieth century, only 40 percent of women lived through all the stages of a normal life course—growing up, marrying, having children, and surviving with a spouse to the age of 50 (Uhlenberg, 1980).

Demographic and economic change has had a profound effect on women's lives. Women today are living longer and having fewer children. When infant and child mortality rates fall, women no longer have to have five or seven or nine children to make sure that two or three will survive to adulthood. After rearing children, the average woman can look forward to three or four decades without maternal responsibilities. Because traditional assumptions about women are based on the notion that they are constantly involved with pregnancy, child rearing, and related domestic concerns, the current ferment about women's roles may be seen as a way of bringing cultural attitudes in line with existing social realities.

As people live longer, they can stay married longer. Actually, the biggest change in contemporary marriage is not the proportion of marriages disrupted through divorce, but the potential length of marriage and the number of years spent without children in the home. By the 1970s the statistically average couple spent only 18 percent of their married lives raising young children, compared with 54 percent a century ago (Bane, 1976). As a result, marriage is becoming defined less as a union between parents raising a brood of children and more as a personal relationship between two individuals.

A Psychological Revolution

The third major transformation is a set of psychocultural changes that might be described as "psychological gentrification" (Skolnick, 1991). That is, cultural advantages once enjoyed only by the upper classes—in particular, education—have been extended to those lower down on the socioeconomic scale. Psychological gentrification also involves greater leisure time, travel, and exposure to information, as well as a general rise in the standard of living. Despite the persistence of poverty, unemployment, and economic insecurity in the industrialized world, far less of the population than in the historical past is living at the level of sheer subsistence.

Throughout Western society, rising levels of education and related changes have been linked to a complex set of shifts in personal and political attitudes. One of these is a more psychological approach to life—greater introspectiveness and a yearning for warmth and intimacy in family and other relationships (Veroff, Douvan, and Kulka, 1981). There is also evidence of an increasing preference on the part of both men and women for a more companionate ideal of marriage and a more democratic family. More broadly, these changes in attitude have been described as a shift to "postmaterialist values," emphasizing self-expression, tolerance, equality, and a concern for the quality of life (Inglehart, 1990).

The multiple social transformations of our era have brought both costs and benefits: Family relations have become both more fragile and more emotionally rich; mass longevity has brought us a host of problems as well as the gift of extended life. Although change has brought greater opportunities for women, persisting gender inequality means women have borne a large share of the costs of these gains. But we cannot turn the clock back to the family models of the past.

Paradoxically, after all the upheavals of recent decades, the emotional and cultural significance of the family persists. Family remains the center of most people's lives and, as numerous surveys show, a cherished value. Although marriage has become more fragile, the parent–child relationship—especially the mother–child relationship—remains a core attachment across the life course (Rossi and Rossi, 1990). The family, however, can be both "here to stay" and beset with difficulties. There is widespread recognition that the massive social and economic changes we have lived through call for public and private-sector policies in support of families. Most European countries have recognized for some time that governments must play a role in supplying an array of supports to families—health care, children's allowances, housing subsidies, support for working parents and children (such as child care, parental leave, and shorter work days for parents), as well as an array of services for the elderly.

Each country's response to these changes, as we've noted earlier, has been shaped by its own political and cultural traditions. The United States remains embroiled in a cultural war over the family; many social commentators and political leaders have promised to reverse the recent trends and restore the "traditional" family. In contrast, other Western nations, including Canada and the other English-speaking countries, have responded to family change by establishing policies aimed at mitigating the problems brought about by economic and social transformations. As a result of these policies, these countries have been spared much of the poverty and other social ills that have plagued the United States in recent decades.

Looking Ahead

The world at the beginning of the twenty-first century is vastly different from what it was at the beginning, or even the middle, of the twentieth century. Families are struggling to adapt to new realities. The countries that have been at the leading edge of family change still find themselves caught between yesterday's norms, today's new realities, and an uncertain future. As we have seen, changes in women's lives have been a pivotal factor in recent family trends. In many countries there is a considerable difference between men's and women's attitudes and expectations of one another. Even where both partners accept a more equal division of labor in the home, there is often a gap between beliefs and behavior. In no country have employers, the government, or men fully caught up to the changes in women's lives.

But a knowledge of family history reveals that the solution to contemporary problems will not be found in some lost golden age. Families have always struggled with outside circumstances and inner conflict. Our current troubles inside and outside the family are genuine, but we should never forget that many of the most vexing issues confronting us derive from benefits of modernization few of us would be willing to give up—for example, longer, healthier lives, and the ability to choose how many children to have and when to have them. There was no problem of the aged in the past, because most people never aged; they died before they got old. Nor was adolescence a difficult stage of the life cycle when children worked, education was a privilege of the rich, and a person's place in society was determined by heredity rather than choice. And when most people were hungry illiterates, only aristocrats could worry about sexual satisfaction and self-fulfillment.

In short, there is no point in giving in to the lure of nostalgia. There is no golden age of the family to long for, nor even some past pattern of behavior and belief that would guarantee us harmony and stability if only we had the will to return to it. Family life is bound up with the social, economic, and ideological circumstances of particular times and places. We are no longer peasants, Puritans, pioneers, or even suburbanites circa 1955. We face conditions unknown to our ancestors, and we must find new ways to cope with them. . . .

REFERENCES

Amato, P. R. 1994. Life span adjustment of children to their parents' divorce. *The Future of Children* 4, no 1. (Spring).

Bane, M. J. 1976. *Here to Stay*. New York: Basic Books.

Bernard, J. 1982. *The Future of Marriage*. New York: Bantarn.

Blake, J. 1978. Structural differentiation and the family: A quiet revolution. Presented at American Sociology Association, San Francisco.

Cherlin, A. J. 1981. *Marriage, Divorce, Remarriage*. Cambridge, Mass.: Harvard University Press.

Demos, John. 1986. *Past, Present, and Personal*. New York: Oxford University Press.

Emde, R. N., and R. J Harmon, eds. 1984. *Continuities and Discontinuities in Development*. New York: Plenum Press.

Farrell, Christopher. 1994. Twenty-first century capitalism: The triple revolution. *Business Week* (November 18): 16–25.

Featherstone, J. 1979. Family matters. *Harvard Educational Review* 49, no. 1: 20–52.

Geertz, G. 1965. The impact of the concept of culture on the concept of man. In *New Views of the Nature of Man*, edited by J. R. Platt. Chicago: University of Chicago Press.

Han, S.-K., and P. Moen. 1999. Work and family over time: A life course approach. *The Annals of the American Academy of Political and Social Sciences* 562: 98–110.

Hrdy, Sarah, B. 1999. *Mother Nature*. New York: Pantheon Books.

Inglehart, Ronald. 1990. *Culture Shift*. New Jersey: Princeton University Press.

Keller, S. 1971. Does the family have a future? *Journal of Comparative Studies*, Spring.

Macfarlane, J. W. 1964. Perspectives on personality consistency and change from the guidance study. *Vita Humana* 7: 115–126.

Masnick, G., and M. J. Bane. 1980. *The Nation's Families: 1960–1990*. Boston: Auburn House.

Mellman, A., E. Lazarus, and A. Rivlin. 1990. Family time, family values. In *Rebuilding the Nest*, edited by D. Blankenhorn, S. Bayme, and J. Elshtain. Milwaukee: Family Service America.

Rossi, A. S., and P. H. Rossi. 1990. *Of Human Bonding: Parent-Child Relations Across the Life Course*. Hawthorne, New York: Aldine de Gruyter.

Rubin, L. 1996. *The Transcendent Child*. New York: Basic Books.

Ryan, M. 1981. *The Cradle of the Middle Class*. New York: Cambridge University Press.

Schnaiberg, A., and S. Goldenberg. 1989. From empty nest to crowded nest: The dynamics of incompletely launched young adults: *Social Problems* 36, no. 3 (June): 251–269.

Skolnick, A. 1991. *Embattled Paradise: The American Family in an Age of Uncertainty*. New York: Basic Books.

Stephens, W. N. 1963. *The Family in Cross-Cultural Perspective*. New York: World.

Tannen, D. 1998. *The Argument Culture*.

Uhlenberg, P. 1980. Death and the family. *Journal of Family History* 5, no. 3: 313–320.

Veroff, J., E. Douvan, and R. A. Kulka. 1981. *The Inner American: A Self-Portrait from 1957 to 1976*. New York: Basic Books.

38

Understanding the Effects of the Internet on Family Life

ROBERT HUGHES, JR. AND JASON D. HANS

When we think of the many new technologies that have emerged in our lifetime, it is usually in terms of the benefits these changes have brought to our physical lives. Rarely do we consider the fact that these technologies also affect our social lives. The authors of this reading want you to understand how one instance of relatively new technology—the computer—is altering the social life of family members.

As you read, consider the following questions as guides:

1. *Have the changes in family life mentioned by the authors affected your family similarly? If not, why not?*
2. *Were you surprised by some of the effects mentioned? Which ones? Why?*
3. *What changes would you suggest to make it easier for your family to adapt to this new technology and help in your interfamily relations?*

Less than two decades ago, computers were primarily used in science, engineering, and business, and the Internet was the province of the military. Yet in 2001, 57% of all households had a computer, and 51% had direct access to the Internet (National Telecommunications and Information Administration [NTIA], 2002). Social critics and technologists have been active in discussing the implications of these changes for individuals, families, work, and society. There are those who see computers and the Internet as a positive force that will foster greater communication and better access to education, promote global understanding, and make the world a better place to live (Rheingold, 1993). Some also believe that the Internet will lead to better social relationships because people will be freed from the constraints of time and place (Katz & Aspden, 1997). Other critics suggest that computer technology will impoverish relationships, isolate family members from each other, and distance families from the outside world (Stoll, 1995)....

In contrast, Burgess (1928) cautioned against the view that families are shaped solely by environmental factors: "Only through research can the necessary basis of fact be found for any practical program to meet the problems of the changing American family" (p. 415).

In keeping with the advice of Burgess (1928), we examine research on the effects of computers and the Internet on families to bring about a better understanding of how this technology influences

SOURCE: In *Handbook of Contemporary Families: Considering The Past, Contemplating The Future* by Coleman, pp. 506–520. Reprinted with permission of Sage Publications.

family life. We begin with an overview of the extent to which computers and the Internet have become part of the landscape of family life. To provide perspective, we also consider information about other communication technologies. Then we look at five questions regarding the effects of computers and the Internet on families: (a) How has the Internet affected romantic relationships? (b) How has the Internet affected family relationships? (c) How does the Internet affect family ties to social networks? (d) How is the intersection of work and family altered by the Internet? and (e) How can the Internet be used to help families?

THE INFORMATION TECHNOLOGY CONTEXT

Much has been made of the rapid introduction of computers and the Internet into the private realm of family life. However, this is just the latest wave of information technology to become commonplace in households. The older technologies—radios, telephones, and televisions—are in well over 90% of U.S. households, and although the proportion of households with these devices has remained steady over the last three decades, an increasing number have multiple radios and televisions (Newburger, 1999).

The prevalence of personal computers and Internet use in the home has grown rapidly in recent years, but ownership of computers and access to the Internet vary on the basis of income, education, household composition, and ethnicity. Low-income and single-parent households and households headed by individuals with little education are far less likely to have Internet access in their homes than affluent, two-parent households composed of adults with more years of formal education (NTIA, 2002). However, the growth rate in Internet access is much greater among the former, leading to a gradual diminishing of the gap. Ethnic differences follow a similar pattern. Asians (60.4%) and whites (59.9%) are more likely to use the Internet than blacks (24.7%) or Hispanics (20.1%), but between 2000 and 2001 Internet use grew the most rapidly among blacks and Hispanics. Personal computers and the Internet may eventually become commonplace in families across a broader spectrum of socioeconomic and racial strata, but for now, there are large differences in who has in-home computer capabilities and who does not.

HOW HAS THE INTERNET AFFECTED ROMANTIC RELATIONSHIPS?

Dating and Intimacy

Although interactive Internet applications (e.g., electronic mail, newsgroups, chat rooms) were first developed in the early 1970s, the role personal computers played in romantic relations went largely unnoticed until the number of Internet users mushroomed in the mid-1990s. Early on, there was a stigma attached to seeking and finding love online, but the number of people engaging in online dating has grown rapidly. Jupiter Research reported that in 2002, 16.3 million people visited online dating sites; there are now more paying subscribers to online dating services than any other content area on the Internet (cited in O'Connell, 2003).

Models have been hypothesized to explain the lure of online romantic/sexual relationships, such as Cooper's (1998) Triple-A Engine (Access, Affordability, Anonymity) and Young's (1999) ACE Model of Cyber-sexual Addiction (Anonymity, Convenience, Escape). Although empirical validation of these models is needed, early research has demonstrated that certain types of people prefer Internet-facilitated courtship. For example, those who are socially anxious or lonely are more likely to form intimate relationships via the Internet (McKenna, Green, & Gleason, 2002), and shy individuals are able to overcome many relationship-initiation barriers (Scharlott & Christ, 1995). It is likely that other barriers to relationship formation (e.g., proximity, social class, some physical traits) also become less important at the outset of computer-mediated relationships (Cooper & Sportolari, 1997;

McKenna at al., 2002), although participants in a study on attitudes toward online relationships indicated discomfort with meeting potential partners online when their physical appearance is not known (Donn & Sherman, 2002).

There is evidence that online relationship development is different from traditional courting. Online romantic relationships have much higher levels of self-disclosure and intimacy early in the relationship (Clark, 1998; Gerlander & Takala, 1997). McKenna et al. (2002) found that the stability of online relationships over a 2-year period compared favorably to that of traditional relationships. If an online interaction is going well, the relationship often evolves into a conventional face-to-face relationship (McKenna et al., 2002; Parks & Roberts, 1998), or at least occasional contacts for those who are geographically distant. Concern has been expressed because one can easily deceive another in computer-mediated relationships (Cooper & Sportolari, 1997), a concern that is especially pertinent for relationships that have not resulted in face-to-face meetings. However, due to the relative anonymity and discreetness of e-mail, many lies (e.g., marital status) may still be quite easily maintained even when meetings occur. . . .

Extramarital Relationships

Infidelity has usually been defined as sexual relationships outside marriage; however, the emergence of computer-mediated relationships and virtual sex has raised new questions about boundaries of intimacy (Merkle & Richardson, 2000). Anecdotal evidence suggests that many people know of a friend, family member, or acquaintance whose face-to-face romantic relationship was threatened by an online relationship. Thus, intimate online relationships, even if not physical, may become common sources of tension in existing face-to-face romantic relationships (Young, Griffin-Shelley, Cooper, O'Mara, & Buchanan, 2000). Indeed, Schneider (2000) found that among 94 respondents whose marital relationships were seriously and adversely affected by a partner's cybersex activities, more than 60% indicated that the online relationship never progressed beyond computer-mediated interaction. Clearly, research is needed that investigates the changing boundaries of intimacy brought about by computer-mediated interaction.

HOW HAS THE INTERNET AFFECTED FAMILY RELATIONSHIPS? . . .

One area of speculation is whether computer technology strengthens or damages relationships among family members. One of the earliest studies on the role of the Internet in family life (Kraut, Mukophadyay, Szczypula, Kiesler, & Scherlis, 2000) monitored a group of parents and their teenagers over their first 1 to 2 years of Internet use. Parents and adolescents used the Internet more often to interact (e.g., sending and receiving e-mail) with non-household members than to seek information or entertainment. They also spent less time communicating in the household with family members than they did before gaining Internet access. These results give credibility to the fears that Internet use damages family relationships; however, a follow-up study 2 to 3 years later found that these initial declines in family communication did not persist (Kraut et al., 1998).

In one of the few observational studies about computers and family relationships, Orleans and Laney (2000) observed 32 children between the ages of 8 and 17 on at least three occasions each for an hour or more while they did computer work on their own or with others at home. Children and their parents seldom talked to each other while the children were using the computer. Generally, children used the computers independently and were more likely to talk with siblings or peers for help regarding computer problems than they were to ask their parents. About 65% of the time that the children were online, they sent and received e-mail, visited chat rooms, and played interactive games. Boys and girls used the computers in different ways:

> The girls were more likely to be serious about using the computer. They were more focused

> on using the computer for particular purposes, and their demeanor while using [the computer] was more somber than the boys. The boys seemed more likely to view the computer as a multipurpose toy that was itself fun to use and integrated it into their social lives. (Orleans & Laney, 2000, p. 67)

Another area of interest to family scientists has been the ways in which parents manage the use of technology. . . . Livingstone (2002) found that only 6% of parents were concerned about their children's use of computers and the Internet. Parents were far more concerned about illegal drugs (51%), crime (39%), and educational standards (38%). These data suggest that when viewed in the context of other hazards children face, parents perceive that there are more serious threats to children's well-being than their children's computer and Internet use. However, 50% of the parents in Livingstone's (2002) study reported having rules about children's use of the Internet. In contrast, children reported about half as many restrictions as their parents. The inconsistency between reports of parents and of their children points to a need for a better understanding of computers and Internet use in family contexts on a day-to-day basis. This may require observational and longitudinal data in addition to self-reports by children and parents.

The contextual nature of parents' Internet concerns compared with their concerns about other aspects of life illustrates the importance of studying the Internet in context to provide a more complete understanding of how the technology fits with other aspects of family life. . . .

Another important question related to the Internet concerns aggressive behavior. There is much research about the contribution of television and video games to aggressive behavior in children and adults (e.g., Johnson, Cohen, Smailes, Kasen, & Brook, 2002). The Internet not only provides additional opportunities for family members to be exposed to violent images and activities but makes it possible to be in contact with people who are engaged with violent material and activities. Given the level of violence in the world, understanding how family members are affected by these potentially harmful opportunities is critical.

Another important direction for research is to focus, not on computer technology itself, but rather on computer technology in the context of family issues such as intergenerational relationships, post-divorce relationships, social network processes, and work. For example, the Internet may provide new ways for older family members to communicate with distant family members, and there are many unanswered questions regarding ways this technology may serve as a bridge for homebound elderly. Similarly, using the Internet to maintain relationships with non-residential children or parents would have quite different effects than participating in online games for recreation. The Internet can become an important way for family members to stay connected after divorce. . . .

HOW DOES THE INTERNET AFFECT FAMILY TIES TO SOCIAL NETWORKS?

Another early concern regarding the Internet was that people would abandon face-to-face relationships and live their lives online. . . .

An early study reporting on new Internet users seemed to confirm the idea that the Internet could lead to withdrawal from social involvements (Kraut et al., 1998). New users who spent more time on the Internet reported less social involvement with both geographically close and distant friends. However, over the next 2 to 3 years, social support and interaction with close and distant network members returned to pre-Internet levels. This study took place when Internet technology was newer and people were less familiar with it than they are now. Thus, participants may have withdrawn from social ties because of the novelty of this new technology and the time needed to master it. Few members of their social network would have had access to the Internet, so they would have been less able to use it to maintain existing social ties than current Internet users would be. In a second study with a new sample, Kraut et al. (2002) added more control variables and a wider range of social network measures. In this study, they found that Internet use was related to *increases* in the number of close and distant social contacts and face-to-face communication with family and friends, indicating that the Internet had a positive impact on development and maintenance of social networks.

Other investigators have begun to provide evidence that the Internet may help maintain social ties. Among home Internet users, 96.6% of women and 93.6% of men reported using the Internet to communicate with friends and family (NTIA, 2002). In fact, the primary reason why people send and receive e-mail messages is to maintain interpersonal relationships (Stafford, Kline, & Dimmick, 1999). Almost twice as many people reported interpersonal reasons for using e-mail (42%) as those who reported using e-mail for business (25%) or information (23%). About 60% of Internet users reported that they communicated more with family and friends now that they had e-mail access (Howard et al., 2001). Franzen (2000) found that, over time, e-mail had a positive effect on the maintenance of social ties. Nearly half of online seniors were persuaded to get Internet access by family members, and a majority reported that the Internet enhanced communication with family members (Fox et al., 2001).

In a unique study of social networks, Hampton and Wellman (2000) surveyed a neighborhood in which all the residents had free access to a high-speed Internet connection. The wired residents recognized almost three times as many neighbors, talked with nearly twice as many, and had been invited or had invited, one and a half times as many neighbors into their homes as had residents of a nearby neighborhood that was not wired. The authors suggested that rather than replacing face-to-face ties, computer-mediated ties supported and strengthened neighborhood social ties by providing new opportunities for social relationships and engagement in community. In a large study of Internet users, Wellman, Haase, Witte, and Hampton (2001) found that online activity supplemented rather than replaced or diminished offline social contacts. Overall, these findings suggest that the Internet has positive effects on family members' ability to maintain real-world social ties outside the immediate family.

HOW IS THE INTERSECTION OF WORK AND FAMILY ALTERED BY THE INTERNET?...

A recent review of telework, defined as work performed at an offsite location and most typically within one's home, suggests that many of the forecasts of changing work environments are unlikely to occur (Ellison, 1999). Although there are many optimistic reports about telework, findings from a recent U.S. Bureau of Labor Statistics (2002) report indicated that only about 15% of employees work at home at least 1 day per week. Taken at face value, telecommuting sounds like a solution to work-family strain, child care, and numerous other family dilemmas. However, studies of telecommuting suggest that there may be significant limitations to overcome with regard to working at home.

In one survey, although 88% of workers preferred telecommuting, only 11% were doing it (Mokhtarian & Salomon, 1996). The constraints against telecommuting have little to do with technology but rather are related to supervisor unwillingness, concern about lack of visibility to management, household distractions, and a lack of self-discipline to do the work (Mokhtarian, Bagley, & Salomon, 1998). Women with children in particular are more likely to cite household distractions as a constraint when compared to women without children.

Reflecting on studies of the impact of technology on work, Kraut (1987) commented, "Office structure has remained virtually unchanged since the late 19th century, despite... major changes in office technology" (p. 130). He suggested that predictors of changes in work as the result of technology have often failed to understand the importance of socializing as a source of worker satisfaction and the importance of co-workers in conducting many work assignments.

Another hypothesis about the effect of technology on work and family life is that the availability of computers and the Internet at home leads to more people taking work home from the office. Surprisingly, the trend has been in the opposite direction. Between 1991 and 1997, there was a modest decline (from 12.2 million to 11.1 million) in the number of workers taking work home (U.S. Bureau of Labor Statistics, 1998).

Although there may not be dramatic changes in the work-family relationship due to computers and the Internet, there have been subtle shifts. Hill and his colleagues (Hill, Hawkins, Ferris, & Weitzman, 2001; Hill, Hawkins, & Miller, 1996) have chronicled the implementation of flextime and flexplace in the IBM Corporation and reports by employees on work-family balance. In one study, workers assigned to virtual offices reported no better ability to balance work and family (Hill et al., 1996); however, recent findings indicate that workers report being better able to achieve work-life balance when they have more flexibility, either flextime or flexplace (Hill et al., 2001). Additionally, some surveys of family members suggest that the boundary between work and home is blurring: About 10% of Internet users who have access only on the job do something unrelated to work almost daily; about 66% report some use of the Internet for home-related activities while at work (Howard et al., 2001). Likewise, almost 25% of Internet users with access only at home report doing something for work at home. These data indicate that work-related tasks are performed in some homes and that some personal tasks are completed in the workplace but that the effects of technology on work and family life appear to be subtle. . . .

HOW CAN THE INTERNET BE USED TO HELP FAMILIES?

The Internet has been used to create new ways of providing peer support, family life education, and family therapy. For example, there are numerous news groups online devoted to family issues such as divorce, death, or children with special needs. Additionally, family life educators and family therapists have begun to create online opportunities to provide help to families. It is important to understand more about the effectiveness of these activities.

Peer Support

Peer support through news groups was one of the first Internet developments. There are groups devoted to a wide range of family issues. Some groups have small readerships, and some have thousands of participants. Participants in these self-help activities find them beneficial (King & Moreggi, 1998). Online groups may be especially important to individuals whose face-to-face social relationships are inadequate or for groups that feel stigmatized, such as parents of special-needs children (King & Moreggi, 1998). Several studies have examined these issues.

Miller and Gergen (1998) concluded, from a content analysis of helping strategies offered on one self-help site, that the help provided differed from change strategies used by trained family therapists. They speculated that although participants may feel like they are getting help through these online groups, the help may not be as effective as that provided by skilled therapists.

Those in online groups, unlike those in face-to-face groups, can participate in three ways: reading messages, posting messages to the group, or sending private e-mail to selected group members (Mickelson, 1997). Different patterns of social support are evident in these three styles of interacting. Mickelson found that although merely reading newsgroup messages was not related to any of the social support behaviors, posting public messages was related to fears of rejection, and private e-mail requests were related to lack of perceived support from spouses. Thus, the Internet may provide an alternative source of social relationships for those who have difficulty developing social ties face to face. Similarly, Cummings, Sproull, and Kiesler (2002) found that members of an online hearing-loss support group were more likely to participate if they lacked real-world social support. Additionally, these researchers found that in online support groups, unlike face-to-face groups, friends and family members can also participate, and participants whose real-world social network participated reported benefiting the most from online help. Cummings and his colleagues concluded that the paths through which social support may benefit individuals may differ in online support groups as opposed to face-to-face groups. It would be important to examine whether the paths found in a hearing-loss group would be similar to those found in family-issue support groups, such as groups related to divorce, single parenting, new parents, or stepfamilies. This function may be especially important for individuals whose face-to-face social relationships among friends and family are inadequate or for groups that feel stigmatized.

Family Life Education and Family Therapy

Family life educators and family therapists have begun to explore the ways in which the Internet can be used to help individuals and families. It has been suggested that the Internet provides a valuable medium through which to teach families (Hughes, Ebata, & Dollahite, 1999) and that the Web may be especially suitable for reaching fathers (Grant, Hawkins, & Dollahite, 2001) because Web-based methods are more instrumental, thereby tending to be a better fit for men's learning style. . . .

Online family therapy poses many of the challenges of family life education, with added concerns about ethics and hazards of these techniques. More has been written about the promise of online therapeutic approaches than about their effectiveness. For example, Jedlicka and Jennings (2001) described their clinical experiences in treating married couples through e-mail, provided insight into their techniques, and shared their clinical judgments about effectiveness, but they did not make comparisons to other treatment approaches.

The ethics of online family therapy remain an important consideration. Until there is evidence that the exclusive use of e-mail, chat rooms, and virtual therapy is effective, online therapy cannot be considered an ethically viable substitute for empirically validated approaches. . . .

Theoretical Issues . . .

The sociology of technology provides some overarching perspective on how to consider the effects of computers on social life. Fischer (1992) described two general approaches to considering the effects of technology on social life. One is a deterministic approach that treats technology as an external force. The other assumes that technology embodies cultural values that shape history. Fischer argued that both of these approaches are problematic because they fail to take into account the ways in which people actively shape the use and influence of technology. For example, it was not inevitable that telephones would be used primarily as private two-way communication devices; early in their development, they were used as a broadcast medium, much as televisions are now used. Thus, the telephone did not determine how people used it; rather, people's use of the telephone shaped how it influenced them. Fischer suggests a social constructivist approach to studying the impact of technology on social life. Research guided by this perspective would examine the ways in which computers get used and the meanings attached to those uses. Researchers should focus their attention on the ways in which the Internet is used in the context of family life. For example, family scientists will obtain a better understanding of the role of the Internet in courtship by studying both online and offline romantic behaviors rather than focusing only on the online behaviors in the absence of broader social interactions.

Methodological Issues . . .

The Internet presents researchers with new methods of data collection that may be appealing to those who study computer technology (e.g., savvy computer and Internet users). Online data collection via e-mail or the Internet can save time and reduce error through automated data entry. Dynamic generation of response options and skip patterns invisible to respondents can allow for complex and personalized survey designs, and printing and postage costs can be avoided. An important disadvantage is that variations in computer hardware and software may result in respondents' experiencing the same survey in different ways (e.g., based on software used, monitor size and resolution), although a skilled programmer and careful planning can minimize (but not eliminate) this problem. Another concern is that representative samples of the general population cannot yet be achieved online due to the socioeconomic bias in Internet users, but samples of specific populations, such as those that are likely to be sought when studying computers and the Internet, can be obtained. Finally, some initial guidelines for online survey design have been outlined, but reliability, validity, and ethical issues of online data collection need further investigation.

CONCLUSION

The major conclusion from this review is that for the most part family scientists are not engaged in exploring the role of computer technology in family life. Much of the debate about the effects of computers on families has been left to social commentators who often have limited access to empirical data or to technologists who predict use on the basis of the capacity of computers. Past approaches to studying technology and families that have assumed that people are passively affected by technology are problematic. It is essential that we develop conceptual models about families in context and study the ways in which families adapt to technological developments.

REFERENCES

Burgess, E. W. (1928). The changing American family. *Religious Education, 23*, 408–415.

Clark, L. S. (1998). Dating on the Net: Teens and the rise of "pure" relationships. In S. G. Jones (Ed.), *Cybersociety 2.0: Revisiting computer-mediated communication and community* (pp. 159–183). Thousand Oaks, CA: Sage.

Cooper, A. (1998). Sexuality on the Internet: Surfing into the new millennium. *CyberPsychology and Behavior, 1*, 181–187.

Cooper, A., & Sportolari, L. (1997). Romance in cyberspace: Understanding online attraction. *Journal of Sex Education and Therapy, 22*, 7–14.

Cummings, J. N., Sproull, L., & Kiesler, S. B. (2002). Beyond hearing: Where real-world and online support meet. *Group Dynamics: Theory, Research, and Practice, 6*, 78–88.

Donn, J. E., & Sherman, R. C. (2002). Attitudes and practices regarding the formation of romantic relationships on the Internet. *CyberPsychology and Behavior, 5*, 107–123.

Ellison, N. B. (1999). Social impacts: New perspectives on telework. *Social Science Computer Review, 17*, 338–356.

Fischer, C. S. (1992). *America calling.* Berkeley: University of California Press.

Fox, S., Rainie, L., Larsen, E., Horrigan, J., Lenhart, A., Spooner, T., et al. (2001). Wired seniors: A fervent few, inspired by family ties. Washington, DC: Pew Internet & American Life Project. Retrieved March 30, 2003, from www.pewinternet.org/reports/pdfs/PIP_Wired_Seniors_Report.pdf.

Franzen, A. (2000). Does the Internet make us lonely? *European Sociological Review, 16*, 427–438.

Gerlander, M., & Takala, E. (1997). Relating electronically: Interpersonality in the Net. *Nordicom Review, 18*, 77–81.

Grant, T. R., Hawkins, A. J., & Dollahite, D. C. (2001). Web-based education and support for fathers: Remote but promising. In J. Fagan & A. J. Hawkins (Eds.), *Clinical and educational interventions for fathers* (pp. 143–170). New York: Haworth.

Hampton, K. N., & Wellman, B. (2000). Examining community in the digital neighborhood: Early results from Canada's wired suburb. In T. Ishida & K. Isbister (Eds.), *Digital cities: Technologies, experiences, and future perspectives* (pp. 194–208). New York: Springer-Verlag.

Hill, E. J., Hawkins, A. J., Ferris, M., & Weitzman, M. (2001). Finding an extra day a week: The positive influence of perceived job flexibility on work and family life balance. *Family Relations, 50*, 49–58.

Hill, E. J., Hawkins, A. J., & Miller, B. C. (1996). Work and family in the virtual office: Perceived influences of mobile telework. *Family Relations, 45*, 293–301.

Howard, P. E. N., Raine, L., & Jones, S. (2001). Days and nights on the Internet: The impact of a diffusing technology. *American Behavioral Scientist, 45*, 383–404.

Hughes, R., Jr., Ebata, A., & Dollahite, D. C. (1999). Family life in the information age. *Family Relations, 48*, 5–6.

Jedlicka, D., & Jennings, G. (2001). Marital therapy on the Internet. *Journal of Technology in Counseling, 2*, 1. Retrieved March 30, 2003, from http://jtc.colstate.edu/vol2_1/Marital.htm

Johnson, J. G., Cohen, P., Smailes, E. M., Kasen, S., & Brook, J. S. (2002). Television viewing and aggressive behavior during adolescence and adulthood. *Science, 295*, 2468–2471.

Katz, J. E., & Aspden, P. (1997). A nation of strangers. *Communications of the ACM, 40(12)*, 81–86.

King, S. A., & Moreggi, D. (1998). Internet therapy and self-help groups: The pros and cons. In J. Gackenbach (Ed.), *Psychology and the Internet: Intrapersonal, interpersonal, and transpersonal implications* (pp. 77–109). San Diego: Academic Press.

Kraut, R. E. (1987). Predicting the use of technology: The case of telework. In R. E. Kraut (Ed.), *Technology and the transformation of white-collar work* (p. 113–133). Hillsdale, NJ: Lawrence Erlbaum.

Kraut, R., Kiesler, S., Boneva, B., Cummings, J., Helgeson, V., & Crawford, A. (2002). Internet paradox revisited. *Journal of Social Issues, 58*, 49–74.

Kraut, R., Mukophadyay, T., Szczypula, J., Kiesler, S., & Scherlis, B. (2000). Information and communication: Alternative uses of the Internet in households. *Information Systems Research, 10*, 287–303.

Kraut, R., Patterson, M., Lundmark, V., Kiesler, S., Mukophadyay, T., & Scherlis, W. (1998). Internet paradox: A social technology that reduces social involvement and psychological well-being. *American Psychologist, 53*, 1017–1031.

Livingstone, S. (2002). *Young people and new media.* Thousand Oaks, CA: Sage.

McKenna, K. Y. A., Green, A. S., & Gleason, M. E. J. (2002). Relationship formation on the Internet: What's the big attraction? *Journal of Social Issues, 58*, 9–31.

Merkle, E. R., & Richardson, R. A. (2000). Digital dating and virtual relating: Conceptualizing computer mediated romantic relationships. *Family Relations, 49*, 187–192.

Mickelson, K. D. (1997). Seeking social support: Parents in electronic support groups. In S. Kiesler (Ed.), *Culture of the Internet* (pp. 157–178). Mahwah, NJ: Lawrence Erlbaum.

Miller, J. K., & Gergen, K. J. (1998). Life on the line: The therapeutic potentials of computer-mediated conversation. *Journal of Marital and Family Therapy, 24*, 189–202.

Mokhtarian, P. C., Bagley, M. N., & Salomon, I. (1998). The impact of gender, occupation, and presence of children on telecommuting motivations and constraints. *Journal for the American Society for Information Science, 49*, 1115–1134.

Mokhtarian, P. C., & Salomon, I. (1996). Modeling the choice of telecommuting 2: A case of the preferred impossible alternative. *Environment and Planning, 28*, 1859–1876.

National Telecommunications and Information Administration. (2002). A nation online: How Americans are expanding their use of the Internet. Retrieved March 30, 2003, from www.ntia.doc.gov/ntiahome/dn.

Newburger, E. C. (1999). *Computer use in the United States* (Current Population Reports, Series P20, No. 522). Washington, DC: U.S. Bureau of the Census. Retrieved March 30, 2003, from www.census.gov/prod/99pubs/p20-522.pdf.

O'Connell, P. L. (2003, February 13). Love clicks. *New York Times*. Retrieved February 21, 2003, from www.nytimes.com.

Orleans, M., & Laney, M. C. (2000). Children's computer use in the home: Isolation or sociation? *Social Science Computer Review, 18*, 56–72.

Parks, M. R., & Roberts, L. D. (1998). "Making MOOsic": The development of personal relationships on line and a comparison to their off-line counterparts. *Journal of Social and Personal Relationships, 15*, 517–537.

Putnam, R. D. (2000). *Bowling alone*. New York: Simon & Schuster.

Rheingold, H. (1993). *The virtual community: Homesteading on the electronic frontier.* Reading, MA: Addison Wesley.

Scharlott, B. W., & Christ, W. G. (1995). Overcoming relationship-initiation barriers: The impact of a computer-dating system on sex role, shyness, and appearance inhibitions. *Computers in Human Behavior, 11*, 191–204.

Schneider, J. P. (2000). Effects of cybersex addiction on the family: Results of a survey. *Sexual Addiction and Compulsivity*, 7, 31–58.

Stafford, L., Kline, S. L., & Dimmick, J. (1999). Home e-mail: Relational maintenance and gratification opportunities. *Journal of Broadcasting and Electronic Media, 43*, 659–669.

Stoll, C. (1995). *Silicon snake oil.* New York: Doubleday.

U.S. Bureau of Labor Statistics. (1998). Work at home in 1997. Retrieved March 30, 2003, from www.bls.gov/news.release/homey.toc.htm.

U.S. Bureau of Labor Statistics. (2002). Work at home in 2001. Retrieved March 30, 2003, from www.bls.gov/news.release/homey.toc.htm.

Wellman, B., Haase, A. Q., Witte, J., & Hampton, K. (2001). Does the Internet increase, decrease, or supplement social capital? *American Behavioral Scientist, 45*, 436–455.

Young, K. S. (1999). Cybersexual addiction. Retrieved March 30, 2003, from www.netaddiction.com/cybersexual_addiction.htm.

Young, K. S., Griffin-Shelley, E., Cooper, A., O'Mara, J., & Buchanan, J. (2000). Online infidelity: A new dimension in couple relationships with implications for evaluation and treatment. *Sexual Addiction and Compulsivity*, 7, 59–74.

39

An Overview of the Nature, Causes, and Consequences of Abusive Family Relationships

ROBERT E. EMERY AND LISA LAUMANN-BILLINGS

The authors note that the development of abusive relationships has a long and mostly overlooked history.

As you read, ask yourself the following questions:

1. *What facts about family abuse surprised you the most?*
2. *What would you add to the means for dealing with family abuse?*

GLOSSARY **Battered children** Physical, mental, or sexual abuse or neglect of a child constitutes child abuse.

SOURCE: From Emery, R. E., & Laumann-Billings, L. (1998). An overview of the nature, causes, and consequences of abusive family relationships toward differentiating maltreatment and violence. *American Psychologist*, 53*(2), 121–135.*

From the vantage point of the present, the public outcry against family violence is shockingly recent. The Society for the Prevention of Cruelty to Children was formed only in 1875 (nine years after the formation of the Society for the Prevention of Cruelty to Animals), and the arousal of professional concern about battered children dates only to 1962 with the publication of Kempe's seminal article (Kempe, Silverman, Steele, Broegemueller, & Silver, 1962; more pervasive professional concern is far more recent). Spousal assault was made a crime in all states only in recent decades, and there still are no laws against spousal rape in some states. Professional concern about the children of battered women is less than two decades old, and sibling and elder abuse are just now beginning to receive clinical and research attention. . . .

THE DEVELOPMENT OF ABUSIVE RELATIONSHIPS

Many factors contribute to the development of family violence, including individual personality factors, family interaction patterns, poverty and social disorganization, acute stressors, and the cultural context in which the family lives (Bronfenbrenner, 1979). Consistent with their comorbidity, a number of the same factors increase the risk for child and spouse abuse, which are the types of abuse that are the focus of our overview. . . .

. . . Understanding Family Violence . . .

Individual Characteristics Personality factors such as low self-esteem, poor impulse control, external locus of control, negative affectivity, and heightened response to stress all increase the likelihood that an individual will perpetrate family violence (Pianta, Egeland, & Erickson, 1989). Alcohol or drug dependence also plays a role both as a background risk factor and as an immediate precipitant of family violence. (Kantor & Straus, 1990). Although the evidence is less consistent, some research has also suggested that the victims of family violence share some common characteristics, including poor physical or mental health, behavioral deviancies, and difficult temperament or personality features (Belsky, 1993). For child victims, age also seems to play a role, as younger children are more prone to be seriously injured as a result of family violence (Lung & Daro, 1996).

Immediate Context Characteristics of the immediate social context, especially of, the family system, have important implications for the etiology or perpetuation of violence in that family (Emery, 1989). Studies have examined structure and size, acute stressors such as the loss of a job or a death in the family, and characteristic styles of resolving conflicts or parenting.

Broader Ecological Context Violence in the family also is related to qualities of the community in which the family is embedded, such as poverty, absence of family services, social isolation, and the lack of social cohesion in the community. As Garbarino and Kostelny (1992) noted, family violence is a social as well as a psychological indicator (p. 463).

Of course, most poor parents do not maltreat their children (Bronfenbrenner, 1979); thus, the relation between poverty and abuse is not simple or direct. Garbarino and colleagues have consistently shown that the principal difference between poor families who do and those who do not maltreat their children lies in the degree of social cohesion and mutual caring found in their communities (Garbarino & Kostelny, 1992).

Societal or Cultural Context A number of commentators contend that family violence is perpetuated by broad cultural beliefs and values, such as the use of physical punishment, extremes in family privacy, and violence in the popular media (Garbarino, 1977). As noted earlier, our societal concern with violence in the family is historically recent, and our continued reluctance to intervene forcefully is evident in the reluctance to terminate parental rights even in egregious cases of child abuse (Besharov, 1996b). Societal policies may not cause family violence, but many of our practices appear to condone it.

CONSEQUENCES FOR VICTIMS . . .

Physical Injuries Physical injury is a clear consequence of many instances of family violence, but information on the physical consequences of family violence is surprisingly incomplete. As noted earlier, an estimated 1,200 to 1,500 children, most of whom are under the age of five years, die each year as a result of either physical abuse (48%), neglect (37%), or both (15%) by a parent or parent figure (Lung & Daro, 1996). In addition, almost one-third of the 4,967 women murdered in 1995 were killed by a boyfriend or husband (Federal Bureau of Investigation, 1996). Official data may somewhat underestimate actual deaths due to family violence, however, because of misclassification of causes of death (McClain, Sacks, & Frohlke, 1993).

Information on nonlethal injuries is much more sketchy. According to NIS-3 (Sedlak & Broadhurst, 1996), nearly 50,000 children were victims of serious physical abuse in 1993, that is, they suffered from life-threatening injuries, long-term physical impairment, or required professional treatment to prevent long-term physical impairment. It has also been estimated that over 18,000 children become severely disabled each year as a result of severe child abuse (Baladerian, 1991). We could not locate epidemiological data on more specific physical outcomes, such as the rate of nonorganic failure to thrive (resulting from gross neglect) among infants, specific injuries such as broken bones, or pregnancies or sexually transmitted diseases resulting from sexual abuse.

The threat of injury often is devastating; however, we need more specific information on injury and disease for several reasons. First, physical safety is the foremost concern about family violence; thus, clear evidence of harm is essential for a number of purposes ranging from basic knowledge to justifying intervention. Second, specific physical injuries provide one relatively unambiguous (if limited) definition of family violence; thus, data would allow investigators to clearly chart the scope of the problem and changes over time. Third, clear information on the physical consequences of family violence should allow researchers to more adequately investigate the psychological consequences of physical abuse, especially given evidence for a dose-response relation, which we review shortly.

Psychological Consequences All types of family violence are linked with diverse psychological problems ranging from aggression, to anxiety, to depression. Still evidence on more specific psychological outcomes may be clouded by the need to consider (a) risk factors correlated with family violence; (b) clusters of symptoms (e.g., disorders like posttraumatic stress disorder [PTSD]) in addition to specific symptoms; (c) subtle psychological consequences that are difficult to document empirically, particularly among children; and (d) psychological processes (not just psychological outcomes) set into motion or disrupted by the experience of family violence.

Correlated Risk Factors Family violence is associated with a number of factors known to place children and adults at risk for psychological problems, for example, poverty, troubled family environments, genetic liability, and so on. These risk factors may account for the apparent relation between family violence and specific psychological outcomes, or their effects may interact with the consequences of abuse. Thus, for example, research comparing abused and nonabused children may actually reflect the psychological effects of anxious attachments, social isolation, or general family stress rather than the consequences of violence per se (National Research Council, 1993). It is essential to consider this possibility both in research and in clinical interventions, which appropriately may focus on risk factors in addition to, or even instead of, the abuse, particularly when abuse is less serious.

Subtle Psychological Effects on Victims The diagnoses of PTSD and acute stress disorder include symptoms such as reexperiencing and dissociation; thus, they offer further impetus for studying some of the more subtle psychological consequences of abuse, particularly among children. The vehement controversy over recovered memories of abuse provides another rationale for research on more subtle psychological reactions (Loftus, 1994). A review of the recovered memories controversy is beyond the scope of his article, but we do note that one longitudinal study found that documented sexual abuse among girls ages 10 months to 12 years was not reported by nearly 40% of the victims when they were asked about their history as young adults (Williams, 1994a). Our present concern is not how supportive or unsupportive these data are of the recovered memories controversy (Williams, 1994b) but the need to better understand the emotional and cognitive processes that may contribute to the forgetting of (or the dissociation from) abuse in the 17 years between victimization and reinterview.

PREVENTION AND INTERVENTION . . .

Child Abuse Reporting

By all accounts, child protective services are driven by—and overwhelmed with—the investigation of child abuse reports. As we have noted, over three million reports are made every year, but less than one-third are substantiated. Moreover, social service workers are so swamped with reports and investigations that about 40% of substantiated cases receive no services at all (McCurdy & Daro, 1993). Finally, only a small proportion of substantiated cases, less than 20%, involve any type of formal court action. Rather, child protective workers encourage the great majority of those involved in substantiated cases to enter treatment voluntarily (Melton et al., 1995).

In order to facilitate the support of families under stress, we would exempt mental health professionals from reporting less serious cases of abuse, whether known or suspected, when a family is actively engaged in treatment. We do not attempt to define these cases here, but we do note the need to define less serious cases (what we have called maltreatment) clearly in the law, as has been done in some experimental programs. Such a change would be a first step toward the broader goal of refocusing the child protection system on supporting rather than policing families under stress, while simultaneously pursuing more vigorous, coercive intervention with cases of serious family violence.

Supportive Interventions

A number of supportive interventions have been developed in an attempt to reduce violent behavior within families, including individual and group therapies for both victims and perpetrators, couples therapy for victims of domestic violence, parent-training and family therapy, and home-visiting programs for the prevention of child abuse. In general, the more serious and chronic the nature of the abuse, the less success these programs have in changing behaviors (National Research Council, 1993). With problems of mild to moderate abuse, however, multilevel programs, which combine behavioral methods, stress management, and relationship skills (parent-child or spousal), lower stress in families and may reduce the likelihood of continued aggression in both child and spouse abuse cases (National Research Council, 1993).

Some interventions show promise, but the need for early intervention and especially prevention is underscored by the difficulty of changing entrenched family violence and the stressful life circumstances that promote abuse (Garbarino & Kostelny, 1992). Home-visitor programs for new parents living in difficult circumstances are one especially promising form of prevention. Home-visitor programs simultaneously assist with material needs (e.g., cribs, child care, transportation), psychological needs (e.g., parenting education and support), and educational needs (e.g., job skills) and many both improve general family well-being and reduce child maltreatment.

Coercive Intervention

As a final point, we note that distinguishing between levels of abuse also may help to improve intervention in cases of serious and extreme family violence. One benefit would be to free investigators and police to focus on the most serious cases. Shockingly, between 35% and 50% of all fatalities that are due to child abuse or neglect occur in cases that have already been brought to the attention of law enforcement and child protection agencies (Lung & Daro, 1996). Another benefit might be to help to clarify when coercive legal intervention is and is not appropriate. For example, debates have errupted over the once promising policy of mandatory arrest for violent partners (even over the objection of victims), because the reduced recidivism found in early research (Sherman & Berk, 1984) has not been replicated in subsequent studies. As another example, many commentators have questioned the overriding goal of family reunification following child abuse; especially in cases like these, the termination of parental rights and early adoption may be the appropriate intervention, especially given the many problems with the overwhelmed foster-care system (Tatara, 1994).

In both of these examples, distinguishing between levels and abuse should help to resolve controversy and thereby clarify the appropriate use of coercive legal intervention. The idea of termination of parental rights or of arrest is far more threatening when our definitions of abuse include relatively minor acts, as they currently do. Clearly, it is difficult to draw a line between cases that should or should not lead to arrest or termination of parental rights (Azar, Benjet, Fuhrmann, & Cavallero, 1995), but the challenge of distinguishing between levels of abuse should not deter us from the task. What we have called maltreatment may be on a continuum of what we have termed violence, and both acts of abuse may differ from normal family aggression only by a matter of degree. As we have argued throughout this article, however, drawing distinctions between levels of abuse seems consistent with the state of our knowledge about the prevalence of abuse, its development, its consequences, and appropriate intervention.

NOTE

Editors' Note: For documentation, see article source on page 286.

40

An Agenda for Family Policy in the United States

GEORGE T. MARTIN, JR.

The author claims that the U.S. has no family policy.

As you read, ask yourself the following questions:

1. *Typically, family policies have four cornerstones. What would you take out as unneeded and what would you add?*
2. *Changes in the family are not due to America's supposed "moral decline." Why do you agree/disagree?*
3. *The family is an obsolete institution: Do you agree? With what would you replace it?*

GLOSSARY **Proletarian** Belonging to the lowest or poorest class of the people. **Welfare state** Such matters as social security, health, education, housing, and working conditions are the responsibility of the government.

Because of its unique importance to society and because of its increased dependence on forces outside its control, the family has attracted growing attention from government policy in recent years. Virtually all societies have comprehensive policies that support family functioning. However, there is a wide range among nations in the level and extent of these family policies. National family policies typically have four cornerstones: cash benefits to supplement the incomes of adults who raise children, comprehensive health services for pregnant women and for children, paid work leaves for parents to care for newborns and ill family members, and a child care program. The nations with the most comprehensive and effective policies are the developed welfare states such as Sweden. The United States, the focus of this analysis, is a welfare state laggard and is one of the few developed nations that lacks a family policy. . . .

Ultimately, the lack of a family policy is a liability to the society as a whole to the extent that it increases the social costs of inadequate attention to family needs and of political conflict between the "haves" and the "have nots." . . .

Societal Change

Specific societal changes are the driving forces in the current transition in family form and purpose (Martin, 1990). For most of human history, the family fulfilled production tasks. Capitalism, especially capitalist industrialization, and urbanization steadily encroached on this purpose of the family. Products formerly made for use in the family now are produced by corporations for profit. In the not-too-distant past, families produced much of their own food, clothing, and shelter. At home, they churned butter, raised buildings, baked bread,

SOURCE: From *Contemporary Parenting* by Terry Arendelled. Thousand Oaks, CA: Sage Publications, 1997, pp. 289–324. Reprinted by permission.

sewed clothing, cultivated gardens, and preserved vegetables for winter. This no longer is the case. Today, even farmers buy their bread.

In losing its production function, the family lost something else. The enormous labor required for production required the services of all the family members; it was an important focus of social life. Children were an economic asset because they could work. Production required cooperation and interaction; it was an activity around which intra- and interfamily communal life was organized. Thus the family lost a basis of its solidarity along with its productive function—the cooperative dependence of its members. Additionally, children have become economic liabilities. . . .

Women's Entry into the Paid Labor Force

Industrial capitalism transformed peasants into proletarians, and women were among the first proletarians. However, because of their caring and domestic responsibilities as well as other reasons, women have lagged behind men in entering the paid labor force. The first group of women to become proletarians were poor and working class, for whom protective policies were created during the Progressive Era of the early 20th century in the United States. . . .

Since the 1960s, economic and political changes have led the last group of women—married women whose husbands are present and who have small children—to enter the paid labor force. . . .

Such a major change as mothers going outside the home to work was bound to have serious ramifications for the family. The impact was heightened by other major social changes that were occurring simultaneously—the transition to a service economy and the post-1960s cultural sexual liberalization and pluralization. As a result of all these changes, the family has undergone an unprecedented transition, the effects of which policy is only beginning to try to accommodate. . . .

Although the U.S. family has undergone major change in the past quarter century, it also is significant that similar changes have been going on in comparable nations. This fact points to structural changes engendered by transformations in the global capitalist economy as being the underlying cause of the changes in the family. This casts considerable doubt on the argument that family changes are due to the fact that the United States has been experiencing moral decline or decay. It may be the case that changes in the family are more *noticeable*—even alarming—in the United States because the nation lacks a comprehensive family policy that could mitigate some of the most negative effects of the global changes. At the least, the current changes in the family make the absence of a family policy in the United States more conspicuous.

Despite the fact that it lacks a comprehensive package of policies that are addressed to families per se, the United States does have a range of social programs that bear on the family. Even though they cannot fully compensate for the lack of a comprehensive family policy, current programs—in the areas of income security, work, and health—provide some support to families. . . .

Welfare state income security programs are decidedly oriented to the labor market, not to the family. This is consistent with the focus in social policy on the traditional family with one breadwinner—the father and husband. Social rights accrue primarily to earners, whereas carers are left largely to fend for themselves, perpetuating a gendering of citizenship. Working women are obliged to make a choice because "full access to social rights is ensured only if their caring commitments are organized so that they do not interfere with formal employment" (Leira, 1992, p. 172).

The overarching purpose of income security programs is to prevent poverty—what Shaw (1907/1958) called "the worst of crimes"—and they have been the prime components of social policy since its inception in Elizabethan England in the 16th century (Martin, 1990, p. 43). Compared to similar nations, the United States has a persistently high poverty rate, and children comprise a group that has experienced large increases in poverty. A 1995 comparative study of 18 nations found that poor children in the United States were poorer than those in 15 other nations. The United States ranked below nations such as Italy that are considerably less affluent than the United States (Bradsher, 1995b). In 1975, when the overall poverty rate stood at 12% of the population, it was 17% for children under age 18. In 1993, the poverty rate for all Americans had risen to 15%; for children, the rise was to 23%. In 1993, there were 15.7 million children under age 18 living below the poverty level. . . .

For the United States to have such a high number and rate of poor children seriously calls into question the effectiveness of its income security programs. These questions relate to the income security programs themselves and to phenomena external to the programs. The most important external phenomenon that is family related is the issue of child support. . . .

Child Support

. . . . In 1995 in the United States, there were 11.5 million custodial parents, of whom 86% were mothers. Lack of financial support from absent parents is a major reason why single-parent families have relatively high poverty rates. . . .

There are three major obstacles to securing adequate support payments from absent fathers: courts awarding inadequate sums, a widespread lack of compliance, and the fact that some fraction of all absent fathers are financially unable to support their children. . . . However, even in the face of the elimination of this obstacle, many single mothers and their children will remain poor because of the other two problems. What may help is a policy shift away from courts and legal contention between mothers and absent fathers to administrative enforcement of child support. . . .

. . . Using such an administrative system eliminates parental jockeying over the children and ensures the government's interest that children not be raised in poverty. Despite the high costs entailed by establishing an assured benefit administrative system, much of the cost could be retrieved by the savings from substantial reductions in public aid and from the saved costs of tending to the social problems engendered by poverty (e.g., poor health).

Public Aid

A piece of erroneous conventional wisdom is that welfare creates family dissolution. The fact is that, as noted by Ellwood (1988), "Virtually every careful social science study that has investigated this issue has found that the welfare system has had little effect on the structure of families" (p. 22). Thus welfare cannot be blamed for the rise either in female-headed households or in out-of-wedlock births. The welfare system responds to changes in the family caused by social changes; it does not create changes in the family. It is important, too, to recognize that households headed by women are not poor because of changes in family composition. For example, even though African American poverty is concentrated in female-headed households, in only about 17% of cases did poverty begin with family changes such as divorce or separation (Bane, 1986). . . .

A comparative study of eight welfare states by Kamerman and Khan (1987) found that the United States stood last in providing income to single-mother families. . . .

Family-Oriented Income Security

The incomes of those raising children can be supported by building a minimum income floor for families. Such a floor could be cobbled together in the United States with an integrated approach to reforms in child support, the minimum wage, child care, health care, and the Earned Income Tax Credit (EITC). . . .

WORK

Short of a comprehensive full-employment policy, there are a number of meaningful reforms that could be made to improve the lot of working families in the United States in addition to expanding the EITC. The working poor would be helped substantially by raising the minimum wage. . . .

Workers who raise families have social needs other than adequate income. Child care at work sites and flexible working hours ("flextime") are the supports that employed parents most desire (Galinsky, 1992). Flextime allows workers to put in their required hours at flexible times and so permits them to arrange their hours for child care. Both child care and flextime provide considerable benefits to employers in the form of reduced employee turnover, improved employee recruitment, higher employee morale, and reductions in employee tardiness and absenteeism. Although there are costs involved for providing child care, the employer costs for implementing flextime are quite low. Yet only a minority of private employers extend these necessary supports to working parents. Research indicates that about one out of five or six unemployed women cannot work because she is unable to make satisfactory child care arrangements (U.S. Commission on Civil Rights, 1984, p. 96)....

PARENTAL LEAVE

In addition to flextime and child care, other steps need to be taken to help families meet their child care needs. One is parental leave. All developed nations except Australia and the United States now provide paid and job-protected maternity leave for working women....

The U.S. Family and Medical Leave Act (FMLA) of 1993 allows a worker to take up to 12 weeks of unpaid leave for the birth of a child or an adoption; to care for a child, spouse, or parent with a serious health condition; or for the worker's own health condition that makes it impossible to work....

Health

The family is at the center of a number of health policy concerns in the United States including maternity care, teenage pregnancy, and domestic violence.

MATERNAL, INFANT, AND CHILD HEALTH

Health care is a vital area of family functioning, in large part because families are the nexus of pregnancy, childbirth, and infant and child development—all of which present unique challenges to health. The health challenges are twofold—to the mother and to the child. In both areas, the United States lags behind comparable nations, sometimes considerably behind. Because of the lack of national health care insurance in the United States, many families are exposed to health risks for financial reasons. Although adopting a national health care plan would be of great assistance to many families, especially poor ones, there is much to be done short of that goal....

Teenage Pregnancy

There is considerable health risk associated with teen parenting ("children who have children"); it is magnified if the teen mother is unwed. These risks include lower exposure to prenatal care, higher incidences of low birthweight, maternal and infant mortality, and higher exposure to poverty....

The most effective and efficient vehicle for reducing teen pregnancy is sex education in the public schools (Gilchrist & Schinke, 1983) and elsewhere. Officials attributed the 1992 decline in teen birthrates in the United States to better contraceptive use among teenagers and to a leveling off of teenage sexual activity (Vobejda, 1994). Although the overwhelming majority of Americans favor sex education in public schools, only four states require it in all schools. An additional problem is that sex education often is not contraception education and is, therefore, not effective as a way in which to prevent pregnancy. Abstinence is the only message commonly given, and it is not convincing to many teenagers. Government leadership for a thorough and honest sex education policy is needed....

Violence

Violence is a major problem in the lives of U.S. families. In 1992, there were nearly 1 million substantiated dispositions of maltreatment of children by their adult caretakers. Of these dispositions, 48% involved neglect, 22% involved physical abuse, 13% involved sexual abuse, and 5% involved emotional abuse. Child maltreatment is a serious matter; child abuse and neglect is the leading cause of trauma death for children age 4 or under, ranking ahead of death from motor vehicle accidents, fires, or other traumas (Action Alliance for Children, 1995, p. 4)....

Spousal abuse also is a problem for U.S. families, and it is primarily abuse of wives by their husbands. About one third of all female homicide victims are killed by their husbands or boyfriends. Elder abuse—the abuse of live-in elderly relatives—has been the newest addition to the abysmal domestic violence problem in the United States....

FRONTIERS...

Technology

...New reproductive technologies (NRTs) carry potential challenges to the nature of the family as well as to the parameters of family policy by altering the nature of parenthood and by introducing new actors such as surrogates. "People are creating new relations that separate genetic, gestational and social parentage" (Russell, 1994, p. 287). The surrogate (gestational) mother creates a new form of blood kinship, a relationship that can exist only through technology. "By this is ... breached the important symbolic conjunction of conjugal and procreative activity, and the equally important opposition between acts undertaken for 'love' and those undertaken for money" (Franklin, 1993, p. 545)....

NRTs have further widened the gap in modern society between the biological act of reproduction and the social activity of parenting....

NRTs have exploded onto the scene at a time of increasing conservative hegemony in social policy. This accents the normal gap between technological innovation and socio-cultural lag. This growing gap is a source of contradiction in societies. "Certainly it is striking to contrast the flourishing of NRTs, whilst infant mortality rates amongst inner-city blacks in the USA remain shockingly high" (McNeil, 1993, p. 503). Such contradictions reflect the growing polarization in U.S. society, in which the "haves" maximize their access to more benefits and services, whereas the "have-nots" experience worsening conditions. Conservative social policy only deepens this polarization because in its reliance on market solutions, it reinforces the already commanding position of the haves in society. Additionally, market-based policy extends the process of commodification into unprecedented areas—with unknown risks....

In 1995, the U.S. Food and Drug Administration proposed federal regulations for sperm and ova banks that set guidelines on testing for HIV, syphilis, and hepatitis as well as on donor screening, retrieval, processing, labeling, storage, and distribution of reproductive tissue (Dieges, 1995). This kind of regulation will address the quality control issues that are currently paramount in dealing with the NRTs. However, around the corner looms the more contentious and muddled arena of social, political, and moral questions posed by NRTs....

Parenting

Biological reproduction through new technologies and the focus on custody in divorce proceedings have contributed to an increasing gap between the biological and social dimensions of being a parent. Leach (1994) defined this gap in the following way:

> We know much more about the reproductive biology and genetics of parenthood than we know about the social, emotional and psychological impacts of parenting, and we

> devote far greater research resources to producing physically healthy babies than to rearing emotionally stable children. Indeed, while family planning, artificial baby foods and a host of childcare aids have dramatically reduced the burdens of traditional mothering roles, those roles themselves have been invalidated and have not been replaced with a workable restructuring of gender roles and relationships. What is needed now is something that cannot be produced by further scientific advance or a new technical fix: a reappraisal of the importance of parenting and fresh approaches to the continuing care and education of children in, and for, changing societies. (pp. xiv–xv)

. . . There are two dangers associated with the upsurge in interest in parenting programs. Unless consciously guided to do otherwise, the tendency is for such programs to reinforce the traditional perception that the responsibility—sole or major—for child rearing lies with mothers. Substantial effort is needed to ensure that both fathers and mothers are involved in such programs. A second danger presented by the current programs is that they single out one group of mothers (perhaps single, young, and poor) for special attention—and blame. The result may be that parenting programs become as stigmatized as welfare is. This produces two negatives: Other parents (married, middle class) may avoid the programs and the programs may deflect needed attention away from other, more basic and material problems of single, young, and poor mothers such as their poverty. . . .

Fathering

In the contemporary discussion of female-headed families, poverty, and welfare, the subject of fatherhood is virtually absent (at least beyond child support maintenance issues). In large part, this is because many fathers also are absent from their families; in the United States in 1994, 27% of children under age 18 were in families in which the fathers were absent. Even when fathers are present, despite the effects of feminism and the fact that more mothers are working in the paid labor force, women still carry a highly disproportionate burden of child care (and housework), even when the women's incomes are higher than the men's (Sidel, 1990, pp. 202–203). . . .

. . . Conceptions and behaviors about fathering are shaped from the earliest experiences of children. Education for fathering (and parenting) is best begun in elementary schools:

> We could begin with parenting and child-care courses in middle school. Both boys and girls need to learn how to care effectively for children and that child care is mother's and father's responsibility. Boys especially need to learn that caring for a baby does not undermine masculinity. (Berry, 1993, p. 219)

Caring for Parents

In addition to their responsibilities for the maintenance of personal life in the family, women face an increasing burden of elder care. The population of aged parents in the United States, especially those who need care, is increasing, and only about 5% of them are in nursing homes at any given time (McLeod, 1995). According to a 1989 study by the Older Women's League, women spend an average of 17 years of their lives caring for children and 18 years aiding aged parents. . . .

There are incremental policy changes that could help parents who care for their parents. Social security could be reformed to meet their special needs. Schorr (1986, p. 69) proposed the following reforms of social security that would help women: (a) upgrading benefits for the "old old," usually defined as people age 85 or older, perhaps by revising the benefits to increase with age rather than decrease as they currently do; (b) liberalizing maximum benefits for families; and (c) restoring a general minimum benefit. These reforms would be especially helpful for those families who care for their elderly members, a growing segment of all families.

CONCLUSION

... In conclusion, we live at a time when families with dependent children in the United States face increasing economic pressures and other stresses, many of them caused by social changes out of the control of individual families. Although many of these families receive some help from private and public sources, much more is needed in the form of policy that addresses their needs. At a time of growing government decentralization and downsizing, it is even more critical that the needs for government intervention and support be rank ordered. Thus far, despite the political rhetoric that is aimed at family functioning, it remains a relatively low priority for policy attention. This is a regrettable fact, one that potentially carries with it quite perilous outcomes for all of society.

A comprehensive family policy is required to meet both the new changes and the recurrent problems that contemporary families face. This family policy must take into consideration mothers, fathers, children, and the family as well as work needs of mothers and fathers. Such a policy could stand as a true testament to this society's devotion to valuing its families.

REFERENCES

Action Alliance for Children. (1995, July/August). A nation's shame: Fatal child abuse and neglect in the U.S. *Children's Advocate*, p. 4.

Bane, M. J. (1986). Household composition and poverty. In S. Danziger & D. Weinberg (Eds.), *Fighting poverty: What works and what doesn't* (pp. 209–231). Cambridge, MA: Harvard University Press.

Berry, M. F. (1993). *The politics of parenthood: Child care, women's rights, and the myth of the good mother.* New York: Penguin Books.

Bradsher, K. (1995b, August 14). Poor children in U.S. are among worst off in study of 18 industrialized countries. *New York Times*, pp. A7, A9.

Dieges, J. (1995, May/June). Sperm and ova banks come into the regulatory fold. *Children's Advocate*, pp. 4–5.

Ellwood, D. T. (1988). *Poor support: Poverty in the American family.* New York: Basic Books.

Franklin, S. (1993). Postmodern procreation; Representing reproductive practice. *Society as Culture, 3*, 522–561.

Galinsky, E. (1992). Work and family: 1992, *Report, 11* (*2*), 2–3. (Family Resource Coalition, Chicago).

Gilchrist, L. D., & Schinke, S. P. (1983). Teenage pregnancy and public policy. *Social Service Review, 57*, 307–322.

Kamerman, S. B., & Kahn, A. J. (1987). Universalism and income testing in family policy: New perspectives on an old debate. *Social Work, 32*, 277–280.

Leach, P. (1994). *Children first.* New York: Vintage Books.

Leira, A. (1992). *Welfare states and working mothers: The Scandinavian experience.* Cambridge, UK: Cambridge University Press.

Martin, G. T., Jr. (1990). *Social policy in the welfare state.* Englewood Cliffs, NJ: Prentice Hall.

McLeod, B. W. (1995, August 21). Elder care, prime social issue of the 21st century. *San Francisco Examiner*, p. A13.

Russell, K. (1994). A value-theoretic approach to childbirth and reproductive engineering. *Science and Society, 58*, 287–314.

Schorr, A. L. 1986. *Common decency: Domestic policies after Reagan.* New Haven, CT: Yale University Press.

Shaw, B. (1958). *Major Barbara.* New York: Longman. (Originally published 1907)

Sidel, R. (1990). *On her own: Growing up in the shadow of the American Dream.* New York: Penguin Books.

U.S. Commission on Civil Rights. (1984). Equal opportunity and the need for child care. In R. Genovese (Ed.), *Cities and change: Social needs and public policies* (pp. 92–105). New York: Praeger.

Vobejda, B. 1994. October 26. Birthrate among teenage girls declines slightly. *Washington Post*, p. A3.

CHAPTER 9

Education

Institution in the Crossfire

Alan Mc Evoy

Alan Mc Evoy

Much of our time up to the age of 18, and often well beyond, is spent learning the roles necessary for survival in society. All societies are concerned with socializing their young to develop skills for knowledge deemed necessary and inculcating loyalty to the social system. In some societies, this education takes place informally through imitation of elders; in others, such as the United States, formal schools have become key mechanisms for the transmission of knowledge and culture. Schools function to perpetuate dominant societal values, prepare young people for roles in society, and allocate societal positions to them.

It is within these functions that controversy arises. For instance, whose culture should be transmitted? Vocal pressure groups argue for curricula to be "inclusive," to represent all groups. Such voices influence textbook content and selection, courses offered, and course content. A comparison of your course catalog today with one of 10 to 20 years ago would probably reflect some of these changes.

Sociologists interested in the institution of education focus on the structure of the system, including the functions performed, the roles played by the various groups and individuals involved, the processes operating within the system, and the pressures from other parts of society that influence the system. Of primary concern is the role education plays in providing the opportunities we have in society; thus, sociologists of education study questions of equality of educational opportunity. Many argue that opportunity is not equal for all who pursue education, and that we lose many bright students in the process of education. Several of our readings focus on this important issue of equal opportunity in education.

Likewise, questions of who gets what training and positions creates controversy. Education in most countries is viewed as key to a better life and opportunities. Competition for entrance into the best schools and universities is keen.

Some assume that if we change certain aspects of the system we can remedy the problems, whereas others see a need to change the larger societal structure, not just education, to reduce inequality in society. Issues of equality of educational opportunity are relevant to most minority groups. The question is whether the dominant group in society has an educational advantage over minority groups, and therefore an advantage in the job market and in long-range earning capacity.

Moving through the levels of the school system, we start with kindergarten; Harry Gracey discusses how children initially are socialized into the rules of society by learning the rules of the classroom—routines, obedience, and discipline. Thus, schools are important tools in preparing children for their later roles in the work world.

In the second reading Jonathan Kozol dramatically illustrates the unequal opportunity that can exist in schools. We present three excerpts from his book *Savage Inequalities*: The first excerpt describes a school in East St. Louis in which the defects of the physical plant lead to the need for periodic school closings. The second describes a poor school in New York City in which children face large classes, lack of facilities, and violence. The third excerpt, about a school located in a wealthy suburb of Chicago, provides an example of a contrasting situation, illustrating the severe inequalities that exist in school quality.

In an effort to address equal opportunity concerns, school districts and courts have required desegregation. This has been an ongoing issue plaguing the American education system, and there is now a trend toward resegregation. The reading by Gary Orfield and colleagues analyzes recent trends and efforts toward achieving equal opportunity in education. Although schools are expected to provide equal opportunity and are criticized for not doing so, Richard Rothstein points out that factors other than schools influence opportunity for children and additional solutions are needed. Moving from kindergarten through high school to college, the last reading addresses the personal and social development of college students as revealed in fraternity hazing. Stephen Sweet provides a theoretical approach to understanding this group behavior.

As you read this chapter, consider the following questions: What are some causes of inequality in education and society? Some consequences? What solutions do the readings propose?

41

Learning the Student Role

Kindergarten as Academic Boot Camp

HARRY L. GRACEY

Schools are agents of socialization, preparing students for life in society. In preparing children, they teach expectations and demand conformity to society's norms. They help teach children the attitudes and behaviors appropriate to the society. Gracey points out that this process of teaching children to "fit in" begins early. His focus is the routines children in kindergarten are expected to follow.

As you read the article, think about the following:

1. *What are key elements of the learning experience of kindergarteners?*
2. *How does the kindergarten routine encourage conformity and help prepare children for later life?*
3. *What might happen in later schooling and life if children do not receive this early childhood socialization?*
4. *How did your schooling prepare you for what you are doing and plan to do?*

GLOSSARY **Educational institution/system** The structure in society that provides systematic socialization of young. **Student role** Behavior and attitudes regarded by educators as appropriate to children in schools.

Education must be considered one of the major institutions of social life today. Along with the family and organized religion, however, it is a "secondary institution," one in which people are prepared for life in society as it is presently organized. The main dimensions of modern life, that is, the nature of society as a whole, is determined principally by the "Primary institutions," which today are the economy, the political system, and the military establishment. Education has been defined by sociologists, classical and contemporary, as an institution which serves society by socializing people into it through a formalized, standardized procedure. At the beginning of this century Emile Durkheim told student teachers at the University of Paris that education "consists of a methodical socialization of the younger generation." He went on to add:

> It is the influence exercised by adult generations on those that are not ready for social life. Its object is to arouse and to develop in the child a certain number of physical, intellectual, and moral states that are demanded of him by the political society as a whole and by the special milieu for which he is specifically destined.... To the egotistic and asocial being that has just been born, [society] must, as rapidly as possible, add another, capable of

SOURCE: In Dennis Wrong and Harry L. Gracey (eds.), *Reading in Introductory Sociology*. New York: Macmillan, 1967. Reprinted with permission.

> leading a moral and social life. Such is the work of education.[1]

The education process, Durkheim said, "is above all the means by which society perpetually recreates the conditions of its very existence."[2] The contemporary educational sociologist, Wilbur Brookover, offers a similar formulation in his recent textbook definition of education:

> Actually, therefore, in the broadest sense education is synonymous with socialization. It includes any social behavior that assists in the induction of the child into membership in the society or any behavior by which the society perpetuates itself through the next generation.[3]

The educational institution is, then, one of the ways in which society is perpetuated through the systematic socialization of the young, while the nature of the society which is being perpetuated—its organization and operation, its values, beliefs, and ways of living—are determined by the primary institutions. The educational system, like other secondary institutions, *serves* the society which is *created* by the operation of the economy, the political system, and the military establishment.

Schools, the social organizations of the educational institution, are today for the most part large bureaucracies run by specially trained and certified people. There are few places left in modern societies where formal teaching and learning is carried on in small, isolated groups, like the rural, one-room schoolhouses of the last century. Schools are large, formal organizations which tend to be parts of larger organizations, local community School Districts. These School Districts are bureaucratically organized and their operations are supervised by state and local governments. In this context, as Brookover says:

> The term education is used . . . to refer to a system of schools, in which specifically designated persons are expected to teach children and youth certain types of acceptable behavior. The school system becomes a . . . unit in the total social structure and is recognized by the members of the society as a separate social institution. Within this structure a portion of the total socialization process occurs.[4]

Education is the part of the socialization process which takes place in the schools; and these are, more and more today, bureaucracies within bureaucracies.

Kindergarten is generally conceived by educators as a year of preparation for school. It is thought of as a year in which small children, five or six years old, are prepared socially and emotionally for the academic learning which will take place over the next twelve years. It is expected that a foundation of behavior and attitudes will be laid in kindergarten on which the children can acquire the skills and knowledge they will be taught in the grades. A booklet prepared for parents by the staff of a suburban New York school system says that the kindergarten experience will stimulate the child's desire to learn and cultivate the skills he will need for learning in the rest of his school career. It claims that the child will find opportunities for physical growth, for satisfying his "need for self-expression," acquire some knowledge, and provide opportunities for creative activity. It concludes, "The most important benefit that your five-year-old will receive from kindergarten is the opportunity to live and grow happily and purposefully with others in a small society." The kindergarten teachers in one of the elementary schools in this community, one we shall call the Wilbur Wright School, said their goals were to see that the children "grew" in all ways: physically, of course, emotionally, socially, and academically. They said they wanted children to like school as a result of their kindergarten experiences and that they wanted them to learn to get along with others.

None of these goals, however, is unique to kindergarten; each of them is held to some extent by teachers in the other six grades at Wright School. And growth would occur, but differently, even if the child did not attend school. The children already know how to get along with others, in their families and their play groups. The unique job of the kindergarten in the educational division of labor seems rather to be teaching children the student role. The student role is the repertoire of behavior and attitudes regarded by educators as appropriate to children in school. Observation in the kindergartens of the Wilbur Wright School revealed a great variety of activities through which children are shown and then drilled in the behavior and attitudes defined as appropriate for school and thereby induced to learn the role of student. Observations of the kindergartens and interviews with the teachers both pointed to the teaching and learning of classroom routines as the main element of the student role. The teachers expended most of their efforts, for the first half of the year at least, in training the children to follow the routines which teachers created. The children were, in a very real sense, *drilled* in tasks and activities created by the teachers for their own purposes and beginning and ending quite arbitrarily (from the child's point of view) at the command of the teacher. One teacher remarked that she hated September, because during the first month "everything has to be done rigidly, and repeatedly, until they know exactly what they're supposed to do." However, "by January," she said, "they know exactly what to do [during the day] and I don't have to be after them all the time." Classroom routines were introduced gradually from the beginning of the year in all the kindergartens, and the children were drilled in them as long as was necessary to achieve regular compliance. By the end of the school year, the successful kindergarten teacher has a well-organized group of children. They follow classroom routines automatically, having learned all the command signals and the expected responses to them. They have, in our terms, learned the student role. The following observation shows one such classroom operating at optimum organization on an afternoon late in May. It is the class of an experienced and respected kindergarten teacher.

AN AFTERNOON IN KINDERGARTEN

At about 12:20 in the afternoon on a day in the last week of May, Edith Kerr leaves the teachers' room where she has been having lunch and walks to her classroom at the far end of the primary wing of Wright School. A group of five- and six-year-olds peers at her through the glass doors leading from the hall cloakroom to the play area outside. Entering her room, she straightens some material in the "book corner" of the room, arranges music on the piano, takes colored paper from her closet and places it on one of the shelves under the window. Her room is divided into a number of activity areas through the arrangement of furniture and play equipment. Two easels and a paint table near the door create a kind of passageway inside the room. A wedge-shaped area just inside the front door is made into a teacher's area by the placing of "her" things there: her desk, file, and piano. To the left is the book corner, marked off from the rest of the room by a puppet stage and a movable chalkboard. In it are a display rack of picture books, a record player, and a stack of children's records. To the right of the entrance are the sink and clean-up area. Four large round tables with six chairs at each for the children are placed near the walls about halfway down the length of the room, two on each side, leaving a large open area in the center for group games, block building, and toy truck driving. Windows stretch down the length of both walls, starting about three feet from the floor and extending almost to the high ceilings. Under the windows are long shelves on which are kept all the toys, games, blocks, paper, paints, and other equipment of the kindergarten. The left rear corner of the room is a play store with shelves, merchandise, and cash register; the right rear corner is a play kitchen with stove, sink, ironing board, and bassinette with baby dolls in it. This area is partly shielded from the rest of the room by a large standing display rack for posters and children's art work. A sandbox is found against the back wall between these two areas. The room is light, brightly colored and filled with things adults feel five- and six-year-olds will find interesting and pleasing.

At 12:25 Edith opens the outside door and admits the waiting children. They hang their sweaters on hooks outside the door and then go to the center of the room and arrange themselves in a semi-circle on the floor, facing the teacher's chair, which she has placed in the center of the floor.

Edith follows them in and sits in her chair checking attendance while waiting for the bell to ring. When she has finished attendance, which she takes by sight, she asks the children what the date is, what day and month it is, how many children are enrolled in the class, how many are present, and how many are absent.

The bell rings at 12:30 and the teacher puts away her attendance book. She introduces a visitor, who is sitting against the wall taking notes, as someone who wants to learn about schools and children. She then goes to the back of the room and takes down a large chart labeled "Helping Hands." Bringing it to the center of the room, she tells the children it is time to change jobs. Each child is assigned some task on the chart by placing his name, lettered on a paper "hand," next to a picture signifying the task—e.g., a broom, a blackboard, a milk bottle, a flag, and a Bible. She asks the children who wants each of the jobs and rearranges their "hands" accordingly. Returning to her chair, Edith announces, "One person should tell us what happened to Mark." A girl raises her hand, and when called on says, "Mark fell and hit his head and had to go to the hospital." The teacher adds that Mark's mother had written saying he was in the hospital.

During this time the children have been interacting among themselves, in their semi-circle. Children have whispered to their neighbors, poked one another, made general comments to the group, waved to friends on the other side of the circle. None of this has been disruptive, and the teacher has ignored it for the most part. The children seem to know just how much of each kind of interaction is permitted—they may greet in a soft voice someone who sits next to them, for example, but may not shout greetings to a friend who sits across the circle, so they confine themselves to waving and remain well within understood limits.

At 12:35 two children arrive. Edith asks them why they are late and then sends them to join the circle on the floor. The other children vie with each other to tell the newcomers what happened to Mark. When this leads to a general disorder Edith asks, "Who has serious time?" The children become quiet and a girl raises her hand. Edith nods and the child gets a Bible and hands it to Edith. She reads the Twenty-third Psalm while the children sit quietly. Edith helps the child in charge begin reciting the Lord's Prayer; the other children follow along for the first unit of sounds, and then trail off as Edith finishes for them. Everyone stands and faces the American flag hung to the right of the door. Edith leads the pledge to the flag, with the children again following the familiar sounds as far as they remember them. Edith then asks the girl in charge what song she wants and the child replies, "My Country." Edith goes to the piano and plays "America," singing as the children follow her words.

Edith returns to her chair in the center of the room and the children sit again in the semi-circle on the floor. It is 12:40 when she tells the children, "Let's have boys' sharing time first." She calls the name of the first boy sitting on the end of the circle, and he comes up to her with a toy helicopter. He turns and holds it up for the other children to see. He says, "It's a helicopter." Edith asks, "What is it used for?" and he replies, "For the army. Carry men. For the war." Other children join in, "For shooting submarines." "To bring back men from space when they are in the ocean." Edith sends the boy back to the circle and asks the next boy if he has something. He replies "No" and she passes on to the next. He says "Yes" and brings a bird's nest to her. He holds it for the class to see, and the teacher asks, "What kind of bird made the nest?" The boy replies, "My friend says a rain bird made it." Edith asks what the nest is made of and different children reply, "mud," "leaves," and "sticks." There is also a bit of moss woven into the nest, and Edith tries to describe it to the children. They, however, are more interested in seeing if anything is inside it, and Edith lets the boy carry it around the semi-circle showing the children its insides. Edith tells the children of some baby robins in a nest in her yard, and some of the children tell about baby birds they have seen. Some children are asking about a small object in the nest which they say looks like an egg, but all have seen the nest now and Edith calls on the next boy. A number of children say, "I know what Michael has, but I'm not telling." Michael brings a book to the teacher and then goes back to his place in the circle of children. Edith reads the last page of the book to the class. Some children tell of books which they have at home. Edith calls the next boy, and three children call out, "I know what David has." "He always has the same thing." "It's a bang-bang." David goes to his table and gets a box which he brings to Edith. He opens it and shows the teacher a scale-model of an old-fashioned dueling pistol. When David does not turn around to the class, Edith tells him, "Show it to the children" and he does. One child says, "Mr. Johnson [the principal] said no guns." Edith replies, "Yes, how many of you know that?" Most of the children in the circle raise their hands. She continues, "That you aren't supposed to bring guns to school?" She calls the next boy on the circle and he brings two large toy soldiers to her which the children enthusiastically identify as being from "Babes in Toyland." The next boy brings an American flag to Edith and shows it to the class. She asks him what the stars and stripes stand for and admonishes him to treat it carefully. "Why should you treat it carefully?" she asks the boy. "Because it's our flag," he replies. She congratulates him, saying, "That's right."

"Show and Tell" lasted twenty minutes and during the last ten one girl in particular announced that she knew what each child called upon had to show. Edith asked her to be quiet each time she spoke out, but she was not content, continuing to offer her comment at each "show." Four children from other classes had come into the room to bring something from another teacher or to ask for something from Edith. Those with requests were asked to return later if the item wasn't readily available.

Edith now asks if any of the children told their mothers about their trip to the local zoo the previous day. Many children raise their hands. As Edith calls on them, they tell what they liked in the zoo. Some children cannot wait to be called on, and they call out things to the teacher, who asks them to be quiet. After a few of the animals are mentioned, one child says, "I liked the spooky house," and the others chime in to agree with him, some pantomiming fear and horror. Edith is puzzled, and asks what this was. When half the children try to tell her at once, she raises her hand for quiet, then calls on individual children. One says, "The house with nobody in it"; another, "The dark little house." Edith asks where it was in the zoo, but the children cannot describe its location in any way which she can understand. Edith makes some jokes but they involve adult abstractions which the children cannot grasp. The children have become quite noisy now, speaking out to make both relevant and irrelevant comments, and three little girls have become particularly assertive.

Edith gets up from her seat at 1:10 and goes to the book corner, where she puts a record on the player. As it begins a story about the trip to the zoo, she returns to the circle and asks the children to go sit at the tables. She divides them among the tables in such a way as to indicate that they don't have regular seats. When the children are all seated at the four tables, five or six to a table, the teacher asks, "Who wants to be the first one?" One of the noisy girls comes to the center of the room. The voice on the record is giving directions for imitating an ostrich and the girl follows them, walking around the center of the room holding her ankles with her hands. Edith replays the record, and all the children, table by table, imitate ostriches down the center of the room and back. Edith removes her shoes and shows that she can be an ostrich too. This is apparently a familiar game, for a number of children are calling out, "Can we have the crab?" Edith asks one of the children to do a crab "so we can all remember how," and then plays the part of the record with music for imitating crabs by. The children from the first table line up across the room, hands and feet on the floor and faces pointing toward the ceiling. After they have "walked" down the room and back in this posture they sit at their table and the children of the next table play "crab." The children love this; they run from their tables, dance about on the floor waiting for their turns and are generally exuberant. Children ask for the "inch worm," and the game is played again with the children squirming down the floor. As a conclusion Edith shows them a new animal imitation, the "lame dog." The children all hobble down the floor on three "legs," table by table to the accompaniment of the record.

At 1:30 Edith has the children line up in the center of the room: she says, "Table one, line up in front of me," and children ask, "What are we going to do?" Then she moves a few steps to the side and says, "Table two over here; line up next to table one," and more children ask, "What for?" She does this for table three and table four, and each time the children ask, "Why, what are we going to do?" When the children are lined up in four lines of five each, spaced so that they are not touching one another, Edith puts on a new record and leads the class in calisthenics, to the accompaniment of the record. The children just jump around every which way in their places instead of doing the exercises, and by the time the record is finished, Edith, the only one following it, seems exhausted. She is apparently adopting the President's new "Physical Fitness" program for her classroom.

At 1:35 Edith pulls her chair to the easels and calls the children to sit on the floor in front of her, table by table. When they are all seated she asks, "What are you going to do for worktime today?" Different children raise their hands and tell Edith what they are going to draw. Most are going to make pictures of animals they saw in the zoo. Edith asks if they want to make pictures to send to Mark in the hospital, and the children agree to this. Edith gives drawing paper to the children, calling them to her one by one. After getting a piece of paper, the children go to the crayon box on the righthand shelves, select a number of colors, and go to the tables, where they begin drawing. Edith is again trying to quiet the perpetually talking girls. She keeps two of them standing by her so they won't disrupt the others. She asks them "Why do you feel you have to talk all the time?" and then scolds them for not listening to her. Then she sends them to their tables to draw.

Most of the children are drawing at their tables, sitting or kneeling in their chairs. They are all working very industriously and, engrossed in their work, very quietly. Three girls have chosen to paint at the easels, and having donned their smocks, they are busily mixing colors and intently applying them to their pictures. If the children at the tables are primitives and neo-realists in their animal depictions, these girls at the easels are the class abstract-expressionists, with their broad-stroked, colorful paintings.

Edith asks of the children generally, "What color should I make the cover of Mark's book?" Brown and green are suggested by some children "because Mark likes them." The other children are puzzled as to just what is going on and ask, "What book?" or "What does she mean?" Edith explains what she thought was clear to them already, that they are all going to put their pictures together in a "book" to be sent to Mark. She goes to a small table in the play-kitchen corner and tells the children to bring her their pictures when they are finished and she will write their message for Mark on them.

By 1:50 most children have finished their pictures and given them to Edith. She talks with some of them as she ties the bundle of pictures together—answering questions, listening, carrying on conversations. The children are playing in various parts of the room with toys, games, and blocks which they have taken off the shelves. They also move from table to table examining each other's pictures, offering compliments and suggestions. Three girls at the table are cutting up colored paper for a collage. Another girl is walking about the room in a pair of high heels with a woman's purse over her arm. Three boys are playing in the center of the room with the large block set, with which they are building walk-ways and walking on them. Edith is very much concerned about their safety and comes over a number of times to fuss over them. Two or three other boys are pushing trucks around the center of the room, and mild altercations occur when they drive through the block constructions. Some boys and girls are playing at the toy store, two girls are serving "tea" in the play kitchen and one is washing a doll baby. Two boys have elected to clean the room, and with large sponges they wash the movable blackboard, the puppet stage, and then begin on the tables. They run into resistance from the children who are working with construction toys on the tables and do not want to dismantle their structures. The class is like a room full of bees, each intent on pursuing some activity, occasionally bumping into one another, but just veering off in another direction without serious altercation. At 2:05 the custodian arrives pushing a cart loaded with half-pint milk containers. He places a tray of cartons on the counter next to the sink, then leaves. His coming and going is unnoticed in the room (as, incidentally, is the presence of the observer, who is completely ignored by the children for the entire afternoon).

At 2:15 Edith walks to the entrance of the room, switches off the lights, and sits at the piano and plays. The children begin spontaneously singing the song, which is "Clean up, clean up. Everybody clean up." Edith walks around the room supervising the clean-up. Some children put their toys, the blocks, puzzles, games, and so on back on their shelves under the windows. The children making a collage keep right on working. A child from another class comes in to borrow the 45-rpm adapter for the record player. At more urging from Edith the rest of the children shelve their toys and work. The children are sitting around their tables now, and Edith asks, "What record would you like to hear while you have your milk?" There is some confusion and no general consensus, so Edith drops the subject and begins to call the children, table by table, to come get their milk. "Table one," she says, and the five children come to the sink, wash their hands and dry them, pick up a carton of milk and a straw, and take it back to their table. Two talking girls wander about the room interfering with the children getting their milk and Edith calls out to them to "settle down." As the children sit, many of them call out to Edith the name of the record they want to hear. When all the children are seated at tables with milk, Edith plays one of these records called "Bozo and the Birds" and shows the children pictures in a book which go with the record. The record recites, and the book shows the adventures of a clown, Bozo, as he walks through a woods meeting many different kinds of birds who, of course, display the characteristics of many kinds of people or, more accurately, different stereotypes. As children finish their milk, they take blankets or pads from the shelves under the windows and lie on them in the center of the room, where Edith sits on her chair showing the pictures. By 2:30 half the class is lying on the floor on their blankets, the record is still playing, and the teacher is turning the pages of the book. The child who came in previously returns the 45-rpm adapter, and one of the kindergartners tells Edith what the boy's name is and where he lives.

The record ends at 2:40. Edith says, "Children, down on your blankets." All the class is lying on blankets now. Edith refuses to answer the various questions individual children put to her because, she tells them, "it's rest time now." Instead she talks very softly about what they will do tomorrow. They are going to work with clay, she says. The children lie quietly and listen. One of the boys raises his hand and when called on tells Edith, "The animals in the zoo looked so hungry yesterday." Edith asks the children what they think about this and a number try to volunteer opinions, but Edith accepts only those offered in a "rest-time tone," that is, softly and quietly. After a brief discussion of animal feeding, Edith calls the names of the two children on milk detail and has them collect empty milk cartons from the tables and return them to the tray. She asks the two children on clean-up detail to clean up the room. Then she gets up from her chair and goes to the door to turn on the lights. At this signal, the children all get up from the floor and return their blankets and pads to the shelf. It is raining (the reason for no outside play this afternoon) and cars driven by mothers clog the school drive and line up along the street. One of the talkative little girls comes over to Edith and pointing out the window says, "Mrs. Kerr, see my mother in the new Cadillac?"

At 2:50 Edith sits at the piano and plays. The children sit on the floor in the center of the room and sing. They have a repertoire of songs about animals, including one in which each child sings a refrain alone. They know these by heart and sing along through the ringing of the 2:55 bell. When the song is finished, Edith gets up and coming to the group says, "Okay, rhyming words to get your coats today." The children raise their hands and as Edith calls on them, they tell her two rhyming words, after which they are allowed to go into the hall to get their coats and sweaters. They return to the room with these and sit at their tables. At 2:59 Edith says, "When you have your coats on, you may line up at the door." Half of the children go to the door and stand in a long line. When the three o'clock bell rings, Edith returns to the piano and plays. The children sing a song called "Goodbye," after which Edith sends them out.

TRAINING FOR LEARNING AND FOR LIFE

The day in kindergarten at Wright School illustrates both the content of the student role as it has been learned by these children and the processes by which the teacher has brought about this learning, or "taught" them the student role. The children have learned to go through routines and to follow orders with unquestioning obedience, even when these make no sense to them. They have been disciplined to do as they are told by an authoritative person without significant protest. Edith has developed this discipline in the children by creating and enforcing a rigid social structure in the classroom through which she effectively controls the behavior of most of the children for most of the school day. The "living with others in a small society" which the school pamphlet tells parents is the most important thing the children will learn in kindergarten can be seen now in its operational meaning, which is learning to live by the routines imposed by the school. This learning appears to be the principal content of the student role.

Children who submit to school-imposed discipline and come to identify with it, so that being a "good student" comes to be an important part of their developing identities, *become* the good students by the school's definitions. Those who submit to the routines of the school but do not come to identify with them will be adequate students who find the more important part of their identities elsewhere, such as in the play group outside school. Children who refuse to submit to the school routines are rebels, who become known as "bad students" and often "problem children" in the school, for they do not learn the academic curriculum and their behavior is often disruptive in the classroom. Today schools engage clinical psychologists in part to help teachers deal with such children.

In looking at Edith's kindergarten at Wright School, it is interesting to ask how the children learn this role of student—come to accept school-imposed routines—and what, exactly, it involves in terms of behavior and attitudes. The most prominent features of the classroom are its physical and social structures. The room is carefully furnished and arranged in ways adults feel will interest children. The play store and play kitchen in the back of the room, for example, imply that children are interested in mimicking these activities of the adult world. The only space left for the children to create something of their own is the empty center of the room, and the materials at their disposal are the blocks, whose use causes anxiety on the part of the teacher. The room, being carefully organized physically by the adults, leaves little room for the creation of physical organization on the part of the children.

The social structure created by Edith is a far more powerful and subtle force for fitting the children to the student role. This structure is established by the very rigid and tightly controlled set of rituals and routines through which the children are put during the day. There is first the rigid "locating procedure" in which the children are asked to find themselves in terms of the month, date, day of the week, and the number of the class who are present and absent. This puts them solidly in the real world as defined by adults. The day is then divided into six periods whose activities are for the most part determined by the teacher. In Edith's kindergarten the children went through Serious Time, which opens the school day, Sharing Time, Play Time (which in clear weather would be spent outside), Work Time, Clean-up Time, after which they have their milk, and Rest Time after which they go home. The teacher has programmed activities for each of these Times.

Occasionally the class is allowed limited discretion to choose between proffered activities, such as stories or records, but original ideas for activities are never solicited from them. Opportunity for free individual action is open only once in the day, during the part of Work Time left after the general class assignment has been completed (on the day reported the class assignment was drawing animal pictures for the absent Mark). Spontaneous interests or observations from the children are never developed by the teacher. It seems that her schedule just does not allow room for developing such unplanned events. During Sharing Time, for example, the child who brought a bird's nest told Edith, in reply to her question of what kind of bird made it, "My friend says it's a rain bird." Edith does not think to ask about this bird, probably because the answer is "childish," that is, not given in accepted adult categories of birds. The children then express great interest in an object in the nest, but the teacher ignores this interest, probably because the object is uninteresting to her. The soldiers from "Babes in Toyland" strike a responsive note in the children, but this is not used for a discussion of any kind. The soldiers are treated in the same way as objects which bring little interest from the children. Finally, at the end of Sharing Time the child-world of perception literally erupts in the class with the recollection of "the spooky house" at the zoo. Apparently this made more of an impression on the children than did any of the animals, but Edith is unable to make any sense of it for herself. The tightly imposed order of the class begins to break down as the children discover a universe of discourse of their own and begin talking excitedly with one another. The teacher is effectively excluded from this child's world of perception and for a moment she fails to dominate the classroom situation. She reasserts control, however, by taking the children to the next activity she has planned for the day. It seems never to have occurred to Edith that there might be a meaningful learning experience for the children in re-creating the "spooky house" in the classroom. It seems fair to say that this would have offered an exercise in spontaneous self-expression and an opportunity for real creativity on the part of the children. Instead, they are taken through a canned animal imitation procedure, an activity which they apparently enjoy, but which is also imposed upon them rather than created by them.

While children's perceptions of the world and opportunities for genuine spontaneity and creativity are being systematically eliminated from the kindergarten, unquestioned obedience to authority and rote learning of meaningless material are being encouraged. When the children are called to line up in the center of the room they ask "Why?" and "What for?" as they are in the very process of complying. They have learned to go smoothly through a programmed day, regardless of whether parts of the program make any sense to them or not. Here the student role involves what might be called "doing what you're told and never mind why." Activities which might "make sense" to the children are effectively ruled out, and they are forced or induced to participate in activities which may be "senseless," such as calisthenics.

At the same time the children are being taught by rote meaningless sounds in the ritual oaths and songs, such as the Lord's Prayer, the Pledge to the Flag, and "America." As they go through the grades children learn more and more of the sounds of these ritual oaths, but the fact that they have often learned meaningless sounds rather than meaningful statements is shown when they are asked to write these out in the sixth grade; they write them as groups of sounds rather than as a series of words, according to the sixth grade teachers at Wright School. Probably much learning in the elementary grades is of this character, that is, having no intrinsic meaning to the children, but rather being tasks inexplicably required of them by authoritative adults. Listening to sixth grade children read social studies reports, for example, in which they have copied material from encyclopedias about a particular country, an observer often gets the feeling that he is watching an activity which has no intrinsic meaning for the child. The child who reads, "Switzerland grows wheat and cows and grass and makes a lot of cheese" knows the dictionary meaning of each of these words but may very well have no conception at all of this "thing" called Switzerland. He is simply carrying out a task assigned by the teacher *because* it is assigned, and this may be its only "meaning" for him.

Another type of learning which takes place in kindergarten is seen in children who take advantage of the "holes" in the adult social structure to create activities of their own, during Work Time or out-of-doors during Play Time. Here the children are learning to carve out a small world of their own within the world created by adults. They very quickly learn that if they keep within permissible limits of noise and action they can play much as they please. Small groups of children formed during the year in Edith's kindergarten who played together at these times, developing semi-independent little groups in which they created their own worlds in the interstices of the adult-imposed physical and social world. These groups remind the sociological observer very much of the so-called "informal groups" which adults develop in factories and offices of large bureaucracies.[5] Here, too, within authoritatively imposed social organizations people find "holes" to create little subworlds which support informal, friendly, unofficial behavior. Forming and participating in such groups seems to be as much part of the student role as it is of the role of bureaucrat.

The kindergarten has been conceived of here as the year in which children are prepared for their schooling by learning the role of student. In the classrooms of the rest of the school grades, the children will be asked to submit to systems and routines imposed by the teachers and the curriculum. The days will be much like those of kindergarten, except that academic subjects will be substituted for the activities of the kindergarten. Once out of the school system, young adults will more than likely find themselves working in large-scale bureaucratic organizations, perhaps on the assembly line in the factory, perhaps in the paper routines of the white collar occupations, where they will be required to submit to rigid routines imposed by "the company" which may make little sense to them. Those who can operate well in this situation will be successful bureaucratic functionaries. Kindergarten, therefore, can be seen as preparing children not only for participation in the bureaucratic organization of large modern school systems, but also for the large-scale occupational bureaucracies of modern society.

NOTES

1. Emile Durkheim, *Sociology and Education* (New York: The Free Press, 1956), pp. 71–72.
2. *Ibid.*, p. 123.
3. Wilbur Brookover, *The Sociology of Education* (New York: American Book Company, 1957), p. 4.
4. *Ibid.*, p. 6.
5. See, for example, Peter M. Blau, *Bureaucracy in Modern Society* (New York: Random House, 1956), Chapter 3.

42

Savage Inequalities

Children in America's Schools

JONATHAN KOZOL

In addition to differences in structure and curricula in schools, rich and poor neighborhoods have different schools. The following excerpts from Jonathan Kozol's Savage Inequalities *illustrate the gap between schools in three different communities.*

As you read, consider the following:

1. *In what ways do the schools in these three communities differ?*
2. *What is the impact of differences between schools on children in those schools?*
3. *What are some causes of differences between schools and what might be done about these differences?*
4. *Are you aware of differences between schools in your area?*

GLOSSARY **Savage inequalities** Differences between schools for children from poor families in inner cities or poor suburbs compared to children from rich communities.

The problems of the streets in urban areas, as teachers often note, frequently spill over into public schools. In the public schools of East St. Louis this is literally the case.

"Martin Luther King Junior High School," notes the *Post-Dispatch* in a story published in the early spring of 1989, "was evacuated Friday afternoon after sewage flowed into the kitchen. . . . The kitchen was closed and students were sent home." On Monday, the paper continued, "East St. Louis Senior High School was awash in sewage for the second time this year." The school had to be shut because of "fumes and backed-up toilets." Sewage flowed into the basement, through the floor, then up into the kitchen and the students' bathrooms. The backup, we read, "occurred in the food preparation areas."

School is resumed the following morning at the high school, but a few days later the overflow recurs. This time the entire system is affected, since the meals distributed to every student in the city are prepared in the two schools that have been flooded. School is called off for all 16,500 students in the district. The sewage backup, caused by the failure of two pumping stations, forces officials at the high school to shut down the furnaces.

At Martin Luther King, the parking lot and gym are also flooded. "It's a disaster," says a legislator. "The streets are underwater; gaseous fumes are being emitted from the pipes under the schools," she says, "making people ill."

In the same week, the schools announce the layoff of 280 teachers, 166 cooks and cafeteria workers, 25 teacher aides, 16 custodians, and 18 painters, electricians, engineers and plumbers. The president of the teachers' union says the cuts, which will bring the size of kindergarten and primary classes up to 30 students, and the size of fourth to twelfth grade classes up to 35, will have "an unimaginable impact" on the students. "If you have a high school teacher with five classes each day and between 150 and 175 students..., it's going to have a devastating effect." The school system, it is also noted, has been using more than 70 "permanent substitute teachers," who are paid only $10,000 yearly, as a way of saving money.

Governor Thompson, however, tells the press that he will not pour money into East St. Louis to solve long-term problems. East St. Louis residents, he says, must help themselves. "There is money in the community," the governor insists. "It's just not being spent for what it should be spent for."

The governor, while acknowledging that East St. Louis faces economic problems, nonetheless refers dismissively to those who live in East St. Louis. "What in the community," he asks, "is being done right?" He takes the opportunity of a visit to the area to announce a fiscal grant for sewer improvement to a relatively wealthy town nearby.

In East St. Louis, meanwhile, teachers are running out of chalk and paper, and their paychecks are arriving two weeks late. The city warns its teachers to expect a cut of half their pay until the fiscal crisis has been eased.

The threatened teacher layoffs are mandated by the Illinois Board of Education, which, because of the city's fiscal crisis, has been given supervisory control of the school budget. Two weeks later the state superintendent partially relents. In a tone very different from that of the governor, he notes that East St. Louis does not have the means to solve its education problems on its own. "There is no natural way," he says, that "East St. Louis can bring itself out of this situation." Several cuts will be required in any case—one quarter of the system's teachers, 75 teacher aides, and several dozen others will be given notice—but, the state board notes, sports and music programs will not be affected.

East St. Louis, says the chairman of the state board, "is simply the worst possible place I can imagine to have a child brought up.... The community is in desperate circumstances." Sports and music, he observes, are, for many children here, "the only avenues of success." Sadly enough, no matter how it ratifies the stereotype, this is the truth; and there is a poignant aspect to the fact that, even with class size soaring and one quarter of the system's teachers being given their dismissal, the state board of education demonstrates its genuine but skewed compassion by attempting to leave sports and music untouched by the overall austerity.

Even sports facilities, however, are degrading by comparison with those found and expected at most high schools in America. The football field at East St. Louis High is missing almost everything—including goalposts. There are a couple of metal pipes—no crossbar, just the pipes. Bob Shannon, the football coach, who has to use his personal funds to purchase footballs and has had to cut and rake the football field himself, has dreams of having goalposts someday. He'd also like to let his students have new uniforms. The ones they wear are nine years old and held together somehow by a patchwork of repairs. Keeping them clean is a problem, too. The school cannot afford a washing machine. The uniforms are carted to a corner laundromat with fifteen dollars' worth of quarters.

Other football teams that come to play, according to the coach, are shocked to see the field and locker rooms. They want to play without a halftime break and get away. The coach reports that he's been missing paychecks, but he's trying nonetheless to raise some money to help out a member of the team whose mother has just died of cancer.

"The days of the tight money have arrived," he says. "It don't look like Moses will be coming to this school."

He tells me he has been in East St. Louis 19 years and has been the football coach for 14 years. "I was born," he says, "in Natchez, Mississippi. I stood on the courthouse steps of Natchez with Charles Evers. I was a teen-age boy when Michael Schwerner and the other boys were murdered. I've been in the struggle all along. In Mississippi, it was the fight for legal rights. This time, it's a struggle for survival.

"In certain ways," he says, "it's harder now because in those days it was a clear enemy you had to face, a man in a hood and not a statistician. No one could persuade you that you were to blame. Now the choices seem like they are left to you and, if you make the wrong choice, you are made to understand you are to blame. . . .

"Night-time in this city, hot and smoky in the summer, there are dealers standin' out on every street. Of the kids I see here, maybe 55 percent will graduate from school. Of that number, maybe one in four will go to college. How many will stay? That is a bigger question.

"The basic essentials are simply missing here. When we go to wealthier schools I look at the faces of my boys. They don't say a lot. They have their faces to the windows, lookin' out. I can't tell what they are thinking. I am hopin' they are saying, 'This is something I will give my kids someday.' "

Tall and trim, his black hair graying slightly, he is 45 years old.

"No, my wife and I don't live here. We live in a town called Ferguson, Missouri. I was born in poverty and raised in poverty. I feel that I owe it to myself to live where they pick up the garbage."

In the visitors' locker room, he shows me lockers with no locks. The weight room stinks of sweat and water-rot. "See, this ceiling is in danger of collapsing. See, this room don't have no heat in winter. But we got to come here anyway. We wear our coats while working out. I tell the boys, 'We got to get it done. Our fans don't know that we do not have heat.' "

He tells me he arrives at school at 7:45 A.M. and leaves at 6:00 P.M.—except in football season, when he leaves at 8:00 P.M. "This is my life. It isn't all I dreamed of and I tell myself sometimes that I might have accomplished more. But growing up in poverty rules out some avenues. You do the best you can." . . .

On the following morning I visit P.S. 79, an elementary school in the same district. "We work under difficult circumstances," says the principal, James Carter, who is black. "The school was built to hold one thousand students. We have 1,550. We are badly overcrowded. We need smaller classes but, to do this, we would need more space. I can't add five teachers. I would have no place to put them."

Some experts, I observe, believe that class size isn't a real issue. He dismisses this abruptly. "It doesn't take a genius to discover that you learn more in a smaller class. I have to bus some 60 kindergarten children elsewhere, since I have no space for them. When they return next year, where do I put them?

"I can't set up a computer lab. I have no room. I had to put a class into the library. I have no librarian. There are two gymnasiums upstairs but they cannot be used for sports. We hold more classes there. It's unfair to measure us against the suburbs. They have 17 to 20 children in a class. Average class size in this school is 30.

"The school is 29 percent black, 70 percent Hispanic. Few of these kids get Head Start. There is no space in the district. Of 200 kindergarten children, 50 maybe get some kind of preschool."

I ask him how much difference preschool makes.

"Those who get it do appreciably better. I can't overestimate its impact but, as I have said, we have no space."

The school tracks children by ability, he says. "There are five to seven levels in each grade. The highest level is equivalent to 'gifted' but it's not a full-scale gifted program. We don't have the funds. We have no science room. The science teachers carry their equipment with them."

We sit and talk within the nurse's room. The window is broken. There are two holes in the ceiling. About a quarter of the ceiling has been patched and covered with a plastic garbage bag.

"Ideal class size for these kids would be 15 to 20. Will these children ever get what white kids in the suburbs take for granted? I don't think so. If you ask me why, I'd have to speak of race and social class. I don't think the powers that be in New York City understand, or want to understand, that if they do not give these children a sufficient education to lead healthy and productive lives, we will be their victims later on. We'll pay the price someday—in violence, in economic costs. I despair of making this appeal in any terms but these. You cannot issue an appeal to conscience in New York today. The fair-play argument won't be accepted. So you speak of violence and hope that it will scare the city into action."

While we talk, three children who look six or seven years old come to the door and ask to see the nurse, who isn't in the school today. One of the children, a Puerto Rican girl, looks haggard. "I have a pain in my tooth," she says. The principal says, "The nurse is out. Why don't you call your mother?" The child says, "My mother doesn't have a phone." The principal sighs. "Then go back to your class." When she leaves, the principal is angry. "It's amazing to me that these children ever make it with the obstacles they face. Many *do* care and they *do* try, but there's a feeling of despair. The parents of these children want the same things for their children that the parents in the suburbs want. Drugs are not the cause of this. They are the symptom. Nonetheless, they're used by people in the suburbs and rich people in Manhattan as another reason to keep children of poor people at a distance."

I ask him, "Will white children and black children ever go to school together in New York?"

"I don't see it," he replies. "I just don't think it's going to happen. It's a dream. I simply do not see white folks in Riverdale agreeing to cross-bus with kids like these. A few, maybe. Very few. I don't think I'll live to see it happen."

I ask him whether race is the decisive factor. Many experts, I observe, believe that wealth is more important in determining these inequalities.

"This," he says—and sweeps his hand around him at the room, the garbage bag, the ceiling—"would not happen to white children."

In a kindergarten class the children sit cross-legged on a carpet in a space between two walls of books. Their 26 faces are turned up to watch their teacher, an elderly black woman. A little boy who sits beside me is involved in trying to tie bows in his shoelaces. The children sing a song: "Lift Every Voice." On the wall are these handwritten words: "Beautiful, also, are the souls of my people."

In a very small room on the fourth floor, 52 people in two classes do their best to teach and learn. Both are first grade classes. One, I am informed, is "low ability." The other is bilingual.

"The room is barely large enough for one class," says the principal.

The room is 25 by 50 feet. There are 26 first graders and two adults on the left, 22 others and two adults on the right. On the wall there is the picture of a small white child, circled by a Valentine and a Gainsborough painting of a child in a formal dress. . . .

Children who go to school in towns like Glencoe and Winnetka do not need to steal words from a dictionary. Most of them learn to read by second or third grade. By the time they get to sixth or seventh grade, many are reading at the level of the seniors in the best Chicago high schools. By the time they enter ninth grade at New Trier High, they are in a world of academic possibilities that far exceed the hopes and dreams of most schoolchildren in Chicago.

"Our goal is for students to be successful," says the New Trier principal. With 93 percent of seniors going on to four-year colleges—many to schools like Harvard, Princeton, Berkeley, Brown, and Yale—this goal is largely realized.

New Trier's physical setting might well make the students of Du Sable High School envious. The *Washington Post* describes a neighborhood of "circular driveways, chirping birds and white-columned homes." It is, says a student, "a maple land of beauty and civility." While Du Sable is sited on one crowded city block, New Trier students have the use of 27 acres. While Du Sable's science students have to settle for makeshift equipment, New Trier's students have superior labs and up-to-date technology. One wing of the school, a physical education center that includes three separate gyms, also contains a fencing room, a wrestling room, and studios for dance instruction. In all, the school has seven gyms as well as an Olympic pool.

The youngsters, according to a profile of the school in *Town and Country* magazine, "make good use of the huge, well-equipped building, which is immaculately maintained by a custodial staff of 48."

It is impossible to read this without thinking of a school like Goudy, where there are no science labs, no music or art classes and no playground—and where the two bathrooms, lacking toilet paper, fill the building with their stench.

"This is a school with a lot of choices," says one student at New Trier; and this hardly seems an overstatement if one studies the curriculum. Courses in music, art and drama are so varied and abundant that students can virtually major in these subjects in addition to their academic programs. The modern and classical language department offers Latin (four years) and six other foreign languages. Elective courses include the literature of Nobel winners, aeronautics, criminal justice, and computer languages. In a senior literature class, students are reading Nietzsche, Darwin, Plato, Freud, and Goethe. The school also operates a television station with a broadcast license from the FCC, which broadcasts on four channels to three counties.

Average class size is 24 children; classes for slower learners hold 15. This may be compared to Goudy—where a remedial class holds 39 children and a "gifted" class has 36.

Every freshman at New Trier is assigned a faculty adviser who remains assigned to him or her through graduation. Each of the faculty advisers—they are given a reduced class schedule to allow them time for this—gives counseling to about two dozen children. . . .

The ambience among the students at New Trier, of whom only 1.3 percent are black, says *Town and Country*, is "wholesome and refreshing, a sort of throwback to the Fifties." It is, we are told, "a preppy kind of place." In a cheerful photo of the faculty and students, one cannot discern a single nonwhite face.

New Trier's "temperate climate" is "aided by the homogeneity of its students," *Town and Country* notes. ". . . Almost all are of European extraction and harbor similar values."

"Eighty to 90 percent of the kids here," says a counselor, "are good, healthy, red-blooded Americans."

The wealth of New Trier's geographical district provides $340,000 worth of taxable property for each child; Chicago's property wealth affords only one-fifth this much. Nonetheless, *Town and Country* gives New Trier's parents credit for a "willingness to pay enough . . . in taxes" to make this one of the state's best-funded schools. New Trier, according to the magazine, is "a striking example of what is possible when citizens want to achieve the best for their children." Families move here "seeking the best," and their children "make good use" of what they're given.

43

Deepening Segregation in American Public Schools

GARY ORFIELD, MARK D. BACHMEIER, DAVID R. JAMES, AND TAMELA EITLE

This excerpt from a report by the Harvard Project on School Desegregation provides a brief history of desegregation in public schools, along with recent statistics and trends relating to the status of race segregation in public schools. The issue of resegregation of public school is of increasing concern in this country. This selection provides some background surrounding desegregation, including the court cases and legislation that formed its core. In addition, current statistics are included so you can see how far we have come.

Questions to consider as you read this selection:

1. *How did the desegregation of schools begin? What were the key decisions that made desegregation possible? Was it successful?*
2. *How does segregation affect the school experiences of students from different ethnic groups? Consider the article "Savage Inequalities" by Kozol. What might be the effects of resegregation?*
3. *What is the status of school segregation in your community. What changes, if any, are needed? Are taking place?*

GLOSSARY ***Brown v. Board of Education*** U.S. Supreme Court decision judging separate school *facilities* to be unequal. **De jure segregation** Separation required by law. **Segregation** To separate groups, usually by racial or ethnic differences.

Decades of legal and political struggle were required to end the apartheid system of mandated segregation in the schools of 17 states and to transform the South from an area of absolute segregation for black students to the most integrated region of the country. We often celebrate this accomplishment as if it were a permanent reversal of a history of segregation and inequality. From the 1950s through the late 1980s, African American students experienced declining segregation, particularly in the southern and border states.

The changes begun by the 1954 Supreme Court decision in *Brown v. Board of Education*, however, are now coming undone. The statistics analyzed for this article show that segregation is increasing for blacks, particularly in the states that once mandated racial separation. For Latinos, an even more severe level of segregation is intensifying across the nation.

The trends reported here are the first since the Supreme Court, in the 1990s, approved a return to segregated neighborhood schools under some conditions. A number of major cities have recently received court approval for such changes and others are in court. The segregation changes reported here are most striking in the southern and border states, but segregation is spreading across the nation, particularly affecting our rapidly growing Latino communities in the West. This report shows that the racial and ethnic segregation of African American and Latino students has produced a deepening isolation from middle-class students and from successful schools. It also highlights a little noticed but extremely important expansion of segregation to the suburbs, particularly in larger metropolitan areas. Expanding segregation is a mark of a polarizing society that lacks effective policies for building multiracial institutions.

Latino students, who will soon be the largest minority group in American public schools, were granted the right to desegregated education by the Supreme Court in 1973, but new data show they now are significantly more segregated than black students, with clear evidence of increasing isolation across the nation. Part of this trend is caused by the very rapid growth in the number of Latino students in several major states. Regardless of the reasons, Latino students now experience more isolation from whites and more concentration in high poverty schools than any other group of students.[1,2]

Desegregation is not just sitting next to someone of another race. Economic class and family and community educational background are also critically important for educational opportunity. School segregation effects go beyond racial separation. Segregated black and Latino schools are fundamentally different from segregated white schools in terms of the background of the children and many things that relate to educational quality. This report shows that only a twentieth of the nation's segregated white schools face conditions of concentrated poverty among their children, but more than 80% of segregated black and Latino schools do. A child moving from a segregated African American or Latino school to a white school will very likely exchange conditions of concentrated poverty for a middle-class school. Exactly the opposite is likely when a child is sent back from an interracial school to a segregated neighborhood school, as is happening under a number of recent court orders that end busing or desegregation choice plans.

The Supreme Court concluded in 1954 that intentionally segregated schools were "inherently unequal," and contemporary evidence indicates that this remains true today. Thus, it is very important to continuously monitor the extent to which the nation is realizing the promise of equal educational opportunity in schools that are now racially segregated. Education was vital to the success of the black tenth of the U.S. population when *de jure* segregation was declared unconstitutional—it is far more important today, when millions of good, low-education jobs have vanished, and when one-third of public school students are non-white.[3]

With the stakes for educational opportunity much higher today, this report shows that we are moving backward toward greater racial separation, rather than pressing gradually forward as we were between the 1950s and the mid-1980s. It shows a delayed impact of the Reagan administration campaign to reverse desegregation orders, which made no progress while Reagan was president, but now has had a substantial impact through appointments that transformed the federal courts. The 1991–95 period following the Supreme Court's first decision authorizing resegregation witnessed the continuation of the largest backward movement toward segregation for blacks in the 43 years since *Brown*.

During the 1980s, the courts rejected efforts to terminate school desegregation, and the level of desegregation actually increased, although the Reagan and Bush administrations advocated reversals. Congress rejected proposals for major steps to reverse desegregation, and there has been no trend toward increasing hostility to desegregation in public opinion. In fact, opinion is becoming more favorable.[4] The policy changes have come from the courts. The Supreme Court, in decisions from 1991 to 1995, has given lower courts discretion to approve resegregation on a large scale, and it is beginning to occur.

The statistics reported here show only the first phase of what is likely to be an accelerating trend. These statistics for the 1994–95 school year do not reflect post-1994 decisions that terminated desegregation plans in metropolitan Wilmington, Broward County (Florida), Denver, Buffalo, Mobile, Cleveland, and a number of other areas. Important cases in several other cities are pending in court now. These decisions are virtually certain to accelerate the trend toward increased racial and economic segregation of African American and Latino students. Thus, the trends reported today should be taken as portents of larger changes now under way.

BACKGROUND OF DESEGREGATION

In 1954 the Supreme Court began the process of desegregating American public education in its landmark decision, *Brown v. Board of Education.* Congress took its most powerful action for school desegregation with the passage of the 1964 Civil Rights Act. In 1971, the great national battle over urban desegregation began with the Supreme Court's decision in the Charlotte, North Carolina, busing case, *Swann v. Charlotte-Mecklenberg Board of Education.*[5] With *Swann*, there was a comprehensive set of policies in place for massive desegregation in the South.

No similar body of law ever developed in the North and West. The Supreme Court first extended some desegregation requirements to the cities of the North and recognized the rights of Hispanic as well as black students from illegal segregation in 1973.[6] In the early 1970s Congress enacted legislation to help pay for the training and educational changes (but not the busing) needed to make desegregation more effective. These last major initiatives intended to foster desegregation took place more than two decades ago.

Since 1974 almost all of the federal policy changes have been negative, even while the nation's non-white population has dramatically increased, particularly its school age children. In what is rapidly becoming a society dominated by suburbia, only a small fraction of white middle-class children are growing up in central cities. The key Supreme Court decision of *Milliken v. Bradley*[7] in 1974 reversed lower court plans to desegregate metropolitan Detroit and provided a drastic limitation on the possibility of substantial and lasting city-suburban school desegregation. That decision ended significant movement toward less segregated schools and made desegregation virtually impossible in many metropolitan areas when the non-white population was concentrated in central cities. (It is not, therefore, surprising that the state of Michigan ranks second in the nation in segregation of black students two decades after the Supreme Court confined desegregation efforts within the boundaries of a largely black and economically declining city.)[8]

The Supreme Court ruled that the courts could try to make segregated schools more equal in its second Detroit decision in 1977, *Milliken v. Bradley II.*[9] The Court authorized an order that the State of Michigan pay for some needed programs in Detroit which were aimed at repairing the harms inflicted by segregation in schools that would remain segregated because of the 1974 decision blocking city-suburban desegregation. Unfortunately, there was little serious follow-up by the courts on the educational remedies, and the Supreme Court severely limited such remedies in the 1995 *Missouri v. Jenkins*[10] decision.

The government turned actively against school desegregation in 1981 under the Reagan administration, with the Justice Department reversing policy on many pending cases and attacking urban desegregation orders. Congress accepted the administration's proposal to end the federal desegregation assistance program in the 1981 Omnibus Budget Reconciliation Act. Twelve years of active efforts to reverse desegregation orders and remake the federal courts followed. The Clinton administration in its first term defended some orders but developed no coherent policy and took no significant initiatives for desegregation.

By far the most important changes in policy have come from the Supreme Court. The appointment of Justice Clarence Thomas in 1991 consolidated a majority favoring cutting back civil rights remedies requiring court-ordered changes in racial patterns. In the 1991 *Board of Education of Oklahoma City v. Dowell*[11] decision, the Supreme Court ruled that a school district that had complied with its court order for several years could be allowed to return to segregated neighborhood schools if it met specific conditions. In the 1992 *Freeman v. Pitts*[12] decision, the Court made it easier to end student desegregation even when the other elements of a full desegregation order had never been accomplished. Finally, in its 1995 *Jenkins* decision, the Court's majority ruled that the court-ordered programs designed to make segregated schools more equal educationally and to increase the attractiveness of the schools to accomplish desegregation through voluntary choices were temporary and did not have to work before they could be discontinued.

In other words, desegregation was redefined from the goal of ending schools defined by race to a temporary and limited process that created no lasting rights and need not overcome the inequalities growing out of a segregated history. These decisions stimulated efforts in a number of cities to end the court orders, sometimes even over the objection of the school district involved.

RACIAL COMPOSITION OF AMERICAN SCHOOLS

As the courts were cutting back on desegregation requirements, the proportion of minority students in public schools was growing rapidly and becoming far more diverse. In the fall of 1994, American public schools enrolled more than 43 million students, of whom 66% were white, 17% African American, 13% Latino, 4% Asian, and 1% Indian and Alaskan. The proportion of Latinos in the United States was higher than that of blacks at the time desegregation began in 1954, and the proportion of whites was far lower. The two regions with the largest school enrollments, the South and the West, were 58% and 57% white, foreshadowing a near future in which large regions of the United States would have white minorities. Table 1 shows that there has been a huge growth (178%) in the number of Latino students during the 26 years since 1968, when data was first available nationally, to 1994. Meanwhile, the number of white (Anglo) students declined 9%, and the number of black students rose 14%.

TABLE 1 Public School Enrollment Changes, 1968–94 (In Millions)

	1968	1980	1994	Change 1968–94
Hispanics	2.00	3.18	5.57	+3.57 (178%)
Anglos	34.70	29.16	28.46	−6.24 (−18%)
Blacks	6.28	6.42	7.13	+0.85 (14%)

SOURCE: DBS Corp., 1982, 1987; Gary Orfield, Rosemary George, and Amy Orfield, "Racial Change in U.S. School Enrollments, 1968–1984," paper presented at National Conference on School Desegregation, University of Chicago, 1968. 1994–95 NCES Common Core of Data.

On a regional level, African Americans remained the largest minority group in the schools of all regions except the West and Alaska and Hawaii. The proportion of black students in the South was, however, about twice the proportion in the Northeast and Midwest and more than four times the level in the West. Latinos, on the other hand, made up more than a fourth of the enrollment in the West but only about a 50th in the Border region and a 25th in the Midwest.

The dramatic changes in the composition of American school enrollment is most apparent in five states that already have a majority of non-white students statewide. These include the nation's two most populous states, California and Texas, which enroll 8.8 million students and are both moving rapidly toward a Latino majority in their school systems.

NATIONAL INCREASE IN SEGREGATION

In the fall of 1972, after the Supreme Court's 1971 busing decision that led to new court orders for scores of school districts, 63.6% of black students were

TABLE 2 Percentage of U.S. Black and Latino Students in Predominantly Minority and 90–100% Minority Schools, 1968–94

	50–100% Minority		90–100% Minority	
	Blacks	Latinos	Blacks	Latinos
1968–69	76.6	54.8	64.3	23.1
1972–73	63.6	56.6	38.7	23.3
1980–81	62.9	68.1	33.2	28.8
1986–87	63.3	71.5	32.5	32.2
1991–92	66.0	73.4	33.9	34.0
1994–95	67.1	74.0	33.6	34.8

SOURCE: U.S. Department of Education Office for Civil Rights data in Orfield, *Public School Desegregation in the United States, 1968–1980,* tables 1 and 10; 1991–92 and 1994–95 NCES Common Core of Data.

TABLE 3 Percentage of White Students in Schools Attended by Typical Black or Latino Students, 1970–94

	Blacks	Latinos
1970	32.0	43.8
1980	36.2	35.5
1986	36.0	32.9
1991	34.4	31.2
1994	33.9	30.6

SOURCE: 1994–95 NCES Common Core of Data Public School Universe.

in schools with less than half white enrollment. Fourteen years later, that percentage was virtually the same, but it rose to 67.1% by 1994–95 (Table 2). Desegregation remained at its high point until about 1988 but then began to fall significantly on this measure.

A second measure of segregation, calculated as the number of students experiencing intense isolation in schools with less than one-tenth whites (i.e., 90–100% minority enrollment), shows that the proportion of black students facing extreme isolation dropped sharply with the busing decisions, declining from 64.3% in 1968 to 38.7% in 1972 and continuing to decline slightly through the mid-1990s (Table 2). This isolation increased gradually from 1988 to 1991 but actually declined slightly from 1991 to 1994. This is the only measure that does not show increased black segregation.

The third measure of desegregation used in this study, the exposure index—which calculates the percentage of white students in a school attended by typical black students—shows the level of contact almost as low as it was before the busing decisions in the early 1970s: 32%, down from its 1980 level of 36.2% (Table 3). Overall, the level of black segregation in U.S. schools is increasing slowly, continuing an historic reversal first apparent in the 1991 enrollment statistics.

. . .

TRENDS FOR LATINO STUDENTS

Latino segregation has become substantially more severe than African American segregation by each of the measures used in this study. In the Northeast, the West, and the South, more than three-fourths of all Latino students are in predominantly non-white schools, a level of isolation found for African American students only in the Northeast. . . . We have been reporting these trends continuously for two decades. They are clearly related to inferior education for Latino students.[13] Although data are limited, the surveys that have been done tend to show considerable interest in desegregated education among Latinos and substantial support for busing if there is no other way to achieve integration.[14]

All three measures of segregation reported in Tables 2 and 3 show a continuing gradual increase in segregation for Latino students nationally. The most significant change comes in the proportion of students in intensely segregated schools, which rose to 34.8% in 1994. In 1968, only 23.1% of Latino students were in these isolated and highly impoverished schools, compared to 64.3% of black students (Table 2). Now the percentage of Latino students in such schools is up by almost half and is slightly higher than the level of intense segregation for black students.

...

RACE AND POVERTY

The relationship between segregation by race and segregation by poverty in public schools across the nation is exceptionally strong. The correlation between the percentage of black and Latino enrollments and the percentage of students receiving free lunches is an extremely high .72. This means that racially segregated schools are very likely to be segregated by poverty as well.

There is strong and consistent evidence from national and state data from across the United States as well as from other nations that high poverty schools usually have much lower levels of educational performance on virtually all outcomes. This is not all caused by the school; family background is a more powerful influence. Schools with concentrations of low income children have less prepared children. Even better prepared children can be harmed academically if they are placed in a school with few other prepared students and, in some cases, in a social setting where academic achievement is not supported.

School achievement scores in many states and in the nation show a very strong relation between poverty concentrations and low achievement.[15] This is because high poverty schools are unequal in many ways that affect educational outcomes. The students' parents are far less educated—a very powerful influence—and the child is much more likely to be living in a single parent home that is struggling with multiple problems. Children are much more likely to have serious developmental and untreated health problems. Children move much more often, often in the middle of a school year, losing continuity and denying schools sufficient time to make an impact on their learning.

High poverty schools have to devote far more time and resource to family and health crises, security, children who come to school not speaking standard English, seriously disturbed children, children with no educational materials in their homes, and many children with very weak educational preparation. These schools tend to draw less qualified teachers and to hold them for shorter periods of time. They tend to have to invest much more heavily in remediation and much less adequately in advanced and gifted classes and demanding materials. The levels of competition and peer group support for educational achievement are much lower in high poverty schools. Such schools are viewed much more negatively in the community and by the schools and colleges at the next level of education as well as by potential employers. In those states that have implemented high stakes testing, which denies graduation or flunks students, the high poverty schools tend to have by far the highest rates of sanctions.[16]

None of this means that the relationship between poverty and educational achievement is inexorable, or that there are not exceptions. Many districts have one or a handful of high poverty schools that perform well above the normal pattern. Students from the same family background may perform at very different levels of achievement, and there are some highly successful students and teachers in virtually every school. The overall relationships, however, are very powerful. Students attending high poverty schools face a much lower level of competition regardless of their own interests and abilities.

This problem is intimately related to racial segregation . . . [o]f the schools in the United States, 60.7% have less than one-fifth black and Latino students, while 9.2% . . . have 80–100% black and Latino students. At the extremes, only 5.4% of the schools with 0–10% black and Latino students have more than half low income students; 70.5% . . . of them have less than one-fourth poor students. Among schools that are 90–100% black and/or Latino, on the other hand, almost nine-tenths (87.7%) are predominantly poor, and only about 3% . . . have less than one-fourth poor children. A student in a segregated minority school is 16.3 times more likely to be in a concentrated poverty school than a student in a segregated white school.[17]

WHERE IS SEGREGATION CONCENTRATED? THE LONG-TERM EFFECTS OF THE SUPREME COURT'S DECISION AGAINST SUBURBAN DESEGREGATION

Blacks living in rural areas and in small and medium-sized towns or the suburbs of small metropolitan areas are far more likely to experience substantial school desegregation than those living in the nation's large cities; they attend schools with an average of about 50% white students.... In contrast, blacks in large cities attend schools that have an average of only 17% white students and those in smaller cities attend schools with an average of 38% white students. Black students in the suburbs of large cities attend schools with an average of 41% white students.... Considering the small proportion of minority students in many suburban rings, this level of segregation is a poor omen for the future of suburbs that will become more diverse.

The nation's non-white population is extremely concentrated in metropolitan areas. Outside the South, this concentration tends to be in the largest metropolitan areas with the largest ghettos and barrios. Many of the small cities and towns in Illinois and Michigan, for example, have few African American students, and the vast majority of the nation's white students live in suburbs divided into scores of separate school districts, all laid over extremely segregated metropolitan housing markets. This means that the central city school districts become extremely isolated by race and poverty and are critical only for non-white students. Since the minority communities are constantly expanding along their boundaries, and virtually all-white developments are continuously being constructed on the outer periphery of suburbia, the central cities have a continual increase in their proportion of black and Latino students.

The suburbs are now the dominant element of our society and our politics. As the nation's population changes dramatically in the coming decades, suburbs are destined to become much more diverse. What kind of access black and Latino children will have to mainstream suburban society will be affected by the racial characteristics of suburban schools. It raises serious concerns to realize that by 1994, blacks were in schools that averaged only 41% white students, and Latinos were in schools that averaged just 36% white students, in the suburbs of the largest cities. Whites in those suburban rings were in schools with an average of only 14%... combined black and Latino enrollments. Latino students, but not blacks, were almost as segregated in the suburbs of smaller metropolitan areas.... If these patterns intensify as the suburban African American and Latino population grows, we may be facing problems that are as serious as those that led to desegregation conflicts in many cities.

It would be profoundly ironic if the Supreme Court decision that meant to protect suburban boundary lines (*Milliken v. Bradley*) ended up making it impossible for suburban communities in the path of racial change to avoid rapid resegregation. Individual suburban school districts are often so small that they can go through racial change much more rapidly and irreversibly than a huge city. A suburb will often have only the enrollment of a single high school attendance area in a city and has little hope of stabilizing its enrollment, once a major racial change begins, without drawing on students from a broader geographic area. This means that in areas with many fragmented school districts, not only the city but also substantial portions of suburban rings may face high levels of segregation. Since non-white suburbanization began in earnest in the 1970s, the cities also have been losing many of their minority middle-class families, leaving the cities with an escalating concentration of poverty.

... These districts contain 18% black students, 23% Latino students, 13% Asian students, but only 2% of the whites. About a fifth of black and Latino students depend on districts that do not matter to 98% of white families. Most of these systems have faced recurrent fiscal and political crises for years and have low levels of educational achievement. Desegregation has become virtually impossible in some of these systems since the *Milliken* decision. The trends of metropolitan racial change since World War II suggest that segregation will become worse in the future.

Consequences of Smaller Districts

Different parts of the country traditionally have very different patterns of organizing school districts, depending in part on local traditions and, in part, on whether or not the districts were organized back when the horse and buggy meant that units of local government had to be very small. In much of the South, counties have traditionally been more important and municipalities less important. In New England the towns existed long before they became part of suburban rings in large metropolitan areas. School districts in the South were often countywide; in the Northeast and the older parts of the Midwest, they were often defined by the structure of local town government set generations in the past. When the Supreme Court decided in 1974 to make it very difficult to desegregate across school district boundary lines, it virtually guaranteed that the regions with large districts would be far less segregated than those with small districts.

The nation's largest school districts are in Maryland, Florida, Louisiana, North Carolina, West Virginia, and Delaware, all states in the Southern and Border regions. Illinois, New York, New Jersey, and Michigan, consistently the most segregated states for black students, have much smaller districts as do Texas and California, where most Latino students are concentrated. There has been no absolute relationship between district size and segregation, of course, because of the widely varying proportions of black and Latino students within various states and because of different types of desegregation plans in place in different areas. Nonetheless, the most segregated states tend to be fragmented into a great many small districts, and the most integrated states tend to have very different patterns, although there are significant exceptions to this pattern. The Maryland and Louisiana statistics show, for example, that it is possible to have very large districts with very high levels of segregation, while the Indiana and Ohio statistics show that small districts in states with multiple desegregation orders and a small black population are compatible with a lower level of intense segregation.

No state with small districts and a substantial African American population has come anywhere near the level of desegregation achieved in the most successful states with large systems. North Carolina and Tennessee, which already have relatively large districts, have seen city and suburban districts consolidated in recent years to create more countywide systems.

. . .

Segregation of Whites at the State Level

In a nation where whites are destined to become one of several minorities in the schools if the existing trends continue, it is important not only to consider the isolation of non-white students from whites but also the isolation of whites from the growing parts of the population.

Except in the historic *de jure* states for blacks and the states taken from Mexico in the war in the 1840s, most white students have not yet experienced substantial desegregation. Although they are growing up in a society where the U.S. Census Bureau predicts that more than half of schoolage children will be non-white in a third of a century, many are being educated in overwhelmingly white schools with little contact with black or Latino students.[18] Those students may be ill prepared as the American workforce changes and skills in race relations become increasingly valuable in many jobs.

. . .

CONCLUSION

In American race relations, the bridge from the twentieth century may be leading back into the nineteenth century. We may be deciding to bet the future of the country once more on "separate but equal." There is no evidence that separate but equal today works any more than it did a century ago.

The debate that has been stimulated by recent Supreme Court decisions is a debate about how and when to end desegregation plans. The most basic need now is for a serious national examination of the cost of resegregation and the alternative solutions to problems with existing desegregation plans. Very few Americans prefer segregation, and most believe that desegregation has had considerable value, but most whites are still opposed to plans that involve mandatory transportation of students. During the last 15 years, plans have been evolving to include more educational reforms and choice mechanisms to try to achieve desegregation and educational gains simultaneously. A stronger fair housing law, a number of settlements of housing segregation cases, and federal initiatives to change the operation of subsidized housing—as well as the very rapid creation of brand new communities in the sunbelt—all offer opportunities to try to change the pattern of segregated housing that underlies school segregation. Policies that would help move the country toward a less polarized society include:

1. Resumption of serious enforcement of desegregation by the Justice Department and serious investigation of the degree to which districts have complied with all Supreme Court requirements by the Department of Education. Such requirements could be appropriately specified in a federal regulation.
2. Creation of a new federal education program to train students, teachers, and administrators in human relations, conflict resolution, and multi-ethnic education techniques and to help districts devise appropriate plans and curricula for successful multiracial schools.
3. Serious federal research on multiracial schools and the comparative success of segregated and desegregated schools.
4. A major campaign to increase non-white teachers and administrators through a combination of employment discrimination enforcement and resources for recruitment and education of potential teachers.
5. Incorporation of successful desegregation into the national educational goals.
6. Federal and state efforts to expand the use of integrated two-way bilingual programs from the demonstration stage to a major technique for improving second language acquisition for both English speakers and other language speakers and for building successful ethnic relationships.
7. Additional Title IV resources to expand state education department staffs working on desegregation and racial equity in the schools.
8. Federal, state, and local plans to coordinate housing policy with school desegregation policy.
9. Examination of choice and charter school plans to assure that they are not increasing segregation and to reinforce their potential contribution to desegregation.
10. Examination of high stakes state testing programs to ensure that they are not punishing the minority students who must attend inferior segregated school under existing state and local policies.

NOTES

1. Distribution of Latinos by ethnicity and state is reported in M. Beatriz Arias, "The Context of Education for Hispanic Students: An Overview," *American Journal of Education 95*(1) (November 1986): 26–57.
2. In this report "white" means non-Hispanic whites. Hispanic or Latino is treated as part of the non-white population although many Latinos define themselves as whites in racial terms. These definitions are used to avoid the awkward and confusing language that would otherwise be necessary and is not an attempt to define Latinos as a race.
3. U.S. Bureau of Census Projections, and Steven A. Holmes, "Census Sees a Profound Ethnic Shift in U.S.," *New York Times*, March 14, 1996; *Education Week*, March 27, 1996, p. 3.
4. Gary Orfield, "Public Opinion and School Desegregation," *Teachers College Record 96*(4) (Summer 1995); Gallup Poll in *USA Today*, May 12, 1994; Gallup Poll in *Phi Delta Kappan*, September 1996. The

1996 survey reported that "the percentages who say integration has improved the quality of education for blacks and for whites have been increasing steadily since these questions were first asked in 1971" (p. 48). The report also showed that 83% of the public believed that interracial schools were desirable.

5. 402 U.S. 1 (1971).
6. *Keyes v. Denver School District No. 1*, 413 U.S. 189 (1973).
7. 418 U.S. 717 (1974).
8. Calculations of metropolitan segregation from 1992 Common Core data.
9. 433 U.S. 267 (1977).
10. 115 S.Ct. 2038 (1995).
11. 498 U.S. 237 (1991).
12. 503 U.S. 467 (1992).
13. Ruben Espinosa and Alberto Ochoa, "Concentration of California Hispanic Students in Schools with Low Achievement: A Research Note," *American Journal of Education 95*(1), (1986): 77–95; Ruben Donato, Martha Menchaca, and Richard R. Valencia, "Segregation, Desegregation, and the Integration of Chicano Students: Problems and Prospects" in Richard R. Valencia, ed., *Chicano School Failure and Success: Research and Policy Agendas for the 1990s* (London: Falmer Press, 1991): 27–63.
14. A national survey by the *Boston Globe* in 1992 reported that 82% of Latinos said that they favored busing if there was no other way to achieve integration and that Latinos expressing an opinion said that they would be "willing to have your own children go to school by bus so the schools would be integrated" ("Poll Shows Wide Support Across U.S. for Integration," *Boston Globe*, January 5, 1992: 15); for evidence of more closely divided opinion earlier see Gary Orfield, "Hispanic Education: Challenges, Research, and Policies," *American Journal of Education 95*(1)(1986): 11-12.
15. See, for example, discussion of the relationship of disadvantaged school status and educational opportunity in Jeannie Oakes, *Multiplying Inequalities: The Effects of Race, Social Class, and Tracking on Opportunities to Learn Mathematics and Science* (Santa Monica: RAND, 1990), figure 2.3.; 1988 NAEP data reported in Educational Testing Service, *The State of Inequality*; Peter Scheirer, "Poverty not Bureaucracy," Working Paper, Metropolitan Opportunity Project, Univ. of Chicago, 1989; Samuel S. Peng, Margaret C. Wang, and Herbert J. Walberg, "Demographic Disparities of Inner-City Eighth Graders," *Urban Education 26*(4) (January 1992): 441–459; David M. Cutler and Edward L. Glaeser, "Are Ghettos Good or Bad?" (Cambridge: National Bureau of Economic Research, 1995); Gary G. Wehlage, "Social Capital and the Rebuilding of Communities," Center on Organization and Restructuring of Schools, Issue Report No. 5, Fall 1993: 3–5; Raymond Hernandez, "New York City Students Lagging in Mastery of Reading and Math," *New York Times*, January 3, 1997: Al, B4; a 1997 study in the suburban Washington district of Prince George's County reported that "with each 10 percent increase in the number of students who qualified for free or reduced-price lunches, a school's score on the state test dropped by an average of 1.8 percentage points in reading, 4.1 percentage points in math and 2.2 percentage points in science" (*Washington Post*, May 15, 1997: A17).
16. A study of 1992 test scores in greater Cleveland showed that district poverty level differences "explained as much as 39 percent of the differences in school district passing rates" on the state proficiency test (*Cleveland Plain Dealer*, Oct. 8, 1995: 3–C).
17. Asian students, who are having far greater success in U.S. schools, attend schools that average only about one-fourth African American and Latino students and they are far less likely to be in high poverty schools, factors that help account for their mobility (*Asian Students and Multiethnic Desegregation*, Harvard Project on School Desegregation, October 1994).
18. A recent Gallup Poll showed that 93% of the public and 96% of school parents thought it was very important to teach students "acceptance of people of different races and ethnic backgrounds" (*Phi Delta Kappan*, October 1993: 139, 145).

44

A Wider Lens on the Black-White Achievement Gap

RICHARD ROTHSTEIN

The achievement gap between white, middle-class students and minority, lower-class students is often thought to be one of the biggest problems schools face. Rothstein argues that relying solely on school reforms and ignoring social class characteristics will not result in improved schools. He points out areas for concern and policy considerations.

As you read the selection, consider the following questions:

1. *What are the variables cited by Rothstein that make a difference in school achievement?*
2. *Why do reformers need to focus on more than just schools to reduce the achievement gap?*
3. *Does this reading present a middle-class bias? If so, is that a problem?*
4. *Based on your reading of this selection, what would you do to reduce the achievement gap?*

GLOSSARY **Achievement gap** Difference between achievement test scores (or other available measures) of two or more groupings of students. **Noncognitive skills** Character traits such as perseverance, self-confidence, self-discipline, punctuality, communication skills, social responsibility, ability to resolve conflicts, etc.

The 50th anniversary of the Supreme Court's school desegregation order in *Brown* v. *Board of Education* has intensified public awareness of the persistent gap in academic achievement between black students and white students. The black-white gap is made up partly of the difference between the achievement of all lower-class students and that of middle-class students, but there is an additional gap between black students and white students—even when the blacks and whites come from families with similar incomes.

The American public and its political leaders, along with professional educators, have frequently vowed to close these gaps. Americans believe in the ideal of equal opportunity, and they also believe that the best way to ensure that opportunity is to enable all children, regardless of their parents' stations, to leave school with skills that position them to compete fairly and productively in the nation's democratic governance and occupational structure. The fact that children's skills can so clearly be predicted by their race and family economic status is a direct challenge to our democratic ideals.

Policy makers almost universally conclude that these existing and persistent achievement gaps must be the result of wrongly designed school policies—either expectations that are too low, teachers who are insufficiently qualified, curricula that are badly

SOURCE: Originally printed in *Phi Delta Kappan*, Vol. 86, Issue 2 (October 2004), pp. 105–110.

designed, classes that are too large, school climates that are too undisciplined, leadership that is too unfocused, or a combination of these factors.

Americans have come to the conclusion that the achievement gap is the fault of "failing schools" because common sense seems to dictate that it could not be otherwise. After all, how much money a family has or the color of a child's skin should not influence how well that child learns to read. If teachers know how to teach reading—or math or any other subject—and if schools emphasize the importance of these tasks and permit no distractions, children should be able to learn these subjects, whatever their family income or skin color.

This commonsense perspective, however, is misleading and dangerous. It ignores how social-class characteristics in a stratified society such as ours may actually influence learning in school. It confuses social class, a concept that Americans have historically been loath to consider, with two of its characteristics: income and, in the U.S., race. For it is true that low income and skin color themselves don't influence academic achievement, but the collection of characteristics that define social-class differences inevitably influences that achievement.

SOCIAL CLASS AND ITS IMPACT ON LEARNING

Distinctly different child-rearing patterns are one mechanism through which class differences affect the academic performance of children. For example, parents of different social classes often have different ways of disciplining their children, different ways of communicating expectations, and even different ways of reading to their children. These differences do not express themselves consistently or in the case of every family; rather, they influence the average tendencies of families from different social classes.

That there would be personality and child-rearing differences, on average, between families in different social classes makes sense when you think about it. If upper-middle-class parents have jobs in which they are expected to collaborate with fellow employees, create new solutions to problems, or wonder how to improve their contributions, they are more likely to talk to their children in ways that differ from those of lower-class parents whose own jobs simply require them to follow instructions without question. Children who are reared by parents who are professionals will, on average, have more inquisitive attitudes toward the material presented by their teachers than will children who are reared by working-class parents. As a result, no matter how competent the teacher, the academic achievement of lower-class children will, on average, almost inevitably be less than that of middle-class children. The probability of such reduced achievement increases as the characteristics of lower-social-class families accumulate.

Many social and economic manifestations of social class also have important implications for learning. Health differences are among them. On average, lower-class children have poorer vision than middle-class children, partly because of prenatal conditions and partly because of how their eyes are trained as infants. They have poorer oral hygiene, more lead poisoning, more asthma, poorer nutrition, less adequate pediatric care, more exposure to smoke, and a host of other problems. Each of these well-documented social-class differences is likely to have a palpable effect on academic achievement, and the combined influence of all of these differences is probably huge.

The growing unaffordability of adequate housing for low-income families is another social-class characteristic that has a demonstrable effect on average achievement. Children whose families have difficulty finding stable housing are more likely to be mobile, and student mobility is an important cause of low student achievement. Urban rents have risen faster than working-class incomes. Even families in which parents' employment is stable are more likely to move when they fall behind in rent payments. In some schools in minority neighborhoods, this need to move has boosted mobility rates to more than 100%: for every seat in the school, more than two children were enrolled at some time during the year.[1] It is hard to imagine how teachers, no matter how well trained, could be as effective for children who move in and out of their classrooms as they can be for children whose attendance is regular.

Differences in wealth between parents of different social classes are also likely to be important determinants of student achievement, but these differences are usually overlooked because most analysts focus only on annual income to indicate disadvantage. This practice makes it hard to understand, for example, why black students, on average, score lower than white students whose family incomes are the same. It is easier to understand this pattern when we recognize that children can have similar family *incomes* but be ranked differently in the social-class structure, even in economic terms. Black families with low income in any particular year are likely to have been poor for longer than white families with similar income in that year. White families are also likely to own far more assets that support their children's achievement than are black families at the same level of current income.

I use the term "lower class" here to describe the families of children whose achievement will, on average, be predictably lower than the achievement of middle-class children. American sociologists were once comfortable with this term, but it has fallen out of fashion. Instead, we tend to use such euphemisms as "disadvantaged" students, "at-risk" students, "inner-city" students, or students of "low socioeconomic status." None of these terms, however, can capture the central characteristic of lower-class families: a collection of occupational, psychological, personality, health, and economic traits that interact, predicting performance—not only in schools but in other institutions as well—that, on average, differs from the performance of families from higher social classes.

Much of the difference between the average performance of black children and that of white children can probably be traced to differences in their social-class characteristics. But there are also cultural characteristics that are likely to contribute a bit to the black-white achievement gap. These cultural characteristics may have identifiable origins in social and economic conditions—for example, black students may value education less than white students because a discriminatory labor market has not historically rewarded black workers for their education—but values can persist independently and outlast the economic circumstances that gave rise to them.

Some lower-class children do achieve at high levels, and many observers have falsely concluded from this that therefore all lower-class children should be able to succeed with appropriate instruction. One of the bars to our understanding of the achievement gap is that most Americans, even well-educated ones, are not expert in discussions of statistical distributions. The achievement gap is a phenomenon of averages, a difference between the average achievement level of lower-class children and the average achievement level of middle-class children. In human affairs, every average characteristic is a composite of many widely disparate characteristics.

For example, we know that lead poisoning has a demonstrable impact on young children's I.Q. scores. Children with high exposure to lead—from fumes or from ingesting paint or dust—have I.Q. scores that, on average, are several points lower than those of children who are not so exposed. But this does not mean that every child with lead poisoning has a lower I.Q. Some children with high lead levels in their blood have higher I.Q. scores than typical children with no lead exposure. When researchers say that lead poisoning seems to affect academic performance, they do not mean that every lead-exposed child performs less well. But the high performance of a few lead-exposed children does not disprove the conclusion that lead exposure is likely to harm academic achievement.

This kind of reasoning applies to each of the social-class characteristics that I discuss here, as well as to the many others that, for lack of space or my own ignorance, I do not discuss. In each case, class differences in social or economic circumstances probably cause differences in the average academic performance of children from different social classes, but, in each case, some children with lower-class characteristics perform better than typical middle-class children.

SCHOOL REFORMS ALONE ARE NOT ENOUGH

The influence of social-class characteristics is probably so powerful that schools cannot overcome it, no matter how well trained their teachers and no matter how well designed their instructional programs and climates. But saying that a social-class achievement gap should be expected is not to make a logical statement. The fact that social-class differences are associated with, and probably cause, a big gap in academic performance does not mean that, in theory, excellent schools could not offset these differences. Indeed, today's policy makers and educators make many claims that higher standards, better teachers, more accountability, better discipline, or other effective practices can close the achievement gap.

The most prominent of these claims has been made by the Heritage Foundation (conservative) and the Education Trust (more liberal), by economists and statisticians who claim to have shown that better teachers do in fact close the gap, by prominent educators, and by social critics. Many (though not all) of the instructional practices promoted by these commentators are well designed, and these practices probably do succeed in delivering a better education to some lower-class children. But a careful examination of each claim that a particular school or practice has closed the race or social-class achievement gap shows that the claim is unfounded.

In some cases, a claim may fail because it reflects a statistical fluke—a school successful for only one year, in only one subject, or in only one grade—or because it reports success only on tests of the most basic skills. In other cases, a claim may fail because the successful schools identified have selective student bodies. Remember that the achievement gap is a phenomenon of averages—it compares the average achievement of lower- and middle-class students. In both social classes, some students perform well above or below the average performance of their social-class peers. If schools can select (or attract) a disproportionate share of lower-class students whose performance is above average for their social class, those schools can appear to be quite successful. Many such schools are excellent and should be commended. But their successes provide no evidence that their instructional approaches would close the achievement gap for students who are average for their social-class groups.

LIMITATIONS OF THE CURRENT TESTING REGIME

Whether efforts to close the social-class achievement gap involve in-school reforms or socioeconomic reforms, it is difficult to know precisely how much any intervention will narrow the gap. We can't estimate the effect of various policies partly because we don't really know how big the achievement gap is overall or how big it is in particular schools or school systems.

This lack of knowledge about the size of the gap or the merits of any particular intervention might surprise many readers because so much attention is devoted these days to standardized test scores. It has been widely reported that, on average, if white students score at around the 50th percentile on a standardized math or reading test, black students typically score around the 23rd percentile. (In more technical statistical terms, black students score, on average, between 0.5 and 1.0 standard deviations below white students.)

But contrary to conventional belief, this may not be a good measure of the gap. Because of the high stakes attached to standardized tests in recent years, schools and teachers are under enormous pressure to raise students' test scores. The more pressure there has been, the less reliable these scores have become. In part, the tests themselves don't really measure the gap in the achievement of high standards because high standards (such as the production of good writing and the development of research skills and analysis) are expensive to test, and public officials are reluctant to spend the money. Instead, schools have tended to use inexpensive standardized tests that mostly, though not entirely, assess more basic skills. Gaps that show up on tests of basic skills may be quite different from the gaps that would show up on tests of higher standards of learning. And it is not the case that students acquire a hierarchy of skills sequentially. Thus truly narrowing the achievement gap would not require children to learn "the basics" first. Lower-class children cannot produce typical middle-class academic achievement unless they learn basic and more advanced skills simultaneously, with each reinforcing the other. This is, in fact, how middle-class children who come to school ready to learn acquire both basic and advanced skills.

The high stakes recently attached to standardized tests have given teachers incentives to revise the priorities of their instruction, especially for lower-class children, so that they devote greater time to drill on basic skills and less time to other, equally important (but untested) learning areas in which achievement gaps also appear. In a drive to raise test scores in math and reading, the curriculum has moved away not only from more advanced mathematical and literary skills, but also from social studies, literature, art, music, physical education, and other important subjects that are not tested for the purpose of judging school quality. We don't know how large the race or social-class achievement gaps are in these subjects, but there is no reason to believe that gaps in one domain are the same as the gaps in others or that the relationships between gaps in different domains will remain consistent at different ages and on different tests.

For example, educational researchers normally expect that gaps in reading will be greater than gaps in math, probably because social-class differences in parental support play a bigger role for reading than for math. Parents typically read to their very young children, and middle-class parents do so more and in more intellectually stimulating ways, but few parents do math problems with their young children. Yet, on at least one test of entering kindergartners, race and social-class gaps in math exceed those in reading.

THE IMPORTANCE OF NONCOGNITIVE SKILLS

We also don't know the extent of the social-class gaps in noncognitive skills—such character traits as perseverance, self-confidence, self-discipline, punctuality, the ability to communicate, social responsibility, and the ability to work with others and resolve conflicts. These are important goals of public education. In some respects, they may be more important than academic outcomes.

Employers, for example, consistently report that workers have more serious shortcomings in these noncognitive areas than in academic areas. Econometric studies show that noncognitive skills are a stronger predictor of future earnings than are test scores. In public opinion surveys, Americans consistently say they want schools to produce good citizens and socially responsible adults first and high academic proficiency second. Yet we do a poor job—actually, no job at all—of assessing whether schools are generating such noncognitive outcomes. And so we also do a poor job of assessing whether schools are successfully narrowing the social-class gap in these traits or whether social and economic reform here, too, would be necessary to narrow these gaps.

There is some evidence that the noncognitive social-class gaps should be a cause for concern. For very young children, measures of antisocial behavior mirror the gaps in academic test scores. Children of lower social classes exhibit more antisocial behavior than children of higher social classes, both in early childhood and in adolescence. It would be reasonable to expect that the same social and economic inequalities that seem likely to produce gaps in academic test scores also produce differences in noncognitive traits.

In some areas, however, it seems that noncognitive gaps may be smaller than cognitive ones. In particular, analyses of some affirmative action programs in higher education find that, when minority students with lower test scores than white students are admitted to colleges, the lower-scoring minority students may exhibit more leadership, devote more serious attention to their studies, and go on to make greater community contributions. This evidence reinforces the importance of measuring noncognitive student characteristics, something that few elementary or secondary schools attempt. Until we begin to measure these traits, we will have no insight into the extent of the noncognitive gaps between lower- and middle-class children.

MOVING FORWARD

Three tracks should be pursued vigorously and simultaneously if we are to make significant progress in narrowing the achievement gap. The first track is school improvement efforts that raise the quality of instruction in elementary and secondary schools. The second track is expanding the definition of schooling to include crucial out-of-school hours in which families and communities now are the sole influences. This means implementing comprehensive early childhood, after-school, and summer programs. And the third track is social and economic policies that will enable children to attend school more equally ready to learn. These policies include health services for lower-class children and their families, stable housing for working families with children, and the narrowing of growing income inequalities in American society.

Many of the reforms in curriculum and school organization that are promoted by critics of education have merit and should be intensified. Repairing and upgrading the scandalously decrepit school facilities that serve some lower-class children, raising salaries to permit the recruitment of more qualified teachers for lower-class children, reducing class sizes for lower-class children (particularly in the early grades), insisting on higher academic standards that emphasize creativity and reasoning as well as basic skills, holding schools accountable for fairly measured performance, having a well-focused and disciplined school climate, doing more to encourage lower-class children to intensify their own ambitions—all of these policies and others can play a role in narrowing the achievement gap. These reforms are extensively covered in a wide range of books, articles, and public discussions of education, so I do not dwell on them here. Instead, my focus is the greater importance of reforming social and economic institutions if we truly want children to emerge from school with equal preparation.

Readers should not misinterpret this emphasis as implying that better schools are not important or that school improvement will not make a contribution to narrowing the achievement gap. Better school practices can no doubt narrow the gap. However, school reform is not enough.

In seeking to close the achievement gap for low-income and minority students, policy makers focus inordinate attention on the improvement of instruction because they apparently believe that social-class differences are immutable and that only schools can improve the destinies of lower-class children. This is a peculiarly American belief—that schools can be virtually the only instrument of social reform—but it is not based on evidence about the relative effectiveness of economic, social, and educational improvement efforts.

While many social-class characteristics are impervious to short-term change, many can easily be affected by public policies that narrow the social and economic gaps between lower- and middle-class children. These policies can probably have a more powerful impact on student achievement (and, in some cases, at less cost) than an exclusive focus on school reform. But we cannot say so for sure, because social scientists and educators have devoted no effort to studying the relative costs and benefits of nonschool and school reforms. For example, establishing an optometric clinic in a school to improve the vision of low-income children could have a bigger impact on their test scores than spending the same money on instructional improvement.[2] Greater proportions of low-income than middle-class children are distracted by the discomfort of untreated dental cavities, and dental clinics can likewise be provided at costs comparable to what schools typically spend on less effective reforms. We can't be certain if this is the case, however, because there have been no experiments to test the relative benefits of these alternative strategies. Of course, proposals to improve all facets of the health of lower-class children, not just their vision and oral health, should be evaluated for their academic impacts.

A full array of health services will be costly, but that cost cannot be avoided if we are truly to embrace the goal of raising the achievement of lower-class children. Some of these costs are not new, of course, and some can be recouped by school clinics by means of reimbursements from other underutilized government programs, such as Medicaid.

Other social reforms—for example, an increase in the number of Section 8 housing vouchers to increase the access of lower-class families to stable housing—also could have a significant educational impact.

Incomes have become more unequally distributed in the United States in the last generation, and this inequality contributes to the academic achievement gap. Proposals for a higher minimum wage or increases in earned income tax credits, which are designed to help offset some of this inequality, should be considered education policies as well as economic ones, for they would be likely to result in higher academic performance from children whose families are more secure.

Although conventional opinion is that "failing" schools contribute mightily to the achievement gap, the evidence indicates that schools already do a great deal to combat it. Most of the social-class difference in average academic potential exists by the time children are 3 years old. This difference is exacerbated over the years that children spend in school, but during these years, the growth in the gap occurs mostly in the after-school hours and during the summertime, when children are not actually in classrooms.[3]

So in addition to school improvement and broader reforms to narrow the social and economic inequalities that produce gaps in student achievement, investments should be made to expand the definition of schooling to cover those crucial out-of-school hours. Because the gap is already huge at 3 years old, the most important focus of this investment should probably be early childhood programs. The quality of these programs is as important as the existence of the programs themselves. To narrow the gap, early childhood care, beginning with infants and toddlers, should be provided by adults who can offer the kind of intellectual environment that is typically experienced by middle-class infants and toddlers. This goal probably requires professional care givers and low child/adult ratios.

Providing after-school and summer experiences to lower-class children that are similar to those middle-class children take for granted would be likely to play an essential part in narrowing the achievement gap. But these experiences should not be restricted only to remedial programs in which lower-class children get added drill in math and reading. Certainly, remedial instruction should be part of an adequate after-school and summer program—but only a part. The advantage that middle-class children gain after school and in the summer probably comes mostly from the self-confidence they acquire and the awareness they develop of the world outside their homes and immediate communities and from organized athletics, dance, drama, museum visits, recreational reading, and other activities that develop their inquisitiveness, creativity, self-discipline, and organizational skills. After-school and summer programs can be expected to have a chance of narrowing the achievement gap only by attempting to duplicate such experiences.

For nearly half a century, the association of social and economic disadvantage with a student achievement gap has been well known to economists, sociologists, and educators. However, most have avoided the obvious implications of this understanding: raising the achievement of lower-class children requires the amelioration of the social and economic conditions of their lives, not just school reform. Perhaps we are now ready to reconsider this needlessly neglected opportunity.

NOTES

1. David Kerbow, "Patterns of Urban Student Mobility and Local School Reform," *Journal of Education for Students Placed at Risk*, vol. 12, 1996, pp. 147–69; and James Bruno and Jo Ann Isken, "Inter- and Intraschool Site Student Transiency: Practical and Theoretical Implications for Instructional Continuity at Inner-City Schools," *Journal of Research and Development in Education*, vol. 29, 1996, pp. 239–52.
2. Paul Harris, "Learning-Related Visual Problems in Baltimore City: A Long-Term Program," *Journal of Optometric Vision Development*, vol. 33, 2002, pp. 75–115; and Marge Christensen Gould and Herman Gould, O.D., "A Clear Vision for Equity and Opportunity," *Phi Delta Kappan*, December 2003, pp. 324–29.
3. See Meredith Phillips, "Understanding Ethnic Differences in Academic Achievement: Empirical Lessons from National Data," in David W. Grissmer and J. Michael Ross, eds., *Analytic Issues in the Assessment of Student Achievement* (Washington, D.C.: U.S. Department of Education, NCES 2000-050, 2000), pp. 103–32, available at http://nces.ed.gov/pubs2000/2000osoa.pdf; Richard L. Allington and Anne McGill-Franzen, "The Impact of Summer Setback on the Reading Achievement Gap," *Phi Delta Kappan*, September 2003, pp. 68–75; and Doris Entwisle and Karl L. Alexander, "Summer Setback: Race, Poverty, School Composition, and Mathematics Achievement in the First Two Years of School," *American Sociological Review*, February 1992, pp. 72–84.

45

Fraternity Hazing: Insights from the Symbolic Interactionist Perspective

STEPHEN SWEET

Finally, we consider higher education and the student role. Applying the symbolic interaction theoretical approach to student experiences, Sweet considers an issue that affects learning on campus and the socialization of students: fraternity hazing.

Using the symbolic interactionist approach, Sweet asks some penetrating questions about why hazing rituals take place, and why young men and women seem willing to submit to horrendous acts. Sweet considers the entire rite of passage—hazing, the phases through which pledges are initiated, how participants define the situation and their social selves, the looking-glass process, role taking, and finally strategies to counter dangerous hazing by applying sociological knowledge.

As you read the selection, consider the following questions:

1. *What purpose does hazing serve for the individual and the group?*
2. *Why do pledges or recruits submit to abuse? Why is it difficult for them to object?*
3. *How does the symbolic interaction perspective help understand hazing?*
4. *What can or should be done to reduce dangerous hazing situations?*

GLOSSARY **Hazing** Ritual to initiate new members to an organization or club. **Rite of passage** Ritual marking the symbolic transition from one social position to another. **Symbolic interaction perspective** A major theoretical approach in sociology that focuses on microlevel interactions. **Definitions of situations** How individuals perceive situations based on their background experiences and social cues. **Looking-glass process** Self as a product of social interactions with others as we see ourselves reflected in responses others make toward us.

Less than a mile from my former office sits the Theta Chi fraternity house. On February 10, 1997, seventeen-year-old Clarkson University freshman. Binaya "Bini" Oja, along with twenty other students, began pledging Theta Chi fraternity. As part of their initiation, the pledges gathered in a semicircle around a bucket and were instructed by fraternity members to take turns drinking hard liquor. If the pledges did not drink the liquor fast enough so that bubbles were seen rising in the bottles, they were also instructed to guzzle a full glass of beer. The point of the game was simple—each pledge was expected to drink until he vomited.

Bini drank a lot and was carried upstairs. The next morning he was discovered dead with his feet

SOURCE: From *College and Society: An Introduction to the Sociological Imagination* by Stephen Sweet. Boston: Allyn & Bacon, 2001, pp. 19–36.

up on a couch and his face on the floor next to a garbage can. An autopsy determined that he died as a consequence of inhaling his own vomit. Six months later, on the same day that the Clarkson University task force on fraternities announced its programs to counter hazing, a similar incident happened at Louisiana State University. Benjamin Wynne, a pledge at Sigma Alpha Epsilon fraternity, died during a fraternity initiation ritual. His autopsy revealed a blood alcohol content of .588, a level six times that of the legal limit. . . .

Especially common in fraternity hazing are situations in which pledges are pressured to drink shots of alcohol in rapid succession. In other circumstances pledges are blindfolded and dropped off in remote areas, sometimes during the middle of winter, sometimes naked. Pledges have been compelled to perform raids on sorority houses, stealing kisses, groping, and ripping clothes. Some fraternities require pledges to perform rigorous calisthenics, sometimes in heavy clothing (Nuwer 1990; Sanday 1990). In the context of discussions with current and former members of Greek organizations, I learned of toes being broken by hammers in games of fear, beatings with paddles, sleep deprivation, submersions in vats of filth, and drinking "games."

It is difficult to provide an accurate assessment of the frequency of hazing. One literature review found over 400 documented hazing incidents resulting in serious injury and death from 1900 to 1990 (Nuwer 1990). Certainly this underestimates how often it occurs. A survey of 283 fraternity advisors revealed that over half of these advisors believed that hazing existed in some of their groups (Shaw and Morgan 1990). Hazing is not limited to college campuses, of course. A report to the Senate Committee on Armed Services found that hazing of "plebes," the incoming recruits to military academies, was also very common (U.S. Government Accounting Office 1992). There are a variety of reasons why hazing is underreported. Fraternities are secretive organizations and pledges are required to keep oaths of silence concerning initiation procedures (Leemon 1972). Students often do not recognize when they are being hazed or abused (Moffatt 1989). Colleges and universities sometimes avoid publicizing hazing incidents for fear of damaging institutional reputations or incurring financial liability to victims (Curry 1989; Nuwer 1990).

The concern of this chapter is not to indict fraternities. In fact, these organizations can be characterized by many positive features, not least of which are community service work and the sense of belonging they provide for their members. The concern is to understand why individuals . . . would willingly participate in the events leading up to their own injury and death. It is also to understand why hazing is integrated into fraternity subculture and why brothers inflict psychological and physical pain on their recruits.

On the surface, these acts appear at worst, sadistic, and at best, stupid. These terms, however, are flawed because they attribute the source of these problems as being within individuals rather than stemming from the interactional processes that happen between people. Sadism, for example, implies the existence of a psychological or moral abnormality in the character of the fraternity brothers or pledges. Although published accounts indicate that some hazed pledges had difficulty adjusting to college life, most appear to be normal, healthy young men (Nuwer 1990). Sadism also suggests that fraternity brothers harbor hostility toward the pledges. In fact, the opposite is true. Fraternity brothers tend to care very deeply for their pledges and feel great regret when a pledge is seriously injured, as was the case with the members of Theta Chi.

The term *stupidity* implies that fraternity members and pledges are of below average intelligence. Data do not seem to support the contention that hazing is a product of low intelligence or ignorance. . . . It is apparent that the source of hazing problems is not the result of personality or intellectual shortcomings of pledges or fraternity members.

These observations suggest that fraternity hazing is not so much a problem of individuals as it is a problem of social relationships. To gain further understanding of the power social relationships have on shaping personal behavior, this chapter reframes observations surrounding fraternity hazing within the symbolic interactionist perspective. From this perspective, rather than focusing on the personality quirks of the victims or perpetrators of hazing, a higher priority is placed on the social context in which hazing events take place.

PLEDGING AS A RITE OF PASSAGE

The typical college student only resides in the college community for a few short years, before moving on to other endeavors. This presents fraternities and sororities with an ongoing problem of keeping their organizations thriving and intact in an environment where almost all potential members are transitory. Within a very limited time span, fraternities and sororities need to be able to select groups of nonmembers (independents) and recruit them to full membership status (Greeks). Once independents are recruited and initiated to full membership status, the continued survival of fraternities and sororities rests on their members' abilities and motivations to replenish themselves with the initiation of a new group of recruits.

Fraternities and sororities are not the only types of organization or institution to face this type of concern. For example, the institution of marriage rests on the creation of a sense of strong loyalty between spouses. In western society, this loyalty does not emerge overnight, but builds through courtship rituals. Training programs for some professions, such as physicians, are structured to create a strong sense of commitment to the group as well. The same holds true for the military, which, in very short order, creates loyal soldiers out of mere citizens. Anthropological studies reveal that all of these groups, fraternities and sororities included, adopt initiation rites that are sociologically similar to the rites of passage of traditional tribal groups (Leemon 1972).

Initiation rites, according to the classic study by Arnold van Gennep (1909) *Les Rites de Passage*, occur in three phases: (1) separation, (2) transition, (3) incorporation. The first step in an initiation rite involves a selection of candidates to undergo the rites and garnering their commitment to a process that has yet to ensue. According to van Gennep, this occurs during the *separation phase*, a brief period when a large pool of potential candidates becomes reduced to a select few. Not every college student is a suitable candidate for fraternity life, just as not every available man is a suitable marriage partner for every available woman. The separation phase requires accomplishing two complementary concerns. First, the fraternity presents itself as attractive, so that the desired candidates become interested in joining the organization. Second, the fraternity screens out undesirable candidates.

This selection process is performed during "rushing," a period when prospective pledges are asked to visit the fraternity houses. During rushing, fraternity and sorority members extol the virtues of fraternity or sorority membership, informing candidates that membership offers instant belonging in a social network of "brothers" and "sisters" who "look out for one another." For college students lacking strong social networks, this can make Greek life particularly appealing, especially if it also offers assistance in generating romantic encounters or access to parties. Prospective members are also shown the advantages of off-campus housing and the increased freedom it can offer. For the right type of student, fraternities are very attractive organizations.

Rushing also enables Greek organizations to screen out people who have lower prospects of becoming committed members. One way fraternities and sororities do this is by selecting candidates that fit consistent standards based on race, class, gender, and physical attractiveness (Strombler 1994). This first phase in fraternity initiation rites has very strong parallels with the earliest phases of courtship, the period when dating partners screen one other during casual encounters, keeping their options open for a suitable match for a long-term relationship. While engaged in this screening, both parties try to make themselves appear as appealing as possible, thereby increasing their chances for attracting the most suitable match.

At the end of rushing, independents are formally asked if they will pledge the fraternity. If candidates agree to pledge, they enter into the transition phase of the initiation rites. The *transition phase* marks a status shift, when the candidates can no longer consider themselves independent of the group, but nor can they consider themselves members either. This quasi-membership status in fraternities is comparable to the status of fiancé in a courtship relationship. After the acceptance of a marriage proposal, for instance, an engagement ring is worn to signify one's change in status. Upon the acceptance of a bid, pledges are given pledge pins and other symbols of the organization, and are asked to swear oaths of loyalty and secrecy. In both cases, pledges and fiancés are aligned and joined with their partners, but still lack the rights and responsibilities that are accorded their fully initiated counterparts.

Entry into the transition phase is a positive one, but it also marks the onset of a period of ongoing tests, whereby the potential members are evaluated as to whether they will serve as faithful and capable members. Fraternities commonly assign a pledge master or "whip" to supervise the pledge class and to create a sense of loyalty to their fellow pledges and the fraternity (Leemon 1972). One way that the whip generates this group loyalty is by assigning tasks, which can range from comparatively trivial assignments (i.e., learning the Greek alphabet) to more difficult activities. Paradoxically, as the fraternity places more severe expectations upon pledges, pledges tend to respond by perceiving the fraternity membership as ever more exclusive and desirable.

Hazing tends to occur during the transition phase, when pledges' loyalties are being tested. For example, Sanday (1990) documents one fraternity that required its pledges to strip from the waist down and to tie a thick string around their genitals. They were then told to tie the other end of the string around a rock. Pledges were then blindfolded and instructed to throw the rock with all their might. Unbeknownst to them, however, the fraternity members cut the strings after the pledges were blindfolded. On the surface, this act appears simply cruel, but in the context of initiation rites, it becomes apparent that it is designed to evaluate the pledges' willingness to submit to group expectations.

Finally, according to van Gennep's model, pledges are incorporated into the fraternity in a final swearing in. The *incorporation phase* is akin to the wedding ceremony, where the individuals' status officially changes to full membership in the group. Few people outside of Greek organizations actually know how these final initiation rites take place. In fact, in one of the most detailed analyses of fraternity initiation rites (Leemon 1972), the investigator was allowed to witness all aspects of the pledging procedure except this final ceremony. He could only report that the initiates returned from the room looking simultaneously drained and elated.

Initiation rites are important to consider when analyzing fraternity hazing, because they are responsible for changing pledges' identities and loyalties. At each phase of the initiation rite, pledges are sent messages to think of themselves not in relation to their college, their parents, or their friends, but rather in relation to the fraternity. This does not occur by accident; it happens by design. The pledges' prospects for becoming persons with new values and loyalties are altered once the initiation rites begin during the separation phase. During the transition phase they are reshaped, molded, and sculpted into fraternity men. Finally, at the incorporation phase they are initiated as full members. By the end of the initiation, students have been molded into fraternity members, adopt the values and codes of conduct expected of these members, and are capable of recruiting and shaping new members in the next year's pledge class. Building on these insights, the symbolic interactionist perspective reveals how these rites transform the ways in which individuals understand their relationships with others and their selves.

SYMBOLIC INTERACTIONIST THEORY

Probably more than any other individual, Herbert Blumer (1969) is responsible for introducing the symbolic interactionist perspective to mainstream sociology in a series of articles, later published in his book *Symbolic Interactionism: Perspective and Method.* Although not the first symbolic interactionist, Blumer was the first sociologist to use the term *symbolic interactionism* as a way of linking a number of sociological studies into a theoretical whole. The research interests of symbolic interactionists vary, and their studies address a wide range of issues, but there are a set of common premises that link symbolic interactionists with one another. These premises include:

1. Human beings act toward things on the basis of the meanings these things have for them.
2. The meanings of things arise out of the social interaction people have with each other.
3. People engage in interpretation when dealing with the things they encounter (Blumer 1969).

Symbolic interactionists challenge many of the taken-for-granted understandings of human behavior, particularly the notion that people are made up of enduring personality traits. It is commonly believed that society is made up of particular types of persons, and that these people are consistent in their values and responses from one situation to the next. In contrast, symbolic interactionists assert that the self is highly malleable and can be shaped and reshaped. The self is anything but stable, and it has great potential to become something new and different. In fact, to be more accurate, according to the symbolic interactionist perspective, the self is better characterized as a process than an object. It is something that is constructed through an ongoing process of interpretation and interaction.

Shifting the imagination in accordance with the symbolic interactionist perspective can be challenging. However, once the perspective is adopted, it opens new horizons and ways of explaining behavior not otherwise sensible. Concepts central to symbolic interactionism can be a useful means of understanding fraternity hazing. They also provide a means to rethink strategies of curtailing abuses within fraternity initiation rites.

Definitions of Situations

Symbolic interactionists believe that people act towards things in accordance with the meanings things have for them. This insight was introduced to social psychologists by W. I. Thomas (1928), who offered a maxim now commonly referred to as the *Thomas theorem*: *if people define situations as real, they are real in their consequences.* The Thomas theorem is important not only to understand fraternity hazing but also to understand a wide range of social behavior. The Thomas theorem imposes two questions of concern. First, how is a definition of a situation created? Second, how is a person's behavior influenced by that definition?

Sometimes the definitions exist because of longstanding traditions. For example, in western society, many believe that there will be one true love in their lives, and as a consequence seek a long-term monogamous relationship with a person identified as being their fated match. In other circumstances, definitions of situations are intentionally created by individuals or organizations seeking to advance their interests. This is what politicians call "spin." One way in which spin is accomplished is through the strategic use of language to redefine situations (Lakoff and Johnson 1980). For example, tobacco companies define cigarette smoking as a "habit" and a "lifestyle," and describe their products as having "flavor." In contrast, critics of cigarette companies use terms such as "addiction" and "cancer sticks." Once the words are shifted, the perception of the same object shifts as well.

Fraternities do not tell their pledges that they are going to be "hazed," "beaten," or "abused," but pledges anticipate some degree of abuse during "hell week." Initiation rites are framed as a "tradition" involving "discipline" that is "character-building" and reveals "loyalty" and "commitment." The belief that hazing is part of a tradition is one of the most serious barriers getting fraternity members to reform their initiation procedures. Colleges are also reluctant to frame hazing as a serious problem on their campuses....

How do fraternities package hazing in the context of initiation rites? If fraternity members seek to define a situation as one of solemn importance, rooms are darkened, candles are lit, robes are worn, and fraternity symbols are displayed (Nuwer 1990). Any event that accompanies these cultural symbols, such as having initiates lay down in a coffin, will likely be defined as solemn simply by being coupled with those symbols. Situations can also be defined as festive by using bright lights and music. Once a situation is defined as a party, or in Bini Oja's case "a game," initiates can enjoy the act of drinking to the point of vomiting....

DISCUSSION: SOCIOLOGICALLY INFORMED STRATEGIES TO COUNTER FRATERNITY HAZING

The symbolic interactionist perspective reveals that hazing is not simply the result of psychologically or morally flawed individuals; it is the result of a confluence of symbols, manipulated identities, and definitions of situations that are organized in the context of fraternity initiation rites. Eliminating fraternity hazing is a challenging proposition because it is so strongly embedded in the subculture of some fraternity houses. The symbolic interactionist perspective offers alternate understandings of pledges' and fraternity brothers' actions and leads to very different approaches in comparison to some "commonsense" solutions....

Symbolic interactionists suggest that reform hinges on the examination of meaning systems. Because individuals respond to the world in accordance with the definitions they hold, it is the definitions that need shifting, not so much the rewards and punishments. Fraternity hazing results from definitions of situations that compel fraternity members to believe that abuse of recruits is a necessary part of entry into the fraternity. In fact, Greek organizations use the same type of strategy other groups use to cultivate strong commitment and loyalty (Leemon 1972). Doctors, for instance, are commonly hazed (although this term is not used) during residency, when they are expected to work incredibly long shifts that necessarily involve sleep deprivation (Becker, Geer, Hughes, and Straus 1961). As discussed above, hazing occurs because fraternities define it as a necessary part of their initiation rites and package it carefully to pledges so as to produce compliance.

Pledges and fraternity members are susceptible to losing perspective on what constitutes a reasonable expectation during pledging rites. Friends and teachers outside of fraternities and sororities, however, may be able to identify problems in the initiation rites and offer some means of redefining situations. For example, a couple of years ago a young woman came into one of my classes with a large paper clock tied around her neck. On the clock was written the phrase "ask me what time it is." The student's head repeatedly dropped during the class and she obviously was struggling to keep her eyes open. At the end of class, I tactfully asked her to stay behind for a moment. With the other students gone, I told her that I knew she was being asked to stay awake by her sorority sisters. She did not deny this. I told her that I didn't expect her to break her vow of secrecy, but that I wanted her to recognize that she was being hazed. She responded that "it would be all over soon." I responded again, "I want you to recognize that you are being hazed." From her nonverbal response, I saw that this registered and a new reality was created. I then followed this statement by saying "I want you to recognize that right now it is a choice for you to submit to this abuse and that you can choose to go to sleep if you want to." This interaction only took about five minutes and demonstrated that definitions of situations can be changed.

Hazing is embedded in Greek culture, and changing a culture is not an easy matter. Studies reveal that Greek culture places a high value on secrecy and autonomy (Leemon 1972). This aspect of Greek society is especially problematic for college authorities, who are likely perceived as "outsiders," and as a consequence have few opportunities to learn about the positive and negative things that occur in fraternities. Fraternities are also not receptive to intrusion into what they perceive as internal affairs. Because college authorities do not constitute a salient reference group for Greek members, well-intentioned advice from advisors or administrators can go unheeded.

Given that college authorities are not a reference group, one possible approach to countering fraternity hazing is to use groups that will be accorded immediate respect by fraternities. National fraternity organization representatives are one such group, and can offer reflections and advice that would otherwise not be well received. There may be a very different reaction, for example, to an instructional program on safe use of alcohol during pledging if it is presented by the student life administrator or if it is presented by an alumnus with a Greek affiliation.

Another possibility is to dramatize the potential consequences of hazing. This approach has been used by the Committee to Halt Useless College Killings (CHUCK) (Gose 1997; Nuwer 1990). CHUCK was founded in 1978 by Eileen Stevens, following the death of her son in a fraternity hazing incident. Rather than college authorities laying down rules of what can and cannot be done in initiations, Eileen Stevens comes to fraternities as an outsider and organizes dramatic representations of what happens following a hazing death. One especially effective means of teaching these lessons involves a mock courtroom, where members of the fraternity are expected to take the stand in defense of a hypothetical pledging rite that resulted in a pledge's death at their fraternity.

Another direct application of symbolic interactionism relates to the insight of the social nature of the self. The symbolic interactionist perspective reveals that pledges' willingness to submit to abuse is linked with their inability to think of themselves outside of their status as future fraternity members. Pledges literally lose their "old self" during the pledging process, as they are given new identity kits, social relations, and definitions of self and are shifted to a new reference group. As pledges construct new identities to correspond with fraternity membership, they have greater difficulties envisioning alternate paths of action that contradict the desires of their reference group. Left unchecked, fraternities and sororities will likely exploit the advantages of socially isolating their pledges. Although this will not necessarily lead to hazing, it increases the likelihood that pledges will submit to being hazed.

Advisors and administrators can curtail this potential by limiting the power of fraternities and sororities to isolate pledges from other social groups. Student affairs professionals can help prevent hazing by structuring policies to maintain pledges' connections with other students outside of the Greek subculture. For example, following a recent hazing death, the Masschusetts Institute of Technology now restricts pledging to students who have already completed their freshman year. Possibly this policy will enable new students to build stronger relationships outside of fraternities, thereby making resistance to hazing a more tenable alternative.

Social problems such as fraternity hazing have no easy solutions. Symbolic interactionism offers some useful insights, though, in explaining this problematic social behavior. Maintaining sensitivity to the ways in which fraternities understand hazing, and the ways fraternities shape pledges' abilities to define their selves, may be the best way to construct programs and policies to prevent hazing.

REFERENCES

Adler, Patricia and Peter Adler. 1989. "The Glorified Self." *Social Psychology Quarterly*. 52:299–310.

Arthur, Linda Boynton. 1997. "Role Salience, Role Embracement, and the Symbolic Self Completion of Sorority pledges." *Sociological Inquiry*. 67:364–379.

Becker, Howard, B. Geer, E. C. Hughes & A. Straus. 1961. *The Boys in White: Student Culture in the Medical World*. Chicago: University of Chicago Press.

Blumer, Herbert. 1969. *Symbolic Interactionism: Perspective and Method*. Englewood Cliffs, NJ: Prentice Hall.

Cooley, Charles Horton. 1970. *Human Nature and the Social Order*. New York: Schocken Books.

Coser, Lewis. 1974. *Greedy Institutions: Patterns of Undivided Commitment*. New York: Free Press.

Curry, Susan. 1989. "Hazing and the 'Rush' Toward Reform: Responses from Universities, Fraternities, State Legislatures, and the Courts." *Journal of College and University Law*. 16:93–117.

Goffman, Erving. 1959. *The Presentation of Self in Everyday Life*. New York: Anchor.

Goffman, Erving. 1961. *Asylums*. Chicago: Aldine.

Gose, Ben. 1997. "Efforts to End Fraternity Hazing Have Largely Failed, Critics Charge." *Chronicle of Higher Education*. April 18: pp. A37, A38.

James, William. 1983. *The Principles of Psychology*. Cambridge, MA: Harvard University Press.

Katz, F. 1993. *Ordinary People and Extraordinary Evil: A Report on the Beguilings of Evil*. Albany, NY: SUNY Press.

Kohn, Alfie. 1993. *Punished by Rewards*. New York: Houghton Mifflin.

Lakoff, George & Mark Johnson. 1980. *Metaphors We Live By*. Chicago: University of Chicago Press.

Leemon, Thomas. 1972. *Rites of Passage in a Student Culture*. New York: Teachers College.

Martin, Patricia Yancey and Robert Hummer. 1989. "Fraternities and Rape on Campus." *Gender and Society*. 3:457–473.

Mead, George Herbert. 1934. *Mind, Self, and Society*. Chicago: University of Chicago Press.

Moffatt, Michael. 1989. *Coming of Age in New Jersey*. New Brunswick, NJ: Rutgers University Press.

Nurius, Paula, Jeanette Norris, Linda Dimeff, and Thomas Graham. 1996. "Expectations Regarding Acquaintance Sexual Aggression Among Sorority and Fraternity Members." *Sex Roles*. 35:427–444.

Nuwer, Hank. 1990. *Broken Pledges: The Deadly Rite of Hazing*. Atlanta: Longstreet Press.

Rhoads, Robert. 1995. "Whale Tales, Dog Piles, and Beer Goggles: An Ethnographic Case Study of Fraternity Life." *Anthropology and Education Quarterly*. 26:306–323.

Sanday, Peggy Reeves. 1990. *Fraternity Gang Rape: Sex, Brotherhood, and Privilege on Campus*. New York: New York University Press.

Shaw, Deborah Lee & Thomas Morgan. 1990. "Greek Advisor's Perceptions of Sorority Hazing." *NASPA Journal*. 28:60–64.

Shibutani, Tamatsu. 1961. *Society and Personality: An Interactionist Approach to Social Psychology*. Englewood Cliffs, NJ: Prentice Hall.

Snow, David and Leon Anderson. 1993. *Salvaging the Self from Homelessness. Down on Their Luck: A Study of Homeless Street People*. Berkeley: University of California Press.

Strombler, Mary. 1994. "'Buddies' or 'Slutties': The Collective Sexual Reputation of Fraternity Little Sisters." *Gender & Society*. 8:297–323.

Thomas, W. I. 1928. *The Child in America*. Chicago: University of Chicago Press.

U.S. Government Accounting Office. 1992. *DOD Service Academies: More Changes Needed to Eliminate Hazing*. Washington, D.C.: GAO/NSIAD-93–36.

van Gennep, Arnold. 1960 [1909]. *The Rites of Passage*, translated by Monika Vizedom and Gabrielle Caffee. Chicago: University of Chicago Press.

Wechsler, Henry. 1996. Alcohol and the American College Campus: A Report from the Harvard School of Public Health. *Change*. 28:20–25.

Zimbardo, P. G. 1972. "Pathology of Imprisonment." *Transaction/Society* 9:4–8.

CHAPTER 10

Religion

The Supernatural and Society

Alan Mc Evoy

Alan Mc Evoy

Alan Mc Evoy

It should come as no surprise that the first reading in this chapter is by Emile Durkheim, an early and major contributor to sociology. For him, the important role that religion plays in society is reflected in his insistence that religion is, in actuality, a reflection of society and thus is connected to and impacts on all other social institutions. This is not a radical claim because all institutions in a given society are interrelated and interdependent. The true impact of religion is better understood once we understand what it actually is. Durkheim defined religion as a more or less coherent system of beliefs (such as monotheism and polytheism) and practices (such as fasts and feasts) that concern a supernatural order of beings (gods, goddesses, angels), places (heaven, hell, purgatory), and forces (such as mana). This rather simplistic definition of religion explains what religion is but not why it is so important to us that it is a universal phenomenon.

In the first reading in this chapter, Emile Durkheim notes that religion allows us to transcend human existence, and this in turn gives people a means of dealing with conditions of uncertainty. Thus, religion helps people overcome their feelings of powerlessness, and in this manner it not only helps them control the conditions that affect their lives but also helps them cope with their unfulfilled psychological and economic needs. Perhaps, then, it is not surprising that for believers religious beliefs transcend national loyalties. Perhaps it is also not surprising that those people who belong to organized religious congregations tend to give their time and money more generously to these organizations than to any other organization outside the family and its circle of friends. It seems obvious that religion is too important a social institution to disappear, but that does not mean it will not undergo change as it, like the other institutions of society, adapts to the impact of changing conditions.

Peter Berger confirms both of these claims—the importance of religion and its unlikely disappearance—in the second reading. Thus, despite the impact of modernization and globalization, religion remains strong; Berger even claims that secularism is in retreat.

James Aho confirms the conclusions of Durkheim and Berger by noting the political extremes that Christians will go to in the name of their religious beliefs. The U.S. Constitution declares that there is a separation of church and state in this country. What this statement means is that the citizen has the right to religious expression but that no religion will be imposed or favored by the government. It is this rule that allows religious institutions to escape paying taxes of any kind because upholders of religious freedom believe that the imposition of taxes infringes on religious freedom. Of course, opponents of this view claim that not imposing taxes on religious organizations means favoring them over other organizations and also encourages participation in religious endeavors. Because their funds are tax free, religious organizations are not allowed to use those funds for political uses. Yet political activism by religious organizations has always been a constant in American politics; in fact, in many situations this activism in the name of a particular religious belief goes to extreme measures. James Aho brings to our attention these extreme measures that are being wrought and the government's reluctance to intrude on the activities of religious organizations, political or otherwise. As the examination of terrorism in Chapter 12 will show, this participation in political affairs by U.S. religious organizations coincides with the activities of other religious groups in other countries. The difference is that only the United States has a law specifically banning such participation.

The conflict over moral issues as well as conflict in other societies for religious hegemony alerts us to the relationship between religious and political institutions. As the abortion issue in the United States shows, religious beliefs can lead to political conflict, and vice versa. This is seen in the second reading by Dallas Blanchard on the many beliefs and goals of the anti-abortion movement.

In the final reading, Gordon Clanton and Shoon Lio seem to disagree with the conclusions made or implied in the preceding selections. America, they say, has witnessed a marked decline both in religion and in its sister, civil religion (in this country, an appreciation of transcendent religious reality as revealed through American experience). This discovery would seem to contradict Berger's claim about the decline of secularism. What do you think? How do the Clanton and Lio findings compare with your own knowledge of these matters?

46

The Elementary Forms of the Religious Life

EMILE DURKHEIM

The author claims that religious beliefs and rituals are real and reflect the societies in which they exist. Note his reasons for this claim. If you agree with his thesis, ask yourself what effect it has on your personal religious beliefs.

As you read, ask yourself the following questions:

1. *Why does the author say that religious beliefs and rituals are real and reflect society?*
2. *How do your religious beliefs and rituals reflect American society?*
3. *Considering religious impact, should religious institutions continue to receive tax breaks? Why?*

GLOSSARY **Profane** Not devoted to religious purposes; secular.
Sui generis In itself.

The theorists who have undertaken to explain religion in rational terms have generally seen in it before all else a system of ideas, corresponding to some determined object. This object has been conceived in a multitude of ways: nature, the infinite, the unknowable, the ideal, etc.; but these differences matter but little. In any case, it was the conceptions and beliefs which were considered as the essential elements of religion. As for the rites, from this point of view they appear to be only an external translation, contingent and material, of these internal states which alone pass as having any intrinsic value. This conception is so commonly held that generally the disputes of which religion is the theme turn about the question whether it can conciliate itself with science or not, that is to say, whether or not there is a place beside our scientific knowledge for another form of thought which would be specifically religious.

But the believers, the men who lead the religious life and have a direct sensation of what it really is, object to this way of regarding it, saying that it does not correspond to their daily experience. In fact, they feel that the real function of religion is not to make us think, to enrich our knowledge, nor to add to the conceptions which we owe to science others of another origin and another character, but rather, it is to make us act, to aid us to live. The believer who has communicated with his god is not merely a man who sees new truths of which the unbeliever is ignorant; he is a man who is *stronger*. He feels within him more force, either to endure the trials of existence, or to conquer them. It is as though he were raised above

SOURCE: From *The Elementary Forms of Religious Life: A Study in Religious Sociology*, by Emile Durkheim, Routledge, 1915. Translated from the French by Joseph Ward Swain. London: Allen & Urwin, 1915. Reprinted with the permission of Simon and Schuster.

the miseries of the world, because he is raised above his condition as a mere man; he believes that he is saved from evil, under whatever form he may conceive this evil. The first article in every creed is the belief in salvation by faith. But it is hard to see how a mere idea could have this efficacy. An idea is in reality only a part of ourselves; then how could it confer upon us powers superior to those which we have of our own nature? Howsoever rich it might be in affective virtues, it could add nothing to our natural vitality; for it could only release the motive powers which are within us, neither creating them nor increasing them. From the mere fact that we consider an object worthy of being loved and sought after, it does not follow that we feel ourselves stronger afterwards; it is also necessary that this object set free energies superior to these which we ordinarily have at our command and also that we have some means of making these enter into us and unite themselves to our interior lives. Now for that, it is not enough that we think of them; it is also indispensable that we place ourselves within their sphere of action, and that we set ourselves where we may best feel their influence; in a word, it is necessary that we act, and that we repeat the acts thus necessary every time we feel the need of renewing their effects. From this point of view, it is readily seen how that group of regularly repeated acts which form the cult get their importance. In fact, whoever has really practised a religion knows very well that it is the cult which gives rise to these impressions of joy, of interior peace, of serenity, of enthusiasm which are, for the believer, an experimental proof of his beliefs. The cult is not simply a system of signs by which the faith is outwardly translated; it is a collection of the means by which this is created and recreated periodically. Whether it consists in material acts or mental operations, it is always this which is efficacious.

Our entire study rests upon this postulate that the unanimous sentiment of the believers of all times cannot be purely illusory. . . . We admit that these religious beliefs rest upon a specific experience whose demonstrative value is, in one sense, not one bit inferior to that of scientific experiments, though different from them. We, too, think that "a tree is known by its fruits," and that fertility is the best proof of what the roots are worth. But from the fact that a "religious experience," if we choose to call it this, does exist and that it has a certain foundation—and, by the way, is there any experience which has none?—it does not follow that the reality which is its foundation conforms objectively to the idea which believers have of it. The very fact that the fashion in which it has been conceived has varied infinitely in different times is enough to prove that none of these conceptions express it adequately. If a scientist states it as an axiom that the sensations of heat and light which we feel correspond to some objective cause, he does not conclude that this is what it appears to the senses to be. Likewise, even if the impressions which the faithful feel are not imaginary, still they are in no way privileged intuitions; there is no reason for believing that they inform us better upon the nature of their object than do ordinary sensations upon the nature of bodies and their properties. In order to discover what this object consists of, we must submit them to an examination and elaboration analogous to that which has substituted for the sensuous idea of the world another which is scientific and conceptual.

This is precisely what we have tried to do, and we have seen that this reality, which mythologies have represented under so many different forms, but which is the universal and eternal objective cause of these sensations *sui generis* out of which religious experience is made, is society. We have shown what moral forces it develops and how it awakens this sentiment of a refuge, of a shield, and of a guardian support which attaches the believer to his cult. It is that which raises him outside himself; it is even that which made him. For that which makes a man is the totality of the intellectual property which constitutes civilization, and civilization is the work of society. Thus is explained the preponderating role of the cult in all religions, whichever they may be. This is because society cannot make its influence felt unless it is in action, and it is not in action unless the individuals who compose it are assembled together and act in common. It is by common action that it takes consciousness of itself and realizes its position;

it is before all else an active cooperation. The collective ideas and sentiments are even possible only owing to these exterior movements which symbolize them, as we have established. Then it is action which dominates the religious life, because of the mere fact that it is society which is its source....

Religious forces are therefore human forces, moral forces. It is true that since collective sentiments can become conscious of themselves only by fixing themselves upon external objects, they have not been able to take form without adopting some of their characteristics from other things: They have thus acquired a sort of physical nature; in this way they have come to mix themselves with the life of the material world, and then have considered themselves capable of explaining what passes there. But when they are considered only from this point of view and in this role, only their most superficial aspect is seen. In reality, the essential elements of which these collective sentiments are made have been borrowed by the understanding. It ordinarily seems that they should have a human character only when they are conceived under human forms,[1] but even the most impersonal and the most anonymous are nothing else than objectified statements....

Some reply that men have a natural faculty for idealizing, that is to say, of substituting for the real world another different one, to which they transport themselves by thought. But that is merely changing the terms of the problem; it is not resolving it or even advancing it. This systematic idealization is an essential characteristic of religions. Explaining them by an innate power of idealization is simply replacing one word by another which is the equivalent of the first; it is as if they said that men have made religions because they have a religious nature. Animals know only one world, the one which they perceive by experience, internal as well as external. Men alone have the faculty of conceiving the ideal, of adding something to the real. Now where does this singular privilege come from? Before making it an initial fact or a mysterious virtue which escapes science, we must be sure that it does not depend upon empirically determinable conditions.

The explanation of religion which we have proposed has precisely this advantage, that it gives an answer to this question. For our definition of the sacred is that it is something added to and above the real: Now the ideal answers to this same definition; we cannot explain one without explaining the other. In fact, we have seen that if collective life awakens religious thought on reaching a certain degree of intensity, it is because it brings about a state of effervescence which changes the conditions of psychic activity. Vital energies are overexcited, passions more active, sensations stronger; there are even some which are produced only at this moment. A man does not recognize himself; he feels himself transformed and consequently he transforms the environment which surrounds him. In order to account for the very particular impressions which he receives, he attributes to the things with which he is in most direct contact properties which they have not, exceptional powers and virtues which the objects of everyday experience do not possess. In a word, above the real world where his profane life passes he has placed another which, in one sense, does not exist except in thought, but to which he attributes a higher sort of dignity than to the first. Thus, from a double point of view it is an ideal world.

The formation of the ideal world is therefore not an irreducible fact which escapes science; it depends upon conditions which observation can touch; it is a natural product of social life. For a society to become conscious of itself and maintain at the necessary degree of intensity the sentiments which it thus attains, it must assemble and concentrate itself. Now this concentration brings about an exaltation of the mental life which takes form in a group of ideal conceptions where is portrayed the new life thus awakened; they correspond to this new set of psychical forces which is added to those which we have at our disposition for the daily tasks of existence. A society can neither create itself nor recreate itself without at the same time creating an ideal. This creation is not a sort of work of supererogation for it, by which it would complete itself, being already formed; it is the act by which it is periodically made and remade. Therefore when some oppose the ideal society to the real society, like two antagonists which would lead us in opposite directions, they materialize and oppose abstractions. The ideal society is not outside of the real society; it is a part of it. Far from being divided between them as between two poles which mutually repel each other, we cannot hold to one without holding to the other. For a society is not made up merely of the mass of individuals who compose it, the ground which they occupy, the things which they use and the movements which they perform, but above all is the idea which it forms of itself. It is undoubtedly true that it hesitates over the manner in which it ought to conceive itself; it feels itself drawn in divergent directions. But these conflicts which break forth are not between the ideal and reality, but between two different ideals, that of yesterday and that of today, that which has the authority of tradition and that which has the hope of the future. There is surely a place for investigating whence these ideals evolve; but whatever solution may be given to this problem, it still remains that all passes in the world of the ideal.

Thus the collective ideal which religion expresses is far from being due to a vague innate power of the individual, but it is rather at the school of collective life that the individual has learned to idealize. It is in assimilating the ideals elaborated by society that he has become capable of conceiving the ideal. It is society which, by leading him within its sphere of action, has made him acquire the need of raising himself above the world of experience and has at the same time furnished him with the means of conceiving another. For society has constructed this new world in constructing itself, since it is society which this expresses. Thus both with the individual and in the group, the faculty of idealizing has nothing mysterious about it. It is not a sort of luxury which a man could get along without, but a condition of his very existence. He could not be a social being, that is to say, he could not be a man, if he had not acquired it. It is true that in incarnating themselves in individuals, collective ideals tend to individualize themselves. Each understands them after his own fashion and marks them with his own stamp; he suppresses certain elements and adds others. Thus the personal ideal disengages itself from the social ideal in proportion as the individual personality develops itself and becomes an autonomous source of action. But if we wish to understand this aptitude, so singular in appearance, of living outside of reality, it is enough to connect it with the social conditions upon which it depends.

Therefore it is necessary to avoid seeing in this theory of religion a simple restatement of historical materialism: That would be misunderstanding our thought to an extreme degree. In showing that religion is something essentially social, we do not mean to say that it confines itself to translating into another language the material forms of society and its immediate vital necessities. It is true that we take it as evident that social life depends upon its material foundation and bears its mark, just as the mental life of an individual depends upon his nervous system and in fact his whole organism. But collective consciousness is something more than a mere epiphenomenon of its morphological basis, just as individual consciousness is something more than a simple efflorescence of the nervous system. In order that the former may appear, a synthesis *sui generis* of particular consciousnesses is required. Now this synthesis has the effect of disengaging a whole world of sentiments, ideas, and images which, once born, obey laws all their own. They attract each other, repel each other, unite, divide themselves, and multiply, though these combinations are not commanded and necessitated by the conditions of the underlying reality. The life thus brought into being even enjoys so great an independence that it sometimes indulges in manifestations with no purpose or utility of any sort, for the mere pleasure of affirming itself. We have shown that this is often precisely the case with ritual activity and mythological thought. . . .

That is what the conflict between science and religion really amounts to. It is said that science denies religion in principle. But religion exists; it is a system of given facts; in a word, it is a reality. How could science deny this reality? Also, insofar as religion is action, and insofar as it is a means of making men live, science could not take its place, for even if this expresses life, it does not create it; it may well seek to explain the faith, but by that very act it presupposes it. Thus there is no conflict except upon one limited point. Of the two functions which religion originally fulfilled, there is one, and only one, which tends to escape it more and more: That is its speculative function. That which science refuses to grant to religion is not its right to exist, but its right to dogmatize upon the nature of things and the special competence which it claims for itself for knowing man and the world. As a matter of fact, it does not know itself. It does not even know what it is made of, nor to what need it answers. It is itself a subject for science, so far is it from being able to make the law for science! And from another point of view, since there is no proper subject for religious speculation outside that reality to which scientific reflection is applied, it is evident that this former cannot play the same role in the future that it has played in the past.

However, it seems destined to transform itself rather than to disappear. . . .

NOTE

1. It is for this reason that Frazer and even Preuss set impersonal religious forces outside of, or at least on the threshold of religion, to attach them to magic.

47

Secularism in Retreat

PETER L. BERGER

According to the author, the belief that modernization lessens the need or desire for religion is not true.

As you read this article, consider the following questions as guides:

1. *Were you surprised by the conclusion that in this supposed age of enlightenment there is still a need for religion? Why or why not?*
2. *Do you believe that the need for religion is good or bad news? Why?*
3. *What is the future effect of insurgencies considering that the Islamic resurgence has a negative view of modernity?*

GLOSSARY **Secularism** A social philosophy that rejects all forms of religious faith and worship. **Fundamentalism** In Christianity, the belief that stresses the inerrancy of the Bible in matters of faith and morals as well as a literal historical record holding to the doctrines of the Christian faith; a reaction against modernism. **Pentecostalism** Christian fundamentalism that stresses holiness of living.

SOURCE: Reprinted with permission from *The National Interest*, Winter 1996/97, pp. 3–12.

The world today, with some exceptions attended to below, is as furiously religious as it ever was, and in some places more so than ever. This means that a whole body of literature written by historians and social scientists over the course of the 1950s and '60s, loosely labeled as "secularization theory," was essentially mistaken. . . .

The key idea of secularization theory is simple and can be traced to the Enlightenment: Modernization necessarily leads to a decline of religion, both in society and in the minds of individuals. It is precisely this key idea that has turned out to be wrong. To be sure, modernization has had some secularizing effects, more in some places than in others. But it has also provoked powerful movements of counter-secularization. Also, secularization on the societal level is not necessarily linked to secularization on the level of individual consciousness. Thus, certain religious institutions have lost power and influence in many societies, but both old and new religious beliefs and practices have nevertheless continued in the lives of individuals, sometimes taking new institutional forms and sometimes leading to great explosions of religious fervor. Conversely, religiously-identified institutions can play social or political roles even when very few people believe or practice the religion supposedly represented by these institutions. To say the least, the relation between religion and modernity is rather complicated.

REJECTION AND ADAPTATION

The proposition that modernity necessarily leads to a decline of religion is, in principle, "value-free." That is, it can be affirmed both by people who think it is good news and by people who think that it is very bad news indeed. Most Enlightenment thinkers and most progressive-minded people ever since have tended toward the idea that secularization is a good thing, at least insofar as it does away with religious phenomena that are "backward," "superstitious," or "reactionary" (a religious residue purged of these negative characteristics may still be deemed acceptable). But religious people, including those with very traditional or orthodox beliefs, have also affirmed the modernity/secularity linkage, and have greatly bemoaned it. Some have defined modernity as the enemy, to be fought whenever possible. Others have, on the contrary, seen modernity as an invincible worldview to which religious beliefs and practices should adapt themselves. In other words, *rejection* and *adaptation* are two strategies open to religious communities in a world understood to be secularized. As is always the case when strategies are based on mistaken perception of the terrain, both strategies have had very doubtful results.

It is possible, of course, to reject any number of modern ideas and values theoretically, but to make this rejection stick in the lives of people is much more difficult. To do that, one can try to take over society as a whole and make one's counter-modern religion obligatory for everyone—a difficult enterprise in most countries in the contemporary world. Franco tried in Spain, and failed; the mullahs are still at it in Iran and a couple of other places; in most of the world such exercises in religious conquest are unlikely to succeed. And this unlikelihood *does* have to do with modernization, which brings about very heterogeneous societies and a quantum leap in intercultural communication, two factors favoring pluralism and *not* favoring the establishment (or re-establishment) of religious monopolies. Another form of rejection strategy is to create religious subcultures so designed as to exclude the influences of the outside society. That is a more promising exercise than religious revolution, but it too is fraught with difficulty. Where it has taken root, modern culture is a very powerful force, and an immense effort is required to maintain enclaves with an airtight defense system. Ask the Amish in eastern Pennsylvania, or a Hasidic rabbi in the Williamsburg section of Brooklyn.

Notwithstanding the apparent power of modern secular culture, secularization theory has been falsified even more dramatically by the results of adaptation strategies attempted by religious institutions. If we really lived in a highly secularized world, then religious institutions could be expected to survive to the degree that they manage to adapt to secularity.

That, indeed, has been the empirical assumption of adaptation strategies. What has in fact occurred is that, by and large, religious communities have survived and indeed flourished to the degree that they have *not* tried to adapt themselves to the alleged requirements of a secularized world. Put simply, experiments with secularized religion have generally failed; religious movements, with beliefs and practices dripping with "reactionary supernaturalism" (the kind utterly beyond the pale at self-respecting faculty parties) have widely succeeded.

The struggle with modernity in the Roman Catholic Church nicely illustrates the difficulties of various rejection and adaptation strategies. In the wake of the Enlightenment and its multiple revolutions, the initial response by the Church was militant and then defiant rejection. Perhaps the most magnificent moment of that defiance came in 1870, when the First Vatican Council solemnly proclaimed the infallibility of the Pope and the immaculate conception of Mary, literally in the face of the Enlightenment about to occupy Rome in the shape of the army of Victor Emmanuel I. The disdain was mutual: The Roman monument to the Bersaglieri, the elite army units that occupied the Eternal City in the name of the Italian Risorgimento, places the heroic figure in his Bersaglieri uniform so that he is positioned with his behind pointing exactly toward the Vatican. The Second Vatican Council, almost a hundred years later, considerably modified this rejectionist stance, guided as it was by the notion of *aggiornamento*—literally, bringing the church "up-to-date" with the modern world. (I remember a conversation I had with a Protestant theologian, whom I asked what he thought would happen at the Council, this before it had actually convened; he replied that he didn't know, but that he was sure that they would not read the minutes of the first Council meeting.)

The Second Vatican Council was supposed to open windows, specifically the windows of the anti-secular Catholic subculture that had been constructed when it became clear that the overall society could not be reconquered. (In the United States this Catholic subculture was quite impressive right up to the very recent past.) The trouble with opening windows is that you cannot control what comes in through them, and a lot has come in—indeed, the whole turbulent world of modern culture—that has been very troubling to the Church. Under the current pontificate the Church has been steering a nuanced course in between rejection and adaptation, with mixed results in different countries.

If one looks at the international religious scene objectively, that of the Roman Catholics as well as virtually all others, one must observe that it is conservative or orthodox or traditionalist movements that are on the rise almost everywhere. These movements, whatever adjective one may choose for them, are precisely those that rejected an *aggiornamento* as defined by progressive intellectuals. Conversely, religious movements and institutions that have made great efforts to conform to a perceived modernity are almost everywhere on the decline. In the United States this has been a much commented-upon fact, exemplified by the decline of so-called mainline Protestantism and the concomitant rise of Evangelicalism; but the United States is by no means unusual in this. Nor is Protestantism.

The conservative thrust in the Roman Catholic Church under John Paul II has borne fruit in both the number of converts and in the renewed enthusiasm among native Catholics, especially in non-Western countries. Following the collapse of the Soviet Union, too, there occurred a remarkable revival of the Orthodox Church in Russia. The most rapidly growing Jewish groups, both in Israel and in the diaspora, are Orthodox groups. There have been similarly vigorous upsurges of conservative religion in all the other major religious communities—Islam, Hinduism, Buddhism—as well as revival movements in smaller communities (such as Shinto in Japan and Sikhism in India).

Of course, these developments differ greatly, not only in religious content (which is obvious), but in their social and political implications. What they have in common, though, is their unambiguously *religious* inspiration. In their aggregate they provide a massive falsification of the idea that modernization and secularization are cognate phenomena. Minimally, one must note that *counter*-secularization is at least as important a phenomenon in the contemporary world as secularization.

TWO REVIVALS . . .

Both in the media and in scholarly publications these religious movements are often subsumed under the category of "fundamentalism." It suggests a combination of several features—great religious passion, a defiance of what others have defined as the *Zeitgeist*, and a return to traditional sources of religious authority. These are indeed common features across cultural boundaries. And they do reflect the presence of secularizing forces, since they must be understood as a reaction *against* them. . . . Clearly, one of the most important topics for a sociology of contemporary religion is precisely this interplay of secularizing and counter-secularizing forces. This is because modernity, for fully understandable reasons, undermines all the old certainties; uncertainty in turn is a condition that many people find very hard to bear; therefore, any movement (not only a religious one) that promises to provide or to renew certainty has a ready market.

While the aforementioned common features are important, an analysis of the social and political impact of the various religious upsurges must take full account of their differences. This becomes clear when one looks at what are arguably the two most dynamic religious upsurges in the world today, the Islamic and the Evangelical ones. Comparison also underlines the weakness of the category "fundamentalism" as applied to both.

The Islamic upsurge, because of its more immediately obvious political ramifications, is the better known of the two. Yet it would be a serious error to see it only through a political lens. It is an impressive revival of emphatically *religious* commitments. And it is of vast geographical scope, affecting every Muslim country from North Africa to Southeast Asia. It continues to gain converts, especially in sub-Saharan Africa, where it is often in head-on competition with Christianity. It is becoming very visible in the burgeoning Muslim communities in Europe and, to a much lesser extent, in North America. Everywhere it is bringing about a restoration not only of Islamic beliefs, but of distinctively Islamic lifestyles, which in many ways directly contradict modern ideas—such as the relation of religion and the state, the role of women, moral codes of everyday behavior and, last but not least, the boundaries of religious and moral tolerance.

An important characteristic of the Islamic revival is that it is by no means restricted to the less modernized or "backward" sectors of society, as progressive intellectuals still like to think. On the contrary, it is very strong in cities with a high degree of modernization, and in a number of countries it is particularly visible among people with Western-style higher education; in Egypt and Turkey, for example, it is often the daughters of secularized professionals who are putting on the veil and other accoutrements expressing so-called Islamic modesty.

Yet there are also very great differences. Even within the Middle East, the Islamic heartland, there are both religiously and politically important distinctions to be made between Sunni and Shi'a revivals—Islamic conservatism means very different things in, say, Saudi Arabia and Iran. As one moves away from the Middle East, the differences become even greater. Thus in Indonesia, the most populous Muslim country in the world, a very powerful revival movement, the Nahdatul-Ulama, is avowedly pro-democracy and pro-pluralism, the very opposite of what is commonly viewed as Muslim "fundamentalism." Where the political circumstances allow it, there is a lively discussion about the relationship of Islam to various modern realities, and there are sharp disagreements between individuals who are equally committed to a revitalized Islam. Still, for reasons deeply grounded in the core of the tradition, it is probably fair to say that, on the whole, Islam has had a difficult time coming to terms with key modern institutions—such as pluralism, democracy, and the market economy.

The Evangelical upsurge is just as breathtaking in scope. Geographically that scope is even wider than that of the Islamic revival. It has gained huge numbers of converts in East Asia—in all the Chinese communities (including, despite severe persecution, in mainland China) and in South Korea, the Philippines, across the South Pacific, throughout sub-Saharan Africa (where it is often synthesized with elements of traditional African religion), and apparently in parts of ex-communist Europe. But the most remarkable success has occurred in Latin America; it is estimated that there are now between forty and fifty million Evangelical Protestants south of the U.S. border, the great majority of them first-generation Protestants.

The most numerous component within the Evangelical upsurge is Pentecostal, combining Biblical orthodoxy and a rigorous morality with an ecstatic form of worship and an emphasis on spiritual healing. Especially in Latin America, conversion to Protestantism brings about a cultural transformation—new attitudes toward work and consumption, a new educational ethos, a violent rejection of traditional *machismo* (women play a key role in the Evangelical churches). The origins of this worldwide Evangelical upsurge are in the United States, from where the missionaries were first dispatched. But it is very important to understand that virtually everywhere, and emphatically in Latin America, the new Evangelicalism is thoroughly indigenous and is no longer dependent on support from U.S. fellow-believers. Indeed, Latin American Evangelicals have been sending missionaries to the Hispanic community in this country, where there has been a comparable flurry of conversions.

Needless to say, the religious contents of the Islamic and Evangelical revivals are totally different. So are the social and political consequences (of which more below). But the two developments also differ in that the Islamic movement is occurring primarily in countries that are already Muslim or among Muslim emigrants (as in Europe); by contrast, the Evangelical movement is growing dramatically throughout the world in countries where this type of religion was previously unknown or very marginal.

. . . AND TWO EXCEPTIONS

The world today, then, is massively religious, and it is anything but the secularized world that had been predicted (be it joyfully or despondently) by so many analysts of modernity. There are two exceptions to this proposition, one somewhat unclear, the other very obvious.

The first apparent exception is in Western Europe, where, if nowhere else, the old secularization theory seems to hold. With increasing modernization there has been an increase in the key indicators of secularization: on the level of expressed beliefs (especially such as could be called orthodox in Protestant or Catholic terms), and dramatically on the level of church-related behavior (attendance at services of worship, adherence to church-dictated codes of personal behavior—especially with regard to sexuality, reproduction, and marriage), and finally, with respect to recruitment to the clergy. These phenomena had been observed for a long time in the northern countries of the continent; since the Second World War they have quickly engulfed the south. Thus Italy and Spain have experienced a rapid decline in church-related religion—as has Greece (thus undercutting the claim of Catholic conservatives that Vatican II is to be blamed for the decline). There is now a massively secular Euro-culture and what has happened in the south can be simply described (though not thereby explained) as the invasion of these countries by that culture. It is not fanciful to predict that there will be similar developments in Eastern Europe, precisely to the degree that these countries too will be integrated into the new Europe.

While these facts are not in dispute, a number of recent works in the sociology of religion (notably in France, Britain, and Scandinavia) have questioned the term "secularization" as applied to these developments. There is now a body of data indicating strong survivals of religion, most of it generally Christian in nature, despite the widespread alienation from the organized churches. If the data hold up to scrutiny, a shift in the institutional location of religion, rather than secularization, would then be a more accurate description of the European situation. All the same, Europe stands out as quite different from other parts of the world. It certainly differs sharply from the religious situation in the United States. One of the most interesting puzzles in the sociology of religion is why Americans are so much more religious as well as more churchly than Europeans.

The other exception to the desecularization thesis is less ambiguous: There exists an international subculture composed of people with Western-type higher education, especially in the humanities and social sciences, which is indeed secularized by any measure. This subculture is the principal "carrier" of progressive, Enlightenment beliefs and values. While the people in this subculture are relatively thin on the ground, they are very influential, as they control the institutions that provide the "official" definitions of reality (notably the educational system, the media of mass communication, and the higher reaches of the legal system). They are remarkably similar all over the world today as they have been for a long time (though, as we have seen, there are also defectors from this subculture, especially in the Muslim countries). Why it is that people with this type of education should be so prone to secularization is not entirely clear, but there is, without question, a globalized *elite* culture. It follows, then, that in country after country religious upsurges have a strongly populist character: Over and beyond the purely religious motives, these are movements of protest and resistance *against* a secular elite. The so-called "culture war" in the United States emphatically shares this feature.

QUESTIONS AND ANSWERS

This somewhat breathless *tour d'horizon* of the global religious scene raises several questions: What are the origins of the worldwide resurgence of religion? What is the likely future course of this religious resurgence? Do resurgent religions differ in their critique of the secular order? How is religious resurgence related to a number of issues not ordinarily linked to religion? Let us take these questions in turn.

As to the origins of the worldwide resurgence of religion, two possible answers have already been mentioned. The first is that modernity tends to undermine the taken-for-granted certainties by which people lived throughout most of history. This is an uncomfortable state of affairs, for many an intolerable one, and religious movements that claim to give certainty have great appeal by easing that discomfort. The second is that a purely secular view of reality has its principal social location in an elite culture that, not surprisingly, is resented by large numbers of people who are not part of it but who nevertheless feel its influence (most troublingly, as their children are subjected to an education that ignores or even directly attacks their own beliefs and values). Religious movements with a strongly anti-secular bent can therefore appeal to people with resentments that sometimes have quite non-religious sources.

But there is yet another answer, which recalls my opening story about certain American foundation officials worrying about "fundamentalism." In one sense, there is nothing to explain here. Strongly felt religion has always been around: what needs explanation is its absence rather than its presence. Modern secularity is a much more puzzling phenomenon than all these religious explosions.... In other words, at one level the phenomena under consideration simply serve to demonstrate *continuity* in the place of religion in human experience.

As to the likely future course of this religious resurgence, it would make little sense to venture a prognosis with regard to the entire global scene, given the considerable variety of important religious movements in the contemporary world. Predictions, if one dares to make them at all, will be more useful if applied to much narrower situations. One, though, can be made with some assurance: There is no reason to think that the world of the twenty-first century will be any less religious than the world is today....

Nonetheless, one will have to speculate very differently regarding different sectors of the religious scene. For example, the most militant Islamic movements will have difficulty maintaining their present stance *vis-á-vis* modernity should they succeed in taking over the governments of their countries (as, it seems, is already happening in Iran). It is also unlikely that Pentecostalism, as it exists today among mostly poor and uneducated people, will retain its present religious and moral characteristics unchanged as many of these people experience upward social mobility (this has already been observed extensively in the United States). Generally, many of these religious movements are linked to non-religious forces of one sort or another, and the future course of the former will be at least partially determined by the course of the latter. Thus in the United States, for instance, the future course of militant Evangelicalism will be different if some of its causes succeed—or continue to be frustrated—in the political and legal arenas.

Finally, in religion as in every other area of human endeavor, individual personalities play a much larger role than most social scientists and historians are willing to concede. Thus there might have been an Islamic revolution in Iran without the Ayatollah Khomeini, but it would probably have looked quite different. No one can predict the appearance of charismatic figures who will launch powerful religious movements in places where no one expects them. Who knows—perhaps the next religious upsurge in America will occur among disenchanted postmodernist academics!

Do the resurgent religious differ in their critique of the secular order?...

The religious impulse, the quest for meaning that transcends the restricted space of empirical existence in this world, has been a perennial feature of humanity.... It would require something close to a mutation of the species to finally extinguish this impulse. The more radical thinkers of the Enlightenment, and their more recent intellectual descendants, hoped for something like such a mutation, of course. Thus far this has not happened and it is unlikely to happen anytime in the foreseeable future. The critique of secularity common to all the resurgent movements is that human existence bereft of transcendence is an impoverished and finally untenable condition.

To the extent that secularity today has a specifically modern form ... the critique of secularity also entails a critique of at least these aspects of modernity. Beyond that, however, different religious movements differ in their relation to modernity.

As noted, an argument can be made that the Islamic resurgence has a strong tendency toward a negative view of modernity; in places it is downright anti-modern or counter-modernizing... By contrast, the Evangelical resurgence is positively modernizing in most places where it occurs, clearly so in Latin America. The new Evangelicals throw aside many of the traditions that have been obstacles to modernization ... and their churches encourage values and behavior patterns that contribute to modernization. Just to take one important case in point: In order to participate fully in the life of their congregations, Evangelicals will want to read the Bible and to be able to join in the discussion of congregational affairs that are largely in the hands of lay persons (indeed, largely in the hands of women). The desire to read the Bible encourages literacy, and, beyond this, a positive attitude toward education and self-improvement. The running of local churches by lay persons necessitates training in various administrative skills, including the conduct of public meetings and the keeping of financial accounts. It is not fanciful to suggest that in this way Evangelical congregations serve (inadvertently, to be sure) as schools for democracy and for social mobility....

RELIGION IN WORLD POLITICS

To assess the role of religion in international politics, it would be useful to distinguish between political movements that are genuinely inspired by religion and those that use religion as a convenient legitimation for political agendas based on non-religious interests. Such a distinction is difficult but not impossible. Thus there is no reason to doubt that the suicide bombers of the Islamic Hamas movement truly believe in the religious motives they avow. By contrast, there is good reason to doubt that the three parties involved in the Bosnian conflict, which is commonly represented as a clash between religions, are really inspired by religious ideas. I think it was P. J. O'Rourke who observed that these three parties are of the same race, speak the same language, and are distinguished only by their religion—in which none of them believe....

It would be very nice if one could say that religion is everywhere a force for peace. Unfortunately, this is not the case. While it is difficult to pinpoint a frequency distribution, very probably religion much more often fosters war, both between and within nations, rather than peace. If so, that is hardly new in history. Religious institutions and movements are fanning wars and civil wars on the Indian subcontinent, in the Balkans, in the Middle East, and in Africa. Occasionally, religious institutions do try to resist warlike policies or to mediate between conflicting parties. The Vatican mediated successfully in some international disputes in Latin America. There have been religiously inspired peace movements in several countries (including the United States, during the Vietnam War). Both Protestant and Catholic clergy have tried to mediate the conflict in Northern Ireland, with notable lack of success. But it is probably a mistake to focus simply on the actions of formal religious institutions or groups. There may be a diffusion of religious values in a society that could have peace-prone consequences even in the absence of formal actions by church bodies. For example, some analysts have argued that the wide diffusion of Christian values played a mediating role in the process that ended the apartheid regime in South Africa, despite the fact that the churches themselves were mostly polarized between the two sides of the conflict (at least until the last few years of the regime, when the Dutch Reformed Church reversed its position on apartheid).

Relatedly, a religious resurgence may well have important implications for economic development.... Something like Weber's "Protestant ethic" is probably functional in an early phase of capitalist growth—an ethic, whether religiously inspired or not, that values personal discipline, hard work, frugality, and a respect for learning.

The new Evangelicalism in Latin America exhibit these values in virtually crystalline purity. Conversely, Iberian Catholicism, as it was well established in Latin America, clearly does *not* foster such values. But religious traditions can change. Spain experienced a remarkably successful period of economic development beginning in the waning years of the Franco regime, and one of the important factors was the influence of Opus Dei, which combined rigorous theological orthodoxy with market-friendly openness in economic matters. Islam, by and large, has difficulties with a modern market economy—especially with modern banking—yet Muslim emigrants have done remarkably well in a number of countries (for instance, in sub-Saharan Africa), and there is a powerful Islamic movement in Indonesia—the aforementioned Nahdatul-Ulama—that might yet play a role analogous to that of Opus Dei in the Catholic world. For years now, too, there has been an extended debate over the part played by Confucian-inspired values in the economic success stories of East Asia; if one is to credit the "post-Confucian thesis" (and also allow that Confucianism is a religion), then here would be a very important religious contribution to economic development.

One morally troubling aspect of this matter is that values functional at one period of economic development may not be functional at another. The values of the "Protestant ethic," or a functional equivalent thereof, are probably essential during the phase that Walt Rostow called "the take-off." It is not at all clear that this is the case in a later phase. Much *less* austere values may be *more* functional in the so-called post-industrial economies of Europe, North America, and East Asia. Frugality, however admirable from a moral viewpoint, may now actually be a vice, economically speaking. Undisciplined hedonists have a hard time climbing out of primitive poverty but, if they are bright enough, they can do very well in the high-tech, knowledge-driven economies of the advanced societies.

Finally, there is the effect of the religious resurgence on human rights and social justice worldwide. Religious institutions have, of course, made many statements on human rights and social justice. Some of these have had important political consequences, as in the civil rights struggle in the United States or in the collapse of communist regimes in Europe. But, as has already been mentioned, there are different religiously articulated views about the nature of human rights. The same goes for ideas about social justice; what is justice to some groups is gross injustice to others. Sometimes it is very clear that positions taken by religious groups on such matters are based on a religious rationale, as with the principled opposition to abortion and contraception by the Roman Catholic Church. At other times, though, positions on social justice, even if legitimated by religious rhetoric, reflect the location of the religious functionaries in this or that network of non-religious social classes and interests. To stay with the same example, most of the positions taken by American Catholic institutions on social justice issues other than those relating to sexuality and reproduction fall into this category.

This mixed analysis is emblematic of what must be our general conclusion. Both those who have great hopes for the role of religion in the affairs of this world and those who fear this role must be disappointed by the factual evidence, which, in the final analysis, points in not just one but several directions simultaneously. In assessing this role, there is no alternative to a nuanced, case-by-case approach. But one statement can be made with great confidence: Those who neglect religion in their analyses of contemporary affairs do so at great peril.

48

Popular Christianity and Political Extremism in the United States

JAMES AHO

Many believe that American political extremism is something that took place in the past and that it is unrelated to religion. According to the author of this article, neither of these beliefs is true.

As you read this selection, consider the following questions as guides:

1. *According to the author, extremists differ from others only in their biographies and that they belong to an extremist organization. Does this claim seem illogical to you? Why or why not?*
2. *Why do you believe or not believe in the role of political extremism in Christianity?*
3. *Think about a number of current movements in the United States and consider whether they are extremist. What would you do about them? Would you eliminate their tax-free status?*

SOURCE: From *Disruptive Religion: The Force of Faith in Social Movement Activism*, Christian Smith, ed. New York: Routledge, 1996, pp. 189–204.

EXTREMISM DEFINED

The word "extremism" is used rhetorically in everyday political discourse to disparage and undermine one's opponents. In this sense, it refers essentially to anyone who disagrees with me politically. In this chapter, however, "extremism" will refer exclusively to particular kinds of behaviors, namely, to nondemocratic actions, regardless of their ideology—that is, regardless of whether we agree with the ideas behind them or not (Lipset and Raab 1970: 4–17). Thus, extremism includes: (1) efforts to deny civil rights to certain people, including their right to express unpopular views, their right to due process at law, to own property, etc.; (2) thwarting attempts by others to organize in opposition to us, to run for office, or vote; (3) not playing according to legal constitutional rules of political fairness: using personal smears like "Communist Jew-fag" and "nigger lover" in place of rational discussion; and above all, settling differences by vandalizing or destroying the property or life of one's opponents. The test is not the end as such, but the means employed to achieve it.

CYCLES OF AMERICAN RIGHT-WING EXTREMISM . . .

American political history has long been acquainted with Christian-oriented rightist extremism. As early as the 1790s, for example, Federalist Party activists, inspired partly by Presbyterian and Congregationalist preachers, took-up arms against a mythical anti-Christian cabal known as the Illuminati—Illuminati = bringers of light = Lucifer, the devil.

The most notable result of anti-Illuminatism was what became popularly known as the "Reign of Terror": passage of the Alien and Sedition acts (1798). These required federal registration of recent immigrants to America from Ireland and France, reputed to be the homes of Illuminatism lengthened the time of naturalization to become a citizen from five to fourteen years, restricted "subversive" speech and newspapers—that is, outlets advocating liberal Jeffersonian or what were known then as "republican" sentiments—and permitted the deportation of "alien enemies" without trial.

The alleged designs of the Illuminati were detailed in a three hundred-page book entitled *Proofs of a Conspiracy Against All the Religions and Governments of Europe Carried on in the Secret Meetings of . . . Illuminati* (Robison 1967 [1798]). Over two hundred years later *Proofs of a Conspiracy* continues to serve as a source-book for right-wing extremist commentary on American social issues. Its basic themes are: (1) *manichaenism:* that the world is divided into the warring principles of absolute good and evil; (2) *populism:* that the citizenry naturally would be inclined to ally with the powers of good, but have become indolent, immoral, and uninformed of the present danger to themselves; (3) *conspiracy:* that this is because the forces of evil have enacted a scheme using educators, newspapers, music, and intoxicants to weaken the people's will and intelligence; (4) *anti-modernism*: that the results of the conspiracy are the very laws and institutions celebrated by the unthinking masses as "progressive": representative government, the separation of church and State, the extension of suffrage to the propertyless, free public education, public-health measures, etc.; and (5) *apocalypticism:* that the results of what liberals call social progress are increased crime rates, insubordination to "natural" authorities (such as royal families and property-owning Anglo-Saxon males), loss of faith, and the decline of common decency—in short, the end of the world.

Approximately every thirty years America has experienced decade-long popular resurrections of these five themes. While the titles of the alleged evil-doers in each era have been adjusted to meet changing circumstances, their program is said to have remained the same. They constitute a diabolic *Plot Against Christianity* (Dilling 1952). In the 1830s, the cabal was said to be comprised of the leaders of Masonic lodges: in the 1890s, they were accused of being Papists and Jesuits; in the 1920s, they were the Hidden Hand; in the 1950s, the Insiders or Force X; and today they are known as Rockefellerian "one-world" Trilateralists or Bilderbergers.

Several parallels are observable in these periods of American right-wing resurgence. First, while occasionally they have evolved into democratically-organized political parties holding conventions that nominate slates of candidates to run for office—the American Party, the Anti-Masonic Party, the People's Party, the Prohibition Party—more often, they have become secret societies in their own right, with arcane passwords, handshakes, and vestments, plotting campaigns of counter-resistance behind closed doors. That is, they come to mirror the fantasies against which they have taken up arms. Indeed, it is this ironic fact that typically occasions the public ridicule and undoing of these groups. The most notable examples are the Know Nothings, so-called because under interrogation they were directed to deny knowledge of the organization; the Ku Klux Klan, which during the 1920s had several million members; the Order of the Star Spangled Banner, which flourished during the 1890s; the Black Legion of Michigan, circa 1930; the Minutemen of the late 1960s; and most recently, the *Bruders Schweigen*, Secret Brotherhood, or as it is more widely known, The Order.

Secondly, the thirty-year cycle noted above evidently has no connection with economic booms and busts. While the hysteria of the 1890s took place during a nation-wide depression, McCarthyism exploded on the scene during the most prosperous era in American history. On close view, American right-wing extremism is more often associated with economic good times than with bad, the 1920s, the 1830s, and the 1980s being prime examples. On the contrary, the cycle seems to have more to do with the length of a modern generation than with any other factor.

Third, and most important for our purposes, Christian preachers have played pivotal roles in all American right-wing hysterias. The presence of Dan Gayman, James Ellison, and Bertrand Comparet spear-heading movements to preserve America from decline today continues a tradition going back to Jedidiah Morse nearly two centuries ago, continuing through Samuel D. Burchard, Billy Sunday, G. L. K. Smith, and Fred Schwarz's Christian Anti-Communist Crusade.

In the nineteenth century, the honorary title "Christian patriot" was restricted to white males with Protestant credentials. By the 1930s, however, Catholic ideologues, like the anti-Semitic radio priest Father Coughlin, had come to assume leadership positions in the movement. Today, somewhat uneasily, Mormons are included in the fold. The Ku Klux Klan, once rabidly anti-Catholic and misogynist, now encourages Catholic recruits and even allows females into its regular organization, instead of requiring them to form auxiliary groups.

CHRISTIANITY: A CAUSE OF POLITICAL EXTREMISM?

The upper Rocky Mountain region is the heartland of American right-wing extremism in our time. Montana, Idaho, Oregon, and Washington have the highest per capita rates of extremist groups of any area in the entire country (Aho 1994: 152–153). Research on the members of these groups show that they are virtually identical to the surrounding population in all respects but one (Aho 1991: 135–163)—they are not less formally educated than the surrounding population. Furthermore, as indicated by their rates of geographic mobility, marital stability, occupational choice, and conventional political participation, they are no more estranged from their local communities than those with whom they live. And finally, their social status seems no more threatened than that of their more moderate neighbors. Indeed, there exists anecdotal evidence that American right-wing extremists today are drawn from the more favored, upwardly-mobile sectors of society. They are college-educated, professional suburbanites residing in the rapidly-growing, prosperous Western states (Simpson 1983).

In other words, the standard sociological theories of right-wing extremism—theories holding, respectively, that extremists are typically undereducated, if not stupid, transient and alienated from ordinary channels of belonging, and suffer inordinately from status insecurity—find little empirical support. Additionally, the popular psychological notion that right-wing extremists are more neurotic than the general population, perhaps paranoid to the point of psychosis, can not be confirmed. None of the right-wing political murderers whose psychiatric records this author has accessed have been medically certified as insane (Aho 1991: 68–82; Aho 1994: 46–49). If this is true for right-wing murderers, it probably also holds for extremists who have not taken the lives of others.

The single way in which right-wing extremists *do* differ from their immediate neighbors is seen in their religious biographies. Those with Christian backgrounds generally, and Presbyterians, Baptists and members of independent fundamentalist Protestant groups specifically, all are overrepresented among intermountain radical patriots (Aho 1991: 164–182). Although it concerns a somewhat different population, this finding is consistent with surveys of the religious affiliations of Americans with conservative voting and attitudinal patterns (Lipset and Raab 1970: 229–232, 359–361, 387–392, 433–437, 448–452; Shupe and Stacey 1983: Wilcox 1992).

Correlations do not prove causality. Merely because American extremists are members of certain denominations and sects does not permit the conclusion that these religious groups compel their members to extremism. In the first place, the vast majority of independent fundamentalists, Baptists, and Presbyterians are not political extremists, even if they are inclined generally to support conservative causes. Secondly, it is conceivable that violently-predisposed individuals are attracted to particular religions because of what they hear from the pulpit; and what they hear channels their *already* violent inclinations in political directions. . . .

The point is not that every extremist is a violent personality searching to legitimize criminality with religion. Instead, the example illustrates the subtle ways in which religious belief, practice, and organization all play upon individual psychology to produce persons prepared to violate others in the name of principle. Let us look at each of these factors separately, understanding that in reality they intermesh in complicated, sometimes contradictory ways that can only be touched upon here.

BELIEF

American right-wing politics has appropriated from popular Christianity several tenets: the concept of unredeemable human depravity, the idea of America as a specially chosen people, covenant theology and the right to revolt, the belief in a national mission, millennialism, and anti-Semitism. Each of these in its own way has inspired rightist extremism.

The New Israel

The notion of America as the new Israel, for example, is the primary axiom of a fast-growing religiously-based form of radical politics known as Identity Christianity. Idaho's Aryan Nations Church is simply the most well-known Identity congregation. The adjective "identity" refers to its insistence that Anglo-Saxons are in truth the Israelites. They are "Isaac's-sons"—the Saxons—and hence the Bible is *their* historical record, not that of the Jews (Barkun 1994). The idea is that after its exile to what today is northern Iran around seven hundred B.C., the Israelites migrated over the Caucasus mountains—hence their racial type, "caucasian"—and settled in various European countries. Several of these allegedly still contain mementos of their origins: the nation of Denmark is said to be comprised of descendants from the tribe of Dan; the German-speaking Jutland, from the tribe of Judah; Catalonia, Scotland, from the tribe of Gad.

Covenant Theology

Identity Christianity is not orthodox Christianity. Nevertheless, the notion of America as an especially favored people, or as Ronald Reagan once said, quoting Puritan founders, a "city on a hill," the New Jerusalem, is widely shared by Americans. Reagan and most conservatives, of course, consider the linkage between America and Israel largely symbolic. Many right-wing extremists, however, view the relationship literally as an historical fact and for them, just as the ancient Israelites entered into a covenant with the Lord, America has done the same. According to radical patriots America's covenant is what they call the "organic Constitution." This refers to the original articles of the Constitution plus the first ten amendments, the Bill of Rights. Other amendments, especially the 16th establishing a federal income tax, are considered to have questionable legal status because allegedly they were not passed according to constitutional strictures.

The most extreme patriots deny the constitutionality of the 13th, 14th, and 15th amendments—those outlawing slavery and guaranteeing free men civil and political rights as full American citizens. Their argument is that the organic Constitution was written by white men exclusively for themselves and their blood descendents (Preamble 1986). Non-caucasians residing in America are considered "guest peoples" with no constitutional rights. Their continued residency in this country is entirely contingent upon the pleasure of their hosts, the Anglo-Saxon citizenry. According to some, it is now time for the property of these guests to be confiscated and they themselves exiled to their places of origin (Pace 1985).

All right-wing extremists insist that if America adheres to the edicts of the organic Constitution, she, like Israel before her, shall be favored among the world's nations. Her harvests shall be bountiful, her communities secure, her children obedient to the voices of their parents, and her armies undefeated. But if she falters in her faith, behaving in ways that contravene the sacred compact, then calamities, both natural and human-made, shall follow. This is the explanation for the widespread conviction among extremists today for America's decline in the world. In short, the federal government has established agencies and laws contrary to America's divine compact: these include the Internal Revenue Service; the Federal Reserve System; the Bureau of Alcohol, Tobacco and Firearms; the Forest Service; the Bureau of Land Management; Social Security; Medicare and Medicaid; the Environmental Protection Agency; Housing and Urban Development; and the official apparatus enforcing civil rights for "so-called" minorities.

Essentially, American right-wing extremists view the entire executive branch of the United States government as little more than "jack-booted Nazi thugs," to borrow a phrase from the National Rifle Association fund-raising letter: a threat to freedom of religion, the right to carry weapons, freedom of speech, and the right to have one's property secure from illegal search and seizure.

Clumsy federal-agency assaults, first on the Weaver family in northern Idaho in 1992, then on the Branch Davidian sect in Waco, Texas, in 1993, followed by passage of the assault weapons ban in 1994, are viewed as indicators that the organic Constitution presently is imperiled. This has been the immediate impetus for the appearance throughout rural and Western America of armed militias since the summer of 1994. The terrorists who bombed a federal building in Oklahoma City in the spring of 1995, killing one hundred sixty-eight, were associated with militias headquartered in Michigan and Arizona. One month after the bombing, the national director of the United States Militia Association warned that after the current government falls, homosexuals, abortionists, rapists, "unfaithful politicians," and any criminal not rehabilitated in seven years will be executed. Tax evaders will no longer be treated as felons; instead they will lose their library privileges (Sherwood 1995).

Millennialism

Leading to both the Waco and Weaver incidents was a belief on the victims' parts that world apocalypse is imminent. The Branch Davidians split from the Seventh-Day Adventists in 1935 but share with the mother church its own millenarian convictions. The Weavers received their apocalypticism from *The Late Great Planet Earth* by fundamentalist lay preacher Hal Lindsey (1970), a book that has enjoyed a wide reading on the Christian right.

Both the Davidians and the Weavers were imbued with the idea that the thousand-year-reign of Christ would be preceded by a final battle between the forces of light and darkness. To this end both had deployed elaborate arsenals to protect themselves from the anticipated invasion of "Babylonish troops." These, they feared, would be comprised of agents from the various federal bureaucracies mentioned above, together with UN troops stationed on America's borders awaiting orders from Trilateralists. Ever alert to "signs" of the impending invasion, both fired at federal officers who had come upon their property; and both ended up precipitating their own martyrdom. Far from quelling millenarian fervor, however, the two tragedies were immediately seized upon by extremists as further evidence of the approaching End Times.

Millenarianism is not unique to Christianity, nor to Western religions; furthermore, millenarianism culminating in violence is not new—in part because one psychological effect of end-time prophesying is a devaluation of worldly things, including property, honors, and human life. At the end of the first Christian millennium (A.D. 1000) as itinerant prophets were announcing the Second Coming, their followers were taking-up arms to prepare the way, and uncounted numbers died (Cohn 1967). It should not surprise observers if, as the second millennium draws to a close and promises of Christ's imminent return increase in frequency, more and more armed cults flee to the mountains, there to prepare for the final conflagration.

Anti-Semitism

Many post-Holocaust Christian and Jewish scholars alike recognize that a pervasive anti-Judaism can be read from the pages of the New Testament, especially in focusing on the role attributed to Jews in Jesus' crucifixion. Rosemary Ruether, for example, argues that anti-Judaism constitutes the "left-hand of Christianity," its archetypal negation (Ruether 1979). Although pre-Christian Greece and Rome were also critical of Jews for alleged disloyalty, anti-Semitism reached unparalleled heights in Christian theology, sometimes relegating Jews to the status of Satan's spawn, the human embodiments of Evil itself.

During the Roman Catholic era, this association became embellished with frightening myths and images. Jews—pictured as feces-eating swine and rats—were accused of murdering Christian children on high feast days, using their blood to make unleavened bread, and poisoning wells. Added to these legends were charges during the capitalist era that Jews control international banking and by means of usury have brought simple, kind-hearted Christians into financial ruin (Hay 1981 [1950]). All of this was incorporated into popular Protestant culture through, among other vehicles, Martin Luther's diatribe, *On the Jews and Their Lies*, a pamphlet that still experiences brisk sales from patriotic bookstores. This is one possible reason for a survey finding by Charles Glock and Rodney Stark that created a minor scandal in the late 1960s. Rigidly orthodox American Christians, they found, displayed far higher levels of Jew-hatred than other Christians, regardless of their education, occupation, race, or income (Glock and Stark 1966).

In the last thirty years there has been "a sharp decline" in anti-Semitic prejudice in America, according to Glock (1993: 68). Mainline churches have played some role in this decline by facilitating Christian-Jewish dialogue, de-emphasizing offensive scriptural passages, and ending missions directed at Jews. Nevertheless, ancient anti-Jewish calumnies continue to be raised by leaders of the groups that are the focus of interest in this reading. Far from being a product of neurotic syndromes like the so-called Authoritarian (or fascist) Personality, the Jew-hatred of many right-wing extremists today is directly traceable to what they have absorbed from these preachments, sometimes as children.

Human Depravity . . .

One of the fundamentals of Calvinist theology, appropriated into popular American Christianity, is this: a transcendent and sovereign God resides in the heavens, relative to whom the earth and its human inhabitants are utterly, hopelessly fallen. True, Calvin only developed a line of thought already anticipated in Genesis and amplified repeatedly over the centuries. However, with a lawyer's penetrating logic, Calvin brought this tradition to its most stark, pessimistic articulation. It is this belief that accompanied the Pilgrims in their venture across the Atlantic, eventually rooting itself in the American psyche.

From its beginnings, a particular version of the doctrine of human depravity has figured prominently in American right-wing extremist discourse. It has served as the basis of its perennial misogyny, shared by both men and women. The female, being supposedly less rational and more passive, is said to be closer to earth's evil. Too, the theology of world devaluation is the likely inspiration for the right-wing's gossipy preoccupation with the body's appetites and the "perilous eroticism of emotion," for its prudish fulminations against music, dance, drink, and dress, and for its homophobia. Here, too, is found legitimation for the right-wing's vitriol against Satanist ouiji boards, "Dungeons and Dragons," and New Age witchcrafters with their horoscopes and aroma-therapies, and most recently, against "pagan-earth-worshippers" and "tree hugging idolaters" (environmentalists). In standing tall to "Satan's Kids" and their cravenness, certain neo-Calvinists in Baptist, Presbyterian, and fundamentalist clothing accomplish their own purity and sanctification.

Conspiratorialism

According to Calvin, earthquakes, pestilence, famine, and plague should pose no challenge to faith in God. We petty, self-absorbed creatures have no right to question sovereign reason. But even in Calvin's time, and more frequently later, many Christians have persisted in asking: if God is truly all-powerful, all-knowing, and all-good, then how is evil possible? Why do innocents suffer? One perennial, quasi-theological response is conspiratorialism. In short, there are AIDS epidemics, murderous holocausts, rampant poverty, and floods because counter-poised to God there exists a second hidden force of nearly equal power and omniscience: the Devil and His human consorters—Jews, Jesuits, Hidden Hands, Insiders, Masons, and Bilderbergers.

By conspiratorialism, we are not referring to documented cases of people secretly scheming to destroy co-workers, steal elections, or run competitors out of business. Conspiracies are a common feature of group life. Instead, we mean the attempt to explain the entirety of human history by means of a cosmic Conspiracy, such as that promulgated in the infamous *Protocols of the Learned Elders of Zion*. This purports to account for all modern institutions by attributing them to the designs of twelve or thirteen—one representing each of the tribes of Israel—Jewish elders (Aho 1994, 68–82). *The Protocols* enjoys immense and endless popularity on the right; and has generated numerous spin-offs: *The International Jew, None Dare Call It Conspiracy*, and the *Mystery of [Jewish] Iniquity*, to name three.

To posit the existence of an evil divinity is heresy in orthodox Christianity. But, theological objections aside, it is difficult indeed for some believers to resist the temptation of intellectual certitude conspiratorialism affords. This certainty derives from the fact that conspiratorialism in the cosmic sense can not be falsified. Every historical event can, and often is, taken as further verification of conspiracies. If newspapers report a case of government corruption, this is evidence of government conspiracy; if they do not, this is evidence of news media complicity in the conspiracy. If the media deny involvement in a cover-up, this is still further proof of their guilt; if they admit to having sat on the story, this is surely an admission of what is already known.

PRACTICE

Christianity means more than adhering to a particular doctrine. To be Christian is to live righteously. God-fearing righteousness may either be understood as a *sign* of one's salvation, as in orthodox Christianity or, as in Mormonism, a way to *earn* eternal life in the celestial heavens.

Nor is it sufficient for the faithful merely to display righteousness in their personal lives and businesses, by being honest, hardworking, and reliable. Many Christians also are obligated to witness to, or labor toward, salvation in the political arena; to work with others to remake this charnel-house world after the will of God; to help establish God's kingdom on earth. Occasionally this means becoming involved in liberal causes—abolitionism, civil rights, the peace and ecological movements; often it has entailed supporting causes on the right. In either case it may require that one publicly stand up to evil. For, as Saint Paul said, to love God is to hate what is contrary to God.

Such a mentality may lead to "holy war," the organized effort to eliminate human fetishes of evil (Aho 1994: 23–34). For some, in cleansing the world of putrefaction their identity as Christian is recognized, it is re-known. This is not to argue that holy war is unique to Christianity, or that all Christians participate in holy wars. Most Christians are satisfied to renew their faith through the rites of Christmas, Easter, baptism, marriage, or mass. Furthermore, those who *do* speak of holy war often use it metaphorically to describe a private spiritual battle against temptation, as in "I am a soldier of Christ, therefore I am not permitted to fight" (Sandford 1966). Lastly, even holy war in the political sense does not necessarily imply the use of violence. Although they sometimes have danced tantalizingly close to extremism (in the sense defined earlier), neither Pat Robertson nor Jerry Falwell, for example, have advocated non-democratic means in their "wars" to avert America's decline.

Let us examine the notion of Christian holy war more closely. The sixteenth-century father of Protestant reform, Martin Luther, repudiated the concept of holy war, arguing that there exist two realms: holiness, which is the responsibility of the Church, and warfare, which falls under the State's authority (Luther 1974). Mixing these realms, he says, perverts the former while unnecessarily hamstringing the latter. This does not mean that Christians may forswear warfare, according to Luther. In his infinite wisdom, God has ordained princes to quell civil unrest and protect nations from invasion. . . . To this day, Lutherans generally are less responsive to calls for holy wars than many other Christians.

John Calvin, on the other hand, rejected Luther's proposal to separate church from State. Instead, his goal was to establish a Christocracy in Geneva along Roman Catholic lines, and to attain this goal through force, if need be, as Catholicism had done. Calvin says that not only is violence to establish God's rule on earth permitted, it is commanded. "Good brother, we must bend unto all means that give furtherance to the holy cause" (Walzer 1965: 17, 38, 68–87, 90–91, 100–109; see Troeltsch 1960: 599–601, 651–652, 921–922 n. 399). . . . And it was the Calvinist ethic, not that of Luther, that was imported to America by the Puritans, informing the politics of Presbyterians and Congregationalists—the immediate heirs of Calvinism—as well as some Methodists and many Baptists. Hence, it is not surprising that those raised in these denominations are often overrepresented in samples of "saints" on armed crusades to save the world for Christ.

Seminal to the so-called pedagogic or educational function of holy war are two requirements. First, the enemy against whom the saint fights must be portrayed in terms appropriate to his status as a fetish of evil. Second, the campaign against him must be equal to his diabolism. It must be terrifying, bloodthirsty, uncompromising.

"Prepare War!" was issued by the now defunct Covenant, Sword and the Arm of the Lord, a fundamentalist Christian paramilitary commune headquartered in Missouri. A raid on the compound in the late 1980s uncovered one of the largest private arms caches ever in American history. . . . When the Lord God has delivered these enemies into our hands, warns the pamphlet quoting the Old Testament, "thou shalt save alive nothing that breatheth: but thou shalt utterly destroy them" (CSA n.d.: 20; see Deuteronomy 20: 10–18).

The 1990s saw a series of State-level initiatives seeking to deny homosexuals civil rights. Although most of these failed by narrow margins, one in Colorado was passed (later to be adjudged unconstitutional), due largely to the efforts of a consortium of fundamentalist Christian churches. . . . Acknowledging that the title of their pamphlet "Death Penalty for Homosexuals" would bring upon them the wrath of liberals, its authors insist that "such slanderous tactics" will not deter the anti-homosexual campaign. "For truth will ultimately prevail, no matter how many truth-bearers are stoned." And what precisely is this truth? It is that the Lord Himself has declared that "if a man also lie with mankind, as he lieth with a woman, both of them have committed an abomination: they shall surely be put to death; their blood shall be upon them" (Peters 1992: i; see Leviticus 20:13). . . .

What should Christians do in the face of this looming specter, asks the pamphlet? "We, today, can and should have God's Law concerning Homosexuality and its judgment of the death penalty." For "they which commit such things," says the apostle Paul, "are worthy of death" (CSA n.d.: 15; see Romans 1:27–32). Extremism fans the flames of extremism.

ORGANIZATION

Contrary to popular thinking, people rarely join right-wing groups because they have a prior belief in doctrines such as those enumerated above. Rather, they come to believe because they have first joined. That is, people first affiliate with right-wing activists and only then begin altering their intellectual outlooks to sustain and strengthen these ties. The original ties may develop from their jobs, among neighbors, among prison acquaintances, or through romantic relationships. . . .

The point of this story is the sociological truth that the way in which some people become right-wing extremists is indistinguishable from the way others become vegetarians, peace activists, or members of mainline churches (Lofland and Stark 1965; Aho 1991: 185–211). *Their affiliations are mediated by significant others already in the movement.* It is from these others that they first learn of the cause; sometimes it is through the loaning of a pamphlet or videotape; occasionally it takes the form of an invitation to a meeting or workshop. As the relationship with the other tightens, the recruit's viewpoint begins to change. At this stage old friends, family members, and cohorts, observing the recruit spending inordinate time with "those new people," begin their interrogations: "What's up with you, man?" In answer, the new recruit typically voices shocking things: bizarre theologies, conspiracy theories, manichaeistic worldviews. Either because of conscious "disowning" or unconscious avoidance, the recruit finds the old ties loosening, and as they unbind, the "stupidity" and "backwardness" of prior acquaintances becomes increasingly evident.

Pushed away from old relationships and simultaneously pulled into the waiting arms of new friends, lovers, and comrades, the recruit is absorbed into the movement. Announcements of full conversion to extremism follow. To display commitment to the cause, further steps may be deemed necessary: pulling one's children out of public schools where "secular humanism" is taught; working for radical political candidates to stop America's "moral decline"; refusing to support ZOG with taxes; renouncing one's citizenship and throwing away social security card and driver's license; moving to a rugged wilderness to await the End Times. Occasionally it means donning camouflage, taking up high-powered weaponry, and confronting the "forces of satan" themselves.

There are two implications to this sociology of recruitment. First and most obviously, involvement in social networks is crucial to being mobilized into right-wing activism. Hence, contrary to the claims of the estrangement theory of extremism mentioned above, those who are truly isolated from their local communities are the last and least likely to become extremists themselves. My research (Aho 1991, 1994) suggests that among the most important of these community ties is membership in independent fundamentalist, Baptist, or Presbyterian congregations.

Secondly, being situated in particular networks is largely a matter of chance. None of us choose our parents. Few choose their co-workers, fellow congregants, or neighbors, and even friendships and marriages are restricted to those available to us by the happenstance of our geography and times. What this means is that almost any person could find themselves in a Christian patriot communications network that would position them for recruitment into right-wing extremism.

As we have already pointed out, American right-wing extremists are neither educationally nor psychologically different from the general population. Nor are they any more status insecure than other Americans. What makes them different is how they are socially positioned. This positioning includes their religious affiliation. Some people find themselves in churches that expose them to the right-wing world. This increases the likelihood of their becoming right-wingers.

CONCLUSION

Throughout American history, a particular style of Christianity has nurtured right-wing extremism. Espousing doctrines like human depravity, white America as God's elect people, conspiratorialism, Jews as Christ killers, covenant theology and the right to revolt, and millennialism, this brand of Christianity is partly rooted in orthodox Calvinism and in the theologically questionable fantasies of popular imagination. Whatever its source, repeatedly during the last two centuries, its doctrines have served to prepare believers cognitively to assume hostile attitudes toward "un-Christian"—hence un-American—individuals, groups, and institutional practices.

This style of Christianity has also given impetus to hatred and violence through its advocacy of armed crusades against evil. Most of all, however, the cults, sects, and denominations wherein this style flourishes have served as mobilization centers for recruitment into right-wing causes. From the time of America's inception, right-wing political leaders in search of supporters have successfully enlisted clergymen who preach these principles to bring their congregations into the fold in "wars" to save America for Christ.

It is a mistake to think that modern Americans are more bigoted and racist than their ancestors were. Every American generation has experienced right-wing extremism, even that occasionally erupting into vigilante violence of the sort witnessed daily on the news today. What is different in our time is the sophistication and availability of communications and weapons technology. Today, mobilizations to right-wing causes has been infinitely enhanced by the availability of personal computer systems capable of storing and retrieving information on millions of potential recruits. Mobilization has also been facilitated by cheap shortwave radio and cable-television access, the telephone tree, desktop publishing, and readily available studio-quality recorders. Small coteries of extremists can now activate supporters across immense distances at the touch of a button. Add to this the modern instrumentality for maiming and killing available to the average American citizen: military-style assault weaponry easily convertible into fully automatic machine guns, powerful explosives manufacturable from substances like diesel oil and fertilizer, harmless in themselves, hence purchasable over-the-counter. Anti-tank and aircraft weapons, together with assault vehicles, have also been uncovered recently in private-arms caches in the Western states.

Because of these technological changes, religious and political leaders today have a greater responsibility to speak and write with care regarding those with whom they disagree. Specifically, they must control the temptation to demonize their opponents, lest, in their declarations of war they bring unforeseen destruction not only on their enemies, but on themselves.

REFERENCES

Aho, J. 1991. *The Politics of Righteousness: Idaho Christian Patriotism.* Seattle: University of Washington Press.

———. 1994. *This Thing of Darkness: A Sociology of the Enemy.* Seattle: University of Washington Press.

Barkun, M. 1994. *Religion and the Racist Right: The Origins of the Christian Identity Movement.* Chapel Hill: North Carolina University Press.

Cohn, N. 1967. *The Pursuit of the Millennium.* New York: Oxford University Press.

Dilling, E. 1952. *The Plot Against Christianity,* n.p.

Glock, C. 1993. "The Churches and Social Change in Twentieth-Century America." *Annals of the American Academy of Political and Social Science.* 527: 67–83.

Glock, C. and R. Stark. 1966. *Christian Beliefs and Anti-Semitism*. New York: Harper & Row.

Hay, M. 1981 (1950). *The Roots of Christian Anti-Semitism*. New York: Anti-Defamation League of B'nai B'rith.

Lindsey, H. 1970. *The Late Great Planet Earth*. Grand Rapids: Zondervan.

Lipset, S. M. and E. Raab. 1970. *The Politics of Unreason: Right-Wing Extremism in America*, 1790–1970. New York: Harper & Row.

Lofland, J. and R. Stark. 1965. "Becoming a World-Saver: A Theory of Conversation to a Deviant Perspective." *American Sociological Review* 30: 862–875.

Luther, M. 1974. *Luther: Selected Political Writings*, J. M. Porter, ed. Philadelphia: Fortress Press.

Mannheim, K. 1952. "The Problem of Generations," in *Essays in the Sociology of Knowledge*. London: Routledge and Kegan Paul.

Nisbet, R. 1953. *The Quest for Community*. New York: Harper and Brothers.

Pace, J. O. 1985. Amendment to the Constitution. Los Angeles: Johnson, Pace, Simmons and Fennel.

Peters, P. 1992. *Death Penalty for Homosexuals*. LaPorte, Colorado: Scriptures for America.

Preamble. 1986. "Preamble to the United States Constitution: Who Are the Posterity?" Oregon City, Oregon: Republic vs. Democracy Redress.

Robison, J. 1967 (1798). *Proofs of a Conspiracy....* Los Angeles: Western Islands.

Ruether, R. 1979. *Faith and Fratricide: The Theological Roots of Anti-Semitism*. New York: Seabury.

Sandford, F. W. 1966. *The Art of War for the Christian Soldier*. Amherst, New Hampshire: Kingdom Press.

Schlesinger, A. 1986. *The Cycles of American History*. Boston: Houghton Mifflin.

Sherwood, "Commander" S. 1995. Quoted in *Idaho State Journal*. May 21.

Shupe, A. and W. Stacey. 1983. "The Moral Majority Constituency," in *The New Christian Right*, R. Liebman and R. Wuthnow, eds. New York: Aldine.

Simpson, J. 1983. "Moral Issues and Status Politics," in *The New Christian Right*, R. Liebman and R. Wuthnow, eds. New York: Aldine.

Solt, L. 1971. *Saints in Arms: Puritanism and Democracy in Cromwell's Army*. New York: AMS Press.

Stark, R. and William Bainbridge. 1985. The *Future of Religion: Secularization, Revival and Cult Formation*. Berkeley: University of California Press.

Stouffer, S. A. 1966. *Communism, Conformity and Civil Liberties*. New York: John Wiley.

Troeltsch, E. 1960. *Social Teachings of the Christian Churches*. Trans. by O. Wyon. New York: Harper & Row.

Watzer, M. 1965. *The Revolution of the Saints*. Cambridge, MA: Harvard University Press.

Wilcox, C. 1992. *God's Warriors: The Christian Right in Twentieth Century America*. Baltimore, MD: Johns Hopkins University Press.

49

Motivation and Ideology

What Drives the Anti-Abortion Movement

DALLAS A. BLANCHARD

Blanchard evaluates the anti-abortion movement, who joins it and why, and the belief systems that drive this and other fundamentalist causes. The anti-abortion movement involves many belief systems and activities, even a way of life for some participants.

As you read, ask yourself the following questions:

1. *Who are those most likely to join the anti-abortion movement, and why?*
2. *What religious and social factors stimulate people to join movements?*
3. *What do fundamentalists from different religions have in common?*
4. *Have you ever joined a social movement? What motivated you to join?*

GLOSSARY **Social movement** Organized activities to encourage or oppose some aspect of change (in this case oppose abortion). **Fundamentalism** Adherence to traditional religious or cultural norms such as respect for authority.

WHY AND HOW PEOPLE JOIN THE ANTI-ABORTION MOVEMENT

Researchers have posited a variety of explanations for what motivates people to join the anti-abortion movement. As with any other social movement, the anti-abortion movement has within it various subgroups, or organizations, each of which attracts different kinds of participants and expects different levels of participation. It might in fact be more appropriate to speak of anti-abortion *movements*.

Those opposing abortion are not unified. Some organizations have a single-issue orientation, opposing abortion alone, while others take what they consider to be a "pro-life" stance on many issues, opposing abortion as well as euthanasia, capital punishment, and the use of nuclear and chemical arms. . . .

Researchers have identified a number of pathways for joining the anti-abortion movement. Luker, in her 1984 study of the early California movement, found that activists in the initial stages of the movement found their way to it through professional associations. The earliest opponents of abortion liberalization were primarily physicians and attorneys who disagreed with their professional associations' endorsement of abortion reform. It is my hypothesis that membership in organizations that concentrate on the education of the public or religious constituencies and on political lobbying is orchestrated primarily through professional networks. With the passage of the California reform

SOURCE: From *The Anti-Abortion Movement and the Rise of the Religious Right: From Polite to Fiery Protest* by Dallas A. Blanchard. New York: Twayne, 1994. Reprinted by permission of the Gale Group.

bill and the increase in abortion rates several years later, many recruits to the movement fell into the category Luker refers to as "self-selected"; that is, they were not recruited through existing networks but sought out or sometimes formed organizations through which to express their opposition.

Himmelstein (1984), in summarizing the research on the anti-abortion movement available in the 10 years following *Roe v. Wade*, concluded that religious networks were the primary source of recruitment. Religious networks appear to be more crucial in the recruitment of persons into high-profile and/or violence prone groups (Blanchard and Prewitt 1993)—of which Operation Rescue is an example—than into the earlier, milder activist groups (although such networks are generally important throughout the movement). Such networks were also important, apparently, in recruitment into local Right to Life Committees, sponsored by the National Right to Life Committee and the Catholic church. The National Right to Life Committee, for example, is 72 percent Roman Catholic (Granberg 1981). It appears that the earliest anti-abortion organizations were essentially Catholic and dependent on church networks for their members; the recruitment of Protestants later on has also been dependent on religious networks (Cuneo 1989, Maxwell 1992).

Other avenues for participation in the anti-abortion movement opened up through association with other issues. Feminists for Life, for example, was founded by women involved in the feminist movement. Sojourners, a socially conscious evangelical group concerned with issues such as poverty and racism, has an anti-abortion position. Some anti-nuclear and anti-death-penalty groups have also been the basis for the organization of anti-abortion efforts.

Clearly, preexisting networks and organizational memberships are crucial in initial enlistment into the movement. Hall (1993) maintains that individual mobilization into a social movement requires the conditions of attitudinal, network, and biographical availability. My conclusions regarding the anti-abortion movement support this contention. Indeed, biographical availability—the interaction of social class, occupation, familial status, sex, and age—is particularly related to the type of organization with which and the level of activism at which an individual will engage.

General social movement theory places the motivation to join the anti-abortion movement into four basic categories: status defense; anti-feminism; moral commitment; and cultural fundamentalism, or defense.

The earliest explanation for the movement was that participants were members of the working class attempting to shore up, or defend, their declining social status. Clarke, in his 1987 study of English anti-abortionists finds this explanation to be inadequate, as do Wood and Hughes in their 1984 investigation of an anti-pornography movement group.

Petchesky (1984) concludes that the movement is basically anti-feminist—against the changing status of women. From this position, the primary goal of the movement is to "keep women in their place" and, in particular, to make them suffer for sexual "libertinism." Statements by some anti-abortion activists support this theory. Cuneo (1989), for example, finds what he calls "sexual puritans" on the fringe of the anti-abortion movement in Toronto. Abortion opponent and longtime right-wing activist Phyllis Schlaffley states this position: "It's very healthy for a young girl to be deterred from promiscuity by fear of contracting a painful, incurable disease, or cervical cancer, or sterility, or the likelihood of giving birth to a dead, blind, or brain-damaged baby (even ten years later when she may be happily married)" (Planned Parenthood pamphlet, no title, n.d. [1990]).... A number of researchers have concluded that sexual moralism is the strongest predictor of anti-abortion attitudes.

The theory of moral commitment proposes that movement participants are motivated by concern for the human status of the fetus. It is probably as close as any explanation comes to "pure altruism." Although there is a growing body of research on altruism, researchers on the abortion issue have tended to ignore this as a possible draw to the movement, while movement participants almost exclusively claim this position: that since the fetus is incapable of defending itself, they must act on its behalf.

In examining and categorizing the motivations of participants in the anti-abortion movement in Toronto, Cuneo (1989:85ff.) found only one category—civil rights—that might be considered altruistic. The people in this category tend to be nonreligious and embarrassed by the activities of religious activists; they feel that fetuses have a right to exist but cannot speak for themselves. Cuneo's other primary categories of motivation are characterized by concerns related to the "traditional" family, the status of women in the family, and religion. He also finds an activist fringe composed of what he calls religious seekers; sexual therapeutics, "plagued by guilt and fear of female sexual power" (115); and punitive puritans, who want to punish women for sexual transgressions. All of Cuneo's categories of participant, with the exception of the civil rights category, seek to maintain traditional male/female hierarchies and statuses.

The theory of cultural fundamentalism, or defense, proposes that the anti-abortion movement is largely an expression of the desire to return to what its proponents perceive to be "traditional culture." This theory incorporates elements of the status defense and anti-feminist theories.

It is important to note that a number of researchers at different points in time (Cuneo 1989; Ginsburg 1990; Luker 1984; Maxwell 1991, 1992) have indicated that (1) there have been changes over time in who gets recruited into the movement and why, (2) different motivations tend to bring different kinds of people into different types of activism, and (3) even particular movement organizations draw different kinds of people with quite different motivations. At this point in the history of the anti-abortion movement, the dominant motivation, particularly in the more activist organizations such as Operation Rescue, appears to be cultural fundamentalism. Closely informing cultural fundamentalism are the tenets of religious fundamentalism, usually associated with certain Protestant denominations but also evident in the Catholic and Mormon faiths. . . .

RELIGIOUS AND CULTURAL FUNDAMENTALISM DEFINED

Cultural fundamentalism is in large part a protest against cultural change: against the rising status of women; against the greater acceptance of "deviant" life-styles such as homosexuality; against the loss of prayer and Bible reading in the schools; and against the increase in sexual openness and freedom. Wood and Hughes (1984) describe cultural fundamentalism as "adherence to traditional norms, respect for family and religious authority, asceticism and control of impulse. Above all, it is an unflinching and thoroughgoing moralistic outlook on the world; moralism provides a common orientation and common discourse for concerns with the use of alcohol and pornography, the rights of homosexuals, 'pro-family' and 'decency' issues." The theologies of Protestants and Catholics active in the anti-abortion movement—many of whom could also be termed fundamentalists—reflect these concerns. . . .

There are at least six basic commonalities to what can be called Protestant, Catholic, and Mormon fundamentalisms: (1) an attitude of certitude—that one may know the final truth, which includes antagonism to ambiguity; (2) an external source for that certitude—the Bible or church dogma; (3) a belief system that is at root dualistic; (4) an ethic based on the "traditional" family; (5) a justification for violence; and, therefore, (6) a rejection of modernism (secularization). . . .

Taking those six commonalities point by point:

1. The certitude of fundamentalism rests on dependence on an external authority. That attitude correlates with authoritarianism, which includes obedience to an external authority, and, on that basis, the willingness to assert authority over others.
2. While the Protestant fundamentalists accept their particular interpretation of the King James Version of the Bible as the authoritative source, Mormon and Catholic fundamentalists tend to view church dogma as authoritative.

3. The dualism of Catholics, Protestants, and Mormons includes those of body/soul, body/mind, physical/spiritual. More basically, they see a distinction between God and Satan, the forces of good and evil. In the fundamentalist worldview, Satan is limited and finite; he can be in only one place at one time. He has servants, however, demons who are constantly working his will, trying to deceive believers. A most important gift of the Spirit is the ability to distinguish between the activities of God and those of Satan and his demons.
4. The "traditional" family in the fundamentalist view of things has the father as head of the household, making the basic decisions, with the wife and children subject to his wishes. Obedience is stressed for both wives and children. Physical punishment is generally approved for use against both wives and children.

   This "traditional" family with the father as breadwinner and the mother as homemaker, together rearing a large family, is really not all that traditional. It arose on the family farm, prior to 1900, where large numbers of children were an economic asset. Even then, women were essential in the work of the farm. . . . In the urban environment, the "traditional" family structure was an option primarily for the middle and upper classes, and they limited their family size even prior to the development of efficient birth control methods. Throughout human history women have usually been breadwinners themselves, and the "traditional" family structure was not an option.
5. The justification for violence lies in the substitutionary theory of the atonement theology of both Protestants and Catholics. In this theory, the justice of God demands punishment for human sin. This God also supervises a literal hell, the images of which come more from Dante's *Inferno* than from the pages of the Bible. Fundamentalism, then, worships a violent God and offers a rationale for human violence (such as Old Testament demands for death when adultery murder, and other sins are committed). The fundamentalist mindset espouses physical punishment of children, the death penalty, and the use of nuclear weapons; fundamentalists are more frequently wife abusers, committers of incest, and child abusers.
6. Modernism entails a general acceptance of ambiguity contingency, probability (versus certitude), and a unitary view of the universe that is, the view that there is no separation between body and soul, physical and spiritual, body and mind (when the body dies, the self is thought to die with it). Rejection of modernism and postmodernism is inherent in the rejection of a unitary worldview in favor of a dualistic worldview. The classic fundamentalist position embraces a return to religion as the central social institution, with education, the family, economics, and politics serving religious ends, fashioned after the social structure characteristic of medieval times.

Also characteristic of Catholic, Protestant, and Mormon fundamentalists are beliefs in individualism (which supports a naive capitalism); pietism; a chauvinistic Americanism (among some fundamentalists) that sees the United States as the New Israel and its inhabitants as God's new chosen people; and a general opposition to intellectualism, modern science, the tenets of the Social Gospel, and communism. (Some liberal Christians may share some of these views.) Amid this complex of beliefs and alongside the opposition to evolution, interestingly, is an underlying espousal of social Darwinism, the "survival of the fittest" ethos that presumes American society to be truly civilized, the pinnacle of social progress. This nineteenth-century American neocolonialism dominates the contemporary political views held by the religious right. It is also inherent in their belief in individualism and opposition to social welfare programs.

Particular personality characteristics also correlate with the fundamentalist syndrome: authoritarianism, self-righteousness, prejudice against minorities, moral absolutism (a refusal to compromise on perceived moral issues), and anti-analytical, anti-critical thinking. Many fundamentalists refuse to accept ambiguity as a given in moral decision making and tend to arrive at simplistic solutions to complex problems. For example, many hold that the solution to changes in the contemporary family can be answered by fathers' reasserting their primacy, by forcing their children and wives into blind obedience. Or, they say, premarital sex can be prevented by promoting abstinence. One popular spokesman, Tim LaHaye, asserts that the antidote to sexual desire, especially on the part of teenagers, lies in censoring reading materials (LaHaye 1980). Strict parental discipline automatically engenders self-discipline in children, he asserts. The implication is that enforced other-directedness by parents produces inner-directed children, while the evidence indicates that they are more likely to exchange parental authoritarianism for that of another parental figure. To develop inner-direction under such circumstances requires, as a first step, rebellion against the rejection of parental authority—the opposite of parental intent.

One aspect of fundamentalism, particularly the Protestant variety, is its insistence on the subservient role of women. The wife is expected to be subject to the direction of her husband, children to their father. While Luker (1984) found that anti-abortionists in California supported this position and that proponents of choice generally favored equal status for women, recent research has shown that reasons for involvement in the anti-abortion movement vary by denomination. That is, some Catholics tend to be involved in the movement more from a "right to life" position, while Protestants and other Catholics are more concerned with sexual morality. The broader right to life position is consistent with the official Catholic position against the death penalty and nuclear arms, while Protestant fundamentalists generally support the death penalty and a strong military. Thus, Protestant fundamentalists, and some Catholic activists, appear to be more concerned with premarital sexual behavior than with the life of the fetus.

Protestants and Catholics (especially traditional, ethnic Catholics) however, are both concerned with the "proper," or subordinate, role of women and the dominant role of men. Wives should obey their husbands, and unmarried women should refrain from sexual intercourse. Abortion, for the Protestants in particular, is an indication of sexual licentiousness (see, for example, LaHaye 1980). Therefore, the total abolition of abortion would be a strong deterrent to such behavior helping to reestablish traditional morality in women. Contemporary, more liberal views of sexual morality cast the virgin female as deviant. The male virgin has long been regarded as deviant. The fundamentalist ethic appears to accept this traditional double standard with its relative silence on male virginity.

Another aspect of this gender role ethic lies in the home-related roles of females. Women are expected to remain at home, to bear children, and to care for them, while also serving the needs of their husbands. Again, this is also related to social class and the social role expectations of the lower and working classes, who tend to expect women to "stay in their place."

Luker's (1984) research reveals that some women in the anti-abortion movement are motivated by concern for maintaining their ability to rely on men (husbands) to support their social roles as mothers, while pro-choice women tend to want to maintain their independent status. Some of the men involved in anti-abortion violence are clearly acting out of a desire to maintain the dependent status of women and the dominant roles of men. Some of those violent males reveal an inability to establish "normal" relationships with women, which indicates that their violence may arise from a basic insecurity with the performance of normal male roles in relationships with women. This does not mean that these men do not have relationships with women. Indeed, it is in the context of relationships with women that dominance-related tendencies become more manifest. It is likely that insecurity-driven behaviors are characteristic of violent males generally, but psychiatric data are not available to confirm this, even for the population in question.

THE COMPLEX OF FUNDAMENTALIST ISSUES

The values and beliefs inherent to religious and cultural fundamentalism are expressed in a number of issues other than abortion. Those issues bear some discussion here, particularly as they relate to the abortion question.

1. *Contraception.* Fundamentalists, Catholic, Protestant, and Mormon, generally oppose the use of contraceptives since they limit family size and the intentions of God in sexuality. They especially oppose sex education in the schools and the availability of contraceptives to minors without the approval of their parents. (See *Nightline*, 21 July 1989.) This is because control of women and sexuality are intertwined. If a girl has knowledge of birth control, she is potentially freed of the threat of pregnancy if she becomes sexually active. This frees her from parental control and discovery of illegitimate sexual intercourse.
2. *Prenatal testing, pregnancies from rape or incest, or those endangering a woman's life.* Since every pregnancy is divinely intended, opposition to prenatal testing arises from its use to abort severely defective fetuses and, in some cases, for sex selection. Abortion is wrong regardless of the origins of the pregnancy or the consequences of it.
3. *In vitro fertilization, artificial fertilization, surrogate motherhood.* These are opposed because they interfere with the "natural" fertilization process and because they may mean the destruction of some fertilized embryos.
4. *Homosexuality.* Homophobia is characteristic of fundamentalism, because homosexual behavior is viewed as being "unnatural" and is prohibited in the Bible.
5. *Uses of fetal tissue.* The use of fetal tissue in research and in the treatment of medical conditions such as Parkinson's disease is opposed, because it is thought to encourage abortion. (See *New York Times*, 16 August 1987; *Good Morning America*, 25 July 1991; *Face to Face with Connie Chung*, 25 November 1989; and *Nightline*, 6 January 1988.)
6. *Foreign relations issues.* Fundamentalists generally support aid to Israel and military funding (Diamond 1989). Indeed, as previously mentioned, they commonly view the United States as the New Israel. Protestant fundamentalists tend to be premillennialists, who maintain that biblical prophecies ordain that the reestablishment of the State of Israel will precede the Second Coming of Christ. Thus, they support aid to Israel to hasten the Second Coming, which actually, then, has an element of anti-Semitism to it, since Jews will not be among the saved.
7. *Euthanasia.* So-called right to life groups have frequently intervened in cases where relatives have sought to remove a patient from life-support systems. Most see a connection with abortion in that both abortion and removal of life-support interfere with God's decision as to when life should begin and end.

The most radical expression of cultural fundamentalism is that of Christian Reconstructionism. . . . The adherents of Christian Reconstructionism, while a distinct minority, have some congregations of up to 12,000 members and count among their number Methodists, Presbyterians, Lutherans, Baptists, Catholics, and former Jews. They are unabashed theocratists. They believe every area of life—law, politics, the arts, education, medicine, the media, business, and especially morality—should be governed in accordance with the tenets of Christian Reconstructionism. Some, such as Gary North, a prominent reconstructionist and son-in-law of Rousas John Rushdoony, considered the father of reconstructionism, would deny religious liberty—the freedom of religious expression—to "the enemies of God," whom the reconstructionists, of course, would identify.

The reconstructionists want to establish a "God-centered government," a Kingdom of God on Earth, instituting the Old Testament as the Law of the Land. The goal of reconstructionism is to reestablish biblical, Jerusalemic society. Their program is quite specific. Those criminals which the Old Testament condemned to death would be executed, including homosexuals, sodomites, rapists, adulterers, and "incorrigible" youths. Jails would become primarily holding tanks for those awaiting execution or assignment as servants indentured to those whom they wronged as one form of restitution. The media would be censored extensively to reflect the views of the church. Public education and welfare would be abolished (only those who work should eat), and taxes would be limited to the tithe, 10 percent of income, regardless of income level, most of it paid to the church. Property, Social Security, and inheritance taxes would be eliminated. Church elders would serve as judges in courts overseeing moral issues, while "civil" courts would handle other issues. The country would return to the gold standard. Debts, including, for example, 30-year mortgages, would be limited to six years. In short, Christian Reconstructionists see democracy as being opposed to Christianity, as placing the rule of man above the rule of God. They also believe that "true" Christianity has its earthly rewards. They see it as the road to economic prosperity, with God blessing the faithful.

REFERENCES

Editors' Note: The original chapter from which this selection was taken has extensive footnotes and references that could not be listed in their entirety here. For more explanation and documentation of sources, please see the source note on page 368.

Blanchard, Dallas A. and Terry J. Prewitt. 1993. *Religious Violence Abortion: The Gideon Project*. Gainesville: University Press of Florida.

Clarke, Alan. 1987. "Collective Action against Abortion Represents a Display of, and Concern for, Cultural Values, Rather than an Expression of Status Discontent." *British Journal of Sociology* 38:235–53.

Cuneo, Michael. 1989. *Catholics against the Church: Anti-Abortion Protest in Toronto, 1969–1985*. Toronto: University of Toronto Press.

Diamond, Sara. 1989. *Spiritual Warfare: The Politics of the Christian Right*. Boston: South End Press.

Ginsburg, Faye. 1990. *Contested Lives: The Abortion Debate in an American Community*. Berkeley: University of California Press.

Granberg, Donald. 1981. "The Abortion Activists." *Family Planning Perspectives* 18:158–61.

Hall, Charles. 1993. "Social Networks and Availability Factors: Mobilizing Adherents for Social Movement Participation" (Ph.D. dissertation, Purdue University).

Himmelstein, Jerome L. 1984. "The Social Basis of Anti-Feminism: Religious Networks and Culture." *Journal for the Scientific Study of Religion* 25:1–25.

50

The Decline of the American Civil Religion?

GORDON CLANTON AND SHOON LIO

The last election had politicians praising religion and patriotism in equal measures, yet the authors claim that both religion and "civil" religion are in decline. The authors argue that both traditional religion and "civil" religion are on the decline.

As you read this article, consider the following questions as guides:

1. *Do you believe that church religion is in decline in this country? What evidence supports your answer?*
2. *Do you believe that American civil religion is on the decline in this country? Why?*
3. *The myths of the civil religion are drawn from the main events of American history. Think of these events and decide which have myths enshrining them.*

GLOSSARY **Transcendent** Having to do with the highest concerns and value **Metanarrative** Grand stories of the nations, its heroes, its myths, its universal truths; stories that answer ultimate questions about the origins, purposes and goals of the societies and their Institutions.

INTRODUCTION

In an influential 1967 essay, "Civil Religion in America," Robert Bellah (1970) proposed that although church religion and traditional beliefs are in decline, American society continues to be held together by a more general, nondenominational civil religion with its own myths, beliefs, and rituals. What constitutes civil religion is the creation of common memories or collective representations that endow successive generations with a sense of a common heritage that provides temporal and social integration (Durkheim, 1915; Schudson, 1989; Schwartz, 1996; Schwartz, 1998).

Since the publication of that essay, there has been considerable debate over the health of civic life in America (Skocpol and Fiorina, 1999; Hall and Lindholm, 1999; Schudson, 1998; Putnam, 1995; Bellah et al., 1985). This debate has focused on the level of civic participation/apathy (Verba, Schlozman and Brady, 1995; Eliasoph, 1998), civic competence (Elkin and Soltan, 1999), and civic knowledge (Delli Carpini and Keeter, 1991; 1996)....We will argue that not only church religion but also the civil religion is in decline. Most of our students are generally ignorant of the U.S. history, politics, and culture that form the base of the American civil religion.

SOURCE: From "The Decline of the American Civil Religion?" by Gordon Clanton and Shoon Lio. Paper presented at the American Sociological Association, 2001. Reprinted with permission of the authors.

CIVIL RELIGION

As formulated by Bellah and further developed by other scholars, civil religion consists of the widely shared myths, beliefs, and rituals of a society (Mathisen, 1989). Civil religion is an expression of the cohesion of the nation. Civil religion is not just public piety or a civil theology (Bellah and Hammond, 1980). Rather, civil religion provides the common language for speaking of the highest aspirations and deepest values of a people (Bellah, 1970). These values are transcendent ethical standards with which to construct lines of social action and are constituted within metanarratives of national identity. The possible meanings of national identity are constituted by the past, which in turn frames social action for the present (Barton and Levstik, 1998; Olick and Robbins, 1998; Schwartz, 1996). Thus, all communities are communities of meaning (Olick and Robbins, 1998; Anderson, 1991; Bellah et al., 1985). American civil religion can be used by various groups to articulate their special values and agendas (Demerath and Williams, 1985:160).

The myths of the civil religion are drawn in part from the main events in American history: the English Puritan colonization of North America, the American Revolution, the Civil War, the Depression, World War II, the Cold War, and what Bellah (1970: 184) calls "the third time of trial," the contemporary "problem of responsible action in a revolutionary world, a world seeking to attain many of the things, material and spiritual, that we have already attained."

The beliefs of the American civil religion are similar to those Plato saw as essential to society: the existence of God, the immortality of the soul, and the moral government of the world (Bellah, 1970: 216). Rousseau, who coined the phrase "civil religion," summarized the core beliefs as: the existence of God, the life to come, the reward of virtue and the punishment of vice, and the exclusion of religious intolerance (Bellah, 1970: 173). Benjamin Franklin, typical of the founding fathers, writes that the essentials of every religion include belief in the existence of the Deity, the obligation of doing good to others, and the punishment of crime and the rewarding of virtue, either here or hereafter. In his 1960 inaugural address, John Kennedy reiterated the American belief that human rights "come not from the generosity of the state but from the hand of God," and he reminded Americans that "here on earth God's work must truly be our own" (Bellah, 1970: 169).

The rituals of civil religion are those constitutive practices that socially generate and regenerate through interaction the contexts in which the meaning of the civil myths and beliefs are lived and understood. They renarrate our collective memories in order to make our lives meaningful. Rituals of American civil religion include presidential inaugural addresses and other state occasions when reference is made to the "God" who blesses America and who also judges America (and Americans) against a high ethical standard. Although almost 90 percent of Americans are associated with some branch of Christianity, the God of the civil religion is not explicitly the Christian God and Jesus typically is not mentioned. The civil religion has its own sacred places (the Capitol, the national shrines and memorials), its own holy days (July 4, Memorial Day, presidential birthdays), and its own saints (presidents, folk heroes, military heroes). The speeches of Abraham Lincoln clearly invoke God as the higher standard by which to judge all human action. Lincoln referred to law as the political religion of the nation. According to Bellah, Lincoln, who never joined any church, was one of the nation's greatest civil theologians (Bellah and Hammond, 1980). Through his assassination, he became a martyr.

More recently, the assassinations of John Kennedy (1963), Robert Kennedy (1968), and Martin Luther King, Jr. (1968), and the ceremonies and remembrances associated with those events, have become part of the civil religion—with especially powerful resonances for Americans of a certain age. . . .

CIVIL RELIGION AGNOSTICS

Based on surveys of our students and our interactions with them, we suggest that not only church religion but also the civil religion is in decline.

Most of our students are generally ignorant not only of traditional religious history, but also of the U.S. history, politics, and culture that form the base of the American civil religion. Most of our students are civil religion agnostics.

Data

Our data about what students know and do not know are drawn from an ongoing series of extra-credit quizzes (six each semester in each course) on current events, cultural literacy (U.S. history and politics, Western civilization) and new and recent course material. Cultural literacy is the common knowledge or collective memory that makes it possible for people to cooperate and upon which all learning depends. These informal surveys provide rough measures of student knowledge in areas of relevance to both traditional religion and the American civil religion. . . .

These data suffer from being the by-products of classroom exercises. The findings are suggestive, not conclusive. We have no longitudinal data and thus offer no formal measure of a decline of the civil religion over the decades, only an intuitive argument about American society that, we believe, has important pedagogic implications. The findings from the extra-credit quizzes are supplemented with information from true/false quizzes from our courses, other surveys, and impressions from conversations with students both in and outside of class. . . .

Despite considerable overlap, different professors have different views of what constitutes the core of basic knowledge a college student should command. We do not claim that ours is comprehensive or definitive. . . .

Findings

Although extra-credit quizzes are composed mostly of items that college students should know, the average student score is 54 percent. A score of about 70 percent usually places a student in the top 10 percent of those tested. More important and more troublesome than this rather low average score, the score on many items relevant to the civil religion were very low. . . .

Student Knowledge of Religion

The surveys reveal general ignorance of religious history. Only about 40 percent can name the three great monotheistic religions of the West (Judaism, Christianity, and Islam) and fewer still know the order in which they emerged. Fewer than half can name the founding prophet of Christianity (Jesus). Ironically, more than half know that Mohammed was the founding prophet of Islam. Many erroneously believe that Judaism is a branch of Christianity. Indeed, one ray of hope from these otherwise bleak findings is that anti-Semitism will continue to decline because most young adults of Christian background don't know who the Jews are. . . .

Many students are surprised to learn that Catholicism is part of Christianity because they associate the term "Christian" with fundamentalist or evangelical Christianity, which they know is quite different from the Roman Catholic Church. Most students are unclear about the religious right, unaware of its political objectives, and hazy about how it is different from other religious groups. For all our tendency to think of the United States as being under the sway of a Judeo-Christian ethic, our data suggest a general ignorance of Jewish and Christian history and teachings. Likewise, a 1984 Gallup poll found that a majority of U.S. adults could not name more than half of the Ten Commandments.

Student Knowledge of World History

Student ignorance of religious history reflects a more general ignorance of world history. Only about half the students know from what event our calendar is dated (the birth of Jesus) and fewer than half understand that 500 B.C. is 1000 years earlier than A.D. 500. Only about 40 percent could name the year ending in two zeroes following the death of Jesus (100 or A.D. 100). Very few are familiar with C.E. (common era) as a secular substitute for A.D. And no more than 10 percent know the meaning of the word "secular."

Most students are unclear about what is meant by the term "Western history." They do not understand that the United States and its political culture were established by the English and other Europeans who in turn were influenced by the Jews, Greeks, Romans, and the other ancient peoples of the Mediterranean basin and the Middle East.

Only about one-third can name the year ending in two zeroes nearest the discovery of the New World by Europeans (1500). More than half give no answer; most of the rest say 1400, indicating that they have remembered 1492 but apparently do not understand that 1492 is much closer to 1500 than to 1400. Some students complain that this is an unfair question because "this is not a math class."

Of special relevance for sociology, most of our students have no clear idea of what is meant by "industrialization," "Industrial Revolution," or "modernization." Only a few can name the defining characteristic of an industrial society (machine power). Very few can name the first machine (the steam engine); for some reason, those who offer an answer most often say the cotton gin. Very few know when (late 1700s) or in what country (England) the factory system began.

Student Knowledge of U.S. History

The extra-credit quizzes reveal that our students are generally ignorant of the U.S. history that Bellah sees as the source of the civil religion. Only a handful of students can name the year ending in two zeroes nearest the beginning of English colonization in the New World (1600). Fewer than half know the year in which the United States became an independent nation. When told that the answer is 1776, some students complain that this is a "trick question." ...

Although most students know that the war that began in 1812 is called the War of 1812, fewer than 10 percent know whom the U.S. fought in that war (England or Britain). Fewer than 10 percent know the intent of the Monroe Doctrine (to prevent colonization of the New World by European powers). Although our students live in California, fewer than half know from whom the U.S. took the Southwest (Mexico)—and fewer than 10 percent know about when this happened (1840s). Only about half know that the Civil War was fought between the North and the South. About half can name the issue over which the war was fought (slavery) and fewer than 10 percent know about when this happened (1860s).

Student Knowledge of Recent U.S. History and Politics

Knowledge of more recent U.S. history is no better. Fewer than half of our students know who was president during the Great Depression and World War II (Franklin D. Roosevelt). One student thought this might be Lincoln; another guessed Henry Ford. Fewer than half can name the major allies (Britain, USSR) and opponents (Germany, Japan) of the U.S. in World War II. Fewer still know that World War II took place in the 1940s. No more than 10 percent know that World War I was fought in the second decade of the twentieth century....

Only a few students know which war was going on in 1950 (Korean) and fewer still know who the opponents of the United States were (North Korea and China). Fewer than half can name the president at the time of the Cuban missile crisis (Kennedy). A narrow majority know that the war that peaked in the late 1960s was the Vietnam War, but only about one-third can name the opponent of the U.S. (North Vietnam or Viet Cong). Only about 60 percent can name the major opponent of the United States in the Cold War (Soviet Union or USSR or Russia) and most are hazy about key differences between communism, socialism, and capitalism.

Fewer than half of the students can name one of the two presidents (Kennedy and Johnson) who supported the Civil Rights and Voting Rights Acts—and fewer still know in which decade these laws were passed (1960s). Almost all can name the main leader of the black civil rights movement (Martin Luther King, Jr.) but fewer than one-third can name the Supreme Court decision that outlawed segregated schools (*Brown v. Board of Education*) and fewer still know that this decision was rendered in the 1950s.

Fewer than two-thirds of our students know which party's platform explicitly wants to outlaw abortion (Republican). Fewer than 40 percent know which party established Social Security (Democratic). Fewer still know what Social Security is. Most students are unaware that most advanced industrial societies have national health insurance systems—and only a few students, most over 30, know anything about Bill Clinton's unsuccessful first-term effort to establish a national health plan....

Other surveys confirm a low level of knowledge of U.S. history and politics. A 1998 poll of American teenagers for the National Constitution Center found that nearly 95 percent could name the actor who played the Fresh Prince of Bel Air (Will Smith) on television, but only 2 percent could name the chief justice of the Supreme Court (William Rehnquist). About 90 percent could name the star of the television series *Home Improvement* (Tim Allen), but less than one-third could name the Speaker of the House (Newt Gingrich). In our class surveys, about 40 percent correctly named the Speaker. A number of other surveys conducted on similar subjects corroborate our finding....

THE GRIM CONCLUSION

Based on informal in-class surveys of our students and our discussions with them both in and outside of class, we suggest that not only church religion but also the civil religion is in slow decline. Most of our students are generally ignorant not only of traditional religious history, but also of the U.S. history, politics, and culture that form the base of the American civil religion. Thus, they are deprived of the ancient language of moral discourse and the metanarratives of the Reformation, the Enlightenment, the City on a Hill, the Declaration of Independence, the lessons of the Civil War, the New Deal, the New Frontier, the civil rights movement, and other social movements with which they can contextualize current events and possible lines of social action....

Our students, like many Americans, cope with the overwhelming complexities of public issues by defining them as being of no concern because we cannot do anything about them, thus justifying the retreat into the private realm of the family or everyday life (Flacks, 1988). As Bellah (1987) has noted, both religion and politics appear to be increasingly relegated to the private sphere, so the modern state increasingly is legitimized by neither religious nor political discourse, but rather is "managed" by bureaucratic and technological expertise.

Postmodernity and Metanarratives

According to Carl Boggs (2000: 210), the crisis of modernity that is central to theories about the post-modern condition is reflective of the pessimism and cynicism about the future and about the very efficacy of collective social action. Much of the discussion on the postmodern condition sees the incredulity towards metanarratives as the defining characteristic of this era (Lyotard, 1979; Lemert, 1997; Schwartz, 1998). Thus, postmodern culture places little trust in the grand narratives of nation, great heroes, myths, universal truths—narratives that answer ultimate questions about the origins, purposes, and fate of societies and their institutions (Schwartz, 1998:64)....

Thus, people see their own lives as a series of unrelated events and history is disconnected from the lived experience of the individual. As Rosenzweig and Thelen (1998) point out, most people's understanding of the past tends to center around their immediate families. Postmodernity encourages a radical individualism, which undermines the civil religion crucial to a participatory culture.

QUALIFYING THE GRIM CONCLUSION....

The grim conclusion we have drawn about the decline of civil religion must be qualified. Some students know much more than others; the range from best to worst scores is enormous. (In a small-group map exercise, one group of four students correctly named all 50 states, but another group of four correctly named only five states.) There is considerable variation in average scores from one class to another.

A few students, probably no more than one in 20, can answer 80 percent of the questions—but the student culture at our university (and, we suspect, in public higher education generally) does not support, expect, and reward knowledge beyond that needed for test-taking and a narrowly-defined careerism. Most students who do well are over 30 and have learned much of what they know by virtue of their longer life experience.

Some students are motivated by the extra-credit quizzes to fill the gaps in their general educations—and most get better over the course of a semester. Some will learn some of these things later in life. Many will find their way into work in which general knowledge is not required, not valued, not rewarded. Perhaps, sadly, ignorance of history, politics, and culture is functional in an increasingly bureaucratic society in which knowledge of such matters is not required and, indeed, may be suspect.

Despite the absence of historical knowledge reported here, some students show evidence of involvement with the civil religion through their participation in volunteer work, their endorsement of the idea of "giving something back," their sense of the importance of making one's own choices and standing up for one's values, and their concern about the degradation of the natural environment. These students have imbibed a version of the civil religion, but a thin and ahistorical one (Swidler, 1999)....

The Future of the American Civil Religion

It is hard to be hopeful about the future of the American civil religion. Most of the undergraduate students we teach, including seniors, are largely lacking the knowledge on which informed participation in democracy depends. Most are without the concepts and the language necessary to discuss and debate issues that bear on the common good.

Equally troubling, for most students learning is not cumulative. Most carry along very little from one course to the next or from one stage of schooling to the next. Even what is "learned" for the final exam in one course is likely to be forgotten quickly, as one student told me, "to make room for the new stuff you need to know to do well in the next course." Pretests of our students reveal that very few have retained sociological ideas, methods, and findings from earlier courses—and sociology majors do only marginally better than other students. Knowledge is increasingly fragmented as people are bombarded with information and trivia characteristic of postmodern culture.

The findings of this research suggest that, at the least, there appears to be a decline in the number and proportion of *practitioners* of the American civil religion. Just as, over time, fewer and fewer people are devoutly religious, so too fewer and fewer people apparently are knowledgeable about the fundamentals underlying the civil religion and familiar with the common language of moral discourse. There have always been some who did not know and did not participate. It is unclear just how small the number of practitioners can become without serious damage to the social fabric of our democracy.

At the somber end of Ray Bradbury's anti-utopian novel *Fahrenheit 451*, when all books have been destroyed by an authoritarian government, knowledge of the great books is kept alive only by a community of outcasts, each of whom has memorized one book and each of whom, before dying, teaches that book to a younger person. This cautionary tale reminds us of the fragility of our institutions and of the knowledge by which they are sustained....

The Social Context

The consciousness of our students is shaped by their social situations. More than 80 percent of students in the California State University system commute to campus. One-third of them work more than 30 hours per week. One-quarter have the responsibility for the care of children or other dependents.

The average age is 26. For many students English is the second language. Even though the CSU system draws its incoming freshman from the top third of California's high school graduates, almost half arrive needing remedial work in English and mathematics—and remedial education costs the CSU system about $10 million per year plus time lost by students and faculty (Reed, 2001). Few students have the free time to experience college life outside the classroom—plays, films, music, clubs, debates, speeches, the campus culture, the community of learning. As Richard Flacks (1988) argues, Americans are constrained by their material conditions in that they are consumed by living their everyday lives rather than making history. . . .

SOCIOLOGY AND THE AMERICAN CIVIL RELIGION

We have argued that most of our students are generally ignorant of the U.S. history, politics, and culture that form the base of the American civil religion and the foundation of a strong participatory culture. We feel that sociology can cultivate the civic practices essential to a democratic culture. These democratic constitutive practices can serve to help people integrate their personal narratives or memories to that of the larger society. In other words, these practices are crucial to the development of a sociological imagination, which in turn serves as a vibrant basis for civil religion. We who teach sociology can make a small but real difference here – and thus help slow the decline of the American civil religion or perhaps help reconstitute it in our postmodern world.

We argue that the quizzes enable us to construct a pedagogy that facilitates civic practices that foster a civil religion for a postmodern society. This pedagogy aims to help our students develop what C. Wright Mills (1959) called "sociological imagination," the ability to comprehend the connection between the personal troubles of individuals and the larger political and economic context. This imagination is especially needed in a postmodern world in which people can be overwhelmed by the amount of trivial information and the speed at which it is communicated.

REFERENCES

Anderson, Benedict. (1991) *Imagined Communities*. London: Verso.

Barton, Keith C., and Linda S. Levstik. (1998) "'It Wasn't a Good Part of History': National Identity and Students' Explanations of Historical Significance." *Teacher College Record*, 99(3) 478–513.

Bellah, Robert N. (1970) *Beyond Belief: Essays on Religion in a Post-traditional World*. New York: Harper & Row.

———. (1987) "Legitimation Processes in Politics and Religion." *Current Sociology*, 35:2, 89–99.

———. (1999) "Freedom, Coercion, and Responsibility," *Academe*, 85:1 (January/February), 16–21.

Bellah, Robert N., and Phillip Hammond. (1980) *Varieties of Civil Religion*. San Francisco: Harper and Row.

Bellah, Robert N., Richard Madsen, William M. Sullivan, Ann Swidler, and Stephen M. Tipton. (1985/1996) *Habits of the Heart: Individualism and Commitment in American Life*. Berkeley: University of California Press.

Boggs, Carl. (2000) *The End of Politics: Corporate Power and the Decline of the Public Sphere*. New York: Guilford Press.

Brownstein, Ronald. (1999) "Words that Inspire: This Century's Top 10 Gems of Political Eloquence." *Los Angeles Times*, May 10, A–5.

Delli Carpini, Michael X., and Scott Keeter. (1991) "Stability and Change in the U.S. Public's Knowledge of Politics." *Public Opinion Quarterly*, 55:583–612.

———. (1996) *What Americans Know about Politics and Why It Matters*. New Haven, CN: Yale University Press.

Demerath, N. J., and Rhys H. Williams. (1985) "Civil Religion in an Uncivil Society," ANNALS, AAPSS, 480: 154–166.

Durkheim, Emile (1915/1995) *The Elementary Forms of the Religious Life*. New York: Free Press.

Eliasoph, Nina. (1998) *Avoiding Politics: How Americans Produce Apathy in Everyday Life*. Cambridge: Cambridge University Press.

Elkin, Stephen L., and Karol Edward Soltan. (1999) *Citizen Competence and Democratic Institutions*. University Park: Pennsylvania State University Press.

Flacks, Richard. (1988)/*Making History: The Radical Tradition in American Life*. New York: Columbia University Press.

Hall, John A., and Charles Lindholm. (1999) *Is America Breaking Apart?* Princeton: Princeton University Press.

Lemert, Charles. (1997) *Postmodernism Is Not What You Think*. Malden: Blackwell.

Lyotard, Jean-Francois. (1979/1984) *The Postmodern Condition*. Minneapolis: University of Minnesota.

Mathisen, James A. (1989) "Twenty Years after Bellah: Whatever Happened to American Civil Religion?" *Sociological Analysis*. 50(2):129–146.

Mills, C. Wright. (1959) *The Sociological Imagination*. New York: Oxford University Press.

Olick, Jeffrey K., and Joyce Robbins. (1998) "Social Memory Studies: From 'Collective Memory' to the Historical Sociology of Mnemonic Practices." *Annual Review of Sociology*, 24:105–140.

Putnam, Robert D. (2000) *Bowling Alone: The Collapse and Revival of American Community*. New York: Simon & Schuster.

Rosenzweig, Roy, and David P. Thelen. (1998) *The Presence of the Past*. New York: Columbia University Press.

Schudson, Michael. (1989) "The Present in the Past versus the Past in the Present." *Communication*, 11:105–113.

———. (1998) *The Good Citizen: A History of American Civic Life*. New York: Martin Kessler Books.

Schwartz, Barry. (1991) "Social Change and Collective Memory: The Democratization of George Washington." *American Sociological Review*, 56 (April): 221–236.

———. (1996) "Memory as a Cultural System: Abraham Lincoln in World War II." *American Sociological Review*, 61(October): 908–927.

———. (1998) "Postmodernity and Historical Reputation: Abraham Lincoln in Late Twentieth-Century American Memory." *Social Forces*, 77, 10: 63–103.

Skocpol, Theda, and Morris P. Fiorina, (1999) *Civic Engagement in American Democracy*. New York: Russell Sage Foundation.

Swidler, Ann, 1999. Personal correspondence.

Verba, Sidney, Kay Lehman Schlozman, and Henry E. Brady. (1995) *Voice and Equality: Civic Voluntarism in American Politics*. Cambridge: Harvard University Press.

CHAPTER 11

Economics

Necessities for Survival

Alan Mc Evoy

Alan Mc Evoy

Each society must have a system for the production and distribution of goods and services. An economic system fulfills these functions. The economic system is interrelated with the other social institutions, a connection that is most easily understood in the relationship between the economic and political institutions. As an illustration, let us examine this relationship between the political systems of feudalism, colonialism, and capitalism and their related economic systems.

When the strong central government of ancient Rome came to an end with the fall of the Roman empire, its monetary system and civic safeguards ended as well. People had to return to cultivating land to raise their own food instead of buying it and were forced to submit to marauding ex-soldiers for protection. Eventually, these ex-soldiers confiscated the land and set up a system of government by electing a strong individual from their ranks as their leader—a king. This system of government was called *feudalism*. As the feudal European states grew and merged, willingly or not, they became large enough to journey overseas and impose their will on other less developed but desirable landholdings. This system of government was called *colonialism*. Because the indigenous people usually had their own land on which to live, grow food and hunt, they saw no reason to work the mines and large plantations of the conquerors. Accordingly, the conquerors developed a political system to come to the aid of the economic system: the best native lands were seized, the natives were required to pay in cash rather than in goods for needed products, and many were arrested on newly made-up laws and sentenced to work on the plantations or in the mines.

The development of *capitalism* led to the growth of large corporations to operate the vast industries that developed. Through mergers, these corporations became so large that they dominated various industries and stifled the economic benefits of competition. Unchecked capitalism led to two political developments on the part of the dominating colonizing societies: regulation to ensure competition and subsidies to ensure that the corporations stayed in business, thus protecting jobs. Although the corporations liked the government subsidies, they did not like the regulations that prevented them from exerting their considerable economic clout to charge whatever they desired for goods and services. The large corporate world soon learned that economic power guarantees political power. A measure of this political power wielded by large corporations is the constant deregulation of laws that limit corporations' activities. Of course, political leaders hide the fact that they are submitting to corporations by claiming that deregulation increases competition and lower prices. As Robert Worth notes in the first reading, this has not been an accurate prediction of the outcome of deregulation.

The major goal of all economic organizations is to increase profits, and their desire to lessen regulation of their industries fits in with this goal. Another means of increasing profits is to reduce costs while maintaining higher margins of profits. In the second reading Jeremy Rifkin notes that this corporate goal will lead to a decline in the global labor force and the birth of a post-market era. He sees the age-old dream of abundance and leisure becoming even more distant at the dawn of the information age. Not taken into account are the changes that will have to be made when work is no longer central in our lives.

If Rifkin's prediction is correct, then measures must be taken by both the economic and political institutions to deal with the resulting system of fewer jobs overall and fewer jobs that pay a livable wage. Both outcomes forecast lower purchasing power to buy corporate products. The next reading, by Diana Zuckerman, deals with this need to increase the public's purchasing power. Rifkin also forecasts that more people will be forced onto welfare. Zuckerman discusses the clashes between politicians and welfare recipients in this highly charged area.

One way to promote economic growth and save jobs is to increase the spending power of the ordinary citizen. In the fourth reading George Ritzer notes how the current economic system has promoted an easy spending philosophy by making it easier both to obtain credit cards and to maintain this available credit by reducing the minimum payment due each month. The promotion of the credit card has changed America's economic philosophy from "pay-as-you-go" to one of living for today.

A final question is raised in this brief look at the economic-political power of corporations. How are they able to get away with this exercise of power? In the final reading, Christopher Kollmeyer attempts to answer this question. He examines the news media's portrayal of economic news and discovers that it is slanted in the corporate interest.

As you read this chapter, think of some current economic issues and how they might be addressed. What can be done about such current issues as individual and public deficits, oligopoly, unemployment, and wage stagnation? Consider whether the ideas proposed in the selections are viable options for dealing with these issues. What techniques would you propose for dealing with these issues? Are the options given in these articles possible in the current political situation?

51

Why Deregulation Has Gone Too Far

ROBERT WORTH

The author explains that government deregulation in several areas has created a number of problems. Yet our politicians tell us that government regulations are bad and want to deregulate even more.

As you read this selection, consider the following questions as guides:

1. *As indicated, there is a contradiction between the problems of deregulating and politicians' desire to deregulate even more. Which side do you agree with? Why?*
2. *It may be easier to answer question 1 by thinking about industries that have either been deregulated or are not subject to strong enforcement of regulations. What are these areas? What have been some outcomes of deregulation and lax enforcement?*
3. *How would you deal with the growing threat of monopoly as a result of corporate mergers?*

GLOSSARY **Oligopoly** a market condition with few sellers.

Twenty years ago this spring, the Senate voted to abolish the Civil Aeronautics Board. Where the government had maintained an elaborate and often unwieldy scaffolding of routes and fares, the free market now came crashing in like a mighty river. Within a few years the flood would carry a host of other regulated industries, including telephones, trucking, and banking, along with it. The Age of Deregulation had begun. . . .

Deregulation became an article of faith, "espoused more or less automatically, even unthinkingly, by a wide range of officeholders and their critics," wrote Martha Derthick and Paul Quirk in their classic study *The Politics of Deregulation*. . . .

Deregulation has definitely made the wheels of the economy spin faster. But the competition it has brought has often been unfair and destructive, and the alleged price cuts often haven't trickled down to the average American.

Take airline deregulation. Sure, the CAB was a clumsy bureaucracy, notorious for its "procedural spaghetti" and the bizarre "kabuki dance" hearings where it decided who could fly where, and for how much. By 1976 even the Board's officers were in favor of some kind of decontrol. But airports aren't like restaurants; you can't just round up some friends, buy a little property, and set up shop. Big carriers control access to airport gates, arrival and departure slots, and even to computer booking systems. A few years after the abolition of the CAB, the major airlines were muscling the little guys out of the sky. And that wasn't so good for customers. Long-distance fares dropped, but smaller communities often got fewer flights for higher prices. Today, the same thing is true. According to the Transportation Department, which is considering punitive action, the big airlines' bullying practices "can hurt consumers in the long run by depriving

SOURCE: From *The Washington Monthly*, July/August 1998 pp. 10–14.

them of the benefits of competition," and leave "much of the demand for low-fare service in many local hub markets unserved."

It could get even worse. On April 3, a passenger on a US Airways jet landing at New York's LaGuardia airport looked out his window and saw another plane coming straight at him. He screamed as the pilot pulled up in an evasive maneuver, avoiding a head-on collision by a mere 20 feet. Panicked traffic controllers breathed a sigh of relief. But they were so busy that the incident didn't even get reported until two months later. Why? Part of the answer is that deregulation has led to a 50 percent increase in air traffic since 1981, with proportionately far fewer controllers and less supervision. In-flight gaffes like the one at LaGuardia are up 19 percent this year, and ground errors are up 49 percent.

"What's gotten left out is somebody still has to manage the system," says MIT economist Lester Thurow. And like it or not, the best way to do that is with tougher, smarter regulation.

MARKET FAILURE

"I strongly oppose regulation," said Sen. John McCain at a hearing in February, "but I don't oppose regulation as much as I oppose unregulated monopolies." He was talking about cable TV, where deregulation was supposed to bring competition from satellite companies to the old cable monopolies. Instead, they've been banding together—witness the recent merger announcement of American Sky Broadcasting, a major satellite company, with Primestar, a cable group. "That's bad for competition and bad for consumers," said Joel Klein, the Justice Department's top antitrust enforcer. It's also one of the reasons cable prices have been rising at five times the rate of inflation.

McCain and Klein could have been talking about any one of a number of deregulated industries. On May 19 members of the Senate's antitrust subcommittee asked Ed Whittaker, CEO of the telecom giant SBC, why his company was proposing to merge with its potential competitor Ameritech, reducing the number of Bell operating companies to four from an original seven. Whittaker insisted that there was no need to worry; the 1996 Telecom Act was working its magic and competition would soon arrive. Then he said he foresaw a mere handful of companies dominating the global phone market, including just *one* of the current Bell operating companies. . . .

Conservatives often counter that the Justice Department is quite capable of enforcing the antitrust laws—witness, for instance, Joel Klein's decision in May to block the American Sky/Primestar merger. They often add that Justice would do an even better job if only the regulators would step further out of the way. Take the current railroad slowdown in the South, where nine men have died and billions of dollars have been lost in what's being called the worst rail melt down in U.S. history. That mess is a direct result of the 1996 merger of the Union Pacific and Southern Pacific railroads—a merger the Justice Department *wanted* to block, but couldn't, because the Surface Transportation Board, a regulatory agency, has jurisdiction over the rails. "One mistake we made with deregulation," says Cliff Winston of the Brookings Institution, "is that we've left a specific regulatory body to handle antitrust issues, instead of just saying 'now you're subject to the Justice Department.' " Defenders of telecom deregulation say the same thing. "[F]or decades, the FCC has legitimized telecom practices that antitrust courts would never have tolerated in its absence," writes Peter Huber of the Manhattan Institute. The remedy, he adds, is to abolish the FCC and hand over the reins to Joel Klein and his trusty knights.

As it happens, the *Monthly* made a very similar case back in 1969, when liberal skepticism of the regulatory bureaucracy was at an all-time high. But the past 30 years have made it clear that antitrust law can't substitute for regulation. For one thing, there's a practical limit to what Klein and his fellow trust-busters can do. Their division is working at historic levels and facing a virtual avalanche of cases. Meanwhile, the number of lawyers on staff has declined from 456 in 1980 to 343 today.

Besides, anyone who still thinks the Justice Department can stand in for an entire regulatory bureaucracy has obviously forgotten all about Charles Keating. It was Keating, after all, who recognized that after the passage of the Garn St.-Germain Act of 1982, which deregulated the savings and loan industry, there was no one to stop him from taking over an S&L, lending out its taxpayer-insured deposits, and gambling them away with his friend Michael Milken the junk-bond king. In the end Keating was caught. But most of his fellow con artists got off scot free, and collectively bilked U.S. taxpayers of $130 billion and counting.

Where was the Justice Department? Standing by, powerless to help. And if you think Congress won't get fooled again, think twice. On May 13 the House voted to abolish Glass-Steagall, the 60-year-old law that prevents banks from gambling away their federally insured deposits by getting into the securities and insurance businesses. Supporters of the measure claim it won't pose any danger, since new technologies and creeping deregulation by the Federal Reserve already allow banks to get around Glass-Steagall. But is that really a good thing? After all, Nationsbank has agreed to pay nearly $37 million in settlements and fines over the past few years for selling high-risk securities to customers—many of them senior citizens—who didn't understand what they were being sold. . . .

Another unsettling consequence of deregulation has been the tendency of competition in newly opened markets to hurt those who can least afford it. In telecom, for instance, the demise of AT&T as a regulated monopoly has brought plenty of competition, and lower rates, to the long-distance market. (It has also brought some of the downsides of competition; if you've ordered phone service lately, you've probably been "slammed"—billed by a phone company you never chose. Complaints about the practice rose 56 percent last year.) But at the local level deregulation has been a disaster, because the regional Bell companies have neither regulators nor real competitors to rein them in. As a result, service is both deteriorating and becoming more expensive.

The deregulation of electricity—the last of the regulated monopolies—poses similar threats to the average consumer. Federal Trade Commission Chairman Robert Pitovsky says the potential for savings with deregulation is "enormous." Maybe so—if you're a Nissan factory. As new power companies compete to sweeten the deal for high-volume business users, they'll be forced to make up the difference on the backs of residential ratepayers. "And transaction costs will go up," says Mark Cooper, research director of the Consumer Federation of America, as the centralized power grid gives way to a crazy quilt of private distributors.

Free-marketeers will counter that all these downsides are a small price to pay compared with regulation, which costs the economy far more every year than the S&L debacle ever did. "Federal regulations cost $1.3 trillion in economic output to be lost each year," according to the Heritage Foundation's voluminous website on the topic. "When a business devotes resources to adhering to regulatory mandates, it is using those resources less efficiently."

But that misses the point. Left to its own devices, the economy—no matter how large or small—always inflicts some unfairness on those who don't own a lot of stock. The purpose of regulation in areas like telecom, electricity, and banking was to defray the cost of service for ordinary people at the expense of the high-volume users—businesses, mostly—who could better afford it. These "cross-subsidies" may have distorted the economy a little, but they maintained a vital safety net for people who might otherwise have fallen through.

In the absence of many of these regulations, there's already plenty of evidence that lower-income people are losing out. The recent wave of banking mergers, for instance, has been hailed by brokers and economists alike—perhaps because most brokers and economists earn enough to profit from it. For those who don't, several recent studies by the Federal Reserve have shown that large interstate banks charge higher service fees than small banks do, and that they're much less likely to make loans to small businesses. Meanwhile, the number of bank branches in inner-city neighborhoods is declining—down 21 percent from 1975 to 1995, according to another Fed study.

The buyout frenzy that deregulation triggered has also eroded some of the other institutions that used to help define a sense of community for middle and lower-income people. In the two years since Congress eased ownership limits on radio stations, 4,000 have been sold, and more than half of big-city stations are in the hands of just five companies. The result has been a rising flood of bland pop songs dictated by big companies that don't want to take risks. So what? Not much, if you have a 2,000 CD collection of your favorite music. If you don't, you could be out of luck.

That doesn't mean we should call in the Quality of Life police on every merger that would close down a neighborhood deli or bookstore. It does mean we should think carefully about who benefits, and how, before we unleash the bulls and bears of the market on our everyday lives.

POISON PILLS

Of course, there are plenty of silly or overburdensome regulations on the books. The Equal Employment Opportunity Commission has been after the Hooters restaurant chain for years because Hooters doesn't employ male waiters (if you don't know why, God bless you.) The EEOC is also pursuing Joe's Stone Crab restaurant in Miami for failing to hire enough women waiters, even though most of the restaurant's managers, and its owner, are women.

Yet too often gaffes like these become an excuse for cutting back on all government regulations, even in areas that ensure our health and safety. For years, Congressional Republicans have targeted the FDA as the country's "leading job killer," a mammoth bureaucracy that takes too long to approve drugs and places intolerable burdens on medical and pharmaceutical companies. Recently they got at least part of what they wanted with last fall's passage of the FDA Modernization Act. But before you accept their claims that this law merely "streamlined" the FDA drug approval process, consider the story of Mary Yount. Last year Mary was on vacation in Florida when she began to gasp for breath. "I felt like my heart was going to pump out of my chest," she says. After a number of tests, doctors found that her heart (which had been tested and found healthy only two years before) had been irreparably damaged by the combination of diet drugs her doctor had prescribed for her ten months earlier, now known as Fen-Phen.

The drugs were soon off the market, but not before causing serious heart valve damage to 285,000 Americans. Next time it could be much worse, because Congress's new FDA "modernization" allows drug and medical companies to do "off-label" marketing—promoting their products for unapproved purposes. "If this law had been passed before the Fen-Phen mess," says Mary, "the drug companies would have been advertising it to thousands of doctors." The Act also lowers the number of clinical investigations required to establish a drug or medical device's safety. And it allows medical device manufacturers to hire private for-profit firms to review the health and safety of their products—even when those "independent" reviewers are then offered jobs with the manufacturer. Dr. Arnold Relman, an emeritus professor at the Harvard Medical School, calls the Act "an invitation to disaster and chaos."

Perhaps worst of all, the law burdens the FDA with new responsibilities in the review process for drugs and medical devices without giving it the money to do the job, just as a spate of recent failures suggest that the agency is *already* incapable of handling its duties. Nick Valestrino found that out the hard way a few months ago when he went to the doctor for a routine injection of intravenous immune globulin, a drug that he and tens of thousands of other Americans depend on for their health. The drug was unavailable, because the FDA had failed to respond promptly to a shortage. Why? One reason, says a top FDA official, is that "[w]e've had a series of annual cuts for five straight years." He added that the new law will only make that worse. "Any time you've already got some measures of program failure, a new law saying 'Go do this, too' will have an impact. The whole agency is kind of stressed out."

If that doesn't scare you, think about what happens when the Department of Agriculture's meat and poultry inspection system fails. Just before Christmas 1992, Lauren Rudolph felt sick to her stomach after eating a hamburger for lunch. "Daddy, I'm going to die," she told her father, who soothed her and told her she'd be fine. But Lauren was right. After a week of unspeakable suffering, her tiny body seized up in a massive heart attack, her breathing stopping and her lips turning blue. Lauren was killed by a microbe known as E. coli O157:H57, which is found in ground beef and anything else that has come in contact with manure. (Vegetarians: that means fruits and vegetables.) Her death was the "index case" in an outbreak that caused 732 illnesses and ultimately killed four children, an outbreak that "might have been identified and stopped in San Diego, where it apparently began, if only California had a reporting requirement and encouraged testing," writes Nicols Fox, an expert on food-borne illness. The outbreak helped pressure Congress into adopting a stricter meat inspection system—a system that the National Academy of Sciences had been urgently recommending for eight years.

But even with the new system, USDA inspection is woefully inadequate. "Do not stop the line," wrote one supervisor last summer at a slaughter-house where up to 50 percent of the carcasses were visibly smeared with feces, bile, or hair, "unless the contamination is dripping." It's not uncommon for inspectors to supervise 300 carcasses an hour—far too many to do a good job. And the number of inspectors is way down, from 12,000 20 years ago to 7,500 today. When they do find something, inspectors cannot order a recall; "they have to rely on voluntary cooperation from companies to get known contaminated meat off the market," says Elizabeth Dahl, a staff attorney for the Center for Science in the Public Interest. So it's no surprise that outbreaks of E. coli and other deadly food-borne illnesses are on the rise. (They've increased almost fivefold since 1988.)

The FDA's food inspectors are even farther behind. According to a recent GAO report, the agency only inspects 1.7 percent of the 2.7 million shipments of fruit, vegetables, seafoods, and processed foods under its jurisdiction. Clinton has repeatedly tried to help the FDA keep up by giving it broad new inspection and enforcement powers, but Congressional Republicans have thus far refused to give him a hearing.

FIX IT

When you look at the human damage weak regulation can do, the question becomes not whether to regulate at all but how to do it better. Sadly, there's plenty of evidence that our regulatory process could use a good kick in the pants. U.S. environmental law, for instance, looks pretty good on the books. But the EPA's inspector general recently detailed a list of leaky sewage plants, belching factories, and dump sites that are in violation of the law. "If one state clearly has more relaxed standards, then businesses might flee to those states," Phillip Wardwell, a state environmental official in New Mexico, told *The New York Times*. "That's exactly the function of the federal agency, to head off that race to the bottom."

This touches on a crucial truth about regulation: No one, particularly in America, wants to be the one sucker who's playing by the rules. Unfortunately, our federal agencies often don't enforce them. Eighteen years after the landmark Superfund legislation was passed for cleaning up toxic waste sites, 63 percent of the nation's 1,359 official Superfund sites haven't been touched. And if you think that's just an issue of old zinc smelters and mines that no one ever goes near, think again. For years, managers at Connecticut's Millstone nuclear power broke federal safety rules on the handling of fuel rods, cooling pipes, gauges, and filters, and punished workers who tried to warn them about the dangers involved. Ultimately the government had to spend $750 million to bring the plant up to speed and re-educate its managers—money that would have been saved if everyone had respected the regulations in the first place.

In some cases, regulators fail to do their job because they're too friendly with the industry they're supposed to be policing. In the past few years the number of deaths in charter and commercial bus accidents has nearly doubled. That's partly because the Department of Transportation does a poor job of inspecting the buses. But it may also have something to do with the fact that in 24 states, charter bus companies are allowed to inspect their own buses, as "20/20" recently reported.

You might answer that bus companies have a pretty strong incentive to keep their buses in good shape. After all, they look bad when even one bus goes careening off the road and kills a group of schoolchildren. But the fact is that bus lines, like all private companies, are competing for market share, and struggling to get as much use as they can out of every tire, axle, and muffler. This is where the remorseless logic of deregulation kicks in: There will always be people willing to pay a little less for a cheaper bus ride. And there will always be someone willing to serve that market. If you doubt it, remember the word "Valujet."

The Valujet crash and others like it might never have happened if the Federal Aviation Administration were not so eager to please the airline industry. In fact, until recently the FAA was officially charged with promoting aviation as well as safety. This dual mission has contributed to countless air disasters, including the crash of Flight 800, which exploded in midair in July 1996 when a spark in its center-wing fuel tank ignited. Investigators at the National Transportation Safety Board (which has never suffered from the FAA's identity crisis) had been pushing the FAA to develop a fuel-tank design that would eliminate the danger for years. The FAA had resisted them, not because it wasn't technologically feasible, but because it would cost money. Ironically, the charges leveled at the FAA in the wake of Flight 800 echoed those made against the CAB in 1978: They had become yes-men for the industry.

But this time the remedy is to make the regulators tougher, not to kick them out. Let's hope the conservative and libertarian policy wonks who are celebrating deregulation's 20th anniversary this summer remember that it, too, has a cost—even when that cost is measured in human suffering rather than dollars and cents.

52

The End of Work

JEREMY RIFKIN

The author notes that we are at the dawn of the informational age and that it is already having a worldwide impact in the loss of employment.

As you read this selection, consider the following questions as guides:

1. *The author paints a grim picture of the future of employment. Do you agree with his assessment? Why?*
2. *Can you think of areas of employment that will escape huge layoffs?*
3. *Think of possible new areas for future employment.*
4. *What, if anything, is the government doing about current and future unemployment?*

From the beginning, civilization has been structured, in large part, around the concept of work. From the Paleolithic hunter/gatherer and Neolithic farmer to the medieval craftsman and assembly line worker of the current century, work has been an integral part of daily existence. Now, for the first time, human labor is being systematically eliminated from the production process. Within less than a century, "mass" work in the market sector is likely to be phased out in virtually all of the industrialized nations of the world. A new generation of sophisticated information and communication technologies is being hurried into a wide variety of work situations. Intelligent machines are replacing human beings in countless tasks, forcing millions of blue and white collar workers into unemployment lines, or worse still, breadlines.

Our corporate leaders and mainstream economists tell us that the rising unemployment figures represent short-term "adjustments" to powerful market-driven forces that are speeding the global economy into a Third Industrial Revolution. They hold out the promise of an exciting new world of high-tech automated production, booming global commerce, and unprecedented material abundance.

Millions of working people remain skeptical. Every week more employees learn they are being let go. In offices and factories around the world, people wait, in fear, hoping to be spared one more day. Like a deadly epidemic inexorably working its way through the marketplace, the strange, seemingly inexplicable new economic disease spreads, destroying lives and destabilizing whole communities in its wake. In the United States, corporations are eliminating more than 2 million jobs annually.[1] In Los Angeles, the First Interstate Bankcorp, the nation's thirteenth-largest bank holding company, recently restructured its operations, eliminating 9,000 jobs, more than 25 percent of its workforce. In Columbus, Indiana, Arvin Industries streamlined its automotive components factory and gave out pink slips to nearly 10 percent of its employees. In Danbury, Connecticut, Union Carbide reengineered its production, administration, and distribution systems to trim excess fat and save $575

SOURCE: From *The End of Work* by Jeremy Rifkin. New York: G. P. Putnam's Sons, 1995, pp. XV–XVIII, 3–19.

million in costs by 1995. In the process, more than 13,900 workers, nearly 22 percent of its labor force, were cut from the company payroll. The company is expected to cut an additional 25 percent of its employees before it finishes "re-inventing" itself in the next two years.[2]

Hundreds of other companies have also announced layoffs. GTE recently cut 17,000 employees. NYNEX Corp said it was eliminating 16,800 workers. Pacific Telesis has riffed more than 10,000. "Most of the cuts," reports *The Wall Street Journal*, "are facilitated, one way or another, by new software programs, better computer networks and more powerful hardware" that allow companies to do more work with fewer workers.[3]

While some new jobs are being created in the U.S. economy, they are in the low-paying sectors and generally temporary employment. In April of 1994, two thirds of the new jobs created in the country were at the bottom of the wage pyramid. Meanwhile, the outplacement firm of Challenger, Gray and Christmas reported that in the first quarter of 1994, layoffs from big corporations were running 13 percent over 1993, with industry analysts predicting even steeper cuts in payrolls in the coming months and years.[4]

The loss of well-paying jobs is not unique to the American economy. In Germany, Siemens, the electronics and engineering giant, has flattened its corporate management structure, cut costs by 20 to 30 percent in just three years, and eliminated more than 16,000 employees around the world. In Sweden, the $7.9 billion Stockholm-based food cooperative, ICA, re-engineered its operations, installing a state-of-the-art computer inventory system. The new laborsaving technology allowed the food company to shut down a third of its warehouses and distribution centers, cutting its overall costs in half. In the process, ICA was able to eliminate more than 5,000 employees, or 30 percent of its wholesale workforce, in just three years, while revenues grew by more than 15 percent. In Japan, the telecommunications company NTT announced its intentions to cut 10,000 employees in 1993, and said that, as part of its restructuring program, staff would eventually be cut by 30,000—15 percent of its workforce.[5]

The ranks of the unemployed and underemployed are growing daily in North America, Europe, and Japan. Even developing nations are facing increasing technological unemployment as transnational companies build state-of-the-art high-tech production facilities all over the world, letting go millions of laborers who can no longer compete with the cost efficiency, quality control, and speed of delivery achieved by automated manufacturing. In more and more countries the news is filled with talk about lean production, re-engineering, total quality management, post-Fordism, decruiting, and downsizing. Everywhere men and women are worried about their future. The young are beginning to vent their frustration and rage in increasing antisocial behavior. Older workers, caught between a prosperous past and a bleak future, seem resigned, feeling increasingly trapped by social forces over which they have little or no control. Throughout the world there is a sense of momentous change taking place—change so vast in scale that we are barely able to fathom its ultimate impact. Life as we know it is being altered in fundamental ways.

SUBSTITUTING SOFTWARE FOR EMPLOYEES

While earlier industrial technologies replaced the physical power of human labor, substituting machines for body and brawn, the new computer-based technologies promise a replacement of the human mind itself, substituting thinking machines for human beings across the entire gamut of economic activity. The implications are profound and far-reaching. To begin with, more than 75 percent of the labor force in most industrial nations engage in work that is little more than simple repetitive tasks. Automated machinery, robots, and increasingly sophisticated computers can perform many if not most of these jobs. In the United States alone, that means that in the years ahead more than 90 million jobs in a labor force of 124 million are potentially vulnerable to replacement by machines. With current surveys showing that less than 5 percent of companies around the world have even begun to make the transition to the new machine culture, massive unemployment of a kind never before experienced seems all but inevitable in the coming decades.[6] Reflecting on the significance of the transition taking place, the distinguished Nobel laureate economist Wassily Leontief has warned that with the introduction of increasingly sophisticated computers, "the role of humans as the most important factor of production is bound to diminish in the same way that the role of horses in agricultural production was first diminished and then eliminated by the introduction of tractors."[7]

Caught in the throes of increasing global competition and rising costs of labor, multinational corporations seem determined to hasten the transition from human workers to machine surrogates. Their revolutionary ardor has been fanned, of late, by compelling bottom-line considerations. In Europe, where rising labor costs are blamed for a stagnating economy and a loss of competitiveness in world markets, companies are hurrying to replace their workforce with the new information and telecommunication technologies. In the United States, labor costs in the past eight years have more than tripled relative to the cost of capital equipment. (Although real wages have failed to keep up with inflation and in fact have been dropping, employment benefits, especially health-care costs, have been rising sharply.) Anxious to cut costs and improve profit margins, companies have been substituting machines for human labor at an accelerating rate. Typical is Lincoln Electric, a manufacturer of industrial motors in Cleveland, which announced plans to increase its capital expenditures in 1993 by 30 percent over its 1992 level. Lincoln's assistant to the CEO, Richard Sobow, reflects the thinking of many others in the business community when he says, "We tend to make a capital investment before hiring a new worker."[8]

Although corporations spent more than a trillion dollars in the 1980s on computers, robots, and other automated equipment, it has been only in the past few years that these massive expenditures have begun to pay off in terms of increased productivity, reduced labor costs, and greater profits. As long as management attempted to graft the new technologies onto traditional organizational structures and processes, the state-of-the-art computer and information tools were stymied, unable to perform effectively and to their full capacity. Recently, however, corporations have begun to restructure the workplace to make it compatible with the high-tech machine culture.

RE-ENGINEERING

"Re-engineering" is sweeping through the corporate community, making true believers out of even the most recalcitrant CEOs. Companies are quickly restructuring their organizations to make them computer friendly. In the process, they are eliminating layers of traditional management, compressing job categories, creating work teams, training employees in multilevel skills, shortening and simplifying production and distribution processes, and streamlining administration. The results have been impressive. In the United States, overall productivity jumped 2.8 percent in 1992, the largest rise in two decades.[9] The giant strides in productivity have meant wholesale reductions in the workforce. Michael Hammer, a former MIT professor and prime mover in the restructuring of the workplace, says that re-engineering typically results in the loss of more than 40 percent of the jobs in a company and can lead to as much as a 75 percent reduction in a given company's workforce. Middle management is particularly vulnerable to job loss from re-engineering. Hammer estimates that up to 80 percent of those engaged in middle-management tasks are susceptible to elimination.[10]

Across the entire U.S. economy, corporate re-engineering could eliminate between 1 million and 2.5 million jobs a year "for the foreseeable future," according to *The Wall Street Journal*.[11] By the time the first stage of re-engineering runs its course, some studies predict a loss of up to 25 million jobs in a private sector labor force that currently totals around 90 million workers. In Europe and Asia, where corporate restructuring and technology displacement is beginning to have an equally profound impact, industry analysts expect comparable job losses in the years ahead. Business consultants like John C. Skerritt worry about the economic and social consequences of re-engineering. "We can see many, many ways that jobs can be destroyed," says Skerritt, "but we can't see where they will be created." Others, like John Sculley, formerly of Apple Computer, believe that the "reorganization of work" could be as massive and destabilizing as the advent of the Industrial Revolution. "This may be the biggest social issue of the next 20 years," says Sculley.[12] Hans Olaf Henkel, the CEO of IBM Deutschland, warns, "There is a revolution underway."[13]

Nowhere is the effect of the computer revolution and re-engineering of the workplace more pronounced than in the manufacturing sector. One hundred and forty-seven years after Karl Marx urged the workers of the world to unite, Jacques Attali, a French minister and technology consultant to socialist president François Mitterrand, confidently proclaimed the end of the era of the working man and woman. "Machines are the new proletariat," proclaimed Attali. "The working class is being given its walking papers."[14]

The quickening pace of automation is fast moving the global economy to the day of the workerless factory. Between 1981 and 1991, more than 1.8 million manufacturing jobs disappeared in the U.S.[15] In Germany, manufacturers have been shedding workers even faster, eliminating more than 500,000 jobs in a single twelve-month period between early 1992 and 1993.[16] The decline in manufacturing jobs is part of a long-term trend that has seen the increasing replacement of human beings by machines at the workplace. In the 1950s, 33 percent of all U.S. workers were employed in manufacturing. By the 1960s, the number of manufacturing jobs had dropped to 30 percent, and by the 1980s to 20 percent. Today, less than 17 percent of the workforce is engaged in blue collar work. Management consultant Peter Drucker estimates that employment in manufacturing is going to continue dropping to less than 12 percent of the U.S. workforce in the next decade.[17]

For most of the 1980s it was fashionable to blame the loss of manufacturing jobs in the United States on foreign competition and cheap labor markets abroad. Recently, however, economists have begun to revise their views in light of new in-depth studies of the U.S. manufacturing sector. Noted economists Paul R. Krugman of MIT and Robert L. Lawrence of Harvard University suggest, on the basis of extensive data, that "the concern, widely voiced during the 1950s and 1960s, that industrial workers would lose their jobs because of automation, is closer to the truth than the current preoccupation with a presumed loss of manufacturing jobs because of foreign competition."[18]

Although the number of blue collar workers continues to decline, manufacturing productivity is soaring. In the United States, annual productivity, which was growing at slightly over 1 percent per year in the early 1980s, has climbed to over 3 percent in the wake of the new advances in computer automation and the restructuring of the workplace. From 1979 to 1992, productivity increased by 35 percent in the manufacturing sector while the workforce shrank by 15 percent.[19]

William Winpisinger, past president of the International Association of Machinists, a union whose membership has shrunk nearly in half as a result of advances in automation, cites a study by the International Metalworkers Federation in Geneva forecasting that within thirty years, as little as 2 percent of the world's current labor force "will be needed to produce all the goods necessary for total demand."[20] Yoneji Masuda, a principal architect of the Japanese plan to become the first fully computerized information based society, says that "in the near future, complete automation of entire plants will come into being, and during the next twenty to thirty years there will probably emerge . . . factories that require no manual labor at all."[21]

While the industrial worker is being phased out of the economic process, many economists and elected officials continue to hold out hope that the service sector and white collar work will be able to absorb the millions of unemployed laborers in search of work. Their hopes are likely to be dashed. Automation and re-engineering are already replacing human labor across a wide swath of service related fields. The new "thinking machines" are capable of performing many of the mental tasks now performed by human beings, and at greater speeds. Andersen Consulting Company, one of the world's largest corporate restructuring firms, estimates that in just one service industry, commercial banking and thrift institutions, re-engineering will mean a loss of 30 to 40 percent of the jobs over the next seven years. That translates into nearly 700,000 jobs eliminated.[22]

Over the past ten years more than 3 million white collar jobs were eliminated in the United States. Some of these losses, no doubt, were casualties of increased international competition. But as David Churbuck and Jeffrey Young observed in *Forbes*, "Technology helped in a big way to make them redundant." Even as the economy rebounded in 1992 with a respectable 2.6 percent growth rate, more than 500,000 additional clerical and technical jobs simply disappeared.[23] Rapid advances in computer technology, including parallel processing and artificial intelligence, are likely to make large numbers of white collar workers redundant by the early decades of the next century.

Many policy analysts acknowledge that large businesses are shedding record numbers of workers but argue that small companies are taking up the slack by hiring on more people. David Birch, a research associate at MIT, was among the first to suggest that new economic growth in the high-tech era is being led by very small firms—companies with under 100 employees. At one point Birch opined that more than 88 percent of all the new job creation was taking place in small businesses, many of whom were on the cutting edge of the new technology revolution. His data was cited by conservative economists during the Reagan Bush era as proof positive that new technology innovations were creating as many jobs as were being lost to technological displacement. More recent studies, however, have exploded the myth that small businesses are powerful engines of job growth in the high-tech era. Political economist Bennett Harrison, of the H. J. Heinz III School of Public Policy and Management at Carnegie-Mellon University, using statistics garnered from a wide variety of sources, including the International Labor Organization of the United Nations and the U.S. Bureau of the Census, says that in the United States "the proportion of Americans working for small companies and for individual establishments . . . has barely changed at all since at least the early 1960s." The same holds true, according to Harrison, for both Japan and West Germany, the other two major economic superpowers.[24]

The fact is that while less than 1 percent of all U.S. companies employ 500 or more workers, these big firms still employed more than 41 percent of all the workers in the private sector at the end of the last decade. And it is these corporate giants that are re-engineering their operations and letting go a record number of employees.[25]

The current wave of job cuts takes on even greater political significance in light of the tendency among economists continually to revise upward the notion of what is an "acceptable" level of unemployment. As with so many other things in life, we often adjust our expectations for the future, on the basis of the shifting present circumstances we find ourselves in. In the case of jobs, economists have come to play a dangerous game of accommodation with steadily rising unemployment figures, sweeping under the rug the implications of an historical curve that is leading inexorably to a world with fewer and fewer workers.

A survey of the past half-century of economic activity discloses a disturbing trend. In the 1950s the average unemployment for the decade stood at 4.5 percent. In the 1960s unemployment rose to an average of 4.8 percent. In the 1970s it rose again to 6.2 percent, and in the 1980s it increased again, averaging 7.3 percent for the decade. In the first three years of the 1990s, unemployment has averaged 6.6 percent.[26]

As the percentage of unemployed workers edged ever higher over the postwar period, economists have changed their assumptions of what constitutes full employment. In the 1950s, 3 percent unemployment was widely regarded as full employment. By the 1960s, the Kennedy and Johnson administrations were touting 4 percent as a full employment goal.

In the 1980s, many mainstream economists considered 5 or even 5.5 percent unemployment[27] as near full employment. Now, in the mid-1990s, a growing number of economists and business leaders are once again revising their ideas on what they regard as "natural levels" of unemployment. While they are reluctant to use the term "full employment," many Wall Street analysts argue that unemployment levels should not dip below 6 percent, lest the economy risk a new era of inflation.[28]

The steady upward climb in unemployment, in each decade, becomes even more troubling when we add the growing number of part-time workers who are in search of full-time employment and the number of discouraged workers who are no longer looking for a job. In 1993, more than 8.7 million people were unemployed, 6.1 million were working part-time but wanted full-time employment, and more than a million were so discouraged they stopped looking for a job altogether. In total, nearly 16 million American workers, or 13 percent of the labor force, were unemployed or underemployed in 1993.[29]

The point that needs to be emphasized is that, even allowing for short-term dips in the unemployment rate, the long-term trend is toward ever higher rates of unemployment. The introduction of more sophisticated technologies, with the accompanying gains in productivity, means that the global economy can produce more and more goods and services employing an ever smaller percentage of the available workforce.

A WORLD WITHOUT WORKERS...

When the first wave of automation hit the industrial sector in the late 1950s and early 1960s, labor leaders, civil rights activists, and a chorus of social critics were quick to sound the alarm. Their concerns, however, were little shared by business leaders at the time who continued to believe that increases in productivity brought about by the new automated technology would only enhance economic growth and promote increased employment and purchasing power. Today, however, a small but growing number of business executives are beginning to worry about where the new high-technology revolution is leading us. Percy Barnevik is the chief executive officer of Asea Brown Boveri, a 29-billion-dollar-a-year Swiss Swedish builder of electric generators and transportation systems, and one of the largest engineering firms in the world. Like other global companies, ABB has recently re-engineered its operations, cutting nearly 50,000 workers from the payroll, while increasing turnover 60 percent in the same time period. Barnevik asks, "Where will all these [unemployed] people go?" He predicts that the proportion of Europe's labor force employed in manufacturing and business services will decline from 35 percent today to 25 percent in ten years from now, with a further decline to 15 percent twenty years down the road. Barnevik is deeply pessimistic about Europe's future: "If anybody tells me, wait two or three years and there will be a hell of a demand for labor, I say, tell me where? What jobs? In what cities? Which companies? When I add it all together, I find a clear risk that the 10% unemployed or underemployed today could easily become 20 to 25%."[30]

Peter Drucker, whose many books and articles over the years have helped facilitate the new economic reality, says quite bluntly that "the disappearance of labor as a key factor of production" is going to emerge as the critical "unfinished business of capitalist society."[31]

For some, particularly the scientists, engineers, and employers, a world without work will signal the beginning of a new era in history in which human beings are liberated, at long last, from a life of back-breaking toil and mindless repetitive tasks. For others, the workerless society conjures up the notion of a grim future of mass unemployment and global destitution, punctuated by increasing social unrest and upheaval. On one point virtually all of the contending parties agree. We are, indeed, entering into a new period in history—one in which machines increasingly replace human beings in the process of making and moving goods and providing services. This realization led the editors of *Newsweek* to ponder the unthinkable in a recent issue dedicated to technological unemployment. "What if there were really no more jobs?" asked *Newsweek*.[32] The idea of a society not based on work is so utterly alien to any notion we have about how to organize large numbers of people into a social whole, that we are faced with the prospect of having to rethink the very basis of the social contract.

Most workers feel completely unprepared to cope with the enormity of the transition taking place. The rash of current technological breakthroughs and economic restructuring initiatives seem to have descended on us with little warning. Suddenly, all over the world, men and women are asking if there is a role for them in the new future unfolding across the global economy. Workers with years of education, skills, and experience face the very real prospect of being made redundant by the new forces of automation and information. What just a few short years ago was a rather esoteric debate among intellectuals and a small number of social writers around the role of technology in society is now the topic of heated conversation among millions of working people. They wonder if they will be the next to be replaced by the new thinking machines. In a 1994 survey conducted by *The New York Times*, two out of every five American workers expressed worry that they might be laid off, required to work reduced hours, or be forced to take pay cuts during the next two years. Seventy-seven percent of the respondents said they personally knew of someone who had lost his or her job in the last few years, while 67 percent said that joblessness was having a substantial effect on their communities.[33]

In Europe, fear over rising unemployment is leading to widespread social unrest and the emergence of neo-fascist political movements. Frightened, angry voters have expressed their frustration at the ballot box, boosting the electoral fortunes of extreme-right-wing parties in Germany, Italy, and Russia. In Japan, rising concern over unemployment is forcing the major political parties to address the jobs issue for the first time in decades.

We are being swept up into a powerful new technology revolution that offers the promise of a great social transformation, unlike any in history. The new high-technology revolution could mean fewer hours of work and greater benefits for millions. For the first time in modern history, large numbers of human beings could be liberated from long hours of labor in the formal marketplace, to be free to pursue leisure-time activities. The same technological forces could, however, as easily lead to growing unemployment and a global depression. Whether a utopian or dystopian future awaits us depends, to a great measure, on how the productivity gains of the Information Age are distributed. A fair and equitable distribution of the productivity gains would require a shortening of the workweek around the world and a concerted effort by central governments to provide alternative employment in the third sector—the social economy—for those whose labor is no longer required in the marketplace. If, however, the dramatic productivity gains of the high-tech revolution are not shared, but rather used primarily to enhance corporate profit, to the exclusive benefit of stockholders, top corporate managers, and the emerging elite of high-tech knowledge workers, chances are that the growing gap between the haves and the have-nots will lead to social and political upheaval on a global scale.

All around us today, we see the introduction of breathtaking new technologies capable of extraordinary feats. We have been led to believe that the marvels of modern technology would be our salvation. Millions placed their hopes for a better tomorrow on the liberating potential of the computer revolution. Yet the economic fortunes of most working people continue to deteriorate amid the embarrassment of technological riches. In every industrial country, people are beginning to ask why the age-old dream of abundance and leisure, so anticipated by generations of hardworking human beings, seems further away now, at the dawn of the Information Age, than at any time in the past half century. The answer lies in understanding a little-known but important economic concept that has long dominated the thinking of both business and government leaders around the world.

53

Welfare Reform in America

A Clash of Politics and Research

DIANA M. ZUCKERMAN

All institutions are interrelated. Zuckerman deals with the impact of politics on welfare reform, illustrating the close relationship of the political and economic institutions.

As you read this article, ask yourself the following questions:

1. *Many people see welfare as "something for nothing." Do you agree that welfare is necessary? Why?*
2. *If welfare is "something for nothing," would not subsidies to various industries also fit this claim? Why?*
3. *How would you change the welfare system?*

The 1996 Personal Responsibility and Work Opportunity Reconciliation Act (PRWORA) radically changed welfare as we knew it, and data on its impact on the most vulnerable Americans are just becoming available. . . .

The controversial passage of PRWORA was the culmination of many years of debate as well as concerns expressed across the political spectrum about the extent to which the welfare system should be considered a failure or an essential safety net. On the right, there were many years of anecdotes about "welfare queens" driving fancy cars, buying steaks with their food stamps, and teaching their children that welfare made working unnecessary. . . . On the left, there was an assumption that welfare saved innocent lives but also a growing concern about the deteriorating conditions in the inner city, where welfare dependence was sometimes a way of life being handed down from generation to generation, often accompanied by drug abuse, violence, teen pregnancy, and other social ills.

Research should have provided essential information to help determine how the welfare system should be changed, but instead, research was used as an ideological weapon to support conflicting points of view. . . . They could be used to prove both that most families on welfare were on it for short periods of time and that there was a hard-core group of families that stay on welfare for many years (Pavetti, 1996). Similarly, the statistics could be used to show that teen pregnancies were statistically significant predictors of long-term welfare and poverty (GAO, 1994, pp. 94–115) or that many welfare recipients were the victims of a crisis and just needed help for a few months before they were able to support themselves (Greenberg, 1993). . . .

This article describes how political and public pressures and compelling anecdotes overpowered the efforts of progressive public policy organizations and researchers, resulting in legislation focused on getting families off welfare rather than getting families out of poverty. . . .

SOURCE: A longer version of this article was published in the *Journal of Social Issues*, Winter 2000, Vol. 56, No. 4.
Dr. Zuckerman is president of the National Center for Policy Research (CPR) for Women & Families and can be reached at dz@center4policy.org.

WELFARE REFORM AS A PRESIDENTIAL ISSUE

As a presidential candidate, Bill Clinton made it clear that he would not defend the welfare system but instead would work to change it. . . . As President, he quickly appointed experts to work on welfare reform, with particular focus on toughening child support laws and requiring welfare recipients to prepare for self-sufficiency (Koppelman, 1993). Although the public and policymakers were ready for major changes in the welfare system, many did not realize that a law passed in 1988, the Family Support Act, already had strengthened child support collection and ordered states to require able-bodied recipients to enter remedial education or job training projects. President Clinton had been instrumental in negotiating that legislation as an officer of the National Governors' Association, and the bill was expected to help single mothers and to encourage more women to move from welfare to work. However, the weakened economy of the late 1980s led to a surge in the welfare rolls instead, with the size of those rolls increasing 25% from 1989 to 1992. As a result of the recession, states were unable to provide matching funds to claim their full share of the federal funding for Job Opportunities and Basic Skills (JOBS), a training program. The law did not seem to be working, and the pressure to "do something about welfare" grew.

As a candidate, Clinton had promised to reform health care as well as welfare, and as President, he decided to focus on health care reform legislation first. This was a logical choice, because the lack of affordable health care was a major disincentive for single mothers who wanted to move from welfare to work. The Clinton administration strategy was to first pass a law that would make health care affordable for the working poor, so that it would be easier to make the other legislative changes necessary to reduce the welfare rolls. Improving access to health care would help those families that would lose Medicaid when they left welfare for low-level jobs that did not offer health insurance as a benefit.

While health care reform took center stage in the public eye, a welfare reform plan was being quietly developed by Clinton administration officials. As Assistant Secretary for Children and Families at HHS, Mary Jo Bane wanted to refocus the welfare system on work and to give states more flexibility regarding welfare policies (Bane, 1997). States were already submitting welfare reform proposals to the federal government to request waivers from the federal requirements. . . .

Meanwhile, polls were showing tremendous public support for requiring all "able-bodied" welfare recipients, including mothers with young children, to get education or training for up to 2 years and then to work (Ellwood, 1996). Assistant HHS Secretaries Bane and David Ellwood wanted the federal law to set consistent national criteria and restrictions, in order to avoid a race to the bottom by the states. According to Ellwood, their proposal to reform welfare had four major goals:

1. *Make work pay*. Low-income workers need a living wage, health care, and child care to make working make sense instead of welfare.
2. *Two-year limits*. Transform the welfare system from a handout to a hand up, with clear requirements to work after 2 years of training or education.
3. *Child support enforcement*. Require absent parents to pay, whether or not they were married.
4. *Fight teen pregnancy*. The plan offered grants to high-risk schools that proposed innovative initiatives to lower rates of teen pregnancy and also supported a national clearinghouse and a few intensive demonstration projects.

The public seemed ready for these kinds of changes, but the Clinton administration was concerned about the cost. In order to move single mothers from welfare to work, it would be necessary to provide job training, child care, and other services for many of them. The cost of those programs and services would initially be high, and it was expected that the savings as families moved off welfare would not be great enough to make up for those extra costs. Since President Clinton had also promised to cut the country's enormous budget deficit, and since a major goal of welfare reform was to save money and make the federal government smaller, it did not seem politically feasible to expect taxpayers to pay more for welfare reform than they were paying for the existing welfare system, even in the short term. To save money, the Clinton plan included a slow phase-in of the program, which caused some conservative critics to accuse the administration of not being serious about reform (Ellwood, 1996).

CONGRESS AND WELFARE REFORM

For a variety of reasons, President Clinton had difficulty obtaining the support of his own party on welfare, health care, and other issues. As a moderate "New Democrat," his positions often seemed too conservative for the more liberal Democrats and too liberal for the more conservative ones, most of whom were southern and concerned about reelection as the Republican Party gained support in the South. Conservative Democrats developed their own welfare reform plan with a faster phase-in. In addition, conservatives from both parties were concerned that the Clinton plan did not sufficiently address out-of-wedlock childbearing and provided too many federal requirements instead of giving the states the autonomy to decide about welfare reform (Ellwood, 1996).

Why did the Clinton plan fail? Ellwood (1996) speculates that they should have tried to pass welfare reform before health care reform instead of afterward.... As someone involved in health care reform legislation in the Senate at the time, I believe that the problem with the timing was not lack of attention but rather the weakened credibility of the Clinton administration because of the barrage of criticism aimed at its health care proposal. Ellwood also speculates that the President's promise to "end welfare as we know it" was a potent sound bite but did not address the concerns of many Democrats about whether the new system would be better than the old. Ellwood believes that the phrase "2 years and you're off" was even more destructive, because it implied no help at all after 2 years, which he says is "never what was intended." I agree that these were problems, and in addition, the tension and lack of trust between the congressional Democrats and the Clinton administration contributed to the view of many Democrats that the Clinton welfare plan was too controversial and would make them politically vulnerable....

The 1994 election, which resulted in the Republican takeover of the House and Senate, changed the political dynamics. The result was that every congressional committee was chaired by a Republican instead of a Democrat and composed primarily of Republicans rather than Democrats. Instead of being chaired by the most liberal Democrats, the committees that would vote on welfare reform were now chaired by conservative Republicans. Although the Republicans had only a small majority in the House, they had the control of committee chairmanships and disproportionate membership on the committees. This meant that the Republican leadership had tremendous power to control the welfare reform bill.

Even more important than the leadership of the congressional committees was the leadership of the House of Representatives, which changed from a liberal Democrat to Newt Gingrich, an outspoken critic of "big government," especially for social programs. Gingrich had been a major architect of the Contract With America, which specified that welfare should be available for only 2 years and that benefits should not be available to minor mothers or for children born to mothers on welfare. The shellacking that the Democrats experienced in the election was perceived as a mandate for the Contract With America (Merida, 1994), including a more punitive welfare reform plan than the one the Clinton administration had proposed.

Meanwhile, state governors demanded more say about how money in the new welfare program would be spent, resulting in a bill that looked more like a block grant than a social program. This had the political benefit of getting Congress off the hook: It would not have to make the difficult political decisions about restrictions and instead could put those decisions in the hands of the state governments.

The media were also influential and tended to focus on the shortcomings of the welfare program that was in place, Aid for Families with Dependent Children (AFDC)....

As the ideological battle continued around it, Congress was under tremendous pressure to pass a welfare reform bill....

Throughout 1995 and 1996, the Republican majority controlled the policy agenda and welfare reform was their cause. In that political climate, it would have been very difficult for progressive organizations to succeed in their public education and lobbying efforts, regardless of research results. It was especially difficult for liberal legislators to suggest that welfare mothers should be able to stay home and care for their children when nationwide 55% of mothers with children under the age of 3 were employed, most of them full time (Pavetti, 1997). To make matters worse, the progressive advocates did not have solid research findings to support their opposition to the Republican leadership's welfare reform bill, other than frightening but questionable statistics describing the number of children who would fall into poverty if their mothers were thrown off welfare. This left the most progressive members of Congress with little ammunition to use against the most conservative proposals. The slim Republican majority in Congress was joined by enough Democrats to create a substantial majority for the welfare reform bill that passed in 1996.

The bill that Congress eventually passed is lengthy and complicated, but the introductory section includes "findings" that explicitly show the ideology behind the legislation. There are 10 findings, and the first three set the tone:

1. Marriage is the foundation of a successful society.
2. Marriage is an essential institution of a successful society which promotes the interests of children.
3. Promotion of responsible fatherhood and motherhood is integral to successful child rearing and the well-being of children. (U.S. Congress, 1996, p. 6)

The bill's focus on marriage and responsible parenthood reflects the Republican Party's ties to the Christian right as well as the growing disenchantment with government programs throughout the country in the early and mid-1990s. The rest of the findings, however, use research data to support this focus, and many of the statistics are ones that have been employed by experts across the ideological spectrum to support public policies aimed at reducing teen pregnancy, reducing poverty, and other goals that are as popular among liberal Democrats as conservative Republicans, as well as everything in between. For example, the next finding pointed out that "only 54 percent of single-parent families with children had a child support order established and, of that 54 percent, only about one-half received the full amount due" (U.S. Congress, 1996, p. 6).

Another major concern was the growth of welfare. The fifth finding points out that the number of individuals on welfare had more than tripled since 1965 and that more than two-thirds were children. The number of children on welfare every month increased from 3.3 million in 1965 to 9.3 million in 1992, although the number of children in the United States declined during those years. The legislation also points out that 89% of the children receiving AFDC were living in homes without fathers and that the percentage of unmarried women nearly tripled between 1970 and 1991.

The next finding includes details regarding pregnancies among unmarried teens and concludes "if the current trend continues, 50 percent of all births by the year 2015 will be out-of-wedlock" (U.S. Congress, 1996, p. 7). The findings recommended that strategies to combat teenage pregnancy "must address the issue of male responsibility, including statutory rape culpability and prevention" and points out that most teen mothers "have histories of sexual and physical abuse, primarily with older adult men."

Despite the clear ideological underpinnings of the bill, some of the findings could be embraced by both political parties. For example, the bill correctly points out that unmarried teenage mothers are more likely to go on welfare and to spend more years on welfare. It also points out that babies of unwed mothers are at risk for very low or moderately low birth weight, for growing up to have lower cognitive attainment and lower educational aspirations, and for child abuse and neglect. They are more likely to grow up to be teen parents and to go on welfare and less likely to have an intact marriage.

The problems of single parenting were also described, showing how they set in motion a cycle of welfare and poverty:

- Mothers under 20 years of age are at the greatest risk of bearing low birth weight babies.
- The younger the single-parent mother, the less likely she is to finish high school.
- Young women who have children before finishing high school are more likely to receive welfare assistance for a longer period of time.
- Children of teenage single parents have lower cognitive scores, lower educational aspirations, and a greater likelihood of becoming teenage parents themselves.
- Children of single-parent homes are three times more likely to fail and repeat a year in grade school and almost four times more likely to be expelled or suspended from school than are children from intact two-parent families.
- Of those youth held for criminal offenses within the State juvenile justice system, only 30% lived primarily in a home with both parents, compared to 74% of the general population of children.

The costs of teen parenting were delineated, with an estimate that "between 1985 and 1990, the public cost of births to teenage mothers under the Aid to Families with Dependent Children program, the food stamp program, and the Medicaid program" (U.S. Congress, 1996, p. 8) was $120 billion. In a climate of deficit reduction and support for smaller government, that estimate was extremely compelling.

The welfare reform bill includes nine sections, referred to as "titles," and several are more generous than would have been expected given the Republican control of Congress and the Republicans' opposition to "big government." For example, Title VI on child care authorizes increased federal money so that child care is more available and affordable. President Clinton, however, expressed considerable concerns about Title IV, which banned most legal immigrants from most federal benefit programs, and Title VIII, which cut the food stamp program across the board and also restricted food stamps to unemployed adults without disabilities or dependents to 3 months out of every 36.

The welfare reform law had a great deal of support from governors around the country, because it gave them enormous flexibility regarding the spending of federal funds. The rationale was that states could experiment with new approaches, under the assumption that what works in a rural state, for example, might not work in a more urban environment. Heavy subsidies for day care or job training might be useful in some areas, for example, but unnecessary in others.

WHY NOT WAIT FOR DATA?

Although data were quoted in the welfare reform bill, there were no convincing data to predict what would actually happen if the bill passed. A "natural experiment" was taking place, however, that could have answered those questions. Between January 1987 and August 1996, 46 states had received approval for waivers to experiment with AFDC and welfare-to-work programs (GAO, 1997). Since welfare reform represented such a dramatic change in policies, with many lives at stake, it would have been logical to delay a federal welfare reform law until the data were analyzed from those programs. For example, by May 1997, the General Accounting Office, which is a research branch of Congress, had published a report entiled *Welfare Reform: States' Early Experiences with Benefits Termination*, based on a study conducted at the request of Senator Pat Moynihan. The study found that the benefits of 18,000 families were terminated under waivers through December 1996, most of them in

Iowa, Massachusetts, and Wisconsin. More than 99% of these families failed to comply with program requirements; for example, some wanted to stay home with their children or were unwilling to do community service or work for low wages (GAO, 1997). These findings could have been used to design a welfare reform process that protected some of these families, but the bill was passed before the data were available.

Instead, the little research that was already completed was used to push welfare reform forward. For example, Vermont had applied for waivers to the welfare requirements, and its welfare restructuring project was the nation's first statewide demonstration of time-limited welfare (Zengerle, 1997). In 2 years, Vermont raised the number of welfare parents with jobs from 20% to 26% and increased their average monthly earnings from $373 to $437. Of course, it was also important to note that Vermont increased its social services budget by 50%, in part to create new jobs. Welfare recipients who were unable to find jobs in the private sector were eligible for 10 months' employment in public jobs or working for nonprofit organizations. It also would have been logical to note that Vermont has relatively few welfare recipients and that its experience is likely to be different from that of most other states. Nevertheless, this study was used to show the success of welfare reform. . . .

THE REALITY OF WELFARE REFORM

Although the welfare reform bill that passed in 1996 was radical and potentially devastating to poor families, the booming economy during the next four years and political compromises resulted in a law that no longer seemed as extreme or partisan as it once did. For example, in the short term, almost every state received more federal funds under welfare reform that it did prior to reform (Nightingale & Brennan, 1998). Perhaps most important, the Clinton administration was able to influence the bill through regulations. The final regulations, announced in April 1999, contain exceptions to the rules that make the bill less rigid. For example, states may continue to provide welfare benefits for longer than 60 months to up to 20% of the welfare caseload based on hardship or domestic violence, and the 20% limit can be exceeded if there are federally recognized "good cause domestic violence waivers" (Schott, Lazere, Goldberg, & Sweeney, 1999). In addition, states will be penalized if they sanction a single parent caring for a child under age 6 if the parent can demonstrate his or her inability to obtain child care.

Research results are finally coming in from across the country. The Urban Institute (Loprest, 1999) reports that most adults leaving welfare between 1995 and 1997 got a new job or increased earnings. They also report that former welfare recipients work more hours than employed near-poor mothers who were not previously receiving welfare. Nevertheless, many families are doing poorly since welfare reform was implemented, and the number of children in extreme poverty (less than half the poverty level) has increased (Sherman, 1999). Between 1995 and 1997, the average income of the poorest 20% of female-headed households fell an average of $580 per family, primarily due to the loss of food stamps and other government benefits (Primus, Rawlings, Larin, & Porter, 1999).

Research had little impact on the passage of welfare reform in 1996, although statistics were used by both sides in the national debate. Is it possible that new research on the impact of welfare reform will influence welfare policies in the future? The studies in this issue provide very useful information about the barriers to success for welfare mothers who attempt to move into the workforce on a permanent basis, with important implications for how policies can maximize success and minimize tragedy for vulnerable families. Now that the costs of the welfare program have been drastically reduced and the national annual budget deficit has turned into a surplus, these studies may manage to attract the attention of policymakers. Unfortunately, welfare reform has made national policy changes in this area even more difficult, because decisions are now made in 50 different states, rather than by the federal government.

REFERENCES

Bane, M. J. (1997). Welfare as we might know it. *American Prospect, 30*, 47–53.

Elwood, and D. T. (1996). Welfare reform as I knew it: When bad things happen to good policies. *American Prospect, 26*, 22–29.

General Accounting Office (GAO). (1994). *Teenage mothers least likely to become self-sufficient.* GAO/HEHS-94-115. Washington, DC: GAO.

General Accounting Office (GAO). (1997). *Welfare reform: States' early experience with benefit termination.* GAO/HEHS-97-74. Washington, DC: GAO.

Greenberg, M. (1993). *What state AFDC studies on length of stay tell us about welfare as a "way of life."* Washington, DC: Center for Law and Social Policy.

Koppelman, J. (1993). Helping AFDC children escape the cycle of poverty: Can welfare reform be used to achieve this goal? *National Health Policy Forum, 627*, 2–9.

Krauthammer, C. (1994, November 11). Republican mandate. *Washington Post*, p. A31.

Loprest, P. (1999). *How families that left welfare are doing: A national picture.* Washington, DC: Urban Institute.

Merida, K. (1994, December 28). Last rites for liberalism? Democrats' legacy now symbolizes their woes. *Washington Post*, p. A1.

Nightingale, D. S., & Brennan, K. (1998). *The welfare-to-work grants program: A new link in the welfare reform chain.* Washington, DC: Urban institute.

Pavetti, L. (1996, May 23). *Time on welfare and welfare dependency.* Testimony before the House of Representatives Committee on Ways and Means, 104th Congress.

Pavetti, L. (1997), *How much more can they work? Setting realistic expectations for welfare mothers.* Washington, DC: Urban Institute.

Primus, W., Rawlings, L., Larin, K., & Porter, K. (1999). *The initial impacts of welfare reform on the incomes of single-mother families.* Washington, DC: Center on Budget and Policy Priorities.

Schott, L., Lazere, E., Goldberg, H., & Sweeney, E. (1999). *Highlights of the final TANF regulations.* Washington, DC: Center on Budget and Policy Priorities.

Sherman, A. (1999). *Extreme child poverty rises sharply in 1997.* Washington, DC: Children's Defense Fund.

U.S. Congress. H.R. 3734: Personal Responsibility and Work Opportunity Reconciliation Act of 1996 (Enrolled Bill Sent to President) (1996) [On-line]. Available: http://thomas.loc.gov/cgi-bin/query/D?cl04:1:./temp/~c104aiBNdl:e19928

Zengerle, J. S. (1997). Welfare as Vermont knows it. *American Prospect*, 30, 54–55.

54

The Credit Card

Private Troubles and Public Issues

GEORGE RITZER

The credit card is seen as a personal part of the individual's economy. Ritzer reveals that anything this ubiquitous has many ramifications for the society in general.

As you read, ask yourself the following questions:

1. *According to the author, what are some good and bad things that can be said about credit cards?*
2. *How does your personal experience tie in with the author's claims about credit cards?*
3. *What restrictions would you put on credit card companies? Why?*

GLOSSARY **Machinations** Crafty schemes.

The credit card has become an American icon. It is treasured, even worshipped, in the United States and, increasingly, throughout the rest of the world. The title of the book [from which this selection was taken], *Expressing America*, therefore has a double meaning: The credit card expresses something about the essence of modern American society and, like an express train, is speeding across the world's landscape delivering American (and more generally consumer) culture. . . .

The credit card is not the first symbol of American culture to play such a role, nor will it be the last. Other important contemporary American icons include Coca-Cola, Levi's, Marlboro, Disney, and McDonald's. What they have in common is that, like credit cards, they are products at the very heart of American society, and they are highly valued by, and have had a profound effect on, many other societies throughout the world. However, the credit card is distinctive because it is a means that can be used to obtain those other icons, as well as virtually anything else available in the world's marketplaces. It is because of this greater versatility that the credit card may prove to be the most important American icon of all. If nothing else, it is likely to continue to exist long after other icons have become footnotes in the history of American culture. When the United States has an entirely new set of icons, the credit card will remain an important means for obtaining them. . . .

THE ADVANTAGES OF CREDIT CARDS

. . . The most notable advantage of credit cards, at least at the societal level, is that they permit people to spend more than they have. Credit cards thereby allow the economy to function at a much higher (and faster) level than it might if it relied solely on cash and cash-based instruments.

Credit cards also have a number of specific advantages to consumers, especially in comparison to using cash for transactions:

- Credit cards increase our spending power, thereby allowing us to enjoy a more expansive, even luxurious lifestyle.
- Credit cards save us money by permitting us to take advantage of sales, something that might not be possible if we had to rely on cash on hand.
- Credit cards are convenient. They can be used 24 hours a day to charge expenditures by phone, mail, or home computer. Thus, we need no longer be inconvenienced by the fact that most shops and malls close overnight. Those whose mobility is limited or who are housebound can also still shop.
- Credit cards can be used virtually anywhere in the world, whereas cash (and certainly checks) cannot so easily cross national borders. For example, we are able to travel from Paris to Rome on the spur of the moment in the middle of the night without worrying about whether we have, or will be able to obtain on arrival, Italian lire.
- Credit cards smooth out consumption by allowing us to make purchases even when our incomes are low. If we happen to be laid off, we can continue to live the same lifestyle, at least for a time, with the anticipation that we will pay off our credit card balances when we are called back to work. We can make emergency purchases (of medicine, for example) even though we may have no cash on hand.
- Credit cards allow us to do a better job of organizing our finances, because we are provided each month with a clear accounting of expenditures and of money due.
- Credit cards may yield itemized invoices of tax-deductible expenses, giving us systematic records at tax time.
- Credit cards allow us to refuse to pay a disputed bill while the credit card company investigates the transaction. Credit card receipts also help us in disputes with merchants.
- Credit cards give us the option of paying our bills all at once or of stretching payments out over a length of time.
- Credit cards are safer to carry than cash is and thus help to reduce cash-based crime. . . .

A KEY PROBLEM WITH CREDIT CARDS

In the course of the twentieth century, the United States has gone from a nation that cherished savings to one that reveres spending, even spending beyond one's means. . . .

At the level of the national government, our addiction to spending is manifest in a once-unimaginable level of national debt, the enormous growth rate of that debt, and the widespread belief that the national debt cannot be significantly reduced, let alone eliminated. As a percentage of gross national product (GNP),* the federal debt declined rather steadily after World War II, reaching 33.3% in 1981. However, it then rose dramatically, reaching almost 73% of GNP in 1992. In dollar terms, the federal debt was just under $1 trillion in 1981, but by September 1993, it had more than quadrupled, to over $4.4 trillion. There is widespread fear that a huge and growing federal debt may bankrupt the nation and a near consensus that it will adversely affect future generations.

Our addiction to spending is also apparent among the aggregate of American citizens. Total

* While the term *GNP* is still used for historical purpose it should be noted that the term *GDP* (gross domestic product) is now preferred.

personal savings was less in 1991 than in 1984, in spite of the fact that the population was much larger in 1991. Savings fell again in the early 1990s from about 5.2% of disposable income in late 1992 to approximately 4% in early 1994. A far smaller percentage of families (43.5%) had savings accounts in 1989 than did in 1983 (61.7%). And the citizens of many other nations have a far higher savings rate. At the same time, our indebtedness to banks, mortgage companies, credit card firms, and so on is increasing far more dramatically than similar indebtedness in other nations. . . .

WHO IS TO BLAME?

The Individual

In a society that is inclined to "psychologize" all problems, we are likely to blame individuals for not saving enough, for spending too much, and for not putting sufficient pressure on officials to restrain government expenditures. We also tend to "medicalize" these problems, blaming them on conditions that are thought to exist within the individual. . . . Although there are elements of truth to psychologistic and medicalistic perspectives, there is also a strong element of what sociologists term "blaming the victim." That is, although individuals bear some of the responsibility for not saving, for accumulating mounting debt, and for permitting their elected officials to spend far more than the government takes in, in the main individuals have been victimized by a social and financial system that discourages saving and encourages indebtedness.

Why are we so inclined to psychologize and medicalize problems like indebtedness? For one thing, American culture strongly emphasizes individualism. We tend to trace both success and failure to individual efforts, not larger social conditions. For another, large social and financial systems expend a great deal of time, energy, and money seeking, often successfully, to convince us that they are not responsible for society's problems. Individuals lack the ability and the resources to similarly "pass the buck." Of perhaps greatest importance, however, is the fact that individual, especially medical, problems appear to be amenable to treatment and even seem curable. In contrast, large-scale social problems (pollution, for example) seem far more intractable. It is for these, as well as many other reasons, that American society has a strong tendency to blame individuals for social problems.

The Government

. . . Since the federal debt binge began in 1981, the government has also been responsible for creating a climate in which financial imprudence seems acceptable. After all, the public is led to feel, if it is acceptable for the government to live beyond its means, why can't individual citizens do the same? If the government can seemingly go on borrowing without facing the consequences of its debt, why can't individuals?

If the federal government truly wanted to address society's problems, it could clearly do far more both to encourage individual savings and to discourage individual debt. For example, the government could lower the taxes on income from savings accounts or even make such income tax-free. Or it could levy higher taxes on organizations and agencies that encourage individual indebtedness. The government could also do more to control and restrain the debt-creating and debt-increasing activities of the credit card industry.

Business

Although some of the blame for society's debt and savings problem must be placed on the federal government, the bulk of the responsibility belongs with those organizations and agencies associated with our consumer society that do all they can to get people to spend not only all of their income but also to plunge into debt in as many ways, and as deeply, as possible. We can begin with American business.

Those in manufacturing, retailing, advertising, and marketing (among others) devote their working hours and a large portion of their energies to figuring out ways of getting people to buy thin

. . . One example is the dramatic proliferation of seductive catalogs that are mailed to our homes. Another is the advent and remarkable growth in popularity of the television home shopping networks. What these two developments have in common is their ability to allow us to spend our money quickly and efficiently without ever leaving our homes. Because the credit card is the preferred way to pay for goods purchased through these outlets, catalogs and home shopping networks also help us increase our level of indebtedness.

Banks and Other Financial Institutions

The historical mission of banks was to encourage savings and discourage debt. Today, however, banks and other financial institutions lead us away from savings and in the direction of debt. Saving is discouraged by, most importantly, the low interest rates paid by banks. It seems foolish to people to put their money in the bank at an interest rate of, say, 2.5% and then to pay taxes on the interest, thereby lowering the real rate of return to 2% or even less. This practice seems especially asinine when the inflation rate is, for example, 3% or 4%. Under such conditions, the saver's money is declining in value with each passing year. It seems obvious to most people that they are better off spending the money before it has a chance to lose any more value.

While banks are discouraging savings, they are in various ways encouraging debt. One good example is the high level of competition among the banks (and other financial institutions) to offer home equity lines of credit to consumers. As the name suggests, such lines of credit allow people to borrow against the equity they have built up in their homes. . . . Banks eagerly lend people money against this equity. Leaving the equity in the house is a kind of savings that appreciates with the value of the real estate, but borrowing against it allows people to buy more goods and services. In the process, however, they acquire a large new debt. And the house itself could be lost if one is unable to pay either the original mortgage or the home equity loan.

The credit card is yet another invention of the banks and other financial institutions to get people to save less and spend more. In the past, only the relatively well-to-do were able to get credit, and getting credit was a very cumbersome process (involving letters of credit, for example). Credit cards democratized credit, making it possible for the masses to obtain at least a minimal amount. Credit cards are also far easier to use than predecessors like letters of credit. Credit cards may thus be seen as convenient mechanisms whereby banks and other financial institutions can lend large numbers of people what collectively amounts to an enormous amount of money.

Normally, no collateral is needed to apply for a credit card. The money advanced by the credit card firms can be seen as borrowing against future earnings. However, because there is no collateral in the conventional sense, the credit card companies usually feel free to charge usurious interest rates.

Credit cards certainly allow the people who hold them to spend more than they otherwise would. . . . Some . . . overwhelmed by credit card debt, take out home equity lines of credit to pay it off. Then, with a clean slate, at least in the eyes of the credit card companies, such people are ready to begin charging again on their credit cards. Very soon many of them find themselves deeply in debt both to the bank that holds the home equity loan and to the credit card companies.

A representative of the credit card industry might say that no one forces people to take out home equity lines of credit or to obtain credit cards; people do so of their own volition and therefore are responsible for their financial predicament. Although this is certainly true at one level, at another level it is possible to view people as the victims of a financial (and economic) system that depends on them to go deeply into debt and itself grows wealthy as a result of that indebtedness. The newspapers, magazines, and broadcast media are full of advertisements offering various inducements to apply for a particular credit card or home equity loan rather than the ones offered by competitors. Many people are bombarded with mail offering all sorts of attractive benefits to those who sign up for yet another card or loan. More generally, one is made to feel foolish, even out of step, if one refuses to be an active part of the debtor society. Furthermore, it has become increasingly difficult to function in our society without a credit card. For example, people who do not have a record of credit card debt and payment find it difficult to get other kinds of credit, like home equity loans, car loans, or even mortgage loans.

AN INDICTMENT OF THE FINANCIAL SYSTEM

The major blame for our society's lack of savings and our increasing indebtedness must be placed on the doorstep of large institutions. We focus on one of those institutions—the financial system, which is responsible for making credit card debt so easy and attractive that many of us have become deeply and perpetually indebted to the credit card firms....

CASE IN POINT: GETTING THEM HOOKED WHILE THEY'RE YOUNG

Before moving on to a more specific discussion of the sociological perspective on the problems associated with credit cards, one more example of the way the credit card industry has created problems for people would be useful: the increasing effort by credit card firms to lure students into possessing their own credit cards. The over 9 million college students (of which 5.6 million are in school on a full-time basis) represent a huge and lucrative market for credit card companies. According to one estimate, about 82% of full-time college students now have credit cards. The number of undergraduates with credit cards increased by 37% between 1988 and 1990. The credit card companies have been aggressively targeting this population not only because of the immediate increase in business it offers but also because of the long-term income possibilities as the students move on to full-time jobs after graduation. To recruit college students, credit card firms are advertising heavily on campus, using on-campus booths to make their case and even hiring students to lure their peers into the credit card world. In addition, students have been offered a variety of inducements. I have in front of me a flyer aimed at a college-age audience. It proclaims that the cards have no annual fee, offer a comparatively low interest rate, and offer "special student benefits," including a 20% discount at retailers like MusicLand and Gold's Gym and a 5% discount on travel.

The credit card firms claim that the cards help teach students to be responsible with money (one professor calls it a "training-wheels operation"). The critics claim that the cards teach students to spend, often beyond their means, instead of saving....

In running up credit card debt, it can be argued, college students are learning to live a lie. They are living at a level that they cannot afford at the time or perhaps even in the future. They may establish a pattern of consistently living beyond their means. However, they are merely postponing the day when they have to pay their debts....

Not satisfied with the invasion of college campuses, credit card companies have been devoting increasing attention to high schools. One survey found that as of 1993, 32% of the country's high school students had their own credit cards and others had access to an adult's card. Strong efforts are underway to greatly increase that percentage. The president of a marketing firm noted, "It used to be that college was the big free-for-all for new customers.... But now, the big push is to get them between 16 and 18." Although adult approval is required for a person under 18 years of age to obtain a credit card, card companies have been pushing more aggressively to gain greater acceptance in this age group....

The motivation behind all these programs is the industry view that about two-thirds of all people remain loyal to their first brand of card for 15 or more years. Thus the credit card companies are trying to get high school and college students accustomed to using their card instead of a competitor's. The larger fear is that the credit card companies are getting young people accustomed to buying on credit, thereby creating a whole new generation of debtors.

A SOCIOLOGY OF CREDIT CARDS

... Sociologists have grown increasingly dissatisfied with having to choose between large-scale, macroscopic theories and small-scale, microscopic theories. Thus, there has been a growing interest in theories that integrate micro and macro concerns. In Europe, expanding interest in what is known there as agency-structure integration parallels the increasing American preoccupation with micro-macro integration.

Mills: Personal Troubles, Public Issues

Of more direct importance here is the now-famous distinction made by Mills in his 1959 work, *The Sociological Imagination,* between micro-level personal troubles and macro-level public issues. Personal troubles tend to be problems that affect an individual and those immediately around him or her. For example, a father who commits incest with his daughter is creating personal troubles for the daughter, other members of the family, and perhaps himself. However, that single father's actions are not going to create a public issue; that is, they are not likely to lead to a public outcry that society ought to abandon the family as a social institution. Public issues, in comparison, tend to be problems that affect large numbers of people and perhaps society as a whole. The disintegration of the nuclear family would be such a public issue....

A useful parallel can be drawn between the credit card and cigarette industries. The practices of the cigarette industry create a variety of personal troubles, especially illness and early death. Furthermore, those practices have created a number of public issues (the cost to society of death and illness traceable to cigarette smoke), and thus many people have come to see cigarette industry practices themselves as public issues. Examples of industry practices that have become public issues are the aggressive marketing of cigarettes overseas.... Similarly, the practices of the credit card industry help to create personal problems (such as indebtedness) and public issues (such as the relatively low national savings rate). Furthermore, some industry practices—such as the aggressive marketing of credit cards to teenagers—have themselves become public issues.

One of our premises is that we need to begin adopting the same kind of critical outlook toward the credit card industry that we use in scrutinizing the cigarette industry....

Mills's ideas give us remarkably contemporary theoretical tools for undertaking a critical analysis of the credit card industry and the problems it generates....

Marx: Capitalist Exploitation

In addition to Mills's general approach, there is the work of the German social theorist Karl Marx (1818–1881), especially his ideas on the exploitation that he saw as endemic to capitalist society....

There have been many changes in the capitalist system, and a variety of issues have come to the fore that did not exist in Marx's day. As a result, a variety of neo-Marxian theories have arisen to deal with these capitalist realities. One that concerns us here is the increasing importance to capitalists of the market for goods and services. According to neo-Marxians, exploitation of the worker continues in the labor market, but capitalists also devote increasing attention to getting consumers to buy more goods and services. Higher profits can come from both cutting costs and selling more products.

The credit card industry plays a role by encouraging consumers to spend more money, in many cases far beyond their available cash, on the capitalists' goods and services. In a sense, the credit card companies have helped the capitalists to exploit consumers. Indeed, one could argue that modern capitalism has come to depend on a high level of consumer indebtedness. Capitalism could have progressed only so far by extracting cash from the consumers. It had to find a way to go further....

Simmel: The Money Economy

...Simmel pointed to many problems associated with a money economy, but three are of special concern:

- The first problem ... is the "temptation to imprudence" associated with a money economy. Simmel argued that money, in comparison to its predecessors, such as barter, tends to tempt people into spending more and going into debt. My view is that credit cards are even more likely than money to make people imprudent. People using credit cards are not only likely to spend more but are also more likely to go deeply into debt....
- Second, Simmel believed that money makes possible many types of "mean machinations" that were not possible, or were more difficult, in earlier economies. For example, bribes for political influence or payments for assassinations are more easily made with money than with barter.... Although bribes or assassinations are generally less likely to be paid for with a credit card than with cash, other types of mean machinations become more likely with credit cards. For example, some organizations associated with the credit card industry engage in fraudulent or deceptive practices in order to maximize their income from credit card users....
- The third problem with a money economy that concerned Simmel was the issue of secrecy, especially the fact that a money economy makes payments of bribes and other types of secret transactions more possible. However, our main concern is the increasing lack of secrecy and the invasion of privacy associated with the growth of the credit card industry....

Weber: Rationalization

... Weber defined rationalization as the process by which the modern world has come to be increasingly dominated by structures devoted to efficiency, predictability, calculability, and technological control. Those rational structures (for example, the capitalist marketplace and the bureaucracy) have had a progressively negative effect on individuals. Weber described a process by which more and more of us would come to be locked in an "iron cage of rationalization." ...The credit card industry has also been an integral part of the rationalization process. By rationalizing the process by which consumer loans are made, the credit card industry has contributed to our society's dehumanization....

Globalization and Americanization

A sociology of credit cards requires a look at the relationship among the credit card industry, personal troubles, and public issues on a global scale. It is not just the United States, but also much of the rest of the world, that is being affected by the credit card industry and the social problems it helps create. To some degree, this development is a result of globalization, a process that is at least partially autonomous of any single nation and that involves the reciprocal impact of many economies. In the main, however, American credit card companies dominate the global market....

The central point is that, in many countries around the world, Americanization is a public issue that is causing personal troubles for their citizens. The credit card industry Americanizes and homogenizes life around the world, with the attendant loss of cultural and individual differences....

OTHER REASONS FOR EXAMINING CREDIT CARDS

Something New in the History of Money

Money in all its forms, especially in its cash form, is part of a historical process. It may seem hard to believe from today's vantage point, but at one time there was no money. Furthermore, some predict that there will come a time in which money, at least in the form of currency, will become less important if not disappear altogether, with the emergence of a "cashless society."...

More important for our purposes, money in the form of currency is being increasingly supplanted by the credit card. Instead of plunking down cash or even writing a check, more of us are saying "Charge it!" This apparently modest act is, in fact, a truly revolutionary development in the history of money. Furthermore, it is having a revolutionary impact on the nature of consumption, the economy, and the social world more generally. In fact, rather than simply being yet another step in the development of money, I am inclined to agree with the contention that credit cards are "an entirely new idea in value exchange." A variety of arguments can be marshaled in support of the idea that in credit cards we are seeing something entirely new in the history of economic exchange, especially relative to cash:

- Credit card companies are performing a function formerly limited to the federal government. That is, they create money ... the Federal Reserve is no longer alone in this ability. The issuing of a new credit card with a $1,000 limit can be seen as creating $1,000. Thus, the credit industry is creating many billions of dollars each year and, among other things, creating inflationary pressures in the process.
- Credit cards do not have a cash or currency form. In fact, they are not even backed by money until a charge is actually made.
- With cash we are restricted to the amount on hand or in the bank, but with credit cards our ceiling is less clear. We are restricted only by the ever-changing limits of each of our credit cards as well as by the aggregate of the limits of all those cards.
- Although we can use our cash anytime we wish, the use of our credit card requires the authorization of another party.
- Unlike cash, which allows for total anonymity, one's name is printed on the front of the credit card and written on the back; a credit card may even have one's picture on it. Furthermore, credit card companies have a great deal of computerized information on us that is drawn on to approve transactions.
- Although cash is simple to produce and use over and over, credit cards require the backing of a complex, huge, and growing web of technologies.
- There is no direct cost to the consumer for using cash, but fees and interest may well accrue with credit card use.
- Although everyone, at least theoretically, has access to cash, some groups (the poor, the homeless, the unemployed) may be denied access to credit cards. Such restrictions sometimes occur unethically or illegally through the "redlining" of certain types of consumers or geographic areas.
- Because of their accordionlike limits, credit cards are more likely than cash to lead to consumerism, overspending, and indebtedness....

A Growing Industry

Another reason for focusing on credit cards is their astounding growth in recent years, which reflects their increasing importance in the social and economic worlds. There are now more than a billion credit cards of all types in the United States. Receivables for the industry as a whole in 1993 were up by almost 16% from the preceding year and by over 400% in a decade. The staggering proliferation of credit cards is also reflected in other indicators of use in the United States:

- Sixty-one percent of Americans now have at least one credit card.
- The average cardholder carries nine different cards.
- In 1992, consumers used the cards to make 5 billion transactions, with a total value of $420 billion.

There has been, among other things, growth in the number of people who have credit cards, the average number of credit cards held by each person, the amount of consumer debt attributable to credit card purchases, the number of facilities accepting credit cards, and the number of organizations issuing cards.... The average outstanding balance owed to Visa and MasterCard increased from less than $400 in the early 1980s to $970 in 1989 and to $1,096 in 1993. The amount of high-interest credit card debt owed by American consumers rose from $2.7 billion in 1969 to $50 billion in 1980 and was approaching $300 billion in 1994....

A Symbol of American Values

A strong case can be made that the credit card is one of the leading symbols of 20th-century America or, as mentioned earlier, that the credit card is an American icon. Indeed, one observer calls the credit card "the twentieth century's symbol par excellence." Among other things, the credit card is emblematic of affluence, mobility, and the capacity to overcome obstacles in the pursuit of one's goals. Thus, those hundreds of millions of people who carry credit cards are also carrying with them these important symbols. And when they use a credit card, they are turning the symbols into material reality....

DEBUNKING CREDIT CARD MYTHS

To most of us, credit cards appear to have near-magical powers, giving us greater access to a cornucopia of goods and services. They also seem to give us something for nothing. That is, without laying out any cash, we can leave the mall with an armload of purchases. Most of us like what we can acquire with credit cards, but some like credit cards so much that they accumulate as many as they can. Lots of credit cards, with higher and higher spending limits, are important symbols of success. That most people adopt a highly positive view of credit cards is borne out by the proliferation of the cards throughout the United States and the world....

A debunking sociology is aimed at revealing the spuriousness of various ideologies. As Berger puts it, "In such analyses the ideas by which men explain their actions are unmasked as self-deception, sales talk, the kind of sincerity ... of a man who habitually believes his own propaganda." From this perspective, the credit card companies can be seen as purveyors of self-deceptive ideologies. They are, after all, in the business of selling their wares to the public, and they will say whatever is necessary to accomplish their goal....

55

Corporate Interests

How the News Media Portray the Economy

CHRISTOPHER J. KOLLMEYER

Despite the fact that most of the news media is owned by large politically conservative organizations, the most frequent complaint is that the media display a liberal political bias. The public does not seem to recognize that a politically conservative bias exists in the coverage of other than political news. This reading reveals the bias shown in economic news.

As you read this selection, ask yourself the following questions:

1. *Considering the bias of economic news, how will you evaluate news about corporations and unions, employers and workers?*
2. *Where will you turn to obtain unbiased economic news, especially during election periods?*
3. *So the news you read and hear is biased—so what?*

The news media's coverage of the economy, along with its attendant political consequences, has been the subject of considerable debate within the social science literature. Although most social scientists and media critics believe that the news media suffer from various shortcomings . . . and that these shortcomings affect the content of the news, the social science literature contains significant disagreements over basic questions about how the news media generally depict the economy. . . .

FOUR PERSPECTIVES ON THE NEWS MEDIA'S COVERAGE OF THE ECONOMY

The following section summarizes four prominent sociological perspectives on the news media. Based on claims associated with each perspective, I derive several hypotheses about the manner in which the news media typically cover the economy. Then, later in the article, I empirically examine these hypotheses with data derived from the *Los Angeles Times'* coverage of the economy between 1997 and 1998.

The News Media Emphasize Bad Economic News

Classic liberals saw the "free press" as an indispensable component of democratic society (see e.g. Mill [1825] 1967; Tocqueville [1840] 1963:181–90). From their perspective, a free and independent press, firmly rooted in a vibrant and pluralistic civil society, can help prevent the rise of tyrannical regimes by critically monitoring the use of state power. When breaches of prescribed authority occur, the free press, according to classic liberals,

SOURCE: From *Social Problems*, Vol. 51, No. 3, pp. 432–452.

could effectively alert the public, who could then mobilize to redress the exposed illegalities, corruption, or incompetence. This capacity, which can be described as the "watchdog" function, makes the news media an important component of democracy's overall system of checks and balances.

Some contemporaries of this tradition, however, believe that the news media have become too zealous in performing their watchdog function (Blood and Phillips 1995; Harrington 1989; Lerner and Rothman 1990; Lichter, Rothman, and Lichter 1986:254–92; Stein 1975; Wattenberg 1984). Advocating what I term the "bad-news-bias perspective," these scholars contend that the news media's coverage of the economy does not accurately reflect the underlying realities of day-to-day events and long-term socioeconomic trends. Instead of "balanced" coverage, the news media purportedly run a disproportionate number of negative stories about the economy. . . . Overall, this body of scholarship suggests the following hypothesis:

> *Hypothesis 1*: Negative accounts of the economy outnumber positive accounts of the economy.

Why would the news media overemphasize negative aspects of the economy? The literature constituting the bad-news-bias perspective contains several explanations, but the most prominent one contends that the journalism profession traditionally attracts individuals who are hostile toward the established political and economic order (Lerner and Rothman 1990; Lichter et al. 1986; Rothman and Black 2001; Rothman and Lichter 1982). Early research on this subject concludes that members of the national press corps, as compared to leaders in the corporate community, are considerably more likely to have left-of-center political views (Lichter et al. 1986; Rothman and Lichter 1982). . . . Recent research on this subject generates similar conclusions. It finds that journalists with the national press corps, on average, hold more liberal political views than the general public (Dautrich and Hartley 1999:95–8) and corporate managers (Rothman and Black 2001). Importantly, these findings . . . suggest that journalists emphasize the economy's shortcomings in order to support liberal political causes. If this is true, and given that the welfare of workers is a central concern of liberal politics, then we would expect the news media to highlight the ways in which the economy performs poorly for the general workforce. This argument suggests the following hypothesis:

> *Hypothesis 2*: Most negative accounts of the economy focus on problems affecting the general workforce rather than problems affecting corporations or investors.

The News Media "De-politicize" the Economy

In stark contrast with the classic liberal tradition, Marxian social theory generally describes civil society as an instrument of class domination. . . . Later in the century, scholars associated with the Frankfurt School continued this line of research. They argued that the mass media, instead of providing a means to rationally debate issues of public importance, most often disseminate a depoliticized discourse that helps elites conceal widespread social and economic inequality (Habermas [1962] 1991; Horkheimer and Adorno 1944; Marcuse 1964).

Often the harshest critics of the news media, contemporaries of this tradition employ what I call the "cultural hegemony perspective." In general, this perspective maintains that the news media, acting on behalf of the powerful societal interests that finance them, portray the economy in a way that lends ideological support to society's existing economic arrangements. . . .

Recent scholarship from this perspective identifies at least three economic and organizational constraints that supposedly prevent journalists from mounting a sustained critique of the corporate community and allied social groups (Bagdikian 2000; Herman 1999; Herman and Chomsky 2002; McChesney 2000; Parenti 1993). First, the ownership structure of the news industry purportedly aligns the news media's coverage of the economy with the overall business interests of the corporate community. Here several scholars contend that, since a handful of corporate conglomerates own the country's major news outlets, and since these corporations strive to maximize profits just like any other for-profit firm, the news media lack the economic and organizational independence needed to critique the economic system from which they and their parent companies benefit (Bagdikian 2000; Herman 1999:11–54; Herman and Chomsky 2002:3–14). Second, certain industry-wide business practices, it is believed, further diminish the news media's capacity to critique the economy. Typically, advertising revenue from large corporations, rather than subscription fees paid by the readers or viewers of the news, represents the primary source of income for most news organizations. Some scholars claim that this business practice makes large corporations, through their purchase of advertising services, the *de facto* "patrons" of the news industry (Bagdikian 2000:105–73; Herman 1999:13–28; Herman and Chomsky 2002:13–8). Importantly, this situation can place the news media in a conflict of interest, one that purportedly impedes their ability to judiciously cover many important economic issues affecting their readership. Finally, according to these same scholars, the norms and practices of the journalism profession place additional constraints on the news media's ability to critique the prevailing economic system. At least tacitly, many scholars studying the news media assume that the political views of journalists significantly affect the content of the news. This presupposition, for example, underlies most of the scholarship constituting the bad-news-bias perspective. By contrast, scholars working from the cultural hegemony perspective typically disagree with this assumption. Instead, they claim that journalistic norms—maintained by the industry's training, editing, and promotional practices—help ensure that journalists, regardless of their personal political views, produce news that is consistent with the business interests of their corporate employers (Reese 1990). In sum, the cultural hegemony perspective maintains that, for the structural reasons mentioned above, the mainstream news media typically obfuscate and ignore economic problems affecting workers and other non-elites, and in doing so, they suppress discourse on serious alternatives to existing economic policies. Based on these arguments, one can reasonably derive the following two hypotheses:

Hypothesis 3: The news media rarely run stories about economic problems affecting workers.

Hypothesis 4: The news media rarely mention economic reforms designed to improve the material well-being of workers.

The News Media Emphasize the Views of Elites

Drawing on Weberian notions of social class, bureaucratic rationality, and mass democracy, C. Wright Mills (1956) developed an influential account of power relations in American society. He argued that during the twentieth century the enlargement and concentration of bureaucratic authority in the United States left a small group of individuals in control of the country's most significant levers of power. This "power elite," as Mills called them, purportedly hold and exercise power through their leadership positions in society's largest private and public bureaucracies, which Mills believed operate with little interference from the electoral process. Importantly, according to his argument, the power elite share common cultural and political views, which along with their mutual economic interests help them function as a *de facto* ruling class.

Scholars drawing on this sociological tradition employ what I call the "power elite perspective" on the news media. This perspective, in general, stresses that the national news media have substantial social and organizational ties with the inner circle of power-holders in the United States (Akhavan-Majid and Wolf 1991; Dreier 1982a; Dreier and Weinberg 1979; Witcover 1990). For example, through interlocking directorates and other associations, the country's largest newspapers have strong institutional affiliations with the corporate community, business policy groups, and prestigious social clubs (Dreier 1982a; Dreier and Weinberg 1979). Similarly, through an extensive two-way flow of personnel, many of the nation's major news organizations have strong social ties to the highest levels of the federal government (Witcover 1990). Conversely, major newspapers rarely, if ever, have these types of affiliations with public interest groups, labor unions, or grass-roots charities.

Due to these social and organizational ties, some scholars maintain that—especially on economic policy issues—the national news media share common political views with the corporate community and other powerful organizations (Croteau 1998; Dreier 1982a, 1982b; Gans 1979).... This centrist political orientation frames the journalism of the country's major newspapers. On one hand, despite structural links to the capitalist economy, and despite strong social ties to the federal government and elite social organizations, major newspapers maintain sufficient autonomy to expose the wrongdoings of corrupt or incompetent corporate executives and government officials. But, on the other hand, since prominent journalists and editors are tangential members of the national elite themselves, major newspapers almost never question the desirability of the country's prevailing economic system or the highly stratified social order that it produces....

In sum, the power elite perspective maintains that the national press corps—due to their social, organizational, and economic ties with national elites—share common viewpoints with corporate leaders about many important economic issues, and that these common viewpoints ultimately influence the manner in which the news media cover the economy. The research supporting this argument reasonably yields the following hypothesis:

> *Hypothesis 5*: When covering the economy, the news media emphasize events and issues affecting corporations and investors and downplay events and issues affecting workers.

The News Media Rely on "Official Sources" of Information...

In general, this perspective emphasizes how the news media's information-gathering practices usually place journalists in contact with a limited range of sources. This occurs because, in an effort to maximize efficiency, the journalism profession has routinized the information-gathering process, matching particular types of stories with particular sources of information.... For instance, national economic correspondents typically gather most of their information from the New York Stock Exchange, Wall Street brokerage houses, the Federal Reserve Bank, the Department of Labor, and other similar sources. This organizational practice, used by both the print and television news media, makes reporters heavily dependent upon "official sources" for much of the information that eventually becomes news. Research on this subject finds that journalists with the national press corps, when gathering information for stories about the economy, say they "nearly always" consult business spokespersons approximately six times more frequently than union representatives (Croteau 1998). Similarly, prominent television news programs, such as ABC's *Nightline* and PBS's *The News Hour*, rely heavily on high-ranking officials from the federal government and corporate community for information and viewpoints on the economy (Croteau and Hoynes 1994; see also Glasgow University Media Group 1980:97–115). These findings suggest the following hypothesis:

> *Hypothesis 6*: Journalists rarely use union leaders, workers, or their spokespersons as sources of information about the economy.

The significance of these findings, of course, rests on the assumption that the social position of the news media's sources affects the content of the news. According to these scholars, this is indeed the case. Under the journalistic norm of "objectivity," sources can express opinions but journalists cannot. Therefore, it logically follows that the social positions of sources, more than the political attitudes of journalists, affect the way the news media portray a broad range of public policy issues.

Extending this logic, other researchers have shown that the news media's portrayal of many important social and economic issues partially reflects the unequal capacity of different social groups to gain the attention of journalists (Dreier 1982b, 1987; Lieberman 2000; Molotch and Lester 1974). The information-gathering process, from this perspective, frequently operates in reverse. Instead of journalists seeking out information, different social groups often vie for opportunities to supply journalists with their viewpoints. Under such circumstances, the ability of corporations and other wealthy organizations to marshal enormous resources gives them a significant advantage over less powerful groups. . . .

But, despite their sizeable advantages, the corporate community's privileged access to the news media can be successfully challenged. . . . During the late 1960s and 1970s, for example, the national news media ran more negative stories about the corporate community and the economy than they had in previous decades. Among other factors, Dreier (1982b, 1987) associates this phenomenon with the strategic actions of several liberal social groups, which enabled them to successfully attract the attention of prominent journalists, something that ultimately manifested as more articles about the economy's shortcomings. But, as Dreier shows, the corporate community, perceiving this upswing in negative news as a political threat to their business interests, responded with a sustained counter-mobilization, which eventually restored their favorable access to the national press corps (see also Vogel 1989, 1996:268–98). Given this body of research, one would expect that the news media's depiction of the economy, at least in part, depends upon the political perspectives and attitudes of their sources. This line of scholarship suggests the final hypothesis:

> *Hypothesis* 7: Articles citing business spokespersons or government officials will report good news about the economy more often than articles citing union leaders, workers, or their spokespersons. . . .

FINDINGS

The discussion of the findings proceeds in two parts. First, drawing on statistical indicators, I describe the growth patterns of California's economy during the 1990s. Then, using these objective indicators of economic performance as a reference, I analyze the more subjective accounts of the economy found in the *Times* during 1997 and 1998.

Objective Indicators of California's Economy

The structure of California's economy has changed significantly over the last few decades. Starting in the 1970s, when technological advancements contributed to a period of worldwide economic restructuring, and culminating after the recession of the early 1990s, when thousands of high-paying manufacturing jobs were lost due to cutbacks in defense spending, California experienced nearly two decades of economic instability. But, by the mid-1990s, the state's economy was growing rapidly, spurred most notably by new innovations and successful entrepreneurship in the high-technology sector. However, unlike the industrial-based economy of earlier decades, where the manufacturing sector created many high-wage jobs, much of the recent employment growth in California has occurred in low-wage industries, especially the consumer services and low-technology manufacturing sectors (Bonacich and Appelbaum 2000; Scott 1993). This pattern of employment growth—coupled with the proliferation of part-time, temporary, and contractual work arrangements—altered the economic foundations that previously supported an expanding middle class in the state. Thus, by the late-1990s, those equipped with the skills or capital needed to participate in the information-based sectors of the economy enjoyed many opportunities for prosperity, while less-educated members of the workforce faced stagnant wages and the vagaries of the contingent labor market.

Statistics on California's economy reflect these diverging trends. . . . Over a seven-year period between 1992 and 1998, corporate profits reported to the California Franchise Board increased nearly

300 percent, even after making downward adjustments for inflation. Moreover, since total tax payments to the state grew by only 20 percent, the average corporation enjoyed substantial increases in net profits during this period. The national economy experienced similar trends, as rising investor confidence, fueled by soaring corporate profits and new market opportunities made possible by globalization and the Internet, generated nearly a decade's worth of unprecedented growth in major U.S. stock markets.

Workers benefited much less from this economic expansion. After adjusting for inflation, wages in the manufacturing sector—which often serve as a proxy for wages across the working class—changed little during the 1990s. And even though unemployment fell substantially—from over nine percent of the workforce in 1992, down to around five percent in 1998—many of these new jobs were part-time positions. Both of these trends, stagnant wages and growth in part-time employment, are traditionally associated with recessions, not long periods of sustained economic growth. So given the uneven distribution of prosperity generated by this economic expansion, how did California's largest newspaper portray the economy's performance?

Journalistic Accounts of California's Economy

Based on an analysis of 201 articles on the economy from the *Los Angeles Times*, I find qualified support for the notion that the news media's coverage of the economy primarily reflects the common interests and concerns of the corporate community and investors. Although more than half of the articles in the data set contained some negative news—thereby supporting the general premise of the bad-news-bias perspective—a detailed analysis of the data reveals a pattern of journalism that downplays economic problems affecting workers. The following paragraphs describe the evidence that supports this conclusion.

Good, Bad, and Mixed Economic News. According to scholarship comprising the bad-news-bias perspective, a disproportionate number of articles on the economy contain negative news (Hypothesis 1). To test this premise, each article in the data set was placed into one of three mutually exclusive categories—good news, bad news, or mixed news—depending upon the way it portrayed the economy's performance. To make this coding process as objective as possible, articles were sorted into these categories based upon the topic covered and the explicit language used in the article. The "good news" category includes any article that focused *solely* on positive aspects of the economy. Many of these articles, for example, discussed favorable macroeconomic conditions, such as falling unemployment and inflation rates, record increases in stock market indices, or the success of the high-technology sector. Conversely, the "bad news" category includes any article that focused *solely* on negative aspects of the economy. Articles in this category, for example, include those that reported on labor disputes, incidences of corporate downsizing, or the hardships of low-skilled workers. Finally, the "mixed news" category includes any article that discussed both positive and negative aspects of the economy. Examples of articles in this category include stories that covered sudden declines in the stock market, but then mentioned that stock prices have been at historic highs throughout the 1990s, or stories that reported robust macroeconomic growth, but then mentioned the possibility that this may cause inflation in the near future.

The results of this coding process partially support the argument that the news media overemphasize negative accounts of the economy. As would be expected given the nature of the economy, the data show that the modal newspaper account, occurring 53.2 percent of the time, reflects a mixed assessment of the economy's performance. However, the data also show that good-news articles outnumber bad-news articles by more than five to one. These results can be interpreted in two ways. With only 7.5 percent of the articles reporting exclusively bad news, as compared to the nearly 40 percent of the articles reporting exclusively good news, the data do not reveal a pronounced inclination toward negative accounts of the economy. But when combining the bad-news and mixed-news articles into one category, the data indicate that over 60 percent of the articles contain at least some criticism of the economy's performance.

Information about the size and placement of articles in the data set helps clarify this ambiguity. During the coding process, the word count and page number of each article were identified and recorded. From this information, two measures of an article's relative prominence were calculated: article size, as measured in number of words, and article placement, as measured by appearance or non-appearance on the front page. Assuming that longer articles are more prominent than shorter articles, and assuming that front-page articles are more prominent than back-page articles, the data show that, by a considerable margin, good-news articles received the most favorable coverage. Articles reporting good news appeared on the front page 77.2 percent of the time, and they had an average length of 1,413 words. Conversely, articles reporting bad news appeared on the front page only 33.3 percent of the time, and they had an average length of only 1,232 words. These findings clearly contradict the assertion that the news media most often emphasize the economy's shortcomings.

Economic Group Affected by Negative News. As discussed earlier, the social science literature suggests competing hypotheses about the news media's coverage of economic problems affecting workers. On one hand, some scholarship predicts that the news media, due to alleged liberal and anti-business attitudes held by journalists, will overemphasize the economy's shortcomings, especially when these shortcomings affect workers (Hypothesis 2). On the other hand, other scholarship anticipates that the news media, since they are major corporations operating within and benefiting from the existing capitalist economy, and since they are embedded within elite social networks, will most often ignore problems affecting workers (Hypotheses 3 and 5). To adjudicate between these competing claims, during the coding process, articles containing negative news about the economy were sorted into one of three categories—corporations and investors, workers, or the economy in general. These mutually exclusive categories represent the economic group, identified by the newspaper, that was most threatened by the problems discussed in the bad- or mixed-news articles. An article was labeled "economy" when the newspaper portrayed the events as threatening the performance of the economy in general rather than the well-being of specific economic actors. After analyzing the articles in this fashion, the data reveal an incongruence between the economic indicators and the journalistic accounts of the economy presented. Although the economic indicators show that workers failed to benefit much, if at all, from the economic prosperity of the mid-1990s, only 14.8 percent of the combined bad- and mixed-news articles discussed problems affecting workers. Conversely, 31.1 percent of the bad- and mixed-news articles discussed problems affecting corporations and investors, even though they were the primary beneficiaries of economic growth during this period. The remaining 53.3 percent of the articles discussed problems with the economy as a whole. Additionally, it should be noted that when the *Times* did focus solely on problems with the economy—meaning those 15 articles coded as bad news—workers received slightly more attention than corporations and investors and the economy in general. However, since the sample size is small, again only 15 articles, the slight variations in the number of articles across the three categories are not statistically significant.

Information about the size and placement of articles in the data set containing criticism of the economy provides further support for the claim that the news media downplay problems affecting workers. Again, using an article's size and appearance on the front page as indicators of its prominence within the newspaper, the data reveal a strong association between an article's prominence and the group affected by the bad or mixed news about the economy. Articles reporting problems threatening corporations and investors received front-page attention 73.7 percent of the time, and they had an average length of 1,447 words. But articles reporting problems threatening workers received front-page attention only 21.1 percent of the time, and they had an average length of 1,315 words. By comparison, articles about problems affecting the economy in general, although they had the smallest average word count, were distributed almost evenly between the front and back pages of the newspaper. Nonetheless, the overall findings portray a pattern of journalism that downplays coverage of economic issues affecting workers.

Articles Mentioning Reforms. Recall that the cultural hegemony perspective suggests that, for several structural reasons, the news media rarely discuss alternatives to existing economic policies, especially when proposed policy alternatives address problems and issues affecting workers (Hypothesis 4). To empirically assess this assertion, articles mentioning one or more potential reforms to existing economic policy were identified. For the purpose of coding the articles, reforms included policies advocating either economic liberalization or new government regulations. This process identified 31 articles—a number equaling 15 percent of the articles in the data set and 25 percent of the articles containing at least some negative news. By itself, this finding does not support the hypothesized outcome. But, after taking two additional factors into consideration, the data more closely support the assertion that the news media rarely mention economic reforms designed to help workers.

First, articles discussing reforms were generally given less-prominent coverage within the newspaper than other articles about the economy. Articles mentioning one or more reforms appeared on the front page only 29 percent of the time, and they had an average length of 1,217 words. By contrast, articles not mentioning reforms appeared on the front page more frequently (67.1 percent of the time), and they were generally longer (1,336 average words). Taken together, this means that articles discussing reforms, as compared to articles not discussing reforms, were half as likely to appear on the front page and almost 9 percent shorter.

Second, when articles mentioned reforms intended to help corporations and investors, they were given more prominent attention, by a considerable margin, than articles mentioning reforms intended to help workers. Articles mentioning reforms intended to help corporations and investors appeared on the front page 83.3 percent of the time, and they had an average length of 1,569 words. But articles mentioning reforms designed to help workers appeared on the front page much less frequently (only 14.3 percent of the time), and they were much shorter on average (only 1,120 words). Or, stated differently, this finding means that articles mentioning reforms designed to help corporations and investors were, on average, almost six times more likely to appear on the front page and approximately 28 percent longer than articles mentioning reforms designed to help workers. Combined, the data presented, while not avoiding the subject altogether, downplays discussions of potential economic reforms, especially when the proposed reforms address problems affecting workers.

Sources of Information. As described earlier, some scholarship on the news media finds that journalists rarely use union leaders, workers, or their spokespersons as sources of information about the economy (Hypothesis 6). To assess this contention, during the coding process, each article's primary source of information, as identified by the journalist, was recorded. To facilitate a quantitative analysis of the resulting information, the identified sources were grouped into one of seven mutually exclusive categories. The results of this coding process provide strong support for the hypothesized outcome. As anticipated, the data show that union leaders, workers, and their spokespersons were rarely used as sources of information about the economy. Specifically, these individuals were used as primary sources for only 7.9 percent of the articles in the data set. This compares unfavorably with most other types of sources—the two exceptions being the category representing *Times* op-ed writers (used in 5 percent of the articles) and the category representing three or more sources (also used in 5 percent of the articles). The most frequently cited source was "business/government," a category for articles citing both business spokespersons and government officials as the primary sources of information. The next most frequently cited sources were business spokespersons (21.3 percent), social scientists and individuals described as authors (19.3 percent), and government officials (18.8 percent). Consistent with the literature of the subject, these findings demonstrate that journalists with the *Times* rely heavily upon individuals representing the corporate community and government for much of the information that eventually becomes news about the economy. In fact, at least 62 percent of the articles in the data set used some combination of business spokespersons and government officials as primary sources—a number nearly 8 times greater than the percentage of articles using union leaders, workers, or their spokespersons as primary sources.

When used as sources of information about the economy, business spokespersons and government officials seemingly received preferential treatment in other ways as well. According to the data there exists a moderate association between an article's source of information and its likelihood of appearing on the front page. Specifically, when journalists used business spokespersons as primary sources—either alone or coupled with government officials—the resulting articles appeared on the front page more than 83 percent of the time. But when journalists used union leaders, workers, or their spokespersons as primary sources, the resulting articles appeared on the front pages less frequently—approximately 56 percent of the time. The least prominent attention, however, went to articles citing social scientists and individuals described as authors. When journalists cited these sources, only 17.9 percent of the resulting articles appeared on the front page. Taken together, the data displayed provide moderate support for the claim that the news media privilege the corporate community and government as sources of information on the economy.

Sources and News Content. Finally, do the news media's choices of sources affect the content of the news? As described earlier, existing scholarship on this subject suggests that articles citing business or government spokespersons will contain favorable accounts of the economy more frequently than articles citing workers or their spokespersons (Hypothesis 7). The data analyzed here, however, provide only limited support for this proposition. First, the association between an article's source of information and its assessment of the economy is ambiguous. On one hand, as anticipated, the data clearly indicate that government officials—whether cited alone or cited along with business spokespersons—were frequently associated with favorable accounts of the economy. When government officials were used as primary sources, 50 percent of the resulting articles reported good news, while only 2.6 percent of articles reported bad news. Similarly, when journalists cited government officials along with business spokespersons as primary sources, 48.9 percent of the resulting articles reported good news, while only 4.4 percent reported bad news. These findings clearly support the hypothesized outcome. But, on the other hand, some findings deviate substantially from the expected results. Most notably, the data indicate that articles citing workers and their spokespersons frequently reported good news about the economy. In fact, articles citing these sources reported good news about the economy more often than articles citing business spokespersons—a finding that clearly contradicts the general thrust of the literature on this subject. Moreover, the data also show that social scientists and authors, not workers and their spokespersons, were the sources most frequently associated with negative accounts of the economy. In particular, when journalists used social scientists and authors as primary sources, over 82 percent of the resulting articles reported either bad or mixed news about the economy. But when journalists used workers or their spokespersons as primary sources, only 56 percent of the resulting articles expressed bad or mixed news about the economy.

Second, in contrast with anticipated results, the data show little correlation between an article's source of information and its likelihood of discussing reforms. The percentage of articles discussing reforms differs insignificantly across the various categories of sources. Although the sample size is small, making this finding statistically unreliable, it clearly does not corroborate the assertion that the news media's choice of sources affects the content of the news. There is, however, a notable exception to this finding. Articles citing social scientists and authors, compared with articles citing other sources of information, were significantly more likely to mention reforms in general and reforms that benefit workers in particular.

CONCLUSION AND DISCUSSION

This study provides new sociological insights into the manner in which the news media portray the economy's performance. Ideally, in a democracy, the news media should help expose weaknesses in society's major institutions and facilitate rational debates on appropriate policy solutions to the identified problems. Scholars and media critics, however, overwhelming agree that the news media in the United States fall far short of this ideal (see review article by Gamson et al. 1992). Yet, despite this unanimity, the social science literature contains highly incompatible arguments about the causes and consequences of the news media's shortcomings. For instance, some media scholars claim that the national press corps hold overly antagonistic views about the corporate community, and that these attitudes, along with a desire to help liberal political causes, manifest as a disproportionate number of articles about the economy's shortcomings. But, on this same issue, other media scholars draw very different conclusions. In general, they contend that the news media—due to their structural links to the capitalist economy, their strong social ties to the corporate community and federal government, and their reliance on elites for much of the information that eventually becomes news—typically cover the economy from a "pro-business" perspective. Using articles that appeared in the *Los Angeles Times* during 1997 and 1998, this study examines these competing claims about the news media. During the late 1990s, rapid economic growth in California produced unequal patterns of prosperity. While corporations and investors enjoyed robust growth in profits and income, a significant portion of the state's workforce experienced stagnant wages and limited opportunities for full-time employment. Given this situation, if the news media's coverage of the economy reflects liberal or "anti-business" political views, as some scholars suggest, then we would expect to find a plurality of articles on economic problems affecting workers. This, however, was not the case. Findings derived from a content analysis show that the *Times* ran relatively few articles about economic problems affecting workers. And when they did cover problems affecting workers, the articles were relatively short, most often placed in the back pages of the newspaper, and rarely discussed alternatives to existing economic policies. Furthermore, the *Times* rarely used union leaders, workers, or their spokespersons as sources of information and viewpoints on the economy. In fact, less than 8 percent of the articles analyzed in this study used this group as the primary source of information.

These results cast considerable doubts over certain claims associated with the bad-news-bias perspective. While the data show that the news media publish a significant amount of negative economic news—and thereby confirm a core premise of this perspective—there is little evidence that the news media cover the economy from a liberal or anti-business viewpoint. This finding, I believe, indicates that the bad-news-bias perspective suffers from several theoretical shortcomings. First and foremost, it typically ignores the class position of those affected by negative economic news. If journalists and editors possess liberal or anti-business attitudes, and if these attitudes affect the content of the news, then the country's major newspapers would emphasize the numerous economic problems facing America's sizeable lower and working classes. But, according to the findings presented here, this is not the case. Even after omitting the business section, the data reveal that the *Los Angeles Times*—supposedly a core member of the "liberal press"—overwhelmingly focused on economic issues and problems important to corporations and investors. This occurred despite the fact that, during the period in question, corporations and investors were enjoying record levels of prosperity, while large segments of the workforce were experiencing economic stagnation. Importantly, these findings imply that the shortcomings of news media are class-specific. On one hand, the news media apparently succeed in drawing the public's attention to current events and economic problems threatening the well-being of corporations and investors. But, on the other hand, the news media apparently provide insufficient scrutiny of the numerous economic problems affecting the general workforce, especially the sizable population of low-wage workers. A second theoretical shortcoming is that scholars working from the bad-news-bias perspective, when analyzing the political attitudes of journalists, often combine an individual's viewpoints on social and economic issues into one overarching political orientation. This practice, I believe, can produce erroneous conclusions about how journalists typically view the economy. Granted, as highly educated and affluent urbanites, it seems likely that most members of the national press corps would have liberal views on social issues. But, even if this is true, it does not logically follow that these same individuals would have liberal views on economic issues. In fact, given their privileged position in the American class structure, it seems more plausible that, as Croteau (1998) argues, many members of the national press corps have center-right views on economic issues. And finally, in my estimation, the entire emphasis on the political orientations of journalists, as the primary determinant of the content of the news, seems misguided. Just like any corporate employee, journalists operate within a workplace explicitly organized around the interests of their employers. For this reason, as well as the other structural constraints discussed above, it seems improbable that the economic news would primarily reflect the personal preferences of journalists and editors rather than the larger business interests of the news industry's executive management and shareholders. This logic does not, however, imply that the news media never criticize corporations or prevailing economic policy—clearly they do on a regular basis—but rather it suggests that powerful social forces, arising from the structural relationship between the news industry and the capitalist economy, help align the norms of economic journalism with the general interests of the corporate community and investors.

Finally, the present study raises two interesting questions about the role sources of information play in shaping the news. First, although existing scholarship suggests otherwise, the data analyzed here reveal little correlation between the news media's choice of sources and the news media's assessment of the economy. Several explanations could plausibly account for the disparity between this finding and the hypothesized results. It could be that previous scholarship on this subject has overestimated the influence sources have on the content of the news. If this is the case, and sources cannot significantly affect coverage of the news, then perhaps journalists most often use sources to merely illustrate and support their preexisting conclusions rather than to gain fresh insights and opinions about their subject matter. But other explanations are equally plausible. It could be that an intervening variable is obscuring an otherwise significant relationship between sources and the content of the news. For example, it is likely that a journalist's specific topic of inquiry, such as a sudden decline in the stock market, plays a larger role in shaping the economic news than a journalist's source of information. A more definitive understanding of this relationship—one that controls for possible intervening variables—will require data and statistical procedures that fall beyond the scope of this research. For this reason, I do not argue that sources have a minimal effect on the content of the news—only that our present understanding of this issue remains unclear. Second, the present study demonstrates that social scientists and authors, not workers or their spokespersons, are most often associated with critical accounts of the economy and discussions of reforms. This finding was not explicitly anticipated by the literature, although in many ways it is consistent with the general notion of cultural hegemony. Again, a more comprehensive understanding of this phenomenon will require additional research. But, despite these two ambiguities, the bulk of the evidence presented here strongly suggests that the news media, when reporting on the economy, privilege the interests of corporations and investors over the interests of the general workforce.

REFERENCES

Akhavan-Majid, Roya and Gary Wolf. 1991. "American Mass Media and the Myth of the Libertarianism: Towards an 'Elite Power Group' Theory." *Critical Studies in Mass Communication* 8:39–151.

Croteau, David. 1998. "Examining the 'Liberal Media' Claim: Journalists' Views on Politics, Economic Policy and Media Coverage." *EXTRA!* Retrieved April 17, 2003 (http://fair.org/reports/journalist-survey.html).

Croteau, David and William Hoynes. 1994. *By Invitation Only: How the Media Limit Political Debate*. Monroe, ME: Common Courage Press.

Dautrich, Kenneth and Thomas H. Hartley. 1999. *How the News Media Fail American Voters: Causes, Consequences, and Remedies*. New York: Columbia University Press.

Dreier, Peter. 1982a. "The Position of the Press in the U.S. Power Structure." *Social Problems* 29:298–310.

———. 1982b. "Capitalists vs. the Media: An Analysis of an Ideological Mobilization Among Business Leaders." *Media, Culture, and Society* 4:111–32.

———. 1987. "The Corporate Complaint Against the Media." Pp. 64–80 in *American Media and Mass Culture: Left Perspectives*, edited by Donald Lazere. Berkeley and Los Angeles: University of California Press.

Dreier, Peter and Steven Weinberg. 1979. "The Ties That Bind: Interlocking Directorates." *Columbia Journalism Review* 17:51–68.

Gans, Herbert J. 1979. *Deciding What's News*. New York: Vintage Books.

———. 1980. *More Bad News*. London: Routledge & Kegan Paul.

Habermas, Jürgen. [1962] 1991. *The Structural Transformation of the Public Sphere: An Inquiry into a Category of Bourgeois Society*. Cambridge, MA: MIT Press.

Herman, Edward S. 1999. *The Myth of the Liberal Media: An Edward Herman Reader*. New York: Peter Lang Publishing.

Herman, Edward S. and Noam Chomsky. 2002. *Manufacturing Consent: The Political Economy of the Mass Media*. 2d ed. New York: Pantheon.

Horkheimer, Max and Theodor Adorno. 1944. *The Dialectic of Enlightenment*. New York: Continuum.

Lerner, Robert and Stanley Rothman. 1990. "Rhetorical Conflict and the Adversarial Media." *International Journal of Group Tensions* 20:203–20.

Lichter, S. Robert, Stanley Rothman, and Linda S. Lichter. 1986. *The Media Elite: America's New Power-brokers*. Bethesda, MD: Alder & Alder.

Lieberman, Trudy. 2000. *Slanting the Story: The Forces that Shape the News*. New York: New Press.

Marcuse, Herbert. 1964. *One-Dimensional Man: Studies in the Ideology of Advanced Industrial Societies*. Boston: Beacon Press.

McChesney, Robert W. 2000. *Rich Media, Poor Democracy: Communication Politics in Dubious Times*. New York: New Press.

Mill, John Stuart [1825] 1967. *Essays on Government, Jurisprudence, Liberty of the Press and Law of Nations*. New York: Kelly.

Mills, C.Wright. 1956. *The Power Elite*. New York: Oxford University Press.

Molotch, Harvey and Marilyn Lester. 1974. "News as Purposive Behavior: On the Strategic Use of Routine Events, Accidents, and Scandals." *American Sociological Review* 39:101–12.

Parenti, Michael. 1993. *Inventing Reality: The Politics of the Mass Media*. 2d ed. New York: St. Martin's Press.

Rothman, Stanley and Amy E. Black. 2001. "Media and Business Elites: Still in Conflict?" *The Public Interest* 143:72–86.

Rothman, Stanley and S. Robert Lichter. 1982. "Media and Business Elites: Two Classes in Conflict?" *The Public Interest* 69:117–25.

Stein, Herbert. 1975. "Media Distortions: A Former Official's View." *Columbia Review of Journalism* 13(6):37–41.

Tocqueville, Alexis Charles de. [1840] 1963. *Democracy in America*. New York: Alfred Knopf.

Wattenberg, Ben J. 1984. *The Good News is the Bad News is Wrong*. New York: Simon and Schuster.

Witcover, Jules. 1990. "Revolving-Door Journalists." *Washington Journalism Review* 12(3):33–8.

CHAPTER 12

Politics

Power and Its Implications

Alan Mc Evoy

In studying the institution of politics, sociologists concern themselves with the groups involved in the political process and the conditions that tend to generate political involvement or apathy. These research interests, in turn, lead to several related areas of inquiry: the problems of a democracy, economic power and its effect on the power structure, the impact of the bureaucracy, factors that thwart the march toward democracy, and the importance of authority. The results of these inquiries are revealed in the readings in this chapter.

As noted, the first of these areas of inquiry is democracy itself. This is an important area of investigation because of the creation of a number of new states since World War II, and the process is continuing as more and more nations and ethnic groups seek their independence from dominating societies. Many of these countries express the desire to become democracies, but this desire is often difficult to achieve and maintain. The difficulty in maintaining a democratic government is best seen in a country considered to be a model of democracy—the United States, which has one of the lowest rates of registration and voter participation of any Western industrial nation. For various reasons, registration seems to be class and age biased; lower-income groups and young people have far lower registration rates than other groups. Yet 11 states have shown that registration can be done at the same time as voting without fraud and a far higher percentage of registration. With voting itself, similarly, the reported turnout of about 50% represents only registered voters; if all possible voters were counted, the actual voting turnout would only be about 30%—one of the lowest figures among Western industrial societies. Perhaps these low voting figures result from the fact that the United States is only one of three of these countries to vote on a workday; the others, with voting on holidays, have a much higher turnout rate. Knowing these facts about low registration and low voting participation, why do you think the situation is not corrected? In the first reading Seymour Martin Lipset deals with this question by examining the conditions necessary for a state to achieve and maintain democracy.

Our examination of the economic institution in the last chapter has revealed not only the requirements of this important institution but also its close relationship to the political elite. In the second reading Dan Clawson and his colleagues examine the role of political action committees (PACs) on the democratic structure. With political campaigns now reduced to a media-run endeavor, candidates need lots of funds to compete. This need has provided the corporate world an opening into political power via its PACs, a situation demonstrated by the fact that about 1% of the wealthiest Americans contribute the bulk of political donations. Does the vast amount of money raised by these political groups from major corporate organizations have an effect on the representation process and if so, how? One study revealed that 20% of members of Congress admitted that campaign contributions affected their voting, another 30% were not sure, and 50% claimed they had no effect. These figures seem to imply that there are now two types of constituents: those who receive bailouts, subsidies, and tax breaks and those who do not. A side benefit of such contributions is that they almost always ensure reelection and that many potential candidates cannot run because of the money gap. In both the 1998 and 2000 elections, over 98% of House incumbents were reelected, as were about 85% of the Senate. A glance at political contributions indicates that incumbents enjoy an 8-to-1 advantage in PAC donations and thus are able to outspend their opponents by over four times. Because incumbents are allowed to roll over unspent political contributions, their accumulated war chest is often enough to scare away legitimate contenders. Some think that U.S. reelection numbers are so high that in reality we have a third party running for office—the incumbent party. What do you think can be done about this situation?

Although terrorism is not a new phenomenon, the large increase in terrorist acts has raised great interest in their impact on democracy. It appears that the rise of religious fundamentalism means that democracy will be denied for both parties, believers and nonbelievers. In the third reading Steven Simon examines the impact of terrorism on the democratic institutions of various nation states.

It would seem obvious that political leadership means the authority to accomplish one's goals. But leaders must have followers, and this fact raises the question of why people submit to leaders. In the final reading of this chapter, Philip Meyer describes a famous experiment testing people's obedience to authority. The study indicated that people tend to obey regardless of who tells them to or the task to be done, an outcome that may explain one of the major difficulties in maintaining a democracy.

In reading this chapter, keep in mind that change is a normative feature of society. There are a number of factors that can encourage, delay, or prevent change. For example, one's socialized beliefs, the media, and/or those with political authority and/or funds may exert change in any direction. This chapter is mostly concerned with political change and its effect on the democratic institution, but that does not mean that sociologists ignore other areas of the political institution. For example, we do not know whether low voter turnout results from the belief that it makes no difference who wins because both parties are more alike than different or from artificial barriers to voting such as advance registration and elections held on workdays. What do you think? There remains one final political factor to be considered. A democratic system implies competition in the realm of ideas, yet some believe that a democratic system cannot function without a strong political party system to articulate a set of principles. What do you think?

56

The Social Requisites of Democracy Revisited

SEYMOUR MARTIN LIPSET

Since the end of World War II, there has been a movement toward the creation of new states. In recent years there has also been a movement toward more democratic states. By looking at the conditions for a democracy, Lipset indicates what these states need if they are truly to achieve democracy. Considering these ideas, what do you see as the future of world affairs?

As you read, ask yourself the following questions:

1. *What are some of the requisites needed to maintain democracy?*
2. *Based on the requisites needed to maintain democracy, how is democracy doing? Why?*
3. *Do the United States have a democracy? Why, or why not?*

GLOSSARY **Efficacy** Producing desired results. **Bourgeois** Member of the middle class. **Facade** Superficial appearance. **Meritocratic** Hiring on the basis of ability rather than patronage. **Totalitarian** Absolute control by the state.

The recent expansion of democracy . . . began in the mid-1970s in Southern Europe. Then, in the early and mid-1980s, it spread to Latin America and to Asian countries like Korea, Thailand, and the Philippines, and then in the late 1980s and early 1990s to Eastern Europe, the Soviet Union, and parts of sub-Saharan Africa. Not long ago, the overwhelming majority of the members of the United Nations had authoritarian systems. As of the end of 1993, over half, 107 out of 186 countries, have competitive elections and various guarantees of political and individual rights—that is more than twice the number of two decades earlier in 1970 (Karatnycky 1994:6; *Freedom Review* 1993:3–4, 10). The move toward democracy is not a simple one.. . . Countries that previously have had authoritarian regimes may find it difficult to set up a legitimate democratic system, since their traditions and beliefs may be incompatible with the workings of democracy.

In his classic work *Capitalism, Socialism, and Democracy*, Schumpeter (1950) defined democracy as "that institutional arrangement for arriving at political decisions in which individuals acquire the power to decide by means of a competitive struggle for the people's vote" (p. 250). This definition is quite broad and my discussion here cannot hope to investigate it exhaustively. Instead, I focus here on . . . the factors and processes affecting the prospects for the institutionalization of democracy.

SOURCE: From the *American Sociological Review*, 1994, Vol. 59 (February), pp. 1–22.

HOW DOES DEMOCRACY ARISE?

Politics in Impoverished Countries

In discussing democracy, I want to clarify my biases and assumptions at the outset. I agree with the basic concerns of the founding fathers of the United States—that government, a powerful state, is to be feared (or suspected, to use the lawyer's term), and that it is necessary to find means to control governments through checks and balances. In our time, as economists have documented, this has been particularly evident in low-income nations. The "Kuznets curve" (Kuznets 1955; 1963; 1976), although still debated, indicates that when a less developed nation starts to grow and urbanize, income distribution worsens, but then becomes more equitable as the economy industrializes.* ... Before development, the class income structure resembles an elongated pyramid, very fat at the bottom, narrowing or thin toward the middle and top (Lipset 1981:51). Under such conditions, the state is a major, usually *the* most important, source of capital, income, power, and status. This is particularly true in statist systems, but also characterizes many so-called free market economies. For a person or governing body to be willing to give up control because of an election outcome is astonishing behavior, not normal, not on the surface a "rational choice," particularly in new, less stable, less legitimate politics.

Marx frequently noted that intense inequality is associated with scarcity, and therefore that socialism, which he believed would be an egalitarian and democratic system with a politically weak state, could only occur under conditions of abundance (Marx 1958:8–9). To try to move toward socialism under conditions of material scarcity would result in sociological abortions and in repression. The Communists proved him correct. Weffort (1992), a Brazilian scholar of democracy, has argued strongly that, although "the political equality of citizens, ... is ... possible in societies marked by a high degree of [economic] inequality," the contradiction between political and economic inequality "opens the field for tensions, institutional distortions, instability, and recurrent violence ... [and may prevent] the consolidation of democracy" (p. 22). Contemporary social scientists find that greater affluence and higher rates of well-being have been correlated with the presence of democratic institutions (Lipset, Seong, and Torres 1993:156–58; see also Diamond 1993a). Beyond the impact of national wealth and economic stratification, contemporary social scientists also agree with Tocqueville's analysis, that social equality, perceived as equality of status and respect for individuals regardless of economic condition, is highly conducive for democracy (Tocqueville 1976: vol. 2, 162–216). ... Weffort (1992) emphasized, "such a 'minimal' social condition is absent from many new democracies, ... [which can] help to explain these countries' typical democratic instability" (p. 18).

The Economy and the Polity

In the nineteenth century, many political theorists noted the relationship between a market economy and democracy (Lipset 1992:2). As Glassman (1991) has documented, "Marxists, classical capitalist economists, even monarchists accepted the link between industrial capitalism and parliamentary democracy" (p. 65). Such an economy, including a substantial independent peasantry, produces a middle class that can stand up against the state and provide the resources for independent groups. ... Schumpeter (1950) held that, "modern democracy is a product of the capitalist process" (p. 297). Moore (1966), noting his agreement with the Marxists, concluded, "No bourgeois, no democracy" (p. 418). ...

Waisman (1992:140–55), seeking to explain why some capitalist societies, particularly in Latin America, have not been democratic, has suggested that ... a free market needs democracy and vice versa.

But while the movement toward a market economy and the growth of an independent middle-class have weakened state power and

*These generalizations do not apply to the East Asian NICS, South Korea, Taiwan, and Singapore.

enlarged human rights and the rule of law, it has been the working class, particularly in the West, that has demanded the expansion of suffrage and the rights of parties. As John Stephens (1993) noted, "Capitalist development is associated with the rise of democracy in part because it is associated with the transformation of the class structure strengthening the working class" (p. 438). . . .

Therefore, a competitive market economy can be justified sociologically and politically as the best way to reduce the impact of nepotistic networks. The wider the scope of market forces, the less room there will be for rent-seeking by elites with privileged access to state power and resources. Beyond limiting the power of the state, however, standards of propriety should be increased in new and poor regimes, and explicit objective standards should be applied in allocating aid, loans, and other sources of capital from outside the state. Doing this, of course, would be facilitated by an efficient civil service selected by meritocratic standards. It took many decades for civil service reforms to take hold in Britain, the United States, and various European countries. To change the norms and rules in contemporary impoverished countries will not be achieved easily. . . .

The Centrality of Political Culture

Democracy requires a supportive culture, the acceptance by the citizenry and political elites of principles underlying freedom of speech, media, assembly, religion, of the rights of opposition parties, of the rule of law, of human rights, and the like. . . . Such norms do not evolve overnight. . . . "Only four of the seventeen countries that adopted democratic institutions between 1915 and 1931 maintained them throughout the 1920s and 1930s. . . . [O]ne-third of the 32 working democracies in the world in 1958 had become authoritarian by the mid-1970s" (Huntington 1991:17–21).

These experiences do not bode well for the current efforts in the former Communist states of Eastern Europe or in Latin America and Africa. And the most recent report by Freedom House concludes: "As 1993 draws to a close, freedom around the world is in retreat while violence, repression, and state control are on the increase. The trend marks the first increase in five years . . ." (Karatnycky 1994:4). A "reverse wave" in the making is most apparent in sub-Saharan Africa, where "9 countries showed improvement while 18 registered a decline" (p. 6). And in Russia, a proto-fascist movement led all other parties, albeit with 24 percent of the vote, in the December 1993 elections, while the Communists and their allies secured over 15 percent.

Almost everywhere that the institutionalization of democracy has occurred, the process has been a gradual one in which opposition and individual rights have emerged in the give and take of politics. . . .

As a result, democratic systems developed gradually, at first with suffrage, limited by and linked to property and/or literacy. Elites yielded slowly in admitting the masses to the franchise and in tolerating and institutionalizing opposition rights. . . . As Dahl (1971:36–37) has emphasized, parties such as the Liberals and Conservatives in nineteenth-century Europe, formed for the purpose of securing a parliamentary majority rather than to win the support of a mass electorate, were not pressed to engage in populist demagoguery.

Comparative politics suggest that the more the sources of power, status, and wealth are concentrated in the state, the harder it is to institutionalize democracy. Under such conditions the political struggle tends to approach a zero-sum game in which the defeated lose all. The greater the importance of the central state as a source of prestige and advantage, the less likely it is that those in power—or the forces of opposition—will accept rules of the game that institutionalize party conflict and could result in the turnover of those in office. Hence, once again it may be noted, the chances for democracy are greatest where, as in the early United States and to a lesser degree in other Western nations, the interaction between politics and economy is limited and segmented. In Northern Europe, democratization let the monarchy and the aristocracy retain their elite status, even though their powers were curtailed. In the United States, the central state was not a major source of privilege for the first half-century or more, and those at the center thus could yield office easily.

Democracy has never developed anywhere by plan, except when it was imposed by a democratic conqueror, as in post–World War II Germany and Japan. From the United States to Northern Europe, freedom, suffrage, and the rule of law grew in a piecemeal, not in a planned, fashion. To legitimate themselves, governmental parties, even though they did not like it, ultimately had to recognize the right of oppositions to exist and compete freely. Almost all the heads of young democracies, from John Adams and Thomas Jefferson to Indira Gandhi, attempted to suppress their opponents. ... Democratic successes have reflected the varying strengths of minority political groups and lucky constellations, as much or more than commitments by new office holders to the democratic process.

Cross-national historical evaluations of the correlates of democracy have found that cultural factors appear even more important than economic ones (Lipset et al. 1993:168–70). ... Dahl (1970:6), Kennan (1977:41–43), and Lewis (1993:93–94) have emphasized that the first group of countries that became democratic in the nineteenth century (about 20 or so) were Northwest European or settled by Northwest Europeans. "The evidence has yet to be produced that it is the natural form of rule for peoples outside these narrow perimeters" (Kennan 1977:41–43).* ...

More particularly, recent statistical analyses of the aggregate correlates of political regimes have indicated that having once been a British colony is the variable most highly correlated with democracy (Lipset et al. 1993:168). ... The factors underlying this relationship are not simple (Smith 1978). In the British/non-British comparison, many former British colonies, such as those in North America before the revolution or India and Nigeria in more recent times, had elections, parties, and the rule of law before they became independent. In contrast, the Spanish, Portuguese, French, Dutch, and Belgian colonies, and former Soviet-controlled countries did not allow for the gradual incorporation of "out groups" into the polity. Hence democratization was much more gradual and successful in the ex-British colonies than elsewhere; their pre-independence experiences were important as a kind of socialization process and helped to ease the transition to freedom.

Religious Tradition

Religious tradition has been a major differentiating factor in transformations to democracy (Huntington 1993:25–29). Historically, there have been negative relationships between democracy and Catholicism, Orthodox Christianity, Islam, and Confucianism; conversely Protestantism and democracy have been positively interlinked. These differences have been explained by (1) the much greater emphasis on individualism in Protestantism and (2) the traditionally close links between religion and the state in the other four religions. Tocqueville (1975) and Bryce (1901) emphasized that democracy is furthered by a separation of religious and political beliefs, so that political stands are not required to meet absolute standards set down by the church. ...

Protestants, particularly the non-state-related sects, have been less authoritarian, more congregational, participatory, and individualistic. Catholic countries, however, have contributed significantly to the third wave of democratization during the 1970s and 1980s, reflecting "the major changes in the doctrine, appeal, and social and political commitments of the Catholic Church that occurred ... in the 1960s and 1970s" (Huntington 1991:281, 77–85). ...

Conversely, Moslem (particularly Arab) states have not taken part in the third wave of democratization. Almost all remain authoritarian. Growth of democracy in the near future in most of these countries is doubtful because "notions of political

*That evidence, of course, has emerged in recent years in South and East Asia, Latin America, and various countries descended from Southern Europe.

freedom are not held in common . . .; they are alien to Islam." As Wright (1992) has stated, Islam "offers not only a set of spiritual beliefs, but a set of rules by which to govern society" (p. 133). . . .

Kazancigil (1991) has offered parallel explanations of the weakness of democracy in Islam with those for Orthodox Christian lands as flowing from their failures "to dissociate the religious from the political spheres" (p. 345). . . .

. . . It is significant that . . . both Confucianism and Maoism in ideological content have explicitly stressed the problems of authority and order" (Pye 1968:16). Though somewhat less pessimistic, He Baogang's (1992) evaluation of cultural factors in mainland China concluded that "evidence reveals that the antidemocratic culture is currently stronger than the factors related to a democratic one" (p. 134). Only Japan, the most diluted Confucian country, "had sustained experience with democratic government prior to 1990, . . . [although its] democracy was the product of an American presence" (Huntington 1991:15). The others—Korea, Vietnam, Singapore, and Taiwan—were autocratic. . . . The situation, of course, has changed in recent years in response to rapid economic growth, reflecting the ways in which economic changes can impact on the political system undermining autocracy.

But India, a Hindu country that became democratic prior to industrialization, is different:

> The most salient feature of Indian civilization, from the point of view of our discussion, is that it is probably the only complete, highly differentiated civilization which throughout history has maintained its cultural identity without being tied to a given political framework. . . . [T]o a much greater degree than in many other historical imperial civilizations politics were conceived in secular forms. . . . Because of the relative dissociation between the cultural and the political order, the process of modernization could get underway in India without being hampered by too specific a traditional-cultural orientation toward the political sphere. (Eisenstadt 1968:32)

These generalizations about culture do not auger well for the future of the third wave of democracy in the former Communist countries. The Catholic Church played a substantial role in Poland's move away from Soviet Communism. But as noted previously, historically deeply religious Catholic areas have not been among the most amenable to democratic ideas. Poland is now troubled by conflicts flowing from increasing Church efforts to affect politics in Eastern Europe even as it relaxes its policies in Western Europe and most of the Americas. Orthodox Christianity is hegemonic in Russia and Belarus. The Ukraine is dominated by both the Catholic and Orthodox Churches. And fascists and Communists are strong in Russia and the Ukraine. Moslems are a significant group in the Central Asian parts of the former Soviet Union, the majority in some—these areas are among the consistently least democratic of the successor Soviet states. Led by the Orthodox Serbians, but helped by Catholic Croats and Bosnian Moslems, the former Yugoslavia is being torn apart along ethnic and religious lines with no peaceful, much less democratic, end in sight. We are fooling ourselves if we ignore the continuing dysfunctional effects of a number of cultural values and the institutions linked to them.

But belief systems change; and the rise of capitalism, a large middle class, an organized working class, increased wealth, and education are associated with secularism and the institutions of civil society which help create autonomy for the state and facilitate other preconditions for democracy. In recent years, nowhere has this been more apparent than in the economically successful Confucian states of East Asia—states once thought of as nearly hopeless candidates for both development and democracy. Tu (1993) noted their totally "unprecedented dynamism in democratization and marketization. Singapore, South Korea, and Taiwan all successfully conducted national elections in 1992, clearly indicating that democracy in Confucian societies is not only possible but also practical" (p. viii). Nathan and Shi (1993), reporting on "the first scientifically valid national sample survey done in China on political behavior and attitudes," stated: "When compared to residents of some of the most stable, long-established democracies in the world, the Chinese population scored lower on the variables we looked at, but not so low as to justify the conclusion that democracy is out of reach" (p. 116). Surveys which have been done in Russia offer similar positive conclusions (Gibson and Duch 1993), but the December, 1993 election in which racist nationalists and pro-Communists did well indicate much more is needed. Democracy is not taking root in much of the former Soviet Union, the less industrialized Moslem states, nor many nations in Africa. The end is not in sight for many of the efforts at new democracies; the requisite cultural changes are clearly not established enough to justify the conclusion that the "third wave" will not be reversed. According to the Freedom House survey, during 1993 there were "42 countries registering a decline in their level of freedom [political rights and civil liberties] and 19 recording gains" (Karatnycky 1994:5).*

INSTITUTIONALIZATION

New democracies must be institutionalized, consolidated, and become legitimate. They face many problems. . . .

Legitimacy

Political stability in democratic systems cannot rely on force. The alternative to force is legitimacy, an accepted systemic "title to rule. . . ."

Weber (1946), the fountainhead of legitimacy theory, named three ways by which an authority may gain legitimacy. These may be summarized:

1. *Traditional*—through "always" having possessed the authority, the best example being the title held in monarchical societies.
2. *Rational-legal*—when authority is obeyed because of a popular acceptance of the appropriateness of the system of rules under which they have won and held office. In the United States, the Constitution is the basis of all authority.
3. *Charismatic*—when authority rests upon faith in a leader who is believed to be endowed with great personal worth, either from God, as in the case of a religious prophet or simply from the display of extraordinary talents. The "cult of personality" surrounding many leaders is an illustration of this (pp. 78–79).

Legitimacy is best gained by prolonged effectiveness, effectiveness being the actual performance of the government and the extent to which it satisfies the basic needs of most of the population and key power groups (such as the military and economic leaders) (Lipset 1979: 16–23; Linz 1988:79–85). This generalization, however, is of no help to new systems for which the best immediate institutional advice is to separate the source and the agent of authority.

The importance of this separation cannot be underestimated. The agent of authority may be strongly opposed by the electorate and may be changed by the will of the voters, but the essence of the rules, the symbol of authority, must remain respected and unchallenged. Hence, citizens obey the laws and rules, even while disliking those who enforce them. . . .

Rational-legal legitimacy is weak in most new democratic systems, since the law had previously operated in the interests of a foreign exploiter or domestic dictator. Efforts to construct rational-legal legitimacy necessarily involve extending the rule of the law and the prestige of the courts, which should be as independent from the rest of the polity as possible. As Ackerman (1992:60–62) and Weingast (1993) note, in new democracies, these requirements imply the need to draw up a "liberal" constitution *as soon as possible*. The constitution can provide a basis for legitimacy, for limitations on state power, and for political and economic rights. . . .

*In the Freedom House survey, a country may move up or down with respect to measures of freedom without changing its status as a democratic or authoritarian system.

To reiterate, if democratic governments which lack traditional legitimacy are to survive, they must be effective, or as in the example of some new Latin American and post-communist democracies, may have acquired a kind of negative legitimacy—an inoculation against authoritarianism because of the viciousness of the previous dictatorial regimes. Newly independent countries that are post-revolutionary, post-coup, or post-authoritarian regimes are inherently low in legitimacy. Thus most of the democracies established in Europe after World War I as a result of the overthrow of the Austro-Hungarian, German, and Czarist Russian empires did not last. . . .

All other things being equal, an assumption rarely achieved, nontraditional authoritarian regimes are more brittle than democratic ones. By definition, they are less legitimate; they rely on force rather than belief to retain power. Hence, it may be assumed that as systems they are prone to be disliked and rejected by major segments of the population. . . .

The record, as in the case of the Soviet Union, seems to contradict this, since that regime remained in power for three-quarters of a century. However, a brittle, unpopular system need not collapse. Repressive police authority, a powerful army, and a willingness by rulers to use brute force may maintain a regime's power almost indefinitely. The breakdown of such a system may require a major catalytic event, a defeat in war, a drastic economic decline, or a break in the unity of the government elite. . . .

In contrast to autocracies, democratic systems rely on and seek to activate popular support and constantly compete for such backing. Government ineffectiveness need not spill into other parts of the society and economy. Opposition actually serves as a communication mechanism, focusing attention on societal and governmental problems. Freedom of opposition encourages a free flow of information about the economy as well as about the polity. . . .

Non-traditional authoritarian regimes seek to gain legitimacy through cults of personality (e.g., Napoleon, Toussaint, Diaz, Mussolini, Hitler). New autocrats lack the means to establish legal-rational legitimacy through the rule of law. Communist governments, whose Marxist ideology explicitly denied the importance of "great men" in history and stressed the role of materialist forces and "the people," were forced to resort to charismatic legitimacy. Their efforts produced the cults of Lenin, Stalin, Mao, Tito, Castro, Ho, Kim, and others. . . .

But charismatic legitimacy is inherently unstable. As mentioned earlier, a political system operates best when the source of authority is clearly separated from the agent of authority. If the ruler and his or her policies are seen as oppressive or exploitive, the regime and its rules will also be rejected. People will not feel obligated to conform or to be honest; force alone cannot convey a "title to rule."

EXECUTIVE AND ELECTORAL SYSTEMS . . .

Executive Systems

In considering the relation of government structure to legitimacy it has been suggested that republics with powerful presidents will, all other things being equal, be more unstable than parliamentary ones in which powerless royalty or elected heads of state try to act out the role of a constitutional monarch. In the former, where the executive is chief of state, symbolic authority and effective power are combined in one person, while in the latter they are divided. With a single top office, it is difficult for the public to separate feelings about the regime from those held toward the policy makers. The difficulties in institutionalizing democracy in the many Latin American presidential regimes over the last century and a half may reflect this problem. The United States presents a special case, in which, despite combining the symbolic authority and power into the presidency, the Constitution has been so hallowed by ideology and prolonged effectiveness for over 200 years, that it, rather than those who occupy the offices it specifies, has become the accepted ultimate source of authority. . . .

Evaluation of the relative worth of presidential and parliamentary systems must also consider the nature of each type. In presidential regimes, the power to enact legislation, pass budgets and appropriations, and make high level appointments are divided among the president and (usually two) legislative Houses; parliamentary regimes are unitary regimes, in which the prime minister and cabinet can have their way legislatively. A prime minister with a parliamentary majority, as usually occurs in most Commonwealth nations and a number of countries in Europe, is much more powerful and less constrained than a constitutional president who can only propose while Congress disposes (Lijphart 1984:4–20). The weak, divided-authority system has worked in the United States, although it has produced much frustration and alienation at times. But, as noted, the system has repeatedly broken down in Latin America, although one could argue that this is explained not by the constitutional arrangements, but by cultural legacies and lower levels of productivity. Many parliamentary systems have failed to produce stable governments because they lack operating legislative majorities. . . . There is no consensus among political scientists as to which system, presidential or parliamentary, is superior, since it is possible to point to many failures for both types.

Electoral Systems

The procedures for choosing and changing administrations also affect legitimacy. Elections that offer the voters an effective way to change the government and vote the incumbents out will provide more stability; electoral decisions will be more readily accepted in those systems in which electoral rules, distribution of forces, or varying party strengths make change more difficult.

Electoral systems that emphasize single-member districts, such as those in the United States and in much of the Commonwealth, press the electorate to choose between two major parties. . . .

In systems with proportional representation, the electorate may not be able to determine the composition of the government. In this type, a representation is assigned to parties which corresponds to their proportions of the vote. . . . Where no party has a majority, alliances may be formed out of diverse forces. . . . Small, opportunistic, or special interest parties may hold the balance of power and determine the shape and policies of post-election coalitions. The tendency toward instability and lack of choice in proportional systems can be reduced by setting up a minimum vote for representation, such as the five percent cut-off that exists in Germany and Russia. In any case, electoral systems, whether based on single-member districts or proportional representation, cannot guarantee particular types of partisan results (Gladdish 1993).

CIVIL SOCIETY AND POLITICAL PARTIES

Civil Society as a Political Base

More important than electoral rules in encouraging a stable system is a strong civil society—the presence of myriad "mediating institutions," including "groups, media, and networks" (Diamond 1993b:4), that operate independently between individuals and the state. These constitute "subunits, capable of opposing and countervailing the state." . . .

Citizen groups must become the bases of—the sources of support for—the institutionalized political parties which are a necessary condition for . . . a modern democracy. . . .

A fully operative civil society is likely to also be a participant one. Organizations stimulate interests and activity in the larger polity; they can be consulted by political institutions about projects that affect them and their members, and they can transfer this information to the citizenry. Civil organizations reduce resistance to unanticipated changes because they prevent the isolation of political institutions from the polity and can smooth over, or at least recognize, interest differences early on. . . .

Totalitarian systems, however, do not have effective civil societies. Instead, they either seek to eliminate groups mediating between the individual and the state or to control these groups so there is no competition. . . .

The countries of Eastern Europe and the former Soviet Union, however, are faced with the consequences of the absence of modern civil society, a lack that makes it difficult to institutionalize democratic polities. . . . Instead, they have had to create parties "from scratch." . . . "Instead of consolidation, there is fragmentation: 67 parties fought Poland's most recent general election, 74 Romania's" (*Economist* 1993a:4). As a result, the former Communists (now "socialists") have either been voted in as the majority party in parliament, as in Lithuania, or have become the largest party heading up a coalition cabinet, as in Poland. In January 1992, the Communist-backed candidate for president in Bulgaria garnered 43 percent of the vote (Malia 1992:73). These situations are, of course, exacerbated by the fact that replacing command economies by market processes is difficult, and frequently conditions worsen before they begin to improve.

Recent surveys indicate other continuing effects of 45 to 75 years of Communist rule. An overwhelming majority (about 70 percent) of the population in nearly all of the countries in Eastern Europe agree that "the state should provide a place of work, as well as a national health service, housing, education, and other services" (*Economist* 1993a:5). . . .

Political Parties as Mediators

Political parties themselves must be viewed as the most important mediating institutions between the citizenry and the state (Lipset 1993). And a crucial condition for a stable democracy is that major parties exist that have an almost permanent significant base of support. That support must be able to survive clear-cut policy failures by the parties. If this commitment does not exist, parties may be totally wiped out, thus eliminating effective opposition. . . .

If, as in new democracies, parties do not command such allegiance, they can be easily eliminated. The Hamiltonian Federalist party, which competed in the early years of the American Republic with the Jeffersonian Democratic-Republicans, declined sharply after losing the presidency in 1800 and soon died out (Dauer 1953). . . . It may be argued then, that having at least two parties with an uncritically loyal mass base comes close to being a necessary condition for a stable democracy. Democracy requires strong parties that can offer alternative policies and criticize each other. . . .

Sources of Political Party Support

. . . These four sources of conflict, *center-periphery, state-church, land-industry*, and *capitalist-worker*, have continued to some extent in the contemporary world, and have provided a framework for the party systems of the democratic polities, particularly in Europe. Class became the most salient source of conflict and voting, particularly after the extension of the suffrage to all adult males (Lipset and Rokkan 1967). Both Tocqueville (1976:vol. 2, 89–93), in the early nineteenth century and Bryce (1901:335), at the end of it, noted that at the bottom of the American political party conflict lay the struggle between aristocratic and democratic interests and sentiments. . . . Given all the transformations in Western society over the first half of the twentieth century, it is noteworthy how little the formal party systems changed. Essentially the conflicts had become institutionalized—the Western party systems of the 1990s resemble those of pre–World War II. . . .

Beginning in the mid-1960s, the Western world appears to have entered a new political phase. It is characterized by the rise of so-called "post-materialistic issues, a clean environment, use of nuclear power, a better culture, equal status for women and minorities, the quality of education, international relations, greater democratization, and a more permissive morality, particularly as affecting familial and sexual issues" (Lipset 1981:503–21). These have been perceived by some social analysts as the social consequences of an emerging third "revolution," the Post-Industrial Revolution, which is introducing new bases of social and political conflict. Inglehart (1990) and others have pointed to new cross-cutting lines of conflict—an *industrial-ecology* conflict—between the adherents of the industrial society's emphasis on production (who also hold conservative positions on social issues) and those who espouse the post-industrial emphasis on the quality-of-life and liberal social views when dealing with ecology, feminism, and nuclear energy. Quality-of-life concerns are difficult to formulate as party issues, but groups such as the Green parties and the New Left or New Politics—all educated middle class groups—have sought to foster them. . . .

The one traditional basis of party differentiation that seems clearly to be emerging in Russia is the center-periphery conflict, the first one that developed in Western society. The second, church-state (or church-secular), is also taking shape to varying degrees. Land-industry (or rural-urban) tension is somewhat apparent. Ironically, the capitalist-worker conflict is as yet the weakest, perhaps because a capitalist class and an independently organized working-class do not yet exist. Unless stable parties can be formed, competitive democratic politics is not likely to last in many of the new Eastern European and Central Asian polities. . . .

THE RULE OF LAW AND ECONOMIC ORDER

Finally, order and predictability are important for the economy, polity, and society. The Canadian Fathers of Confederation, who drew up the newly unified country's first constitution in 1867, described the Constitution's objective as "peace, order, and good government" (Lipset 1990b:xiii). Basically, they were talking about the need for the "rule of law," for establishing rules of "due process," and an independent judiciary. Where power is arbitrary, personal, and unpredictable, the citizenry will not know how to behave; it will fear that any action could produce an unforeseen risk. Essentially, the rule of law means: (1) that people and institutions will be treated equally by the institutions administering the law—the courts, the police, and the civil service; and (2) that people and institutions can predict with reasonable certainty the consequences of their actions, at least as far as the state is concerned. . . .

In discussing "the social requisites of democracy," I have repeatedly stressed the relationship between the level of economic development and the presence of democratic government. . . .

Clearly, socioeconomic correlations are merely associational, and do not necessarily indicate cause. Other variables, such as the force of historical incidents in domestic politics, cultural factors, events in neighboring countries, diffusion effects from elsewhere, leadership, and movement behavior can also affect the nature of the polity. Thus, the outcome of the Spanish Civil War, determined in part by other European states, placed Spain in an authoritarian mold, much as the allocation of Eastern Europe to the Soviet Union after World War II determined the political future of that area and that Western nations would seek to prevent the electoral victories of Communist-aligned forces. Currently, international agencies and foreign governments are more likely to endorse pluralistic regimes. . . .

CONCLUSION

Democracy is an international cause. A host of democratic governments and parties, as well as various non-governmental organizations (NGOs) dedicated to human rights, are working and providing funds to create and sustain democratic forces in newly liberalized governments and to press autocratic ones to change (*Economist* 1993c:46). Various international agencies and units, like the European Community, NATO, the World Bank, and the International Monetary Fund (IMF), are requiring a democratic system as a condition for membership or aid. A diffusion, a contagion, or demonstration effect seems operative, as many have noted, one that encourages democracies to press for change and authoritarian rulers to give in. It is becoming both uncouth and unprofitable to avoid free elections, particularly in Latin America, East Asia, Eastern Europe, and to some extent in Africa (Ake 1991:33). Yet the proclamation of elections does not ensure their integrity. The outside world can help, but the basis for institutionalized opposition, for interest and value articulation, must come from within.

Results of research suggest that we be cautious about the long-term stability of democracy in many of the newer systems given their low level of legitimacy. As the Brazilian scholar Francisco Weffort (1992) has reminded us, "In the 1980s, the age of new democracies, the processes of political democratization occurred at the same moment in which those countries suffered the experience of a profound and prolonged economic crisis that resulted in social exclusion and massive poverty. ... Some of those countries are building a political democracy on top of a minefield of social apartheid ..." (p. 20). Such conditions could easily lead to breakdowns of democracy as have already occurred in Algeria, Haiti, Nigeria, and Peru, and to the deterioration of democratic functioning in countries like Brazil, Egypt, Kenya, the Philippines, and the former Yugoslavia, and some of the trans-Ural republics or "facade democracies," as well as the revival of anti-democratic movements on the right and left in Russia and in other formerly Communist states.

What new democracies need, above all, to attain legitimacy is efficacy—particularly in the economic arena, but also in the polity....

REFERENCES

Editors' Note: The original chapter from which this selection was taken has extensive references that could not be listed here. For more documentation see the source note on page 431.

Ackerman, Bruce. 1992. *The Future of Liberal Revolution.* New Haven, CT: Yale University.

Bryce, James. 1901. *Study in History and Jurisprudence.* New York: Oxford University.

Dahl, Robert. 1970. *After the Revolution: Authority in a Good Society.* New Haven, CT: Yale University.

———. 1971. Polyarchy: *Participation and Opposition.* New Haven, CT: Yale University.

Gibson, James L. and Raymond M. Duch. 1993. "Emerging Democratic Values in Soviet Political Culture." Pp. 69–94 in *Public Opinion and Regime Change,* A. A. Miller, W. M. Reisinger, and V. Hesli, Boulder, CO: Westview.

Gladdish, Ken. 1993. "The Primacy of the Particular." *Journal of Democracy* 4(1):53–65.

Glassman, Ronald. 1991. *China in Transition: Communism, Capitalism and Democracy.* Westport, CT: Praeger.

He Baogang. 1992. "Democratization: Antidemocratic and Democratic Elements in the Political Culture of China." *Australian Journal of Political Science* 27:120–36.

Huntington, Samuel. 1968. *Political Order in Changing Societies.* New Haven, CT: Yale University.

Kohli, Atul. 1992. "Indian Democracy: Stress and Resilience." *Journal of Democracy* 3(1):52–64.

Kuznets, Simon. 1955. "Economic Growth and Income Inequality." *American Economic Review* 45:1–28.

———. 1963. "Quantitative Aspects of the Economic Growth of Nations: VIII, The Distribution of Income by Size." *Economic Development and Cultural Change* 11:1–80.

———. 1976. *Modern Economic Growth: Rate, Structure and Spread.* New Haven, CT: Yale University.

Lewis, Bernard. 1993. "Islam and Liberal Democracy." *Atlantic Monthly.* 271(2):89–98.

Linz, Juan J. 1988. "Legitimacy of Democracy and the Socioeconomic System." Pp. 65–97 in *Comparing Pluralist Democracies: Strains on Legitimacy,* M. Dogan. Boulder, CO: Westview.

Lipset, Seymour Martin and Stein Rokkan. 1967. "Cleavage Structures, Party Systems and Voter Alignments." Pp. 1–64 in *Party Systems and Voter Alignments,* S. M. Lipset and S. Rokkan. New York: Free Press.

Lipset, Seymour Martin, Kyoung-Ryung Seong, and John Charles Torres. 1993. "A Comparative Analysis of the Social Requisites of Democracy." *International Social Science Journal* 45:155–75.

Malia, Martin. 1992. "Leninist Endgame." *Daedalus* 121(2):57–75.

Weffort, Francisco C. 1992. "New Democracies, Which Democracies?" (Working Paper #198). Washington, DC: The Woodrow Wilson Center, Latin American Program.

Weingast, Barry. 1993. "The Political Foundations of Democracy and the Rule of Law." The Hoover Institution, Stanford, CA: Unpublished manuscript.

Wright, Robin. 1992. "Islam and Democracy." *Foreign Affairs* 71(3):131–45.

57

Money Changes Everything

DAN CLAWSON, ALAN NEUSTADTL, AND DENISE SCOTT

The United States claims to be a representative democracy. The implication of this reading is that our political representation is far narrower than trumpeted. What can be done to bring about our democratic claims? Do you think it will happen?

As you read, ask yourself the following questions:

1. *What effects have PACs had on campaign financing?*
2. *How would you change the financing of political campaigns?*

GLOSSARY **Hegemony** Leadership dominance.

In the past twenty years political action committees, or PACs, have transformed campaign finance. . . .

Most analyses of campaign finance focus on the candidates who receive the money, not on the people and political action committees that give it. PACs are entities that collect money from many contributors, pool it, and then make donations to candidates. Donors may give to a PAC because they are in basic agreement with its aims, but once they have donated they lose direct control over their money, trusting the PAC to decide which candidates should receive contributions. . . .

WHY DOES THE AIR STINK?

Everybody wants clean air. Who could oppose it? "I spent seven years of my life trying to stop the Clean Air Act," explained the PAC director for a major corporation that is a heavy-duty polluter. Nonetheless, he was perfectly willing to use his corporation's PAC to contribute to members of Congress who voted for the act:

> How a person votes on the final piece of legislation often is not representative of what they have done. Somebody will do a lot of things during the process. How many guys voted against the Clean Air Act? But during the process some of them were very sympathetic to some of our concerns.

In the world of Congress and political action committees things are not always what they seem. Members of Congress want to vote for clean air, but they also want to receive campaign contributions from corporate PACs and pass a law that business accepts as "reasonable." The compromise solution to this dilemma is to gut the bill by crafting dozens of loopholes inserted in private meetings or in subcommittee hearings that don't receive much (if any) attention in the press. Then the public vote on the final bill can be nearly unanimous: members of Congress can assure their constituents that they voted for the final bill and their corporate PAC contributors that they helped weaken the bill in

SOURCE: In *Money Talks: Corporate PACs and Political Influence*. Basic Books, 1992. Reprinted with permission.

private. We can use the Clean Air Act of 1990 to introduce and explain this process.

The public strongly supports clean air and is unimpressed when corporate officials and apologists trot out their normal arguments: "corporations are already doing all they reasonably can to improve environmental quality"; "we need to balance the costs against the benefits"; "people will lose their jobs if we make controls any stricter." The original Clean Air Act was passed in 1970, revised in 1977, and not revised again until 1990. Although the initial goal of its supporters was to have us breathing clean air by 1975, the deadline for compliance has been repeatedly extended—and the 1990 legislation provides a new set of deadlines to be reached sometime far in the future.

Because corporations control the production process unless the government specifically intervenes, any delay in government action leaves corporations free to do as they choose. Not only have laws been slow to come, but corporations have fought to delay or subvert implementation. The 1970 law ordered the Environmental Protection Agency (EPA) to regulate the hundreds of poisonous chemicals that are emitted by corporations, but as William Greider notes, "in twenty years of stalling, dodging, and fighting off court orders, the EPA has managed to issue regulatory standards for a total of seven toxics."

Corporations have done exceptionally well politically, given the problem they face: the interests of business often are diametrically opposed to those of the public. Clean air laws and amendments have been few and far between, enforcement is ineffective, and the penalties for infractions are minimal. . . .

This corporate struggle for the right to pollute takes place on many fronts. One front is public relations: the Chemical Manufacturers Association took out a two-page Earth Day ad in the *Washington Post* to demonstrate its concern for the environment; coincidentally many of the corporate signers are also on the EPA's list of high-risk producers. Another front is research: expert studies delay action while more information is gathered. The federally funded National Acid Precipitation Assessment Program (NAPAP) took ten years and $600 million to figure out whether acid rain was a problem. Both business and the Reagan administration argued that no action should be taken until the study was completed. The study was discredited when its summary of findings minimized the impact of acid rain—even though this did not accurately represent the expert research in the report. But the key site of struggle has been Congress, where for years corporations have succeeded in defeating environmental legislation. In 1987 utility companies were offered a compromise bill on acid rain, but they "were very adamant that they had beat the thing since 1981 and they could always beat it," according to Representative Edward Madigan (R-III.). Throughout the 1980s the utilities defeated all efforts at change. . . .

The stage was set for a revision of the Clean Air Act when George Bush was elected as "the environmental president" and George Mitchell, a strong supporter of environmentalism, became the Senate majority leader. But what sort of clean air bill would it be? "What we wanted," said Richard Ayres, head of the environmentalists' Clean Air Coalition, "is a health-based standard—one-in-1-million cancer risk." Such a standard would require corporations to clean up their plants until the cancer risk from their operations was reduced to one in a million. "The Senate bill still has the requirement," Ayres said, "but there are forty pages of extensions and exceptions and qualifications and loopholes that largely render the health standard a nullity." Greider reports, for example, that "according to the EPA, there are now twenty-six coke ovens that pose a cancer risk greater than 1 in 1000 and six where the risk is greater than 1 in 100. Yet the new clean-air bill will give the steel industry another thirty years to deal with the problem."

This change from what the bill was supposed to do to what it did do came about through what corporate executives like to call the "access" process. The main aim of most corporate political action committee contributions is to help corporate executives attain "access" to key members of Congress and their staffs. Corporate executives (and corporate PAC money) work to persuade the member of Congress to accept a carefully pre-designed loophole that sounds innocent but effectively undercuts the stated intention of the bill. Representative Dingell (D-Mich.), chair of the House Committee on Energy and Commerce, is a strong industry supporter; one of the people we interviewed called him "the point man for the Business Roundtable on clean air." Representative Waxman (D-Calif.), chair of the Subcommittee on Health and the Environment, is an environmentalist. Observers of the Clean Air Act legislative process expected a confrontation and contested votes on the floor of Congress.

The problem for corporations was that, as one Republican staff aide said, "If any bill has the blessing of Waxman and the environmental groups, unless it is totally in outer space, who's going to vote against it?" But corporations successfully minimized public votes. Somehow Waxman was persuaded to make behind-the-scenes compromises with Dingell so members didn't have to publicly side with business against the environment during an election year. Often the access process leads to loopholes that protect a single corporation, but for "clean" air most special deals targeted entire industries, not specific companies. The initial bill, for example, required cars to be able to use strictly specified cleaner fuels. But the auto industry wanted the rules loosened, and Congress eventually modified the bill by incorporating a variant of a formula suggested by the head of General Motors' fuels and lubricants department.

Nor did corporations stop fighting after they gutted the bill through amendments. Business pressed the EPA for favorable regulations to implement the law: "The cost of this legislation could vary dramatically, depending on how EPA interprets it," said William D. Fay, vice president of the National Coal Association, who headed the hilariously misnamed Clean Air Working Group, an industry coalition that fought to weaken the legislation. An EPA aide working on acid rain regulations reported, "We're having a hard time getting our work done because of the number of phone calls we're getting" from corporations and their lawyers.

Corporations trying to convince federal regulators to adopt the "right" regulations don't rely exclusively on the cogency of their arguments. They often exert pressure on a member of Congress to intervene for them at the EPA or other agency. Senators and representatives regularly intervene on behalf of constituents and contributors by doing everything from straightening out a social security problem to asking a regulatory agency to explain why it is pressuring a company. This process—like campaign finance—usually follows accepted etiquette. In addressing a regulatory agency the senator does not say, "Lay off my campaign contributors, or I'll cut your budget." One standard phrasing for letters asks regulators to resolve the problem "as quickly as possible within applicable rules and regulations." No matter how mild and careful the inquiry, the agency receiving the request is certain to give it extra attention; only after careful consideration will they refuse to make any accommodation.

The power disparity between business and environmentalists is enormous during the legislative process but even larger thereafter. When the Clean Air Act passed, corporations and industry groups offered positions, typically with large pay increases, to congressional staff members who wrote the law. The former congressional staff members who work for corporations know how to evade the law and can persuasively claim to EPA that they know what Congress intended. Environmental organizations pay substantially less than Congress and can't afford large staffs. They are rarely able to become involved in the details of the administrative process or influence implementation and enforcement.

Having pushed Congress for a law, and the Environmental Protection Agency for regulations, allowing as much pollution as possible, business then went to the Quayle Council for rules allowing even more pollution. Vice President J. Danforth Quayle's Council, technically the Council on Competitiveness, was created by President Bush specifically to help reduce regulations on business. Quayle told the *Boston Globe* "that his council has an 'open door' to business groups and that he has a bias against regulations." The Council reviews, and can override, all federal regulations, including those by the BPA setting the limits at which a chemical is subject to regulation. The council also recommended that corporations be allowed to increase their polluting emissions if a state did not object within seven days of the proposed increase. Corporations thus have multiple opportunities to win. If they lose in Congress, they can win at the regulatory agency; if they lose there, they can try again at the Quayle Council. If they lose there, they can try to reduce the money available to enforce regulations, tie up the issue in the courts, or accept a minimal fine.

The operation of the Quayle Council probably would have received little publicity, but reporters discovered that the executive director of the Council, Allan Hubbard, had a clear conflict of interest. Hubbard chaired the biweekly White House meetings on the Clean Air Act. He owns half of World Wide Chemical, received an average of more than a million dollars a year in profits from it while directing the Council, and continues to attend quarterly stockholder meetings. According to the *Boston Globe*, "Records on file with the Indianapolis Air Pollution Control Board show that World Wide Chemical emitted 17,000 to 19,000 pounds of chemicals into the air last year." The company "does not have the permit required to release the emissions," "is putting out nearly four times the allowable emissions without a permit, and could be subject to a $2,500-a-day penalty," according to David Jordan, director of the Indianapolis Air Pollution Board. . . .

The real issue is the system of business-government relations, and especially of campaign finance, that offers business so many opportunities to craft loopholes, undermine regulations, and subvert enforcement. Still worse, many of these actions take place outside of public scrutiny.

The Candidates' Perspective

. . . Money has always been a critically important factor in campaigns, but the shift to expensive technology has made it the dominant factor. Today money is the key to victory and substitutes for everything else—instead of door-to-door canvassers, a good television spot; instead of a committee of respected long-time party workers who know the local area, a paid political consultant and media expert. To be a viable political candidate, one must possess—or be able to raise—huge sums. Nor is this a one-time requirement; each reelection campaign requires new infusions of cash.

The quest for money is never ending. Challengers must have money to be viable contenders; incumbents can seldom predict when they might face a tight race. In 1988 the average winning candidate for the House of Representatives spent $388,000; for the Senate, $3,745,000. Although the Congress, especially the Senate, has many millionaires, few candidates have fortunes large enough to finance repeated campaigns out of their own pockets. It would take the entire congressional salary for 3.1 years for a member of the House, or 29.9 years for a senator, to pay for a single reelection campaign. Most members are therefore in no position to say, "Asking people for money is just too big a hassle. Forget it. I'll pay for it myself." They must raise the money from others, and the pressure to do so never lets up. To pay for an average winning campaign, representatives need to raise $3,700 and senators $12,000 during *every week* of their term of office.

Increasingly incumbents use money to win elections before voters get involved. Senator Rudy Boschwitz (R-Minn.) spent $6 million getting reelected in 1984 and had raised $1.5 million of it by the beginning of the year, effectively discouraging the most promising Democratic challengers. . . .

Fundraising isn't popular with the public, but candidates keep emphasizing it because it works: the champion money raiser wins almost regardless of the merits. *Almost* is an important qualifier here, as Boschwitz would be the first to attest: in his 1990 race he outspent his opponent by about five to one and lost nonetheless. . . .

It is not only that senators leave committee hearings for the more crucial task of calling people to beg for money. They also chase all over the country because reelection is more dependent on meetings with rich people two thousand miles from home than it is on meetings with their own constituents.

. . . Do members of Congress incur any obligations in seeking and accepting these campaign contributions? Bob Dole, Republican leader in the Senate and George Bush's main rival for the 1988 Republican presidential nomination, was quoted by the *Wall Street Journal* as saying, "When the Political Action Committees give money, they expect something in return other than good government." One unusually outspoken business donor, Charles Keating, made the same point: "One question among the many raised in recent weeks had to do with whether my financial support in any way influenced several political figures to take up my cause. I want to say in the most forceful way I can, I certainly hope so." . . .

The Current Law The law, however, regulates fundraising and limits the amount that any one individual or organization may (legally) contribute. According to current law:

1. A *candidate* may donate an unlimited amount of personal funds to his or her *own* campaign. The Supreme Court has ruled this is protected as free speech.
2. Individuals may not contribute more than $1,000 per candidate per election, nor more than $25,000 in total in a given two-year election cycle.
3. Political action committees may contribute up to $5,000 per candidate per election. Since most candidates face primaries, an individual may contribute $2,000 and a PAC $10,000 to the candidate during a two-year election cycle. PACs may give to an unlimited number of candidates and hence may give an unlimited amount of money.
4. Individuals may contribute up to $5,000 per year to a political action committee.
5. Candidates must disclose the full amount they have received, the donor and identifying information for any individual contribution of $200 or more, the name of the PAC and donation amount for any PAC contribution however small, and all disbursements. PACs must disclose any donation they make to a candidate, no matter how small. They must also disclose the total amount received by the PAC and the names and positions of all contributors who give the PAC more than $200 in a year.
6. Sponsoring organizations, including corporations and unions, may pay all the expenses of creating and operating a PAC. Thus a corporation may pay the cost of the rent, telephones, postage, supplies, and air travel for all PAC activities; the salaries of full-time corporate employees who work exclusively on the PAC; and the salaries of all managers who listen to a presentation about the PAC. However, the PAC money itself—the money used to contribute to candidates—must come from voluntary donations by individual contributors. The corporation may not legally take a portion of its profits and put it directly into the PAC.
7. Corporations may establish and control the PAC and solicit stockholders and/or managerial employees for contributions to the PAC. It is technically possible for corporations to solicit hourly (or nonmanagerial) employees and for unions to solicit managers, but these practices are so much more tightly regulated and restricted that in practice cross-solicitation is rare.
8. The Federal Election Commission (FEC) is to monitor candidates and contributors and enforce the rules.

These are the key rules governing fundraising, but the history of campaign finance is that as time goes on, loopholes develop. . . . What is generally regarded as the most important current loophole is that there are no reporting requirements or limits for contributions given to political parties as opposed to candidates. Such money is ostensibly to be used to promote party building and get-out-the-vote drives; in 1988 literally hundreds of individuals gave $100,000 or more in unreported "soft money" donations. Many of these loopholes are neither accidents nor oversights. Three Democrats and three Republicans serve as federal election commissioners, and commissioners are notorious party loyalists. Because it requires a majority to investigate a suspected violation, the FEC not only fails to punish violations, it fails to investigate them.

Corporate Pacs

The Federal Election Commission categorizes PACs as corporate, labor, trade-health-membership, and nonconnected. Nonconnected PACs are unaffiliated with any other organization: they are formed exclusively for the purpose of raising and contributing money. Most subsist by direct-mail fundraising targeted at people with a commitment to a single issue (abortion or the environment) or philosophical position (liberalism or conservatism). Other PACs are affiliated with an already existing organization, and that organization—whether a corporation, union, trade, or membership association—pays the expenses associated with operating the PAC and decides what will happen to the money the PAC collects.

Candidates increasingly rely on PACs because they can easily solicit a large number of PACs, each of which is relatively likely to make a major contribution. "From 1976–88, PAC donations rose from 22 per cent to 40 per cent of House campaign receipts, and from 15 per cent to 22 per cent of Senate receipts." Almost half of all House members (205 of the 435) "received at least 50 per cent of their campaign contributions from PACs." The reliance of PACs is greater in the House than in the Senate: PACs give more to Senate candidates, but Senate races are more expensive than House races, so a larger fraction of total Senate-race receipts comes from individual contributions.

Although other sorts of PACs deserve study, we believe the most important part of this story concerns corporate PACs. We focus on corporate PACs for three interrelated reasons. First, they are the largest concentrated source of campaign money and the fastest growing. In 1988 corporate PACs contributed more than $50 million, all trade-membership-health PACs combined less than $40 million, labor PACs less than $35 million, and nonconnected PACs less than $20 million. Moreover, these figures understate the importance of corporate decisions about money because industry trade associations are controlled by corporations and follow their lead. In addition, corporate executives have high incomes and make many individual contributions; a handful of labor leaders may attempt to do the same on a reduced scale, but rank-and-file workers are unlikely to do so. Second, corporations have disproportionate power in U.S. society, magnifying the importance of the money they contribute. Finally, corporate PACs have enormous untapped fundraising potential. They are in a position to coerce their donors in a way no other kind of PAC can and, if the need arose, could dramatically increase the amount of money they raise. . . .

Corporate PACs follow two very different strategies, pragmatic and ideological. . . .

Pragmatic donations are given specifically to advance the short-run interests of the donor, primarily to enable the corporation to gain a chance to meet with the member and argue its case. Because the aim of these donations is to gain "access" to powerful members of Congress, the money is given without regard to whether or not the member needs it and with little consideration of the member's political stance on large issues. The corporation's only concern is that the member will be willing and able to help them out—and virtually all members, regardless of party, are willing to cooperate in this access process. Perhaps the most memorable characterization of this strategy was by Jay Gould, nineteenth-century robber baron and owner of the Erie Railroad: "In a Republican district I was a Republican; in a Democratic district, a Democrat; in a doubtful district I was doubtful; but I was always for Erie."

Ideological donations, on the other hand, are made to influence the political composition of the Congress. From this perspective, contributions should meet two conditions: (1) they should be directed to politically congenial "pro-free enterprise" candidates who face opponents unsympathetic to business (in practice, these are always conservatives); and (2) they should be targeted at competitive races where money can potentially influence the election outcome. The member's willingness to do the company favors doesn't matter, and even a conservative "free enterprise" philosophy wouldn't be sufficient: if the two opponents' views were the same, then the election couldn't influence the ideological composition of Congress. Most incumbents are reelected: in some years as many as 98 percent of all House members running are reelected. Precisely because incumbents will probably be reelected even without PAC support, ideological corporations usually give to nonincumbents, either challengers or candidates for open seats.

Virtually all corporations use some combinations of pragmatic and ideological strategies. The simplest method of classifying PACs is by the proportion of money they give to incumbents: the higher this proportion, the more pragmatic the corporation.... In 1988 about a third (36 percent) of the largest corporate PACs gave more than 90 percent of their money to incumbents, and another third (34 percent) gave 80 to 90 percent to incumbents. Although roughly a third gave less than 80 percent to incumbents, only eight corporate PACs gave less than 50 percent of their money to incumbents (that is, more than 50 percent to nonincumbents).... The pragmatic emphasis of recent years is a change from 1980, when a large number of corporations followed an ideological approach.

OUR RESEARCH

...Our quantitative analyses concentrate on Democrats and Republicans in general-election contests for congressional seats. We focus on the 309 corporate PACs that made the largest contributions in the period from 1975 to 1988. As might be expected, these are almost exclusively very large corporations: on average in 1984 they had $6.7 billion in sales and 48,000 employees.... Moreover, not all firms with large PACs have huge sales, so our sample includes about twenty-five "small" firms with 1984 revenues of less than $500 million....

On average, in 1988 these PACs gave 52.7 percent of their money to Republicans and 47.3 percent to Democrats. They gave 83.6 percent of their money to incumbents, 10.2 percent to candidates for open seats, and 6.2 percent to challengers.

The PAC officials we interviewed were selected from this set of the 309 largest corporate PACs and were representative of the larger sample in terms of both economic and political characteristics....

A third source of original data supplements our quantitative analyses of the 309 largest corporate PACs and our 38 in-depth interviews. In November and December of 1986 we mailed surveys to a random sample of ninety-four directors of large corporate PACs, achieving a response rate of 58 percent. For the most part, we use this to place our interview comments in context: if a PAC director tells us a story of being pressured by a candidate, how typical is this? How many other PAC directors report similar experiences? Finally, our original data also are supplemented by books, articles, and newspaper accounts about campaign finance....

OVERVIEW AND BACKGROUND

...We argue that corporate PACs differ from other PACs in two ways: (1) as employees, managers can be—and are—coerced to contribute; and (2) corporate PACs are not democratically controlled by their contributors (even in theory)....

We argue that PAC contributions are best understood as gifts, not bribes. They create a generalized sense of obligation and an expectation that "if I scratch your back you scratch mine."...

A corporation uses the member of Congress's sense of indebtedness for past contributions to help it gain access to the member. In committee hearings and private meetings the corporation then persuades the member to make "minor" changes in a bill, which exempt a particular company or industry from some specific provision.

Even some corporations are troubled by this "access" approach, and . . . consider the alternative: donations to close races intended to change the ideological composition of the Congress. In the late 1980s and early 1990s only a small number of corporations used this as their primary strategy, but most corporations make some such donations. In the 1980 election a large group of corporations pursued an ideological strategy. We argue this was one of the reasons for the conservative successes of that period. . . .

Do competing firms or industries oppose each other in Washington, such that one business's political donations oppose and cancel out those of the next corporation or industry? More generally, how much power does business have in U.S. society, and how does its political power relate to its economic activity?. . .

The PAC directors we interviewed are not very worried about reform: they don't expect meaningful changes in campaign funding laws, and they assume that if "reforms" are enacted, they will be easily evaded. . . .

Three interrelated points. . . . First, power is exercised in many loose and subtle ways, not simply through the visible use of force and threats. Power may in fact be most effective, and most limiting, when it structures the conditions for action—even though in these circumstances it may be hard to recognize. Thus PAC contributions can and do exercise enormous influence through creating a sense of obligation, even if there is no explicit agreement to perform a specific service in return for a donation. Second, business is different from, and more powerful than, other groups in the society. As a result, corporations and their PACs are frequently treated differently than others would be. Other groups could not match business power simply by raising equivalent amounts of PAC money. Third, this does not mean that business always wins, or that it wins automatically. If it did, corporate PACs would be unnecessary. Business must engage in a constant struggle to maintain its dominance. This is a class struggle just as surely as are strikes and mass mobilizations, even though it is rarely thought of in these terms.

What Is Power?

Our analysis is based on an understanding of power that differs from that usually articulated by both business and politicians. The corporate PAC directors we interviewed insisted that they have no power. . . .

The executives who expressed these views used the word *power* in roughly the same sense that it is usually used within political science, which is also the way the term was defined by Max Weber, the classical sociological theorist. Power, according to this common conception, is the ability to make someone do something against his or her will. If that is what power means, then corporations rarely have power in relation to members of Congress. As one corporate senior vice president said to us, "You certainly aren't going to be able to buy anybody for $500 or $1,000 or $10,000. It's a joke." In this regard we agree with the corporate officials we interviewed: a PAC is not in a position to say to a member of Congress, "Either you vote for this bill, or we will defeat your bid for reelection." Rarely do they even say, "Vote for this bill, or you won't get any money from us." . . . Therefore, if power is the ability to make someone do something against his or her will, then PAC donations rarely give corporations power over members of Congress.

This definition of power as the ability to make someone do something against his or her will is what Steven Lukes calls a *one-dimensional view of power. A two-dimensional view* recognizes the existence of nondecisions: a potential issue never gets articulated or, if articulated by someone somewhere, never receives serious consideration. . . . A two-dimensional view of power makes the same point: in some situations no one notices power is being exercised—because there is no overt conflict.

Even this model of power is too restrictive, however, because it still focuses on discrete decisions and nondecisions. . . . Such models do not recognize "the idea that the most fundamental use of power in society is its use in structuring the basic manner in which social agents interact with one another."... Similarly, the mere presence of a powerful social agent alters social space for others and causes them to orient to the powerful agent. One of the executives we interviewed took it for granted that "if we go see the congressman who represents [a city where the company has a major plant], where 10,000 of our employees are also his constituents, we don't need a PAC to go see him." The corporation is so important in that area that the member has to orient himself or herself in relation to the corporation and its concerns. In a different sense, the mere act of accepting a campaign contribution changes the way a member relates to a PAC, creating a sense of obligation and need to reciprocate. The PAC contribution has altered the member's social space, his or her awareness of the company and wish to help it, even if no explicit commitments have been made.

Business Is Different

Power therefore is not just the ability to force people to do something against their will; it is most effective (and least recognized) when it shapes the field of action. Moreover, business's vast resources, influence on the economy, and general legitimacy place it on a different footing from other so-called special interests. Business donors are often treated differently from other campaign contributors. When a member of Congress accepts a $1,000 donation from a corporate PAC, goes to a committee hearing, and proposes "minor" changes in a bill's wording, those changes are often accepted without discussion or examination. The changes "clarify" the language of the bill, perhaps legalizing higher levels of pollution for a specific pollutant or exempting the company from some tax. The media do not report on this change, and no one speaks against it....

Even groups with great social legitimacy encounter more opposition and controversy than business faces for proposals that are virtually without public support. Contrast the largely unopposed commitment of more than $500 billion for the bailout of savings and loan associations with the sharp debate, close votes, and defeats for the rights of men and women to take *unpaid* parental leaves. Although the classic phrase for something noncontroversial that everyone must support is to call it a "motherhood" issue, and it would cost little to guarantee every woman the right to an unpaid parental leave, nonetheless this measure generated intense scrutiny and controversy, ultimately going down to defeat. Few people are prepared to publicly defend pollution or tax evasion, but business is routinely able to win pollution exemptions and tax loopholes. Although cumulatively these provisions may trouble people, individually most are allowed to pass without scrutiny. *No* analysis of corporate political activity makes sense unless it begins with a recognition that the PAC is a vital element of corporate power, but it does not operate by itself. The PAC donation is always backed by the wider range of business power and influence.

Corporations are different from other special interest groups not only because business has far more resources, but also because of this acceptance and legitimacy. When people feel that "the system" is screwing them, they tend to blame politicians, the government, the media—but rarely business. Although much of the public is outraged at the way money influences elections and public policy, the issue is almost always posed in terms of what politicians do or don't do. This pervasive double standard largely exempts business from criticism....

Many people who are outraged that members of Congress recently raised their pay to $125,100 are apparently unconcerned about corporate executives' pay. One study calculated that CEOs at the largest U.S. companies are paid an average of $2.8 million a year, 150 times more than the average U.S. worker and 22 times as much as members of Congress. More anger is directed at Congress for delaying new environmental laws than at the companies who fight every step of the way to stall and subvert the legislation. When members of Congress do favors for large campaign contributors, the anger is directed at the senators who went along, not at the business owner who paid the money (and usually initiated the pressure). The focus is on the member's receipt of thousands of dollars, not on the business's receipt of millions (or hundreds of millions) in tax breaks or special treatment. It is widely held that "politics is dirty," but companies getting away with murder—quite literally—generates little public comment and condemnation. This disparity is evidence of business's success in shaping public perceptions. Lee Atwater, George Bush's campaign manager for the 1988 presidential election, saw this as a key to Republican success:

> In the 1980 campaign, we were able to make the establishment, insofar as it is bad, the government. In other words, big government was the enemy, not big business. If the people think the problem is that taxes are too high, and the government interferes too much, then we are doing our job. But, if they get to the point where they say that the real problem is that rich people aren't paying taxes . . . then the Democrats are going to be in good shape.

. . . We argue corporations are so different, and so dominant that they exercise a special kind of power, what Antonio Gramsci called *hegemony*. Hegemony can be regarded as the ultimate example of a field of power that structures what people and groups do. It is sometimes referred to as a world view—a way of thinking about the world that influences every action and makes it difficult to even consider alternatives. But in Gramsci's analysis it is much more than this; it is a culture and set of institutions that structure life patterns and coerce a particular way of life. . . .

Hegemony is most successful and most powerful if it is unrecognized. . . . In some sense gender relations in the 1950s embodied a hegemony even more powerful than that of race relations. Betty Friedan titled the first chapter of *The Feminine Mystique* "The Problem That Has No Name" because women literally did not have a name for and did not recognize the existence of their oppression. Women as well as men denied the existence of inequality or oppression and denied the systematic exercise of power to maintain unequal relations.

We argue that today business has enormous power and exercises effective hegemony, even though (perhaps because) this is largely undiscussed and unrecognized. *Politically* business power today is similar to white treatment of blacks in 1959: business may sincerely deny its power, but many of the groups it exercises power over recognize it, feel dominated, resent this, and fight the power as best they can. *Economically* business power is more similar to gender relations in 1959: virtually no one sees this power as problematic. If the issue is brought to people's attention, many still don't see a problem: "Well, so what? How else could it be? Maybe we don't like it, but that's just the way things are." . . .

Hegemony is never absolute. . . . A hegemonic power is usually opposed by a counter-hegemony. . . .

The Limits To Business Power

We have argued that power is more than winning an open conflict, and business is different from other groups because of its pervasive influence on our society—the way it shapes the social space for all other actors. These two arguments, however, are joined with a third: a recognition of, in fact an insistence on, the limits to business power. We stress the power of business, but business does not feel powerful. . . .

Executives believe that corporations are constantly under attack, primarily because government simply doesn't understand that business is crucial to everything society does but can easily be crippled by well-intentioned but unrealistic government policies. A widespread view among the people we interviewed is that "far and away the vast majority of things that we do are literally to protect ourselves from public policy that is poorly crafted and nonresponsive to the needs and realities and circumstances of our company." These misguided policies, they feel, can come from many sources—labor unions, environmentalists, the pressure of unrealistic public-interest groups, the government's constant need for money, or the weight of its oppressive bureaucracy. Simply maintaining equilibrium requires a pervasive effort: if attention slips for even a minute, an onerous regulation will be imposed or a precious resource taken away. . . . But evidently the corporation agrees . . . since it devotes significant resources to political action of many kinds, including the awareness and involvement of top officials. Chief executive officers and members of the board of directors repeatedly express similar views.

Both of these views—the business view of vulnerability and our insistence on their power—are correct. . . .

Perhaps once upon a time business could simply make its wishes known and receive what it wanted; today corporations must form PACs, lobby actively, make their case to the public, run advocacy ads, and engage in a multitude of behaviors that they wish were unnecessary. From the outside we are impressed with the high success rates over a wide range of issues and with the lack of a credible challenge to the general authority of business. From the inside they are impressed with the serious consequences of occasional losses and with the continuing effort needed to maintain their privileged position.

Business power does not rest *only* on PAC donations, but the PAC is a crucial aspect of business power. A football analogy can be made: business's vast resources and its influence on the economy may be equivalent to a powerful offensive line that is able to clear out the opposition and create a huge opening, but someone then has to take the ball and run through that opening. The PAC and the government relations operation are, in this analogy, like a football running back. When they carry the ball they have to move quickly, dodge attempts to tackle them, and if necessary fight off an opponent and keep going. The analogy breaks down, however, because it implies a contest between two evenly matched opponents. Most of the time the situation approximates a contest between an NFL team and high school opponents. The opponents just don't have the same muscle. Often they are simply intimidated or have learned through past experience the best thing to do is get out of the way. Occasionally, however, the outclassed opponents will have so much courage and determination that they will be at least able to score, if not to win.

58

The New Terrorism

Securing the Nation against a Messianic Foe

STEVEN SIMON

Messianic movements see their actions as performing God's will against opposing ideologies. This apocalyptic vision makes al Qaeda's terrorist actions different from those of other terrorist groups. It also means that traditional deterrents against this group will not work and that different types of offensive and defensive actions alike will be needed.

As you read this article, consider the following questions as guides:

1. *What other offensive steps should the United States take?*
2. *What other defensive steps should the United States take?*
3. *Is the United States in a jihad against various Islamic nations? Why, or why not?*
4. *What do you think about the author's ideas for securing the nation?*

GLOSSARY **Apocalyptic** Predicting widespread devastation or doom. **Messianic** Relating to a belief that a cause or leader is destined to save the world.

In the minds of the men who carried them out, the attacks of September 11 were acts of religious devotion—a form of worship, conducted in God's name and in accordance with his wishes. The enemy was the infidel; the opposing ideology, "Western culture." That religious motivation, colored by a messianism and in some cases an apocalyptic vision of the future, distinguishes al-Qaida and its affiliates from conventional terrorists groups such as the Irish Republican Army, the Red Brigades, or even the Palestine Liberation Organization. Although secular political interests help drive al-Qaida's struggle for power, these interests are understood and expressed in religious terms. Al-Qaida wants to purge the Middle East of American political, military, and economic influence, but only as part of a far more sweeping religious agenda: a "defensive jihad" to defeat a rival system portrayed as an existential threat to Islam.

The explicitly religious character of the "New Terrorism" poses a profound security challenge for the United States. The social, economic, and political conditions in the Arab and broader Islamic world that have helped give rise to al-Qaida will not be easily changed. The maximalist demands of the new terrorists obviate dialogue or negotiation. Traditional strategies of deterrence by retaliation are unlikely to work because the jihadists have no territory to hold at risk, seek sacrifice, and court Western attacks that will validate their claims about Western hostility to Islam. The United States will instead need to pursue a strategy of containment,

SOURCE: From *The New Era of Terrorism: Selected Readings*, Gus Martin, ed. *The Brookings Review*, 211 (Winter 2003). Reprinted with permission of The Brookings Institute.

while seeking ways to redress, over the long run, underlying causes.

THE FABRIC OF NEW TERRORISM

Religiously motivated terrorism, as Bruce Hoffman of the RAND Corporation first noted in 1997, is inextricably linked to pursuit of mass casualties. The connection is rooted in the sociology of biblical religion. Monotheistic faiths are characterized by exclusive claims to valid identity and access to salvation. The violent imagery embedded in their sacred texts and the centrality of sacrifice in their liturgical traditions establish the legitimacy of killing as an act of worship with redemptive qualities. In these narratives, the enemy must be eradicated, not merely suppressed.

In periods of deep cultural despair, eschatology—speculation in the form of apocalyptic stories about the end of history and dawn of the kingdom of God—can capture the thinking of a religious group. History is replete with instances in which religious communities—Jewish, Christian, Islamic—immolated themselves and perpetrated acts of intense violence to try to spur the onset of a messianic era. Each community believed it had reached the nadir of degradation and was on the brink of a resurgence that would lead to its final triumph over its enemies—a prospect that warranted and required violence on a massive scale. . . .

THE DOCTRINAL POTENCY OF AL-QAIDA

Similar thinking can be detected in narrative trends that inform al-Qaida's ideology and actions. Apocalyptic tales circulating on the web and within the Middle East in hard copy tell of cataclysmic battles between Islam and the United States, Israel, and sometimes Europe. Global battles see-saw between infidel and Muslim victory until some devastating act, often the destruction of New York by nuclear weapons, brings Armageddon to an end and leads the world's survivors to convert to Islam.

The theological roots of al-Qaida's leaders hark back to a medieval Muslim jurisconsult, Taqi al Din Ibn Taymiyya, two of whose teachings have greatly influenced Islamic revolutionary movements. The first was his elevation of jihad—not the spiritual struggle that many modern Muslims take it to be, but physical combat against unbelievers—to the rank of the canonical five pillars of Islam (declaration of faith, prayer, almsgiving, self-purification, and pilgrimage to Mecca). The second was his legitimization of rebellion against Muslim rulers who do not enforce *sharia*, or Islamic law, in their domains. . . .

Al-Qaida embodies both the Egyptian and Saudi sides of the jihad movement, which came together in the 1960s when some Egyptian militants sought shelter in Saudi Arabia, which was locked in conflict with Nasserist Egypt. Osama bin Laden himself is a Saudi, and his second-in-command, Ayman al Zawahiri, an Egyptian who served three years in prison for his role in Sadat's assassination.

The jihadist themes in Ibn Taymiyya's teachings are striking an increasingly popular chord in parts of the Muslim world.

AL-QAIDA'S GEOPOLITICAL REACH

Religiously motivated militants have now dispersed widely to multiple "fields of jihad." The social problems that have fueled their discontent are well known—low economic growth, falling wages and increasing joblessness, poor schooling, relentless but unsustainable urban growth, and diminishing environmental resources, especially water. Political alienation and resentment over the intrusion into traditional societies of offensive images, ideas, and commercial products compound these problems and help account for the religious voice given to these primarily secular grievances. The mobilization of religious imagery and terminology further transforms secular issues into substantively religious ones, putting otherwise negotiable political issues beyond the realm of bargaining and making violent outcomes more likely.

The political power of religious symbols has led some pivotal states, in particular Egypt and Saudi Arabia, to use them to buttress their own legitimacy. In so doing they perversely confer authority on the very clerical opposition that threatens state power and impedes the modernization programs that might, over the long haul, materially improve quality of life. Although the jihadists are unable to challenge these states, Islamists nevertheless dominate public discourse and shape the debate on foreign and domestic policy.... In Egypt and Saudi Arabia, Islamists have inextricably intertwined the near and far enemies. The governments' need to cater to the sentiments aroused within mosques and on the Islamist airwaves to keep their regimes secure dictates their tolerance or even endorsement of extreme anti-American views. At the same time, strategic circumstances compel both states to provide diplomatic or other practical support for U.S. policies that offend public sensitivities. It is small wonder that Egyptians and Saudis are the backbone of al-Qaida and that Saudi Arabia spawned most of the September 11 attackers.

The fields of jihad stretch far and wide. In the Middle East, al-Qaida developed ties in Lebanon and Jordan. In Southeast Asia, Indonesians, Malaysians, and Singaporeans trained in Afghanistan, or conspired with those who had, to engage in terror, most horrifically the bombing in Bali. In Central Asia, the Islamic Movement of Uzbekistan became a full-fledged jihadist group. In Pakistan, jihadists with apocalyptic instincts nearly provoked a nuclear exchange between India and Pakistan. Videotapes of atrocities of the Algerian Armed Islamic Group circulate in Europe as recruitment propaganda for the global jihad.

Given its role as a springboard for the September 11 attacks, Europe may be the most crucial field of jihad. Lack of political representation and unequal access to education, jobs, housing, and social services have turned European Muslim youth against the states in which they live. In the United Kingdom, the Muslim prison population, a source of recruits for the radical cause, has doubled in the past decade. Close to a majority of young Muslims in Britain have told pollsters that they feel no obligation to bear arms for England but would fight for bin Laden.

The United States remains al-Qaida's prime target. Suleiman Abu Ghaith, the al-Qaida spokesman, has said that there can be no truce until the group has killed four million Americans, whereupon the rest can convert to Islam.

THE RECALCITRANCE OF THE JIHADISTS

How should the United States respond to the jihadist threat? To the extent one can speak of the root causes of the new terrorism, they defy direct and immediate remedial action. Population in the Middle East is growing rapidly, and the median age is dropping. The correlation between youth and political instability highlights the potential for unrest and radicalization. In cities, social welfare programs, sanitation, transportation, housing, power, and the water supply are deteriorating. In much of the Muslim world, the only refuge from filth, noise, heat, and, occasionally, surveillance is the mosque. Economists agree that the way out of the morass is to develop institutions that facilitate the distribution of capital and create opportunity; how to do that, they are unsure. The West can offer aid but cannot correct structural problems.

Improving public opinion toward the United States is also deeply problematic. Decades of official lies and controlled press have engendered an understandable skepticism toward the assertions of any government, especially one presumed hostile to Muslim interests. Trust is based on confidence in a chain of transmission whose individual links are known to be reliable. Official news outlets or government spokespersons do not qualify as such links. Nor, certainly, do Western news media.

Moreover, highly respected critics of the United States in Saudi Arabia demonstrate an ostensibly profound understanding of U.S. policies and society, while offering a powerful and internally consistent explanation for their country's descent from the all-powerful, rich supplier of oil to the West to a debt-ridden, faltering economy protected by Christian troops and kowtowing to Israel. These are difficult narratives to counter, especially in a society where few know much about the West.

The prominent role of clerics in shaping public opinion offers yet more obstacles. The people who represent the greatest threat of terrorist action against the United States follow the preaching and guidance of Salafi clerics—the Muslim equivalent of Christian "fundamentalists." Although some Salafi preachers have forbidden waging jihad as harmful to Muslim interests, their underlying assumptions are that jihad qua holy war against non-Muslims is fundamentally valid and that Islamic governments that do not enforce *sharia* must be opposed. No authoritative clerical voice offers a sympathetic view of the United States.

The prognosis regarding root causes, then, is poor. The world is becoming more religious; Islam is the fastest-growing faith; religious expression is generally becoming more assertive and apocalyptic thinking more prominent. Weapons of mass destruction, spectacularly suited to cosmic war, are becoming more widely available. Democratization is at a standstill. Governments in Egypt, Saudi Arabia, Pakistan, and Indonesia are unwilling or unable to oppose anti-Western religiously based popular feeling. Immigration, conversion, and inept social policies will intensify parallel trends in Europe.

At least for now, dialogue does not appear to be an option. Meanwhile, global market forces beyond the control of Western governments hasten Western cultural penetration and generate ever-greater resentment. Jihadists could conceivably argue that they have a negotiable program; cessation of U.S. support for Israel, withdrawal from Saudi Arabia, broader American disengagement from the Islamie world. But U.S. and allied conceptions of international security and strategic imperatives will make such demands difficult, if not impossible to accommodate.

REDUCING VULNERABILITY TO NEW TERRORISM

Facing a global adversary with maximal goals and lacking a bargaining option or means to redress severe conditions that may or may not motivate attackers, the United States is confined primarily to a strategy of defense, deterrence by denial, and, where possible and prudent, preemption. Deterrence through the promise of retaliation is impossible with an adversary that controls little or no territory and invites attack.

Adjusting to the new threat entails disturbing conceptual twists for U.S. policymakers. After generations of effort to reduce the risk of surprise attack through technical means and negotiated transparency measures, surprise will be the natural order of things. The problem of warning will be further intensified by the creativity of this adversary, its recruitment of Europeans and Americans, and its ability to stage attacks from within the United States. Thinking carefully about the unlikely—"institutionalizing imaginativeness," as Dennis Gormley has put it—is by definition a paradox, but nonetheless essential for American planners.

With warning scarce and inevitably ambiguous, it will be necessary to probe the enemy both to put him off balance and to learn of his intentions. The United States has done so clandestinely against hostile intelligence agencies, occasionally with remarkable results. Against al-Qaida, a more difficult target, the approach will take time to cohere. Probes could also take the form of military action against al-Qaida-affiliated cantonments, where they still exist. The greater the movement's virtuality, however, the fewer such targets will be available for U.S. action. Preemptive strikes could target sites that develop, produce, or deploy weapons of mass destruction.

A decade of al-Qaida activity within the United States has erased the customary distinction between the domestic and the foreign in intelligence and law enforcement. The relationship between the Central Intelligence Agency and the Federal Bureau of Investigation must change. Only a more integrated organization can adapt to the seamlessness of the transnational arenas in which the terrorists operate.

Civil liberties and security must be rebalanced. How sweeping the process turns out to be will depend largely on whether the nation suffers another attack or at least a convincing attempt. Americans will have to be convinced that curtailing civil liberties is unavoidable and limited to the need to deal with proximate threats. They will need to see bipartisan consensus in Congress and between Congress and the White House and be sure that politicians are committed to keeping the rebalancing to a minimum. . . .

The pursuit of public-private partnership will have to be extended to all potentially vulnerable critical infrastructures by a government that does not yet understand perfectly which infrastructures are truly critical and which apparently dispensable infrastructures interact to become critical. . . .

The United States must also devise ways to block or intercept vehicles that deliver weapons of mass destruction. It cannot do that alone. The cruise missile threat, for instance, requires the cooperation of suppliers, which means an active American role in expanding the remit of the Missile Technology Control Regime (MTCR). . . .

Offensive opportunities will be limited but not impossible. They do, however, require impeccable intelligence, which has been hard to come by. The Afghan nexus in which jihadis initially came together and the cohesion of the groups that constitute the al-Qaida movement have made penetration forbiddingly complicated. But as al-Qaida picks up converts to Islam and Muslims who have long resided in Western countries, penetration may become easier. The more they look like us, the more we look like them. . . .

Without revoking the longstanding executive order prohibiting assassination, the United States should also consider targeted killing, to use the Israeli phrase, of jihadists known to be central to an evolving conspiracy to attack the United States or to obtain weapons of mass destruction. As a practical matter, the intelligence value of such a person alive would generally outweigh the disruptive benefits of his death, assuming that U.S. or friendly intelligence services could be relied on to keep him under surveillance. But this will not always be so. When it is not, from a legal standpoint, targeted killing falls reasonably under the right to self-defense. Such a policy departure is unsavory. But in a new strategic context, with jihadis intent on mass casualties, unsavory may not be a sensible threshold.

ALLIED COOPERATION

As the al-Qaida movement dissolves into virtuality in 60 countries worldwide, international cooperation becomes ever more indispensable to countering the threat.

Many countries that host al-Qaida will cooperate with the United States out of self-interest; they do not want jihadis on their soil any more than Americans do on theirs. A durable and effective counterterrorism campaign, however, requires not just bare-bones cooperation, but political collaboration at a level that tells the bureaucracies that cooperation with their American counterparts is expected. Such a robust, wholesale working relationship is what produces vital large-scale initiatives—a common diplomatic approach toward problem states; a sustainable program of economic development for the Middle East, domestic policy reforms that lessen the appeal of jihadism to Muslim diaspora communities, improved border controls; and tightened bonds among the justice ministries, law enforcement, customs, and intelligence agencies, and special operations forces on the front lines.

Whether this level of burden sharing emerges, let alone endures, depends on the give and take among the players. Since September 11, the United States has fostered allied perceptions that Washington is indifferent to their priorities. Apart from slow progress toward a UN Security Council resolution on Iraq, the United States has not yet paid a serious penalty in terms of allied cooperation. The scale of the attacks and the administration's blend of resolve and restraint in the war on terrorism have offset allies' disappointment in its go-it-alone posture. But as the war grinds on, good will is certain to wear thin. The United States would be wise to forgo some of its own trade- and treaty-related preferences, at least in short term, to ensure allied support in the crises that will inevitably come.

Washington's interests would also be well served by modifying what appears at times to be a monolithic view of terrorist networks that equates the Arafats and Saddams of the world with bin Laden (or his successors). Several European partners regard Arafat and his ilk as considerably more controllable through diplomacy than bin Laden and view countries such as Iran, which has used terrorism against the United States, as amenable to "constructive dialogue." Greater American flexibility may prove essential for ensuring European capitals' military, law-enforcement, and intelligence cooperation. And the fact remains that al-Qaida has killed more Americans than have Iraq, Iran, or Palestinian groups and would use weapons of mass destruction against the United States as soon as it acquired them.

ISRAEL AND THE PALESTINIANS

Since the heyday of the Middle East peace process under Ehud Barak's Labor government, jihadists have exploited the Israeli-Palestinian conflict to boost their popularity. The strategem has worked: jihadists are seen as sticking up for Palestinian rights, while Arab governments do nothing. Direct, energetic U.S. diplomatic intervention in the conflict would lessen the appeal of jihadi claims and make it marginally easier for regional governments to cooperate in the war on terrorism by demonstrating American concern for the plight of Palestinians.

The Bush administration fears becoming entangled in a drawn out, venomous negotiation between irreconcilable parties. They see it distracting them from higher priorities and embroiling them in domestic political disputes over whether Washington should pressure Israel. Still, the administration has been drawn in by degrees and has announced its support for creating a Palestinian state. If the war on terrorism is now the highest U.S. priority, then more vigorous—and admittedly risky—involvement in the Israeli-Palestinian conflict is required. The jihadi argument that the United States supports the murder of Palestinian Muslims must be defanged.

DEMOCRATIZATION IN THE MIDDLE EAST

If it continues to engage with the authoritarian regimes in Cairo and Riyadh, Washington should try to renegotiate the implicit bargain that underpins its relations with both. The current bargain is structured something like this: Egypt sustains its commitment to peace with Israel, Saudi Arabia stabilizes oil prices, and both proffer varying degrees of diplomatic support for American objectives in the region, especially toward Iraq. In return, Washington defers to their domestic policies, even if these fuel the growth and export of Islamic militancy and deflect public discontent onto the United States and Israel. With jihadis now pursuing nuclear weapons, that bargain no longer looks sensible.

Under a new bargain, Cairo and Riyadh would begin to take measured risks to lead their publics gradually toward greater political responsibility and away from Islamist thinking (and action) by encouraging opposition parties of a more secular cast and allowing greater freedom of expression. Saudi Arabia would throttle back on its wahhabiization of the Islamic world by cutting its production and export of unemployable graduates in religious studies and reducing subsidies for foreign mosques and madrassas that propagate a confrontational and intolerant form of Islam while crowding out alternative practices. Both countries would be pushed to reform their school curricula—and enforce standards—to ensure a better understanding of the non-Islamic world and encourage respect for other cultures. With increased financial and technical assistance from the West, regimes governing societies beset by economic problems that spur radicalism would focus more consistently on the welfare of their people. In this somewhat utopian conception, leaders in both countries would use their newly won credibility to challenge Islamist myths about America and the supposed hostility of the West toward Islam. In sum, Cairo and Riyadh would challenge the culture of demonization across the board, with an eye toward laying the groundwork for liberal democracy.

In the framework of this new bargain, the United States would establish contacts with moderate opposition figures in Egypt, Saudi Arabia, and perhaps other countries. The benefit would be twofold. Washington would get a better sense of events on the ground and would also gain credibility and perhaps even understanding on the part of critics. For this effort to bear fruit, however, the United States would have to use regional media efficiently—something for which it has as yet no well-developed strategy. Washington would also have to engage in a measure of self scrutiny and explore ways in which its policies contribute—in avoidable ways—to Muslim anti-Americanism. "Rebranding" is not enough.

Change will be slow. The regimes in Cairo and Riyadh face largely self-inflicted problems they cannot readily surmount without serious risks to stability. Nor is the United States entirely free to insist on the new bargain: it will need Saudi cooperation on Iraq as long as Saddam Hussein is in power, if not longer, given the uncertainties surrounding Iraq's future after Saddam leaves the stage. Egyptian support for a broader Arab-Israeli peace will also remain essential. But change has to start sometime, somewhere. It will take steady U.S. pressure and persistent attempts to persuade both regimes that a new bargain will serve their countries' long-term interests. The sooner the new deals are struck, the better.

Hazardous but Not Hopeless

Western democracies face a serious, possibly transgenerational terrorist threat whose causes are multidimensional and difficult to address. The situation is hazardous, but not hopeless. The United States possesses enormous wealth, has capable allies, and stands on the leading edge of technological development that will be key to survival. A strategy that takes into account the military, intelligence, law-enforcement, diplomatic, and economic pieces of the puzzle will see America through. For the next few years, the objective will be to contain the threat, in much the same way that the United States contained Soviet power throughout the Cold War. The adversary must be prevented from doing his worst, while Washington and its allies wear down its capabilities and undermine its appeal to fellow Muslims. Success will require broad domestic support and a strong coalition abroad.

Prospects are, in many respects, bleak. But the dangers are not disproportionate to those the nation faced in the 20th-century. America's initial reaction to September 11 was and indeed had to be its own self-defense: bolstering homeland security, denying al-Qaida access to failing or hostile states, dismantling networks, and developing a law-enforcement and intelligence network able to better cope with the new adversary. Not all vulnerabilities can be identified and even fewer remedied, and al-Qaida need launch only one attack with a weapon of mass destruction to throw the United States into a profound crisis. Washington and its partners must convince Muslim populations that they can prosper without either destroying the West or abandoning their own traditions to the West's alien culture. That is a long-term project. American and allied determination in a war against apocalyptic—and genocidal—religious fanatics must be coupled with a generous vision about postwar possibilities. Militant Islam cannot be expected to embrace the West in the foreseeable future. But the United States can lay the foundation for a lasting accommodation by deploying its considerable economic and political advantages. It is not too late to begin.

59

The End of War?

GREGG EASTERBROOK

Considering the fact that almost all current news headlines are about conflicts in different parts of the world, the title of this reading must come as a bit of surprise.

As you read this article, ask yourself the following questions as a guide:

1. *Do you believe the claims of the author. Why, or why not?*
2. *How do you think your friends will react to the idea that war is coming to an end?*
3. *How will the end of war affect your political beliefs and activities?*

Daily explosions in Iraq, massacres in Sudan, the Koreas staring at each other through artillery barrels, a Hobbesian war of all against all in eastern Congo—combat plagues human society as it has, perhaps, since our distant forebears realized that a tree limb could be used as a club. But here is something you would never guess from watching the news: War has entered a cycle of decline. Combat in Iraq and in a few other places is an exception to a significant global trend that has gone nearly unnoticed—namely that, for about 15 years, there have been steadily fewer armed conflicts worldwide. In fact, it is possible that a person's chance of dying because of war has, in the last decade or more, become the lowest in human history.

Five years ago, two academics—Monty Marshall, research director at the Center for Global Policy at George Mason University, and Ted Robert Gurr, a professor of government at the University of Maryland—spent months compiling all available data on the frequency and death toll of twentieth-century combat, expecting to find an ever-worsening ledger of blood and destruction. Instead, they found, after the terrible years of World Wars I and II, a global increase in war from the 1960s through the mid-'80s. But this was followed by a steady, nearly uninterrupted decline beginning in 1991. They also found a steady global rise since the mid-'80s in factors that reduce armed conflict—economic prosperity, free elections, stable central governments, better communication, more "peacemaking institutions," and increased international engagement. Marshall and Gurr, along with Deepa Khosla, published their results as a 2001 report, *Peace and Conflict*, for the Center for International Development and Conflict Management at the University of Maryland. At the time, I remember reading that report and thinking, "Wow, this is one of the hottest things I have ever held in my hands." I expected that evidence of a decline in war would trigger a sensation. Instead it received almost no notice.

"After the first report came out, we wanted to brief some United Nations officials, but everyone at the United Nations just laughed at us. They could not believe war was declining, because this went against political expectations," Marshall says. Of course, 2001 was the year of September 11. But, despite the battles in Afghanistan, the Philippines, and elsewhere that were ignited by Islamist terrorism and the West's response, a second edition of

SOURCE: Originally printed in *The New Republic*, May 30, 2005, pp. 18–21.

Peace and Conflict, published in 2003, showed the total number of wars and armed conflicts continued to decline. A third edition of the study, published last week, shows that, despite the invasion of Iraq and other outbreaks of fighting, the overall decline of war continues. This even as the global population keeps rising, which might be expected to lead to more war, not less. . . .

First, the numbers. The University of Maryland studies find the number of wars and armed conflicts worldwide peaked in 1991 at 51, which may represent the most wars happening simultaneously at any point in history. Since 1991, the number has fallen steadily. There were 26 armed conflicts in 2000 and 25 in 2002, even after the al Qaeda attack on the United States and the U.S. counterattack against Afghanistan. By 2004, Marshall and Gurr's latest study shows, the number of armed conflicts in the world had declined to 20, even after the invasion of Iraq. All told, there were less than half as many wars in 2004 as there were in 1991.

Marshall and Gurr also have a second ranking, gauging the magnitude of fighting. This section of the report is more subjective. Everyone agrees that the worst moment for human conflict was World War II; but how to rank, say, the current separatist fighting in Indonesia versus, say, the Algerian war of independence is more speculative. Nevertheless, the *Peace and Conflict* studies name 1991 as the peak post–World War II year for totality of global fighting, giving that year a ranking of 179 on a scale that rates the extent and destructiveness of combat. By 2000, in spite of war in the Balkans and genocide in Rwanda, the number had fallen to 97; by 2002 to 81; and, at the end of 2004, it stood at 65. This suggests the extent and intensity of global combat is now less than half what it was 15 years ago.

How can war be in such decline when evening newscasts are filled with images of carnage? One reason fighting seems to be everywhere is that, with the ubiquity of 24-hour cable news and the Internet, we see many more images of conflict than before. A mere decade ago, the rebellion in Eritrea occurred with almost no world notice; the tirelessly globe-trotting Robert Kaplan wrote of meeting with Eritrean rebels who told him they hoped that at least spy satellites were trained on their region so that someone, somewhere, would know of their struggle. Today, fighting in Iraq, Sudan, and other places is elaborately reported on, with a wealth of visual details supplied by minicams and even camera-enabled cell phones. News organizations must prominently report fighting, of course. But the fact that we now see so many visuals of combat and conflict creates the impression that these problems are increasing: Actually, it is the reporting of the problems that is increasing, while the problems themselves are in decline. Television, especially, likes to emphasize war because pictures of fighting, soldiers, and military hardware are inherently more compelling to viewers than images of, say, water-purification projects. Reports of violence and destruction are rarely balanced with reports about the overwhelming majority of the Earth's population not being harmed. . . .

With war now in decline, for the moment men and women worldwide stand in more danger from cars and highways than from war and combat. World Health Organization statistics back this: In 2000, for example, 300,000 people died in combat or for war-related reasons (such as disease or malnutrition caused by war), while 1.2 million worldwide died in traffic accidents. That 300,000 people perished because of war in 2000 is a terrible toll, but it represents just .005 percent of those alive in that year.

This low global risk of death from war probably differs greatly from most of the world's past. In prehistory, tribal and small-group violence may have been endemic. Steven LeBlanc, a Harvard University archeologist, asserts in his 2003 book about the human past, *Constant Battles*, that warfare was a steady feature of primordial society. LeBlanc notes that, when the aboriginal societies of New Guinea were first observed by Europeans in the 1930s, one male in four died by violence; traditional New Guinean society was organized around endless tribal combat. Unremitting warfare characterized much of the history of Europe, the Middle

East, and other regions; perhaps one-fifth of the German population died during the Thirty Years War, for instance. Now the world is in a period in which less than one ten-thousandth of its population dies from fighting in a year. The sheer number of people who are *not* being harmed by warfare is without precedent.

Next consider a wonderful fact: Global military spending is also in decline. Stated in current dollars, annual global military spending peaked in 1985, at $1.3 trillion, and has been falling since, to slightly over $1 trillion in 2004, according to the Center for Defense Information, a nonpartisan Washington research organization. Since the global population has risen by one-fifth during this period, military spending might have been expected to rise. Instead, relative to population growth, military spending has declined by a full third. In current dollars, the world spent $260 per capita on arms in 1985 and $167 in 2004.

The striking decline in global military spending has also received no attention from the press, which continues to promote the notion of a world staggering under the weight of instruments of destruction. Only a few nations, most prominently the United States, have increased their defense spending in the last decade. Today, the United States accounts for 44 percent of world military spending; if current trends continue, with many nations reducing defense spending while the United States continues to increase such spending as its military is restructured for new global anti-terrorism and peacekeeping roles, it is not out of the question that, in the future, the United States will spend more on arms and soldiers than the rest of the world combined.

Declining global military spending is exactly what one would expect to find if war itself were in decline. The peak year in global military spending came only shortly before the peak year for wars, 1991. There's an obvious chicken-or-egg question, whether military spending has fallen because wars are rarer or whether wars are rarer because military spending has fallen. Either way, both trend lines point in the right direction. This is an extremely favorable development, particularly for the world's poor—the less developing nations squander on arms, the more they can invest in improving daily lives of their citizens.

What is causing war to decline? The most powerful factor must be the end of the cold war, which has both lowered international tensions and withdrawn U.S. and Soviet support from proxy armies in the developing world. Fighting in poor nations is sustained by outside supplies of arms. To be sure, there remain significant stocks of small arms in the developing world—particularly millions of assault rifles. But, with international arms shipments waning and heavy weapons, such as artillery, becoming harder to obtain in many developing nations, factions in developing-world conflicts are more likely to sue for peace. For example, the long, violent conflict in Angola was sustained by a weird mix of Soviet, American, Cuban, and South African arms shipments to a potpourri of factions. When all these nations stopped supplying arms to the Angolan combatants, the leaders of the factions grudgingly came to the conference table.

During the cold war, Marshall notes, it was common for Westerners to say there was peace because no fighting affected the West. Actually, global conflict rose steadily during the cold war, but could be observed only in the developing world. After the cold war ended, many in the West wrung their hands about a supposed outbreak of "disorder" and ethnic hostilities. Actually, both problems went into decline following the cold war, but only then began to be noticed in the West, with confrontation with the Soviet empire no longer an issue.

Another reason for less war is the rise of peacekeeping. The world spends more every year on peacekeeping, and peacekeeping is turning out to be an excellent investment. Many thousands of U.N., NATO, American, and other soldiers and peacekeeping units now walk the streets in troubled parts of the world, at a cost of at least $3 billion annually. Peacekeeping has not been without its problems; peacekeepers have been accused of paying very young girls for sex in Bosnia and Africa, and NATO bears collective shame for refusing support to the Dutch peacekeeping unit that might have prevented the Srebrenica massacre of 1995. But, overall, peacekeeping is working. Dollar for dollar, it is far more effective at preventing fighting than purchasing complex weapons systems. A recent study from the notoriously gloomy RAND Corporation found that most U.N. peacekeeping efforts have been successful.

Peacekeeping is just one way in which the United Nations has made a significant contribution to the decline of war. American commentators love to disparage the organization in that big cereal-box building on the East River, and, of course, the United Nations has manifold faults. Yet we should not lose track of the fact that the global security system envisioned by the U.N. charter appears to be taking effect. Great-power military tensions are at the lowest level in centuries; wealthy nations are increasingly pressured by international diplomacy not to encourage war by client states; and much of the world respects U.N. guidance. Related to this, the rise in "international engagement," or the involvement of the world community in local disputes, increasingly mitigates against war.

The spread of democracy has made another significant contribution to the decline of war. In 1975, only one-third of the world's nations held true multiparty elections; today two-thirds do, and the proportion continues to rise. In the last two decades, some 80 countries have joined the democratic column, while hardly any moved in the opposite direction. Increasingly, developing-world leaders observe the simple fact that the free nations are the strongest and richest ones, and this creates a powerful argument for the expansion of freedom. Theorists at least as far back as Immanuel Kant have posited that democratic societies would be much less likely to make war than other kinds of states. So far, this has proved true: Democracy-against-democracy fighting has been extremely rare. Prosperity and democracy tend to be mutually reinforcing. Now prosperity is rising in most of the world, amplifying the trend toward freedom. As ever-more nations become democracies, ever-less war can be expected, which is exactly what is being observed.

For the great-power nations, the arrival of nuclear deterrence is an obvious factor in the decline of war. The atomic bomb debuted in 1945, and the last great-power fighting, between the United States and China, concluded not long after, in 1953. From 1871 to 1914, Europe enjoyed nearly half a century without war; the current 52-year great-power peace is the longest period without great-power war since the modern state system emerged. Of course, it is possible that nuclear deterrence will backfire and lead to a conflagration beyond imagination in its horrors. But, even at the height of the cold war, the United States and the Soviet Union never seriously contemplated a nuclear exchange. If it didn't happen then it seems unlikely for the future.

In turn, lack of war among great nations sets an example for the developing world. When the leading nations routinely attacked neighbors or rivals, governments and emerging states dreamed of the day when they, too, could issue orders to armies of conquest. Now that the leading nations rarely use military force—and instead emphasize economic competition—developing countries imitate the model. This makes the global economy more turbulent, but reduces war. . . .

Is it possible to believe that war is declining, owing to the spread of enlightenment? This seems the riskiest claim. Human nature has let us down many times before. Some have argued that militarism as a philosophy was destroyed in World War II, when the states that were utterly dedicated to martial organization and violent conquest were not only beaten but reduced to rubble by free nations that initially wanted no part of the fight. World War II did represent the triumph of freedom over militarism. But memories are short: It is unrealistic to suppose that no nation will ever be seduced by militarism again.

Yet the last half-century has seen an increase in great nations acting in an enlightened manner toward one another. Prior to this period, the losing sides in wars were usually punished; consider the Versailles Treaty, whose punitive terms helped set in motion the Nazi takeover of Germany. After World War II, the victors did not punish Germany and Japan, which made reasonably smooth returns to prosperity and acceptance by the family of nations. Following the end of the cold war, the losers—the former Soviet Union and China—have seen their national conditions improve, if fitfully; their reentry into the family of nations has gone reasonably well and has been encouraged, if not actively aided, by their former adversaries. Not punishing the vanquished should diminish the odds of future war, since there are no generations who suffer from the victor's terms, become bitter, and want vengeance. . . .

As recently as the Civil War in the United States and World War I in Europe, it was common to view war as inevitable and to be fatalistic about the power of government to order men to march to their deaths. A spooky number of thinkers even adulated war as a desirable condition. Kant, who loved democracy, nevertheless wrote that war is "sublime" and that "prolonged peace favors the predominance of a mere commercial spirit, and with it a debasing self-interest, cowardice and effeminacy." Alexis De Tocqueville said that war "enlarges the mind of a people." Igor Stravinsky called war "necessary for human progress." In 1895, Oliver Wendell Holmes Jr. told the graduating class of Harvard that one of the highest expressions of honor was "the faith . . . which leads a soldier to throw away his life in obedience to a blindly accepted duty."

Around the turn of the twentieth century, a counter-view arose—that war is usually absurd. One of the best-selling books of late-nineteenth-century Europe, *Lay Down Your Arms!*, was an antiwar novel. Organized draft resistance in the United Kingdom during World War I was a new force in European politics. England slept during the '30s in part because public antiwar sentiment was intense. By the time the U.S. government abolished the draft at the end of the Vietnam War, there was strong feeling in the United States that families would no longer tolerate being compelled to give up their children for war. Today, that feeling has spread even to Russia, such a short time ago a totalitarian, militaristic state. As average family size has decreased across the Western world, families have invested more in each child; this should discourage militarism. Family size has started to decrease in the developing world, too, so the same dynamic may take effect in poor nations.

There is even a chance that the ascent of economics to its pinnacle position in modern life reduces war. Nations interconnected by trade may be less willing to fight each other: If China and the United States ever fought, both nations might see their economies collapse. It is true that, in the decades leading up to World War I, some thought rising trade would prevent war. But today's circumstances are very different from those of the fin de siècle. Before World War I, great powers still maintained the grand illusion that there could be war without general devastation; World Wars I and II were started by governments that thought they could come out ahead by fighting. Today, no major government appears to believe that war is the best path to nationalistic or monetary profit; trade seems much more promising.

The late economist Julian Simon proposed that, in a knowledge-based economy, people and their brainpower are more important than physical resources, and thus the lives of a country's citizens are worth more than any object that might be seized in war. Simon's was a highly optimistic view—he assumed governments are grounded in reason—and yet there is a chance this vision will be realized. Already, most Western nations have achieved a condition in which citizens' lives possess greater economic value than any place or thing an army might gain by combat. As knowledge-based economics spreads throughout the world, physical resources may mean steadily less, while life means steadily more. That's, well, enlightenment.

In his 1993 book, *A History of Warfare*, the military historian John Keegan recognized the early signs that combat and armed conflict had entered a cycle of decline. War "may well be ceasing to commend itself to human beings as a desirable or productive, let alone rational, means of reconciling their discontents," Keegan wrote. Now there are 15 years of positive developments supporting the idea. Fifteen years is not all that long. Many things could still go badly wrong; there could be ghastly surprises in store. But, for the moment, the trends have never been more auspicious: Swords really are being beaten into plowshares and spears into pruning hooks. The world ought to take notice.

60

If Hitler Asked You to Electrocute a Stranger, Would You? Probably

PHILIP MEYER

Many have wondered how a former corporal (Hitler) could manage to influence so many people and get them to commit atrocities such as those that occurred in the concentration camps. This reading reveals the roles of authority and charisma in those decisions.

As you read, ask yourself the following questions:

1. *What were two of the findings of the Milgram study that were surprising?*
2. *In the position of the testees, how would you have reacted? Why?*
3. *Do the findings explain the violence in the world? Some of it? Why?*

GLOSSARY **Pathological** Disordered in behavior. **Macabre** Gruesome or horrible. **Sadistic** Deriving pleasure from inflicting pain.

In the beginning, Stanley Milgram was worried about the Nazi problem. He doesn't worry much about the Nazis anymore. He worries about you and me, and, perhaps, himself a little bit too.

Stanley Milgram is a social psychologist, and when he began his career at Yale University in 1960 he had a plan to prove, scientifically, that Germans are different. The Germans-are-different hypothesis has been used by historians, such as William L. Shirer, to explain the systematic destruction of the Jews by the Third Reich. One madman could decide to destroy the Jews and even create a master plan for getting it done. But to implement it on the scale that Hitler did meant that thousands of other people had to go along with the scheme and help to do the work. The Shirer thesis, which Milgram set out to test, is that Germans have a basic character flaw which explains the whole thing, and this flaw is a readiness to obey authority without question, no matter what outrageous acts the authority commands.

The appealing thing about this theory is that it makes those of us who are not Germans feel better about the whole business. Obviously, you and I

SOURCE: Reprinted by permission from *Esquire*, February 1970. © Esquire, Inc.

are not Hitler, and it seems equally obvious that we would never do Hitler's dirty work for him. But now, because of Stanley Milgram, we are compelled to wonder. Milgram developed a laboratory experiment which provided a systematic way to measure obedience. His plan was to try it out in New Haven on Americans and then go to Germany and try it out on Germans. He was strongly motivated by scientific curiosity, but there was also some moral content in his decision to pursue this line of research, which was, in turn, colored by his own Jewish background. If he could show that Germans are more obedient than Americans, he could then vary the conditions of the experiment and try to find out just what it is that makes some people more obedient than others. With this understanding, the world might, conceivably, be just a little bit better.

But he never took his experiment to Germany. He never took it any farther than Bridgeport. The first finding, also the most unexpected and disturbing finding, was that we Americans are an obedient people: not blindly obedient, and not blissfully obedient, just obedient. "I found so much obedience," says Milgram softly, a little sadly, "I hardly saw the need for taking the experiment to Germany."

There is something of the theater director in Milgram, and his technique, which he learned from one of the old masters in experimental psychology, Solomon Asch, is to stage a play with every line rehearsed, every prop carefully selected, and everybody an actor except one person. That one person is the subject of the experiment. The subject, of course, does not know he is in a play. He thinks he is in real life. The value of this technique is that the experimenter, as though he were God, can change a prop here, vary a line there, and see how the subject responds. Milgram eventually had to change a lot of the script just to get people to stop obeying. They were obeying so much, the experiment wasn't working—it was like trying to measure oven temperature with a freezer thermometer.

The experiment worked like this: If you were an innocent subject in Milgram's melodrama, you read an ad in the newspaper or received one in the mail asking for volunteers for an educational experiment. The job would take about an hour and pay $4.50. So you make an appointment and go to an old Romanesque stone structure on High Street with the imposing name of The Yale Interaction Laboratory. It looks something like a broadcasting studio. Inside, you meet a young, crew-cut man in a laboratory coat who says he is Jack Williams, the experimenter. There is another citizen, fiftyish, Irish face, an accountant, a little overweight, and very mild and harmless-looking. This other citizen seems nervous and plays with his hat while the two of you sit in chairs side by side and are told that the $4.50 checks are yours no matter what happens. Then you listen to Jack Williams explain the experiment.

It is about learning, says Jack Williams in a quiet, knowledgeable way. Science does not know much about the conditions under which people learn and this experiment is to find out about negative reinforcement. Negative reinforcement is getting punished when you do something wrong, as opposed to positive reinforcement which is getting rewarded when you do something right. The negative reinforcement in this case is electric shock. You notice a book on the table titled, *The Teaching-Learning Process*, and you assume that this has something to do with the experiment.

Then Jack Williams takes two pieces of paper, puts them in a hat, and shakes them up. One piece of paper is supposed to say, "Teacher" and the other, "Learner." Draw one and you will see which you will be. The mild-looking accountant draws one, holds it close to his vest like a poker player, looks at it, and says, "Learner." You look at yours. It says, "Teacher." You do not know that the drawing is rigged, and both slips say "Teacher." The experimenter beckons to the mild-mannered "learner."

"Want to step right in here and have a seat, please?" he says. "You can leave your coat on the back of that chair ... roll up your right sleeve, please. Now what I want to do is strap down your arms to avoid excessive movement on your part during the experiment. This electrode is connected to the shock generator in the next room.

"And this electrode paste," he says, squeezing some stuff out of a plastic bottle and putting it on the man's arm, "is to provide a good contact and to avoid a blister or burn. Are there any questions now before we go into the next room?"

You don't have any, but the strapped-in "learner" does.

"I do think I should say this," says the learner. "About two years ago, I was at the veterans' hospital . . . they detected a heart condition. Nothing serious, but as long as I'm having these shocks, how strong are they—how dangerous are they?"

Williams, the experimenter, shakes his head casually. "Oh, no," he says. "Although they may be painful, they're not dangerous. Anything else?"

Nothing else. And so you play the game. The game is for you to read a series of word pairs: for example, blue-girl, nice-day, fat-neck. When you finish the list, you read just the first word in each pair and then a multiple-choice list of four other words, including the second word of the pair. The learner, from his remote, strapped-in position, pushes one of four switches to indicate which of the four answers he thinks is the right one. If he gets it right, nothing happens and you go on to the next one. If he gets it wrong, you push a switch that buzzes and gives him an electric shock. And then you go to the next word. You start with 15 volts and increase the number of volts by 15 for each wrong answer. The control board goes from 15 volts on one end to 450 volts on the other. So that you know what you are doing, you get a test shock yourself, at 45 volts. It hurts. To further keep you aware of what you are doing to that man in there, the board has verbal descriptions of the shock levels, ranging from "Slight Shock" at the left-hand side, through "Intense Shock" in the middle, to "Danger: Severe Shock" toward the far right. Finally, at the very end, under 435- and 450-volt switches, there are three ambiguous X's. If, at any point, you hesitate, Mr. Williams calmly tells you to go on. If you still hesitate, he tells you again.

Except for some terrifying details, which will be explained in a moment, this is the experiment. The object is to find the shock level at which you disobey the experimenter and refuse to pull the switch.

When Stanley Milgram first wrote this script, he took it to fourteen Yale psychology majors and asked them what they thought would happen. He put it this way: Out of one hundred persons in the teacher's predicament, how would their break-off points be distributed along the 15-to-450 volt scale? They thought a few would break off very early; most would quit someplace in the middle, and a few would go all the way to the end. The highest estimate of the number out of one hundred who would go all the way to the end was three. Milgram then informally polled some of his fellow scholars in the psychology department. They agreed that a very few would go to the end. Milgram thought so too.

"I'll tell you quite frankly," he says, "before I began this experiment, before any shock generator was built, I thought that most people would break off at 'Strong Shock' or 'Very Strong Shock.' You would get only a very, very small proportion of people going out to the end of the shock generator, and they would constitute a pathological fringe."

In his pilot experiments, Milgram used Yale students as subjects. Each of them pushed the shock switches one by one, all the way to the end of the board.

So he rewrote the script to include some protests from the learner. At first, they were mild, gentlemanly, Yalie protests, but "it didn't seem to have as much effect as I thought it would or should," Milgram recalls. "So we had more violent protestations on the part of the person getting the shock. All of the time, of course, what we were trying to do was not to create a macabre situation, but simply to generate disobedience. And that was one of the first findings. This was not only a technical deficiency of the experiment, that we didn't get disobedience. It really was the finding: that obedience would be much greater than we had assumed it would be and disobedience would be much more difficult than we had assumed."

As it turned out, the situation did become rather macabre. The only meaningful way to generate disobedience was to have the victim protest with great anguish, noise, and vehemence. The protests were tape-recorded so that all the teachers ordinarily would hear the same sounds and nuances, and they started with a grunt at 75 volts, proceeded through a "Hey, that really hurts," at 125 volts, got desperate with, "I can't stand the pain, don't do that," at 180 volts, reached complaints of heart trouble at 195, an agonized scream at 285, a refusal to answer at 315, and only heart-rending, ominous silence after that.

Still, 65 percent of the subjects, twenty- to fifty-year-old American males, everyday, ordinary people, like you and me, obediently kept pushing those levers in the belief that they were shocking the mild-mannered learner, whose name was Mr. Wallace, and who was chosen for the role because of his innocent appearance, all the way up to 450 volts.

Milgram was now getting enough disobedience so that he had something he could measure. The next step was to vary the circumstances to see what would encourage or discourage obedience. There seemed very little left in the way of discouragement. The victim was already screaming at the top of his lungs and feigning a heart attack. So whatever new impediment to obedience reached the brain of the subject had to travel by some route other than the ear. Milligan thought of one.

He put the learner in the same room with the teacher. He stopped strapping the learner's hand down. He rewrote the script so that at 150 volts the learner took his hand off the shock plate and declared that he wanted out of the experiment. He rewrote the script some more so that the experimenter then told the teacher to grasp the learner's hand and physically force it down on the plate to give Mr. Wallace his unwanted electric shock.

"I had the feeling that very few people would go on at that point, if any," Milgram says. "I thought that would be the limit of obedience that you find in the laboratory."

It wasn't.

Although seven years have now gone by, Milgram still remembers the first person to walk into the laboratory in the newly rewritten script. He was a construction worker, a very short man. "He was so small," says Milgram, "that when he sat on the chair in front of the shock generator, his feet didn't reach the floor. When the experimenter told him to push the victim's hand down and give the shock, he turned to the experimenter, and he turned to the victim, his elbow went up, he fell down on the hand of the victim, his feet kind of tugged to one side, and he said, 'Like this, boss?' ZZUMPH!"

The experiment was played out to its bitter end. Milgram tried it with forty different subjects. And 30 percent of them obeyed the experimenter and kept on obeying.

"The protests of the victim were strong and vehement, he was screaming his guts out, he refused to participate, and you had to physically struggle with him in order to get his hand down on the shock generator," Milgram remembers. But twelve out of forty did it.

Milgram took his experiment out of New Haven. Not to Germany, just twenty miles down the road to Bridgeport. Maybe, he reasoned, the people obeyed because of the prestigious setting of Yale University. If they couldn't trust a center of learning that had been there for two centuries, whom could they trust? So he moved the experiment to an untrustworthy setting.

The new setting was a suite of three rooms in a run-down office building in Bridgeport. The only identification was a sign with a fictitious name: "Research Associates of Bridgeport." Questions about professional connections got only vague answers about "research for industry."

Obedience was less in Bridgeport. Forty-eight percent of the subjects stayed for the maximum shock, compared to 65 percent at Yale. But this was enough to prove that far more than Yale's prestige was behind the obedient behavior.

For more than seven years now, Stanley Milgram had been trying to figure out what makes ordinary American citizens so obedient. The most obvious answer—that people are mean, nasty, brutish, and sadistic—won't do. The subjects who gave the shocks to Mr. Wallace to the end of the board did not enjoy it. They groaned, protested, fidgeted, argued, and in some cases, were seized by fits of nervous, agitated giggling.

"They even try to get out of it," says Milgram, "but they are somehow engaged in something from which they cannot liberate themselves. They are locked into a structure, and they do not have the skills or inner resources to disengage themselves."

Milgram, because he mistakenly had assumed that he would have trouble getting people to obey the orders to shock Mr. Wallace, went to a lot of trouble to create a realistic situation.

There was crew-cut Jack Williams and his grey laboratory coat. Not white, which might denote a medical technician, but ambiguously authoritative grey. Then there was the book on the table, and the other appurtenances of the laboratory which emitted the silent message that things were being performed here in the name of science, and were therefore great and good.

But the nicest touch of all was the shock generator. When Milgram started out, he had only a $300 grant from the Higgins Fund of Yale University. Later he got more ample support from the National Science Foundation, but in the beginning he had to create this authentic-looking machine with very scarce resources except for his own imagination. So he went to New York and roamed around the electronic shops until he found some little black switches at Lafayette Radio for a dollar apiece. He bought thirty of them. The generator was a metal box, about the size of a small footlocker, and he drilled the thirty holes for the thirty switches himself in a Yale machine shop. But the fine detail was left to professional industrial engravers. So he ended up with a splendid-looking control panel dominated by the row of switches, each labeled with its voltage, and each having its own red light that flashed on when the switch was pulled. Other things happened when a switch was pushed. Besides the ZZUMPH-ing noise, a blue light labeled "voltage energizer" went on, and a needle on a dial labeled "voltage" flicked from left to right. Relays inside the box clicked. Finally, in the upper left-hand corner of the control panel was this inscription, engraved in precise block letters:

SHOCK GENERATOR TYPE ZLB
DYSON INSTRUMENT COMPANY
WALTHAM, MASS.
OUTPUT: 15 VOLTS–450 VOLTS

One day a man from the Lehigh Valley Electronics Company of Pennsylvania was passing through the laboratory, and he stopped to admire the shock generator.

"This is a very fine shock generator," he said. "But who is this Dyson Instrument Company?" Milgram felt proud at that, since Dyson Instrument Company existed only in the recesses of his imagination.

When you consider the seeming authenticity of the situation, you can appreciate the agony some of the subjects went through. It was pure conflict. As Milgram explains to his students, "When a parent says, 'Don't strike old ladies,' you are learning two things: the content, and, also, to obey authority. This experiment creates conflicts between the two elements."

Subjects in the experiment were not asked to give the 450-volt shock more than three times. By that time, it seemed evident that they would go on indefinitely. "No one," says Milgram, "who got within five shocks of the end ever broke off. By that point, he had resolved the conflict."

Why do so many people resolve the conflict in favor of obedience?

Milgram's theory assumes that people behave in two different operating modes as different as ice and water. He does not rely on Freud or sex or toilet-training hang-ups for this theory. All he says is that ordinarily we operate in a state of autonomy, which means we pretty much have and assert control over what we do. But in certain circumstances, we operate under what Milgram calls a state of agency (after agent, n. . . . one who acts for or in the place of another by authority from him; a substitute; a deputy.—*Webster's Collegiate Dictionary*). A state of agency, to Milgram, is nothing more than a frame of mind.

"There's nothing bad about it, there's nothing good about it," he says. "It's a natural circumstance of living with other people. . . . I think of a state of agency as a real transformation of a person; if a person has different properties when he's in that state, just as water can turn to ice under certain conditions of temperature, a person can move to the state of mind that I call agency . . . the critical thing is that you see yourself as the instrument of the execution of another person's wishes. You do not see yourself as acting on your own. And there's a real transformation, a real change of properties of the person."

To achieve this change, you have to be in a situation where there seems to be a ruling authority whose commands are relevant to some legitimate purpose; the authority's power is not unlimited.

But situations can be and have been structured to make people do unusual things, and not just in Milgram's laboratory. The reason, says Milgram, is that no action, in and of itself, contains meaning.

"The meaning always depends on your definition of the situation. Take an action like killing another person. It sounds bad.

"But then we say the other person was about to destroy a hundred children, and the only way to stop him was to kill him. Well, that sounds good.

"Or, you take destroying your own life. It sounds very bad. Yet, in the Second World War, thousands of persons thought it was a good thing to destroy your own life. It was set in the proper context. You sipped some saki from a whistling cup, recited a few haiku. You said, 'May my death be as clean and as quick as the shattering of crystal.' And it almost seemed like a good, noble thing to do, to crash your kamikaze plane into an aircraft carrier. But the main thing was, the definition of what a kamikaze pilot was doing had been determined by the relevant authority. Now, once you are in a state of agency, you allow the authority to determine, to define what the situation is. The meaning of your actions is altered."

So, for most subjects in Milgram's laboratory experiments, the act of giving Mr. Wallace his painful shock was necessary, even though unpleasant, and besides they were doing it on behalf of somebody else and it was for science. There was still strain and conflict, of course. Most people resolved it by grimly sticking to their task and obeying. But some broke out. Milgram tried varying the conditions of the experiment to see what would help break people out of their state of agency.

"The results, as seen and felt in the laboratory," he has written, "are disturbing. They raise the possibility that human nature, or more specifically the kind of character produced in American democratic society, cannot be counted on to insulate its citizens from brutality and inhumane treatment at the direction of malevolent authority. A substantial proportion of people do what they are told to do, irrespective of the content of the act and without limitations of conscience, so long as they perceive that the command comes from a legitimate authority. If in this study, an anonymous experimenter can successfully command adults to subdue a fifty-year-old man and force on him painful electric shocks against his protest, one can only wonder what government, with its vastly greater authority and prestige, can command of its subjects."

This is a nice statement, but it falls short of summing up the full meaning of Milgram's work. It leaves some questions still unanswered.

The first question is this: Should we really be surprised and alarmed that people obey? Wouldn't it be even more alarming if they all refused to obey? Without obedience to a relevant ruling authority there could not be a civil society. And without a civil society, as Thomas Hobbes pointed out in the seventeenth century, we would live in a condition of war, "of every man against every other man," and life would be "solitary, poor, nasty, brutish, and short."

In the middle of one of Stanley Milgram's lectures at CUNY recently, some mini-skirted undergraduates started whispering and giggling in the back of the room. He told them to cut it out. Since he was the relevant authority in that time and place, they obeyed, and most people in the room were glad that they obeyed.

This was not, of course, a conflict situation. Nothing in the coeds' social upbringing made it a matter of conscience for them to whisper and giggle. But a case can be made that in a conflict situation it is all the more important to obey. Take the case of war, for example. Would we really want a situation in which every participant in a war, direct, or indirect—from front-line soldiers to the people who sell coffee and cigarettes to employees at the Concertina barbed-wire factory in Kansas—stops and consults his conscience before each action? It is asking for an awful lot of mental strain and anguish from an awful lot of people. The value of having civil order is that one can do his duty, or whatever interests him, or whatever seems to benefit him at the moment, and leave the agonizing to others. When Francis Gary Powers was being tried by a Soviet military tribunal after his U-2 spy plane was shot down, the presiding judge asked if he had thought about the possibility that his flight might have provoked a war. Powers replied with Hobbesian clarity: "The people who sent me should think of these things. My job was to carry out orders. I do not think it was my responsibility to make such decisions."

It was not his responsibility. And it is quite possible that if everyone felt responsible for each of the ultimate consequences of his own tiny contributions to complex chains of events, then society simply would not work. Milgram, fully conscious of the moral and social implications of his research, believes that people should feel responsible for their actions. If someone else had invented the experiment, and if he had been the naive subject, he feels certain that he would have been among the disobedient minority.

"There is no very good solution to this," he admits, thoughtfully. "To simply and categorically say that you won't obey authority may resolve your personal conflict, but it creates more problems for society which may be more serious in the long run. But I have no doubt that to disobey is the proper thing to do in this [the laboratory] situation. It is the only reasonable value judgment to make."

The conflict between the need to obey the relevant ruling authority and the need to follow your conscience becomes sharpest if you insist on living by an ethical system based on a rigid code—a code that seeks to answer all questions in advance of their being raised. Code ethics cannot solve the obedience problem. Stanley Milgram seems to be a situation ethicist, and situation ethics does offer a way out: When you feel conflict, you examine the situation and then make a choice among the competing evils. You may act with a presumption in favor of obedience, but reserve the possibility that you will disobey whenever obedience demands a flagrant and outrageous affront to conscience. This, by the way, is the philosophical position of many who resist the draft. In World War II, they would have fought. Vietnam is a different, an outrageously different, situation.

Life can be difficult for the situation ethicist, because he does not see the world in straight lines, while the social system too often assumes such a God-given, squared-off structure. If your moral code includes an injunction against all war, you may be deferred as a conscientious objector. If you merely oppose this particular war, you may not be deferred.

Stanley Milgram has his problems, too. He believes that in the laboratory situation he would not have shocked Mr. Wallace. His professional critics reply that in his real-life situation he has done the equivalent. He has placed innocent and naive subjects under great emotional strain and pressure in selfish obedience to his quest for knowledge. When you raise this issue with Milgram, he has an answer ready. There is, he explains patiently, a critical difference between his naive subjects and the man in the electric chair. The man in the electric chair (in the mind of the naive subject) is helpless, strapped in. But the naive subject is free to go at any time.

Immediately after he offers this distinction, Milgram anticipates the objection.

"It's quite true," he says, "that this is almost a philosophic position, because we have learned that some people are psychologically incapable of disengaging themselves. But that doesn't relieve them of the moral responsibility."

The parallel is exquisite. "The tension problem was unexpected," says Milgram in his defense. But he went on anyway. The naive subjects didn't expect the screaming protests from the strapped-in learner. But that went on.

"I had to make a judgment," says Milgram. "I had to ask myself, was this harming the person or not? My judgment is that it was not. Even in the extreme cases, I wouldn't say that permanent damage results."

Sound familiar? "The shocks may be painful," the experimenter kept saying, "but they're not dangerous."

After the series of experiments was completed, Milgram sent a report of the results to his subjects and a questionnaire, asking whether they were glad or sorry to have been in the experiment. Eighty-three and seven-tenths percent said they were glad and only 1.3 percent were sorry; 15 percent were neither sorry nor glad. However, Milgram could not be sure at the time of the experiment that only 1.3 percent would be sorry.

Kurt Vonnegut Jr. put one paragraph in the preface to *Mother Night*, in 1966, which pretty much says it for the people with their fingers on the shock-generator switches, for you and me, and maybe even for Milgram. "If I'd been born in Germany," Vonnegut says, "I suppose I would have *been* a Nazi, bopping Jews and gypsies and Poles around, leaving boots sticking out of snow-banks, warming myself with my sweetly virtuous insides. So it goes."

Just so. One thing that happened to Milgram back in New Haven during the days of the experiment was that he kept running into people he'd watched from behind the one-way glass. It gave him a funny feeling, seeing those people going about their everyday business in New Haven and knowing what they would do to Mr. Wallace if ordered to. Now that his research results are in and you've thought about it, you can get this funny feeling too. You don't need one-way glass. A glance in your own mirror may serve just as well.

PART V

Some Processes of Social Life

An early pioneer in sociology, Auguste Comte, divided sociology into two major parts: statics and dynamics. Statics is the study of order, whereas dynamics is the study of social processes. Although the terms utilized by Comte have changed, this basic division of sociology remains in use today as the study of social structure and functions and social change. This section deals with some of the processes or dynamics in society.

In the preceding sections, we noted that individuals are transformed into social beings through group interactions in a specific cultural context. We saw that socialization not only transforms people into social beings but also makes them viable members of society by imbuing them with the culture of the society.

However, the powerful members of societies also develop systems that perpetuate their positions and favor dominant groups over others, and these others are often members of race, class, or gender minorities. But society does not consist just of structure and institutions; social *processes* may bring about change that can either aid or hinder individuals or society in meeting their needs. One outcome of social processes for minority group members is to sometimes be labeled as deviant or, because of circumstances, to engage in deviance.

Chapter 13 considers the status of the elderly and issues surrounding health care. In some societies the process of aging is accompanied by enhanced status. Other societies, such as the United States, do not value the possible contributions of the elderly, but instead concentrate on youth with "new" knowledge. Whatever the status of the elderly, this group has some specific health care needs. Increasing medical expertise allows people to live longer and fight diseases such as AIDS that were formerly a death warrant, but has also raised the cost of medical care and insurance. As the baby boom generation ages, we are likely to see changes in health care policies and options in the dying and death process.

Chapter 14 deals with population processes and the urban scene, two topics that are related in many ways. Population growth has led to the development of urban areas that fulfill necessary functions for the population. But there is another side to population growth. Unchecked growth can mean enormous consumption of food and energy resources which creates numerous problems. As people seek better lives free from war and poverty, and more economic opportunities, they may migrate to other areas of their own countries or to other countries.

Trends in migration patterns reflect economic and social problems as people seek better opportunities. Population movements are most often to urban areas. As the pull of urban possibilities draws people to cities, urban governments are stretched to provide infrastructure and services to meet the needs of the burgeoning population.

The theme of Chapter 15, our final chapter, is change. All societies undergo change, but change can be so rapid that a society may find itself with few norms to guide behavior; the resulting breakdown of norms is a situation referred to as *anomie*. The push for change may result from what sociologists call *collective behaviors*, such as rumor, panic, and even revolution. One common push for change comes from groups of people dissatisfied with the societal system who join together to form social movements directed toward accomplishing desired changes. Several examples of social movements are presented in the final chapter. Another effect of rapid change is the difficulty for societies to keep their beliefs and value systems in line with changes in the material culture, a situation called *cultural lag*.

Some social scientists specialize in predicting the future. Throughout the book we have presented some of their forecasts on technology, aging, health care, religious fundamentalism, terrorism, and other issues facing societies in a globalizing world. By understanding the process of change and studying future predictions, we can better prepare ourselves for rewarding, productive, and useful lives.

Some people fear change and its consequences for their lives, and indeed some change may be disruptive to routines. However, change can be positive rather than dysfunctional. In fact, the possibilities open to us through societal change can be both exciting and fulfilling. Perhaps you have been involved in efforts to bring about change in social situations, maybe even through a social movement. The readings in this book have presented many problems and proposed many solutions. Consider the role you can play locally or nationally in bringing about needed change.

CHAPTER 13

Aging and Health

Alan Mc Evoy

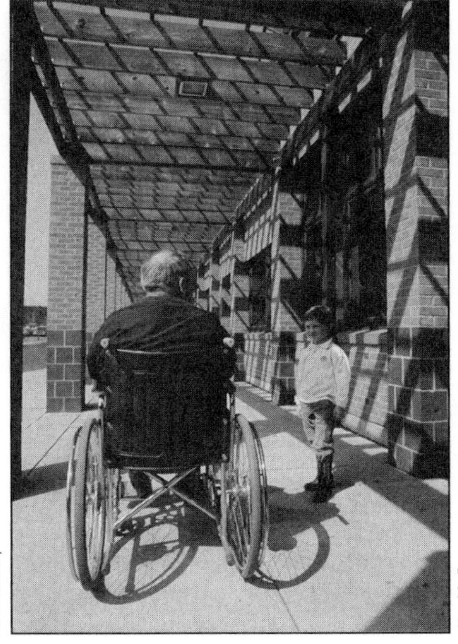

Alan Mc Evoy

Life and death parallel health, disease, and age. The readings in this chapter deal with the status of the elderly, the aging process, death and dying, problems in the health care system, and disease epidemics. Thanks to influences such as modernization, urbanization, and globalization, the aging process has been going through a period of rapid change. People are living longer, more productive lives in much of the world because of improved living standards and health care; the numbers of the elderly are growing rapidly, and in most countries the percentage of the elderly is also increasing. As it ages and reduces its productive capacity, this group becomes dependent on the younger population.

Although workers have contributed to government health care systems in most modern societies and these have provided a safety net for older citizens, the safety net in many societies is under strain or failing because of the increasing numbers of elderly citizens. Add to these pressures the advances in medical care and technology that add significant costs to health care services and many feel a crisis is pending.

The world's population *is* getting old and fat! Outlining some of the issues from an international perspective, Jeremy Seabrook discusses aging around the world. He reviews the changing status of the elderly, the growing number of elderly persons in societies, and the strain this creates on societies that need younger people to carry out the work that will support aging populations. Add to the problem the fact that birthrates have declined dramatically in many modern societies and there will not be enough workers to support the dependent young and old populations. Some have suggested that the political ramifications of the changing status of the elderly include a rise of religious fundamentalism and the influence of aging leaders in traditional societies who are attempting to maintain the status quo and influence of the elders in those societies. Seabrook points out the changing status of the elderly around the world through examples of fulfillment in the senior years versus declining capacity and dependence. He contrasts the elderly who are honored for their wisdom, respected, and given obedience versus those who are discarded because their wisdom is no longer relevant in the modern world, losing power and influence as younger generations take control.

The costs of growing old continue their rapid rise: insurance, health care, the cost of dying. These costs create potential conflicts between generations as different age groups compete for limited governmental funds. Sam Ervin presents fourteen forecasts for an aging society, issues from work and independence to shortages of health and elder care, that reflect rising costs.

Eventually we all face death. Michael Rybarski discusses the large baby boom generation that has rewritten the rules in a number of arenas, influencing every stage of the aging process; he points out that as the boomers face death, here also they will change the American "way of death." Older notions about "how to die" and "what to do with the deceased" are already changing with this generation.

Medical sociology is a major topic. Sociologists view health and medicine as one of the major institutions in society along with family, education, religion, politics, and economics. Because of the growing number of older people in the world, the institution is taking on increasing significance as demand for health services by the elderly rises. In fact, a large and increasing number of sociologists are in the specialty of medical sociology. Among the many issues sociologists study in the field of health and medicine are doctors as professionals, other roles in health care, the patient (or sick) role, hospitals as organizations, national health care systems, and health care for elderly patients, a focus in this chapter.

As people live longer and modern health care becomes increasingly sophisticated—with technology to deal with many conditions that were death sentences in the past—questions of cost and delivery of health care increase. Jim Taylor and Florence Comite explore the health care system in the United States and the impending crisis it faces. Medical technology has progressed to the point where we can diagnose illnesses and keep people alive for much longer periods than were possible in the past. This ability raises many practical and ethical questions. An ethical dilemma occurs, for example, when medical technology can keep individuals alive well beyond the point when they would have died naturally without intervention. The costs of keeping people alive may be depriving others of needed health care. What should be done in such situations?

In the final reading, Edward Kain considers a disease that does not discriminate on the basis of age. However, in the case of HIV/AIDS, younger sexually active individuals are at greater risk of contracting the disease. This disease is putting tremendous stress on health care systems, especially in poor countries that are already pressed to meet the basic health care needs of citizens.

This section raises questions about health, aging, and ways of dying and death in society. As you read, think about issues of access to health care and "right to life," increasing costs of care, and the process of dying and death in society today.

61

A World Growing Old

JEREMY SEABROOK

Once upon a time the elderly held high status in societies around the world. Today the number of societies in which this is true has dwindled, leaving the elderly in many societies feeling left behind, ignored, and disenfranchised. Seabrook analyzes the reasons for this change, the current status of the elderly, and the impact globalization is having on the elderly. One problem he points to is the lack of younger people to offset the growing percentage of elderly persons in most Western societies and the problem this imbalance causes for economies, especially in trying to support their older populations.

As you read about populations growing older, think about the following:

1. *What is producing the increase in the percentage of the elderly in many societies?*
2. *What is the impact on societies of this increase?*
3. *What can or should societies do about the growing number of elderly persons and the issues of support for the elderly?*
4. *Have you noted an increase in the elderly in your community? If so, what changes is this bringing about?*

GLOSSARY **Globalization** Increased economic, political, and social interconnectedness and interdependence among societies in the world. **Ageism** Prejudice and discrimination against the elderly.

RESPONSES TO AGEING

The world has, over time, produced a vast range of responses towards old age. These often contradict one another, as well they might, given the ambiguities surrounding old age itself. Growing old may be regarded as a time of ripeness and fulfilment or a period of declining health and failing powers. The storehouse of human societies has amassed a great variety of ways and means of coming to terms with an experience which remains essentially *that of other people*, until, at last, it catches up with us too.

There are good reasons not to anticipate the decline that comes with ageing, not least the tendency to avoid meeting trouble halfway. Received ideas about ageing are often a means of evasion and denial. "I'll worry about that when the time comes." "I'm not going to live that long." "I believe in living in the present." Our own old age is almost inconceivable until it is upon us. That it is a time of serenity, or that it holds all the terrors associated with standing on the edge of eternity, are beliefs of convenience, a mechanism to distance our younger selves from our own fate.

Throughout most of recorded time old bones were rare, and the great majority of people would have died by the time they reached what we would now consider middle age. In Britain, in

SOURCE: From *A World Growing Old*, by Jeremy Seabrook. London and Sterling, VA: Pluto Press, 2003, pp. 1–13.

1901 8 per cent of the population were over 60. By 1941 this had risen to 14 per cent. In 1991 it was 20 per cent. Today, although the old are present in increasing numbers, they nevertheless suffer a different kind of invisibility. They have become part of the landscape, obstacles on the sidewalk, impediments to the accelerating tempo of life, delaying the swiftly moving crowds in their urgent forward movement. Although they constitute one-fifth of the population, as one elderly woman in North London said, "People look through you. If you are old and a woman, you are doubly invisible. We have become like ghosts before we die."

The present moment inflects the ancient puzzle of old age and its meaning in ways that are historically unprecedented. In Britain, in 2002 it was remarked that for the first time there are more people over 60 than under 16. This ought, in a democracy, to give greater power to the elderly.

Yet the testimony of the old suggests something different. Paradoxically, as they become more numerous, they observe a growing indifference towards them. It seems to them that the rich reservoir of their accumulated experience is a wasting – and often wasted – resource. They find themselves speaking an alien language to those who have little wish to understand. They no longer recognise the world they live in. "We have lived too long" is a recurring theme.

It is remarkable that, now that the elderly are so numerous in the world, they should lament their loss of influence and power. Although in the past there were cultures which exiled or even killed their old, for the most part, when they were comparatively few, they commanded both respect and obedience. It is, perhaps, easier to create myths of wisdom and discernment in hoary heads when these are uncommon; and the nodding of senescence might well frequently have passed for sagacity.

But when life expectancy rises well into the 70s – and in Japan now, for women it is over 80 – the scarcity value of the old is undermined. The growing numbers of elderly in the world, far from representing a precious store of wisdom, are often perceived as a constraint upon the freedom and development of the young. It is not that large numbers of older people are abandoned or institutionalised. The myth of a more caring past persists, even though it has been rare for elderly parents to live with their families. In 1929–30, for instance, less than one-fifth of over-60s lived in extended families, and only 7 per cent lived in three-generation households. It was more common for people to live closer to their elderly parents than is now the case: in the dense mesh of the streets of industrial Britain, relatives often lived a few doors, or a couple of streets, away. The distance between people, which some observe today, is only partly spatial. It is also psychological, since the destinies of individuals diverge more obviously than they did when most people worked in the staple industry of a single town and expected their children to do likewise.

Conflict between the generations is no new thing. All cultures tell of a new generation, eager to play its part in the life of society, excluded and often humiliated by those in positions of power and influence. And that means the old, seniors, chiefs and headmen. Youthful energy, repressed by elders, is a persistent theme.

In many societies, the authority and prestige of elders were often unlimited. In Thailand, traditional law stated that wives and children were liable for the commission of crimes by the (senior, male) head of the family. 'The liability was not due to the fact that they were members of a family, but because their status in the family was property owned by the head of the family. Which was not so different from the manner by which slaves were owned.'[1] In some cultures a child could be given as payment to a creditor. A girl might be given to cancel a debt, and she would become the mistress of the individual to whom she was given.

Feudalism in Europe was a hierarchical system which was believed to reflect on earth the hierarchy of heaven, with its archangels, angels and saints, and an omnipotent God at its apex. Social reconstructions of this belief in the arrangements of religious institutions, and the societies that evolved around them, have shown a remarkable persistence through time.

Veneration of the elderly, especially of men, had an even more direct significance in tribal societies where the hierarchy of the dead and living was blurred. The ancestors were closest to God and had to be propitiated in order to earn their goodwill towards the living. Among the living, the oldest members of the tribe, being close to death, had a privileged relationship to all those who had gone before. Ancestor worship was an extension into the supernatural of existing family structures, in which the older members enjoyed a high level of authority. The family comprised both the material world and the invisible, but no less real, world of the spirits. The family and the tribe transcended mortality, and the oldest were the bridge between the living and the dead.

Nor is this unintelligible to us. Even today, many people in the West think of the dead as "looking down," "watching over" the living, a mixture of guardian angel and moral police. The dead are granted the compensatory privilege of supervising our mortal lives. I was much struck, at the time of the death of Diana, Princess of Wales, by the number of cards and mementoes left by people outside Kensington Palace referring to her caring for people, and her ability to do so now from her place in heaven. Speaking ill of the dead remains a taboo, even if much weakened by a market avid for revelations and the true story of dead celebrities.

The idea of the patriarch, the paterfamilias, the head of family, has been remarkably tenacious in all castes and classes. Their power was not uncontested – the resentment it created in the young may be read in the almost universal severity of the laws against parricide. The next generation must have been often tempted to put an end to the tyranny of those who lived on, denying them their inheritance, land and the power that went with it. This temptation had to be limited by the threat of the most draconian punishments.

Nor was the power of the patriarch curbed by the coming of industrial society. Industrial discipline only strengthened the authority of senior males in all social classes, exemplified by the often tyrannical, though sometimes paternalistic, mill or factory owner. The industrial workers, who were at the mercy of the arbitrary power of employers, visited their own victimhood on those over whom they had control, their wives and children.

STATUS OF THE ELDERLY

Now, everywhere in the world, gerontocracy is dying, although faster in some cultures than others. In certain areas of the world, the weakening powers of the old have called forth a vigorous reaction and a sometimes violent reassertion of authority. This is one possible reading of the emergence of religious fundamentalism: the reclamation of traditional forms of social and spiritual control by priests, imams and all the other – usually aged – intermediaries between this world and the next. A reaffirmation of dominance expresses itself in a hardening of old faiths: fundamentalism, ostensibly "a return to tradition," is a very contemporary phenomenon, a response to a modernisation which robs elders of power and undermines sources of authority.

In Africa, where rural, clan-based societies bestowed social and religious knowledge on elders, and where the main productive resource – land – was controlled by them, these patterns were first disrupted by colonialism. Later, Western-style education discredited ancient patterns of lordship by shamans, traditional healers and priests, and empowered those who had acquired the skills and knowledge appropriate to a new, urbanising and industrial society.

In Asia, joint and extended families are rapidly decaying under the same influences. The knowledge of the old is perceived increasingly as of dwindling use to, and an encroachment upon, the lives of a generation formed for a quite different way of living from anything known to their forebears. That the young should see this as liberation, and the elderly as evidence of deterioration, is scarcely surprising. But contemporary shifts in sensibility go far beyond a familiar cross-generational friction. They are symptomatic of more profound social and economic movements in the world, which have caught up whole cultures and civilisations in the compulsions of globalisation.

These have their origin in convulsive changes that have occurred in the West, where accelerating technological innovation, "de-industrialisation" and economic restructuring have rapidly removed the skills and competences of an older generation in favour of the flexibility and adaptability of the young. The "virtues" of frugality, thrift and self-denial have been eclipsed, since these are an embarrassment to a consumer society where status reflects spending power, and extravagance is a sign of success. Youth has acquired a social supremacy it has hitherto rarely enjoyed. This has been at the expense of the old.

LIFE EXPECTANCY AND GLOBALISATION

The dramatic rise in life expectancy is, to a considerable extent, a result of the application of medical technologies, which have prolonged life far beyond anything foreseen by the introduction of the welfare state in the mid-twentieth century. But in the rich countries, other factors have contributed to the rising proportion of elderly people, some of which are puzzling.

It was not anticipated that populations would fail to replenish themselves in the "developed" world. In Britain, in 2002 the birth rate fell to 1.6, which is just below the level at which the population will maintain itself. Wolfgang Lutz of Austria's International Institute for Applied Systems Analysis estimates that almost half the population of Western Europe and Japan will be over 60 by the end of the twenty-first century.[2] This forecast may, of course, prove false, as demographic extrapolations often have been in the past. (There was, for instance, a scare in Britain in the 1930s about the future depopulation of the country. It was forecast then that the total population of Britain by 2000 would be a mere 35 million. This prediction was swiftly overtaken after the Second World War, when the birth rate rose again, affluence became widespread and, above all, young and healthy migrants from the Caribbean, India and Pakistan came to ease labour shortages, and in the process rejuvenated the population.) In spite of this, however, there is no doubt that a reduction in the proportion of people of working age in relation to the retired is imminent.

The social, economic and moral consequences of these developments are far-reaching, although there is by no means unanimity on their meaning. Some researchers find nothing disturbing in the projections. . . .[3] Optimists argue that with a healthier older population and their desire to go on working longer, with continuing economic growth and improving productivity, there is no reason for excessive concern.

. . . Others argue that globalisation endangers the collective social transfers that are essential to elders in later life, pointing out that work, the family and collective institutions are all jeopardised by the neo-liberal ideology that presently dominates the global economy: work is decreasingly available to older people in the West (despite the current talk of raising the retirement age), as well as in the South, as the informal economy is replacing a "liberalised" formal sector; family support is eroded by growing individualism, while resistance to public spending is part of the global ideological curb on state provision for old age.

REPLACING THE GENERATIONS

The United States is the only industrialised country which has a fertility rate above the replacement level of 2.1 children per woman. The United States has also maintained a fairly steady flow of immigrants from all over the world. About 30 million people in the US were born outside the country, while there are an estimated 6 million undocumented migrants. These factors combine to protect the US against the threat of drastic population decline or a very high proportion of elderly. In spite of this, however, it is estimated that by 2020 23 per cent of the US population will be over 60. After the trauma of September 11, it may be that migration into the country will become more tightly controlled; the effect of this on the population profile and, consequently, on the dynamism and energy of the US is not yet clear.

In the US, the proportion of the population over 65 is expected to double by 2030 to 70 million, while the number of people over 80 will rise from 9.3 million in 2000 to 19.5 million in 2030. This will lead to increased health-care costs. In 1997, the US had the highest per capita health-care spending per person over 65 (US $12,100), by far greater than that of Canada (US $6,800) and the UK (US $3,600.) In the US, nursing home and home health-care spending doubled between 1990 and 2001, when it reached US $132 billion.

In North America, on average individuals between the ages of 65 and 69 have a further life expectancy of about 15 years. Between 75 and 79 it reaches ten years, and even at 80 it is six or seven years. . . . International agencies, governments, national charities and local organisations now routinely commit themselves to policies against ageism. These remain largely declaratory, although legislation against age discrimination in employment has been effective in the US, where the over-60s make up a larger proportion of the workforce than in any other Western country.

However this may be, the *social* power of the elderly shows little sign of being enhanced by their numbers. Youth, as an increasingly scarce commodity, is likely to go on appreciating in the demographic marketplace. If it has traditionally been the destiny of the young to rail against the authoritarianism and tyranny of age, there is little evidence that when the young are in the ascendant they are likely to be more merciful to their elders than these were to those subordinated to *them* in the past.

Nevertheless, the capacity to prolong life yet further, into the tenth and eleventh decade, is constantly advertised by enthusiasts of technological progress. These promises of a provisional immortality are limited only by questions sometimes raised about the purpose and function of superfluous aged populations, their unproductiveness and their dead weight on the declining number of earners of the future. It seems we are likely to hear much more about the desirability – or otherwise – of shortening, rather than extending, the lifespan by a further 20 years. . . .

What is certain is that, within little more than a generation, the population of much of the developed world will be ageing and falling. It will also be fat (more than 20 per cent are expected to be obese). These mutations in European society are unparalleled in modern times, and it is scarcely surprising that the policies to deal with them are both improvised and inadequate.

What does it mean, if rich societies fail to replenish themselves? Have they become too – what? – selfish? frightened? liberated? Does it matter? Are declining populations a blessing to the crowded lands of Japan or the Netherlands? Has child-bearing become too burdensome? Should we celebrate the freedom of women from an ancient cycle of pregnancy and childbirth, subservience and enslavement to the will of men? What are the consequences of elective childlessness for the future structure and cohesion of society?

Or have children simply become too expensive? In the United States, where most aspects of human life have been meticulously costed, the Department of Agriculture estimates that it now costs between US $121,000 and US $241,000 to bring up a child. A baby born today will be even more costly. By the age of 17, these omnivorous infants will have devoured between US $171,000 and US $340,000. It seems that the privileged people of the world are coming to regard children as something of a luxury. The comfort of the present depends not only upon growing inequality in the distribution of the wealth of the world, but is also constructed on the absence of the unborn. How future – and possibly depleted – generations will regard the legacy bequeathed by their begetters scarcely troubles a world which feels the pressing problems of today weigh upon it quite heavily enough without having to think about those of a distant tomorrow.

The Western model of development has now usurped all others and is presented as the sole source of hope and renewal to the whole world. What are the implications of this, when it creates a Japan or an Italy peopled by shadows, whose lives have been prolonged by technology far beyond anything that can be understood as their 'natural term'? What will these people do, sitting in the low-watt penumbra of old-age homes, their hearing ruined by decades of hyper-decibel music, their eyesight dimmed by long years of voyeuristic television, their memories all but erased by the media-crowded images of the day before yesterday? Even in the West such an achievement chills the spirit. Can it be, should it be, exported globally?

In its example to the world of abstention from increasing its population, we have a rare case of Europe and Japan practising what they preach – a birth control so effective that we can see future generations dwindling before our eyes. In the meantime, it is clear that the people of the rich countries will be able to ransack the countries of the South for urgently needed personnel to service our dereliction. Having already extracted maximum profit from their crop lands, forests, seas and mineral riches, and having taken advantage of the cheapness of their labour in the slums of Mexico City, Jakarta and Dhaka, we shall now pluck out the people they depend on most to help their own countries deal with the asperities of globalisation – doctors and nurses, carers for the old and infirm. Of course, people-stealing is not new. It was once known as slavery, but in the transformed circumstances of globalisation, this now appears as privilege.

NOTES

1. *The History of Thai Laws*, Chanvit Kasetsiri and Vikul Pongpanitanondha (eds), Bangkok, n.d.
2. *Guardian*, August 20, 2002.
3. *Guardian*, August 16, 2002.

62

Fourteen Forecasts for an Aging Society

SAM L. ERVIN

A population trend that accompanies modernization is lengthened life spans; in other words, people live longer. And the "oldest-old" groups are expanding most rapidly. This change creates new challenges for societies as they provide for the needs of aging citizens. Ervin addresses forecasts for societies in the future.

As you read, think about the following:

1. *How does an aging population affect society as a whole?*
2. *How are societies likely to change due to an aging population?*
3. *What is the relationship between an aging population and a society's health care system?*
4. *Which of the 14 forecasts have you seen come true? Give examples.*

GLOSSARY **Medicare** U.S. government program to fund health care for the elderly. **HMO** Health maintenance organization, a form of group insurance. **Baby boomers** Members of the large post-World War II cohort, born between 1946 and 1964.

As the baby-boom generation ages and the pool of retirees increases exponentially, a period of great change in elder care looms.

The median age of the U.S. population has been steadily rising. In 1900, one American in 25 was 65 or over. By 2050, that figure will increase to one in five. The U.S. Census Bureau projects that the over-65 population will more than double between 2000 and 2050. The proportion of "oldest-old" Americans, those 85 and over, will grow even more rapidly—quadrupling over the same period. By the year 2020, the ratio of over-65 individuals to the working-age adult population will be about one to four.

As a result, we will see sweeping changes in health care, including Medicare, the government program that currently covers 39 million Americans. There will also be major growth in options for elder care, as well as a flood of new products and services aimed directly at this swelling segment of the population.

Here are 14 forecasts based on recent surveys and studies, many of them conducted by the SCAN Health Plan, a not-for-profit plan serving about 39,000 seniors in Southern California.

SOURCE: Originally published in the November/December 2000 issue of *The Futurist*. Used with permission from the World Future Society, 7910 Woodmont Avenue, Suite 450, Bethesda, Maryland 20814. Telephone: 301/656-8274; Fax: 301/951-0394; http://www.wfs.org

1. THE RETIRED WILL WORK AGAIN

More seniors are likely to reenter the labor force, thanks to new legislation allowing those 65 to 69 to earn without penalizing Social Security benefits. Currently, 23% of the 9.2 million people in this age bracket are in the labor force. That is over 2 million senior workers.

Seniors' job-search success will no doubt be boosted by their increased comfort and proficiency with computers.

In addition to the financial benefits of earning a steady paycheck, seniors might get some health benefits, too. A SCAN Health Plan study among its members found that nonworking seniors are more likely to have significant health and daily living problems.

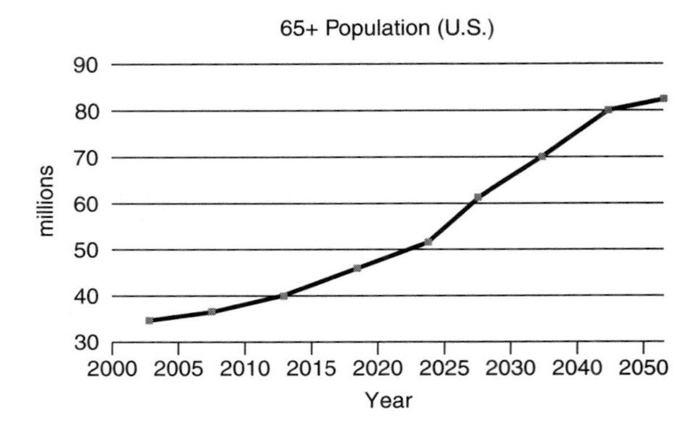

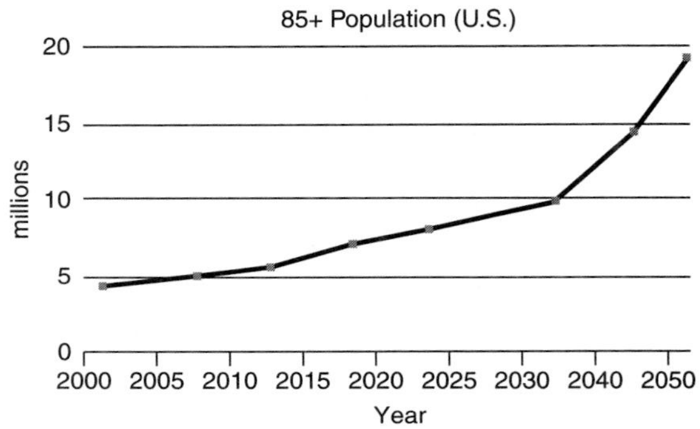

The population of seniors 65 and older will more than double by 2050. Seniors 85 and up, the most in need of elder care, will quadruple.

SOURCE: SCAN Health Plan. Data supplied by U.S. Census Bureau.

2. TECH-SAVVY SENIORS WILL MAINTAIN THEIR INDEPENDENCE

Elder-friendly technology will significantly improve access to resources and information to assist those who are frail and vulnerable. It will also reduce isolation among those living in rural or hard-to-reach areas of the country.

Products such as the multifunctional pager, which alerts seniors when it is time to take a particular medication, will be readily available. The increased use of technology will be a key factor in helping tech-savvy seniors to remain living independently, as it will enhance their ability to communicate and obtain valuable health-care information. Technology will also allow health-care providers to better monitor their patients.

Currently, nearly a quarter of the 20 million seniors aged 60 to 69 own and use a computer, as do 14% of the 16 million seniors 70 to 79, according to AgeLight Institute. Computer ownership among members of SCAN Health Plan is even greater. A survey found that 36% of seniors 65 to 74 reported having a PC in the home. Of those 74 to 89, 34% owned a computer.

While technology will play a wider role for seniors in the twenty-first century, the potential for isolation will increase the need for service fostering human interaction.

3. THE HOTTEST FITNESS BUFFS? SENIORS!

Health plans may begin to offer health club memberships and personal trainers as part of their coverage for seniors because of the proven benefit regular exercise has on seniors' overall health. Health clubs report that seniors are the fastest-growing group of members.

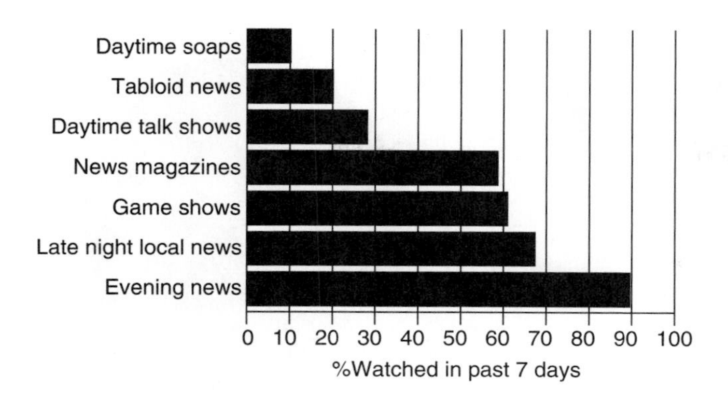

Seniors stay informed. In a SCAN Health Plan survey, seniors reported watching some type of news program more frequently than any other type of programming. Only 29% of respondents said they watched daytime talk shows, and soap operas drew just 10% of the seniors. SOURCE: SCAN Health Plan.

A recent survey finds more than two-thirds of seniors engage in regular physical exercise, double the national average for younger adults.

A SCAN survey of 2,035 seniors aged 65 to 90 found that 68% maintain a regular regimen of exercise ranging from moderate activities like walking to more intense workouts including weight lifting, jogging, cycling, tennis, and even heart-pounding handball. Forty percent of the survey respondents said they spend more than four hours per week on these activities. Another 44% exercise one to four hours each week.

Among a group of seniors in a SCAN-sponsored mall-walking program, an insulin-dependent diabetic was able to reduce insulin intake by four units per day. Others reported improved heart conditions, lower blood pressure, less pain from arthritis, and the elimination of leg cramps.

4. SENIOR-FRIENDLY CARS WILL OFFER INDEPENDENCE

Automakers may one day market cars that are easier and safer for America's growing population of seniors to drive.

The "senior-mobiles" may include such features as higher seats, larger numbers on the speedometer, and slower acceleration.

Retaining the ability to drive is the chief concern among aging seniors, according to SCAN research. For many, losing the ability to drive means losing one's independence and being forced to rely on others for transport.

5. SENIORS WILL BE IMPORTANT VOTERS

Seniors will wield the most power of any demographic group in the voting booth.

Seniors are big-time voters. In the 1996 federal elections, more than two-thirds of those 65 and older voted, according to the Census Bureau. That is 36% better than the 25 to 44 age group.

Seniors not only rank as the top voters, but they are likely to be the most-informed voters. Contrary to the popular belief that seniors are not interested in current events, a SCAN Health Plan survey reveals that they are major consumers of the daily news.

According to the survey, four of the top five television programs most frequently watched by seniors are some type of news program. Only 29% of respondents said they watched daytime talk shows. Soap operas drew just 10% of the seniors. By comparison, 90% said they had watched the evening news.

Seniors are also avid readers. Eighty-seven percent of seniors ranked reading the newspaper among their most favored regular activities. Magazine reading was favored by 75%.

6. MORE ALTERNATIVES TO NURSING HOMES WILL EMERGE

One of the government's biggest tasks in the new millennium will be an extensive education campaign to increase awareness of the wide range of alternatives to nursing homes.

Because of Medicaid's reliance on publicly funded nursing homes and hospital care, most seniors are not aware of alternatives, such as assisted living, independent living, life-care communities, and adult day care. A Harvard School of Public Health survey found that a majority of adults over 50 had never heard or read about six of 10 alternatives to nursing homes listed in the survey.

Nursing-home care cost Medicaid $40.6 billion in 1998 (24% of total outlays), compared to $14.7 billion in 1985. Clearly, more cost-effective options are needed, although these alternatives must provide a high standard of care.

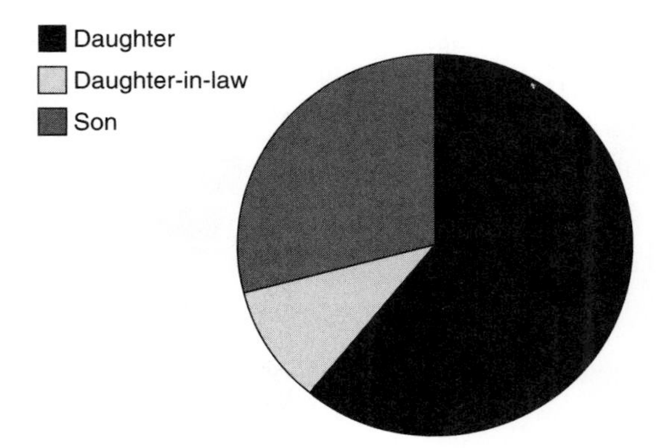

Women's climb up the corporate ladder may be slowed as they remain the predominant caregiver for aging parents. A SCAN Health Plan survey found that, of seniors relying on a child for caregiving, 71% rely on a daughter or daughter-in-law.
SOURCE: SCAN Health Plan.

7. BOOMERS COULD END UP IMPOVERISHED

Many aging baby boomers who thought they would be spending their golden years in relative financial comfort may actually find themselves impoverished because they did not prepare for the costs of long-term care.

A survey by the American Health Care Association found that 68% of baby boomers are not financially prepared for long-term care should they need it later in life. Half of the boomers polled had not even given any thought to how they will pay for long-term-care needs.

Part of the problem may stem from a lack of understanding about how long-term care is paid for.

A separate survey by the National Council on Aging, in conjunction with John Hancock Mutual Life Insurance, found that one-third of baby boomers incorrectly assume that Medicare is the primary source for long-term-care funding. The fact is, Medicare covers only about 53% of all health-care costs. Most people are surprised to find that it does not cover long-term care, prescription drugs, or vision or dental care.

8. ELDER CARE SHORTAGE IS COMING

As the population of those 85 and older doubles to 8.4 million by 2030, the demand of professional home-care aides will skyrocket. By 2008, personal care and home health aide jobs will rank second only to systems analysts in terms of sheer numbers (1.19 million vs. 1.18 million), according to the Bureau of Labor Statistics.

Twenty-nine percent of seniors age 65 and up rely on a daughter or daughter-in-law for caregiving assistance compared to 12% who rely on a son, according to a SCAN survey. However, daughters will become increasingly unable to take on the caregiver's role; more and more women will take on time-consuming managerial and executive positions as they continue to outpace men in earning bachelor's degrees (637,000 vs. 500,000 in 2000).

As the need for in-home elder-care-service providers soars, a major shortage of caregivers could result.

9. AGING BOOMERS WILL FORCE HEALTH-CARE-POLICY CHANGES

Baby boomers can be expected to use their influence to ensure that quality health-care services and programs are available for their parents now, not to mention for themselves in the not-too-distant future.

There is power in numbers among the baby boomers. Approximately 76 million boomers born between 1946 and 1964 will soon join the ranks of older Americans. Many of these boomers are beginning to provide care for their aging parents.

However, they also have a significant vested interest in ensuring that quality health care will be available for themselves. According to the Census Bureau, one in every nine boomers will live at least 90 years.

Perhaps the biggest concern among boomers is long-term care, which is currently not covered by Medicare. In fact, one survey found that long-term care has replaced child care as the number-one concern among baby boomers. Nearly 66% of boomers polled in the 1999 survey by the National Council on the Aging and John Hancock Life Insurance said they would expand Medicare to cover long-term care, even if it means paying higher taxes.

10. ELDER CARE WILL HURT WOMEN'S CAREERS

Women's progress toward finally shattering the glass ceiling will be slowed because of the increasing burden of elder care.

Despite women's advancement up the corporate ladder, a SCAN survey indicates that women are still disproportionately affected by elder care.

Strong evidence of elder care's adverse impact on career advancement comes from a national study conducted by the National Alliance for Caregiving and the American Association of Retired Persons. The study found that 31% of caregivers significantly alter their career paths; some leave the work force altogether.

According to the study, which polled 1,200 people who provided caregiving assistance to someone 50 or older, 11% of caregivers took a leave of absence, 7% opted to work fewer hours, 4% lost job benefits, 3% turned down a promotion, and 10% took early retirement or quit their jobs.

11. TELECOMMUTING WILL ASSIST FAMILY CAREGIVERS

Increased telecommuting will allow working adults to move closer to their aging parent and thus ease the burden of long-distance caregiving.

The National Council on the Aging (NCOA) estimates that nearly 7 million caregivers provide care for someone who lives at least one hour away. As more and more companies offer telecommuting, employees will be able to work from anywhere in the country and will have the flexibility to meet elder-caregiving responsibilities.

The percentage of companies with telecommuting rose steadily during the 1990s. In 1993, just 6% of the companies surveyed by William M. Mercer Inc. offered telecommuting. That figure grew to 14% by 1995 and 33% by 1998.

12. MORE EMPLOYERS WILL OFFER ELDER CARE

Elder-care benefits will become a major issue for workers and their employers as increased elder care-related absences and falling productivity begin to take a toll on the workplace.

The Census Bureau projects that over 40 million people in the United States will be older than 65 by the year 2010, an increase of 19% from 1995. As a result, about 12% of informal caregivers will quit their jobs to provide care full time, estimates the Family Caregiver Alliance. In a tight labor market, employers will offer more elder-care benefits to combat employee turnover.

In order to attract and retain quality employees, as well as strengthen productivity, companies will develop ways to help their employees deal with the burden of caring for aging parents. This trend will mirror the movement among corporations to provide child-care assistance in the 1980s and early 1990s.

In 1999, 47% of companies offered some type of elder care, up from 40% in 1998, according to an annual survey of over 1,000 employers by benefits consultant Hewitt Associates of Lincolnshire, Illinois. By comparison, 90% of companies offered some type of child-care assistance.

13. CAREGIVERS WILL NEED INTERVIEWING SKILLS

The sandwich generation, those caught between raising children and taking care of aging parents, will have to learn a lesson from human-resource executives.

As the senior population grows, there will be a constant stream of new business start-ups offering products and services to this group. Caregivers will be forced to make decisions on whom to hire and which organization or firm to use, much as human-resource professionals do on a daily basis.

Employers, working in concert with their human-resource departments, will set up training sessions for employees to allow them to make better decisions when it comes to hiring home aides or choosing elder-care-service providers. These sessions will stress that employees need to:

- Brush up on interviewing skills. Most of us have been through an interview process, but only as a candidate.
- Ask about availability of services. Twenty-four-hour, seven-day-a-week services are a must.
- Have potential caregivers submit résumés, complete with references.
- Ask, "Will the qualified care manager have a qualified back-up during vacations and time off?"
- View parents and relatives receiving care as "upper management," in that no decision on hiring should be made without upper management's input.

14. WORKING FAMILIES WILL GAIN STATE ALLIES

As the balance between work and family becomes an increasingly major issue, more state legislatures are likely to follow in the footsteps of the four that have already taken steps to ensure that companies accommodate employees' caregiving responsibilities.

While the legislation covers family caregiving, it is especially valuable to workers caring for an aging parent.

Terms of the legislation approved in California, Oregon, Washington, and Minnesota allow workers not covered by union contracts to use up to one-half of their paid sick leave to care for an ill child, spouse, or parent.

Elder care is likely to become an increasingly worrisome issue for employers as aging seniors eschew nursing homes in favor of independent living. Annual growth in nursing home residents has slowed to just 0.4%, down from 4.8% in the mid-1970s, according to the National Bureau of Economic Research.

However, many independently living seniors still need some type of caregiving assistance. SCAN surveyed 1,453 members, averaging age 82: None lived in a nursing home, despite meeting state qualifications. The survey found that 76% rely on a caregiver for assistance with daily activities. In 39.5% of these situations, the primary caregiver is a son or daughter.

THE FACTS OF LONG-TERM LIFE

Consider the following concluding statistics:

- In the twenty-first century, one in five Americans will be 65 or older. One in nine current baby boomers will live to at least age 90. The number of those 85 years old and over will quadruple by 2050.
- About 6.5 million older people need assistance with activities of daily living (e.g., bathing, cooking, cleaning, dressing). That number is expected to double by 2020.
- Women account for 72% (18 million) of the approximately 25 million family caregivers in the United States.
- The U.S. Census Bureau projects that the number of caregivers will drop from 11 for each person needing long-term care in 1990 to four in 2050.
- Collectively, family caregivers spend $2 billion of their own assets each month to assist relatives.
- By 2005, noninstitutionalized people over age 65 may spend an average of $14,000 annually on healthcare.
- Nearly 90% of baby boomers say taking care of their parents is among their top three life priorities.
- Ninety-four percent of seniors believe their health conditions do not affect their adult children's quality of life, but 80% of children say they do.
- Less than one-quarter of seniors expect to move in with their children; more than half of baby boomers anticipate having their parents move in at some point.
- Eighty-one percent of seniors do not believe their children will have to provide a great deal of financial support for their care; one-third of children believe they will.

The number of older Americans is increasing rapidly, while the human and financial resources to care for them are dwindling. Clearly then, elder care will become a major political issue in local, state, and national elections in the decades ahead.

63

Boomers After All Is Said and Done

MICHAEL RYBARSKI

The American way of dying and death has changed dramatically as the baby boom generation ages. Traditions from the past are being replaced with choices and alternatives, such as hospice, that are more cost efficient and personalized. Michael Rybarski discusses these social changes in death and burial customs.

As you read this selection, ask yourself the following questions:

1. *What are some differences in "way of death" between the baby boomer generation and their parents?*
2. *What has brought about changes in the way of death?*
3. *What do these changes mean for the social taboos, attitudes, and behaviors surrounding death?*
4. *Have you seen any of these changes reflected in deaths in your family or among your friends' families?*

GLOSSARY **Boomerization** Social changes brought about by the large baby boom generation, born between 1946 and 1964.

Just as they've reinvented or modified every life stage they've entered, the Baby Boom generation is beginning to rewrite the way America deals with life's final chapter. A major new trend among Boomers is to crack open the taboo, question institutionalized approaches to death and replace them with a more personalized, more humane model. The geometric growth of hospice care and alternative approaches to funerals, including the increase in cremation and casket stores are indications that traditional and institutional approaches to the end of life are now undergoing "Boomerization."

Economics accounts for part of this change. Institutional programs for end of life are just plain expensive—two to 20 times more expensive than alternatives. But, there's more to it than cost. As they plan for the end of life, often for their parents these days, issues of control, increased choice and a new search for meaning and ritual motivate Boomers. The expense and sterility of extended hospital care and "one-size-fits-all" funerals no longer cut it. They want better ways to say so long.

THE LAST TABOO

In the way of that goal is a lack of knowledge. The end of life remains the last great taboo, about which much of our culture remains silent. As they do, Boomers are breaking the taboo, looking for choices and opening up a dialogue about end of life. As with the sexual revolution, the end of life is surrounded by myths and half-truths. For instance, today, in all 50 states, you can do a funeral at home, or in a church, without using a funeral home. For over 20 years federal rules have allowed consumers to purchase caskets and monuments from outside the funeral home, but few know this. Hospice care is entirely paid for by Medicare, and can save families thousands of dollars and a lot of grief, but again, many don't realize they have a choice.

One reason for this is a lingering discomfort with thinking and talking about death. Another is a self-interested unwillingness among health-care and funeral advisers to offer alternative information.

So the death-care industry is becoming polarized. Options are emerging: Home-based hospice care is providing an alternative to hospitals; individualized care is challenging institutional care and customized burial and memorial services are starting to replace prefabricated funeral home offerings. To imagine the amount and pace of change that could surround end-of-life options, think of the U.S. highways before minivans.

BOOMERS CONFRONT PARENTS' DEATH FIRST

Demographics are power, and the cultural impact of the Baby Boom is largely driven by their numbers. When 75 million people confront an issue, it becomes culturally significant. Currently, there are about 2.4 million deaths in this country annually. That figure has remained constant for years. By 2040, the total will double, to about 4.1 million deaths per year, as Baby Boomers themselves die in greater numbers. By then, odds are that the process of dying—and the death-care industry—will have substantially altered to fit the Boomers' needs. It is not the immediate death of the Boomers that will drive that change, but the need for them to address the issues that surround the death of their parents. Psychologically, emotionally, financially and culturally, those changes are occurring today.

Psychographic and socioeconomic forces that have shaped the cohort experience of Boomers and their parents are forging this new face of death. Evident among these forces is the extended life spans of healthy and ill people alike. When the first Boomers were born in 1946, their parents were unlikely to contemplate prolonged end-of-life care for their parents. The average life span for men and for women born in 1900 was 47. While many people did live to an older age, they didn't have the medical resources to prolong life as we do now.

At the same time, most of our grandparents lived as adults within 15 miles of the town in which they were born. Burial rituals were well-established and required little pre-planning or forethought. People who lived close together often shared similar ethnicity, religious beliefs and cultural traditions. So when their parents began to die, there was no confusion—and little choice—for the Greatest Generation cohort about what to do.

END-OF-LIFE CONFUSION

Contrast that with what Boomers face. The most geographically and culturally diverse generation in history, Boomers live in a world where interfaith marriages, divorce and same-sex relationships are not unusual, nor are the revised rituals and traditions that accompany them. Home births, home-schooling, self-written marriage vows, home offices and increased involvement in one's own wellness and treatment are Boomer-initiated trends that have become commonplace.

As cultures, ethnicities, religions and geography blur with the Baby Boom and their parents, there can be no simple guide to end-of-life planning as existed with their grandparents. What are the new rules for planning cross-religious, cross-racial or cross-country funerals? And how do you transform an institutional death into a meaningful end-of-life celebration and ritual?

NEW CHOICES: THE MONEY ADDS UP

Dramatic evidence of the shift in thinking about death is the rise of hospice care. Since the creation of a Medicare Hospice Benefit in 1984, the

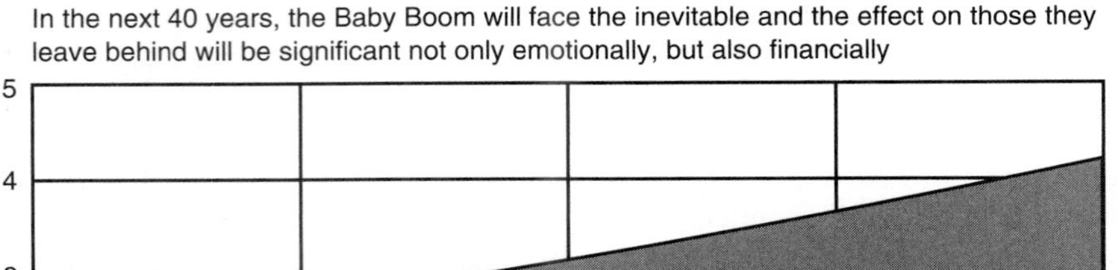

The Doubling of Death

The cultural level of death describes society's expectations based on historical trends in the annual number of deaths. As illustrated in this chart, Baby Boomers arrival at their final years will have a great impact on these levels.

SOURCE: U.S. Census Bureau

number of hospice patients has increased exponentially. In 1984, 154,000 patients were in hospice care. In 2003, the National Hospice and Palliative Care Organization estimates that 855,000 Americans received end-of-life care through a hospice. Dying is increasingly seen as a time for, in the words of Dr. Elisabeth Kubler-Ross, "spiritual, emotional and financial help for final care at home . . .," not isolation, institutionalization and staggering expense.

While compassion and opportunity for closure drive the hospice movement, it is bolstered somewhat by a third set of "c's": cost containment. The Agency for Healthcare Research and Quality estimates that 30 days of acute end-of-life care in a hospital would average about $109,000. By contrast, 30 days of hospice care—which includes far more one-to-one time with the patient and comparable palliative measures—costs about $4,000.

The end of end-of-life care means attending to the dead. In this instance, too, the Baby Boom population is taking a closer look at practices in an industry that were once a given. The business and foibles of funerals is moving into the mainstream with quirky, ironic series like HBO's *Six Feet Under* and A&E's *Family Plots*.

It's little-known that no state requires use of a funeral home to bury someone. Church and home services are legal in every state, and sources exist from which individuals may purchase third-party funeral supplies and monuments. Nontraditional approaches which may include a church funeral and the direct purchase of a casket and monument, can save families thousands of dollars. Cremation, which represents about 28 percent of all final dispositions, is also a way families save. Groups like the Neptune Society perform cremations for very low costs. Still, there are 22,000 funeral homes in America, or one per 100 deaths.

Why are there so many funeral homes? Because few people know their options, which can range from traditional funeral home services to burial in the backyard in a hand-decorated casket and artisan's monument. For Boomers and their parents, the ways to say goodbye will increase as they gain more knowledge about end-of-life issues and the power that goes with it.

POLARIZATION AMONG SIBLINGS

The most significant impact of the passing of the Boomers' parents may be a psychological one—as the Baby Boom generation is compelled to become the new head of the tribe. No matter what age they have been, as long as their parents were alive, Boomers were always the children at the center of their parents' universe. Internalizing that the world does not center on them, may be akin to a new Copernican revolution of the psyche.

Who makes the decisions concerning long-term care, funeral planning, hospital and hospice care for their parents and their spouses, will either polarize these families, or be new opportunities for coming together.

INDIVIDUALISM OR INSTITUTIONALISM?

A growing polarization between the institutional and personal approaches to end-of-life issues seems inevitable as there is a growing awareness of end-of-life choices. At the heart of the issue are decisions about who is in control at the end of life.

Given the extraordinary costs and economic consequences involved, an end-of-life revolution may be our next great cultural milestone, as the last life stage becomes the new center of gravity for the Baby Boom. And with this generational group, one thing is as certain as death and taxes: the status quo no longer applies.

64

Blood Sport

Do-or-Die Time

JIM TAYLOR AND FLORENCE COMITE

The health care system in the United States provides the highest quality of care for those who can pay or who know how to "work the system." As the numbers of older Americans (and individuals worldwide) increases, the costs of health care are soaring. Jim Taylor and Florence Comite address the serious state of health care finance and delivery and what individuals must do to get the care they need in the future.

As you read this selection, think about the following questions:

1. *What problems does the U.S. health care system display?*
2. *What has brought about these problems?*
3. *Is health care a right or a privilege for humans?*
4. *What can individuals do to alleviate these problems and obtain needed health care? Where does this leave those who cannot afford to pay?*

GLOSSARY **Blood sport** Refers here to competition to receive health care.

SOURCE: Originally published in *American Demographics*, Vol. 26, Issue 9 (November 2004), pp. 34–41.

THIS ARTICLE IS A WAKE-UP CALL

As we have been writing it, media attention has suddenly preoccupied itself with the sorry state of the American health-care finance and delivery. The furor reached a new level of consequence on September 8, [2004], as Federal Reserve Chairman Alan Greenspan digressed from his near-term economic analysis for the House Budget Committee to hold prophetic court on the decades ahead: "As a nation, we may have already made promises to coming generations of retirees that we will be unable to fulfill." The reality of health care in America is that we don't have the dollars to foot the bill for today's needs, nor tomorrow's escalating costs.

The current government obligation for taxpayer-funded Social Security and Medicare is said to total more than 10 times the $4.2 trillion national debt. What's more, we know that with age, health-care expenditures increase. As new technologies and pharmaceuticals intervene to keep Americans alive longer, even as they suffer from chronic or acute disease, post-symptomatic life expectancy could grow by leaps and bounds. An irony of living longer, however, is that the price is steep and certain to soar, especially during the last two years of life. America's health-care complex can barely tolerate today's level of demand, let alone serve an exponentially growing patient population.

An analysis of the financial and political options for health-care management and reform today, in light of an almost certain meltdown in the nation's capacity to keep a rapidly aging populace healthy, might well promote a deep sense of unease in every citizen, but especially those of us over age 50. The reality of American health care is that it is woefully unequipped with lifeboats for all.

Let's look at the math for a moment:

- 59 million are over age 55, when people begin to show health disorders.
- 38 million people are ages 45 to 55.
- 23 million more are ages 40 to 45.
- Fast forward to 2020, apply current death rates, and you get a grand total of 110 million people over age 55. Likely, they'll compete among one another for vital health resources.

Moreover, we need to consider the rapid run-up in life span. Nothing like this has ever occurred in human history:

- 35 million Americans today are over 65, receiving Medicare.
- 4.6 million Americans are over 85, consuming health resources at the rate of more than $16,000 per person per year.[1] By 2020, Americans in the over-85 category will at least double, maybe triple, or even get worse as life-prolonging technology advances.
- The typical aging person, suffering a slow death from a chronic disease, burns roughly $150,000 in the last 6 months of life.

According to Lincolnshire, Ill–based global human resources consultancy Hewitt Associates,[2] in 2004, health insurance costs will exceed $7,000 per employee, and this figure continues to rise at a double-digit rate. The heaviest part of this load is being borne by Boomers, but today's over-85 cohort—many of them Boomers' parents—are consuming more than $16,000 in health care, per capita. Current health-care expenditures for a 65-year-old are now four times those for a 40-year-old. U.S. health-care expenditures are projected to increase 25 percent by 2030, because the population will be older and greater in number.[3] We're almost certain that expenditures for Social Security, fixed and variable pensions and disability payments will burst their respective cupboards in the mid-2020s.

In any case, we'd estimate that total 2025 expenditures for the caring and feeding of America's aging populations will be roughly $5 trillion per year when the youngest Baby Boomer turns 65. To service that population, roughly 20 percent of the U.S. workforce will be dedicated to taking care of old people. In a recent forum on the nation's health-care crisis, president of the Hospice of Michigan, Dottie Deremo, cautioned, "Baby Boomers will be like velociraptors eating every health-care dollar in sight."[4] Yikes!

BLOOD SPORT: ACCESS TO CARE

Nothing in the world quite matches the American health-care complex.[5] It is one of seven trillion-dollar industries in the U.S. It can save a blue baby from Rwanda, and deny treatment to an uninsured middle-class baby in Chicago. It can provide the most sophisticated health care available to both an affluent person and an indigent person. Too often, though, it leaves masses of people between the two extremes in a state of arrested risk.

Many believe that insurance and Medicare will provide protection against health liabilities. But, given the supply squeeze, costs are disproportionately rising. In any case, we'd bet that minimum co-pays will continue to rise significantly (as they have this year), service cuts will be inevitable and age limits will creep up as the system becomes loaded with 110 million eligible citizens 65 and over.

At the same time, preventive health is practically nonexistent in our disease-centric medical delivery complex.[6] (While the U.S. spends more on health care than any other industrial nation—13.1 percent of gross domestic product—it ranked as only the No. 37 healthiest nation, according to World Health Organization data.) Traditional annual exams leave much to be desired. Tests are restricted; practitioner time is limited. From an economic perspective, the fact is that it is not a good investment for an insurance company to spend on a bone scan today to prevent a hip fracture in 30 years. By then, a broken hip will be Medicare's problem. Insurance is a transactional business—cash reserves in the present inevitably trump certain risk in the future.

Relatively few people work for a company that offers lifelong health benefits. These companies are under extreme pressure to reduce benefit costs.[7] As a result, post-retirement benefits are being cut or eliminated. Second, those companies that offer post-retirement benefit liabilities are moving the age line of qualification.

But the big bulge factor is the moving of the modal age in America. Today, 1 in 5 in a population just shy of 300 million is over the age of 55. Over 1 million people are over the age of 90, and by 2050, 5 million people or more will likely eclipse age 100. Who knows how long Boomers—birth years 1946 to 1964—and the "greatest Generation"—those born between 1927 and 1944—will last?

An already growing litany of strains puts the care of the greatest Generation at risk. First, we have fewer children compared with our grandparents, averaging less than two per nuclear family. Hence, fewer available caregivers. That grown children are less likely to live in the same town as their parents, due to jobs, relationships, etc. compounds the shortage of familial caregivers. High divorce rates, and the resulting disruption of the family of origin, further intensify the challenge of caring for America's seniors. All told, a weakened sense of kinship and community will exacerbate the pressures on the American health-care complex as social safety nets fail. These issues will only become more problematic over the next couple of decades as Boomers age.

Baby Boomers began turning 50 in 1996, 60 in 2006, and they begin turning 70 in 2016. Never has a generation entered its mature years as strong, as beautiful, as healthy and as aggressive at defending personal interests as the Baby Boom. With what is at stake, they will not go quietly into the sunset of their lives. Trouble lurks on the social horizon.

The magnitude of the problem does not end with demographics. Medical technology and science are heavily invested in extension of life therapies. Personalized gene therapies may provide you the opportunity to purchase (no insurance here!) replacement parts made from your own DNA, grown in a lab for insertion into diseased tissue. The conventional health-care community is beginning to explore and research a broad range of Complementary & Alternative Medicine[8] (CAM) therapies. These therapies are widely available in America for wellness and health maintenance as well as disease management. Molecular science businesses are distilling life-energizing proteins and related complex molecules from food sources, putting therapeutic review outside the scope of the Food and Drug Administration (FDA). If you can afford them, clinical services, nutritional intelligence, naturopathic remedies, physical fitness and attention to your personal spiritual needs may give you the added potential to maintain and sustain physical and mental health well into your 100s.

SITUATIONAL VECTOR

The situation is frighteningly simple: More people will demand more services from a shrinking complex. Some will be in a position to benefit; others will not. The complex will disallow service through queuing strategies, triage strategies, denial of service strategies, delayed entrance strategies. The complex will do whatever it takes to reduce the patient load to that which the system can sustain. We fear that a rising provider bias for cash and a seemingly random access model for everyone else will squeeze the kindness out of American health care.

Triage by inability to pay has already begun. The 20th century expectation that "health care is a right, not a privilege" has already been hard hit, given the sheer numbers of uninsured Americans. Forty-four million working Americans have been excluded, either because of poverty or the inability to acquire insurance from their employer. At the same time, denial of service by insurance companies is growing for health services that range from diagnostic tests to life-extending interventions for patients with cancer, health care becomes a blood sport: to the victor goes a long life. . . .

Interestingly, advantage redounds to those who fall seriously ill at a relatively young age, or who have a loved one who has to deal with current disaggregated health-care delivery. They are better informed because of their exposure to the inconsistencies and gaps in the present complex. However an individual can wise up to the complex, it is essential. There are not enough health resources to go around today. And, because there is little or no chance there will be a sufficiency of resources when you are in your 70s, most individuals will suffer if they fail to build adequate personal health management practices at the time they were in the best position to do so—i.e., *now*!

DENIAL AIN'T A RIVER IN EGYPT

Forewarned, as the coming health implosion unfolds over the next two decades, is forearmed. There are three trends simultaneously threatening the viability of the American health-care complex.

We are all aware of the first trend: aging healthy or healthy aging. Life expectancy hit just under 77 in 2000, but more interestingly, the rate of increase in life expectancy over the prior 10 years averaged a month in life for every year lived. By this calculus, we should have been on track to hit 80 around 2020. But recently, statistics show a sudden surge. Life expectancy rose 3 months in 2001, with women crossing the 80-year mark in 2002. Also, the gap between men and women is narrowing. The guys are only a year or two behind the gals.

As the Canadian government—a group that does behave like a system—has noted in a Ministerial report issued in 2000,[9] the process of life extension is likely to continue to accelerate. Technologies, surgical procedures, dialysis procedures, pharmacogenomics, heart disease innovations, new pharmaceuticals and genetically-specific drug design have already begun to play significant roles in symptomatic disease management. . . .

DOCTORS' RANKS DECLINE . . .

The American Academy of Medical Colleges reports that roughly 18,000 new medical students (both U.S. and foreign medical colleges) annually enter professional schools.[10] Assuming no drop-outs, approximately 270,000 new doctors will be in place by 2020. But, as 425,000 MDs retire (or partially retire) and demand doubles, we are in for some trouble in the health-care supply side economics.

What kind of trouble? If Adam Smith is correct, and supply-demand disequilibrium results in compensating cost changes, the supply shortfall could add as much as 33 percent to the cost burden. In simple terms, this would mean an added cost burden in 2020 equal to the entire Medicare budget today (on top of an estimated 58 percent increase from structural demand issues).

BLIND JUSTICE...

Fairness is the current standard by which life-extending therapies are to be distributed, with the presumption that sufficient supply exists for most therapeutic requirements. But this is already not the case. (And, too, fairness only became the standard in the last 50 years. Historically, health care has been inequitably distributed by income: Wellness was a luxury of the rich.) For example, there is a shortage of replacement organs such as hearts, lungs, kidneys, livers, etc. The fairness remedy is queuing. In the case of organ transplant material, fairness is managed by a system in which each recipient waits in line for a genetically sympathetic organ without respect to class or means. If the material fails to arrive by the moment of terminal denouement, you die. If it arrives in time, you live.

We estimate that by 2020 (based upon a simple doubling of the size of the relevant age group) that demand for coronary artery bypass heart surgeries will increase to over 1,200,000 per year, from the present volume of about 600,000.[11] Assuming that the 50 percent retirement rate applies to coronary specialists, and the same replacement rate applies as for other specialties, we forecast a 30 percent decline in available heart surgeons by 2020. Even if we allow for big improvements in productivity, robotics, surgery and better prevention of coronary disease, 400,000 people will still not be able to get the procedure. The same will occur for people requiring cancer treatments, dialysis, intervention for neurological disorders and the like. If the procedures are unavailable through lack of supply, people who would otherwise live long "post-symptomatic" lives will perish.

The question is how fairness will be adjudicated. In the United Kingdom and Canada, fairness is managed by the creation of age-based triage rules and random queuing management. In the U.K., if you are over age 65, you are at the bottom of the list, as long as a younger patient remains in the queue. Should a procedure become available, you receive it in the order of your diagnosis date, with some forgiveness for criticality. In Canada, except for emergency trauma, it is all queue-based. The result: Canadians come stateside for self-paid procedures.

One Canadian reported that her sister underwent an MRI for possible multiple sclerosis in November of last year, but no one was available to interpret the results for nine months, during which the disease advanced. Under Canadian rules, she was queued until a specialist became available. In Britain, a woman reported that her husband finally made the list for a knee reconstruction last year, three years after he had passed away. Apocryphal or not, these stories suggest that fairness will be rule-governed. Only the clever and those with the resources to buy their way into privatized health practices will be able to avoid a new game of lotto in which you literally bet your life. Nasty. . . .

INTERIM CONSIDERATIONS

Partial fixes may offer a modicum of protection and short-term solutions. For example, a broader utilization of medical professionals such as nurse practitioners and physician assistants as physician "extenders" will help.[12] These clinicians are capable of managing patients independently, in teams with other clinicians and/or with doctor oversight. Licensing and independent practices vary from state to state, and consumers, who are used to seeing "The Doctor" may protest. Extender precedents exist in specialized care, such as for diabetics, who are frequently monitored by a "team" including nurses, educators and nutritionists. At university health centers, students are often screened and treated by a nurse practitioner or PA only. Pregnant women typically see nurse midwives at prenatal care visits, and even for the delivery in some cases.

The reality, however, is that American health care is broken, and perhaps impossible to reconfigure for now, in part due to fragmentation and lack of incentives for current stakeholders. There is no one poised to create or train teams or set up viable networks. And, if this route is taken, it will be an agonizingly slow process. Doctors are not trained to work in teams, they like autonomy. Paraprofessionals may run into difficulty without adequate malpractice insurance coverage. Educating twice the number of docs will take a while; there just aren't enough medical school dollars, professors nor the ability to reorganize to ensure rapid ramp up of doctors that quickly.

Plus, the need to increase the numbers of graduating doctors would not only have an impact on medical schools, but also on residency training and supervision. The structure of medical school classes, and on the wards of hospitals would not lend itself to the short-term nor even medium-term doubling in class size. All these reasons do not even take into account the need for an influx of funds, the tax dollars to support such changes, and the willingness of medical school administrations and faculty to adopt a logarithmic increase in class size. Since a miraculous cure to fix the ills of American health care is unlikely to happen quickly, to help you stay healthy to a ripe old age, what are your options?

STRATEGIES GET PERSONAL

In this blood sport, the wealthy, the informed and the motivated players will be the victors who will outlive the losers. Winners will be the ones who anticipate that access and preventive measures will extend their life and improve the chance of support from the unrelieved ugliness of an overburdened system. While many health problems arise based upon genetics, lifestyle choices, such as smoking, overeating, physical inactivity, are estimated to contribute over 50 percent of an individual's health destiny.[13] In a recent government study reported in the Journal of the American Medical Association (JAMA), researchers analyzed data from 2000 for the leading causes of mortality and for those preventable factors known to contribute to them.[14] Tobacco, obesity and inactivity increase the risks for the top three killers: heart disease, cancer and cerebrovascular disease, including strokes. Obesity and inactivity also strongly increase diabetes risk, the sixth leading cause of death. An editorial accompanying the study in JAMA says national leadership and policy changes are needed to help curb preventable causes of death.

Changes don't happen quickly. If you invest your time and attention in preparing for the inevitable risks of aging and participate in managing those risks, you will reduce your exposure to an overwhelmed health system and, in your own way, reduce the burden on that system. Plus, given all the emerging technologies, you will ensure a quality of life that has never been available to aging people before—just like the childhood you had, that no one ever had before.

A good defense is the best offense. To survive well you need to plan to stay healthy. There are *five* keys to achieve this goal, and they depend on you:

First, you must develop a close relationship with an internist, and in some cases, a family practice physician, who will see to it that you achieve a preferred position in the access queue. . . .

Second, choose to be an active participant in health-care decisions. . . .

Third, consolidate your health records. . . .

Fourth, build a program of mental and physical wellness. . . .

Fifth, engage your spiritual faculties. Whatever belief system you practice, it becomes more and more relevant as the years of ambition are displaced by years of wisdom and contemplation. The objective of the game is happiness and wellness. Without the capacity for reflection neither of these are easily available. . . .

THE BIG BULGE FACTOR

Currently, one American in 5 is over age 55, 4.6 million are over 85, including 1 million over 90. By 2050, it is projected that about 5 million will over 100. This expected growth in the older population, combined with weakened family support, could place excessive pressure on the American health-care complex, with the result that only those at the higher-income level will be able to afford advanced medical care.

NOTES

1. http://www.agingstats.gov/chartbook2000/healthcare.html.
2. Hewitt Associates; press release; October 11, 2004.
3. http://www.cdc.gov/nccdphp/aag/aag%5faging.htm. Healthy Aging: Preventing Disease and Improving Quality of Life Among Older Americans, At A Glance, 2004.
4. Ravenous Baby Boomers set to devour health-care dollars; Rick Haglund, Booth Newspaper, September 22, 2004.
5. We use the word "complex" in lieu of the word "system" because system implies the existence of a self-correcting feedback mechanism; the complex under discussion consists of fiercely competing elements, each determined to protect its interests.
6. Who Will Keep the Public Healthy? Educating Public Health Professionals for the 21st Century (2003), Institute of Medicine, The National Academy of Sciences, National Academy Press, Washington, DC.

   Crossing the Quality Chasm: a new health system for the 21st century. Committee on Quality health-care in America, Institute of Medicine, National Academy of Sciences, 2001. National Academy Press, Washington, D.C.
7. The Benefit Trap, BusinessWeek; July, 19, 2004.
8. Complementary & Alternative Medicine, abbreviated as CAM, includes clinical and non-clinical interventions, a multiplicity of therapeutic (acupuncture), health (rolfing), and wellness (massage, yoga) approaches, as well as complete systems of care, such as Ayurvedic Medicine, Traditional Chinese Medicine, which incorporates acupuncture, herbal therapies and more.
9. Understanding Canada's Health-Care Costs (c.f. Google)
10. American Academy of Medical Colleges, 2004.
11. John Muir/Mt. Diablo Health System, Web Publication 2004.
12. Is There a Doctor in the House? Perhaps not, as nurse practitioners take on many of the roles long played by physicians, Andrew Blackman, WSJ, October 11, 2004.
13. Julie Britt, SHRMOnLine, 5.27.04.
14. Ali H. Mokdad, PhD; et al, Actual Causes of Death in the United States, 2000, JAMA. 2004;291: 1238–1245.

65

Some Sociological Aspects of HIV Disease

EDWARD L. KAIN

At first glance, the problem of AIDS might not appear to be an issue related to sociology. But if the sociologist is "interested in the action of human conduct," then AIDS—a disease spread through human behavior—is an object of sociological interest. Kain explains this relationship.

As you read, ask yourself the following questions:

1. *The social construction of AIDS can be seen from what two perspectives?*
2. *How does the notion of "stigma" relate to AIDS?*
3. *How would you deal with the problem of AIDS in the United States? In Africa?*
4. *What effect is HIV/AIDS having on health care systems around the world?*

GLOSSARY **Disease** biological aspects of sickness. **Illness** Social construct affecting society's reaction to disease. **Role of the other** Label and expectations attached by one individual to another individual or group. **Social construction** Process individuals use to create their reality through social interaction. **Stigma** Negative label that changes an individual's self-concept and social identity.

One of the most crucial social problems emerging in the last two decades of the twentieth century is the epidemic of HIV disease. From an illness that at first was perceived as affecting only a small number of gay and bisexual men in a few U.S. cities, it has grown to a pandemic (a worldwide epidemic). Both the number of AIDS cases and the estimated number of people who are infected with HIV have grown at a staggering rate.[1] In April of 1991, scarcely a decade into the epidemic, the World Health Organization estimated that approximately 9 million people were infected worldwide (World Health Organization, 1991). By 2004, that estimate had reached just under 40 million with over 20 million having died since the first cases were identified in the early 1980's (UNAIDS, 2004).

The epidemic has implications far beyond these numbers. Worldwide, there are nearly 13,500 additional infections per day—resulting in 4.9 million new infections in 2004, with an estimated 3.1 million deaths from AIDS in the same year (UNAIDS,

1. This reading makes a distinction between AIDS (Acquired Immune Deficiency Syndrome) and HIV disease. HIV disease begins when a person is infected with the human immunodeficiency virus. For most adults there is an extended period of as much as 10 to 15 years during which there are very few symptoms of infection. AIDS is the name associated with the end stages of HIV disease.

SOURCE: This reading was rewritten for the 8th edition of *Sociological Footprints*, edited by Leonard Cargan and Jeanne H. Ballantine. Research for the paper was funded, in part, by a Fellowship from the Brown Foundation of Houston, Texas.

2005). When adults die from HIV disease, they often leave behind children who must be cared for by others. Worldwide, the pandemic has left over 13 million children with one or both parents dead; that number will increase to more than 25 million by the year 2010 (UNAIDS, 2002).

Although disease is a biological concept, illness is a social construct. As such, it has an impact upon how well a society is able to combat the disease as well as upon the personal experience of individuals who contract it. Discussion of HIV disease ultimately points to important sociological concepts from a number of areas within the discipline. The next few pages examine some of the ways in which a sociological perspective can help us understand the social impact of AIDS and HIV disease.

A full sociological analysis of the epidemic is beyond the scope of this reading. This discussion focuses upon three topics: (1) the social construction of illness, (2) HIV disease and stigma, and (3) shifting definitions of HIV disease.

THE SOCIAL CONSTRUCTION OF ILLNESS

The current social construction of HIV disease and AIDS must be understood in the context of historical changes in the broader social definition of disease. Rosenberg (1988) suggests that in late eighteenth- and early nineteenth-century America illness was conceived primarily in terms of the individual. "Even epidemic disease was understood to result from an unbalanced state in a particular individual . . . thus the conventional and persistent emphases on regimen and diet in the cause and cure of sickness" (Rosenberg, 1988: 17). Using the example of cholera, Rosenberg points out that certain groups of people were seen as "predisposed" to falling ill—those who had poor nutrition, who were dirty or who were gluttonous. Such an approach had the function of reducing the seeming randomness with which the epidemic struck.

This approach to disease in the West slowly shifted throughout the nineteenth century. The Paris clinical school argued that disease was something "lesion-based" which played itself out in each person who was afflicted. By the end of the century the germ theory of disease gave such a model an explanatory mechanism. At the same time, a broader range of behaviors, previously linked to concepts of sin and deviance, came under the purview of medicine. The medicalization of alcoholism, mental illness, and a variety of sexual behaviors expanded the authority of medicine to deal with behavior previously not thought of as illness.

One result of medicalization is to reduce the amount of individual blame. Defining alcoholism, for example, as a disease transforms its very character. The alcoholic is now someone who has an affliction rather than someone who is of weak moral character. The blame begins to shift from the individual to the disease.

Blame

Blame never shifts entirely away from the individual, however. In the West, although we may be able to trace an historical shift along a continuum from blaming the individual to blaming an outside causal agent of some type (whether it be a germ, a gene, or a virus), this shift has not occurred equally for various types of diseases. Illness or injury resulting from behaviors that are defined by the society as morally wrong receive much greater attribution of individual responsibility than those resulting from behaviors which do not have the same degree of moral censure.

Ultimately, many types of illness can be linked to individual behavior. Smoking and overeating both are behaviors that are "chosen" by individuals. People who develop lung cancer, strokes, or heart attacks related to these behaviors are not, by and large, blamed for their illness. Sports, like football, (or more obviously hot-dog skiing) lead to a large number of serious injuries and deaths each year. Yet these are viewed as "accidents." Typically, those who suffer the results are not defined as deserving their fate.

Sexually transmitted diseases are much more likely to elicit reactions of blame and deservedness because of their linkage to behavior that is defined as morally wrong by the community. Further, there is a long cultural history of dividing the innocent victims[2] (spouses and children who are unknowingly infected by a partner, usually a husband, who strayed) from guilty sufferers who deserve their illness as the wages of sin.

The stigmatization of those who are ill and a search for someone to blame are not contemporary phenomena. During the 1656 outbreak of bubonic plague in Rome, for example, foreigners, the poor, and the Jews were blamed for the epidemic. Similarly, the poor and immigrants were blamed for the 1832 cholera outbreak in New York City. In this case, the moral failings that led to poverty were seen as the root cause of the illness. Indeed, in his examination of past responses to epidemics (1988, p. 57), Guenter Risse concludes that "in the face of epidemic disease, mankind has never reacted kindly . . . the response to disease is a powerful tool to buttress social divisions and prejudices."

The Role of the Other

The cause of a disease is often understood so as to shift blame for an epidemic upon the other. Whether this other is an ethnic or racial group, a religious or social category, or a group stigmatized for behavior which is labeled as deviant, this conception of the other is powerful in shaping the social response to disease.

The most extreme social response is total isolation of those who are sick in an attempt to stop the spread of the illness from one segment of society to another. Historically, quarantine has been tried as a method for coping with a variety of illnesses. With tuberculosis, yellow fever, cholera, and leprosy, all efforts to quarantine large numbers of people have been failures. Rather than being effective public health measures, these mass quarantines have been expressions of fear of the other—attacks upon the civil liberties of groups not accepted by the general public (Musto, 1988).

In the case of HIV disease, the complex interactions among blame, fear, discrimination, and stigma (a concept discussed in the next section) have led to a social construction of the disease in which "us" and "them" play a central role. The worldwide social impact of the epidemic, however, makes it clear that *all* of us are living with HIV disease (see Gilmore and Somerville, 1994, for an excellent discussion of this issue.)

The Role of "Deserving to Have the Disease"

The search for a group to blame for illness, combined with the tendency to blame the "other," often creates the conception that those who are sick deserve to have the disease. This is reflected in early cultural constructions of AIDS in the United States. Those who contracted the disease were often divided into "innocent victims" of AIDS and those who somehow deserved their illness. This ascription of personal responsibility for their illness affects not only people with AIDS but also their care givers (see Sosnowitz and Kovacs, 1992).

HIV DISEASE AND STIGMA

A key to understanding the social construction of HIV disease is the concept of stigma. Erving Goffman defines stigma as "an attribute that is deeply discrediting." He goes on to say that there are three types of stigma—what he calls "abominations of the body" (physical deformities); "blemishes of individual character" (some examples include a weak will, dishonesty, alcoholism and other addictions, homosexuality, radicalism); and "the tribal stigma of race, nation, and religion" (Goffman, 1963).

2. The use of the word *victim* is politically charged because of its implication of powerlessness. Much of the literature on HIV disease and AIDS suggests that *HIV-positive*, *seropositive*, and *persons with AIDS (PWA)* be used rather than *victim*. In this context I use the word *victim* because the popular consciousness clearly separates "innocent victims" from "guilty sufferers."

HIV disease has the potential of developing all three types of stigma described by Goffman. Some of the opportunistic infections associated with end-stage HIV disease (AIDS) can be disfiguring—the skin lesions of Kaposi's sarcoma being a prime example. The extreme weight loss associated with end-stage HIV disease also creates what Goffman called an "abomination of the body." The social and cultural construction of HIV disease involving blame, as noted, further links it to a number of "blemishes in individual character"—drug use, homosexuality, and inability to control one's own behavior in a safe manner. Finally, what Goffman called the "tribal stigma" of race, nation, and religion, also apply to the cultural construction of AIDS and HIV disease. In the early years of the epidemic one of the major "risk groups" was Haitians. A number of authors have suggested that the cultural constructions of AIDS are tinged with racism. In *AIDS and Accusation*, for example, Paul Farmer argues that ethnocentrism and racism in the United States were key factors in the theories about a Haitian origin for AIDS (Farmer, 1992).

Why Is HIV Disease Particularly Stigmatized?

Most of the literature on HIV disease and AIDS talks about the importance of stigma. From this literature we find a number of characteristics that predict whether or not a disease will be particularly stigmatized. One of the first analyses of the stigma associated with AIDS was by Peter Conrad (1986). After a discussion of the public hysteria surrounding this illness, Conrad identifies four social aspects of AIDS which lead to its peculiar status.

First, and foremost, throughout the early years of the epidemic AIDS was associated with "risk groups" which were both marginal and stigmatized. Because early cases of AIDS were found in homosexual men and intravenous drug users, a powerful cultural construction emerged that defined the illness as a gay disease.

Second, because a major mode of transmission involves sexual activity, the disease is thus further stigmatized. In his insightful book on the history of venereal disease in the United States, Allan Brandt suggests that human societies have never been very effective in dealing with sexually transmitted diseases. He begins his book by noting that "the most remarkable change in patterns of health during the last century has been the largely successful conquest of infectious disease." The striking exception to this pattern has been an explosion in sexually transmitted diseases. He asks, "Why, if we have been successful in fighting infectious disease in this century, have we been unable to deal effectively with venereal disease?" His answer lies in an examination of the social and cultural responses to sexuality and sexually transmitted disease. He argues that venereal disease was a social symbol for "a society characterized by a corrupt sexuality" and was used as "a symbol of pollution and contamination." The power of this social construction has rendered efforts to control sexually transmitted diseases ineffective (Brandt, 1987).

Third, Conrad points out that contagion plays a major role in whether or not a disease stigmatizes the individual who is infected. If a disease is contagious, or if it is *perceived* as contagious, then the stigma of the illness increases.

Fourth, Conrad argues that the fact that AIDS is a deadly disease also adds to its stigma. Because of these social characteristics, Conrad concludes that AIDS is a disease with a triple stigma: "it is connected to stigmatized groups, it is sexually transmitted, and it is a terminal disease."

One of the best sociological discussions on the stigma associated with HIV disease is found in Rose Weitz's *Life with AIDS*. Weitz says that stigma is greatest when an illness evokes the strongest blame and dread. She delineates six conditions that increase blame and dread. Like Conrad, Weitz includes (1) linkage to stigmatized groups, (2) an association with sexuality, (3) if the illness is perceived to be contagious and if there is no vaccine available, and (4) if the illness is "'consequential,' producing death or extensive disability and appearing to threaten not just scattered individuals but society as a whole" (Weitz, 1991, pp. 45–48). To Conrad's list she adds two more—if the illness creates dehumanizing or disfiguring changes that "seem to transform the person into something beastly or alien" and "if mysteries remain regarding their natural history."

To this list I would add two more factors. First, industrial societies have not had to deal with fatal infectious diseases in several decades. We have come to expect that young people will not be struck down by disease in the prime of their life. Polio was the last great infectious disease to affect modern industrial societies, and much of the population has grown up in a world with no experience of consequential infectious diseases. This circumstance increases the fear and dread associated with HIV disease.

Finally, if a disease has an impact upon mental functioning, it increases stigma. HIV can directly infect brain cells. As prophylactic measures such as the use of inhalant pentamidine and anti-viral drugs such as AZT increase the time between exposure to HIV and first opportunistic infections, the number of people with HIV disease who have impaired mental functioning because of infection with the virus will increase . . . thus increasing the potential for stigma. Indeed, dementia is now the most common neuropsychiatric problem found among HIV patients (Buckingham, 1994).

The Future of Stigma as It Relates to HIV Disease

Just as the relative stigma of various illnesses can be predicted by examining the eight characteristics just delineated, the stigma of a particular illness varies by culture and will change over time within any particular culture as these variables change. In cultures where sexual behavior is more openly discussed and where attitudes are more tolerant of homosexuality, the stigma associated with HIV disease will be less. Similarly, in cultures where drug use is conceptualized differently (the Netherlands is an instructive example here), the stigma of HIV will be lessened.

The point here, however, is that no other disease has such a high potential for stigma on all eight characteristics, making HIV disease the prototypical example of stigmatized illness in modern times.

Shifting Definitions of HIV Disease

There is no single social definition of HIV disease and AIDS. Public perception of the disease varies among cultures and has changed over time. It also varies considerably from one segment of the population to another and from region to region in a population. Much of this variation is linked to epidemiology.

The social epidemiology of HIV disease in Africa, for example, is radically different from that in our country. Rather than being associated with certain high risk groups, it is more equally distributed among males and females, and appears to be as common in the heterosexual population as it is in the homosexual and bisexual population. The disease has continued to be "feminized" over time, so that by 2004, 57% of all adult cases in sub-Saharan Africa are among women (UNAIDS, 2004). These differences in epidemiology clearly have a major impact upon the social definition of the disease in different parts of the world as well as the treatment and ultimate social consequences of the disease.

It is interesting to note that coverage of the disease by the popular press shifted both in magnitude and in tone when it became clear that AIDS could be contracted by so-called "innocent" people—babies, hemophiliacs, and recipients of blood transfusions (Altman, 1986; Kain and Hart, 1987). Indeed, the social definition of AIDS shifted considerably over the first decade of its existence. Some of the early literature referred to the illness as GRID (Gay-Related Immune Deficiency), and there is evidence that when it was defined as a "gay plague" the scientific community joined the general public in its reluctance to take the disease as seriously as was warranted (Shilts, 1987).

Panic, fear, and rumor were common elements of popular press reports in the early 1980s. Indeed, many scientists and AIDS educators were frustrated by continued misconceptions about modes of HIV transmission. These early responses to AIDS reflect basic principles in research on collective behavior. Rumors are most likely to develop when there is ambiguity about something. They help to clarify the situation when data are unavailable (Shibutani,

1966; Macionis, 2003). The early years of the HIV epidemic fit this description perfectly. In addition, because the disease is fatal, people have a high degree of interest in the topic. Fear of contagion is a very predictable response.

From Acute to Chronic

Before the causal agent of AIDS was understood, and before effective treatments had been developed for some of the opportunistic infections, AIDS was defined as a short-term disease that led to death in a relatively short time. As the etiology of the disease has become more clear, it is becoming defined as a long-term chronic illness (see Fee and Fox, 1992). This shift has implications for the cost of treatment, the stigma associated with the disease, and calculations of the social impact of the epidemic.

One change resulting from the shift to defining HIV disease as a chronic illness is the emergence of new issues for the delivery of health care. Chronic illness has typically been associated with the elderly. Long-term care facilities must rethink their methods of patient care when working with younger persons with HIV (Zablotsky and Ory, 1995). Further, policy makers and clinicians may view this shift to HIV as a chronic disease differently. Thus a simple redefinition of the disease as chronic may be inadequate for planning patient care in the case of HIV (Clarke, 1994).

Changing Epidemiology

As the epidemiology of the disease changes, so will social definitions. In the United States there has been a shift over time in which a larger proportion of new AIDS cases are linked to drug use rather than homosexual and bisexual activity. Further, African Americans and Hispanics are disproportionately affected by the disease. The general trend in this country has been that the epidemic increasingly affects women, the poor, people of color, and the heterosexual population (Karon *et al.*, 2001). As in Africa, the proportion of new cases in the United States found among women has been increasing over time (UNAIDS, 2004). Because HIV has not been defined as a women's disease in the West, and because of women's lower social status, women with HIV have a shorter survival time than men. As the epidemiology changes, approaches to women's health issues must also change (Lea, 1994). The combination of race and gender are also important to examine. By the mid-1990s, women of color made up well over two-thirds of all cases of HIV-infected women in the United States (Land, 1994). By 2002, nearly half of all the HIV-positive people on the planet were women (UNAIDS, 2004).

Higher rates of infection among African Americans and Hispanics in the United States have led researchers to explore ways in which HIV education programs may need to be targeted to specific populations. Different intervention techniques may be more effective with one population than another, and ethnographic analysis can help identify the best programs to use for a particular ethnic or racial group (Goicoechea-Balbona, 1994; Bletzer, 1995).

The history of the epidemic has also seen a shift in the worldwide distribution of reported cases of AIDS as well as projections concerning HIV infection. When the early reported cases were concentrated in North America, AIDS was defined as an American disease. In a number of communist countries; it was defined as a disease of Western capitalism. As data on the epidemic improved, it became clear that large numbers of people were infected in the Third World. Current estimates suggest that sub-Saharan Africa has more than 60% of all worldwide HIV infections. It appears that the locus of the epidemic in the future may be Asia, where the population size holds the potential for large numbers of infections. The diversity of epidemiology in the region has led to a discussion of HIV epidemics rather than a single epidemic. Some areas (including Thailand and Cambodia) have been very successful in bringing the epidemic under control, whereas other areas (some parts of India and southwestern China) have high prevalence and thus far little success in limiting the spread of infection (MAP, 2004).

As the epidemiology of HIV disease shifts, so will our cultural constructions of the disease. As women and heterosexuals become larger proportions of the

HIV-positive population in Western industrialized countries, the stigma will decrease. In addition, infection rates are actually beginning to decline in some of the industrialized West, and treatments using mixtures of "cocktail" drugs have been very promising in slowing (or even reversing) the replication of HIV within infected individuals. As the disease becomes more treatable, it will become less stigmatized. Worldwide, the epidemiology will continue to evolve. Although Asia currently has low infection rates, the rate of spread is rapid in some areas. In Latin America and the Caribbean the overwhelming proportion of infections are found in marginalized groups, which predicts high stigmatization in those regions. In Eastern Europe most cases are associated with drug use—again predicting stigmatization of the disease.

Data on patterns and trends in HIV infection change relatively rapidly. Luckily there are several excellent resources on the Internet that provide up-to-date information both for the United States and other countries. The Centers for Disease Control maintain a number of websites related to HIV/AIDS. For general information, statistics, and other information, go to: http://www.cdc.gov Once at the CDC website, you can use the search function, or look for HIV/AIDS in the alphabetical list of statistics under "FASTSTATS A–Z".

The U.S. Census Bureau website maintains an international database that includes infection rates for various countries around the world. They also have links to maps of the worldwide pandemic. The general URL for the Census Bureau is http://www.census.gov Once you are at the website, search within "Subjects A–Z."

Perhaps the most comprehensive site for reports on the global pandemic is http://www.unaids.org

CONCLUSION

A full sociological analysis of the HIV epidemic would examine a wide variety of issues. The economic and demographic impact of the disease has already been devastating in a number of central African nations. Gains in infant mortality that have taken four decades to achieve have been wiped out in less than a decade. Worldwide, women are particularly vulnerable to HIV disease both because of their lack of power in sexual relationships and because of the relative ease of viral transmission between sexual partners (Panos Institute, 1990). Indeed, issues of social stratification are central to understanding the epidemic. In most societies, race, class, and gender are critical variables in predicting who is more likely to become infected, and once infected, who will receive adequate treatment. Worldwide, the poorest countries are among the hardest hit in the epidemic. Unfortunately, much of the progress made in the developed countries relies upon very expensive drug treatments, which will be unavailable to the majority of those infected with HIV throughout the world.

Although this reading has not covered nearly all the issues involved in such an analysis of the HIV pandemic, it has illustrated how a sociological perspective informs our understanding of one of the greatest social problems facing the world today.

REFERENCES

Altman, Dennis. 1986. *AIDS in the Mind of America.* Garden City, NY: Anchor Press/Doubleday.

Bletzer, Keith V. 1995. Use of ethnography in the evaluation and targeting of HIV/AIDS education among Latino farm workers. *AIDS Education and Prevention* 7(2) 178–91.

Brandt, Allan M. 1987. *No Magic Bullet: A Social History of Venereal Disease in the United States since 1880.* Expanded ed. New York: Oxford.

Buckingham, Stephan L. 1994. HIV-associated dementia: A clinician's guide to early detection, diagnosis, and intervention. *Families in Society* 75(6) 333–45.

Clarke, Aileen. 1994. What is a chronic disease? The effects of a re-definition in HIV and AIDS. *Social Science and Medicine* 39(4) 591–97.

Conrad, Peter. 1986. The social meaning of AIDS. *Social Policy* (Summer 1986): 51–56.

Farmer, Paul. 1992. *AIDS and Accusation: Haiti and the Geography of Blame.* Berkeley: University of California Press.

Fee, Elizabeth, and Daniel M. Fox. 1992. *AIDS: The Making of a Chronic Disease*. Berkeley: University of California Press.

Gilmore, Norbert, and Margaret A. Somerville. 1994. Stigmatization, scapegoating and discrimination in sexually transmitted diseases: overcoming "them" and "us." *Social Science and Medicine* 39(9) 1339–58.

Goffman, Erving. 1963. *Stigma: Notes on the Management of Spoiled Identity*. Englewood Cliffs, NJ: Prentice-Hall.

Goicoechea-Balbona, Anamaria. 1994. Why we are losing the AIDS battle in rural migrant communities. *AIDS and Public Policy Journal* 9(1): 36–48.

Kain, Edward L. and Shannon Hart. 1987. *AIDS and the Family: A Content Analysis of Media Coverage*. Presented to the National Council on Family Relations, Atlanta.

Karon, John M., PhD; Fleming, Patricia L., PhD; Steketee, Richard W., MD; De Cock, Kevin M., MD. 2001. HIV in the United States at the turn of the century: An epidemic in transition. *American Journal of Public Health* 91(7): 1060–68.

Land, Helen. 1994. AIDS and women of color. *Families and Society* 75(6): 355–61.

Lea, Amandah. 1994. Women with HIV and their burden of caring. *Health Care for Women International* 15(6): 489–501.

Macionis, John J. 2003. *Sociology*, 9th ed. Englewood Cliffs, NJ: Prentice Hall.

MAP (Monitoring the AIDS Pandemic). 2004. *AIDS in ASIA: Face the Facts: A Comprehensive Analysis of the AIDS Epidemics in Asia*.

Musto, David F. 1988. Quarantine and the problem of AIDS. Pp. 67–85 in Elizabeth Fee and Daniel M. Fox (eds.), *AIDS: The Burdens of History*. Berkeley: University of California Press.

Panos Institute. 1990. *Triple Jeopardy: Women & AIDS*. London: Author.

Risse, Guenter B. 1988. Epidemics and history: Ecological perspectives and social responses. Pp. 33–66 in Elizabeth Fee and Daniel M. Fox (eds.), *AIDS: The Burdens of history*. Berkeley: University of California Press.

Rosenberg, Charles E. 1988. Disease and social order in America: Perceptions and expectations. Pp. 12–32 in Elizabeth Fee and Daniel M. Fox (eds.), *AIDS: The Burdens of History*. Berkeley: University of California Press.

Shibutani, Tomotsu. 1966. *Improvised News: A Sociological Study of Rumor*. Indianapolis, IN: Bobbs-Merrill.

Shilts, Randy. 1987. *And the Band Played on: Politics, People, and the AIDS Epidemic*. New York: St. Martin's Press.

Sosnowitz, Barbara G., and David R. Kovacs. 1992. From burying to caring: Family AIDS support groups. Pp. 131–44 in Joan Huber and Beth E. Schneider, eds., *The Social Context of AIDS*. Newbury Park, CA: Sage.

UNAIDS. 2002. *Children on the Brink 2002: A Joint Report on Orphan Estimates and Program Strategies*.

UNAIDS. 2004. *2004 Report on the Global HIV/AIDS Epidemic: 4th Global Report*.

UNAIDS. 2005. Global summary of the HIV and AIDS epidemic, December 2004. Accessed at *www.unaids.org*.5/23/2005.

Weitz, Rose. 1991. *Life with AIDS*. New Brunswick, NJ: Rutgers University Press.

World Health Organization. 1991. *Current and Future Dimensions of the HIV/AIDS Pandemic: A Capsule Summary*.

Zablotsky, Diane L., and Marcia G. Ory. 1995. Fulfilling the potential: Modifying the current long-term care system to meet the needs of persons with AIDS. *Research in the Sociology of Health Care* 12: 313–28.

CHAPTER 14

Spaceship Earth

Population, Urbanization, and the Human Environment

Alan Mc Evoy

Population expansion and related ecological problems are with us today and will remain with us for many years to come—if we survive. This is the verdict of many experts, called *demographers*, who study changes in human populations. Three variables are crucial to population change: fertility (the birth rate), mortality (the death rate), and migration (population movement). In studying populations, demographers focus on how these variables affect three different areas:

1. Population growth and decline: size of the population.
2. Population distribution: where the population is located.
3. Population composition and structure: characteristics of the population, such as age, sex, education, and so forth.

Each of these areas must be examined to understand the population in the world today. The readings in this chapter deal with these factors and with the impact of population on the environment.

We must be concerned with the rapid growth rate that many countries are experiencing, both for humanitarian reasons and because of the increasing demand on scarce resources. The Population Reference Bureau estimates that in developing countries population can double every 20 to 35 years. This means that natural resources, food-production capacity, and other essentials must double in the same period in order to maintain present lifestyles. Yet most countries are demanding more food, improved communications, better education, scientific advances, and higher standards of living at the same time that resources and capacities to produce are already severely strained or declining.

The themes running through these readings are (1) factors that change the population of nations, such as birthrates and migration; (2) the stress of increased population on the environment and natural resources, (3) human migration patterns, (4) the results of migration to urban areas, and (5) major events that have affected the world's population and environment in recent years.

As you read these articles, consider the implications of population conditions and urbanization along with the benefits and value of protecting our environment.

A major factor in the depletion of the earth's resources is overconsumption by developed countries. Exploitation of resources by the world's richest countries threatens the well-being of all. Lester Brown raises questions about the relationships between birthrate and population growth, use of the world's resources, and the human condition.

Another population variable, migration, is the subject of William Frey's article. As population increases in many countries and resources are strained, some people migrate to "gateway" cities in the United States to seek greater economic opportunities where resources are more plentiful. Domestic migration takes different routes, but for similar reasons—a better life and improvement of the human condition.

Cities developed centuries ago as agricultural surplus freed people from farming. Those not needed in the vital task of feeding the population were attracted to urban areas by work opportunities and the excitement of city living. This process of migration to cities continues today in many parts of the world.

Urban areas are the focus of the next two articles. Louis Wirth wrote about urbanism in 1934 in an essay still considered a classic today for its insights on variables affecting urban life. However, the rapid expansion of cities has caused strains on communications, transportation, and distribution of needed goods and services. Rapidly growing urban areas struggle to meet the needs of swelling populations. At the same time, those who can afford to move to the suburbs cause increased problems for cities by reducing the urban tax base. Unable to cope with deteriorating infrastructures and the demands of newcomers, cities are decaying in both physical structure and social control. Molly O'Meara Sheehan focuses on urban slums created by the influx of migrants from rural areas and the struggles of many cities to meet the needs of their populations.

Finally, we include a timetable illustrating changes that have taken place in the world's environment, the human condition, and human rights in one recent year.

As you read these selections, consider how population dynamics relate to economic crises, environmental degradation, and city problems. How might we change our consumption patterns to influence and help solve world problems?

66

A Planet Under Stress

Rising to the Challenge

LESTER R. BROWN

Population and the Environment: *The world's nonrenewable resources are finite—they won't last forever. Lester Brown raises questions about growing populations and reduction in resources, leaving much of the world with hunger, illiteracy, disease, water shortages, desertification—the list goes on and on. Brown outlines strategies some countries are pursuing to address environmental problems that affect the well-being of humans and the sustainability of the environment.*

Think about these questions as you read this article:

1. *What problems are intensifying as the world's population increases?*
2. *What solutions are being tried in different parts of the world? Are you aware of any attempts to deal with problems in your community?*
3. *To what is Brown referring when he talks of "honest market," "taxing indirect costs," and "shifting subsidies"?*
4. *Which of the ideas discussed might work in your community?*

GLOSSARY **Honest market** An economic market that reflects ecological/environmental realities. **Tax shifting** Tax alternatives such as lowering income taxes while raising taxes on environmentally destructive activities. **Shifting subsidies** Changing the underwriting (paying) for environmentally destructive activities to subsidies for environmentally sound activities.

Early in this new century, the world is facing many longstanding social challenges, including hunger, illiteracy, and disease. If developing countries add nearly 3 billion people by mid-century, as projected, population growth will continue to undermine efforts to improve the human condition. The gap between the billion richest and the billion poorest will continue to widen, putting even more stress on the international political fabric.

As a species, our failure to control our numbers is taking a frightening toll. Slowing population growth is the key to eradicating poverty and its distressing symptoms, and, conversely, eradicating poverty is the key to slowing population growth. With time running out, the urgency of moving simultaneously on both fronts seems clear.

The challenge is to create quickly the social conditions that will accelerate the shift to smaller families. Among these conditions are universal education, good nutrition, and prevention of infectious diseases. We now have the knowledge and resources

SOURCE: Originally printed in *The Futurist* Vol. 37, Issue 6 (November/December 2003), pp. 18–24.

to reach these goals. In an increasingly integrated world, we also have a vested interest in doing so.

Historically, we have lived off the interest generated by the earth's natural capital assets, but now we are consuming those assets themselves. We have built an environmental bubble economy, one where economic output is artificially inflated by overconsumption of the earth's natural assets. The challenge today is to deflate the bubble before it bursts.

Keeping the bubble from bursting will require an unprecedented degree of international cooperation to stabilize population, climate, water tables, and soils—and at wartime speed. Indeed, in both scale and urgency the effort required is comparable to U.S. mobilization during World War II.

Our only hope now is rapid systemic change—change based on market signals that tell the ecological truth. This means restructuring the tax system: lowering income taxes and raising taxes on environmentally destructive activities, such as fossil fuel burning, to incorporate the ecological costs. Unless we can get the market to send signals that reflect reality, we will continue making faulty decisions as consumers, corporate planners, and government policy makers. Ill-informed economic decisions and the economic distortions they create can lead to economic decline.

Continuing with business as usual offers an unacceptable outcome—continuing environmental degradation and disruption and a bursting of the economic bubble. The warning signals are coming more frequently, whether they be collapsing fisheries, melting glaciers, or falling water tables. Thus far the wake-up calls have been local, but soon they could become global, and time is running out. Bubble economies, which by definition are artificially inflated, do not continue indefinitely. Our demands on the earth exceed its regenerative capacity by a wider margin with each passing day.

DEFLATING THE BUBBLE

Stabilizing world population at about 7.5 billion is central to avoiding economic breakdown in countries with large projected population increases that are already overconsuming their natural capital assets. Some 36 countries, all in Europe except Japan, have essentially stabilized their populations. The challenge now is to create the economic and social conditions and to adopt the priorities that will lead to population stability in all remaining countries. The keys here are extending primary education to all children, providing vaccinations and basic health care, and offering reproductive health care and family-planning services in all countries.

Shifting from a carbon-based to a hydrogen-based energy economy to stabilize climate is now technologically possible. Advances in wind turbine design and in solar cell manufacturing, the availability of hydrogen generators, and the evolution of fuel cells provide the technologies needed to build a climate-benign hydrogen economy. Moving quickly from a carbon-based to a hydrogen-based energy economy depends on getting the price right and on incorporating the indirect costs of burning fossil fuels into the market price.

Iceland is the first country to adopt a national plan to convert its carbon-based energy economy to one based on hydrogen. Denmark now gets 18% of its electricity from wind turbines and plans to increase this to 40% by 2030. Japan leads the world in electricity generation from solar cells. The Netherlands leads the industrial world in exploiting the bicycle as an alternative to the automobile. The Canadian province of Ontario is emerging as a leader in phasing out coal. It plans to replace its five coal-fired power plants with gas-fired plants, wind farms, and efficiency gains. This initiative calls for the first plant to close in 2005 and the last one in 2015. The resulting reduction in carbon emissions is equivalent to taking 4 million cars off the road.

Stabilizing water tables depends on quickly raising water productivity. It is difficult to overstate the urgency of this effort. Failure to stop the fall in water tables by systematically reducing water use will lead to the depletion of aquifers, an abrupt cutback in water supplies, and the risk of a precipitous drop in food production. By pioneering in drip irrigation technology, Israel has become the world leader in the efficient use of agricultural water. This unusually labor-intensive irrigation practice, now being used to produce high-value crops in many countries, is ideally suited where water is scarce and labor is abundant.

With soil erosion, we have no choice but to reduce the loss to the rate of new soil formation or below. The only alternative is a continuing decline in the inherent fertility of eroding soils and cropland abandonment. South Korea, with once denuded mountainsides and hills now covered with trees, has achieved a level of flood control, water storage, and hydrological stability that is a model for other countries. In the United States as well, farmers have reduced soil erosion by nearly 40% in less than two decades thanks to a combination of several programs and practices.

Thus all the things we need to do to keep the bubble from bursting are now being done in at least a few countries. If these highly successful initiatives are adopted worldwide, and quickly, we can deflate the bubble before it bursts, similar to the way U.S. mobilization helped lead Allied forces to victory in less than four years.

In retrospect, the speed of the conversion from a peacetime to a wartime economy at the beginning of World War II was stunning. One month after Pearl Harbor, President Roosevelt announced plans to produce 60,000 planes, 45,000 tanks, 20,000 anti-aircraft guns, and 6 million tons of merchant shipping. The automobile industry went from producing nearly 4 million cars in 1941 to producing 24,000 tanks and 17,000 armored cars in 1942—but only 223,000 cars, and most of which were produced early in the year, before the conversion began. Essentially the auto industry was closed down from early 1942 through the end of 1944. In 1940, the United States produced some 4,000 aircraft. In 1942, it produced 48,000. By the end of the war, more than 5,000 ships were added to the 1,000 that made up the American Merchant Fleet in 1939.

Various other firms likewise converted. A spark-plug factory switched to producing machine guns; a manufacturer of stoves produced lifeboats; a merry-go-round factory made gun mounts; a toy company turned out compasses; a corset manufacturer produced grenade belts; and a pinball machine plant began to make armor-piercing shells.

This mobilization of resources within a matter of months demonstrates that a country and, indeed, the world can restructure its economy quickly if it is convinced of the need to do so.

CREATING AN HONEST MARKET

The key to restructuring the economy is the creation of an honest market, one that tells the ecological truth. The market has three fundamental weaknesses: It does not incorporate the indirect costs of providing goods or services into prices. It does not value nature's services properly. It does not respect the sustainable-yield thresholds of natural systems such as fisheries, forests, rangelands, and aquifers.

As the global economy has expanded and as technology has evolved, the indirect costs of some products have become far larger than the price fixed by the market. The price of a gallon of gasoline, for instance, includes the cost of production but not the expense of treating respiratory illnesses from breathing polluted air or the repair bill from acid rain damage. Nor does it cover the cost of rising global temperature, ice melting, more destructive storms, or the relocation of millions of refugees forced from their homes by sea-level rise.

If we have learned anything over the last few years, it is that accounting systems that do not tell the truth can be costly. Faulty corporate accounting systems that overstate income or leave costs off the books have driven some of the world's largest corporations into bankruptcy, costing millions of people their lifetime savings, retirement incomes, and jobs.

Unfortunately, we also have a faulty economic accounting system at the global level, but with potentially far more serious consequences. Economic prosperity is achieved in part by running up ecological deficits, costs that do not show up on the books, but costs that someone will eventually pay. Some of the record economic prosperity of recent decades has come from consuming the earth's productive assets and from destabilizing its climate.

No one has attempted to assess fully the worldwide costs of rising temperature and then to allocate them by gallon of gasoline or ton of coal. A summary of eight studies done during the 1990s indicates that, if the price were raised enough to make drivers pay some of the indirect costs of automobile use, a gallon of gas would cost anywhere from $3.03 to $8.64, with the variations largely due to how many indirect costs were covered. For example, some studies included the military costs of protecting petroleum supply lines and ensuring access to Middle Eastern oil, while others did not. No studies, unfortunately, incorporated all the costs of using gasoline—including the future inundation of coastal cities, island countries, and rice-growing river floodplains.

Not only are some of the looming costs associated with continued fossil fuel burning virtually incalculable, but the outcome is unacceptable. What is the cost of inundating half of Bangladesh's riceland by a one-meter rise in sea level? How much is this land worth in a country that is the size of New York state and has a population half that of the United States? And what would be the cost of relocating the 40 million Bangladeshis who would be displaced by the one-meter rise in sea level? Would they be moved to another part of the country? Or would they migrate to less densely populated countries, such as the United States, Canada, Australia, or Brazil?

Another challenge in creating an honest market is to get it to value nature's services. For example, after several weeks of flooding in the Yangtze River basin in 1998 inflicted $30 billion worth of damage, the Chinese government announced that it was banning all tree cutting in the basin. It justified the ban by saying that trees standing are worth three times as much as trees cut.

Once we calculate all the costs of a product or service, we can incorporate them into market prices by restructuring taxes. If we can get the market to tell the truth, then we can avoid being blindsided by faulty accounting systems that lead to bankruptcy.

TAXING INDIRECT COSTS

The need for tax shifting—lowering income taxes while raising taxes on environmentally destructive activities—in order to get the market to tell the truth has been widely endorsed by economists. The basic idea is to establish a tax that reflects the indirect costs to society of an economic activity. For example, a tax on coal would incorporate the increased health-care costs associated with breathing polluted air, the costs of damage from acid rain, and the costs of climate disruption.

Among the activities taxed in Europe are carbon emissions, emissions of heavy metals, and the generation of garbage (so-called landfill taxes). The Nordic countries, led by Sweden, pioneered tax shifting at the beginning of the 1990s. By 1999, a second wave of tax shifting was under way, this one including the larger economies of Germany, France, Italy, and the United Kingdom. Tax shifting does not change the level of taxes, only their composition. One of the better known changes was a four-year plan adopted in Germany in 1999 to shift taxes from labor to energy. By 2001, this had lowered fuel use by 5%. A tax on carbon emissions adopted in Finland in 1990 lowered emissions there 7% by 1998.

There are isolated cases of environmental tax reform elsewhere. The United States, for example, imposed a stiff tax on chlorofluorocarbons to phase them out in accordance with the Montreal Protocol of 1987. At the local level, the city of Victoria, British Columbia, adopted a trash tax of $1.20 per bag of garbage, reducing its daily trash flow 18% within one year.

One of the newer taxes gaining in popularity is the so-called congestion tax. City governments are turning to a tax on vehicles entering the city, or at least the inner part of the city where traffic congestion is most serious. In London, where the average speed of an automobile was nine miles per hour—about the same as a horse-drawn carriage—a congestion tax was adopted in early 2003. The £5 ($8) charge on all motorists driving into the center city between 7 a.m. and 6:30 p.m. immediately reduced the number of vehicles by 24%, permitting traffic to flow more freely while cutting pollution and noise.

Environmental tax shifting usually brings a double dividend. In reducing taxes on income, labor becomes less costly, creating additional jobs while protecting the environment. This was the principal motivation in the German four-year shift of taxes from income to energy. The shift from fossil fuels to more energy-efficient technologies and to renewable sources of energy reduces carbon emissions and represents a shift to more labor-intensive industries. By lowering the air pollution from smokestacks and tailpipes, it also reduces respiratory illnesses, such as asthma and emphysema, and healthcare costs—a triple dividend.

When it comes to reflecting the value of nature's services, ecologists can calculate the values of services that a forest in a given location provides. Once these are determined, they can be incorporated into the price of trees as a stumpage tax of the sort that Bulgaria and Lithuania have adopted. Anyone wishing to cut a tree would have to pay a tax equal to the value of the services provided by that tree. The market would then be telling the truth. The effect of this would be to reduce tree cutting, since forest services may be worth several times as much as the timber, and to encourage wood and paper recycling.

Some 2,500 economists, including eight Nobel Prize winners in economics, have endorsed the concept of tax shifts. Former Harvard economics professor N. Gregory Mankiw, chairman of the President's Council of Economic Advisers, wrote in Fortune magazine: "Cutting income taxes while increasing gasoline taxes would lead to more rapid economic growth, less traffic congestion, safer roads, and reduced risk of global warming—all without jeopardizing long-term fiscal solvency. This may be the closest thing to a free lunch that economics has to offer." Mankiw could also have added that it would reduce the military expenditures associated with ensuring access to Middle Eastern oil.

The *Economist* has recognized the advantage of environmental tax shifting and endorses it strongly: "On environmental grounds, never mind energy security, America taxes gasoline too lightly. Better than a one-off increase, a politically more feasible idea, and desirable in its own terms, would be a long-term plan to shift taxes from incomes to emissions of carbon." In Europe and the United States, polls indicate that at least 70% of voters support environmental tax reform once it is explained to them.

SHIFTING SUBSIDIES

Each year the world's taxpayers underwrite $700 billion of subsidies for environmentally destructive activities, such as fossil fuel burning, overpumping aquifers, clear-cutting forests, and overfishing. A 1997 Earth Council study, *Subsidizing Unsustainable Development*, observes that "there is something unbelievable about the world spending hundreds of billions of dollars annually to subsidize its own destruction."

Iran provides a classic example of extreme subsidies when it prices oil for internal use at one-tenth the world price, strongly encouraging the consumption of gasoline. The World Bank reports that if this $3.6 billion annual subsidy were phased out it would reduce Iran's carbon emissions by a staggering 49%. It would also strengthen the economy by freeing up public revenues for investment in the country's economic and social development. Iran is not alone. The Bank reports that removing energy subsidies would reduce carbon emissions in Venezuela by 26%, in Russia by 17%, in India by 14%, and in Indonesia by 11%.

Some countries are eliminating or reducing these climate-disrupting subsidies. Belgium, France, and Japan have phased out all subsidies for coal. Germany reduced its coal subsidy from $5.4 billion in 1989 to $2.8 billion in 2002, meanwhile lowering its coal use by 46%. It plans to phase them out entirely by 2010. China cut its coal subsidy from $750 million in 1993 to $240 million in 1995. More recently, it has imposed a tax on high-sulfur coals. Together these two measures helped to reduce coal use in China by 5% between 1997 and 2001 while the economy was expanding by one-third.

The environmental tax shifting described earlier reduces taxes on wages and encourages investment in such activities as wind electric generation and recycling, thus simultaneously boosting employment and lessening environmental destruction. Eliminating environmentally destructive subsidies reduces both the burden on taxpayers and the destructive activities themselves.

Subsidies are not inherently bad. Many technologies and industries were born of government subsidies. Jet aircraft were developed with military R&D expenditures, leading to modern commercial airliners. The Internet was a result of publicly funded efforts to establish links between computers in government laboratories and research institutes. And the combination of the federal tax incentive and a robust state tax incentive in California gave birth to the modern wind power industry.

But just as there is a need for tax shifting, there is also a need for subsidy shifting. A world facing the prospect of economically disruptive climate change, for example, can no longer justify subsidies to expand the burning of coal and oil. Shifting these subsidies to the development of climate-benign energy sources such as wind power, solar power, and, geothermal power is the key to stabilizing the earth's climate. Shifting subsidies from road construction to rail construction could increase mobility in many situations while reducing carbon emissions.

In a troubled world economy facing fiscal deficits at all levels of government, exploiting these tax and subsidy shifts with their double and triple dividends can help balance the books and save the environment. Tax and subsidy shifting promise both gains in economic efficiency and reductions in environmental destruction, a win-win situation.

A CALL TO GREATNESS

There is a growing sense among the more thoughtful political and opinion leaders worldwide that business as usual is no longer a viable option and that, unless we respond to the social and environmental issues that are undermining our future, we may not be able to avoid economic decline and social disintegration. The prospect of failing states is growing as megathreats such as the HIV epidemic, water shortages, and land hunger threaten to overwhelm countries on the lower rungs of the global economic ladder. Failed states are a matter of concern not only because of the social costs to their people but also because they serve as ideal bases for international terrorist organizations.

We have the wealth to achieve these goals. What we do not yet have is the leadership. And if the past is any guide to the future, that leadership can only come from the United States. By far the wealthiest society that has ever existed, the United States has the resources to lead this effort. Economist Jeffrey Sachs sums it up well: "The tragic irony of this moment is that the rich countries are so rich and the poor so poor that a few added tenths of one percent of GNP from the rich ones ramped up over the coming decades could do what was never before possible in human history: ensure that the basic needs of health and education are met for all impoverished children in this world. How many more tragedies will we suffer in this country before we wake up to our capacity to help make the world a safer and more prosperous place not only through military might, but through the gift of life itself?"

The additional external funding needed to achieve universal primary education in the 88 developing countries that require help is conservatively estimated by the World Bank at $15 billion per year. Funding for an adult literacy program based largely on volunteers is estimated at $4 billion. Providing for the most basic health care is estimated at $21 billion by the World Health Organization. The additional funding needed to provide reproductive health and family planning services to all women in developing countries is $10 billion a year.

Closing the condom gap and providing the additional 9 billion condoms needed to control the spread of HIV in the developing world and eastern Europe requires $2.2 billion—$270 million for condoms and $1.9 billion for AIDS prevention education and condom distribution. The cost per year of extending school lunch programs to the 44 poorest countries is $6 billion per year. An additional $4 billion per year would cover the cost of assistance to preschool children and pregnant women in these countries.

In total, this comes to $62 billion. If the United States offered to cover one-third of this additional funding, the other industrial countries would almost certainly be willing to provide the remainder, and the worldwide effort to eradicate hunger, illiteracy, disease, and poverty would be under way.

This reordering of priorities means restructuring the U.S. foreign policy budget. Stephan Richter, editor of *The Globalist*, notes, "There is an emerging global standard set by industrialized countries, which spend $1 on aid for every $7 they spend on defense. . . . At the core, the ratio between defense spending and foreign aid signals whether a nation is guided more by charity and community—or by defensiveness." And then the punch line: "If the United States were to follow this standard, it would have to commit about $48 billion to foreign aid each year." This would be up from roughly $10 billion in 2002.

The challenge is not just to alleviate poverty, but in doing so to build an economy that is compatible with the earth's natural systems—an eco-economy, an economy that can sustain progress. This means a fundamental restructuring of the energy economy and a substantial modification of the food economy. It also means raising the productivity of energy and shifting from fossil fuels to renewables. It means raising water productivity over the next half century, much as we did land productivity over the last one.

This economic restructuring depends on tax restructuring, on getting the market to be ecologically honest. Hints of what might lie ahead came from Tokyo in early 2003 when Environment Minister Shunichi Suzuki announced that discussions were to begin on a carbon tax, scheduled for adoption in 2005. The benchmark of political leadership in all countries will be whether or not leaders succeed in restructuring the tax system.

It is easy to spend hundreds of billions in response to terrorist threats, but the reality is that the resources needed to disrupt a modern economy are small, and a Department of Homeland Security, however heavily funded, provides only minimal protection from suicidal terrorists. The challenge is not just to provide a high-tech military response to terrorism, but to build a global society that is environmentally sustainable, socially equitable, and democratically based—one where there is hope for everyone. Such an effort would more effectively undermine the spread of terrorism than a doubling of military expenditures.

We can build an economy that does not destroy its natural support systems, a global community where the basic needs of all the earth's people are satisfied, and a world that will allow us to think of ourselves as civilized. This is entirely doable. To paraphrase Franklin Roosevelt at another of those hinge points in history, let no one say it cannot be done.

The choice is ours—yours and mine. We can stay with business as usual and preside over a global bubble economy that keeps expanding until it bursts, leading to economic decline. Or we can be the generation that stabilizes population, eradicates poverty, and stabilizes climate. Historians will record the choice, but it is ours to make.

67

Zooming In on Diversity

WILLIAM H. FREY

Migration Patterns: *The United States is diverse, sometimes dubbed a "tossed salad," sometimes a "melting pot" of diverse peoples, but always a mix of many types of domestic and immigrant groups. Patterns of group movement (migration) and settlement differ by the status of group members in the society. States gaining population fall into one of two categories: those that have gateway cities for immigrants and the warmer southern states below the 35th parallel, the politically "blue states,"* where northerners are migrating for jobs and retirement; population losses are occurring in northern and mid-country states. Frey discusses diversity and migration patterns in the United States.*

As you read this selection, consider the following questions:

1. *What are the differences between cities that are magnets for immigrants versus domestic migrants?*
2. *What are different migration patterns between minority groups (African American, Hispanic, Asian) in the United States?*
3. *What are the prospects for the United States becoming more or less of a "melting pot"?*
4. *What are the migration patterns in your community? Do they involve in-migration or out-migration?*

GLOSSARY **Melting pot** Biological and cultural mixing of racial and ethnic groups. **Magnet communities** Destination communities for immigrants or domestic migrants seeking familiarity and comfort zones. For immigrants these are also known as gateway communities. **Multicultural flight** Movement of diverse peoples from urban to suburban areas and (in the U.S.) from north to south.

America's changing racial and ethnic makeup will profoundly transform the nation's regional landscape for at least the next four decades. Consumer markets, politics and day-to-day personal transactions simply will not go on as they have up to now as this change sweeps the nation. By 2050, only half the population will be non-Hispanic white, the Census Bureau projects. The Hispanic and Asian populations will both triple, the black population will almost double and the white population will barely hold its own.

Yet, what looks to the naked eye like a diversified melting pot at the national level takes on a dramatically different look if you zoom in on

*Jim Taylor, "Manifest Destiny 3.0," *American Demographics* (September 2004): 30.

SOURCE: Originally printed in *American Demographics*, Vol. 26, Issue 6 (July/August 2004), pp. 27–32.

specific regions and metropolitan areas. Why? Mostly because people moving to the United States from other countries pick certain places to settle, whereas people who already live within national borders choose others. The lion's share of immigrating Hispanics and Asians tends to cluster in gateway locales, while domestic migration networks of whites and blacks often follow different paths. And there are reasons for this.

Diverging migration patterns have unevenly distributed racial and ethnic diversity into America's regions. Three states—New Mexico, Hawaii and California—already stand out as the nation's first nonwhite majority states. At the same time, there are the 15 states, where minorities account for less than 15 percent of the population. Each minority group, including Hispanics, Asians and blacks, has tended to cluster in geographic patterns that begin to suggest staying power.

Exactly what is the role of race and ethnicity in ingrained and emerging patterns of migration? How rapidly are Hispanics and Asians dispersing away from the traditional gateway regions? And to what extent is their migration converging with the mainstream? New answers to these questions can be drawn from census migration statistics for 2000–2003, and freshly compiled Census 2000 race migration statistics for 1995–2000. After analysis, we might assert that while America may not become a true national melting pot anytime soon, there is a measure of "simmering" going on.

IMMIGRANT MAGNETS. DOMESTIC MIGRANT MAGNETS

Arriving immigrant minorities tend to cluster geographically because destination communities provide them with a comfort zone of familiarity, while others require greater acclimatization for new residents to survive and thrive. Ethnic enclaves in gateway metropolises like Los Angeles and New York contain already established institutions—churches, community centers, stores, neighbors—that make new arrivals feel at home, and give them social and economic support. Immigration laws also foster clustering. Since family reunification is regarded as a priority in legal immigration, family-related migrant "chains" direct many new arrivals to the nation's gateway cities.

If immigrants choose a U.S. destination based on familiarity and cultural support to get started, domestic migrants tend to move for more pragmatic, hardheaded reasons. Most times, it's economic opportunity in the labor market. What's more, whites and blacks are better represented among domestic migrants. The places in which they choose to resettle are less constrained than those of immigrants, for whom a familiar language and the presence of family mean more than the local unemployment rate in selecting a destination.

A distinction between immigration and domestic migration—not a new phenomenon—still holds (see Table 1). Among the nine leading "magnet metro areas" for immigrants, only one—Dallas—is on the list of the largest metro destinations of domestic migrants. Traditional gateways, New York, Los Angeles, San Francisco, Chicago, Miami and Washington, D.C. attract the greatest number of immigrants. These six areas have led all others since the mid-1960s, when current immigration laws came into effect. They've become permanent beacons for newcomers, despite the fact that geographic labor markets have shifted through the years.

But even as the welcome mat cities stay the same, domestic migrant magnets come and go and change as the pushes and pulls of metro economies shift opportunity from place to place over time. Clearly the past decade has brought population and jobs to western areas such as Phoenix and Las Vegas, in addition to longstanding southern juggernauts, Atlanta and Dallas. Increased appeal among metros in Florida (Tampa, Orlando, West Palm Beach) and North Carolina (Charlotte and Raleigh) is also evident.

A striking fact to take note of: if a place appeals to immigrants, it tends to have the opposite effect on people who choose to move domestically. Seven of nine leading immigrant magnets lose domestic migrants to the rest of the country.

TABLE 1 Metropolitan* Magnets for Immigrants and Domestic Migrants, 1995–2003

Immigrant Magnet Metros				Domestic Migration Magnets			
		Immigrants from Abroad	Net Domestic Migration			Immigrants from Abroad	Net Domestic Migration
1	New York	1,605,530	−1,511,765	1	Phoenix	224,305	387,482
2	Los Angeles	1,196,359	−676,213	2	Las Vegas	98,813	368,434
3	San Francisco	613,037	−556,777	3	Atlanta	258,889	338,015
4	Chicago	527,651	−525,974	4	Dallas–Fort Worth	386,647	212,758
5	Miami	493,056	−162,715	5	Tampa–St. Petersburg	99,097	206,223
6	Washington, D.C.	451,546	−22,018	6	Orlando	112,061	188,480
7	Dallas–Fort Worth	386,647	212,758	7	Sacramento	89,368	155,167
8	Houston	353,738	22,794	8	Austin	83,113	146,412
9	Boston	301,915	−141,665	9	Charlotte	66,159	143,406

*Metro areas are defined as CMSAs, MSAs and NECMAs. Names are abbreviated.

SOURCE: William H. Frey analysis of Census 2000 and U.S. Census Bureau Population Estimates, 2000–2003.

New York, Los Angeles, San Francisco and Chicago lose the most domestic migrants among all metros in the U.S. Domestic migrants are leaving immigrant magnets, not as a response to immigrants per se, but because of the increasing congestion and high costs of living in highly urbanized metro areas. The numbers show that recent immigrants, and by extension, immigrant minorities will continue to dominate these areas' population gains.

While migration patterns among immigrants and domestic movers tend to mirror those reported in the 1990s (see "Immigrant and Native Migrant Magnets" *American Demographics*, November 1996), the gravitational pull among areas that attract immigrants does appear to be losing some power. The nine immigrant magnet areas represent 51 percent of all U.S. destinations among recent arrivals, compared with 57 percent in the late 1980s.

The new data also reveals that immigrants play a significantly larger role in the gains of domestic migrant magnets. Domestic in-migrants create jobs in construction and other services that attract immigrants as well. The greater the dispersal among domestic migrants, the speedier the dispersal of immigrant minorities across the country.

The parallel immigration and domestic migration patterns that characterize metropolitan areas also play out at the state level. Again, six immigrant magnet states, California, Texas, New York, Florida, Illinois and New Jersey attract nearly 3 in 5 migrants from abroad, a slight drop from 3 in 4 in the 1980s. Still, high immigration states house more racially and ethnically diverse populations than the entire rest of the country (see Figure 1). However, especially in the West, a significant dispersal of Hispanics and Asians has begun to develop in states such as Nevada, Arizona and Colorado.

WHITES AND BLACKS

It is whites and blacks who largely define and delineate domestic migration patterns in the U.S. And as you might imagine, there are marked differences in the destinations each gravitates toward. White migration tends to run consistent with

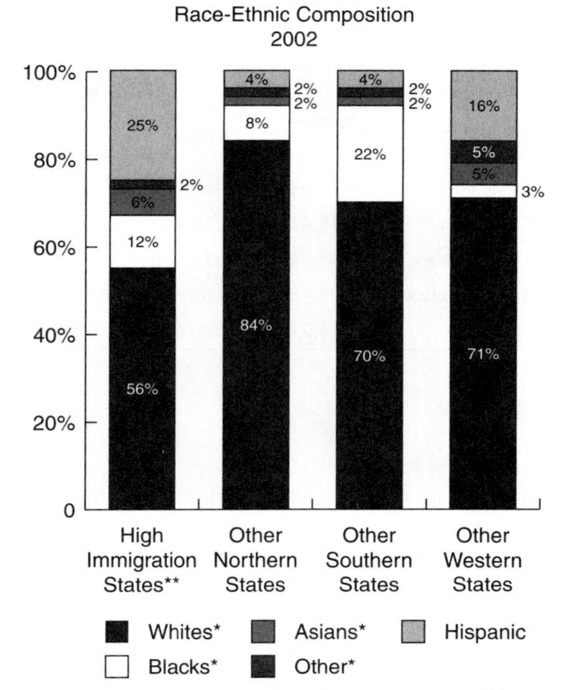

FIGURE 1 High Immigration States Are Melting Pots. Others Are Simmering

*Non-Hispanic members of the racial group.
**California, New York, Texas, Florida, Illinois, New Jersey.
SOURCE: William H. Frey analysis of U.S. Census Bureau Estimates, 2002.

overall domestic migration patterns observed earlier. Most whites exit congested urbanized metropolises such as New York, Chicago, Los Angeles and San Francisco, and several Rust Belt metros. Their destinations tend to be Phoenix and Las Vegas and other Sun Belt "hot spots" (see Table 2).

Blacks have a preference for southern destinations. In fact, the 1990s was the decade in which a surge in a black "return to the South" began, representing a wholesale reversal of the South-to-North migration of earlier decades. For the first time, the South gained blacks in its migration exchanges with each of the other regions of the country. The culture and heritage in the South appear to have a strong appeal among blacks, along with a strong economy. African American Gen Xers and Gen Ys tend to be less concerned by the region's history of racial discrimination, than with available jobs and the chance to network with other middle-class blacks. Young black college graduates lead the way among this group of movers. However, the South is gaining blacks from almost every demographic group, including seniors.

So, while white domestic migrants head to the Sun Belt destinations in both the West and the South, southern destinations are more dominant among black migrants. The top ranking black destination is Atlanta, followed by Orlando, Charlotte and Dallas. At the state level, blacks' preferences are apparent, compared with whites. Georgia, North Carolina, Florida, Maryland and Texas received the largest black gains; while whites show a greater preference for Arizona and Nevada. Still, both racial groups are departing from the same places. Four of the top immigration states, New York, California, Illinois and New Jersey, show the greatest out-migration of both African Americans and Anglos, mirroring the overall domestic migration flight from these areas.

Both the black and white movement patterns help to reinforce the distinct racial and ethnic structures of their destination regions. The out-migration of whites, in particular, serves to increase the minority profiles of the big immigration states. At the same time, the white movement to the "non-California" West tends to give their states and metros a more suburban feel. And the black migration back to the South tends to reinforce the regions longstanding largely white-black demographic profile.

HISPANICS AND ASIANS

There are two ways that Hispanic and Asian movements affect metro areas and states. As we've said, destinations among recent immigrants tend to focus on traditional gateways. However, there is an increasing tendency for both groups to participate in the domestic migration. This seems to be the

TABLE 2 Largest Domestic Migration Gainers & Decliners Whites and Blacks, 1995–2000

	Greatest Gainers					Greatest Decliners			
	Whites		Blacks			Whites		Blacks	
	Metro*	Size	Metro*	Size		Metro*	Size	Metro*	Size
1	Phoenix	169,220	Atlanta	114,478	1	New York	–470,685	New York	–193,344
2	Las Vegas	121,908	Dallas	39,360	2	Chicago	–219,449	Chicago	–59,282
3	Tampa	74,657	Charlotte	23,313	3	Los Angeles	–199,048	Los Angeles	–38,833
4	Austin	70,032	Orlando	20,222	4	San Francisco	–121,180	San Francisco	–30,613
5	Atlanta	66,911	Las Vegas	18,912	5	Detroit	–111,211	Detroit	–15,955

*Metro areas are defined as CMSAs, MSAs and NECMAs. Names are abbreviated.
SOURCE: William H. Frey analysis of Census 2000

TABLE 3 Hispanic Immigrant Destinations & Domestic Migrant Magnets, 1995–2000

Greatest Immigrant Destinations			Greatest Domestic Net In-Migration			Greatest Domestic Net Out-Migration	
	Metro*	Size		Metro*	Size	Metro*	Size
1	Los Angeles	378,858	1	Las Vegas	57,926	Los Angeles	–272,712
2	New York	338,878	2	Phoenix	51,838	New York	–162,246
3	Miami	198,350	3	Dallas	42,853	San Francisco	–60,994
4	Dallas	145,132	4	Orlando	38,173	Chicago	–32,278
5	Chicago	140,069	5	Atlanta	32,831	El Paso	–26,165

*Metro areas are defined as CMSAs, MSAs and NECMAs. Names are abbreviated.
SOURCE: Census 2000

primary vehicle for their dispersion beyond the gateways.

The greatest destinations for Hispanic immigrants include the traditional gateways, led by Los Angeles, New York and Miami. Only two non-immigrant magnet metros—Phoenix and Atlanta—rank in the top 10 recent Hispanic immigrant destinations. Domestic migration among Hispanics contrasts with the destinations of recent immigrants (see Table 3). To a large degree, these Hispanic domestic migration patterns mirror those for whites. Las Vegas and Phoenix represent the top destinations. Similarly, many domestic migrant Hispanics are leaving the traditional immigrant magnet areas of Los Angeles, New York, San Francisco and Chicago. To be sure, the latter metros are gaining many more Hispanics from immigration than are lost through domestic out-migration; but a pattern of dispersal is apparent.

The same pattern is shown at the state level where California, Texas, Florida, New York, Illinois and New Jersey lead all others in gaining Hispanic immigrants.

Yet, four of them are losing Hispanic domestic migrants to a broader swath of states across the country. This dispersion of Hispanic domestic migrants reflects smaller numbers than those of the more concentrated immigrant flows, but the numbers are tending to increase over time.

TABLE 4 Asian Immigrant Destinations & Domestic Migrant Magnets, 1995–2000

	Greatest Immigrant Destinations			Greatest Domestic Net In-Migration		Greatest Domestic Net Out-Migration	
	Metro*	Size		Metro*	Size	Metro*	Size
1	New York	221,132	1	San Francisco	17,881	New York	−36,351
2	Los Angeles	171,806	2	Las Vegas	17,562	Los Angeles	−28,323
3	San Francisco	151,151	3	Dallas	13,605	Honolulu	−12,486
4	Washington, D.C.	64,687	4	Atlanta	13,522	Fresno, CA	−9,266
5	Chicago	58,117	5	Seattle	11,727	Merced, CA	−2,952

*Metro Areas are defined as CMSAs, MSAs and NECMAs. Names are abbreviated.
SOURCE: Census 2000

The Asian patterns are similar to those of Hispanics, wherein immigrant destinations are losing Asians through domestic migration dispersal. However, compared with Hispanics, the dispersion of Asians away from the gateways is much lighter, reflecting their recency of arrival. The exception to this pattern is San Francisco, which draws Asians from both immigration and domestic migration (see Table 4).

MULTICULTURAL FLIGHT

Ten years ago, I wrote an article called "The New White Flight:" (*American Demographics*, April 1994) to highlight the fact that most of the domestic migrants leaving immigrant gateways were non-Hispanic whites. That movement did, and still does, reflect a kind of suburbanization across metropolitan areas and state boundaries toward places that offer affordable housing and improving employment prospects. The difference now is that this domestic out-migration from Los Angeles, New York and other high immigration metros is much more multicultural. There are more Hispanics than whites leaving Los Angeles as a result of domestic migration. In New York, while whites still dominate the exodus, Hispanics and blacks account for a larger share of out-migrants in the late 1990s than 10 years prior. It appears now that blacks and immigrant minorities are responding to the same middle-class crunch as whites, in leaving high immigration metropolitan areas.

The other side of the coin is what is happening in domestic migration magnets. Phoenix and Atlanta provide good examples. In both of these metros, domestic migrating whites and blacks tend to lead the overall gains. But noteworthy in both metros is a phenomenon in the increasing domestic in-migration of Hispanics and, to a lesser degree, Asians. While the overall demographic profiles of these areas do not have nearly the melting pot character of immigrant magnets like Los Angeles or New York, their strong domestic migration flows are creating employment opportunities for new immigrants.

CONCLUSION

Recent race-ethnic migration patterns show that America still has a long way to go before it becomes a coast-to-coast melting pot, where racial and ethnic groups spread evenly across the land. High immigration gateway states and metros continue to stand out as the most racially and ethnically diverse in the nation. They are not, by and large, magnets for white and black domestic migrants who tend to flow to ever-changing constellations of domestic migration magnets. Yet, this analysis also confirms, with migration data from the census, that there exists a true dispersion of Hispanics and Asians as part of a new multicultural flight away from several immigrant gateways. Gradually, these migration patterns will accelerate as they lead to further simmering among our diverse peoples toward a more integrated society.

68

Urbanism as a Way of Life

LOUIS WIRTH

As world population trends show people moving from rural to urban areas, newcomers to cities find different lifestyles. In a classic discussion of the urban way of life, Wirth points out three characteristics of urbanism—size of urban populations, density of the populations, and heterogeneity. These three variables explain many of the differences between urban and rural life and life in different-sized cities.

As you read this classic article about urban life, consider the following:

1. *What key variables describe differences in urban and rural life?*
2. *What elements of urban life make this a logical area of study for sociologists?*
3. *What are some examples of these differences in your community versus the surrounding urban or rural areas?*

GLOSSARY **Population aggregate** Those people who mass in certain areas, in this case urban areas. **Density** Concentration of population in a limited space. **Heterogeneity** A variety of different types of peoples (class, ethnic background, etc.).

A SOCIOLOGICAL DEFINITION OF THE CITY

For sociological purposes a city may be defined as a relatively large, dense, and permanent settlement of socially heterogeneous individuals. On the basis of the postulates which this minimal definition suggests, a theory of urbanism may be formulated in the light of existing knowledge concerning social groups. . . .

The central problem of the sociologist of the city is to discover the forms of social action and organization that typically emerge in relatively permanent, compact settlements of large numbers of heterogeneous individuals. We must also infer that urbanism will assume its most characteristic and extreme form in the measure in which the conditions with which it is congruent are present. Thus the larger, the more densely populated, and the more heterogeneous a community, the more accentuated the characteristics associated with urbanism will be....

To say that large numbers are necessary to constitute a city means, of course, large numbers in relation to a restricted area or high density of settlement. There are, nevertheless, good reasons for treating large numbers and density as separate factors, because each may be connected with significantly different social consequences. Similarly the need for adding heterogeneity to numbers of population as a necessary and distinct criterion of urbanism might be questioned, since we should expect the range of differences to increase with numbers. In defense, it may be said that the city shows a kind and degree of heterogeneity of population which cannot be wholly accounted for by the law of large numbers or adequately represented by means of a normal distribution curve. Because the population of the city does not reproduce itself, it must recruit its migrants from other cities, the countryside, and ... from other countries. The city has thus historically been the melting-pot of races, peoples, and cultures, and a most favorable breeding-ground of new biological and cultural hybrids. It has not only tolerated but rewarded individual differences. It has brought together people from the ends of the earth *because* they are different and thus useful to one another, rather than because they are homogeneous and like-minded.

A number of sociological propositions concerning the relationship between (a) numbers of population, (b) density of settlement, (c) heterogeneity of inhabitants and group life can be formulated on the basis of observation and research.

Size of the Population Aggregate

Ever since Aristotle's *Politics*, it has been recognized that increasing the number of inhabitants in a settlement beyond a certain limit will affect the relationships between them and the character of the city. Large numbers involve, as has been pointed out, a greater range of individual variation. Furthermore, the greater the number of individuals participating in a process of interaction, the greater is the *potential* differentiation between them. The personal traits, the occupations, the cultural life, and the ideas of the members of an urban community may, therefore, be expected to range between more widely separated poles than those of rural inhabitants.

That such variations should give rise to the spatial segregation of individuals according to color, ethnic heritage, economic and social status, tastes and preferences, may readily be inferred. The bonds of kinship, of neighborliness, and the sentiments arising out of living together for generations under a common folk tradition are likely to be absent or, at best, relatively weak in an aggregate the members of which have such diverse origins and backgrounds. Under such circumstances competition and formal control mechanisms furnish the substitutes for the bonds of solidarity that are relied upon to hold a folk society together.

Increase in the number of inhabitants of a community beyond a few hundred is bound to limit the possibility of each member of the community knowing all the others personally. Max Weber, in recognizing the social significance of this fact, explained that from a sociological point of view large numbers of inhabitants and density of settlement mean a lack of that mutual acquaintanceship which ordinarily inheres between the inhabitants in a neighborhood.[1] The increase in numbers thus involves a changed character of the social relationships. As Georg Simmel points out: "[If] the unceasing external contact of numbers of persons in the city should be met by the same number of inner reactions as in the small town, in which one knows almost every person he meets and to each of whom he has a positive relationship, one would be completely atomized internally and would fall into an unthinkable mental condition."[2] The multiplication of persons in a state of interaction under conditions which make their contact as full personalities impossible produces that segmentalization of human relationships which has sometimes been seized upon by students of the mental life of the cities as an explanation for the "schizoid" character of urban personality. This is not to say that the urban inhabitants have fewer acquaintances than rural inhabitants, for the reverse may actually be true; it means rather that in relation to the number of people whom they see and with whom they rub elbows in the course of daily life, they know a smaller proportion, and of these they have less intensive knowledge.

Characteristically, urbanites meet one another in highly segmental roles. They are, to be sure, dependent upon more people for the satisfactions of their life-needs than are rural people and thus are associated with a greater number of organized groups, but they are less dependent upon particular persons, and their dependence upon others is confined to a highly fractionalized aspect of the other's round of activity. This is essentially what is meant by saying that the city is characterized by secondary rather than primary contacts. The contacts of the city may indeed be face to face, but they are nevertheless impersonal, superficial, transitory, and segmental. The reserve, the indifference, and the blasé outlook which urbanites manifest in their relationships may thus be regarded as devices for immunizing themselves against the personal claims and expectations of others.

The superficiality, the anonymity, and the transitory character of urban social relations make intelligible, also, the sophistication and the rationality generally ascribed to city-dwellers. Our acquaintances tend to stand in a relationship of utility to us in the sense that the role which each one plays in our life is overwhelmingly regarded as a means for the achievement of our own ends. Whereas the individual gains, on the one hand, a certain degree of emancipation or freedom from the personal and emotional controls of intimate groups, he loses, on the other hand, the spontaneous self-expression, the morale, and the sense of participation that comes with living in an integrated society. This constitutes essentially the state of *anomie*, or the social void, to which Durkheim alludes in attempting to account for the various forms of social disorganization in technological society.

The segmental character and utilitarian accent of interpersonal relations in the city find their institutional expression in the proliferation of specialized tasks which we see in their most developed form in the professions. The operations of the pecuniary nexus lead to predatory relationships, which tend to obstruct the efficient functioning of the social order unless checked by professional codes and occupational etiquette. The premium put upon utility and efficiency suggests the adaptability of the corporate device for the organization of enterprises in which individuals can engage only in groups. The advantage that the corporation has over the individual entrepreneur and the partnership in the urban-industrial world derives not only from the possibility it affords of centralizing the resources of thousands of individuals or from the legal privilege of limited liability and perpetual succession, but from the fact that the corporation has no soul.

The specialization of individuals, particularly in their occupations, can proceed only, as Adam Smith pointed out, upon the basis of an enlarged market, which in turn accentuates the division of labor. This enlarged market is only in part supplied by the city's hinterland; in large measure it is found among the large numbers that the city itself contains. The dominance of the city over the surrounding hinterland becomes explicable in terms of the division of labor which urban life occasions and promotes. The extreme degree of interdependence and the unstable equilibrium of urban life are closely associated with the division of labor and the specialization of occupations. This interdependence and this instability are increased by the tendency of each city to specialize in those functions in which it has the greatest advantage.

In a community composed of a larger number of individuals than can know one another intimately and can be assembled in one spot, it becomes necessary to communicate through indirect media and to articulate individual interests by a process of delegation. Typically in the city, interests are made effective through representation. The individual counts for little, but the voice of the representative is heard with a deference roughly proportional to the numbers for whom he speaks.

While this characterization of urbanism, in so far as it derives from large numbers, does not by any means exhaust the sociological inferences that might be drawn from our knowledge of the relationship of the size of a group to the characteristic behavior of the members, for the sake of brevity the assertions made may serve to exemplify the sort of propositions that might be developed.

Density

As in the case of numbers, so in the case of concentration in limited space certain consequences of relevance in sociological analysis of the city emerge. Of these only a few can be indicated.

As Darwin pointed out for flora and fauna and as Durkheim noted in the case of human societies,[3] an increase in numbers when area is held constant (i.e., an increase in density) tends to produce differentiation and specialization, since only in this way can the area support increased numbers. Density thus reinforces the effect of numbers in diversifying men and their activities and in increasing the complexity of the social structure.

On the subjective side, as Simmel has suggested, the close physical contact of numerous individuals necessarily produces a shift in the media through which we orient ourselves to the urban milieu, especially to our fellow-men. Typically, our physical contacts are close but our social contacts are distant. The urban world puts a premium on visual recognition. We see the uniform which denotes the role of the functionaries, and are oblivious to the personal eccentricities hidden behind the uniform. We tend to acquire and develop a sensitivity to a world of artifacts, and become progressively farther removed from the world of nature.

We are exposed to glaring contrasts between splendor and squalor, between riches and poverty, intelligence and ignorance, order and chaos. The competition for space is great, so that each area generally tends to be put to the use which yields the greatest economic return. Place of work tends to become dissociated from place of residence, for the proximity of industrial and commercial establishments makes an area both economically and socially undesirable for residential purposes.

Density, land values, rentals, accessibility, healthfulness, prestige, aesthetic consideration, absence of nuisances such as noise, smoke, and dirt determine the desirability of various areas of the city as places of settlement for different sections of the population. Place and nature of work income, racial and ethnic characteristics, social status, custom, habit, taste, preference, and prejudice are among the significant factors in accordance with which the urban population is selected and distributed into more or less distinct settlements. Diverse population elements inhabiting a compact settlement thus become segregated from one another in the degree in which their requirements and modes of life are incompatible and in the measure in which they are antagonistic. Similarly, persons of homogeneous status and needs unwittingly drift into, consciously select, or are forced by circumstances into the same area. The different parts of the city acquire specialized functions, and the city consequently comes to resemble a mosaic of social worlds in which the transition from one to the other is abrupt. The juxtaposition of divergent personalities and modes of life tends to produce a relativistic perspective and a sense of toleration of differences which may be regarded as prerequisites for rationality and which lead toward the secularization of life.[4]

The close living together and working together of individuals who have no sentimental and emotional ties foster a spirit of competition, aggrandizement, and mutual exploitation. Formal controls are instituted to counteract irresponsibility and potential disorder. Without rigid adherence to predictable routines a large compact society would scarcely be able to maintain itself. The clock and the traffic signal are symbolic of the basis of our social order in the urban world. Frequent close physical contact, coupled with great social distance, accentuates the reserve of unattached individuals toward one another and, unless compensated by other opportunities for response, gives rise to loneliness. The necessary frequent movement of great numbers of individuals in a congested habitat causes friction and irritation. Nervous tensions which derive from such personal frustrations are increased by the rapid tempo and the complicated technology under which life in dense areas must be lived.

Heterogeneity

The social interaction among such a variety of personality types in the urban milieu tends to break down the rigidity of caste lines and to complicate the class structure; it thus induces a more ramified and differentiated framework of social stratification than is found in more integrated societies. The heightened mobility of the individual, which brings him within the range of stimulation by a great number of diverse individuals and subjects him to fluctuating status in the differentiated social groups that compose the social structure of the city, brings him toward the acceptance of instability and insecurity in the world at large as a norm. This fact helps to account, too, for the sophistication and cosmopolitanism of the urbanite. No single group has the undivided allegiance of the individual. The groups with which he is affiliated do not lend themselves readily to a simple hierarchical arrangement. By virtue of his different interests arising out of different aspects of social life, the individual acquires membership in widely divergent groups, each of which functions only with reference to a single segment of his personality. Nor do these groups easily permit a concentric arrangement so that the narrower ones fall within the circumference of the more inclusive ones, as is more likely to be the case in the rural community or in primitive societies. Rather the groups with which the person typically is affiliated are tangential to each other or intersect in highly variable fashion.

Partly as a result of the physical footlooseness of the population and partly as a result of their social mobility, the turnover in group membership generally is rapid. Place of residence, place and character of employment, income, and interests fluctuate, and the task of holding organizations together and maintaining and promoting intimate and lasting acquaintanceship between the members is difficult. This applies strikingly to the local areas within the city into which persons become segregated more by virtue of differences in race, language, income, and social status than through choice or positive attraction to people like themselves. Overwhelmingly the city-dweller is not a home-owner, and since a transitory habitat does not generate binding traditions and sentiments, only rarely is he a true neighbor. There is little opportunity for the individual to obtain a conception of the city as a whole or to survey his place in the total scheme. Consequently he finds it difficult to determine what is to his own "best interests" and to decide between the issues and leaders presented to him by the agencies of mass suggestion. Individuals who are thus detached from the organized bodies which integrate society comprise the fluid masses that make collective behavior in the urban community so unpredictable and hence so problematical.

Although the city, through the recruitment of variant types to perform its diverse tasks and the accentuation of their uniqueness through competition and the premium upon eccentricity, novelty, efficient performance, and inventiveness, produces a highly differentiated population, it also exercises a leveling influence. Wherever large numbers of differently constituted individuals congregate, the process of depersonalization also enters. . . .

NOTES

1. *Wirtschaft und Gesellschaft* (Tübingen, 1925), part I, chap. 8, p. 514.
2. "Die Grossstädte und das Geistesleben," *Die Grossstadt*, ed. Theodor Petermann (Dresden, 1903), pp. 187–206.
3. E. Durkheim, *De la division du travail social* (Paris, 1932), p. 248.

4. The extent to which the segregation of the population into distinct ecological and cultural areas and the resulting social attitude of tolerance, rationality, and secular mentality are functions of density as distinguished from heterogeneity is difficult to determine. Most likely we are dealing here with phenomena which are consequences of the simultaneous operation of both factors.

69

Uniting Divided Cities

MOLLY O'MEARA SHEEHAN

People migrate from rural to urban areas and from poor countries to urban centers in wealthier countries (as described in the first reading of this chapter), usually looking for greater economic opportunities or freedom from oppression or conflict. Cities around the world are faced with the task of providing services for the influx of migrants. Often the poor cannot afford urban housing, resulting in squatter settlements that develop at the periphery of cities. Sheehan discusses the rich-poor divide in cities and the problems that slums create for cities trying to offer services for their old and new residents.

Consider the following questions as you read this selection:

1. *What are the divisions within cities discussed by Sheehan, and why do these divisions exist?*
2. *What problems do cities face in trying to deal with the needs of slum dwellers? What are some of these needs?*
3. *What is the "paradox of slums," and how does the author propose dealing with this problem?*
4. *What do you see as the most promising solutions to urban problems?*

GLOSSARY **Slum** Urban area with miserable living conditions that vary from place to place. Also known as *kampong, favela, gecekondu, bidonville,* and *squatter settlement.*

Cities divided into rich and poor, healthy and unhealthy, "legal" and "illegal," are all too common worldwide.[1] In some sense, this is nothing new. Plato observed around 400 BC that "any city, however small, is in fact divided into two, one the city of the poor, the other of the rich." Centuries of technological innovations and social progress have done little to close the gap. Priced out of the "legal" real estate market, hundreds of millions of people seek shelter in the most precarious places,

SOURCE: Originally printed in and Linda Starke et al. (eds.), *State of the World 2003.* Reprinted by permission of Worldwatch Institute Report on Progress toward a Sustainable Society, pp. 130–39. http://www.worldwatch.org/pubs/sow/2003.

on steep hillsides or floodplains, living not only with the constant threat of possible eviction but also more vulnerable to natural disasters, pollution, and disease from lack of water and toilets. More than half the people in Cairo, Nairobi, and Mumbai (formerly Bombay), for example, lack adequate housing—living in slums or even on the pavement.[2]

Slum residents have not gained much from society's intense use of key resources over the last century, a use that has pushed the planet's support systems to their limits. One group of scientists has estimated that people have transformed half of Earth's land surface through agriculture, forestry, and urbanization; contributed to a 30-percent increase in atmospheric carbon dioxide concentration since the beginning of the Industrial Revolution; and today use more than half of all available surface fresh water. The benefits of all this activity, however, have accrued to a relatively wealthy minority. In 2001, 52 percent of the gross world product went to the 12 percent of the world living in industrial nations—the same group responsible for a disproportionate share of industrial timber consumption, paper use, and carbon emissions. These inequities are perhaps most glaring in the world's slums, where the poor are exposed to the worst environmental conditions, including pollution from the wealthy.[3]

While the inequalities of wealth, power, opportunities, and survival prospects that hobble humanity are crystallized in cities, these places will have an important role to play in any shift toward development that does not destroy the environment. At the root of sustainable development—which can be defined as meeting the needs of all today without endangering the prospects of future generations—is the challenge of improving the welfare of billions of people without further undermining Earth's support systems. Cities are where most of the world's people will live and where an even greater share of key planetary resources will be used in the coming decades. Key global environmental problems have their roots in cities—from the vehicle exhaust that pollutes and warms the atmosphere, to the urban demand for timber that denudes forests and threatens biodiversity, to the municipal thirst that heightens tensions over water.[4]

Cities will have to be the building blocks of development that values nature and people—and they do hold enormous potential for both environmental and social progress. When people are concentrated in one place, they ought to be able to use fewer materials, and to recycle them with greater ease, than widely dispersed populations can; at the same time, they should be more easily linked to schools, health care, and other key services. Compared with higher forms of government, city halls are closer to people, so organized citizens theoretically have a better chance of changing the status quo on matters of environmental and social concern. Throughout history, higher levels of health and education have come after periods of urbanization; today, the countries that rank highest in surveys of freedom and human development are also the most urbanized. City-level investments in water infrastructure, waste provision, health, and education match up with national rankings of human development that take into account life expectancy and literacy. Many cities perform better or worse in these measures of "development" than could be explained by income alone, suggesting that municipal policies can make a big difference.[5]

By ensuring that their poorest slum dwellers feel secure in their own homes, can make a living, and are healthy, the world's relatively poorer cities could leapfrog their wealthier counterparts in the North, creating an urban model that values both people and nature. Cities typically are responsible for granting titles to property, providing water and waste disposal, organizing public transportation, and making building codes and land use rules. These activities could be carried out in a way that makes it easier for poor people to survive, while also having environmental benefits for the whole city and the world. Local governments can, for instance, promote metals recycling, organic waste composting, and urban agriculture, can give priority to cheap public transportation, and can allow people to run small businesses out of their homes.

TABLE 1 World's 10 Largest Urban Areas, 1000, 1800, 1900, and 2001

1000		1800		1900		2001	
			(million population)				
Cordova[1]	0.45	Peking[3]	1.10	London	6.5	Tokyo	26.5
Kaifeng	0.40	London	0.86	New York	4.2	São Paulo	18.3
Constantinople[2]	0.30	Canton[4]	0.80	Paris	3.3	Mexico City	18.3
Angkor	0.20	Edo[5]	0.69	Berlin	2.7	New York	16.8
Kyoto	0.18	Constantinople	0.57	Chicago	1.7	Mumbai[7]	16.5
Cairo	0.14	Paris	0.55	Vienna	1.7	Los Angeles	13.3
Baghdad	0.13	Naples	0.43	Tokyo	1.5	Calcutta	13.3
Nishapur	0.13	Hangchow[6]	0.39	St. Petersburg	1.4	Dhaka	13.2
Hasa	0.11	Osaka	0.38	Manchester	1.4	Delhi	13.0
Anhilvada	0.10	Kyoto	0.38	Philadelphia	1.4	Shanghai	12.8

[1]Cordoba today. [2]Istanbul today. [3]Beijing today. [4]Guangzhou today. [5]Tokyo today.
[6]Hangzhou today. [7]Formerly Bombay.

SOURCE: 1000–1900 from Tertius Chandler, *Four Thousand Years of Urban Growth: An Historical Census* (Lewiston, NY: Edwin Mellen Press, 1987); 2001 from U.N. Population Division, *World Urbanization Prospects: The 2001 Revision* (2002).

Such activities have the potential to green cities, create job opportunities, and reduce the demand for materials from logging, mining, and industrial agriculture, all of which take an enormous environmental toll.

Urban centers in the developing South now dominate the ranks of the world's largest cities, so they are well positioned to capture the public's imagination. While most of the world actually lives in smaller cities, towns, and villages, big cities command special attention. Many people either know of or have been to large metropolises, which often serve as national capitals, financial hubs, sites for major airports, and centers of commerce and media. The cities of the industrial North were center stage in this regard for just a brief moment in history, claiming all the slots in the top 10 in 1900. By 2001, however, only Tokyo and New York remained on that list. (See Table 1.) Demographers expect that by 2015, Los Angeles and Shanghai will be bumped from the top 10, as Karachi and Jakarta move up. Why shouldn't some of the cities that lead us toward a more equitable and environmentally friendly model of development be some of these behemoths of the twenty-first century?[6]

In many cases, municipal reforms that benefit the poorest people and nature will be more likely if city halls become more open and accountable. Local governments usually do not boldly address the needs of their poorest people in ways that would yield wide-ranging environmental benefits because people who have more money and influence—from real estate developers to leaders of polluting industries—often push a different agenda. In the last decade, some cities have started to include their poorest citizens in decisionmaking, often with national and international support. From slum dwellers federations worldwide to an innovative budgeting process in many Brazilian cities, poor people's voices are rising in open political arenas. If they are to help unite divided cities, governments will have to work even more closely with large numbers of poor urbanites, many of whom live in slums.[7]

POVERTY AND INEPT GOVERNMENT IN AN URBANIZING WORLD

Slums are an intensely local phenomenon with growing global significance. A neighborhood-by-neighborhood look at the world's cities would reveal that not all poor people live in slums, and that not all slums are uniformly poor. As urban poverty concentrates in slums, however, these neighborhoods offer government officials distinct places on the ground where they could find and work with some of their poorest constituents.

Although "slums" are generally understood to be urban areas with miserable living conditions, they vary dramatically from place to place and are described by a universe of overlapping terms—some of them are colorful; many of them, like "slum," are frankly negative; and few are synonymous. "Squatter settlements" are formed when poor people build shelter on land that does not belong to them. Such settlements may also be called "illegal" or "informal," terms that are often used interchangeably when describing the off-the-books nature of some slums. Other development authorized by landowners that is not in the squatter category may still be illegal or informal because the land is not zoned for building, or because it has been unlawfully subdivided into smaller parcels, or because the dwellings are not up to the standards of building codes.[8]

All these terms can give a false impression of the character of communities without conveying the basic problem of insecurity. Law-abiding people often live in "illegal" housing. Many "squatter settlements" are packed with rent-paying tenants. Neighborhoods settled by squatters decades ago may no longer be slums. And some illegally built or subdivided neighborhoods may be upscale from the outset. As every city has its own history, culture, economy, and real estate peculiarities, each slum has its own look and feel—whether it's a *kampung* in Indonesia, a *favela* in Brazil, a *gecekondu* in Turkey, or a *bidonville* in parts of francophone Africa. Despite the tremendous variation, one common characteristic of slums tends to be the insecurity that residents feel in their own homes, which often thwarts them from improving their living conditions and reaching their full potential.[9]

The United Nations estimates that 712 million people lived in slums in 1993 and that their ranks swelled to at least 837 million by 2001, with slum dwellers accounting for 56 percent of the urban population in Africa, 37 percent in Asia and Oceania, and 26 percent in Latin America and the Caribbean. These rough numbers, drawn from surveys and census data that may be incomplete or out of date, give some sense of the scale of the global slum population, although they may well underestimate it. Another U.N. study suggests that more than 1 billion people worldwide live in slums.[10]

Urban growth is meeting up with poverty and inept governments to fuel the current proliferation of slums. World population increased by some 2.4 billion in the past 30 years, and roughly half of that growth took place in cities. Over the next three decades, the industrial North is not expected to expand in total population very much. In contrast, demographers believe that in many developing countries, urban migration and growth combined with high birth rates will mean that between 2000 and 2030 nearly all of the 2.2 billion people added to world population will end up in urban centers of the developing world. While the size and growth of the urban population in developing nations dominates global population projections, there is always a lag between censuses, and all nations have their own definitions of "urban" that tend to change over time, so these estimates are rough.[11]

Poverty may be even harder to measure on a global basis than population size is, but various studies do point to greater numbers of urban poor. One U.S. dollar will buy far less food in Jakarta or São Paulo than in Dacca or Nairobi—and it will buy even less in New York. For that reason, the international standard of a $1 a day income to denote "extreme poverty" or lack of money to meet basic food needs invariably underestimates poverty in cities. Still, the World Bank suggests that some 1.2 billion people worldwide were extremely poor as of 1998, with rural sub-Saharan Africa and South Asia hardest hit. Martin Ravallion of the World Bank estimates that the urban share of extreme poverty is currently 25 percent worldwide, and likely to reach 50 percent by 2035. By then the urban share of world population is likely to have grown from nearly 50 percent today to more than 60 percent.[12]

While rural people tend to have less access to cash, education, clean water, and sanitation than city dwellers do, the deficits cause more severe problems in an urban setting. People are less able to grow their own food in cities, so they must rely on the cash economy for survival. Urban jobs tend to require higher levels of education. And inadequate sanitation brings infectious disease to more people in cities, where dense populations make it easier for disease to spread. Addressing the World Bank in April 2002, economist Jeffrey Sachs noted that too often the fact that most of the poor live in rural areas is used to argue for only a rural-led growth strategy to end poverty. "We need a better urban-based strategy as well," he pointed out.[13]

Slums take root when local governments fail to serve large numbers of poor people. Many cities in Africa, Asia, and Latin America have housing laws and codes copied from those written in nineteenth- or twentieth-century Europe that make little sense in their current context. Poor people, by building their own shelters, have become the developing world's "most important organizers, builders, and planners," in the words of researchers Jorge Hardoy and David Satterthwaite. Yet most codes are written not for these local builders but for engineers or architects in a different time and place. In Nairobi, for example, Kenyan codes call for the building materials standard in the United Kingdom.[14]

Even if appropriate housing codes were on the books, the larger problem of governments being unable or unwilling to enforce laws and provide needed urban services would remain. In many countries, national governments have given local governments more responsibility for providing services in the last several decades, but have been slower to give cities money from national tax revenue or to allow local governments to raise the needed funds themselves. Moreover, the disparity between the budgets of rich and poor cities is striking. A survey of 237 cities worldwide shows an average municipal revenue per person of just $15.20 in Africa, $248.60 in Asia, $252.20 in Latin America, and $2,763.30 in Western Europe, the United States, Japan, and the rest of the industrial world. The ratio of city budgets in Africa to those of the industrial world, 1:182, is far higher than the 1:51 ratio of per capita income between sub-Saharan Africa and high-income nations.[15]

As money buys political influence virtually everywhere in the world, bribes and kick-backs often keep cash-strapped local officials in developing nations from operating in the interests of their poorest constituents. The nongovernmental organization (NGO) Transparency International, in a ranking of 102 nations, found corruption rampant in many nations with large or growing numbers of urban poor, including Bangladesh, Bolivia, Indonesia, Kenya, Nigeria, and Uganda.[16]

The available data on population, poverty, and corruption, while patchy, thus suggest that the conditions for large and growing slums exist in many parts of the world. The areas of particular concern include sub-Saharan Africa, South Asia, and parts of Latin America. (See Figure 1).[17]

THE PARADOX OF SLUMS

A slum can demonstrate both the very best and the very worst in society, showing the ingenuity of poor people in desperate circumstances as well as the failure of government to make the most of this human energy. People who are not born into informal settlements may find their way there because their other options are far bleaker. While the energy that people in slums may invest in securing a better future for their families shows the resiliency of the human spirit, if government were functioning well, people would not have to try so hard to achieve a decent standard of living. Mtumba, an informal community in Nairobi, is

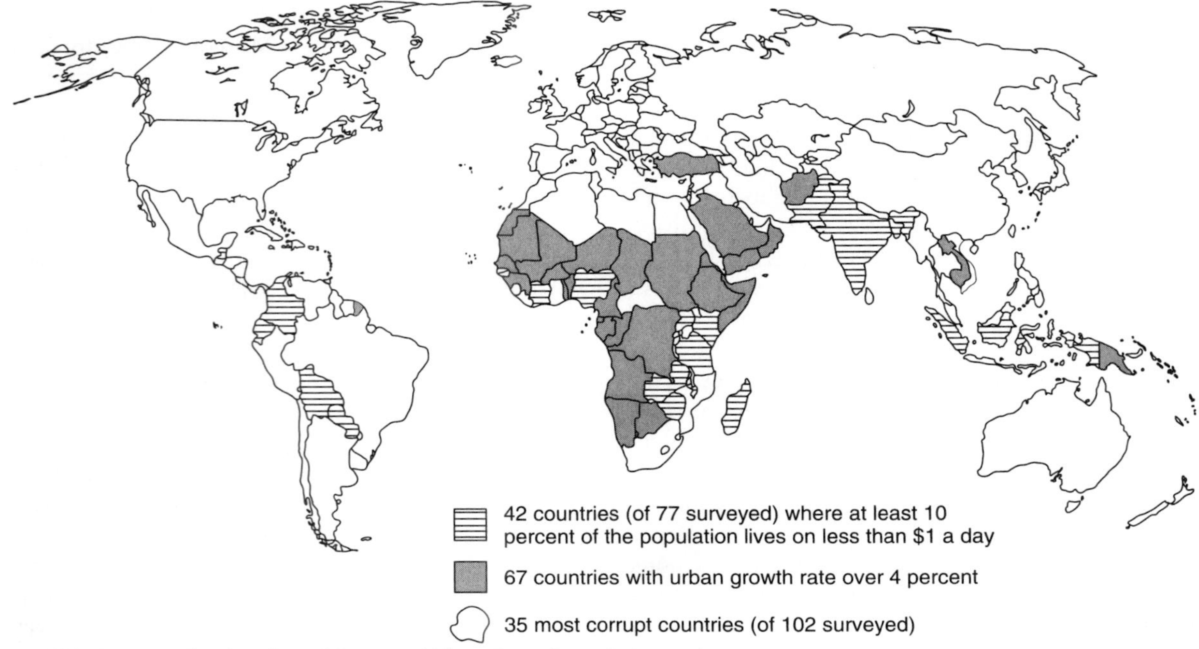

FIGURE 1 The Overlap of Poverty, Urban Growth, and Corruption
SOURCE: World Bank, UNPD, Transparency Intl.

one place where it is easy to see both the good and the bad aspects of life in a slum.[18]

All over the world, people move to new places for better opportunities, and when they choose informal settlements in urban areas it is often because these slums, shanty-towns, or squatter settlements offer the best chance for them to survive. In some cases, slums may offer the most affordable lodging close to jobs, even if the location still requires a very long commute. In general, the "off-the-books" nature of informal communities confers certain advantages. People can skirt zoning laws that separate residences from businesses, and can set up shop inside their home or just outside. Plus they face low short-term costs: low rent and no property taxes.

But the same informality that may help poor people gain a tenuous toehold on the ladder toward economic security can also prevent them from moving up that ladder. As informal settlements legally do not exist, people who live there often lack not only money and political power, but any legal means of solving problems. The owners of slum dwellings can more easily get away with charging exorbitant rents. Although the share of residents who are owners versus renters varies among communities, the proportion of renters is often higher than commonly thought (people assume that most, if not all, inhabitants are recent migrants who have built their own accommodation). The shacks can be lucrative investments, but their owners do not typically reinvest their profits by repairing them or hooking them up to electricity or water, and tenants have no way to hold landlords accountable.[19]

A related irony is that the poorest urban residents often pay the highest price for essential goods and services that are delivered by government at much lower cost to wealthier residents. In some cases, this is because a household without a formal address does not qualify for hookup to the public water system, entry into public schools, or other essentials. In Mumbai, for example, pavement dwellers have trouble obtaining the cards that qualify poor people for food aid and health care. What makes informal settlements so cheap in the short term is that the cost of urban services are not factored in from the beginning, as they would be in the formal construction sector; in the latter case, the government provides streets and services and someone buys the land at the outset, then the buildings are constructed, and only afterwards do people move into finished houses. In contrast, informal settlements begin with people moving onto land, then building homes and businesses over time, and only later, if at all, negotiating connections to urban streets, water systems, and electricity grids and obtaining title to property.[20]

As a result, poor people often end up building their own schools and latrines and purchasing water, at a very high cost, from private vendors. The price of water in the slums may be 7–11 times the tap price of piped water in wealthier areas in Nairobi, 12–25 times the tap price in Dhaka, 16–34 times the tap price in Tegucigalpa, 20–60 times the tap price in Surabaya, and 28–83 times the tap price in Karachi. Toilet stalls, generally operated by governments rather than private vendors, are not as pricey but far less common. Pointing to deficiencies in data supplied by governments, researchers at the UK-based International Institute for Environment and Development (IIED) have estimated that as much as two thirds of the urban population in Africa, Asia, and Latin America has no safe way to dispose of human waste.[21]

Water that is too expensive for slum dwellers to use in adequate amounts, combined with few toilets, leads to disease. When asked at an international conference to sum up the worst environmental health threats to poor people in cities of the developing world, IIED's David Satterthwaite showed a slide with a single word: "shit." At a later meeting, renowned epidemiologist Sir Richard Doll provided a more thorough summary: "bugs and shit." Slum dwellers pay more for each liter of water they consume than wealthier residents hooked up to municipal water mains and sewers, and they disproportionately suffer from water- and waste-borne pathogens—from diarrhea-causing *E. coli* and rota virus to roundworm. "Put bluntly," writes health researcher Carolyn Stephens, "the poor pay more for their cholera." Gullies filled with stagnant water often serve as cesspools in slums, and attract mosquitoes, so slum residents are also more vulnerable to malaria.[22]

Furthermore, the money slum dwellers spend on water, kerosene, or other key items from private vendors does not reach public coffers, where it could then be used to extend public services, from water pipes to health clinics, into slums. In Mumbai, local authorities are beginning to understand that bringing slum dwellers "onto the books" will help the city, and they are now working with the National Slum Dwellers Federation and its partners, other NGOs. "From sanitation to access to policymaking—when poor people are allowed these things, government has an easier job," says Sheela Patel, who directs an NGO in Mumbai, the Society for the Promotion of Area Resource Centres (SPARC), that works with slum dwellers.[23]

Slums can also breed disease that threatens broader public health. While pathogens travel quickly in crowded slum conditions, they do not stop at the gates of wealthier enclaves. By weakening people's immune systems, the AIDS virus makes people more susceptible to other communicable diseases, speeding the transmission of airborne pathogens such as the tuberculosis bacteria. Both HIV and tuberculosis are spreading rapidly in urban centers of the developing world.[24]

Moreover, economic inequality, in the form of glaring disparities between poor slums and posh gated enclaves, may itself be a drag on public health. Researchers comparing U.S. metropolitan areas found a higher level of premature deaths in the places with the highest income inequality, while in 13 industrial countries there were lower levels of premature deaths from certain diseases in more egalitarian countries. One theory to explain these findings is that cities or countries with high levels of inequality may be underinvesting in important physical and social infrastructure, such as education, that could serve to prevent some diseases. Another possibility is that high levels of inequality contribute to social tensions that stress people; as the immune system takes cues from both body and mind, people under stress are more susceptible to illness.[25]

The persistence of slums in an era of unprecedented prosperity may also contribute to tensions that threaten local, national, and even global security. Slums do not create criminals, but the lack of policing in bad neighborhoods allows criminals to victimize a city's poorest people. Following the September 2001 attacks on the United States, *New York Times* columnist Thomas Friedman wrote that in an increasingly interconnected world, it will be impossible to ignore the problems of people living in desperate conditions at home or abroad: "If you don't visit a bad neighborhood, a bad neighborhood will visit you." The educated and relatively wealthy young hijackers who used planes as weapons on September 11 did not come from the slums; however, the contrast of poverty in the Middle East with wealth in the United States and Western Europe appears to have been at least one factor motivating their actions.[26] . . .

NOTES

1. History of apartheid from Keith S. O. Beavon, "Johannesburg: A City and Metropolitan Area in Transformation," in Carole Rakodi, ed., *The Urban Challenge in Africa: Growth and Management of Its Largest Cities* (Tokyo: United Nations University Press, 1997), pp. 150–91; toilets in Johannesburg from and Jo Beall, Owen Crankshaw, and Susan Parnell, "Victims, Villains and Fixers: The Urban Environment and Johannesburg's Poor," *Journal of Southern African Studies*, December 2000, pp. 833–55; Alexandra from Jo Beall, Owen Crankshaw, and Susan Parnell, *Uniting a Divided City: Governance and Social Exclusion in Johannesburg* (London: Earthscan, 2002), pp. 154–59; inequality in public health from Lynn Dalrymple, "Building Healthy Cities and Improving Health Systems for the Urban Poor in South Africa," in Samson James Opolot, ed., *Building Healthy Cities: Improving the Health of Urban Migrants and The Urban Poor in Africa* (Washington, DC: Woodrow Wilson International Center for Scholars, 2002), pp. 121–29; share of Johannesburg's population that is black from Steven Friedman, "A Quest for Control: High Modernism and its Discontents in Johannesburg, South Africa," in Blair A. Ruble et al., eds., *Urban Governance Around the World* (Washington, DC: Woodrow Wilson International Center for Scholars, 2002), p. 32.
2. Stephen Berrisford, "Law and Urban Change in the New South Africa," in Edésio Fernandes and Ann Varley, eds., *Illegal Cities: Law and Urban Change in Developing Countries* (London: Zed Books, 1998), pp. 213–29; Plato, *The Republic*, cited in James Clapp, *The City: A Dictionary of Quotable Thoughts on Cities and Urban Life* (New Brunswick, NJ: Center for Urban Policy Research, Rutgers University, 1984), p. 194; Laila Iskandar Kamel, *The Informal Solid Waste Sector in Egypt: Prospects for Formalization* (Cairo: CID with funding from the Institute of International Educationand the Ford Foundation, January 2001); African Population and Health Research Center, *Population and Health Dynamics in Nairobi's Informal Settlements*, Report of the Nairobi Cross-Sectional Slums Survey (Nairobi: April 2002); Mumbai from Arjun Appadurai, "Deep Democracy: Urban Governmentality and the Horizon of Politics," *Environment and Urbanization*, October 2001, pp. 23–43.
3. Jane Lubchenco, presentation at the State of the Earth conference, Columbia University Earth Institute, 13 May 2002, updating Peter M. Vitousek et al., "Human Domination of Earth's Ecosystems," *Science*, 25 July 1997, pp. 494–99; global world product from David Malin Roodman, "Economic Growth Falters," in Worldwatch Institute, *Vital Signs 2002* (New York: W.W. Norton & Company, 2002), pp. 58–59.
4. Definition of sustainable development from World Commission on Environment and Development, *Our Common Future* (Oxford: Oxford University Press, 1987). Roughly 78 percent of carbon emissions from fossil fuel burning and cement manufacturing, and 76 percent of industrial wood use worldwide, occur in urban areas. Some 60 percent of the planet's water that is tapped for human use goes to cities in one form or another. About half of this water irrigates food crops for urban residents, roughly a third is used by city industry, and the remainder is for drinking and sanitation. These calculations, done by the author in May 1999, are based on urban population from United Nations,

World Urbanization Prospects: The 1996 Revision (New York: 1996), on share of gross domestic product from industry and services from World Bank, *World Development Indicators 1997*, CD-ROM (Washington, DC: 1997), and from World Bank, *World Development Indicators 1998* (Washington, DC: 1998), on carbon emissions from G. Marland et al., "Global, Regional, and National CO_2 Emission Estimates from Fossil Fuel Burning, Cement Production, and Gas Flaring: 1751–1995 (revised March 1999)," Oak Ridge National Laboratory, <cdiac.esd.ornl.gov>, viewed 22 April 1999, on industrial round-wood consumption from U.N. Food and Agriculture Organization, *FAOSTAT Statistics Database*, at <apps.fao.org>, and on water from I.A. Shiklomanov, "Global Water Resources," *Nature and Resources*, vol. 26, no. 3 (1990).

5. Levels of urbanization from U.N. Population Division (UNPD), *World Urbanization Prospects: The 2001 Revision* (New York: 2002); Freedom House, *Freedom in the World: The Annual Survey of Political Rights and Civil Liberties, 2000–2001* (New York: 2001); U.N. Development Programme (UNDP), *Human Development Report 2002* (New York: 2002); Figure 7–1 based on UN-HABITAT, *The State of the World's Cities 2001* (Nairobi: United Nations Centre for Human Settlements, 2001), pp. 116–17.
6. Cities in history from Peter Hall, *Cities in Civilization* (New York: Pantheon Books, 1998); cities today from Joseph Rykwert, *The Seduction of Place: The City in the Twenty-first Century* (New York: Pantheon Books, 2000); cities in 2015 from UNPD, op. cit. note 5.
7. Richard Stren, "Introduction: Toward the Comparative Study of Urban Governance," in Richard Stren and Patricia McCarney, *Urban Governance in the Developing World: Innovations and Discontinuities* (Washington, DC: Woodrow Wilson Center and Johns Hopkins University, in press); Nick Devas with Philip Amis et al., "*Urban Governance and Poverty: Lessons From a Study of Ten Cities in the South* (Birmingham, U.K.: University of Birmingham, June 2001).
8. Jorge E. Hardoy and David Satterthwaite, *Squatter Citizen: Life in the Urban Third World* (London: Earthscan, 1989).
9. The United Nations Human Settlements Programme (UN-HABITAT) uses an index weighted heavily toward lack of secure tenure to estimate the number of people living in slums worldwide; UN-HABITAT, "Millennium Development Goal 7, Target 11, Indicator 31" (Nairobi: 2001), from Christine Auclair, Global Urban Observatory, UN-HABITAT, e-mail to author, 30 October 2002.
10. Estimates for 1993 and 2001 from ibid.; higher estimate of more than 1 billion people living in inadequate housing from UN-HABITAT, op. cit. note 5, p. 30.
11. UNPD, op. cit. note 5, p. 1; David Satterthwaite, *Coping with Rapid Urban Growth*, Leading Edge Series (London: Royal Institution of Chartered Surveyors, September 2002).
12. World Bank, *World Development Report 2000/2001* (New York: Oxford University Press, 2000), p. 23 (the $1 a day estimate is in 1993 purchasing power parity terms); 1.2 billion from Shaohua Chen and Martin Ravallion, *How Did the World's Poorest Fare in the 1990s?* World Bank Group Policy Research Work Paper 2409 (Washington, DC August 2000); Martin Ravallion, "On the Urbanization of Poverty," *Journal of Development Economics*, vol. 857 (2002, in press); Martin Ravallion, e-mail to Elizabeth Bast, World-watch Institute, 15 May 2002.
13. Jeffrey Sachs, "The Economics of Sustainability," Keynote Address for the Distinguished Lecture Series, World Bank, Washington, DC, 16 April 2002.
14. Hardoy and Satterthwaite, op. cit. note 8, p. 15; Kenyan building codes from Richard Stren, University of Toronto, e-mail to author, 23 September 2002.
15. UN-HABITAT, "Global Urban Indicators," at <www.unchs.org/guo/gui/>, data are comparable as of 1993; World Bank, op. cit. note 12, p. 275; Stren, op. cit. note 7.
16. Transparency International, *Transparency International Corruption Perceptions Index 2002* (Berlin: 2002).
17. Figure 1 from the following: urban growth from UNPD, op. cit. note 5; poverty from World Bank, *World Development Indicators 2001* (Washington DC: 2001); corruption from Transparency International, op. cit. note 16, based on surveys from 2000–02.
18. Lawrence Apiyo, Pamoja Trust, Nairobi, and George Ng'ang'a, Tom Werunga, and Isaac Mburu, Mtumba Governing Council, Nairobi, discussions with author, 19 February 2001; F. Nii-Amoo Dodoo, "The Urban Poor and Health Systems in East Africa: Voices from Nairobi'a Slums," in Opolot, op. cit. note 1, pp. 9–19; under-five mortality rates from African Population and Health

Research Center, *Population and Health Dynamics in Nairobi's Informal Settlements*, Report of the Nairobi Cross-Sectional Slums Survey (Nairobi: April 2002), pp. 86–92.

19. UN-HABITAT, *Nairobi Housing Survey* (Nairobi: 2001); Hardoy and Satterthwaite, op. cit. note 8.
20. Richard Stren and Mario Polèse, "Understanding the New Sociocultural Dynamics of Cities: Comparative Urban Policy in a Global Context," in Mario Polèse and Richard Stren, *The Social Sustainability of Cities: Diversity and the Management of Change* (Toronto, Canada: University of Toronto Press, 2000), pp. 26–28.
21. Water prices from Jorge E. Hardoy, Diana Mitlin, and David Satterthwaite, *Environmental Problems in an Urbanizing World* (London: Earthscan, 2001), p. 48; sanitation from ibid., p. 57.
22. Two conferences from David Satterthwaite, International Institute for Environment and Development, e-mail to author, 27 September 2002; related health problems from Hardoy, Mitlin, and Satterthwaite, op. cit. note 21, pp. 39–43; cholera from Carolyn Stephens, "Healthy Cities or Unhealthy Islands: The Health and Social Implications of Urban Inequality," *Environment and Urbanization*, October 1996, p. 16; malaria from Gilbert M. Khadiagala, "Urban Governance and Health in East Africa," in Opolot, p. cit. note 1, pp. 112–14.
23. Sheela Patel, Society for Preservation of Area Resource Centres (SPARC), Mumbai, presentation at UN-HABITAT, Nairobi, 14 February 2001.
24. "Childhood Tuberculosis in an Urban Population in South Africa: Burden and Risk Factor," *Archives of Disease in Childhood*, vol. 80, no. 5 (1999), pp. 433–37; HIV and tuberculosis from Lisa Mastny, "Tuberculosis Resurging Worldwide," in Lester R. Brown et al., *Vital Signs 2000* (New York: W.W. Norton & Company), pp. 148–49.
25. George A. Kaplan et al., "Inequality in Income and Mortality in the United States," in Ichiro Kawachi, Bruce P. Kennedy, and Richard G. Wilkinson, eds., *The Society and Population Health Reader: Income Inequality and Health* (New York: New Press, 1999), pp. 50–59; industrial nations from Sandra J. McIsaac and Richard G. Wilkinson, "Income Distribution and Cause-Specific Mortality," in ibid., pp. 124–36.
26. Thomas Friedman, "Ask Not What?" (op ed), *New York Times*, 9 December 2001; Terry McDermott, "The Plot: How Terrorists Hatched a Simple Plan to Use Planes As Bombs," *Los Angeles Times*, 1 September 2002; Douglas Frantz et al., "Threats and Responses: Pieces of a Puzzle; On Plotters' Path to U.S., A Stop at bin Laden Camp," *New York Times*, 10 September 2002.

70

State of the World

A Year in Review

LORI BROWN

Each year the Worldwatch Institute publishes State of the World, *including "A Year in Review." The following timeline from the 2006 edition covers some significant announcements and reports from October 2004 through September 2005. It is a mix of progress, setbacks, and missed steps around the world that are affecting society's environmental and social goals.*

Timeline events were selected to increase your awareness of the connections between people and the environment. An online version of the timeline with links to Internet resources is available at www.worldwatch.org/features/timeline.

As you study the events listed here, consider the following:

1. *How do these events shed light on the problems discussed in this chapter?*
2. *Which events do you feel are most significant for the future of the planet? Why?*
3. *Which events have affected you directly?*

SOURCE: *State of the World 2006,* compiled by Lori Brown. The Worldwatch Institute, 2005.
http://www.worldwatch.org/pubs/sow/2006.

ACTIVISTS
Kenyan environmental activist Wangari Maathai receives 2004 Nobel Peace Prize for her contributions to sustainable development, democracy, and peace.

HEALTH
International health officials warn that more than 2.5 million people in Asia and the Pacific die each year due to such environmental problems as air pollution, unsafe water, and poor sanitation.

CLIMATE
World Meteorological Organization reports 2004 was the fourth-hottest year on record, extending a trend since 1990 that has registered the 10 warmest years in over a century.

GOVERNANCE
International treaty on wildlife trade upholds 14-year ban on commercial ivory trade and adopts plan to crack down on African ivory markets.

MARINE SYSTEMS
UN report says climate change is the single greatest threat to corals, with 20 percent of all reefs damaged beyond recovery and another 50 percent facing collapse.

NATURAL DISASTERS
Degraded coastlines, reefs, and mongroves contribute to massive destruction and death as an enormous tsunami sweeps across the Indian Ocean, affecting more than 2.4 million people.

OCTOBER NOVEMBER DECEMBER

2004 STATE OF THE WORLD: A YEAR IN REVIEW

2 4 6 8 10 12 14 16 18 20 22 24 26 28 30 2 4 6 8 10 12 14 16 18 20 22 24 26 28 30 2 4 6 8 10 12 14 16 18 20 22 24 26 28 30

URBANIZATION
China reports over 40 percent of its population lives in urban areas, including in 46 megacities of 10 million or more, and an urban economy supplying 70 percent of GDP.

MARINE SYSTEMS
Scientists conducting the first global census of marine life announce the discovery of 106 new marine fish species, raising the total known number to 15,482.

AGRICULTURE
Report says North America leads in biotech crop production but China ranks second in research funding; half of China's farmland is likely to be planted in GMO crops within 10 years.

AGRICULTURE
US government confirms the domestic arrival of soybean rust, a fungal infection that threatens reduced yields and higher production costs.

ENERGY
Finland begins constructing the world's largest nuclear reactor to meet rising energy demand, reduce greenhouse gas emissions, and lower dependence on imported oil.

HUMAN RIGHTS
Burmese villagers and a multinational energy company Unocal agree to a ground-breaking settlement for abuses that occurred during the building of Myanmar's Yadana gas pipeline.

JANUARY FEBRUARY MARCH 2005

CLIMATE
World's largest climate prediction project reports human greenhouse gas emissions could raise global temperatures 2–11 degrees Celsius by century's end, exceeding earlier predictions.

HUMAN RIGHTS
American nun Dorothy Stang is murdered in Brazil because of her outspoken efforts on behalf of landless peasants and wildlife in the Amazon.

WASTE
Four leading global plastics companies voluntarily commit to using biodegradable and compostable polymers to manufacture packaging materials.

INDIGENOUS PEOPLES
Australian Aboriginals win their long battle to halt further development of uranium mining on their traditional lands in Kakadu National Park, Northern Territory.

ENERGY
China passes its first comprehensive renewable energy law in a bid to increase renewable energy capacity to 10 percent by 2020.

AGRICULTURE
India bans a livestock veterinary drug that has caused a rapid 90-percent population crash of three vulture spieces in southern Asia.

GOVERNANCE
Peru creates 2.7-million-hectare Alto Purus National Park and Communal Reserve, one of the largest combined indigenous reserves and protected areas co-managed by local communities and the state.

CLIMATE
World's first mandatory carbon emissions trading scheme to reduce greenhouse gases begins, with participation by 21 European Union nations.

FORESTS
European Commission adds 5,000 boreal and northern woodland forests to its network of sites protecting threatened and vulnerable species and habitats.

CLIMATE
The Kyoto Protocol, the key international agreement aimed at slowing climate change and cutting carbon emissions, enters into force.

HEALTH
Global convention on tobacco control enters into force as 57 countries agree to raise tobacco taxes, ban advertising, expand warning labels, and crack down on smuggling.

MARINE SYSTEMS
UN reports that 7 of the top 10 marine fish species are already fully exploited or overexploited but that world fish consumption may rise by more than 25 percent by 2015.

HUMAN RIGHTS
A protest by rural villagers in China's Zhejiang province over local industrial pollution turns violent, reflecting widening social unrest and a 15-percent rise in grassroots protests nationwide.

ECOSYSTEMS
Key UN assessment reports that 60 percent of Earth's ecosystem services—including fresh water, soil, nutrient cycles, and biodiversity—are being degraded or used unsustainably.

2005 STATE OF THE WORLD: A YEAR IN REVIEW

APRIL · MAY · JUNE

ENERGY
Germany leads in new solar power installation, making it the world's largest photovoltaic market, followed by Japan and the United States.

WASTE
"Rethink," an industry recycling program, is initiated after a survey finds that more than half of US households have working electronics items that are no longer being used.

CLIMATE
Scientists report that 87 percent of 244 glacier fronts on the Antarctic Peninsula have retreated over the past half-century, a trend that matches atmospheric warming patterns.

WATER
Report finds that poor countries with access to improved water and sanitation have a 3.7-percent annual growth in GDP; those without this infrastructure have 0.1-percent growth.

MARINE SYSTEMS
Thai fishers net a 646-pound catfish, the largest freshwater fish ever caught and a critically endangered species due to environmental damage along the Mekong River.

BIODIVERSITY
Scientists report that nearly half of all bird species in the US and Canada rely on the Boreal Forest Region, a massive area increasingly threatened by logging and fragmentation.

FORESTS
Brazil's Amazon deforestation rate rose 6 percent in a year, with 26,000 square kilometers lost—an area the size of Belgium and the second largest loss on record.

CLIMATE
An alliance of institutional investors, managing $3.22 trillion, demands that capital market regulators require rigorous corporate disclosure of climate risks because of growing costs

SECURITY
Researchers report that world military expenditures in 2004 surpassed $1.04 trillion, nearing the historic peak reached during the cold war.

URBANIZATION
Government report finds that more than half of 500 Chinese cities failed to meet minimum air pollution standards, nearly 200 had no wastewater treatment, and only half had clean water.

HUMAN RIGHTS
Guatemalan Sipakapense communities organize a region-wide vote against a gold mining project, demanding that the government respect their sovereign decision and voting rights.

ENERGY
European Environment Agency reports that the rise in coal use for electricity generation has pushed up greenhouse gas emissions across the European Union.

ENERGY
Nepalese win award for implementing a biogas project in 85 percent of the nation's districts that is credited with saving 400,000 tons of firewood and preventing 600,000 tons of greenhouse gas emissions.

ENERGY
China embarks on a quest to more than double its nuclear power generating capacity by 2020 to meet the soaring demands for electricity in its booming economy.

INDIGENOUS PEOPLES
Venezuelan government, for the first time, presents property titles to several indigenous groups, recognizing their ownership of ancestral lands.

BIODIVERSITY
World atlas of great apes is released, presenting the first comprehensive review of the threats and conservation efforts for all species of chimpanzees, orangutans, and gorillas.

GOVERNANCE
World Bank and IMF agree that the debts owed by 18 of the world's poorest countries will be cancelled without conditions.

CLIMATE
Mumbai, India, gets a record 37 inches of rain in one day—the strongest rains ever recorded in India—as monsoons destroy 700,000 hectares (1.7 million acres) of crops.

ENERGY
Crude oil prices to $70 a barrel after hurricanes strike the US oil industry—an increase of nearly 60 percent over the previous year.

OZONE LAYER
Mexico becomes the first developing country to announce the halt in further production of ozone-depleting chlorofluorocarbons.

JULY AUGUST SEPTEMBER

2 4 6 8 10 12 14 16 18 20 22 24 26 28 30 2 4 6 8 10 12 14 16 18 20 22 24 26 28 30 2 4 6 8 10 12 14 16 18 20 22 24 26 28

MARINE SYSTEMS
Canada's government reports that ocean temperatures in the North Atlantic hit an all-time high, raising concerns about the effects of climate change.

BIOTECHNOLOGY
Researchers in South Korea announce they have successfully cloned a dog, a milestone in genetic technology with implications for human cloning.

HEALTH
FAO warns that the deadly strain of avian flu affecting several Asian nations is likely to be carried by wild birds migrating to the Middle East, Europe, South Asia, and Africa.

CLIMATE
Researchers warn that Arctic ice melting is accelerating, with an 8-percent loss in sea ice area over the past 30 years and the possibility of ice-free summers before 2100.

FORESTS
Government satellite surveys show that while India's total tree cover has increased, areas covered by dense forests have shrunk due to mining and industrial development.

NATURAL DISASTERS
Hurricane Katrina wreaks catastrophic damage along the US Gulf Coast, leaving 80 percent of New Orleans underwater and causing estimated $125 billion in economic damage.

HEALTH
Twenty-year review of the Chernobyl nuclear accident—the largest in history—says radiation-induced death and disease is lower than predicted but that damaging psychological fallout is far greater.

CHAPTER 15

Social Movements and Global Change

Society In Flux

Alan Mc Evoy

SOCIAL CHANGE IS THE ALTERATION, OVER time, of a basic pattern of social organization. It involves two related types of changes: changes in folkways, mores, and other cultural elements of the society; and changes in the social structure and social relations of the society. These cultural and structural changes affect all aspects of society, from international relations to our individual lifestyles. An example of a change in cultural elements is women increasingly entering the labor force; an example of structural change is the rise of secondary-group relationships with a corresponding increase in formal organization.

Social change happens through the triple impact of *diffusion* (the speed of ideas across culture), *discovery* (unpremeditated findings), and *invention* (purposeful new arrangements). Each of these actions occurs at geometrically expanding rates—that is to say, the period of time over which an initial invention evolves may require centuries, but once it is established, further changes related to that invention are rapid. For example, it took less than fifty years to advance from the Wright brothers' 12-second flight to supersonic travel to trips to the moon. Similarly, the rates of diffusion and discovery are now greater because of advancements in communication and travel.

Change occurs in all societies, but at different rates. While change is rapid and radical in some societies, such as eastern European countries and the former Soviet Union, and can actually tear societies apart, as in the case of the former country of Yugoslavia, change generally occurs more slowly in stable, traditional societies.

There are times when the norms that guide societal behavior break down. At these times, *collective behavior*—behavior that is both spontaneous and organized—may predominate. These forms are major sources of change in societies.

Many of us have been involved in a form of collective behavior called a social movement, a group of people attempting to resist or bring about change. Recent examples include the civil rights movement; various educational reform movements such as "back to basics"; fundamentalist Christian movements; and the women's movement. A movement may have a short life span, or it may become an institutionalized, stable part of the ongoing society. The reading by Barry Adam discusses a worldwide movement that has been influenced by globalization—the mobilization of gay and lesbian communities.

A debate that arises from globalization is how alike or different countries are becoming. The next two articles address this issue. The first by Ronald Inglehart and Wayne Baker explores world values and reports on a survey that distinguishes "traditional" values from "modern" values. This distinction helps us understand the split between world societies referred to in the next article by Michael Marien and his colleagues on the world after September 11, 2001.

Futurists predict trends that will lead us into the future. Sociologists are active in this field, contributing study methods and an understanding of social systems and the process of change. In some cases, they focus on specific areas of change. The reading by Robin Gunston discusses trends in postmodern society that are influencing the direction of sports.

Finally, David Pearce Snyder discusses five metatrends that futurists have predicted are likely to change the world. These are evolutionary, systemwide developments arising from demographic, economic, and technological changes. Understanding worldwide trends gives us a look into the future and helps individuals, communities, nations, and world organizations prepare for the inevitability of change in a systematic way.

As you read these selections on collective behavior and social change, consider the changes since September 11 and whether they fit the patterns described.

1. What are some current social movements and how are they affecting society?
2. Were the predicted changes to take place, what would the strains be on existing structures and roles?
3. What kind of society and world do we want, and how can we attain it?

71

Globalization and the Mobilization of Gay and Lesbian Communities

BARRY D. ADAM

Change occurs in many ways. Stresses from outside the country or organization can bring about change, just as strains from within the country or organization can. Change can be rapid or slow, planned or unplanned. Social movements—groups of individuals organized to push for or resist change in some part of society—are one means of putting pressure on systems to change. The first reading provides an example of change brought about by international globalization processes that can shape movements.

As you read it, consider the following:

1. *What does Adam mean when he states that "lesbian, gay, bisexual and transgender identities have become 'necessary fictions' ". . .?*
2. *How do modern gay and lesbian relationships to the world differ from historical ones?*
3. *What characteristics of the modern world system affect homosexuality, and what is the effect of globalization on social movements?*
4. *Do you belong to any social movements (women's rights, animal rights, "Greens," etc.)? How do you see your movement bringing about change?*

GLOSSARY **Social movement** Organized effort to encourage or oppose some dimension of change. **Globalization** Economic and political changes spanning many nations of the world.

There has been a strong tendency to treat gay and lesbian communities and their movements as cultural artifacts unrelated to structure, political economy, or the modern world-system (Seidman 1996). This tendency gives gay and lesbian people over to the "globalization-as-culture" school virtually by default. A good deal of contemporary social movement theory has assimilated lesbian and gay mobilization into the "new social movement" camp, and into "identity politics" in particular. New social movements now supposedly "seem to exist independently from . . . structural and cultural conditions" (Eder 1993:44) and gay movements are, somewhat inexplicably, all about affirming and expressing identities (Dudink and Verhaar 1994). What is missing from this view are the ways the changes wrought by the world-system articulate with the historical rise of gay and lesbian forms of same-sex connection and how lesbian, gay, bisexual, and transgender identities have become

SOURCE: From Pierre Hamel, Henri Lustiger-Thaler, Jan Nederveen, Pieterse and Sasha Roseneil, eds., *Globalization and Social Movements 2001*. New York: St. Martin's/Palgrave, pp. 166–179. Reprinted with permission from the author.

"necessary fictions" (Weeks 1993) in contemporary political environments.

. . . The "gay" and "lesbian" categories of the modern world, when compared to cross-cultural and historical forms of same-sex relations, show several specific characteristics:

1. In societies where kinship has declined as a primary organizing principle determining the survival and well-being of their members, homosexual relations have developed autonomous forms apart from dominant heterosexual family structures.
2. Exclusive homosexuality has become increasingly possible for both partners and a ground for household formation.
3. Same-sex bonds have developed relatively egalitarian forms characterized by age and gender "endogamy" rather than involving people in differentiated age and gender classes.
4. People have come to discover each other and form large-scale social networks because of their homosexual interests and not only in the context of pre-existing social relationships (such as households, neighborhoods, schools, militaries, churches, etc.).
5. Homosexuality has come to be a social formation unto itself characterized by self-awareness and group identity (Adam 1995:7).

These characteristics presume a sociological infrastructure characteristic of the modern world-system:

1. The financial "independence" of wage labor. In societies where access to the means of production depends on kin ties and inheritance, homosexuality is either integrated into the dominant kinship order (through gender redesignation or limitation of homosexual ties to a life stage), or subordinated to it as a hidden and "unofficial" activity.
2. Urbanization and personal mobility. The growth of cities, the invention of public spaces, migration away from traditional settings and from the supervisory gaze of families have all created new opportunities, especially at first for men, to form new, nontraditional relationships.
3. Disruption of traditional gender rules. As minorities and women have been entering wage labor in increasing numbers, often occupying newly created locations in the division of labor, more of the choices taken for granted by men have become available.
4. Development of the welfare state. The creation of social services in health, education, welfare, employment insurance, pensions, and so on have supplemented traditional family functions, providing alternatives to reliance on kin.
5. Liberal democratic states. Legal guarantees of basic civil liberties also facilitate the ability of people to love and live with those of their choice, though most liberal democratic states have lengthy histories of violating basic constitutional freedoms of conscience, assembly, and free speech in order to suppress their gay and lesbian citizens; and gay and lesbian communities have also carved out small spaces for themselves in the hostile environments of authoritarian states.

Having enumerated some of these factors, it is important to note that these are structural underpinnings of contemporary gay and lesbian worlds, not of homosexuality itself, which is best conceived as a universal human potential with a wide range of expression across cultures. Even though gay and lesbian worlds (as defined above) have now been documented for at least three centuries in Europe, these social conditions do not in themselves determine that same-sex relationships will necessarily take gay/lesbian forms. Medieval constructions of sodomy (Jordan 1997), Judeo-Christian obsession with narrowing legitimate sexuality to its procreative form, and the Western tendency to force sexuality to "confess truth" about the nature of persons (Foucault 1978) all enter into western constrictions of the "gay" and "lesbian." Japan offers an example of a society with the political economy of an advanced industrial society but none of the history of religious or state persecution of homosexuality. Though Japan has developed gay (and to a very limited degree, lesbian) public spaces like its Western counterparts, the formation of a popular sense of identity and movement has been less evident (Lunsing 2001). Finally, even among the citizens of advanced industrial societies, many more people have homosexual experiences than identify as "gay," "lesbian," or "bisexual," and these identities may, as well, be less widespread among non-white and working class people who nevertheless have homosexual interests and practices (Laumann et al. 1994).

Still, where this sociological infrastructure is lacking, gay and lesbian identities, and the movements built out of a sense of commonality signified by these identities, are also unusual. As Badruddin Khan (1997) remarks about urban Pakistan, widespread practices of sexuality among men tend not to lead to gay identities: where family networks remain major determinants of one's well-being, families make sure their progeny marry regardless of sexual orientation, and very few young people have the financial ability or freedom to form households of their own choosing. Similar conditions apply to much of Latin America with the partial exception of the urban middle classes. The Communist states of eastern Europe, on the other hand, showed a different array of social conditions that both facilitated and inhibited the growth of gay and lesbian cultures. State socialism typically pursued a development model founded on industrialization, resulting in urban migration, conversion of much of the population to wage labor, and improved opportunities for women. As in capitalist societies, state socialism displaced kinship as the primary determinant of people's life chances, disestablished churches, and devolved decisions about family and reproduction to the individual level (Adam 1995:166). It is not surprising, then, that gay and lesbian bars and coffeehouses became part of the urban scene in eastern European capitals well before the fall of Communism and movement organizations came into existence in several countries despite the power of central bureaucracies to control the mass media and administer labor migration, commercial meeting places, and housing (Hauer et al. 1984). When the Communist states collapsed in the early 1990s, gay and lesbian movement organizations rapidly emerged across the region in a pattern similar to western Europe because the sociological infrastructure was already in place....

THE GLOBAL MOVEMENT

Today the International Lesbian and Gay Association federates movement organizations and attempts to monitor civil rights around the world, twin established organizations with new, and link international efforts around asylum, the military, churches, youth, ableism, health, trade unions, AIDS, and prisoners. A primary objective of national movements has been to win legal recourse against discrimination. Beginning with Norway in 1981, sexual orientation had been included in the human rights legislation of Canada, Denmark, France, Iceland, Netherlands, New Zealand, Slovenia, South Africa, Spain, and Sweden, as well as eleven of the fifty United States, two states in Australia, and seventy-three cities in Brazil (as of 1997). Ireland and Israel also legislated workplace protection in the 1990s. Even more recent is legislation recognizing the spousal status of same-sex relationships. Norway, Sweden, Netherlands, and Iceland have followed the precedent set by Denmark in 1989, and a supreme court decision in Hungary has included same-sex couples in common-law spousal status. Initial break-throughs in relationship recognition have also occurred in 1997 in the Canadian province of British Columbia and the U.S. state of Hawaii. All of these laws, however, fall short of full equality, often barring gay and lesbian couples from full-fledged marriage, adoption, or access to artificial insemination.

Gay and lesbian movements have emerged along with (and sometimes in explicit coalition with) democratic movements to oppose authoritarian rule in Spain (after the death of Franco), South Africa (in overcoming apartheid), in eastern Europe (with the fall of Communism), and Brazil, Argentina, Uruguay, and Chile (in ending military dictatorships).

Once these precedents are in place, gay and lesbian communities may take the opportunity to mobilize in neighboring countries. With the African National Congress in power in South Africa and sexual orientation entrenched in the new constitution, groups in Zimbabwe (1990), Botswana (1996), Namibia (1996), and Swaziland (1997) have declared themselves often in the face of virulent homophobia enunciated by local rulers. The struggle against AIDS has created the first opening for gay mobilization in even less hospitable places such as Kenya, Malaysia, and Ecuador.

This record of civil emancipation must be viewed in the context of ongoing homophobic practices of states and their agents, of reactionary civil and religious movements, and of those in control of cultural reproduction from the schools to television programming. Amnesty International documents the incarceration or murder of homosexual people in several countries, most notably, Iran, Turkey, and Romania. Movement groups continue to face repeated police raids on gay gathering places and death squad activity in many cities of Latin America, especially in Colombia, Peru, Mexico, and Brazil. Gay and transgender Argentines were among the targets of the "dirty war" perpetuated by the police and death squads and sanctioned by the military and Roman Catholic hierarchies (Jauregui 1987). Sweeping police powers continued after the fall of the military dictatorship into the 1990s resulting in numerous bar raids and arbitrary arrests. The work of the Comunidad Homosexual Argentina, in collaboration with other democratic organizations, resulted in 1996 in the revocation of police powers and inclusion of sexual orientation as legally protected categories in Buenos Aires and Rosario.

Globalization processes, then, have affected the mobilization of gay and lesbian communities in a wide variety of ways. Insofar as globalization refers to processes of incorporation of local communities into the political and economic networks of the capitalist world-system, globalization has accelerated processes of proletarianization, urbanization, family change, and democratization. These processes have created social conditions where indigenous forms of same-sex adhesiveness (to use Walt Whitman's term) have tended to evolve toward gay and lesbian social forms. Contemporary gay and lesbian communities and movements have also, in turn, provided fertile ground for the defense, growth, and reworking of older, more traditional bisexual and trans-gendered forms of sexual life. Where globalization refers to the faster and easier circulation of cultural practices and ideologies, gay and lesbian movements have been participants as well. Though gay and lesbian cultural forms usually have an uneasy or oppositional relation with the established, institutional ideological circuits of religion, nationalism, and "family values," an international gay and lesbian culture has emerged throughout the metropoles and, increasingly around the world, through personal contacts, the gay press, and now more formal, if fragile, organizations like the International Lesbian and Gay Association. Pride celebrations, which originated in political demonstrations against repression, have become so massive, in such cities as Sydney, Toronto, and San Francisco, that they have changed the cultural landscape and attracted commercial interests intent on making inroads into a new market of consumers.

Globalization, as a political and economic process underlying the transition from the welfare state to neoliberalism, has consequences for the well-being and potential mobilization of numerous social movement constituencies. Neoliberal governments, particularly in the United States and the United Kingdom, have drawn on conservative ideologies of work discipline, delayed gratification, racial and national supremacy, and patriarchal definitions of gender and family to "sell" wage restraints and the withdrawal of social services, as well as to re-attribute blame for the declining quality of life. As Stuart Hall (1988b: 48) remarks, Thatcherism combined "organic Toryism—nation, family, duty, authority, standards, traditionalism—with the aggressive themes of a revived neoliberalism—self-interest, competitive individualism, anti-statism." Neoconservative governments thereby encouraged and exploited the resentment of social classes damaged by the world economy, channeling their anger toward traditional lightning rods of popular prejudice, including lesbians and gay men

(Adam 1995:Ch 6). The global economy has struck hard at some social groups, typically rural people, small business people, and workers located in heavy industry. Declining or beleaguered ethnic and class groups have proven to be especially susceptible to right-wing mobilization, from the antisemites of the early twentieth century to the homophobes and xenophobes of the late twentieth century.

At the same time, gay and lesbian movements, with the other "new social movements," have flourished in the neoliberal age. The welfare state both changed and reinforced family and gender requirements in the 1950s not only by (1) relieving families of some of the functions they could not always fulfill but also by (2) restoring the home as women's sphere and by overtly suppressing gay and lesbian life. The social policies of neoliberal governments of the 1980s tended to press for the "restoration" of traditional gender and family scripts (restoring to some degree a family that never was [Coontz 1992]) while reducing the social services available to them. Gay and lesbian mobilization, like the women's movement, is part of the larger wave of "anti-systemic" movements since 1968 (Wallerstein 1989), building on and expressing some aspects of structural change—such as women's increasing participation in wage labor—while resisting others—such as explicitly conservative social policy. And as capital has gone global, social movements as well have sought unprecedented world-wide networks.

REFERENCES

Editors' Note: Notes appear in original source.

Adam, Barry. 1995. The Rise of a Gay and Lesbian Movement. Revised Edition.New York: Twayne/ Simon & Schuster Macmillan.

Coontz, Stephanie. 1992. The Way We Never Were. New York: Basic Books.

Dudink, Stefan, and Odile Verhaar. 1994. "Paradoxes of identity politics." Homologie 4–94:29–36.

Eder, Klaus. 1993. The New Politics of Class. London: Sage.

Foucault, Michel. 1978. La Volonté de savoir Vol. 1 of Histoire de la sexualité (1976). Trans. Robert Hurley as The History of Sexuality Volume 1: An Introduction. New York: Pantheon.

Hall, Stuart. 1988b. The Hard Road to Renewal. London: Verso.

Hauer, Gudrun, et al. 1984. Rosa Liebe unterm roten Stern. Hamburg: Frühlings Erwachen.

Jauregui, Carlos. 1987. La homosexualidad en la Argentina. Buenos Aires: Tarso.

Jordan, Mark. 1997. The Invention of Sodomy in Christian Theology. Chicago: University of Chicago Press.

Khan, Badruddin. 1997. "Not-so-gay life in Pakistan in the 1980s and 1990s." In Islamic Homosexualities, edited by Stephen Murray and Will Roscoe. New York: New York University Press.

Laumann, Edward, John Gagnon, Robert Michael, and Stuart Michaels. 1994. The Social Organization of Sexuality. Chicago: University of Chicago Press.

Lunsing, Wim. 2001. Beyond Common Sense: Negotiating Constructions of Sexuality and Gender in Contemporary Japan. London and New York: Kegan Paul International.

Seidman, Steven. 1996. Queer Theory/Sociology. Cambridge, MA: Blackwell.

Wallerstein, Immanuel. 1989. The Modern World-System. Volumes 1–3. San Diego: Academic Press.

Weeks, Jeffrey. 1993. "Necessary fictions." In Constructing Sexualities, edited by Jacqueline Murray. Windsor, Ontario: University of Windsor Humanities Research Group.

72

Modernization's Challenge to Traditional Values

Who's Afraid of Ronald McDonald?

RONALD INGLEHART AND WAYNE E. BAKER

Is there such a thing as world *values? We turn to a discussion of the "World Values Survey" to find that world values break down into rich and poor societies' values. These values create stress on all countries—pressure toward change and resistance to change, as seen in the next article on terrorism. Understanding challenges to traditional values helps shed light on terrorism in the world.*

Think about the following questions as you read this article:

1. *How do values shape culture?*
2. *What are major distinctions between rich and poor countries' values?*
3. *When does modernization lead to change, and when is there resistance to change?*
4. *What current political events are fueled by traditional verses modern value conflicts?*

GLOSSARY **Values** A culture's collective ideas about right and wrong, good or bad, desirable or undesirable. **Economic globalization** Influence of dominant economic systems spreading around the world.

The World Values Survey—a two-decade-long examination of the values of 65 societies coordinated by the University of Michigan's Institute for Social Research—is the largest investigation ever conducted of attitudes, values, and beliefs around the world. This study has carried out three waves of representative national surveys: the first in 1981–1982, the second in 1990–1991, and the third in 1995–1998. The fourth wave is being completed in 1999–2001. The study now represents some 80% of the world's population. These societies have per capita GNPs ranging from $300 to more than $30,000. Their political systems range from long-established stable democracies to authoritarian states.

The World Values Survey data have been used by researchers around the world for hundreds of publications in more than a dozen languages. Studies that have been based on the data cover a wide range of topics, including volunteerism in Europe, political partisanship and social class in Ireland, democratization in Korea, liberalization in Mexico, future values in Japan, and the religious vote in Western Europe.

SOURCE: Originally published in the March/April 2001 issue of *The Futurist*. Used with permission from the World Future Society, 7910 Woodmont Avenue, Suite 450, Bethesda, Maryland 20814. Telephone: 301/656-8274; Fax: 301/951-0394; http://www.wfs.org

This article examines the relationship between cultural values and economic globalization and modernization: What impact does economic development have on the values of a culture, and vice versa? Is a future "McWorld" inevitable?

RICH VALUES, POOR VALUES

The World Values Survey data show us that the world views of the people of rich societies differ systematically from those of low-income societies across a wide range of political, social, and religious norms and beliefs. The two most significant dimensions that emerged reflected, first, a polarization between *traditional* and *secular-rational* orientations toward authority and, second, a polarization between *survival* and *self-expression* values. By *traditional* we mean those societies that are relatively authoritarian, place strong emphasis on religion, and exhibit a mainstream version of preindustrial values such as an emphasis on male dominance in economic and political life, respect for authority, and relatively low levels of tolerance for abortion and divorce. Advanced societies, or *secular-rational*, tend to have the opposite characteristics.

A central component of the survival vs. self-expression dimension involves the polarization between materialist and postmaterialist values. Massive evidence indicates that a cultural shift throughout advanced industrial society is emerging among generations who have grown up taking survival for granted. Values among this group emphasize environmental protection, the women's movement, and rising demand for participation in decision making in economic and political life. During the past 25 years, these values have become increasingly widespread in almost all advanced industrial societies for which extensive time-series evidence is available.

Economic development brings with it sweeping cultural change, some modernization theorists tell us. Others argue that cultural values are enduring and exert more influence on society than does economic change. Who's right?

One goal of the World Values Survey is to study links between economic development and changes in values. A key question that we ask is whether the globalization of the economy will necessarily produce a homogenization (or, more specifically, an Americanization) of culture—a so-called "McWorld."

In the nineteenth century, modernization theorists such as Karl Marx and Friedrich Nietzsche made bold predictions about the future of industrial society, such as the rise of labor and the decline of religion. In the twentieth century, non-Western societies were expected to abandon their traditional cultures and assimilate the technologically and morally "superior" ways of the West.

Clearly now, at the start of the twenty-first century, we need to re-think "modernization." Few people today anticipate a proletarian revolution, and non-Western societies such as East Asia have surpassed their Western role models in key aspects of modernization, such as rates of economic growth. And few observers today attribute moral superiority to the West.

On the other hand, one core concept of modernization theory still seems valid: Industrialization produces pervasive social and cultural consequences, such as rising educational levels, shifting attitudes toward authority, broader political participation, declining fertility rates, and changing gender roles. On the basis of the World Values Surveys, we believe that economic development has systematic and, to some extent, predictable cultural and political consequences. Once a society has embarked on industrialization—the central element of the modernization process—certain changes are highly likely to occur. But economic development is not the *only* force at work.

In the past few decades, modernization has become associated with *post*-industrialization: the rise of the knowledge and service-oriented economy. These changes in the nature of work had major political and cultural consequences, too. Rather than growing more materialistic with increased prosperity, postindustrial societies are experiencing an increasing emphasis on quality-of-life issues, environmental protection, and self-expression.

While industrialization increased human dominance over the environment—and consequently created a dwindling role for religious belief—the emergence of postindustrial society is stimulating further evolution of prevailing world views in a different direction. Life in postindustrial societies centers on services rather than material objects, and more effort is focused on communicating and processing information. Most people spend their productive hours dealing with other people and symbols.

Thus, the rise of postindustrial society leads to a growing emphasis on self-expression. Today's unprecedented wealth in advanced societies means an increasing share of the population grows up taking survival for granted. Their value priorities shift from an overwhelming emphasis on economic and physical security toward an increasing emphasis on subjective well-being and quality of life. "Modernization," thus, is not linear—it moves in new directions.

HOW VALUES SHAPE CULTURE

Different societies follow different trajectories even when they are subjected to the same forces of economic development, in part because situation-specific factors, such as a society's cultural heritage, also shape how a particular society develops. Recently, Samuel Huntington, author of *The Clash of Civilizations* (Simon & Schuster, 1996), has focused on the role of religion in shaping the world's eight major civilizations or "cultural zones": Western Christianity, Orthodox, Islam, Confucian, Japanese, Hindu, African, and Latin American. These zones were shaped by religious traditions that are still powerful today, despite the forces of modernization.

Other scholars observe other distinctive cultural traits that endure over long periods of time and continue to shape a society's political and economic performance. For example, the regions of Italy in which democratic institutions function most successfully today are those in which civil society was relatively well developed in the nineteenth century and even earlier, as Robert Putnam notes in *Making Democracy Work* (Princeton University Press, 1993). And a cultural heritage of "low trust" puts a society at a competitive disadvantage in global markets because it is less able to develop large and complex social institutions, Francis Fukuyama argues in *Trust: The Social Virtues and the Creation of Prosperity* (Free Press, 1995).

The impression that we are moving toward a uniform "McWorld" is partly an illusion. The seemingly identical McDonald's restaurants that have spread throughout the world actually have different social meanings and fulfill different social functions in different cultural zones. Eating in a McDonald's restaurant in Japan is a different social experience from eating in one in the United States, Europe, or China.

Likewise, the globalization of communication is unmistakable, but its effects may be overestimated. It is certainly apparent that young people around the world are wearing jeans and listening to U.S. pop music; what is less apparent is the persistence of underlying value differences.

... PREDICTING VALUES

... Cross-cultural variation is highly constrained. That is, if the people of a given society place a strong emphasis on religion, that society's relative position on many other variables can be predicted—such as attitudes toward abortion, national pride, respect for authority, and childrearing. Similarly, survival vs. self-expression values reflect wide-ranging but tightly correlated clusters of values: Materialistic (survival-oriented) societies can be predicted to value maintaining order and fighting inflation, while postmaterialistic (self-expression-oriented) societies can be predicted to value freedom, interpersonal trust, and tolerance of outgroups.

Economic development seems to have a powerful impact on cultural values: The value systems of rich countries differ systematically from those of poor countries. If we superimpose an income "map" over the values map, we see that all 19 societies with an

Two Dimensions of Cross-Cultural Variation

1. Traditional vs. Secular-Rational Values	2. Survival vs. Self-Expression Values
Traditional values emphasize the following:	**Survival** values emphasize the following:
▪ God is very important in respondent's life.	▪ Respondent gives priority to economic and physical security over self-expression and quality of life.
▪ Respondent believes it is more important for a child to learn obedience and religious faith than independence and determination.	▪ Respondent describes self as not very happy.
▪ Respondent believes abortion is never justifiable.	▪ Respondent has not signed and would not sign a petition.
▪ Respondent has strong sense of national pride.	▪ Respondent believes homosexuality is never justifiable.
▪ Respondent favors more respect for authority.	▪ Respondent believes you have to be very careful about trusting people.
Secular-Rational values emphasize the opposite.	**Self-Expression** values emphasize the opposite.

SOURCE: World Values Survey (http://wvs.isr.umich.edu)

annual per capita GNP of over $15,000 rank relatively high on both dimensions.... This economic zone cuts across the boundaries of the Protestant, ex-Communist, Confucian, Catholic, and English-speaking cultural zones.

On the other hand, all societies with per capita GNPs below $2,000 fall into a cluster... in an economic zone that cuts across the African, South Asian, ex-Communist, and Orthodox cultural zones. The remaining societies fall into two intermediate cultural-economic zones. Economic development seems to move societies in a common direction, regardless of their cultural heritage. Nevertheless, distinctive cultural zones persist two centuries after the industrial revolution began.

Of course, per capita GNP is only one indicator of a society's level of economic development. Another might be the percentage of the labor force engaged in the agricultural sector, the industrial sector, or the service sector. The shift from an agrarian mode of production to industrial production seems to bring with it a shift from traditional values toward increasing rationalization and secularization.

But a society's cultural heritage also plays a role: All four of the Confucian-influenced societies (China, Taiwan, South Korea, and Japan) have relatively secular values, regardless of the proportion of their labor forces in the industrial sector. Conversely, the historically Roman Catholic societies (e.g., Italy, Portugal, and Spain) display relatively traditional values when compared with Confucian or ex-Communist societies with the same proportion of industrial workers. And virtually all of the historically Protestant societies (e.g., West Germany, Denmark, Norway, and Sweden) rank higher on the survival/self-expression dimension than do all of the historically Roman Catholic societies, regardless of the extent to which their labor forces are engaged in the service sector.

We can conclude from this that changes in GNP and occupational structure have important influences on prevailing world views, but traditional cultural influences persist.

Religious traditions appear to have had an enduring impact on the contemporary value systems of the 65 societies. But a society's culture reflects its entire historical heritage. A central historical event of the twentieth century was the rise and fall of a Communist empire that once ruled one-third of the world's population. Communism left a clear imprint on the value systems of those who lived under it. East Germany remains culturally close to West Germany despite four decades of Communist rule, but its value system has been drawn toward the Communist zone. And although China is a member of the Confucian zone, it also falls within a broad Communist-influenced zone. Similarly, Azerbaijan, though part of the Islamic cluster, also falls within the Communist superzone that dominated it for decades.

THE DEVIANT U.S.

The World Value Map clearly shows that the United States is a deviant case. We do not believe it is a prototype of cultural modernization for other societies to follow, as some postwar modernization theorists have naively assumed. The United States has a much more traditional value system than any other advanced industrial society.

On the traditional/secular-rational dimension, the United States ranks far below other rich societies, with levels of religiosity and national pride comparable to those found in developing societies. The United States does rank among the most advanced societies along the survival/self-expression dimension, but even here it does not lead the world. The Swedes and the Dutch seem closer to the cutting edge of cultural change than do the Americans.

Modernization theory implies that as societies develop economically their cultures tend to shift in a predictable direction. Our data supports this prediction. Economic differences are linked with large and pervasive cultural differences. But we find clear evidence of the influence of long-established cultural zones.

Do these cultural clusters simply reflect economic differences? For example, do the societies of Protestant Europe have similar values simply because they are rich? No. The impact of a society's historical-cultural heritage persists when we control for GDP per capita and the structure of the labor force. On a value such as *interpersonal trust* (a variable on the survival/self-expression dimension), even rich Catholic societies rank lower than rich Protestant ones.

Within a given society, however, Catholics rank about as high on interpersonal trust as do Protestants. The shared historical experience of given nations, not individual personality, is crucial. Once established, the cross-cultural differences linked with religion have become part of a national culture that is transmitted by the educational institutions and mass media of given societies to the people of that nation. Despite globalization, the nation remains a key unit of shared experience, and its educational and cultural institutions shape the values of almost everyone in that society.

THE PERSISTENCE OF RELIGIOUS AND SPIRITUAL BELIEFS

As a society shifts from an agrarian to an industrial economy and survival comes to be taken for granted, traditional religious beliefs tend to decline. Nevertheless, as the twenty-first century opens, cleavages along religious lines remain strong. Why has religion been so slow to disappear?

History has taken an ironic turn: Communist-style industrialization was especially favorable to secularization, but the collapse of Communism has given rise to pervasive insecurity—and a return to religious beliefs. Five of the seven ex-Communist societies for which we have time-series data show rising church attendance.

Throughout advanced industrial societies we see two contrasting trends: the decline of attendance at religious services on the one hand, and on the other the persistence of religious beliefs and the rise of spirituality. The need for answers to spiritual questions such as why we are here and where we are going does not die out in postindustrial society. Spiritual concerns will probably always be part of the human outlook. In fact, in the three successive waves of the World Values Survey, concern for the meaning and purpose of life became *stronger* in most advanced industrial societies.

CONCLUSION: WHITHER MODERNIZATION?

Economic development is associated with pervasive, and to an extent predictable, cultural changes. Industrialization promotes a shift from traditional to secular-rational values; postindustrialization promotes a shift toward more trust, tolerance, and emphasis on well-being. Economic collapse propels societies in the opposite direction.

Economic development tends to push societies in a common direction, but rather than converging they seem to move along paths shaped by their cultural heritages. Therefore, we doubt that the forces of modernization will produce a homogenized world culture in the foreseeable future.

Certainly it is misleading to view cultural change as "Americanization." Industrializing societies in general are *not* becoming like the United States. In fact, the United States seems to be a deviant case: Its people hold much more traditional values and beliefs than do those in any other equally prosperous society. If any societies exemplify the cutting edge of cultural change, it would be the Nordic countries.

Finally, modernization is probabilistic, not deterministic. Economic development tends to transform a given society in a predictable direction, but the process and path are not inevitable. Many factors are involved, so any prediction must be contingent on the historical and cultural context of the society in question.

Nevertheless, the central prediction of modernization theory finds broad support: Economic development is associated with major changes in prevailing values and beliefs. The world views of rich societies differ markedly from those of poor societies. This does not necessarily imply cultural convergence, but it does predict the general direction of cultural change and (insofar as the process is based on intergenerational population replacement) even gives some idea of the rate at which such change is likely to occur.

In short, economic development will cause shifts in the values of people in developing nations, but it will not produce a uniform global culture. The future may *look* like McWorld, but it won't feel like one.

73

The New Age of Terrorism

Futurists Respond

MICHAEL MARIEN ET AL.

September 11 brought change to the world—change in every institution and structure. In response, the United States declared a "war on terrorism." This selection features a number of noted scholars and futurists predicting what will change as a result of September 11 and what can or should be done to deal with the threat of terrorism.

SOURCE: Originally published in the January/February 2002 issue of *The Futurist*. Used with permission from the World Future Society, 7910 Woodmont Avenue, Suite 450, Bethesda, Maryland 20814. Telephone: 301/656-8274; Fax: 301/951-0394; http://www.wfs.org.

As you read these commentaries, think about the following:

1. *Can you spot themes that run through the following commentaries? How do they relate to changes in societal structures (institutions) and processes?*
2. *What fuels terrorism?*
3. *From what you have read, is terrorism likely to increase or decrease? Where and why?*
4. *What steps would you take to reduce the threat of terrorism in the world?*

GLOSSARY **Futurist** One who specializes in predicting future events from past events and trend data.

INTRODUCTION

Michael Marien

September 11, 2001, is a major turning point into the future—the end of the post–Cold War era and the beginning of a new Age of Terrorism, perhaps a World War III. The horrible attacks on the World Trade Center and the Pentagon, and whatever follows as a result, will change many lives, many organizations, many industries, and many nations.

The terrorism scenario of which many futurists and military analysts have warned has unfolded, although not as expected. It was not as bad as a nuclear bomb or a smallpox epidemic (and such mega-disasters could yet happen), but it was deadly serious enough to command global attention. In its wake, it is quite reasonable to ask, "What next?" And futurists, who specialize in addressing this question, have something distinctive and valuable to say.

As a group, futurists' response differs markedly from the flood of editorials released by the September 11 jolt. In general, they take a longer view; some see the terrorist attack as a portent of darker things to come. Some, by taking a normative view, or including a positive scenario among several possibilities, envision upside developments from this tragic event.

The October 2001 issue of *Future Survey* was devoted to futurists' analyses of the September 11 attacks, providing detailed abstracts of many of the essays received by the World Future Society. Nearly all of these essays, invited or otherwise, have been posted in their entirety on the World Future Society Web site (www.wfs.org/terror.htm). What follows are excerpts from some of them.

WIDER GAPS, WIDER CONFLICTS

W. Warren Wagar

Whatever the United States and its allies do in response to the September 11 attacks, there is little chance that they can eliminate all terrorist networks or prevent the formation of new ones. As the experience of Israel suggests, the struggle against terrorism is never-ending. It cannot be brought to a successful conclusion without removing its root causes, which lie deep within the structure of the modern world system.

In fact, the long-term prospects are for widening conflict between the rich and poor nations of the world, for widening conflict among the poor nations themselves, and for increasing destabilization worldwide. A system that routinely rewards the few who are rich and powerful at the expense of the many who are poor and weak—especially given the range of weapons and terrorist strategies available to the poor and weak—is a system that cannot stand. It is programmed for self-destruction, and its eventual collapse, if my guess is right, will be the principal event of the twenty-first century.

TWO WINNING SOLUTIONS

Hazel Henderson

President Bush now has $40 billion of discretionary funds granted by Congress. He could take $500 million of this and pay what the United States still owes to the United Nations and to our allies for past UN peacekeeping actions. Bush needs to create the very broadest coalition of support for the United States in dealing with terrorism. Only the United Nations can deliver this.

Then, instead of bombing raids, the United States and other countries with large inventories and backlogs of unsold food, clothing, and other consumer goods could air-lift those surplus goods and parachute them all over Afghanistan.

As unsold commodities, consumer goods, radios, and magazines float down from the skies, the Taliban will become further discredited for their economic failures. As unsold inventories are reduced in stagnating economies in Europe, Asia, North and South America, production can be revived and people re-employed. These economies and companies will rediscover the power of bartering—their surplus goods for peace—at a fraction of the cost of weapons and military strikes.

ACTS BEYOND RELIGIOUS BELIEF

Ziauddin Sardar

In an openly declared act of war, the Prophet Muhammad forbade the killing of civilians, women and children, the old and the infirm; the destruction of property, the burning of crops; and the slaughter of animals. Kidnapping, hijacking, and other acts of terror are as contrary to the teachings of Islam as they are to Christianity and Judaism.

The suicide bomber is a special breed. He stands outside normality, beyond reason. He may justify his rage with perverse self-righteousness and twisted religious notions, but his utterances and pieties are as impenetrable to Muslims as they are to anyone else.

Islam cannot explain the actions of the suicide hijackers, just as Christianity cannot explain the gas chambers, Catholicism the bombing at Omagh. They are acts beyond belief, religious belief. The perpetrators had long ago abandoned the path of Islam. But if you bracket all Muslims with the hijackers, we all go into the dark with those so brutally ripped from life in that inferno of twisted metal and concrete.

Are people calling themselves Muslims capable of such atrocities? Are they reading the same Qur'an? Are they the followers of the same Prophet Muhammad?

The Qur'an has very few general commands. Many find it surprising that over one-third of the Qur'an is devoted to extolling the virtues of reason and the pursuit of knowledge and thought.

That's the message that early Muslim communities took to heart. That's the message that terrorists like Osama bin Laden, and fundamentalist regimes like the Taliban, are trying to suppress. Surely, this message of knowledge and thought has to be our main weapon in the fight against the terrorists.

The task before us is to tackle the threat of terrorism.

Already, innumerable voices have been raised to bemoan the failure of intelligence. But ending the evil of terrorism cannot be found in better intelligence. It must be sought through better peacemaking, through engaging with the needs of the neediest, and through forming partnerships with the oppressed and tyrannized so that they are nurtured and helped.

The intelligence we need must light a path to a common edifice of compassion that is the shared inheritance of all people. Unless we have the intelligence to make common cause, through our differences, across communities, we cannot throw into relief or properly eradicate the dark hiding places of the unconscionable. We are all victims of terror. Our common revulsion, our shared horror is the intelligence we must gather.

GLOBALIZATION ISN'T WHAT IT USED TO BE

Walter Truett Anderson

Before September 11, the world was in the process of giving birth to the first global civilization. It was a painful birth, beset by problems and inequities and angry opposition to globalization itself.

Yet, for all the controversy, there were signs of growing cooperative efforts, born of the recognition that we were becoming a "global risk society" in which all people everywhere were potentially threatened by global-scale challenges such as climate change and AIDS.

Now, global society turns out to have been even riskier than we had thought. A new threat to peace and progress has leaped to the top of the list: It's not only about whatever damage terrorists may do, but also about the many grim scenarios of how retaliation efforts might go wrong.

But is globalization itself at risk? Will we shut down the borders, crank up national passions, return to a world of localism and closed systems? I don't think so.

Globalization is a massive process of evolutionary changes—economic, political, cultural, and biological—that have been going on over the entire course of human evolution and have accelerated in recent decades. We can no more reverse or undo them than we can unscramble an egg. Undoubtedly the shape and speed of globalization will be different now—yet the process will continue, and in some ways may even accelerate.

Globalization thrives on mobility—especially the mobility of people and the mobility of symbols. Both of those have been increasing exponentially. What happens now?

In the immediate aftermath of the September attacks, people's mobility declined precipitously: Airports were shut down, travel plans canceled, border guards increased, immigration policies reviewed. But at the same time, the mobility of symbols—particularly through the electronic media—increased. I doubt that there has ever been such an outpouring of news, information, misinformation, speculation, discussion, and dialogue.

The healthiest development I have seen has been the determination of many people to move toward a more mature understanding of precisely who our enemies are, and why they are our enemies.

So it is possible that some good things may come out of the event. One would be an end to the Lone Ranger style of foreign policy that was being demonstrated in the early days of the Bush administration, a seeming willingness to go it alone. Since September 11, there are many signs of a recognition—we have yet to see how deep or how lasting—that there is no going alone, not even for a superpower.

STOPPING WORLD WAR 4

Igor Bestuzhev-Lada

There is no more terrorism in the world. There is World War 4, as different from World War III (the "Cold War," 1946–1989) as the latter was from the second and first World Wars.

World War 4 is a conflict between "Poor South" and "Rich North." The first stage was in the 1980s and 1990s (Lebanon, Afghanistan, Yugoslavia, Israel, Chechnya, and Russia as a whole). The second stage began on September 11 in the United States. The third stage will begin when a weapon of mass destruction is used—nuclear, chemical, biological, or computer destruction of information systems, including finance.

The way to stop World War 4 before its third stage is through the refusal of the remaining participants of the Cold War (e.g., anti-Russian NATO, anti-Byelorussian intrigues, anti-Serbian repression). We need to transition from confrontational South vs. North and create instead a Transnational Antiterrorist Corporation (including both Americas, Europe, Eurasia, Japan, China, India, etc.) against an already existing "shadow transnational corporation" like that of bin Laden.

GLOBALIZATION AND TERRORISM: THE LONG ROAD AHEAD

Victor Ferkiss

The tragedy of September 11 revealed in bloody horror that globalization is a two-edged sword. Bitter arguments exist as to whether or not globalization has on balance improved the lot of the people of the world economically. But many millions in the world are convinced that it is a cultural, political, and religious disaster. As a result, there is now no question but that terrorism, like trade, extends beyond national borders.

What can the United States do about it? Our economy has been struck a major blow. We will need all of our economic strength for the end game if we are to dry up the roots of terrorism in a lasting victory.

Bush has said that we "are not into nation-building." But we must be, because our long-term interest—extirpating terrorism—demands it. We will need to create many Marshall Plans for poor nations and the political equivalent to sustain their benefits. If we defeat the Taliban we must extend economic and political aid to Afghanistan so as to enable it to take the first halting steps toward globalization.

Even more problematic, we must somehow spur democratic governments and equitable economic globalization in such nations as Pakistan, Egypt, and Saudi Arabia, among others.

The choice is ours and must be made very soon.

FORTRESS, VENGEANCE, OR GAIA: THREE SCENARIOS

Sohail Inayatullah

Three scenarios for the near and long-term future can be summarized as, first, Fortress USA/OECD and Fortress Islam, which gives a short-term illusion of security but results in general impoverishment; second, Cowboy War—Vengeance Forever, as George W. Bush has already invoked with the Wanted–Dead or Alive image. This scenario will only exacerbate the deep cleavages in the world.

The third scenario I call Gaian Bifurcation. The problems facing us cannot be resolved by making traditionalists modernist—i.e., turning them from loving land and God to loving money and scientific rationality. We must move from tradition to a transmodernity that is inclusive of multiple, layered realities. Gaian Bifurcation is a world of interdependent civilizations plus a system of international justice. The context would be a new equity-based multicultural globalization, moving to world governance. Specifically, this means:

- Human and animal rights.
- Indexing of wealth of poor and rich on a global level; that is, economic democracy.
- Prama, or meditative, philosophies that create dynamic balances between regions and between rural areas and cities, that see the world economy through the ecological metaphor (i.e., Gaia, the world as a living organism), but without excluding technological innovation.
- Self-reliant, ecological, electronically linked communities (and not states).
- Gender partnership.
- A transformed United Nations, with increased direct democracy, influence of the social movements, and transparency within multinationals.

This entails moving from a strategic discourse to a healing discourse, moving from trauma to transcendence. In the non-West, there is also a desire to move away from feudal structures but to retain spiritual heritage, to be "modern," but in a different way.

Do we have the courage to create this emergent future? We need to choose life.

THE NEW COMMUNISM

Kevin Kelly

Radical Islam will become the new communism, if it isn't that already. It has a deep appeal, even to those subjugated to it. There are aspects about it that even supporters don't like and can't stand, but they will submit to it because they believe it is better overall than the alternative of Western capitalism.

Our chief concern should be that there is nothing we have they want. They don't want recognition. They don't want our trade. They don't want our culture. They don't want our aspirations—democracy, free choice, high technology. They don't want our values. They don't want our wealth.

For the first time since communism, there is a competing destiny in the world. It is not the end of history as far as they are concerned. There were plenty of bombs dropped and people killed in Western capitalism's battle against communism, but little of all that warring made as much difference in the end as the simple fact that the West, through its own improvement, came to offer many things that those in communist countries wanted. And when we had something they wanted, communism collapsed.

I think we need to enlarge Western civilization so that we have something young Islamic believers want. Providing it will be the only way, and the only honest way, to triumph.

A POSITIVE OUTCOME?

Michael Marien

Many commentators have already emphasized that any serious "war on terrorism" must attack the roots of terrorism. To counter the growing divisions of rich and poor, within and between societies, this means extensive action to end poverty and despair, realize human rights for all, enhance human security and civil society, improve education and health care, fight global crime and corruption, develop democracy at national and global levels, ensure safe and abundant drinking water, promote greening of the market system, stabilize currency systems, reconsider damaging government subsidies, pursue sustainable development, and encourage environmentally benign technology.

No one has argued against undertaking these global reforms; rather, the many arguments for doing so are simply ignored. The new focus on terrorism, however, has already quickly and radically changed U.S. foreign policy from unilateralist to multilateralist. This initial step may lead to at least some of the numerous global reforms that are needed. And thus the acceleration of globalization and widespread global reform could very well be the most positive outcome of the September 11 terrorist attack.

The overall long-term benefits from this acceleration could possibly surpass the many costs of the terrorist attack and counterattack, likely to total hundreds of billions—if not trillions—of dollars. Broad reforms toward "a world that works for all" is thus a *futurible* that deserves our serious contemplation, and our heartfelt efforts to realize.

74

Play Ball!

ROBIN GUNSTON

Often trends in one area reflect larger societal trends. Gunston discusses the changes taking place in the arena of sports and trends that predict future directions of sports: big business, individualization of sports, terrorism's impact, drug controversies, and high-tech innovations.

SOURCE: Originally printed in *The Futurist, Vol. 31,* (January-February 2005*), pp. 33–36.*

As you read this article, ask yourself the following questions:

1. *What are the major trends in sports today?*
2. *How will sports change in the twenty-first century?*
3. *What trends and changes can you foresee in sports today and in the future?*
4. *What changes predicted by Gunston do you see already happening in your community?*

GLOSSARY **Postmodern** Life patterns in postindustrial societies. **Sport** Forms of physical activity that aim at expressing or improving physical fitness and mental well-being, forming social relationships, or obtaining results in competition.

Welcome to the twenty-first century's wide world of sports—a rapid-paced world where technology is as much a part of the game as muscle, where sports celebrity rivals religious worship, and where winning at all costs is the name of the game. To understand how sports will evolve during the coming decade, let's take a look backward and examine the ongoing tensions among sports, technology, celebrity, and politics. Those elements and how they mix in years to come will greatly influence the nature of sports and sportsmanship for future generations.

The Council of Europe defined sport as "all forms of physical activity, which, through casual or organized participation, aim at expressing or improving physical fitness and mental well-being, forming social relationships, or obtaining results in competition at all levels." This is a good test to put to any new future of sport.

KEY TRENDS IN SPORTS AND CULTURE

Here are the key trends that I believe will have the most impact and that may lead to different possible futures for sports.

Sports have become an entertainment business. Postmodernism has transformed sports. Sports scholars Bob Stewart and Aaron Smith note, "By the 1990s a number of professional sport leagues had emerged as amateurism lost its snobbish appeal and sport went about building its commercial value." As a result, stadiums became billboards, athletes became celebrities, competitions became sense-bombarding experiences, and fans shifted loyalties from one team to the next, unbound by the parochial tribalism of the ancients. "The Sydney 2000 Olympic Games convincingly demonstrated that Australian sport had become a chaotic mix of ancient ritual, traditional athletic contests, slickly marketed and customized leisure experiences, and ultra-professional sports that combine complex strategy with Hollywood-style showmanship," Stewart and Smith conclude.

Anyone looking for further evidence that sports are big business only needs to pick up a local television guide and look at the sheer variety of choices available to the viewer: poker, lawn-mower racing, bungee-jumping, elephant polo, juggling, and more. Can this trend go on forever? Increasingly, the answer is, "no." Both ESPN and NBC stated more than 10 years ago that their networks no longer had any "must-see" events, including the Olympics, which only survives financially on such network revenue and on the sponsorship of corporations and conglomerates. Many sports organizations, participants, and viewers are also starting to resent the intrusion of the television scheduler and advertisers into the flow of the game.

Team sports versus the individual. An emerging trend that seems set to continue is the demise of team sports and the ascendance of individual sports. This trend seems closely associated with changes to work–life balance and the culture of individualism apparent in most of Generation X. The modern worker is losing the battle to balance participation in organized sports with the demands of the workplace and the home. With the demise of the standard working day and the trend toward holding down multiple jobs, fewer and fewer chances exist for today's busy worker to commit to a regular training schedule for a team sport. Thus, the serious fitness addict or sporting person is turning more to individual pursuits—such as triathlons, marathons, the personal fitness regimen at the gym, and Ironman competitions—to achieve prowess.

Club ownership. The majority of team sports, including baseball, basketball, soccer, and rugby, are in professional leagues and managed as business franchises. Ownership comes in many shapes and forms; baseball management, for example, consists of clubs owned by a business that employs players on salaries. Clubs regularly change hands for colossal sums; sometimes the entire team will change cities when new owners are based elsewhere and want their team closer to control management better.

Business owners, however much they may like a sport or a club, want one thing above all else: a better-than-normal rate of return on their investment. Inevitably, this trend will create demands on coaches and players that create a win-at-any-cost mentality, leading to tragedies such as the infamous 1919 Black Sox scandal, the World Series "game-throwing" incident. The true essence of sport will then be lost.

The impact of terrorism. Terrorism is normally perceived as more of a wild card than a trend, as it usually seeks publicity about a cause by capturing public interest through a dramatic event, such as the 1972 Munich Olympics siege. However, modern terrorism is an international, omnipresent movement seemingly inspired by hatred for all things Western and American. This trend could eventually affect sports at all levels.

The modern terrorist is prepared to take time to accomplish his ends, often planning for events many years in advance. Today, there may already be individuals training not only to excel as athletes, but also to be able to disrupt major events in the future. Imagine a winning World Cup team clutching the Jules Rimet trophy, then blowing themselves up with it on the podium. Such an event may seem far-fetched, but it is plausible given the international attention paid to sports and the desperation of some individuals and groups.

Designer drugs. The trend for sports people to enhance their performance through substances is not a modern one. Records of the ancient Games show athletes selectively feeding on herbs for many weeks before such major events. It was not until 1969 that robust analytical techniques were used for drug detection in sports, and since then, both coaches and pharmaceutical interests have been trying to find drugs that avoid detection. When testers catch up, athletes inevitably end up losing—their records, their medals, their sponsorships, and their reputations.

In 2003, a new human gene—the so-called "speed gene"—was discovered in East Africa. Geneticists hailed the day when a purpose-built athlete could be cloned, and the American Association for the Advancement of Science announced a conference to explore the "potential uses of genetic enhancement in competitive sports from the perspective of athletic organizations, athletes, scientists, and ethicists."

High-technology equipment. We may see completely new forms of artificial-intelligence-based machinery taking over areas of human activity within the next 20 years. Sports are no exception to this trend. You can now go to the local sports store and plug into a machine that will fit each foot with a shoe designed for a particular sport. At the golf range, a computerized swing analyzer will tell you what shaft length you need and what club head and what ball will give the greatest distance. Specially designed track-and-field equipment that aligns the characteristics of the athlete and the stadium can give an individual athlete more than a meter's advantage over competitors. In motor sports and America's Cup yacht racing, technologists sit amid a vast array of computers taking race data, analyzing it, and sending it back to the driver or helmsman to make minute corrections that might win them the race. It may not be long before a humanoid robot sits in the driving seat of a Formula One car and the champion driver sits under the grandstand or in an apartment at a beach in another country driving the car remotely. The only time the fans or officials will suspect the difference is when the champagne flows and the robot shorts out!

The sports industry. Sports are no longer just pastimes. They are *big* business. Over a 20-year period, there has been a 10,000% increase in sports sponsorship, affecting every possible sport imaginable—even sheep shearing. Most professional sports cannot afford to operate without guaranteed television rights payments and commercial sponsorship. Even at the amateur level, club finances rely on the local sportswear store to provide the uniform and equipment, while the local butcher or hardware store may have its name adorning shirts or goalposts, a privilege *someone* is collecting money for.

Will this trend continue? As ethical investors compare the economic, environmental, and social results of a business, many are finding it difficult to account for the benefits that sport sponsorship actually brings in the marketplace. This can be especially problematic when the supported sport or a particular team performs badly or has something of a poor reputation. In Australia, for example, almost an entire rugby league team has been recently accused of pack rape. If they are convicted, expect to see their sponsors distance themselves quickly from the team.

Without sponsors, there will be no teams or individual superstar athletes. Without teams, there will be no leagues. Without leagues, there will be no major competitions. Without major competitions, there will be no sports television, no merchandising, no corporate boxes. And thus it goes on.

FUTURE SCENARIOS FOR SPORTS

To understand where sport is heading, we need to examine five key drivers of change:

1. The clear distinction between work and leisure is growing blurrier and changing the types of sports we play.
2. The drive for instant entertainment will place high demands on sports people and the industry.
3. The drive by media companies and other businesses to own our allegiance to their products and services is dictating increasing control of sporting performance and behavior.
4. As sports bodies such as the International Olympics Committee, the International Cricket Council, and the International Rugby Board become neopolitical entities, their decisions will control key aspects of the destiny and sovereignty of sporting nations.
5. The loss of core values in society due to the waning influence of the church creates a spiritual vacuum into which sports may move.

Based on these drivers of change, we can discern four possible long-term scenarios.

Religiosport could develop as major sports replace conventional religion. Religiosport will have its shrines (stadiums), costumes (uniforms), services (games and events), rituals (chants and songs), high priests (star athletes), and piety (fan loyalty). Religiosport will actively condone violence against rival sects (teams).

Machosport is a future where individual sports people become popular idols, feted wherever they go, promoted by the media, and put on display as being the ideal of modern man or woman. In this scenario, knowing about the sport is incidental to knowing about the person. It is increasingly associated with the worst forms of idolatry and leads to individuals losing their human rights and respect. Supporters will form fan clubs for individuals instead of participating in the sport, and as the heroes grow older, they will pass, with their "sport," out of existence.

Technosport develops when winning is everything and ethics counts for nothing. At this stage, the sport exists and is managed entirely by large businesses that appoint the sports-administration body and control all aspects of the sport's development, rules, and competitions. Individual players, coaches, and managers are only pawns to winning at all costs. In this scenario, only two international sports eventually remain—soccer and basketball—with most nations having only one team.

Valuesport will see an end to the big business of organized team sports and events. This future will be driven largely in response to some wild-card event, such as a terrorist attack at an Olympiad. Another threat to the Games' viability would be a last-minute pullout by the broadcast media covering them, perhaps in a battle over naming rights.

Another driver in this scenario is the obesity crisis. Health specialists, trying to cope with the vastly increased death rates stemming from society-wide lack of fitness, will lobby governments for a return to a different style of sports participation at all levels of society. They will no longer permit advertising to be linked to sports, and all teams participating in healthy competition will be backed by their community and financially supported by additional local taxes on unhealthy products such as alcohol, certain drugs, and tobacco.

GLOBAL COMMUNITY GAMES

Valuesport is, in this author's opinion, the preferred scenario. But how can we turn everything around now and prevent the onset of one of the other possible scenarios? The key will be to strengthen underlying values and continue to reinforce them for young people and new participants. These moral and ethical principles have to be continually reinforced through many avenues, including a more sensitive media, to ensure that sports' positive contributions to society are not undermined.

Fortunately this work has already begun, quietly and without fanfare. For the past five years, an international group of sports people and community workers have been collaborating and bringing about such a change. These dedicated volunteers are using a simple experiential learning model used with sports and games and values derived from Bible stories to get people of all ages involved in improving their physical, moral, and spiritual health. Many sporting champions have also lent their support to the program by appearing at some of the opening and closing ceremonies of KidsGames, TeenGames, EdgeGames, and FamilyGames to encourage people of all ages to reach their goals. Some of these events have been the largest sporting events ever held in their cities or countries.

The future of large sports events may no longer be with the Olympians, but rather with those participating in something much grander: a values-based movement committed to ensuring that sports have a positive role in society. The result will be what the International Sports Coalition has called Global Community Games.

75

Five Meta-Trends Changing the World

DAVID PEARCE SNYDER

In our final reading, Snyder discusses the future—major trends that are affecting the direction of change in the world. These meta-trends *combine demographic, economic, and technological events and changes that add up to predictions for the future.*

SOURCE: Originally printed in *The Futurist*, 38:4 (July-August 2004), pp. 22–27.

As you read this final selection, think about the following:

1. *What are meta-trends and who do they affect?*
2. *What events or social indicators have led to these meta-trends?*
3. *Discuss a scenario of the future in your community based on these meta-trend predictions.*
4. *How might these meta-trends affect you and your family?*

GLOSSARY **Cultural modernization** evolution of education, urbanization, institutional order and tenets of equality, personal freedom, and self-fulfillment that are changing traditional cultures. **Meta-trend** A major, overarching trend that influences the direction of other areas of change.

What follows are five meta-trends I believe are profoundly changing the world. They are evolutionary, system-wide developments arising from the simultaneous occurrence of a number of individual demographic, economic, and technological trends. Instead of each being individual freestanding global trends, they are composites of trends.

TREND 1—CULTURAL MODERNIZATION

Around the world over the past generation, the basic tenets of modern cultures—including equality, personal freedom, and self-fulfillment—have been eroding the domains of traditional cultures that value authority, filial obedience, and self-discipline. The children of traditional societies are growing up wearing Western clothes, eating Western food, listening to Western music, and (most importantly of all) thinking Western thoughts. Most Westerners—certainly most Americans—have been unaware of the personal intensities of this culture war because they are so far away from the "battle lines." Moreover, people in the West regard the basic institutions of modernization, including universal education, meritocracy, and civil law, as benchmarks of social progress, while the defenders of traditional cultures see them as threats to social order.

Demographers have identified several leading social indicators as key measures of the extent to which a nation's culture is modern. They cite the average level of education for men and for women, the percentage of the salaried workforce that is female, and the percentage of population that lives in urban areas. Other indicators include the percentage of the workforce that is salaried (as opposed to self-employed) and the percentage of GDP spent on institutionalized socioeconomic support services, including insurance, pensions, social security, civil law courts, worker's compensation, unemployment benefits, and welfare.

As each of these indicators rises in a society, the birthrate in that society goes down. The principal measurable consequence of cultural modernization is declining fertility. As the world's developing nations have become better educated, more urbanized, and more institutionalized during the past 20 years, their birthrates have fallen dramatically. In 1988, the United Nations forecast that the world's population would double to 12 billion by 2100. In 1992, their estimate dropped to 10 billion, and they currently expect global population to peak at 9.1 billion in 2100. After that, demographers expect the world's population will begin to slowly decline, as has already begun to happen in Europe and Japan.

The effects of cultural modernization on fertility are so powerful that they are reflected clearly in local vital statistics. In India, urban birthrates are similar to those in the United States, while rural birthrates remain unmanageably high. Cultural modernization is the linchpin of human sustainability on the planet.

The forces of cultural modernization, accelerated by economic globalization and the rapidly spreading wireless telecommunications info-structure, are likely to marginalize the world's traditional cultures well before the century is over. And because the wellsprings of modernization—secular industrial economies—are so unassailably powerful, terrorism is the only means by which the defenders of traditional culture can fight to preserve their values and way of life. In the near-term future, most observers believe that ongoing cultural conflict is likely to produce at least a few further extreme acts of terrorism, security measures not withstanding. But the eventual intensity and duration of the overt, violent phases of the ongoing global culture war are largely matters of conjecture. So, too, are the expert pronouncements of the probable long-term impacts of September 11, 2001, and terrorism on American priorities and behavior.

After the 2001 attacks, social commentators speculated extensively that those events would change America. Pundits posited that we would become more motivated by things of intrinsic value—children, family, friends, nature, personal self-fulfillment—and that we would see a sharp increase in people pursuing pro bono causes and public-service careers. A number of media critics predicted that popular entertainment such as television, movies, and games would feature much less gratuitous violence after September 11. None of that has happened. Nor have Americans become more attentive to international news coverage. Media surveys show that the average American reads less international news now than before September 11. Event-inspired changes in behavior are generally transitory. Even if current conflicts produce further extreme acts of terrorist violence, these seem unlikely to alter the way we live or make daily decisions. Studies in Israel reveal that its citizens have become habituated to terrorist attacks. The daily routine of life remains the norm, and random acts of terrorism remain just that: random events for which no precautions or mind-set can prepare us or significantly reduce our risk.

In summary, cultural modernization will continue to assault the world's traditional cultures, provoking widespread political unrest, psychological stress, and social tension. In developed nations, where the great majority embrace the tenets of modernization and where the threats from cultural conflict are manifested in occasional random acts of violence, the ongoing confrontation between tradition and modernization seems likely to produce security measures that are inconvenient, but will do little to alter our basic personal decision making, values, or day-to-day life. Developed nations are unlikely to make any serious attempts to restrain the spread of cultural modernization or its driving force, economic globalization.

TREND 2—ECONOMIC GLOBALIZATION

On paper, globalization poses the long-term potential to raise living standards and reduce the costs of goods and services for people everywhere. But the short-term marketplace consequences of free trade threaten many people and enterprises in both developed and developing nations with potentially insurmountable competition. For most people around the world, the threat from foreign competitors is regarded as much greater than the threat from foreign terrorists. Of course, risk and uncertainty in daily life is characteristically high in developing countries. In developed economies, however, where formal institutions sustain order and predictability, trade liberalization poses unfamiliar risks and uncertainties for many enterprises. It also appears to be affecting the collective psychology of both blue-collar and white-collar workers—especially males—who are increasingly unwilling to commit themselves to careers in fields that are likely to be subject to low-cost foreign competition.

Strikingly, surveys of young Americans show little sign of xenophobia in response to the millions of new immigrant workers with whom they are competing in the domestic job market. However, they feel hostile and helpless at the prospect of competing with Chinese factory workers and Indian programmers overseas. And, of course, economic history tells us that they are justifiably concerned. In those job markets that supply untariffed international industries, a "comparable global wage" for comparable types of work can be expected to emerge worldwide. This will raise workers' wages for freely traded goods and services in developing nations, while depressing wages for comparable work in mature industrial economies. To earn more than the comparable global wage, labor in developed nations will have to perform incomparable work, either in terms of their productivity or the superior characteristics of the goods and services that they produce. The assimilation of mature information technology throughout all production and education levels should make this possible, but developed economies have not yet begun to mass-produce a new generation of high-value-adding, middle-income jobs.

Meanwhile, in spite of the undeniable short-term economic discomfort that it causes, the trend toward continuing globalization has immense force behind it. Since World War II, imports have risen from 6% of world GDP to more than 22%, growing steadily throughout the Cold War, and even faster since 1990. The global dispersion of goods production and the uneven distribution of oil, gas, and critical minerals worldwide have combined to make international interdependence a fundamental economic reality, and corporate enterprises are building upon that reality. Delays in globalization, like the September 2003 World Trade Organization contretemps in Cancun, Mexico, will arise as remaining politically sensitive issues are resolved, including trade in farm products, professional and financial services, and the need for corporate social responsibility. While there will be enormous long-term economic benefits from globalization in both developed and developing nations, the short-term disruptions in local domestic employment will make free trade an ongoing political issue that will be manageable only so long as domestic economies continue to grow.

TREND 3—UNIVERSAL CONNECTIVITY

While information technology (IT) continues to inundate us with miraculous capabilities, it has given us, so far, only one new power that appears to have had a significant impact on our collective behavior: our improved ability to communicate with each other, anywhere, anytime. Behavioral researchers have found that cell phones have blurred or changed the boundaries between work and social life and between personal and public life. Cell phones have also increased users' propensity to "micromanage their lives, to be more spontaneous, and, therefore, to be late for everything," according to Leysia Palen, computer science professor at the University of Colorado at Boulder.

Most recently, instant messaging—via both cell phones and online computers—has begun to have an even more powerful social impact than cell phones themselves. Instant messaging initially tells you whether the person you wish to call is "present" in cyberspace—that is, whether he or she is actually online at the moment. Those who are present can be messaged immediately, in much the same way as you might look out the window and call to a friend you see in the neighbor's yard. Instant messaging gives a physical reality to cyberspace. It adds a new dimension to life: A person can now be "near," "distant," or "in cyberspace." With video instant messaging—available now, and widely available in three years—the illusion will be complete. We will have achieved what Frances Cairncross, senior editor of *The Economist*, has called "the death of distance."

Universal connectivity will be accelerated by the integration of the telephone, cell phone, and other wireless telecom media with the Internet. By 2010, all long-distance phone calls, plus a third of all local calls, will be made via the Internet, while 80% to 90% of all Internet access will be made from Web-enabled phones, PDAs, and wireless laptops. Most important of all, in less than a decade, one-third of the world's population—2 billion people—will have access to the Internet, largely via Web-enabled telephones. In a very real sense, the Internet will be the "Information Highway"—the infrastructure, or infostructure, for the computer age. The infostructure is already speeding the adoption of flexplace employment and reducing the volume of business travel, while making possible increased "distant collaboration," outsourcing, and offshoring.

As the first marketing medium with a truly global reach, the Internet will also be the crucible from which a global consumer culture will be forged, led by the first global youth peer culture. By 2010, we will truly be living in a global village, and cyberspace will be the town square.

TREND 4—TRANSACTIONAL TRANSPARENCY

Long before the massive corporate malfeasance at Enron, Tyco, and WorldCom, there was a rising global movement toward greater transparency in all private and public enterprises. Originally aimed at kleptocratic regimes in Africa and the former Soviet states, the movement has now become universal, with the establishment of more stringent international accounting standards and more comprehensive rules for corporate oversight and record keeping, plus a new UN treaty on curbing public-sector corruption. Because secrecy breeds corruption and incompetence, there is a growing worldwide consensus to expose the principal transactions and decisions of all enterprises to public scrutiny.

But in a world where most management schools have dropped all ethics courses and business professors routinely preach that government regulation thwarts the efficiency of the marketplace, corporate and government leaders around the world are lobbying hard against transparency mandates for the private sector. Their argument: Transparency would "tie their hands," "reveal secrets to their competition," and "keep them from making a fair return for their stockholders."

Most corporate management is resolutely committed to the notion that secrecy is a necessary concomitant of leadership. But pervasive, ubiquitous computing and comprehensive electronic documentation will ultimately make all things transparent, and this may leave many leaders and decision makers feeling uncomfortably exposed, especially if they were not provided a moral compass prior to adolescence. Hill and Knowlton, an international public-relations firm, recently surveyed 257 CEOs in the United States, Europe, and Asia regarding the impact of the Sarbanes-Oxley Act's reforms on corporate accountability and governance. While more than 80% of respondents felt that the reforms would significantly improve corporate integrity, 80% said they also believed the reforms would not increase ethical behavior by corporate leaders.

While most consumer and public-interest watchdog groups are demanding even more stringent regulation of big business, some corporate reformers argue that regulations are often counterproductive and always circumventable. They believe that only 100% transparency can assure both the integrity and competency of institutional actions. In the world's law courts—and in the court of public opinion—the case for transparency will increasingly be promoted by nongovernmental organizations (NGOs) who will take advantage of the global infostructure to document and publicize environmentally and socially abusive behaviors by both private and public enterprises. The ongoing battle between institutional and socioecological imperatives will become a central theme of Web newscasts, Netpress publications, and Weblogs that have already begun to supplant traditional media networks and newspaper chains among young adults worldwide. Many of these young people will sign up with NGOs to wage undercover war on perceived corporate criminals.

In a global marketplace where corporate reputation and brand integrity will be worth billions of dollars, businesses' response to this guerrilla scrutiny will be understandably hostile. In their recently released Study of Corporate Citizenship, Cone/Roper, a corporate consultant on social issues, found that a majority of consumers "are willing to use their individual power to punish those companies that do not share their values." Above all, our improving comprehension of humankind's innumerable interactions with the environment will make it increasingly clear that total transparency will be crucial to the security and sustainability of a modern global economy. But there will be skullduggery, bloodshed, and heroics before total transparency finally becomes international law—15 to 20 years from now.

TREND 5—SOCIAL ADAPTATION

The forces of cultural modernization—education, urbanization, and institutional order—are producing social change in the developed world as well as in developing nations. During the twentieth century, it became increasingly apparent to the citizens of a growing number of modern industrial societies that neither the church nor the state was omnipotent and that their leaders were more or less ordinary people. This realization has led citizens of modern societies to assign less weight to the guidance of their institutions and their leaders and to become more self-regulating. U.S. voters increasingly describe themselves as independents, and the fastest-growing Christian congregations in America are nondenominational.

Since the dawn of recorded history, societies have adapted to their changing circumstances. Moreover, cultural modernization has freed the societies of mature industrial nations from many strictures of church and state, giving people much more freedom to be individually adaptive. And we can be reasonably certain that modern societies will be confronted with a variety of fundamental changes in circumstance during the next five, 10, or 15 years that will, in turn, provoke continuous widespread adaptive behavior, especially in America.

During the decade ahead, information—the automated collection, storage, and application of electronic data—will dramatically reduce paperwork. As outsourcing and off-shoring eliminate millions of U.S. middle-income jobs, couples are likely to work two lower-pay/lower-skill jobs to replace lost income. If our employers ask us to work from home to reduce the company's office rental costs, we will do so, especially if the arrangement permits us to avoid two hours of daily commuting or to care for our offspring or an aging parent. If a wife is able to earn more money than her spouse, U.S. males are increasingly likely to become househusbands and take care of the kids. If we are in good health at age 65, and still enjoy our work, we probably won't retire, even if that's what we've been planning to do all our adult lives. If adult children must move back home after graduating from college in order to pay down their tuition debts, most families adapt accordingly.

Each such lifestyle change reflects a personal choice in response to an individual set of circumstances. And, of course, much adaptive behavior is initially undertaken as a temporary measure, to be abandoned when circumstances return to normal. During World War II, millions of women voluntarily entered the industrial workplace in the United States and the United Kingdom, for example, but returned to the domestic sector as soon as the war ended and a prosperous normalcy was restored. But the Information Revolution and the aging of mature industrial societies are scarcely temporary phenomena, suggesting that at least some recent widespread innovations in lifestyle—including delayed retirements and "sandwich households"–are precursors of long-term or even permanent changes in society.

The current propensity to delay retirement in the United States began in the mid-1980s and accelerated in the mid-1990s. Multiple surveys confirm that delayed retirement is much more a result of increased longevity and reduced morbidity than it is the result of financial necessity. A recent AARP survey, for example, found that more than 75% of baby boomers plan to work into their 70s or 80s, regardless of their economic circumstances. If the baby boomers choose to age on the job, the widely prophesied mass exodus of retirees will not drain the workforce during the coming decade, and Social Security may be actuarially sound for the foreseeable future.

The Industrial Revolution in production technology certainly produced dramatic changes in society. Before the steam engine and electric power, 70% of us lived in rural areas; today 70% of us live in cities and suburbs. Before industrialization, most economic production was home- or family-based; today, economic production takes place in factories and offices. In preindustrial Europe and America, most households included two or three adult generations (plus children), while the great majority of households today are nuclear families with one adult generation and their children.

Current trends in the United States, however, suggest that the three great cultural consequences of industrialization—the urbanization of society, the institutionalization of work, and the atomization of the family—may all be reversing, as people adapt to their changing circumstances. The U.S. Census Bureau reports that, during the 1990s, Americans began to migrate out of cities and suburbs into exurban and rural areas for the first time in the twentieth century. Simultaneously, information work has begun to migrate out of offices and into households. Given the recent accelerated growth of telecommuting, self-employment, and contingent work, one-fourth to one-third of all gainful employment is likely to take place at home within 10 years. Meanwhile, growing numbers of baby boomers find themselves living with both their debt-burdened, underemployed adult children and their own increasingly dependent aging parents. The recent emergence of the "sandwich household" in America resonates powerfully with the multigenerational, extended families that commonly served as society's safety nets in preindustrial times.

LEADERSHIP IN CHANGING TIMES

The foregoing meta-trends are not the only watershed developments that will predictably reshape daily life in the decades ahead. An untold number of inertial realities inherent in the common human enterprise will inexorably change our collective circumstances—the options and imperatives that confront society and its institutions. Society's adaptation to these new realities will, in turn, create further changes in the institutional operating environment, among customers, competitors, and constituents. There is no reason to believe that the Information Revolution will change us any less than did the Industrial Revolution.

In times like these, the best advice comes from ancient truths that have withstood the test of time. The Greek philosopher-historian Heraclitus observed 2,500 years ago that "nothing about the future is inevitable except change." Two hundred years later, the mythic Chinese general Sun Tzu advised that "the wise leader exploits the inevitable." Their combined message is clear: "The wise leader exploits change."